Natural Resources Engineering

Natural Resources Engineering

Ernest W. Tollner, P.E.

Iowa State Press
A Blackwell Publishing Company

Ernest W. Tollner, P.E., is professor of Biological and Agricultural Engineering at the University of Georgia, Athens. His teaching and research interests include management of runoff and nonpoint source pollution in agriculture and urban settings.

A Blackwell Publishing Company

Iowa State Press
2121 State Avenue, Ames, Iowa 50014

Orders:	1-800-862-6657
Office:	1-515-292-0140
Fax:	1-515-292-3348
Web site:	www.iowastatepress.com

First edition, 2002

Library of Congress Cataloging-in-Publication Data

Tollner, Ernest W.
Natural resources engineering / Ernest W. Tollner.-1st ed.
p. cm.
Includes bibliographical references and index.
ISBN 0-8138-1847-8
1. Agricultural engineering. 2. Bioengineering. I. Title.
S675. T54 2002
627–dc21

2001006117

The last digit is the print number: 9 8 7 6 5 4 3 2 1

Contents

Acknowledgments xiii

Preface xv

Chapter 1 Natural Resources Engineering Opportunities 1

Defining Natural Resources Engineering and Related Terms 2

The Hydrologic Cycle 3

Changing Land Uses Due to Societal Forces 4

The Scope of Natural Resources Engineering 6

Outlook 15

Problems and Questions 17

References 18

Chapter 2 Precipitation 20

Precipitation Defined 20

Raindrop Characteristics 21

Precipitation Mechanisms 21

Precipitation Types and Seasonal Effects 22

Precipitation Measurement 23

Precipitation Analyses 25

Introduction to Hydrologic Frequency Analyses:
Return Period Analysis 27

Calculating Storm Magnitudes for Design Purposes 33

Local-Scale Precipitation Analyses 43

Pollution Transport by Precipitation 46

Outlook 46

Problems and Questions 47

Notes 49

References 49

Chapter 3 Infiltration 51

Infiltration, Percolation, and Subsurface Flow Defined 51
Factors Affecting Infiltration 51
Darcy's Law 52
Infiltration and Percolation Prediction 57
Infiltration Measurement 62
Pollutant Transport by Subsurface Flow 64
Outlook 65
Problems and Questions 66
Notes 68
References 68

Chapter 4 Evapotranspiration 69

Factors Affecting Evaporation 70
Evaporation Prediction Approaches 70
Evaporation from Water Surfaces Based Solely on Aerodynamic Effects or Mass Balance 71
Evaporation Prediction from the Energy Balance: The Bowen Ratio Method 73
Evaporation (E_P) Prediction Accounting for Aerodynamic Effects and Energy Balance: The Penman-Monteith Combination Equation Method 74
Empirical and Approximate Methods for E Prediction 82
Physical Analog: The Standard Evaporation Pan 87
Evaporation and Evapotranspiration Measurement 89
Pollutants and Evaporation 90
Outlook 90
Problems and Questions 91
References 92

Chapter 5 Runoff 94

Background 94
Watershed or Catchment Delineation 96
Runoff Volume for a Compact Watershed 98
Peak Runoff Rate with Compact Watersheds 105
Runoff Hydrograph Development 123
Impact of Errors on Runoff Assessment 132
Outlook 133
Problems and Questions 133
Notes 135
References 136

Chapter 6 Erosion by Water 137

Factors Affecting Erosion by Water 138
Soil Erosion versus Sediment Yield 140
Soil Loss Tolerance 140
Types of Water Erosion 140
Erosion Mechanics 142
Predicting Soil Detachment and Upland Erosion 144
The Weighted $R_{rf}C_{prac}$ Factor for Average Annual Erosion Computation 154
Note on Units 158
Prediction of Sediment Yield 159
Modeling Approaches for Sediment Yield Prediction 162
Erosion Control Practices in Agriculture 165
Sediment Control Practices in Construction 168
Erosion and Pollution 168
Future Directions of Water Erosion Prediction 169
Outlook 169
Going from Engineering Science to Engineering Design: Why Some Approaches Are Empirical 170
Problems and Questions 171
Notes 174
References 174

Chapter 7 Water Quality and Upland Nonpoint Pollution 177

Important Concepts and Selected Pollution Measurement Techniques 179
Water Quality Background 183
Scale Effects 186
Some Best Management Practices for Nonpoint Pollution Abatement 189
Quantitative Removal and Renovation of Selected Pollution Constituents 197
Outlook 199
Problems and Questions 201
References 203

Chapter 8 Open Channel Hydraulics 205

Fundamentals of Hydraulics 205
Fluid Statics Review 208
Channel Design and Construction 215
Outlook 231
Problems and Questions 233
Notes 236
References 236

Chapter 9 Vegetated Waterways 238

Background 239
Channel Planning 240
Design Procedures 245
Other Design Approaches 253
Construction and Maintenance 253
Outlook 253
Problems and Questions 255
References 256

Chapter 10 Terraces and Diversions 257

Background 257
Terrace Definitions and Types 259
Function of Terraces 261
Erosion Terrace Spacing for Erosion Prevention 261
Terrace Channel Length, Cross Section, Maximum Grade, and Capacity 264
Broadbase and Bench Erosion Terrace Design Summary 269
Diversion Terraces and Diversions in Agricultural and Urban Applications 270
Diversion Terraces for Construction Site Erosion Control 274
Terrace Outlet Selection 276
Calculations for Tile Outlet Terrace or Sedimentation Pond 276
Parallel Terrace System Planning and Design 281
Outlook 283
Problems and Questions 287
Notes 289
References 290

Chapter 11 Hydraulics of Water Management Structures 291

Structure Types 292
Hydraulic Concepts 294
Stage-Discharge Relationships of Weir Inlets and Flumes 297
Discharge Relations of Orifices and Sluice Gate Inlet Devices 304
Stage-Discharge Curves for Culverts and Spillways 316
Closed Conduit Systems for Urban Storm Water Collection 321
Water Measurement Structures for Irrigation and Aquaculture 322
Outlook 323
Problems and Questions 325
References 327

Chapter 12 Impoundments and Embankments 329

Soil Fundamentals for Embankment Construction 330
Flood Routing through Reservoirs and Related Structures 335
General Pond Design 342
Spillway Design 352
Embankment Safety: Selection of Design Storm 354
Applications 355
Construction and Safety Issues 362
Outlook 363
Problems and Questions 366
Notes 369
References 369

Chapter 13 Shallow Groundwater Management with Drains and Wells 371

Surface Drainage 371
Introduction to Subsurface Flow 375
Pipe Drainage 376
Shallow Wells 390
Uniform Infiltration and Drainage to a Nearby Stream 396
Outlook 398
Problems and Questions 398
References 400

Chapter 14 Introduction to Irrigation Design 402

Irrigation Systems Overview 403
Soil-Water-Plant Relations 406
Soil Intake Rate 413
Water Quality Issues and Leaching Requirement for Crop Production 414
Irrigation Efficiency 416
Irrigation Scheduling 419
Computing *ET* Requirements: Peak *ET* 420
Effective Rainfall for Agricultural Irrigation 420
Computing *ET* Requirements: Seasonal Water Use 421
Irrigation Pumping Rate for Meeting *ET* Requirements 423
Water Rights 423
Sprinkler Lateral Design for Agricultural Applications 426
Center Pivot System Specification 436
Design for Nonagricultural Applications 439

Microirrigation System Design with Pressure-Compensated Emitters 442
Safety and Maintainability 445
Outlook 447
Problems and Questions 447
Notes 449
References 449

Chapter 15 Streamflow Processes and Lake Dynamics 452

Background 452
How Waters Are Biologically Impaired 454
Predominant Physical Processes 454
Biological Processes 474
Predominant Nutrient and Chemical Processes 482
Outlook 487
Problems and Questions 488
Notes 490
References 490

Appendix A Ethics, Biotic Relationships, and Resource Economics 493

Review of Values and Ethics 493
Ethical Concepts and the Environment 497
Principles of Ecological Engineering or Design 498
Resource Economics: Social Traps 500
What Defines a Well-Designed Project? 501
Case Study 1: Pitcher Plant Community Viability 502
Case Study 2: Clean Water Act Compliance, a Hypothetical Case 505
Ethics, Ecological Knowledge, and Natural Resources Problem Solution 506
Notes 507
References 507

Appendix B Solutions for Trapezoidal Channels with Constrained Velocity 509

Appendix C Structure Installation and Soil-Imposed Forces 511

Impoundment and Embankment Volume Calculations 511
Setting Slope Stakes for Cuts and Fills 515
Techniques for Installation of Channels and Other Structures 516
Materials Specifications for Channels and Structures 521

Rural Road Construction 524
Buried Pipe Loading 525
References 529

Appendix D Land-Forming Computations for Surface Drainage 530

References 533

Appendix E Centrifugal Pumps and Pumping System Design 534

Pump Fundamentals 534
Centrifugal Pump Type Selection 536
Centrifugal Pump Performance Curves 537
Friction Losses in Pipes 537
Suction Line Design Considerations 539
Total Dynamic Head and Power 540
Determination of the System Head-Flow Curve 542
Selecting and Matching Pumps to a System 543
Pumps in Series and Parallel 545
Storage Requirements at Pumping Plants 545
Pump and Power Plant Fittings and Protection 546
Problems and Questions 547
Notes 548
References 548

Appendix F Plant Water Potential 550

References 552

Appendix G Unit Conversions 553

Index 555

Acknowledgments

This book is a product of considerable interaction among teaching faculty, research colleagues, and students. I thank Dr. Jerry Chesness and Mr. Carlisle Cobb for use of their excellent slides, and K. C. Das, D. E. Radcliffe (Department of Crop and Soil Sciences, University of Georgia), and Sharyl Walker (University of Illinois) for artwork.

Faculty colleagues Jerry Chesness, K. C. Das, David Gattie, John Goodrum, Todd Rasmussen (University of Georgia forest hydrologist), Mark Risse, Matt Smith, David Stooksbury, Gary VanWicklin, and John Worley reviewed chapters corresponding to their respective areas of expertise. Research colleagues Donald Pfost (University of Missouri), Allen Thompson (University of Missouri), and Neil Pellmann (USDA-NRCS, Columbia, Missouri) likewise contributed excellent comments. Postdoctoral associate Ali Esteghalian provided helpful material relating to bioethics. Mr. Wes Byne, research engineer, provided helpful comments relating to erosion, water quality, and stream processes. Graduate students Mikali Bruton, Dan Hitchcock, Brandy Hunt, K. C. Price, Jordan Smith, and Simon van Donk also provided valuable reviews. Jerry Chesness, Sharyl Walker (University of Illinois), Dan Yoder (University of Tennessee), John Hayes (Clemson University), and Richard Warner (University of Kentucky) reviewed the entire manuscript and contributed constructive review comments related to content analysis and organization. My thanks also go to the many undergraduates from EGR 325 and ENGR 3410 classes for their helpful remarks. The book would not have been possible without the excellent help and advice of Judi Brown, Catherine Conover, and others at Iowa State Press. The help of Pam Bruton, copy editor, is especially appreciated.

As author, I am solely responsible for errors and omissions. Please notify me of errors or significant omissions; I would greatly appreciate constructive suggestions for improving future editions.

My general farm and dairy background, made possible by my parents, Ernest and Ruby Tollner, was of incalculable value to this undertaking. The values gleaned from the farm experience are priceless. Likewise, the general agricultural and biological engineering experiences gained while studying under the guidance of B. J. Barfield, Tom Haan (University of Kentucky), Charlie Busch, and Dave Hill (Auburn University) are much appreciated. Brahm Verma and Dale Threadgill (University of Georgia) encouraged me to challenge assumptions and provided valuable career guidance. I deeply appreciate the patience and encouragement of my wife, Caren. She was always there, enduring many lonely evenings and weekends during the writing of the book. I dedicate this effort to my parents and to Caren.

Preface

In this book, I define *natural resources engineering* as the design of planned activities that are complementary to or opposed to natural or societal forces and that lead to modifications of the soil, water, biota, and/or air environment. The problem space is on the farm or field scale. The purpose of natural resources engineering is resource development and/or environmental management. Thus, I broaden the conventional definition of natural resource use from the usual oil/gas mining extraction activity to include crop production and development activities related to urbanization. Ancillary site development activities associated with bioremediation, bioconversion, and resource extraction are included in this more general definition.

Natural resources engineers solve problems arising from nonagricultural rural population, concentration of animal production, outward advances of the "urban fringe," and waste/residue recycling to productive land. Natural resources engineers also address traditional agricultural production problems. Pressure for improving the quality of water leaving upland agricultural areas by implementing comprehensive nutrient management plans (CNMPs) and best management practices (BMPs) is growing. Regulatory agencies are phasing in storm water discharge requirements and associated storm water BMPs. CNMPs and storm water discharge permits are motivated by a desire to enable continued growth and development without exceeding basin total mass daily loadings (TMDLs). The need to determine TMDLs for watersheds offers many exciting employment opportunities. The engineers meeting these demands in the future are likely to be from nonfarm backgrounds. Many graduates in the natural resources engineering find employment in areas requiring the modeling of environmental problems associated with large-scale agriculture, bioremediation, and the urban fringe.

The text first provides a quantitative overview of the hydrologic cycle, culminating with runoff. Topics then proceed in a "downhill" fashion, beginning with upland erosion and water quality coverage. From upland flow, the text moves to general channel design followed by specific channel types important to natural resources engineering. The hydraulics of common structures is introduced, followed by a discussion of ponds and embankments. The text then presents drainage and irrigation topics. Just as water continues its flow toward the ocean through stream and lake systems, so the text concludes with a discussion of stream and lake quality issues. This book focuses on activities related to water and soil resources. For air quality issues relating to particulates and odors, topics of increasing importance at the urban-suburban fringe, the student should refer to the several excellent texts available that deal with air quality engineering.

Many instructional programs are applying various total quality management (TQM) tools and concepts to engineering design. Use of engineering design methodology is stressed to aid in keeping the client's interest in the forefront. Prompt sheets are presented for open channel design and structures to help students probe customer requirements and identify technical specifications.

Internet addresses are provided so users may obtain current information on the topics discussed. Information provided by the expanding Internet links includes practical tips for construction of channels, ponds, embankments, and irrigation and drainage systems. The use of Internet key word searches is encouraged for finding up-to-date links to governmental and commercial modeling and design software, design methodology, and construction methods.

Courses launching those with environmental interests into more advanced courses should benefit from this text. The overarching goal is to provide the basic background to natural resources engineering for agricultural, biological, and civil engineers who use sophisticated computer models for natural resources engineering and watershed assessments. Students should have completed most basic engineering science courses, especially statics and fluid mechanics. This reference and text will aid students and practicing professionals as they prepare for helping clients achieve environmental "friendliness" and stewardship while engaging in agricultural production or related commercial activities leading to economic well-being.

1

Natural Resources Engineering Opportunities

Water, soil, air, plants, animals, people—these are the primary components of our environment. For millennia, men and women have been cultivating food and fiber, domesticating animals, and developing resources. The provision of tools to facilitate "dressing and keeping" while engaging in economic activities defines an overarching vision for the natural resources engineer.

Where do we begin? The study of natural resources engineering begins with the study of the primary physical forces operating in the environment. These forces are manifest in the *hydrologic cycle*. According to Biswas (1970), philosophers have documented natural forces since the time of Hammurabi (c. 1700 B.C.E.). Solomon (c. 900 B.C.E.) made one of the more elegant statements pointing toward the hydrologic cycle in Ecclesiastes (1:4–7):

> A generation goes and a generation comes, but the earth remains forever. Also, the sun rises and the sun sets; and, hastening to its place, it rises there again. Blowing toward the south, then turning toward the north, the wind continues swirling along; on its circular courses the wind returns. All the rivers flow into the sea, yet the sea is not full. To the place where the rivers flow, there they flow again.

Humankind establishes boundaries in time and space not respected by water, soil, and air. Neither can people totally control the biotic component. Nearly all of humankind's activities on the soil have affected and will affect not only the soil but the water, biota, and air at the site. Impacts spread across human boundaries, spatial and temporal. Impacts become manifest in the face of population increases, the increasing desires of developing nations to achieve a standard of living comparable to that of developed nations, and increasing environmental awareness in developed nations. Society looks for the path of moderation between development and conservation. Engineers, along with the practitioners of many other scientific and liberal arts disciplines, play important roles in identifying the moderate path.

Goals

- Identify fundamental terms such as *environment, natural resources engineering, urbanization, intensive agriculture, extensive agriculture,* and *farm and field scale.*

- Recognize the basic components of the hydrologic cycle and how the hydrologic cycle both drives natural resources engineering activities and requires engineering activities to meet societal needs.
- Evaluate how land use is changing in response to societal forces and visualize the resulting opportunities for natural resources engineering.
- Comprehend the vast scope of natural resources engineering by defining and solving problems affected by the hydrologic cycle.
- Determine the similarities in agriculture- and urbanization-related problems with respect to potential environmental impacts, enabling one to visualize the opportunities and challenges created as public and societal needs interact with natural realities.

Defining Natural Resources Engineering and Related Terms

Natural resources engineering is the design of planned activities that complement or oppose natural and societal forces and that lead to modifications of the soil, water, biota, and/or air environment. The natural forces arise from the hydrologic cycle. Societal forces stem from the desires of people. The problem space is the farm or field scale. The purpose is resource development and/or environmental management.

This definition broadens the scope of natural resource utilization from the usual oil/gas extraction activity to include crop production and urban development activities. Ancillary site development activities associated with bioremediation, bioconversion, and resource extraction are also covered by the more general definition. The following terms often appear in the literature of natural resources engineering:

Bioconversion: Biological conversion; the biologically mediated physical and chemical conversion of municipal, agricultural, and industrial organics and other by-products into useful products.

Bioremediation: Biological remediation; the application of plants or microbial organisms, or other biological amendments, to sequester or remove chemical or biological toxins from a soil body.

Environment (ISO 14001): The surroundings in which organisms operate, including air, water, land, natural resources, flora, fauna, and humans.

Extensive agriculture: Traditional agricultural enterprises, with production measured by yield units per hectare (e.g., field scale) and with minimal environmental controls. Examples include forage production, row crops, forest systems, and some animal production systems.

Farm, field, and factory scale: Refers to typical problem size; different from industrial, regional, and global scales, on the one hand, and greenhouse, room, and microbial scales, on the other. Agricultural engineers manage soil and water at the farm and field scales. Though many biological engineers will work at the room (e.g., greenhouse) scale and microbial scale in the future, they will also find remediation and waste conversion opportunities at the field and farm (or factory) scale.

Hydrology: The scientific study of water, its properties and distribution in the atmospheric, surface, and subsurface realms.

Intensive agriculture: Agricultural enterprises with production or yield measured in units per square meter (e.g., room or greenhouse scales) and with considerable environmental controls. Examples include greenhouse production systems and concentrated animal production systems. Future natural resources engineers will work in this domain.

Urbanization: The transformation of land from natural or agricultural use to intensive commercial, residential, mining, or industrial uses. For example, urbanization occurs when natural forest, managed forest, or grasslands are developed for housing. One may make the case that problems arising from intensive agricultural enterprises and from urbanization are similar.

The Hydrologic Cycle

Water and wind are the driving forces for production and pollution. Thus, one must be concerned with the hydrologic cycle. For example, consider the continental United States. The equivalent depth of water passing over the United States in the atmosphere is 300 in. Average precipitation over the U.S. landmass is about 30 in (762 mm), partitioned as follows:

26 in (660 mm) as rain
4 in (102 mm) as snow, sleet, hail
9 in (229 mm) percolates to groundwater or runs off
21 in (533 mm) returns to the atmosphere
0.73 in (18.5 mm) is consumptively used

Agriculture consumes 83% of the consumptively used water; competition is increasing. Agricultural irrigation requires about 40% of consumptively used water; of this, 40% is lost due to inefficient irrigation and due to seepage below the root zone.

Corresponding partitions for the world are (Maidment 1993)

31 in (800 mm) falls as precipitation
12 in (320 mm) runs off to land
5 in (130 mm) runs off directly to oceans

Figure 1.1 schematically shows the hydrologic cycle. In the hydrologic cycle:

- Water falls to earth as precipitation: rain, drizzle, snow, sleet, hail. Water also forms directly by condensation (dew).
- Plants may intercept precipitation reaching the earth's surface; water may infiltrate and percolate into the soil, run off the surface, or evaporate.
- Evaporation may occur directly from the precipitation, from plant leaves (transpiration), from the soil surface, from storage structures, from streams and other water bodies, and from the ocean.
- Winds transport moisture, particulates (wind eroded and those emitted by human activities), and odors.
- Plants use infiltrated water (transpiration).

Water may seep to groundwater or seep slowly to streams and surface water bodies. It may also move laterally after infiltrating and reappear on the surface at some point downslope (interflow). A more complete development of the wind erosion and air quality aspects is left to other texts.

Common hydrologic measurements include

- precipitation, by rain and snow gauges
- accumulated snow, by snow surveys
- runoff, by gauging streams

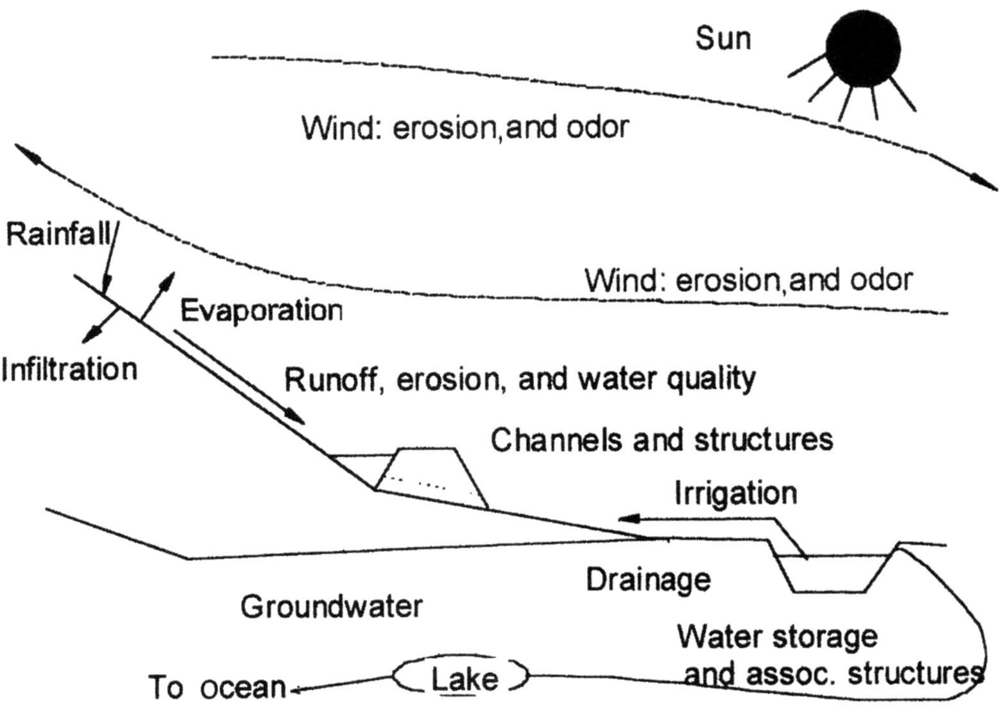

Figure 1.1. Schematic of the hydrologic cycle superimposed on natural resources engineering scope and processes. Topics emphasized in this text are in black.

- evaporation, by evaporation pans; evapotranspiration, by lysimeters
- groundwater level, by monitoring wells
- wind speed and direction (wind may predominantly come from one or multiple directions, depending on location); fully equipped weather stations also contain instrumentation for measuring humidity and solar radiation

A visit to the U.S. Geological Survey (USGS) home page at www.usgs.gov will provide much information relating to runoff and groundwater levels at sites around the United States.

Changing Land Uses Due to Societal Forces

Throughout the world, social forces consisting of increasing population pressure, the desire of developing countries to better their standard of living, growing environmental awareness in developed countries, and people's efforts to improve their living situation combine in various ways to drive land use change. The trends listed in Table 1.1 suggest opportunities for natural resources engineers (NRCS 1985, 1998). These trends include outward expansion of the "urban fringe," the increasing concentration of animal production, and the applications of agricultural and municipal wastes to land. An evaluation of land use and trends motivated the U.S. Department of Agriculture (USDA) Soil Conservation Service to change its name and focus. The Soil Conservation Service is now the Natural Resource Conservation Service (NRCS).

In the United States, we are diverting forests and prime farmlands to nonagricultural uses (urban areas, reservoirs, etc.). Change in land use implies change in land ownership. Change in land ownership or use frequently triggers legally mandated assessments related

Table 1.1. Synopsis of U.S. land use changes over the last 40 years.

Land use category	Change from 1958 to 1977	Trend from 1977 to 1997
Cropland	−8%	Continued decrease
Pastureland	+12%	Leveled off
Forest land	−18%	Some increase
Urban land	+76%	Accelerating increases
Open water	+27%	Leveled off

to environmental appropriateness for the intended use and, in some instances, financial assistance for making the change. A visit to the USDA National Resources Inventory at www.NRCS.USda.gov will yield up-to-date information on trends in land use change.

The application of soil conservation practices to cropland in the United States is increasing. Between 1996 and 1997, the use of field border and filter strips increased by 5100 acres. Grassed waterways increased by 20,700 acres. A total of 1.6 million acres are currently managed with border strips and waterways. Contour farming increased by 278,100 acres. Terraces, diversions, hillside ditches, and water/sediment control basins increased by 72,300 acres. Strip-cropping for wind erosion control decreased. However, the area protected by wind erosion control buffers increased by 17.9% to 320,900 acres.

In agriculture, we collectively must be concerned with the loss of prime farmland to nonagricultural uses. Prime farmland is the most productive land, the least erodible. The use of other types of land for crops requires increased inputs with possible detrimental environmental impacts. Increased human populations will require increased production of food and fiber. Urbanization intensifies air quality and water quality problems, intensifying the need for air and water quality control measures. Thus air, water, and land quality issues will continue to gain in importance.

Urbanization results in cover removal and exposure to erosive rainfalls. Urbanization also increases the population segment having heightened sensitivities to clogged streams and other pollution. The change of land use in and around another owner's real estate provides opportunities for natural resources engineers. One must anticipate—in the problem space—outcomes of management changes on soil, water, air, and biotic state. The alteration may be simply a change in crop rotations. The alteration may also involve moving from an agricultural use to a residential, commercial, waste treatment, or remediation application. Our knowledge is at best approximate. Cooperation among disciplines and decision makers is critical.

Accelerated land use change is a worldwide phenomenon due to the 2% annual increase in population and to the efforts of developing nations to catch up with developed countries. In addition to all the U.S. land use change motivators, developing nations around the world face major problems related to the creation or expansion of infrastructure (e.g., roadways), production agriculture, and associated postharvest handling facilities such as grain elevators, fruit/vegetable packing plants, port facilities, and other appropriate storage and shipping facilities. Environmentally conscious production and manufacturing are commonly not high priorities. This contributes to water and air quality problems. Developing countries often employ agricultural and manufacturing technologies and approaches that were common in the United States some 30, 50, or as much as 100 years ago, though many are struggling valiantly to catch up even while undergoing tremendous population growth. Overgrazing in Haiti, loss of tropical forests in Brazil and Malaysia, urban encroachment in Europe, and development in China are having or shortly will have repercussions on these societies and, indirectly, the world.

Governmental regulatory agencies are the primary vehicle for coupling land use changes with opportunities for the natural resources engineer. In the United States, the Clean Water Act and the Safe Drinking Water Act regulate point source pollution discharges. Agencies are now requiring municipalities requesting permission to increase discharge volumes to demonstrate lack of impairment of receiving streams. This requirement entails consideration of nonpoint pollution loadings (e.g., from agriculture and storm water runoff). The determination of the total maximum daily loading (TMDL) for a stream requires an accounting and modeling of both point and nonpoint loadings. Sophisticated modeling approaches are required to evaluate best management practices (BMPs). Engineers will be heavily involved in evaluating BMPs and in modeling their effectiveness in managing pollution loadings. Similar approaches may be adopted in developing countries as their pollution problems manifest.

The Scope of Natural Resources Engineering

In this section we will discuss problem areas portraying the scope of natural resources engineering. The purpose of this section is to clearly establish the relevance of diverse topics to agricultural, biological, and civil engineers. Each example listed below ties in with one or more components of the hydrological cycle.

Flooding

Figure 1.2 shows severe flooding. Flooding in upstream watersheds is a major natural phenomenon. Its consequences become more significant as urbanization occurs. It causes loss of life, property damage, crop loss, health hazards, loss of services, and lack of access to remote areas. Flood damage in rural areas such as shown in Figure 1.2 costs an estimated

Figure 1.2. Flooding in an urbanized rural area.

$2 billion annually in the United States. Losses may increase with continued development. The federal government funds most flood control projects. The Georgia floods of 1994 and the midwestern floods of 1993 are recent examples of the devastation caused by flooding. Floods of this magnitude are regional in scale and thus beyond the scope of this text. The text addresses channel design and management issues.

Flood losses are catastrophic in terms of human life and property around the world. As an extreme example, 70% of Bangladesh was inundated by the flooding of the Ganges River in 1998. Tens of millions of people lost their homes, and thousands died from drowning and disease. Deforestation and mountainside cultivation in the Himalayas have increased the amount of sediment carried by the rivers to 3 billion tonnes (metric tons) annually, raising the riverbeds and increasing the likelihood of flooding. Deforestation due to population pressure often precedes catastrophic flooding events.

Erosion Control

Erosion occurs when rainfall dislodges soil particles and excess rainfall (runoff) transports the particles off-site. Figure 1.3 shows an example of catastrophic erosion near a construction site. Terracing, conservation tillage, and application of soil stabilizers are erosion control strategies appropriate to the urban or rural setting. Control of wind and water erosion is necessary to maintain high levels of production. Wind erosion control reduces sediment and chemical pollution. Croplands in the United States produce approximately 4800 tons of eroded soil per square mile annually. The annual erosion from agricultural fields and pastures is about 1 m from one thousand 200 ha farms (3,800,000,000,000 kg at a density of 1900 kg/m^3, or 4,189,500,000 tons). Construction on a site may increase erosion from 2 to 40,000 times (Goldman et al. 1986). Erosion from cropland and construction is declining due to control measures (NRCS 1998). The eroded material contains higher proportions of fine particles, organic matter, and chemicals than the original soil. This affects soil productivity and runoff water quality. Erosion by wind can also create air quality problems.

Figure 1.3. Runoff and erosion damage on a construction site.

Worldwide, of the 6 billion hectares of arable land, water erosion affects approximately 1.1 billion hectares and wind erosion affects 0.5 billion hectares. Estimates of global soil loss rates range from 0.09 mm/yr to 0.3 mm/year (Lal 1994). A loss rate of 0.1 mm/yr from 6 billion hectares is 1.14×10^{14} kg. Federal legislation in 1978 mandated that states must develop plans for controlling sediment pollution from nonpoint sources. The 1985 Federal Farm Security Act encouraged farmers to adopt conservation practices.

Thirty years ago, much time would have been devoted to mechanistic approaches to soil and water conservation engineering. In many parts of the world, the mechanistic approaches continue to be appropriate. Chemical-based weed controls and accompanying tillage systems effectively address the erosion problem on many field crops. Challenges remain for specialty crops such as vegetables. Mining reclamation and forestry production have unique problems germane to biological, agricultural, and civil engineers.

Water Quality Renovation and Management

Poor water quality is the consequence of the transportation of surface or subsurface pollutants by excess rainfall, or runoff. Water quality suffers when pesticides, nutrients, organics, industrial pollutants, or soils become suspended or dissolved in water. Pollution constituents are chemical elements in undesirable places. Pollution is a sign of inefficiency (Risse and Cheadle 1996). The development and adoption of best management practices and environmentally conscious manufacturing practices have improved water quality in the United States. Nevertheless, researchers (Leeden et al. 1990) have detected at least 16 different pesticides in the groundwater in 26 states. Water quality is a continuing problem in much of the developing world.

We must continue to reduce the adverse impacts of agricultural and other practices on water quality. Wetlands, riparian zones, and vegetation strips trap sediments and remove nutrients and organics. Careful planning and management of large livestock facilities concentrating large quantities of animal manure are essential. Waste lagoons and storage structures are an important component of waste handling in concentrated animal-feeding operations. Figure 1.4 shows a waste lagoon for a poultry production operation. Erosion due to urban construction is an issue in water quality. The common approach for dealing with these problems is to extend the data base for agriculture to include these activities.

Research identifying technologies that reduce inputs, conserve resources, or lead to useful by-products is a high priority. Pollution from animal waste constituents and pesticides and other water quality problems will continue to grow due to urbanization and concentrated animal production. Comprehensive nutrient management and storm water management plans are increasingly important in rural and urban areas. Best management practices often involve engineered structures such as channels, lagoons, and catch basins.

Irrigation

Crops and turf must be irrigated for reliable growth even in humid areas. The timing and distribution of rainfall often impose stress at critical times. Irrigation compensates for a lack of timely rainfall. Managed irrigation increases crop yields, enhances germination and emergence, moderates air temperatures, and applies nutrients and pesticides. Drip

Figure 1.4. Waste treatment lagoon for a poultry operation. (Courtesy of Georgia Soil and Water Conservation Commission.)

irrigation for intensive specialty-crop production and sprinkler irrigation for turf are two important types. Figure 1.5 shows a center pivot system using sprinklers. Irrigation requires large quantities of water to overcome high evapotranspiration to avoid salinity buildup.

In the United States, 8% of cropland is irrigated, and that 8% produces 25% of the total value of farm crops. Most of the irrigated acreage is in the West. The irrigated area of the world, expanding at 1% per year, is about 260 million hectares, with approximately 19 million in the United States (Hoffman et al. 1990). Engineering support is

Figure 1.5. Center pivot sprinkler irrigation system irrigating soybeans. (Courtesy of Georgia Soil and Water Conservation Commission.)

Figure 1.6. Poor stand of corn due to prolonged high soil moisture levels.

needed in developing countries, where less than an estimated 2% of potentially productive land is currently irrigated. Most irrigation statistics do not include nonstandard irrigation such as bioconversion applications, crop cooling, etc.

Drainage

Drainage is important both in crop production and in many remediation projects. The need for drainage results from excess rainfall. Drainage can be surface or subsurface. Agricultural drainage is often necessary (see Figures 1.6 and 1.7) for high levels of production and timely field operations. Modern surface drainage systems are water management systems for providing drainage and supplemental irrigation water.

Figure 1.7. Poor drainage hinders timeliness of tillage and planting operations.

In arid areas, salinity remediation demands subsurface drainage. Worldwide, of 260 million irrigated hectares, 60 million hectares suffer from salinization (Jensen 1993).

In 1985, 96% of all cropland drained in the United States was concentrated in 23 states. Sixty percent of land drained was in organized drainage districts. Thirty-four percent of drainage involved subsurface (pipe) drainage. Schwab et al. (1993) estimated that 22 million hectares of cropland could benefit from drainage. They also estimated that 75% of the land currently drained will need replacement drainage systems due to system failure. Failures occur due to poor design, outdated design life of older technologies, and lack of maintenance. Responding to a growing ecological appreciation of wetlands, however, the "swamp buster" provisions of the 1985 Food Security Act severely restricted additional U.S. drainage efforts.

Bioremediation

Novel bioremediation techniques are now in use in many parts of the developed world. Bioremediation is the application of plant, microbial, or other biological organisms to sequester or remove chemically or biologically toxic materials from a soil body. Organic petroleum distillates and other solvents and pesticide residues are candidates for treatment.

Groundwater pollution, like surface pollution, starts as point source pollution. It becomes a nonpoint remediation problem due to transport. Many bioremediation projects occur on a farm scale with respect to area. The bioremediation method shown in Figure 1.8 is an extension of simple groundwater hydraulic analyses.

The in situ bioremediation technique, shown schematically in Figure 1.8, uses microbes to convert contaminants to innocuous constituents. The technique may involve the application of

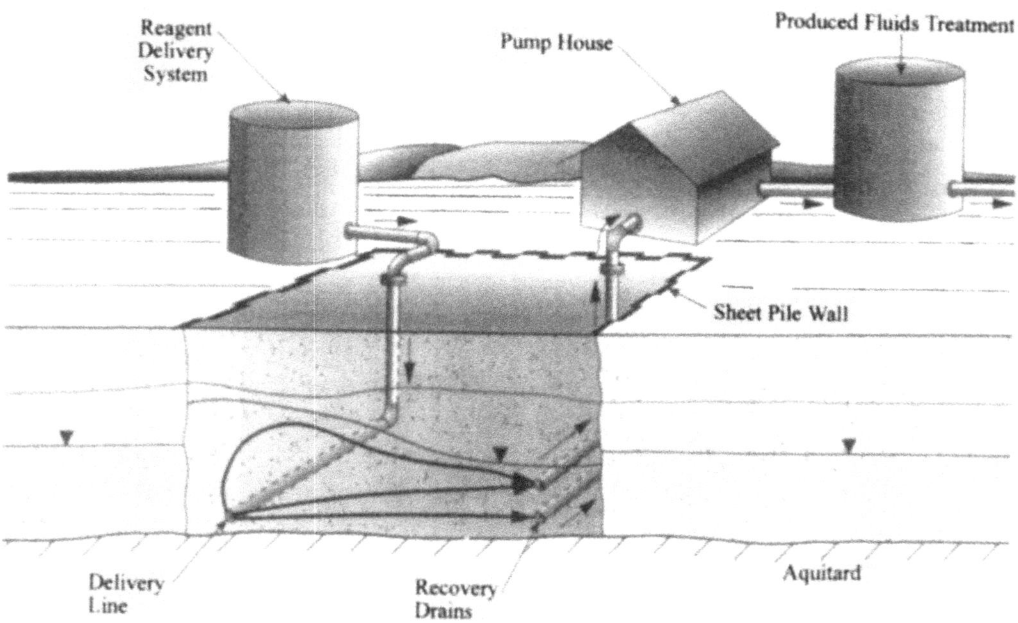

Figure 1.8. Cutaway view of a bioremediation system where indigenous microbes within the sheet piling boundary are stimulated with a nutrient mixture to metabolize soil organic pollutants. Drainage and well flow hydraulics explain the flow movement through the soil. Pipe flow hydraulics explains the fluid movement through the piping. (From Nyer et al. 1996; used with permission of Lewis Publishers.)

nutrients and electron acceptors. It may involve adjustment of pH, temperature, and/or other microbial growth factors. Selected petroleum hydrocarbons are successfully treated using microbial bioremediation. Polyaromatic hydrocarbons present in heavier petroleum blends degrade at slower rates. Critical to the design is knowledge of the biochemical flow paths and rates. The groundwater flow must be sufficient to deliver the nutrients at required rates.

Other texts, such as Hillel 1998, Tindall and Kunkel 1999, Schwarzenbach et al. 1993, and Odum 1983, provide excellent background on biological sustainability and interactions between soil, water, and air. Excellent Internet resources for natural resources engineering topics may be found on the World Wide Web.

Air Quality

As the scale of concentrated agricultural enterprises increases, air quality becomes a larger issue. Odor emission problems arise with large poultry and swine operations in the southeastern United States. Beef feedlots create odor problems in the Midwest. Composting and bioconversion operations are concerned with odor and particulate emissions. Buffer strips are one technique for managing odor and particulate problems. The discipline of industrial air quality engineering is highly developed. Natural resources engineers can help to modify appropriate industrial air quality control strategies to make them applicable to the farm and field scale. Detailed air quality engineering is beyond the scope of this text. Works such as Cooper and Alley 1994 and Rafson 1998 provide strategies for industrial air quality management, many of which are transferable to the farm and field scale.

Watershed-Stream Assessment

Many aspects of natural resources engineering are driven by the requirement to prevent various impending disasters. Most emphasis in the past has been on moderating the effects of too much or too little rainfall. Issues of water quality are now coming to the forefront. Sophisticated modeling techniques provide the capability to model the response of entire watersheds to various climatic events and imposed land uses, and thereby prevent disasters associated with environmental degradation. Engineers, along with ecologists, geologists, and other professionals, are devoting more time to developing whole-farm comprehensive nutrient management plans wherein enterprise suitability is evaluated based on the capacity of the land to receive nutrients. Suitability depends upon such factors as slope, depth to bedrock, and proximity to surface water bodies and wells (see Figure 1.9). Engineers are involved with quantifying, describing, and restoring natural streams (see Figure 1.10).

Natural resources engineers are working with municipalities to develop comprehensive agricultural nutrient management plans and industry/municipality storm water discharge management plans. Natural resources engineers will take a lead role in the management of nonpoint source pollution in the coming decades. Modeling approaches applicable to watershed assessment are presented throughout the text. Adequate assessment is a precursor to ecological engineering.

Ecological Engineering

Changing land uses, increasing environmental awareness, increasing competition for land and water, and increasing human populations motivate society to demand the best solutions.

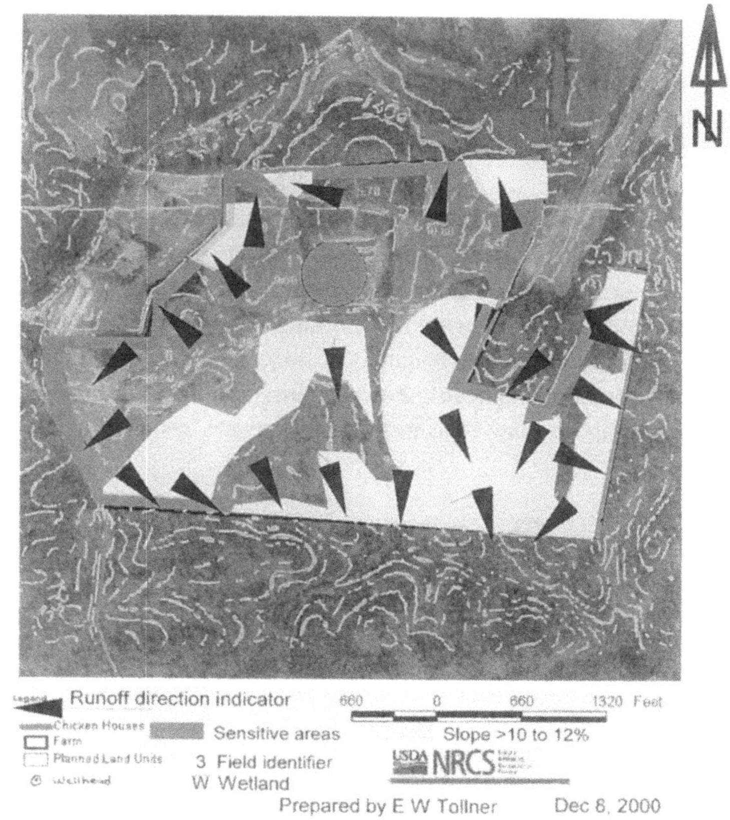

Figure 1.9. Aerial photograph showing a broiler production operation and surrounding fields in north Georgia. A topographic map has been superimposed enabling delineation of slopes. A soils map (not shown) may reveal additional unsuitable zones for waste application. (Courtesy of Dr. Mark Risse, Department of Biological and Agricultural Engineering, University of Georgia.)

Figure 1.10. Data collection for a physical, biological, and chemical stream assessment. (Courtesy of Dr. David Gattie, Department of Biological and Agricultural Engineering, University of Georgia.)

The ecological engineer is distinguished from the conventional civil and agricultural engineer by a focus on holistic, as opposed to linear, approaches. The ecological engineer concentrates on ways to foster system self-design (see below), whereas the conventional civil and agricultural engineer concentrates on the linear, unit operations approach. The natural resources engineer works with ecologists and other professionals to develop environmentally sustainable systems. The ecologist tends to view field-scale environmental problems through the lens of a macroscopic control volume with somewhat indistinct boundaries, such as shown in Figure 1.11. Consider a control volume with the bottom boundary at some soil depth, the top boundary at some aerial height, and the other boundaries facing north, east, south, and west. Property lines or problem extent dictate the problem space. Our inability to draw clear distinctions between resource management issues

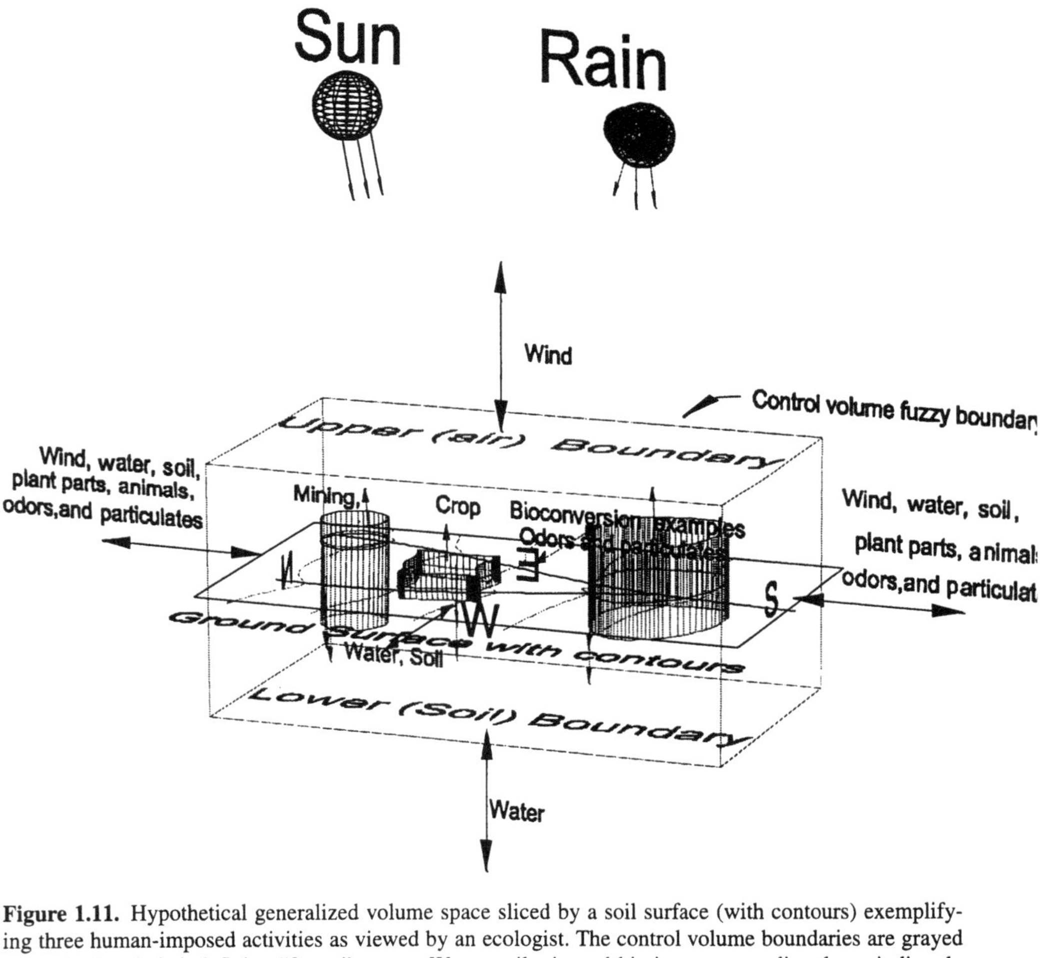

Figure 1.11. Hypothetical generalized volume space sliced by a soil surface (with contours) exemplifying three human-imposed activities as viewed by an ecologist. The control volume boundaries are grayed to symbolize their indefinite, "fuzzy," nature. Water, soil, air, and biotic states are directly or indirectly managed at the fuzzy side boundaries as suggested by the fluxes (not all of which are shown). The lower fuzzy boundary is the depth at which soil removes all mass except possibly water. The upper fuzzy boundary marks where odors and particulates are ignored. Box boundaries are fuzzy because of the lack of crisp distinctions between environmental management and resource management. The system within the indicated volume is both manipulated by the forcing activities and reacting to the forcing functions according to various relationships describing the system. Boundaries become less fuzzy as the subsystem within the box becomes smaller and more controlled. This text will be concerned mainly with surface processes but will give introductory overviews and some discussion of atmospheric and subsurface processes.

and environmental management issues may make the problem boundaries fuzzy. Air, water, soil, and biotic material represent states within the volume.

Key concepts of ecology are "self-design" and a holistic viewpoint, as opposed to imposed design and highly mechanistic viewpoints. Nomenclature appropriate for describing a given situation may depend on the scale of the problem. Engineers and quantitative ecologists are familiar with the use of differential-equation nomenclature to describe the natural resources engineering space. The initial condition represents the state of the water-soil-air-biota system at time zero, the time we engage the problem. Sites require preliminary assessment (e.g., watershed and stream) to define the relevant initial physical, biological, and chemical states. Fluxes through any of the boundaries result from various gradients or levels of the state variables existing around the boundaries. A suite of generalized, yet unspecified, differential equations or other chaotic functional forms describes the constituent movements within the volume. This represents the self-design approach associated with ecological engineering. Much effort is under way to better understand the paradigms associated with self-design.

The description in Figure 1.11 is of limited quantitative value for current problem solutions because the differential-equation forms are unknown for biotic components. Figure 1.11 is a vision of how ecological, mathematical, and engineering research may lead to workable, ecologically sound designs/management at the field and watershed scale. Self-design in the problem space begins with each natural or artificial perturbation in the problem space. The ecological engineer controls and manipulates the boundaries of the problem space, allowing self-design to affect desired transformations in the problem space. A discussion of ethics, natural resources economics, and ecological principles can be found in Appendix A.

Ecological engineering is by nature less intrusive than other engineering approaches. Our knowledge of how to adequately practice ecological engineering is in its infancy. The less intrusive approach of ecological engineering will become more predominant in the practice of natural resources engineering as the application of ecological engineering principles is better understood. The ecological engineer must have an excellent grasp of the holistic problem space and at the same time must have excellent command of various unit operations within that space.

Outlook

Population increases, the desire for a better standard of living, and environmental awareness are among the societal forces that put increasing pressure on existing land resources for sustainable production. Societal forces operate in the context of natural forces, stemming mainly from the hydrologic cycle. Shortsighted environmental remedies invariably lead to serious pollution in developed countries. Wrong remedies for economic problems may lead to devastating deforestation and overgrazing in undeveloped countries, creating the potential for serious flooding and erosion and consequent further loss of production capability. Unplanned, unmanaged urban development often leads to similar effects. Truly sustainable production requires systems approaches involving ecological input, public education, and appropriate technology introduction. *Truly sustainable agricultural and urban development requires systems-based planning involving understanding of the hydrological cycle* (Hes and van Lier 1999).

The current state of the art of natural resources engineering is rapidly moving forward along with our knowledge of ecological processes. Pollution represents inefficiency and excess cost (see Puri 1996; Tibor and Feldman 1997). Many agricultural and manufacturing sectors accept that pollution is inefficient and costly. This understanding led to ISO 14000, an environmental standard that was developed by the International Standards Organization (ISO) and that outlines a method for achieving continuous improvement in environmental impacts of manufacturing practices. ISO 14000 sets protocols for tracking all material inflows, product outflows, and transformations within the manufacturing operation. The ecological paradigm outlined above reflects a similar view for natural resources management and development. Natural resources engineers will find employment with firms that implement environmental management systems, risk assessments, and quality audits.

Figure 1.12 presents types of problems encountered in natural resources engineering. Agriculture and urban-related activity impose some similar conditions in a field-scale space. Environmental impacts are similar. The hydrological cycle affects both cases. Both cases produce runoff affecting downstream users. Each requires either irrigation or water application engineering. Environmental preservation and management will call upon specialists in both types of production. One can easily visualize interfaces developing between agricultural and urban activities since these often occur in close proximity. This has led to the coining of the phrase *rural-urban interface*. Shown in Figure 1.12 are only two of many project types requiring the skills of the natural resources engineer. The natural resources engineer contributes to process definition, process maintenance, and flux anticipation and manipulations at the problem boundaries, all to achieve satisfactory self-design within the problem space.

The immediate goal of this text is to provide the background for intelligent use of sophisticated, commercially available software packages for natural resources engineering design and watershed assessment. The natural resources engineer must learn the unit operations with an eye toward holistic systems applications. Most topics shown in Figure 1.1

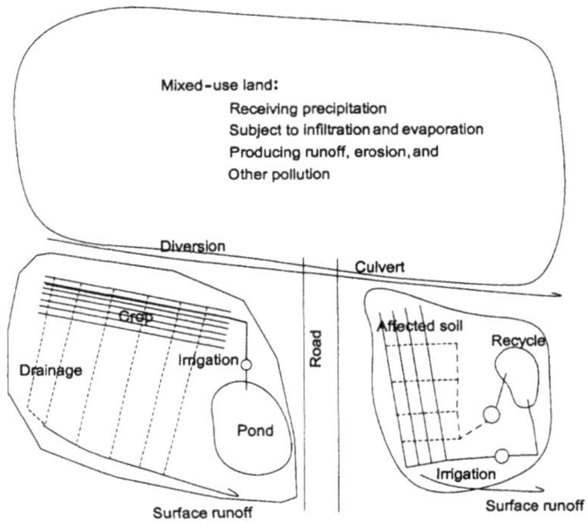

Figure 1.12. Sketch showing similarities between conservation resource management and bioremediation on the farm/field scale. Each involves water, soil, and air components of the natural environment, resulting in similarity in environmental impact. Although crisp boundaries are conveyed in a drawing, they are fuzzy in reality.

are discussed in the following chapters, which have specific goals and numerous hand calculation examples. Some commercial and agency software models are mentioned throughout the text, not for endorsement purposes but as examples.

Problems and Questions

1. Define the main components of the hydrologic cycle. Describe how drainage, irrigation, soil erosion, bioconversion, and bioremediation each relates to the hydrologic cycle.
2. Why should the biological engineer be concerned with the macroscopic (e.g., nonmicroscopic) approaches emphasized in this text?
3. Describe what is meant by the rural-urban interface. Do a web search to find your discussion resources.
4. What is meant by *farm/field scale*? Compare and contrast the farm/field scale with the global, room, and microbial scales. Start the comparison by defining *global, room,* and *microscopic scales.*

Note: Consult the Appendix A discussion on ethics before answering the ethics-related questions below. See a text such as Underwood 1998 or *National Agricultural Biotechnology Council Bio-Ethics Institute Handbook* (1999) for a collection of environmental case studies. These may be helpful in answering some of the questions below.

5. Describe your views on the need to conserve trees versus the desirability of maintaining a viable forest products industry. Discuss this topic in the light of current changes in land use trends.
6. Describe your views on the need to protect endangered species versus economic development (e.g., the spotted owl controversy). Identify the views forming the foci of the selected controversy. Discuss in the light of bioresource ethics.
7. Why should the agricultural engineer be concerned with trying to satisfy "city folks," whose noses may be "hypersensitive" to odors associated with animal production? Discuss in the light of current changes in land use trends and bioresource ethics.
8. What does the term *bioresource ethics* mean to you?
9. Why should one bother (or not bother) with saving endangered species? Consider this question using views listed in Table A.2 of Appendix A. Add views to this table if you find yours missing.
10. Visit the website and work your way through maps and other resources available. Give a comprehensive discussion of the rural-urban interface issues touched upon by these resources.
11. Compare and contrast the ethics associated with a forest products industry versus the ethics associated with cutting old- or new-growth trees.
12. Discuss the pros and cons of the suggested resolution to the dilemma of the city of Blissful in case study 2 of Appendix A.
13. Summarize in a paragraph the intent of ISO 14000 using references such as Puri 1996 and Tibor and Feldman 1997.
14. Select an endangered species in your area and analyze the impact of a hydrological modification (your instructor will pick the relevant modification) on the viability of the species. What have we lost if the species goes extinct? Does a plant or an animal

have legal standing? Define the problem boundaries. Define the ecological "forcing functions." Define the control measures. Define the boundary fluxes and the sources/sinks.

15. Write a 200- to 300-word essay describing your concept of natural resources engineering.

References

Biswas, A. K. 1970. *History of hydrology.* North-Holland Publishing Co., Amsterdam, Holland.

Cooper, C. D., and F. C. Alley. 1994. *Air pollution control: A design approach.* 2d ed. Waveland Press, Prospect Heights, IL.

Epstein, E. 1997. *The science of composting.* Technomic Press, Lancaster, PA.

Goldman, S. J., K. Jackson, and T. A. Bursztynsky. 1986. *Erosion and sediment control handbook.* McGraw-Hill, New York, NY.

Hes, M. B. A., and H. N. van Lier. 1999. Land and water use planning. In *CIGR handbook of agricultural engineering: Land and water engineering.* American Society of Agricultural Engineers, St. Joseph, MI.

Hillel, D. 1998. *Environmental soil physics.* Academic Press, New York, NY.

Hoffman, G. J., T. A. Howell, and K. H. Solomon, eds. 1990. *Management of farm irrigation systems.* American Society of Agricultural Engineers, St. Joseph, MI.

Jensen, M. E. 1993. The impacts of irrigation and drainage on the environment. Fifth N. D. Gulhati Memorial Lecture. International Commission on Irrigation and Drainage, The Hague, Netherlands.

Jorgensen, S. E., and W. J. Mitsch. 1989. Ecological engineering principles. In *Ecological engineering: An introduction to ecotechnology,* ed. W. J. Mitsch and S. E. Jorgensen. John Wiley & Sons, New York, NY.

Lal, R. 1994. Soil erosion by wind and water: Problems and prospects. In *Soil erosion research methods,* 2d ed., ed. R. Lal. Soil and Water Conservation Society, Ankeny, IA.

Leeden, F. van der, F. L. Troise, and D. K. Todd. 1990. *The water encyclopedia.* Lewis Publishers, Chelsea, MI.

Maidment, D. R. 1993. Hydrology. In *Handbook of hydrology,* ed. D. R. Maidment. McGraw-Hill, New York, NY.

National Agricultural Biotechnology Council Bio-Ethics Institute handbook. 1999. Notes distributed at the North Carolina State University Bio-Ethics Institute Workshop, 22–27 May 1999, Raleigh, NC.

NRCS. 1985. *America's soil and water: Conditions and trends.* U.S. Department of Agriculture, Washington, DC.

———. 1998. *Stream corridor restoration: Principles, processes, and practices.* USDA Natural Resource Conservation Service, http://www.usda.nrcs.gov, Washington, DC.

Nyer, E. K., D. F. Kidd, P. L. Palmer, T. L. Crossman, S. Fam, F. J. Johns II, G. Boettcher, and S. S. Suthersan. 1996. *In situ treatment technology.* Lewis Publishers, Boca Raton, FL.

Odum, E. P. 1983. *Basic ecology.* CBS College Publishing, Philadelphia, PA.

Puri, S. 1996. *Stepping up to ISO 14000: Integrating environmental quality with ISO 9000 and TQM.* Productivity Press, Portland, OR.

Rafson, H. J. 1998. *Odor and VOC control handbook.* McGraw-Hill, New York, NY.

Risse, L. M., and S. A. Cheadle. 1996. Pollution prevention in agricultural livestock production. Report submitted to the Georgia Pollution Prevention Assistance Division, Georgia Department of Natural Resources, Atlanta, GA.

Schwab, G. O., D. D. Fangmeier, W. J. Elliot, and R. K. Frevert. 1993. *Soil and water conservation engineering*. 4th ed. John Wiley & Sons, New York, NY.

Schwarzenbach, R. B., P. M. Gschwend, and D. M. Imboden. 1993. *Environmental organic chemistry*. John Wiley & Sons, New York, NY.

Tibor, T., and I. Feldman. 1997. *Implementing ISO 14000*. McGraw-Hill, New York, NY.

Tindall, J. A., and J. R. Kunkel. 1999. *Unsaturated zone hydrology for scientists and engineers*. Prentice Hall, Upper Saddle River, NJ.

Underwood, L. 1998. *Case studies in environmental science*. Saunders College Press (Harcourt Brace), New York, NY.

2

Precipitation

Precipitation—rain, snow, drizzle, sleet, hail—is crucial to life on earth. Too much or too little precipitation and poor distribution of precipitation amount and form have major undesirable consequences for the production of food, feed, and fiber for humanity's needs. Along with the study of heat and air movement, the study of precipitation belongs to the discipline of meteorology. The distribution of water from precipitation forms the core of the discipline of hydrology. Precipitation is important to natural resources engineers since it determines runoff volume and flow rate. These are determining factors in sizing channels and structures. Precipitation is the catalyst for many runoff and drainage engineering projects. Because most precipitation occurs as rainfall, this text regards the terms *precipitation* and *rain* or *rainfall* as synonymous.

Goals

- Appreciate the wide variety of precipitation types in various climate regions and lay a foundation for how precipitation affects upcoming topics such as runoff and erosion.
- Obtain and analyze precipitation data using hydrological frequency analysis on a regional scale.
- Quantitatively use important concepts such as return period and storm duration.
- Compute precipitation amount and intensity for specified design frequencies and durations.
- Simulate a storm hyetograph useful for runoff and erosion analysis.
- Analyze precipitation data from a densely gauged area.

Precipitation Defined

Precipitation is the primary input to the hydrologic cycle. Atmospheric moisture condenses via mechanisms discussed below to form rain, snow, hail, drizzle, sleet, fog, and dew. Other climatic factors such as wind, temperature, and atmospheric pressure influence the form and quantity of precipitation. Snow forms when the atmosphere is below freezing and moisture moves from the vapor phase directly to the solid phase. Hail forms when lower-level convection currents (e.g., 1000 m) transport water droplets to the upper atmosphere (e.g., 10,000 m), where they freeze. Sleet forms when rain moves through a cold lower atmosphere and freezes before impact. Fog forms when water vapor condenses near the ground. Dew results from water vapor condensation near cold surfaces. Drizzle is precipitation at an intensity of less than 1 mm/hr (0.04 in/hr). Precipitation intensities greater than 1 mm/hr define rain. Reversal of seasonal prevailing winds during the year defines the monsoon. The monsoon rain may persist for days at a time. Monsoon rains occur in Southeast Asia, India, and East Africa. *Rain* is often a synonym for *precipitation* since most precipitation is rain.

Precipitation Characteristics

Raindrops may be as large as 7 mm (0.25 in) in diameter. Uniform droplets of less than 0.5 mm (0.02 in) in diameter define drizzle. Raindrop sizes vary within a storm and the size distribution varies with intensity. For a "typical" storm, Smith (1993) relates the distribution of drop size diameters and kinetic energy to precipitation amounts.

Friction from air resistance deforms raindrops. A droplet more than 5 mm in diameter is unstable. The velocity of fall is dependent upon the size of the drop. Terminal velocities of raindrops range from 5 to 9 m/s ($\approx$ 16–30 ft/s) for droplets from 1 to 5 mm (0.04–0.2 in) in diameter. Terminal velocity V_{term} (m/s) may be approximated by $V_{term} = 3.86D^{0.67}$ where D is the droplet diameter in millimeters (Smith 1993).

Precipitation Mechanisms

Condensation is primarily responsible for rain and other precipitation. Sleet occurs if rain falls through subfreezing air before impact. Snow forms if the atmosphere is below freezing in the zone of condensation. Fog forms when warm, moist air is cooled near the ground surface. Dew forms when moist air contacts a cold surface. The following processes cause condensation: dynamic, or adiabatic, cooling; mixing of air masses having different temperatures; contact cooling; and cooling by radiation.

Dynamic Cooling

Dynamic cooling is responsible for most precipitation. Condensation occurs on small, hygroscopic (i.e., water-loving, or hydrophilic) particles called *condensation nuclei* as turbulent eddies transport the particles up and down within the atmosphere. Recall from thermodynamics that a control volume of air compressed adiabatically (i.e., without heat gain or loss) warms. Conversely, decompression leads to cooling. Pressure decreases with increasing elevation. As turbulent eddies and terrain features such as mountains cause movement of small volumes of air up and down in the atmosphere, condensation occurs on condensation nuclei, leading to precipitation. Particles of ocean salt and combustion products containing sulfurous and nitrous acids are effective condensation nuclei. Water droplets grow due to coalescence or ice crystallization processes until they reach sufficient size (500–4000 microns; 0.0127–0.1 mm) to overcome air resistance and fall to the earth as precipitation. The converse of dynamic cooling, dynamic heating, occurs on the lee side of mountains and causes deserts. Dynamic heating occurs as winds move down mountainsides, causing hot, dry conditions. The Santa Ana winds of California are an example of dynamic heating.

Mixing of Air Masses

The meeting of continental and maritime air masses causes condensation. The final temperature causes the moisture to supersaturate, leading to atmospheric condensation. Air mass mixing is important on the U.S. west coast. Frontal systems mix air masses.

Contact Cooling

Contact-cooling condensation occurs when warm, moist air contacts a cold surface. Dew is the common result of contact cooling. Radiation heat transfer (with clear evening skies) or rapid passage of a cold front over warm surfaces results in condensation at or near the surface (e.g., fog). Although dew is a minor contributor to total precipitation in most regions, some desert areas receive significant precipitation as dew. Some areas, such as the Canadian Maritimes, receive much precipitation from fog.

PRECIPITATION TYPES AND SEASONAL EFFECTS

Precipitation mechanisms usually work in combination to produce various precipitation types. Of the mechanisms discussed above, dynamic cooling is the primary cause of condensation and is responsible for most precipitation. Thus, vertical transport of air masses is a requirement for most precipitation. Three major categories of precipitation follow from the mechanisms causing vertical air motion. The meeting of warm and cool air masses often results in the moisture supply for precipitation.

Convective

Heating of air near the ground surface causes convective precipitation. The heated air expands and rises. As the air expands, it takes up increased amounts of moisture. The warm, moisture-laden air rises, and dynamic cooling occurs, causing condensation and precipitation. The resultant precipitation may have extremely high intensity. This mechanism is important in the southeastern and midwestern United States.

Orographic

Mechanical lifting of moist, horizontal air currents over natural barriers such as mountain ranges causes orographic precipitation. This mechanism is prevalent along the Pacific Ocean coastline.

Cyclonic

Movement of air masses from high- to low-pressure regions results in cyclonic storms. Zones of the atmosphere containing significant moisture differences cause pressure differentials. The displacement of air (molecular weight MW = 28) by water (MW = 18) results in lower regional air pressure.

Meteorologists classify cyclonic precipitation as either frontal or nonfrontal. Precipitation associated with low-pressure areas over land is an example of frontal precipitation. Winds move clockwise around high-pressure areas and counterclockwise around low-pressure areas. Air sinks near the center of the high-pressure area. Air rises in the center of a low-pressure area. The falling air in a high-pressure cell and the rising air in a low-pressure cell interact with the earth's rotation via the Coriolis force, resulting in the respective air mass rotations. Fronts often form at air mass boundaries.

Hurricanes are examples of nonfrontal cyclonic storms. Any low barometric pressure area can produce precipitation as frontal convergence lifts inflowing air. Frontal precipitation results from lifting warm air over cold air. This mechanism is important over continental regions.

Seasonal Effects

The location affects the distribution of precipitation throughout the year. In a Mediterranean climate, most rain will fall in the winter months. The coast of California is an example of the Mediterranean climate type. In a continental climate, precipitation is more nearly uniformly distributed, with more falling in the summer months. Summer precipitation exceeds winter precipitation because rainfall provides more precipitation than snowfall. One exception is in the western mountain ranges of the United States, where up to 90% of precipitation falls as snow. Precipitation distribution can affect construction timing and the selection of cropping practices for reducing soil loss due to water erosion.

Precipitation Measurement

A precipitation gauge measures the depth and, sometimes, the intensity of precipitation falling on a flat surface. Hydrologists classify precipitation gauges as recording or nonrecording. Common recording precipitation gauges include weighing gauges and tipping-bucket gauges.[1] Recording precipitation gauges require consistent service. The trace of cumulative precipitation amount versus time is the starting point for analyses. Precipitation per unit time as a function of time is the slope of the cumulative trace. Hydrologists call precipitation per unit time versus time a precipitation hyetograph. Figure 2.1 shows a weather station containing instrumentation for continuously recording precipitation.

We must properly site gauges to reduce adverse impacts of topography, vegetation, etc. Locate gauges such that the distance from an obstruction is twice the obstruction height. Wind also causes errors in precipitation measurement. One should avoid windswept slopes and should not put gauges on roofs. Gauges should be mounted parallel with the horizontal to prevent cosine errors. Gauges should be checked regularly to remove insect debris and ensure functionality.

The number of gauges per unit area becomes an issue when characterizing regional precipitation. Experience has shown that precipitation data from a single location within an area are not representative of the average depth over an area. The average storm depth is, on average, less than the depth recorded at a single location. Figure 2.2 models the recommended reduction in effective rainfall to be applied when applying a single gauge reading over an area (Hershfield 1961). The depth-area-time relationship reflects the nonuniform reality of area precipitation. The variation increases with short-duration storms. The true storm maximum precipitation is illusory since one can never be certain the maximum occurs over a gauge. Hershfield (1961) prepared Figure 2.2 for use on small areas. Holtan et al. (1962) suggested one gauge per 30 acres for watersheds up to 100 acres. They recommended decreased gauge density as the watershed size increases.

Figure 2.1. Precipitation gauge recording weight versus time.

Equation 2.1, developed using a regression procedure (Hintze 2001), describes Figure 2.2 with an R^2 of 0.96:

$$\text{percent of point rainfall for a given area} = \{(98.86415) + (8.851868^{-2}) \times \text{Area} + (7.626413^{-5}) \times \text{Area}^2 + (53.17964) \times [\text{LN(Durat)}] + (4.856125^{-2}) \times \text{Area} \times [\text{LN(Durat)}] + (5.010615) \times [\text{LN(Durat)}]^2)\}/\{1 + (2.811765^{-3}) \times \text{Area} + (1.137505^{-7}) \times \text{Area}^2 + (.5230235) \times [\text{LN(Durat)}] + (2.521296^{-4}) \times \text{Area} \times [\text{LN(Durat)}] + (5.282063^{-2}) \times [\text{LN(Durat)}]^2\} \quad \textbf{(2.1)}$$

Area in the above relationship is in square kilometers and duration (durat) is in hours. For small watersheds and areas of less than, say, 10 km^2, one or two gauges are often adequate.

Example 2.1: The precipitation gauge located within a 625 km^2 watershed recorded a depth of 76 mm over a 6 hr period. What was the average depth received over the watershed from this storm?

The above relationship (equation 2.1) gave a point-area percentage of 0.84. Thus the effective precipitation was 76 mm (0.84) = 63.88 mm.

Measurement of snowfall is important for predicting future water supplies in many areas. Snowpack measurements are used in predicting possible spring flooding. The water content of freshly fallen snow can vary from 40 mm to 400 mm per meter of snow. A commonly

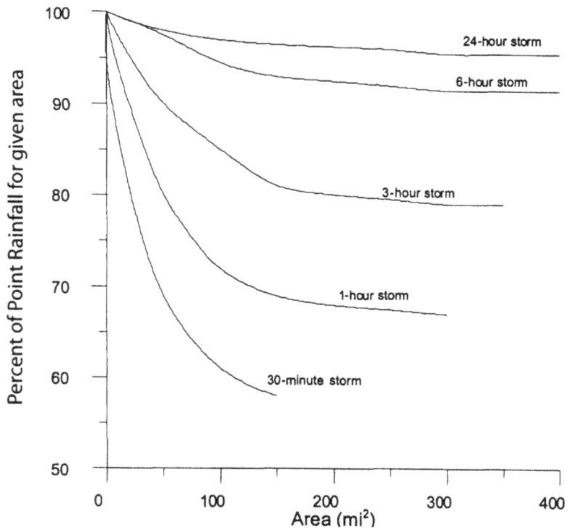

Figure 2.2. Area-depth curves in relation to point rainfall for use in cases where regional gauge density is very low. Note that 1 mi^2 = 2.591 km^2 (From Hershfield 1961.)

accepted value is 100 mm per meter (10%), or 1 in per 10 in of snow. Water yields in compacted snow may reach as high as 30–50%. Apply this percentage factor to the measured snowfall to arrive at the water yield.

One method of measuring snowfall is to place antifreeze in a standard precipitation gauge. The snow melts upon entrance and the water equivalent of the snow is measured. Hydrologists also sample the snowpack to assess snowfall.

Precipitation Analyses

Precipitation data analyses are important in research applications where local precipitation trends are of interest. Studies correlating precipitation to runoff and other variables require accurate local analyses. Figure 2.3 is a hypothetical recording precipitation gauge output. The graph in Figure 2.3 suggests the definition of precipitation intensity.

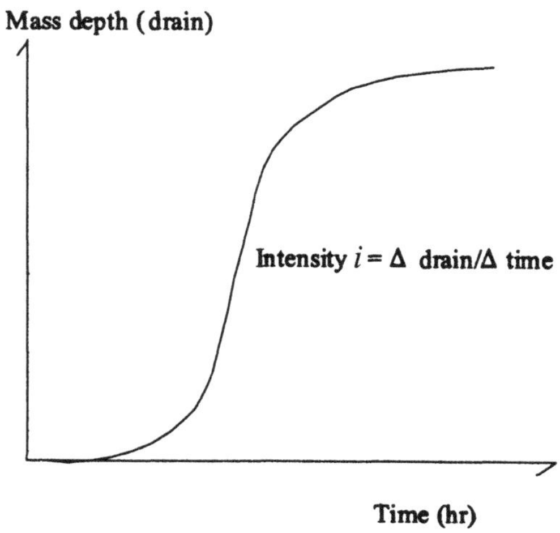

Figure 2.3. Hypothetical rainfall amount vs. time curve.

Example 2.2: Given the precipitation data shown in columns A and B of the spreadsheet in Figure 2.4, compute precipitation intensity as a function of time.

A Time	B Rain gauge clicks	C Cum. rain (in)	D Cum. rain (mm)	E Intensity (mm/hr)	F 30 min intensity (mm/hr)
6:00	0	0	0	—	—
05	2	0.02	0.508	6.096	—
10	2	0.02	0.508	0	—
15	3	0.03	0.762	3.048	15.24
20	7	0.07	1.778	12.192	24.384
25	20	0.2	5.08	39.624	26.924
30	30	0.3	7.62	30.48	28.956
35	50	0.5	12.7	60.96	28.956
40	55	0.55	13.97	15.24	23.368
45	60	0.6	15.24	15.24	18.288
50	64	0.64	16.256	12.192	—
55	66	0.66	16.764	6.096	—
7:00	66	0.66	16.764	0	—

Col A is the time scale.
Col B is the logged rain gauge clicks.
Cols C and D are the cumulative rainfall based on 0.01 in (0.254 mm) per click.
Col E is the computed intensity computed by differencing the cumulative rain and dividing by 5/60 hr.
Col F is the 30 min intensity found by differencing the cumulative rainfall over a 30 min interval, centered about the respective times, and dividing by 0.5 hr. The significance of this number will become apparent in the erosion discussion. (see Chapter 6)

Figure 2.4. Rainfall intensity calculations for Example 2.2 in spreadsheet form.

Recording precipitation gauges give cumulative precipitation versus time. Intensity is the slope of the cumulative mass curve. For example, the 20 min intensity is the 20 min duration of the storm with the most rainfall, divided by 20 min. The 24 hr intensity of the storm is the total volume divided by 24 hr. The slope, or precipitation intensity as a function of time, defines the precipitation hyetograph. The hyetograph may be uniform, advanced, delayed, or intermediate (Schwab et al. 1993). Examples are shown in Figure 2.5.

The intermediate pattern in Figure 2.5 usually results from storms spawned by warm-front passage. The uniform pattern is an idealized distribution. The uniform distribution provides the basis for theoretical precipitation analyses. The uniform pattern roughly approximates the pattern from orographic lifting in the western United States. The advanced pattern usually results from cold-front passage. The delayed pattern (not shown in Figure 2.5) is not associated with a common weather pattern. Any pattern can occur at any given location.

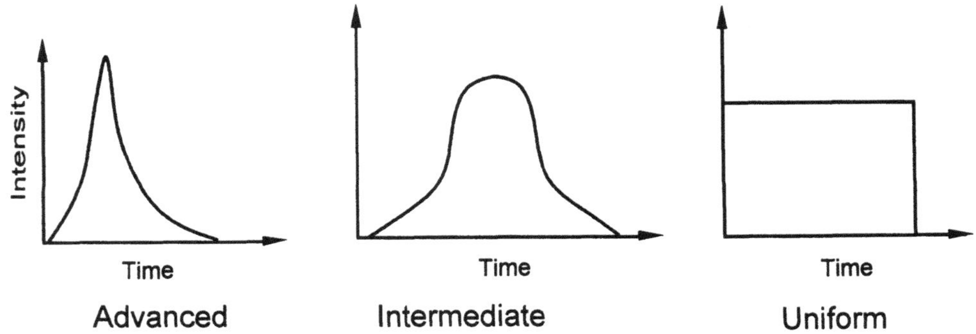

Figure 2.5. Hypothetical rainfall hyetograph types.

Introduction to Hydrologic Frequency Analyses: Return Period Analysis

Hydrologic frequency analysis connects discrete precipitation event analyses to design storm prediction. Hydrologic frequency analysis was the technique used for predicting storms of specified frequency and duration for a given location. Data from gauges distributed across the United States went into precipitation design maps. The techniques of hydrologic frequency analysis are useful for many other analyses (e.g., runoff) also.

Important characteristics of precipitation include the intensity (mm/hr), duration, and frequency of expected events. The concept of return period is fundamental. *The return period (recurrence interval) is the period within which the depth of precipitation for a specified duration will be equaled or exceeded once on the average.* The 100 yr–24 hr precipitation depth for Athens, Georgia, is 8 in (203 mm). Thus, Athens receives 8 in (203 mm) or more of precipitation within a 24 hr period an average of once every 100 yr over many 100 yr periods.

A common analysis method for arriving at design precipitations is the *synthetic storm approach.* We can statistically generate a hypothetical storm—called a design or synthetic storm—that will serve as a basis for design and characterization. Design precipitation determination at a location begins with selecting storm duration, then finding the maximum intensity for the selected duration for each rainfall event in a year. Keep the maximum for the year. Repeat this for as many years as records exist for the location. Do the analyses for the annual maximums following the procedure. Redo the entire analysis for each desired storm duration. Understanding return period analyses is crucial for specifying storms for channel, structure, and embankment design.

The essence of return period analyses is to fit precipitation data to statistical distributions. The statistical distribution most often used in fitting hydrologic data is the lognormal distribution. The lognormal distribution is in effect the fitting of the logarithms of data to the normal distribution. The lognormal distribution is more appropriate than the normal. The normal distribution models data from minus infinity to plus infinity, whereas the lognormal distribution models data from zero to infinity. Because most hydrological data are positive, the lognormal distribution makes more sense.

Return period analyses consist of plotting the data on cumulative probability paper and then plotting a best-fit statistical distribution. After taking logarithms, use standard cumulative normal paper for plotting purposes. Follow these steps:

1. Begin with all events (e.g., total annual precipitation, heating degree days, or other data type) for as many years as possible, one duration, one location.
2. Pick the extreme event for each year.
3. Rank the maximum precipitations for each year,[2] for the selected duration, in descending order. Rank = 1 for largest value.
4. Calculate the plotting return period (T) from equation 2.2a:

$$T_{RtPd} = \frac{\max rank + 1}{rank}; \qquad plot\ prob. = \frac{1}{T_{RtPd}} \tag{2.2a}$$

Plotting probability P is $1/T$. Equation 2.2a is the Weibull plotting position equation (Stedinger et al. 1993) and is also frequently referred to as the Gumbel equation. A spreadsheet is helpful for these calculations.

5. Develop a plot of calculated probability P versus the ranked variable (e.g., precipitation intensity, depth). Plot the data on lognormal probability paper (other distributions, such as extreme value or the log Pearson Type III distribution, are also available [Haan et al. 1994]).

Fitting a probability distribution to the plotted data begins with the following statistical concepts: the mean, the standard deviation, and the coefficient of variation. The mean refers to the simple arithmetic average given by

$$\overline{V} = \frac{\sum_{i=1}^{i=N} V_i}{N} \tag{2.2b}$$

where

V_i is the value of event I, and
N is the total number of events V averaged.

Parameter V_i may represent the log of a particular value. The standard deviation (s.d.) may be found using

$$s.d. = \sqrt{\frac{(V - V)^2}{N - 1}} = \sqrt{\frac{1}{N - 1}\left[\sum_{i=1}^{i=N} V_i^2 - \frac{1}{N}\left(\sum_{i=1}^{i=N} V_i\right)^2 \right]} \tag{2.2c}$$

The coefficient of variation (C_V) is the standard deviation divided by the mean. These concepts are necessary to fit statistical distributions to precipitation or other data types. We will be mainly concerned with the lognormal distribution in this text. The steps for fitting a lognormal distribution are

1. Calculate the mean.
2. Calculate the standard deviation (s.d.).
3. Compute the coefficient of variation (C_V) = s.d./mean.
4. Go to the frequency factor table (given in Table 2.1 for lognormal; otherwise consult Haan et al. 1994) and obtain the K_T factor corresponding to the C_V and each desired theoretical return period T_t. See Table 2.1 for an abbreviated table of K_{RtPd} factors for the lognormal distribution.
5. Compute the corresponding theoretical lognormal magnitude X_{RtPd} for a given theoretical probability or theoretical duration (prob. $= 1/T_{RtPd}$) from equation 2.3:

$$X_{RtPd} = X(1 + C_V K_{RtPd}) \tag{2.3}$$

6. Plot the theoretical lognormal points (X_{RtPd}, prob.) on the probability plot (theoretical prob. $= 1/X_{RtPd}$).
Caveats: Data used in a return period analysis must be independent. Dependence implies serial correlation, and the analysis assumes no serial correlation. Serial correlation implies that data for time $x + I$ are related to those at time x. For example, if it is warm today, it will probably be reasonably warm tomorrow. These analyses do not account for long-term drift in data caused by phenomena such as global climate change.

In equation 2.3, X_{RtPd} is the theoretical value at return period T, 0 is the mean, C_V is the coefficient of variation (s.d./mean), and K_{RtPd} is the frequency factor from Table 2.1.

Is the record length adequate? Especially for small projects, the 100 yr storm is usually the maximum storm of interest. If you are confident in the 100 yr storm, then confidence

Table 2.1. Frequency factors (K_{RtPd}) for the lognormal probability distribution.

Return Period = T_{RtPd} (probability = $1/T_{RtPd}$)					C_v
1.01 (0.99)	2 (0.5)	5 (0.2)	20 (0.05)	100 (0.01)	
−2.33	0	0.84	1.64	2.33	0
−2.18	−0.04	0.83	1.7	2.47	0.067
−2.11	−0.06	0.82	1.72	2.55	0.1
−2.04	−0.07	0.81	1.75	2.62	0.136
−1.91	−0.1	0.79	1.79	2.77	0.197
−1.79	−0.11	0.78	1.81	2.84	0.230
−1.74	−0.14	0.76	1.84	2.97	0.292
−1.63	−0.16	0.73	1.86	3.15	0.351
−1.54	−0.18	0.71	1.88	3.21	0.409
−1.45	−0.20	0.68	1.89	3.31	0.462
−1.41	−0.21	0.67	1.89	3.36	0.490
−1.34	−0.22	0.64	1.89	3.44	0.544
−1.28	−0.24	0.61	1.89	3.52	0.596
−1.22	−0.25	0.59	1.89	3.59	0.643
−1.17	−0.26	0.57	1.88	3.65	0.691
−1.10	−0.27	0.54	1.87	3.72	0.755
−1.06	−0.27	0.52	1.86	3.76	0.796
−0.98	−0.29	0.47	1.83	3.84	0.895
−0.9	−0.29	0.42	1.78	3.91	1.0

Source: Adapted from Haan et al. 1994.

in the return periods for less than 100 years is justified. The question then becomes: Is the length of record long enough to result in a reasonable estimate of the 100 yr storm? Mockus (1960) suggested the following approach based on the t distribution to answer this question. The Mockus (1960) test uses equation 2.4, requiring an iterative solution:

$$Ye_{\min} = \left[4.3\ Ts_{(Ye_{\min}-6),90\%}\log_{10}\left(\frac{X_{RtPd(100)}}{X_{RtPd(2)}}\right)\right]^2 + 6 \quad \textbf{(2.4)}$$

where

Ye_{min} is the minimum acceptable number of years of record;
$Ts_{(Ye[min]-6),90\%}$ is the Student's Ts value at the 90% significance level with Ye_{min} −6 degrees of freedom (see Table 2.2);
$X_{RtPd(i)}$ denotes the 100 yr (i = 100) and 2 yr (i = 2) storms.

Use the Mockus (1960) equation as follows: Determine the number of years N of record and set $Ye_{min} = N$. Next, go to the $Ts_{(Ye[min]-6),90\%}$ table (Table 2.2) and obtain the $T_{(Ye[min]-6),90\%}$ factor. Compute $Ye_{min}\rightarrow$. If $Ye_{min}\rightarrow$ is less than Ye_{min}, adjust $Ye_{min}=$ downward, determine a new $Ts_{(Ye[min]=-6),90\%}$, and recalculate. If $Ye_{min}\rightarrow$ is greater than $Ye_{min}=$, adjust $Ye_{min}=$ upward, obtain a new $Ts_{(Y=-6),90\%}$, and recalculate. Continue adjustments on Y until Ye_{min} $Ye_{min}\rightarrow$ $Ye_{min\ final}$. If $Ye_{min\ final}$ is less than or equal to N, then the record is adequate by the Mockus (1960) criterion. If $Ye_{min\ final}$ is greater than N, then the record is not adequate.

Table 2.2. *Ts* table for use with the Mockus (1960) record length evaluation procedure.

Degrees of freedom ($Y - 6$)	$Ts(Ye - 6)$, 90% (sign ignored)
1	6.31
2	2.90
3	2.35
4	2.13
5	2.01
6	1.94
7	1.85
8	1.86
9	1.83
10	1.81
11	1.80
12	1.78
13	1.77
14	1.76
15	1.75
20	1.73
40	1.68
120	1.65
∞	1.645

Source: Excerpted from Steel and Torrie 1960. Two-sided *Ts* values are shown.

In brief, the larger the ratio of the 100 yr to the 2 yr storm, the more years of record required to give a reasonable degree of confidence in the 100 yr storm. Because the $T_{(df),90\%}$ is used, one should realize that 90% of the time the computed years of adequacy should be reasonable; 10% of the time the computed years of adequacy will be insufficient.

Estimating a 100 yr event with less than 100 yr of data places one in the realm of statistical vagueness. Beware of using a statistical distribution to extrapolate beyond the years of available data. These analyses do not account for phenomena such as global climate change.

Probability plots and calculated confidence intervals are replacing the Mockus procedure. Haan et al. (1994) discuss a method for calculating confidence intervals. Hintze (2001) and others provide software to generate the plot.

Example 2.3: Determine the 20 yr–1 hr storm values with the data shown in Figure 2.6.

Organize the data in columns 1 and 2. Copy column 2 to column 3 and sort in descending order. Use the spreadsheet "data fill" or its equivalent to place the order in column 4. Calculate the return period in column 5. For column 6, take the inverse of column 5, giving the probability of being equaled or exceeded. We can plot column 3 versus column 6 on probability paper. Calculations involving the lognormal distribution plotting factors appear at the bottom of the spreadsheet. The probability shown in the body of the table is the probability of *being equaled or exceeded.* Alternatively, we can use a statistical package such as Hintze 1996. Figure 2.7 shows a probability plot for

Year	Intensity (in/hr)	Rnk. int.	Order	Ret. per. (yr)	Prob.
1933	1.9	5.1	1	22	0.04
1934	2	5	2	11	0.09
1935	2.5	4.8	3	7.33	0.14
1936	2.5	4.5	4	5.5	0.18
1937	2.2	4.05	5	4.4	0.23
1938	3	4	6	3.67	0.27
1939	2.8	3.7	7	3.14	0.32
1940	1.99	3.6	8	2.75	0.36
1941	1.5	3.3	9	2.44	0.41
1942	5	3.2	10	2.2	0.45
1943	4.5	3.1	11	2	0.5
1944	4	3	12	1.83	0.55
1945	3.6	2.8	13	1.69	0.59
1946	3.2	2.7	14	1.57	0.64
1947	3.3	2.5	15	1.47	0.68
1948	2.7	2.5	16	1.38	0.73
1949	5.1	2.2	17	1.29	0.77
1950	4.8	2	18	1.22	0.82
1951	4.05	1.99	19	1.16	0.86
1952	3.7	1.9	20	1.1	0.91
1953	3.1	1.5	21	1.05	0.95

Avg. Int. ***3.211429 in/hr (probability is probability of being equaled or exceeded)***
s.d. ***1.04 in***
Cv ***0.32***

The maximum 1 hr intensity to be equaled or exceeded in 20 yrs is:
Theoretical magnitude at 20 yr storm: $X_{T(20)}$ = ***3.21 in/hr [(1+0.324*1.85)] = 5.13 in/hr***

Figure 2.6. Intensity data set and analysis results for Example 2.3.

the data of Figure 2.6. The probability in the plot is the probability of *not being equaled or exceeded.* The probability of not being equaled or exceeded is 1 minus the probability of being equaled or exceeded. The statistical package computes confidence intervals for the plot as shown in Figure 2.7. Precipitation maps have been developed using many frequency analyses from weather stations around the United States.

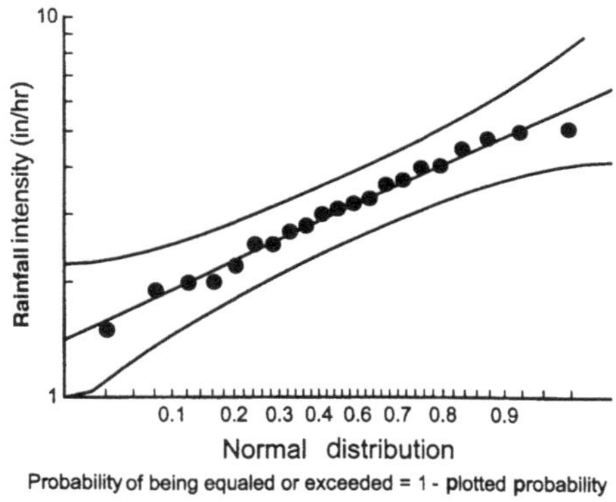

Figure 2.7. Probability plot of the precipitation intensity data of Figure 2.6. If all points are within the confidence intervals, the adequacy of the lognormal distribution for describing the data is confirmed.

Example 2.4: Determine the annual precipitation for return periods of 1, 2, 5, 20, and 100 yr at Griffin, Georgia, by the log-probability law based on 20 yr of annual precipitation. The data are from Kinard and Tippins 1982. The procedure used in Example 2.3 yields Figure 2.8.

Organize the precipitation by year in the first two spreadsheet columns, as shown in Figure 2.8. Convert to SI precipitation units (mm) in the third column. Copy the precipitation data to the fourth column and sort in descending order. Use "data fill" to create the ranking shown in the fifth column. Calculate the return period in the sixth column using Gumbel's formula and take the reciprocal of the return period to obtain the plotting probability (seventh column). Calculations for the log distribution are shown at the bottom of the spreadsheet. Figure 2.9 shows the probability plot (Hintze 1996) for the data of Figure 2.8.

Annual Precipitation at Griffin, GA (35 Mi South of Atlanta)

1962	44.05	1118.87	1748.28	1	21	0.05
1963	53.36	1355.34	1728.72	2	10.5	0.1
1964	68.83	1748.28	1557.53	3	7	0.14
1965	48.09	1221.49	1531.62	4	5.25	0.19
1966	59.2	1503.68	1503.93	5	4.2	0.24
1967	52.06	1322.32	1503.68	6	3.5	0.29
1968	48.69	1236.73	1368.81	7	3	0.33
1969	45.73	1161.54	1355.34	8	2.63	0.38
1970	53.89	1368.81	1337.06	9	2.33	0.43
1971	68.06	1728.72	1322.32	10	2.1	0.48
1972	61.32	1557.53	1322.07	11	1.91	0.52
1973	48.4	1229.36	1272.29	12	1.75	0.57
1974	59.21	1503.93	1271.27	13	1.62	0.62
1975	60.3	1531.62	1236.73	14	1.5	0.67
1976	50.05	1271.27	1229.36	15	1.4	0.71
1977	52.05	1322.07	1228.34	16	1.32	0.76
1978	42.65	1083.31	1221.49	17	1.24	0.81
1979	52.64	1337.06	1161.54	18	1.17	0.86
1980	50.09	1272.29	1118.87	19	1.11	0.90
1981	48.36	1228.34	1083.31	20	1.05	0.95
Year	In	mm	Rank Prec	Rank	Ret Per	Plot Prob
(1)	(2)	(3)	(4)	(5)	(6)	(7)

AVG 1355.1 SD 186.1 CV 0.14

Computed theoretical magnitudes for selected return periods (Equation 2.3)

P (%)	T (Year)	K	$X_{T(T)}$
99	1.01	-2.04	975.494
50	2	-0.07	1342.1
20	5	1.75	1680.79
5	20	1.88	1704.99
1	100	2.62	1842.7

Figure 2.8. Analyses spreadsheet for annual precipitation data, with lognormal distribution predictions for indicated probabilities.

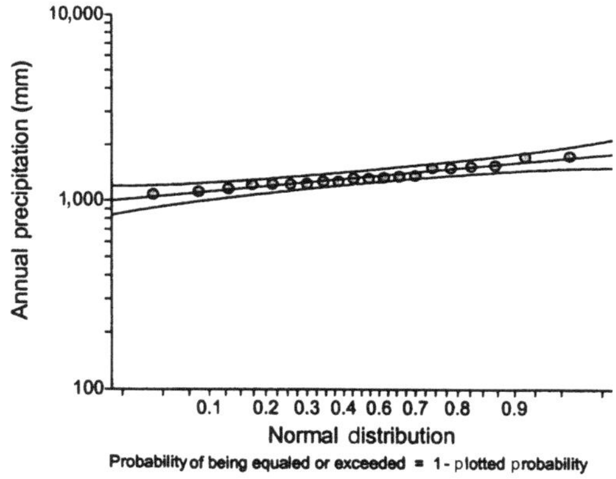

Figure 2.9. Probability plot of annual rainfall data in Example 2.4.

Is the length of record adequate in Example 2.4 according to the Mockus (equation 2.4) criterion?

The Mockus 100 yr/2 yr ratio is $X_{RtPd(100)}/X_{RtPd(2)} = 1842.7/1342.1 = 1.373$. The $Ts_{90(20-6=14)} = 1.76$.

$Ye_{min} = [4.3(1.76)(\log_{10} 1.373)]^2 + 6 = 7.1$ yr. Change Ye_{min} to 8. Then $Ts_{90(8-6)}$ is 2.9. After recalculating, $Ye_{min} = 8.9$. $Ts_{90(3)}$ is 2.35. After recalculating, $Ye_{min} = 8$. This is close enough: $Ye_{min} = 8$ or 9, converging well below 20. The record is adequate by the Mockus criterion.

Evaluation of confidence intervals is probably the best approach for determining adequacy of data set length. The confidence intervals shown in the probability plots of Figures 2.7 and 2.9 suggest adequate fits since no points lie outside the confidence intervals. However, the confidence intervals are wider apart than desirable. More years of record will narrow the gap between the confidence intervals. *As a rule, climatologists like to have four times the recurrence interval years of record.*

Climatic data for Class A and other weather stations around the United States are commercially available. For example, we obtained the following annual precipitation data from Earth Info—National Climatic Data Center (NCDC) Summary of the Day[3] for Atlanta, Georgia:

Record years:	1930–95, or 66 yr
Average annual precipitation:	49.25 in (1250.95 mm)
Standard deviation:	8.08 in (205.23 mm)
Coefficient of variation:	0.164
Skew coefficient:	0.8

One could choose a distribution such as the lognormal to model the probability of equaling or exceeding some value X_i of average annual precipitation. The lognormal usually works well; however, one should generate a probability plot when major projects are involved. The good lognormal fit of the Griffin data bodes well for a good lognormal fit of the Atlanta data.

Calculating Storm Magnitudes for Design Purposes

We will discuss design storms from a U.S. perspective. Precipitation data for other countries are sometimes available from respective climatological agencies or from UN agencies. One should consult these agencies for precipitation prediction in various locations around the world.

Table 2.3. Factors for the x_r value of the precipitation equation.

Return period in years	Variate x_r value
1	−6.93
2	0
5	9.2
10	16.1
25	25.3
50	32.1
100	39.1

Source: Adapted from Weiss 1962.

The Weiss Method

The Weiss method enables computation of duration–return period combinations using four types of storms: the 2 yr–1 hr, 2 yr–24 hr, 100 yr–1 hr, and 100 yr–24 hr storms. The method represents the Hershfield 1961 data numerically in four maps. We can estimate the expected precipitation depth for a given location and duration using equation 2.5 (Weiss 1962):

$$P = 0.0256(C_r - A_r)x_r + 0.000256[(D_r - C_r) - (B_r - A_r)]x_r y_r + 0.01(B_r - A_r)y_r + A_r \tag{2.5}$$

where

P = precipitation depth in inches;
x_r = return period variate (random variable) from Table 2.3;
y_r = duration variate from Table 2.4;
A_r = 2 yr–1 hr precipitation (in) from Figure 2.10;

Table 2.4. Factors for the y_r value of the precipitation equation.

Duration in hours	Duration in minutes	Linearized rainfall duration variate y_r value
0.083	5	−50
0.17	10	−37.0
0.33	20	−24.0
0.3	30	−15.6
0.67	40	−9.4
1	60	0
2	120	17.6
3	180	28.8
6	360	49.9
12	720	73.4
24	1440	100

Source: Adapted from Weiss 1962, except for the 0.083 hr (5 min) value, which was extrapolated.

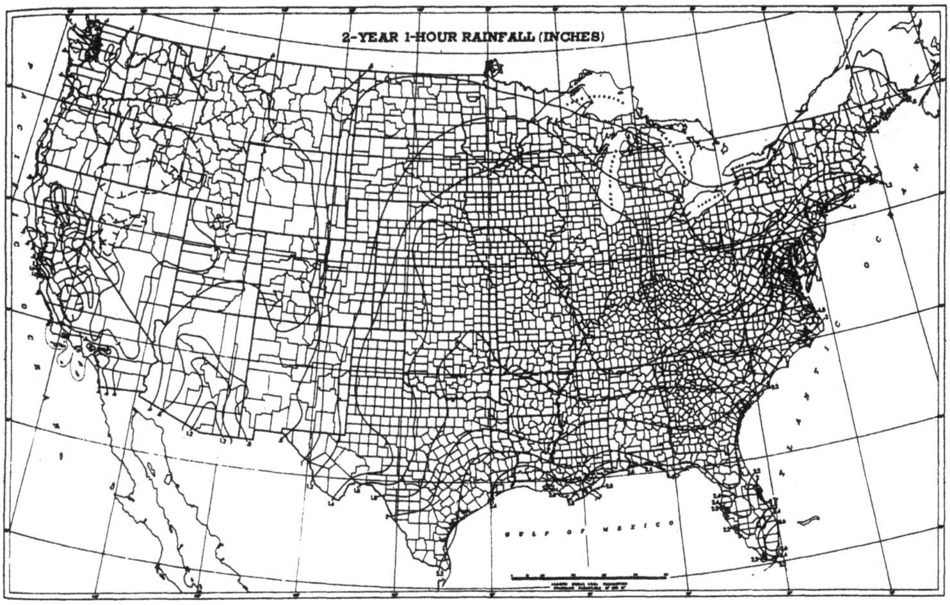

Figure 2.10. Map of 2 yr–1 hr rainfall (in) for the continental United States. (From Hershfield 1961.)

B_r = 2 yr–24 hr precipitation (in) from Figure 2.11;
C_r = 100 yr–1 hr precipitation (in) from Figure 2.12;
D_r = 100 yr–24 hr precipitation (in) from Figure 2.13.

Multiply amounts in inches by 25.4 to obtain amounts in millimeters. The charts in Figures 2.10–2.13 are approximate only.

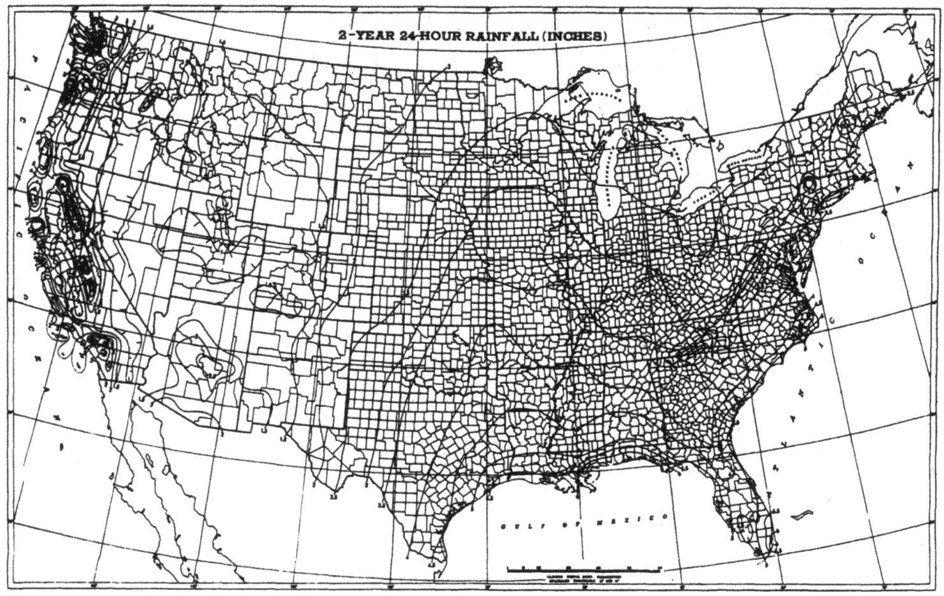

Figure 2.11. Map of 2 yr–24 hr rainfall (in) for the continental United States. (From Hershfield 1961.)

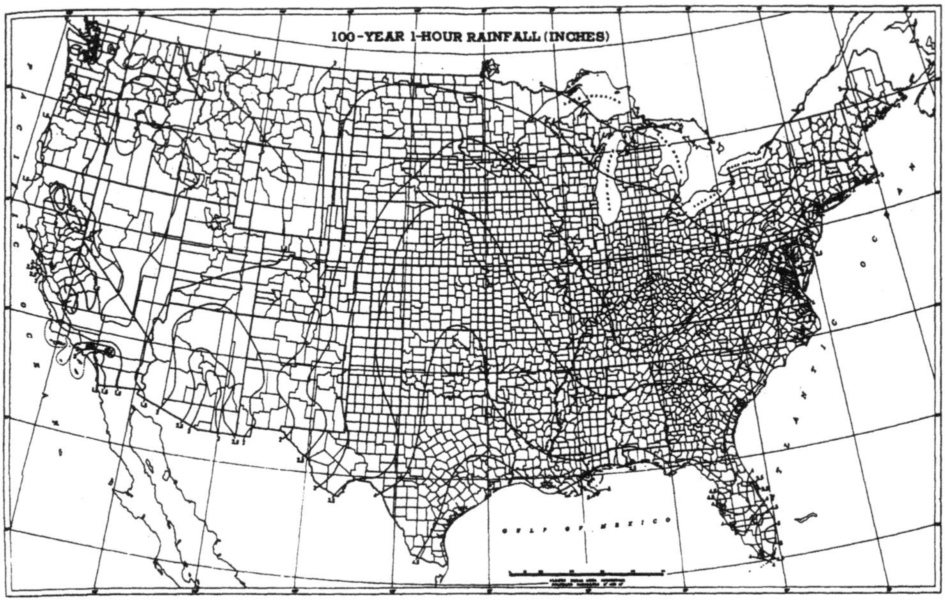

Figure 2.12. Map of 100 yr–1 hr rainfall (in) for the continental United States. (From Hershfield 1961.)

The Weiss (1962) equation provides easy estimates of precipitation. It is convenient to use over the humid eastern regions of the United States. Local trends in data even in the eastern United States are not modeled by the Weiss equation. One may obtain more localized data from state-level geological survey offices.

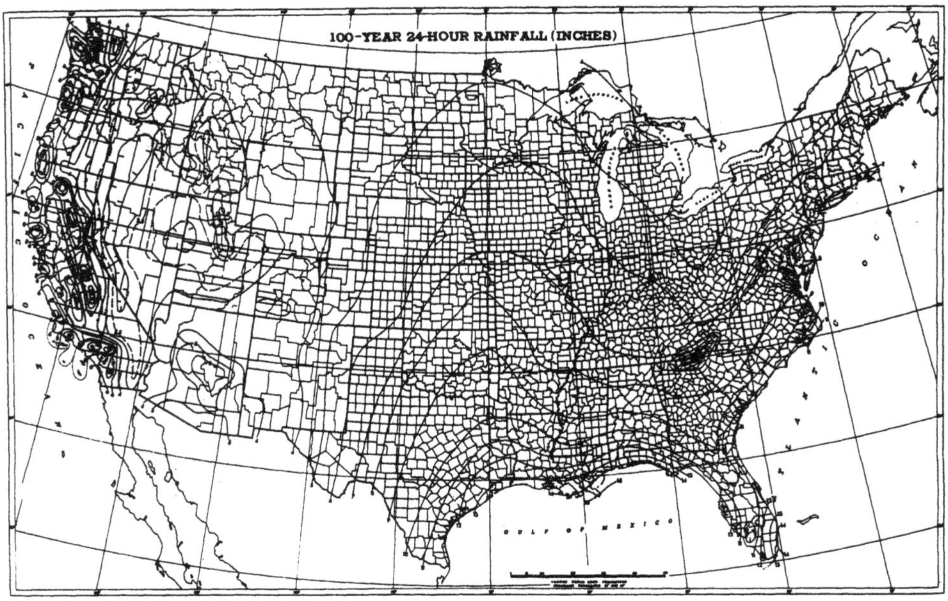

Figure 2.13. Map of 100 yr–24 hr rainfall (in) for the continental United States. (From Hershfield 1961.)

Example 2.5: What is the depth of precipitation and intensity of a 10 yr–12 hr storm in Atlanta, Georgia?

From Table 2.3 for a 10 yr return period, read $x_r = 16.1$.

From Table 2.4 for a 12 hr duration, read $y_r = 73.4$.

From Figure 2.10, read $A_r = 1.8$ in.

From Figure 2.11, read $B_r = 4$ in.

From Figure 2.12, read $C_r = 3.5$ in.

From Figure 2.13, read $D_r = 8$ in.

From equation 2.5,

$$P = 0.0256(C_r - A_r)x_r + 0.000256[(D_r - C_r) - (B_r - A_r)]x_r y_r + 0.01(B_r - A_r)y_r + A_r$$
$$= 0.0256(1.7 \text{ in})(16.1) + 0.000256(4.5 \text{ in} - 2.2 \text{ in})(16.1)(73.4) + 0.01(2.2 \text{ in})(73.4) + 1.8 \text{ in}$$
$$= 4.81 \text{ in (122 mm)}$$

Intensity, $i = P/\text{duration} = 4.81 \text{ in}/12 \text{ hr} = 0.4 \text{ in/hr}$ (10.2 mm/hr)

Assuming no variation in the X and Y coefficients, a variation of about 2–3% is typical, depending on how well one reads the precipitation maps. A 5% variation (see Fredericks et al. 1977) in both X and Y causes 10% or greater variation in predicted precipitation variation.

Intensity-Duration-Frequency (IDF) Maps

Plots of precipitation intensity-duration-frequency (IDF) have been created for all major U.S. cities. These plots are available in publications such as Hershfield 1961. We can compute estimates of IDF plots for any location in the United States using the Hershfield atlas. Example 2.6 demonstrates the computations. Figure 2.14 shows typical plots from the Hershfield atlas. Many engineering offices use these plots to estimate precipitation intensities for runoff calculations.

Example 2.6: Compute the intensity and total depth of a 5 yr–1 hr storm in Baltimore, Maryland.

Enter the IDF curve for Baltimore (Figure 2.14) at duration = 1 hr and go up to the 5 year curve. Read from the scale on the left $I = 2$ in/hr.

The total depth of the storm is P = intensity * duration:

$$P = 2 \text{ in/hr} * 1 \text{ hr} = 2 \text{ in (51 mm)}$$

The relationship between return period and probability of occurrence is expressed by equation 2.2a.

NRCS (1986) Method

After evaluating 8400 station records, Hershfield (1961) prepared precipitation maps that included durations from 30 min to 24 hr and return periods from 1 to 100 yr. Frederick et al. (1977) analyzed 5–60 min storms for the eastern and central United States. The

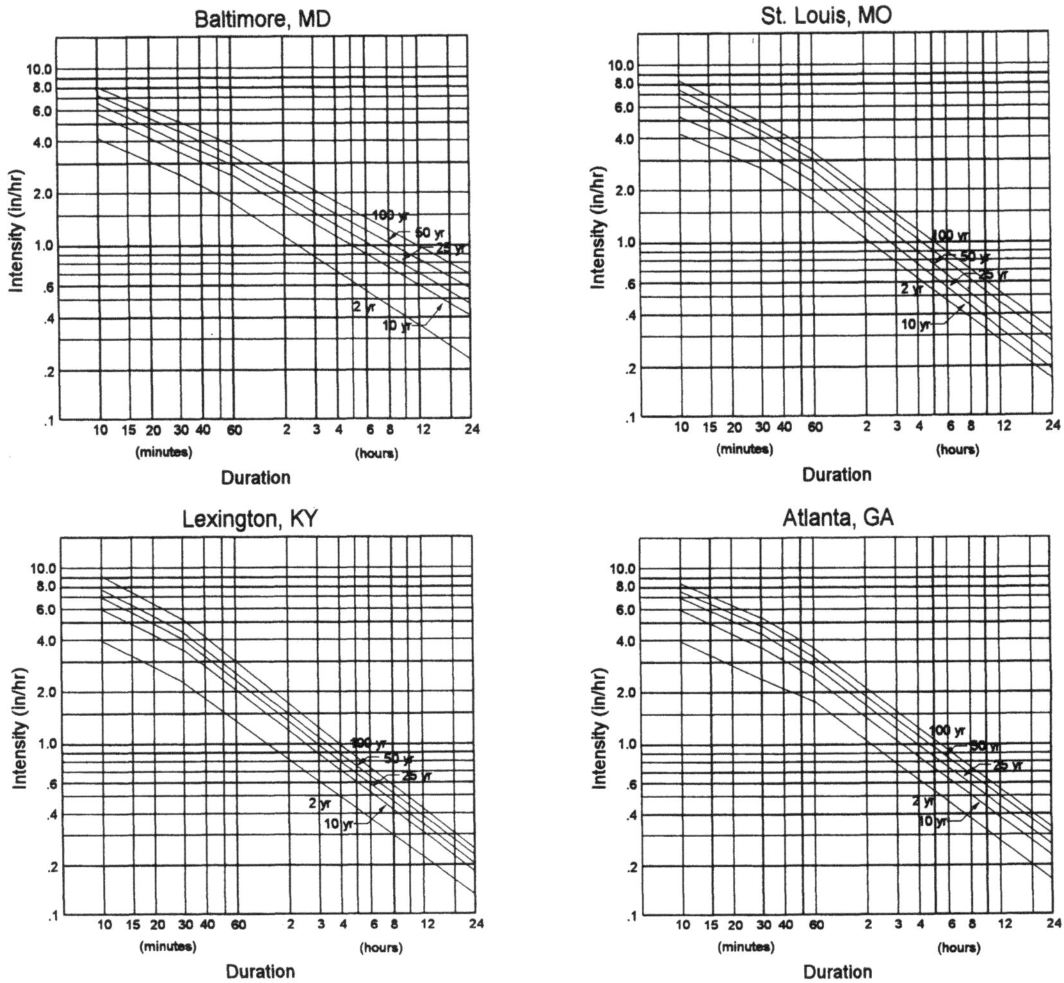

Figure 2.14. Intensity-duration-frequency curves for indicated locations (data from Hershfield 1961.)

NRCS (1986) developed a method for computing storms of various durations from a 24 hr storm. The method comprises the following steps:

1. Determine storm depth for the desired return period and 24 hr duration as given in precipitation maps (e.g., Hershfield 1961; NRCS 1986; see Figs. 2.11, 2.13, 2.15–2.18).
2. Determine patterns from the pattern distribution map shown in Figure 2.19. A loose relation exists between the storm types and the uniform, advanced, intermediate, and delayed hyetographs. The type II (continental) storm is often an advanced storm. The type I and Ia storms associated with the west coast are often approximately uniform. The type III storm associated with the Gulf coast may be considered intermediate. Delayed storms are possible in any region. The normalized coordinates shown in Table 2.5 best fit the storms of each type.
3. Locate the 12 hr line in the coordinates table (Table 2.5) and extend away from the 12 hr line by 2 times the desired duration. For example, for a 3 hr storm, extend back 3/2 hr to 10.5 and record this as the zero point. Record the coordinate. Record the coordinate for each point forward through 12 + 3/2 by 2 hr intervals. Get the storm

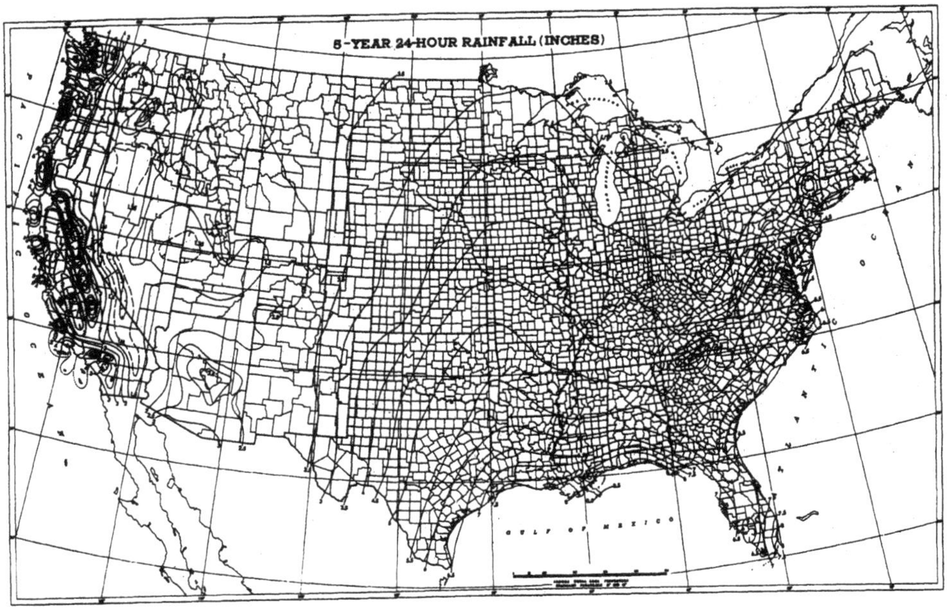

Figure 2.15. 5 yr–24 hr storms for the eastern United States. (From NRCS 1986.)

coordinates by noting zero at the first entry and subtracting the former entry from the current entry. Find all storm coordinates. Multiply each storm entry by the total storm precipitation as read from the appropriate 24 hr map. One may use equations 2.6a–d (see Table 2.5) to compute the ordinates of the respective storms. Equations 2.6a–d were developed using nonlinear regression techniques (Hintze 2001) building on the equation form shown in Haan et al. 1994. Including the time offset in the

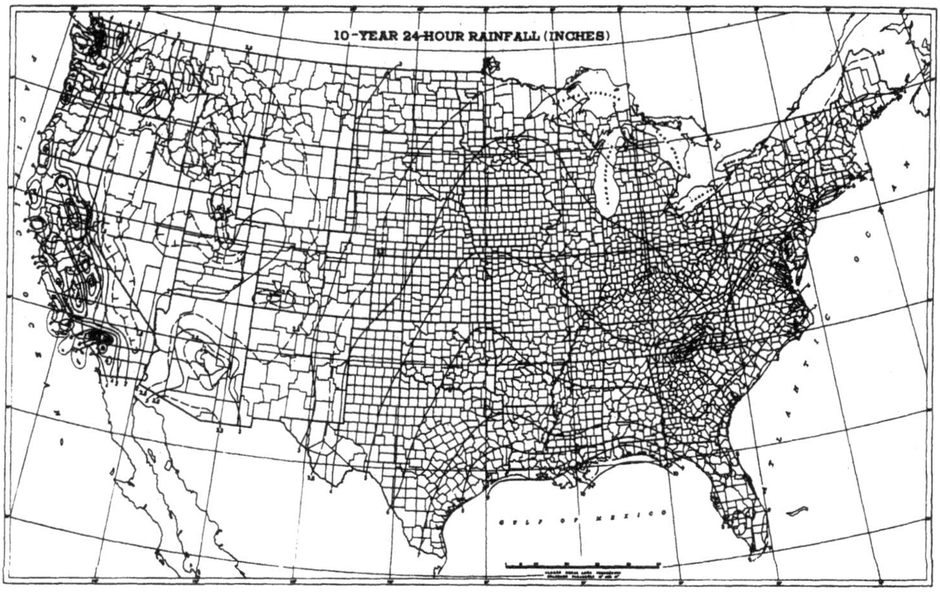

Figure 2.16. 10 yr–24 hr storms for the eastern United States (From NRCS 1986.)

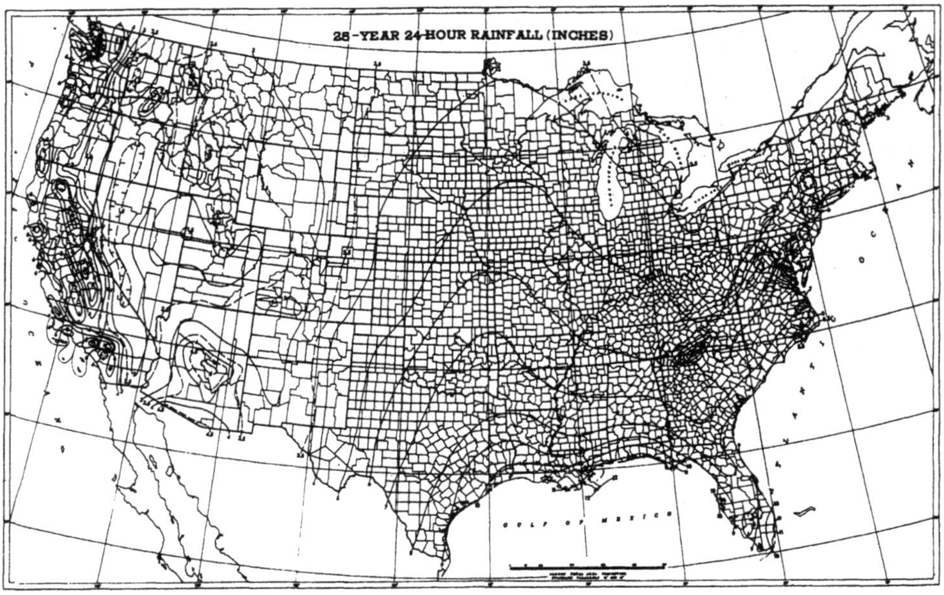

Figure 2.17. 25 yr–24 hr storms for the eastern United States. (From NRCS 1986.)

optimized parameter set enabled usable models not suggested in Haan et al. 1994 for types I and Ia. The equation of Haan et al. 1994 was confirmed for the type II and type III storms. Maximum absolute residual differences less than 0.03 were obtained with each model. Computed upper and lower confidence intervals for each estimated parameter were within 5% of the estimates with each model. Convergence occurred within 10 iterations for each model, suggesting excellent robustness.

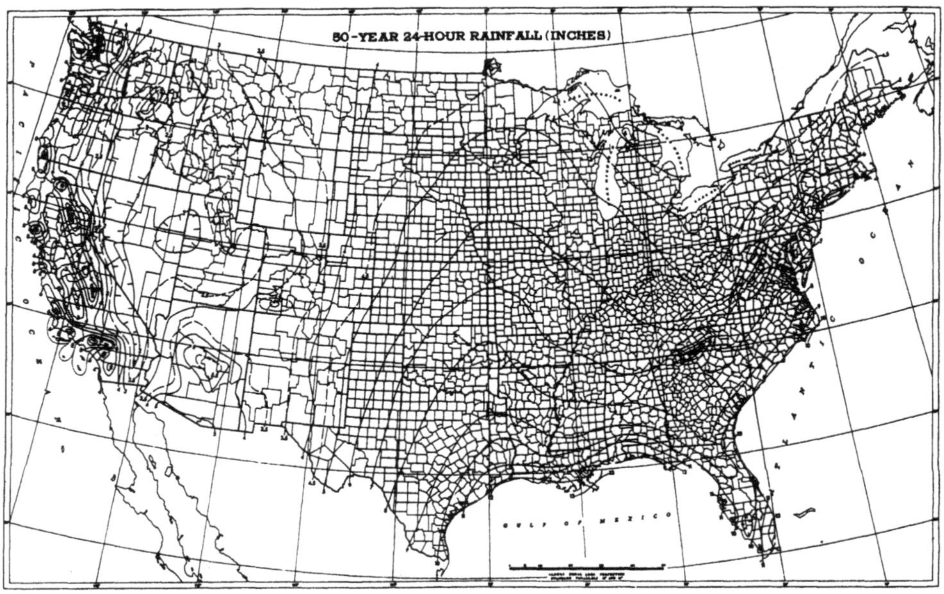

Figure 2.18. 50 yr–24 hr storms for the eastern United States. (From NRCS 1986.)

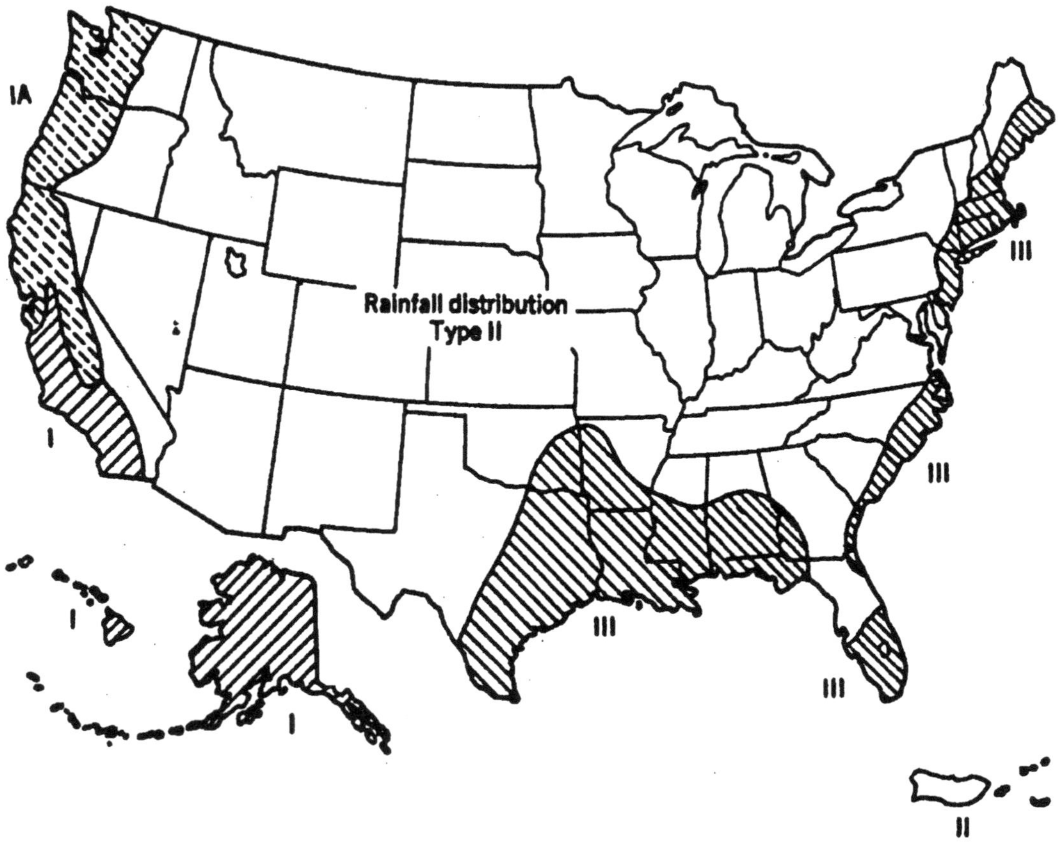

Figure 2.19. NRCS storm type distribution over the United States. (From NRCS 1986.)

$$P(\text{time})/P_{24} = [(\text{time} - 9.93)/24]\{24.58/[2\text{ABS}(\text{time} - 9.93) + 0.04]\}^{0.59} + 0.45 \quad (R^2 = 0.999) \text{ (Type 1)} \tag{2.6a}$$

$$P(\text{time})/P_{24} = [(\text{time} - 7.91)/24]\{25.80/[2\text{ABS}(\text{time} - 7.91) + 0.04]\}^{0.427} + 0.39 \quad (R^2 = 0.999) \text{ (Type 1a)} \tag{2.6b}$$

$$P(\text{time})/P_{24} = [(\text{time} - 11.845)/24]\{23.94/[2\text{ABS}(\text{time} - 11.845) + 0.04]\}^{0.76} + 0.497 (R^2 = 0.999) \text{ (Type 2)} \tag{2.6c}$$

$$P(\text{time})/P_{24} = [(\text{time} - 12.0)/24]\{24.54874/[2\text{ABS}(\text{time} - 12.) + 0.04]\}^{0.734} + 0.5 (R^2 = 0.999) \text{ (Type 3)} \tag{2.6d}$$

Table 2.5. NRCS storm coordinates where the entries are $P(t)/P(24)$.

Row no	Time	Type 1	Type 1a	Type 2	Type 3
1	0	0	0	0	0
2	0.5	.008	.01	.005	.005
3	1	.017	.02	.011	.01
4	1.5	.026	.035	.016	.015
5	2	.035	.05	.022	.02
6	2.5	.045	.066	.028	.025

(Continued)

Table 2.5. (*Continued*)

Row no	Time	Type 1	Type 1a	Type 2	Type 3
7	3	.055	.082	.034	.031
8	3.5	.065	.098	.041	.037
9	4	.076	.116	.048	.043
10	4.5	.087	.135	.055	.05
11	5	.099	.156	.063	.057
12	5.5	.111	.18	.071	.064
13	6	.125	.206	.08	.072
14	6.5	.14	.237	.089	.081
15	7	.156	.268	.098	.091
16	7.5	.173	.31	.109	.102
17	8	.194	.425	.12	.114
18	8.5	.219	.48	.133	.128
19	9	.254	.52	.147	.146
20	9.5	.303	.55	.163	.166
21	10	.515	.577	.181	.189
22	10.5	.583	.601	.204	.217
23	11	.624	.624	.235	.25
24	11.5	.654	.645	.283	.298
25	12	.682	.664	.663	.5
26	12.5	.706	.683	.735	.702
27	13	.727	.701	.772	.75
28	13.5	.748	.719	.799	.783
29	14	.767	.736	.82	.811
30	14.5	.785	.753	.838	.834
31	15	.801	.769	.854	.854
32	15.5	.816	.785	.868	.872
33	16	.83	.8	.88	.886
34	16.5	.843	.815	.892	.898
35	17	.856	.83	.903	.91
36	17.5	.868	.844	.913	.919
37	18	.879	.858	.922	.928
38	18.5	.891	.871	.93	.936
39	19	.902	.884	.938	.943
40	19.5	.914	.896	.945	.95
41	20	.926	.908	.952	.957
42	20.5	.936	.92	.958	.963
43	21	.946	.932	.964	.969
44	21.5	.955	.944	.97	.975
45	22	.964	.956	.976	.981
46	22.5	.974	.967	.982	.986
47	23	.982	.978	.988	.991
48	23.5	.991	.989	.994	.996
49	24	1	1	1	1

Source: Adapted from Haan et al. 1994. Equations for storms other than Type 2 were developed using nonlinear regression techniques in Hintze 1996.

Example 2.7: The 100 yr–24 hr storm for Athens has a total depth of 8 in (203 mm). Estimate the 100 yr–2 hr storm.

From Figure 2.19, Athens is in the type II storm region. For the 2 hr storm, from Table 2.5 read the 11 and 13 points as 0.235 and 0.772, respectively (or calculate them using equation 2.6b). For example, using the above relationship to get the 11 hr ordinand yields

$$P_{11}/P_{24} = [(11 - 11.845)/24]\{23.94/[2.\mathrm{ABS}(11 - 11.845) + 0.04]\}^{0.76} + 0.497 = 0.238$$

The corresponding calculation for hour 13 is 0.779, compared to 0.772. Then the 100 yr–2 hr storm is 8 in (0.772 − 0.235) = 4.3 in (109 mm). Compare this with the Weiss predicted value of approximately 4.29 in (109 mm). The precipitation hyetograph is found as follows: Rain for 1st increment = 8(0.283 − 0.235) = 0.384 in; 2d increment = 8(0.663 − 0.283) = 3.04 in; 3d increment = 8(0.735 − 0.663) = 0.576 in; 4th increment = 8(0.772 − 0.735) = 0.296 in. This accounts for the 2 hr duration, given the 30 min increments in Table 2.5. The hyetograph is found by dividing each incremental amount by the time interval (0.5 hr). This yields 0.77 in/hr, 6.08 in/hr, 1.15 in/hr, and 0.59 in/hr (20, 154, 29, and 15 mm/hr, respectively). We may then plot the hyetograph versus time starting with the first time increment beginning at zero hours.

Probable Maximum Precipitation (PMP)

The probable maximum precipitation (PMP) for an area is, in the judgment of the nation's top climatologists, the amount likely to fall in a region. The assigned PMP for a location is based on climate records and climate modeling. Some have criticized the use of PMP because the values are neither probable nor maximum but instead are based on expert judgment and worst-case analyses. The use of PMP is relevant when structure failure might endanger human life, disrupt vital community services, or cause unusual economic loss. Regulators have embedded the PMP in many regulations. The U.S. Weather Bureau (1947, 1960) and the National Research Council give mappings of PMP for the United States (contiguous 48 states).

Local-Scale Precipitation Analyses

Our look at precipitation has been regional thus far. Information addressing farm- and field-scale precipitation is scarce. It is often known (or thought to be known) by the local populace. Urbanizing activities can have significant impact on precipitation received by a given location and thus introduce farm/field-scale variations. Gauging networks need to be much denser than one per square kilometer to resolve local effects. The correction of Figure 2.2 is not applicable to defining local effects. The Thiessen and hyetograph approaches for determining average precipitation with dense gauging networks are applicable to scales smaller than regional. Holtan et al. (1962) evaluates the effects of gauge density on measurement accuracy in the midwestern United States, and Yoo and Boyd (1994) provide similar analyses for the Alabama piedmont region.

There are many different types of local-scale effects. Large lakes may increase or decrease precipitation received in a year. Rivers and mountains influence precipitation received. The precipitation maps of Figures 2.10–2.13 are not reliable in the mountains because of local topographic variation. Urbanization, which reduces grassed and forested areas, often raises temperatures due to increased heat absorption by asphalt and buildings. These heat islands often experience somewhat more precipitation than surrounding areas (Wanielista et al. 1997). Many believe nutrient constituents found in precipitation vary with proximity to surrounding industries or waste treatment facilities emitting the nutrients. Documentation of local effects is a continuing research need.

Precipitation assessment is useful for research purposes when studying runoff as a function of various management practices. When more than one gauge is used for precipitation estimation in a watershed, one may employ Thiessen polygons or the isohyetal method for computing the representative average.

Simple Average Method

We can take a simple average; however, if several gauges are close together, unusual localized storm events may distort the average.

Thiessen Polygon Method

Suppose the goal is to arrive at a weighted average of some component weighted with respect to area. Precipitation distributed over a watershed is such an example. This method is fast and accurate for many applications. Hydrologists call it the Thiessen polygon, polygraph, or network (Figure 2.20). A watershed is a stream plus all the land area draining into the stream. The core assumption of the method is that the gauge surrounded by the Thiessen polygon equally reflects the area of the surrounding polygon.

The procedure for determining average precipitation using a Thiessen network is as follows:

1. Define the watershed and show gauging stations.
2. Connect adjacent points (stations) with a dashed line and "grow" a bisector on each connecting line.

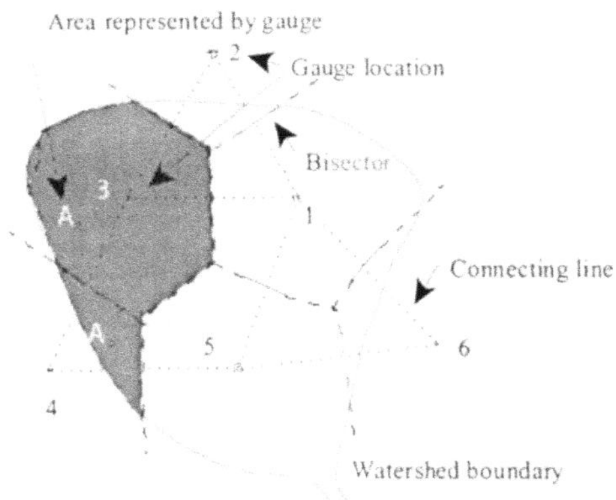

Figure 2.20. Watershed with Thiessen network shown. Develop the network by (1) indicating gauge sites; (2) connecting the gauge sites by lines and bisecting each line; (3) "growing" the bisectors into lines and points of equidistance between gauges; (4) measuring the area of each zone; and (5) computing the area-weighted value.

3. "Grow" the bisectors to approximate polygons in a preliminary way. The goal is to define boundaries equidistant from gauge points. This step comes with practice. Remember, the polygons are approximate.
4. Some polygons may not overlap any watershed area. They can therefore be neglected. Find the area of each polygon as bounded by the lines of step 3 or the watershed boundary.
5. Transfer the final map to the precipitation analysis report.

Compute the weighted area using equation 2.7:

$$drain_{wtd} = \frac{\sum_0^N (A_i drain_i)}{\sum_0^N A_i} \qquad \textbf{(2.7)}$$

where

$drain_{wtd}$ is the area-weighted precipitation depth;
N is the total number of polygons;
A_i is the area of polygon i;
$drain_i$ is the rain depth in the ith gauge.

Isohyetal Method

This method consists of plotting the precipitation depths at the gauge locations and plotting isohyets (lines of equal precipitation) using topographic mapping techniques. Contouring packages used in representing survey elevations are useful for developing the plot. We then determine areas represented by contour bands and compute a weighted average. Figure 2.21 shows an isohyetal plot for a test watershed. We may compute the average precipitation as follows: Determine the area of the watershed, compute the product of precipitation amount and area of each isohyet (within the watershed boundary), sum the respective amount-isohyet products, then divide the sum by the total watershed area. Software packages such as Surfer, version 7 (1994)[4], and NCSS (Hintze 2001) are useful for isohyetal analysis. The isohyetal method allows several gauges to influence the reading at a point, whereas the Thiessen polygon method allows influence from only one gauge.

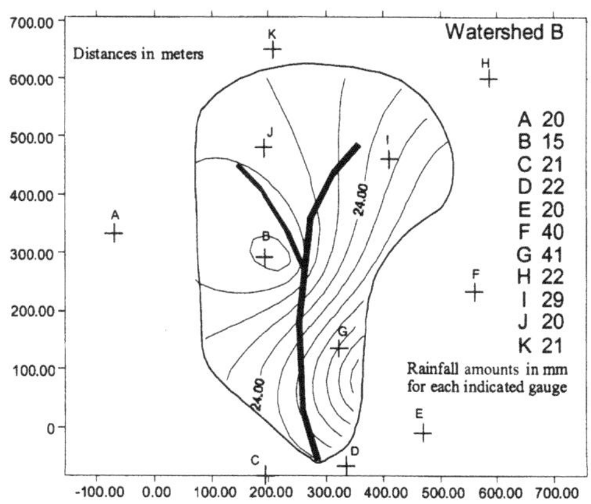

Figure 2.21. Isohyetal plot of the test watershed.

Pollution Transport by Precipitation

Nonpoint pollution begins with precipitation. Precipitation is not reagent-grade distilled water. Precipitation is something more than H_2O as it falls from the sky onto the earth. Precipitation has many other constituents, depending on location and time. The mix of industries in a given region appears to play a part in determining water constituents. Bouwer (1978) presented data for the Lake Ontario basin for 1970–71, some of which are shown in Table 2.6. Table 2.6 is not exhaustive. Further background on physical, chemical, and biological factors affecting water quality is available (e.g., see Tchobanoglous and Schroeder 1985).

Animals, wind, and water distribute natural and human-induced nonpoint pollution over surfaces. Physical or chemical adsorption is the primary nutrient retention mechanism. Volatilization, biologically induced decomposition, and wind and water transport are the primary nutrient transport mechanisms. Constituents are subjected to involved dynamics such as the nitrogen cycle and other biochemical cyclical processes. Virtually every factor affecting physical, chemical, and biological processes affects nonpoint pollution generation or attenuation. Table 2.6 shows relative contributions of selected pollutants attributable to nonpoint pollution. The remainder is due to point pollution. Nonpoint pollution analyses require answers to questions such as: What is the effect of an existing natural or manmade structure (e.g., wetland) on water quality? What is the water quality impairment associated with urbanization in a region? Are rainfalls purifying water via the soil, or does the soil impart nutrients and other chemicals to the incoming water? Surface and groundwater hydrology is critical to answering these questions. Organic and inorganic chemical and biological processes interact with hydrological processes to result in both constituent attenuation and constituent enrichment.

Outlook

Precipitation—snow, rain, drizzle, hail—initiates runoff events. Precipitation and its prediction are major branches of hydrology. Techniques of computing watershed precipitation averages with multiple gauges are equally applicable to computing weighted

Table 2.6. Relative contributions of selected freshwater pollutants attributable to nonpoint pollution.

Pollutant	Estimated percentage from nonpoint sources
Chemical oxygen demand	70
Total P	66
Total Kjeldahl N	90
Fecal coliforms	90
Lead	57
Cadmium	16
Chromium	50
Zinc	70
Arsenic	5
Mercury	2

Source: Adopted from Novotny 1995.

averages of other spatially distributed variables. The return period approach for determining design storms, having broad applicability to any type of data where extremes of noncorrelated data are of interest, is used to quantify annual precipitation amounts. The return period method says in effect: *The storm will be equaled or exceeded at least once in that period, in the long term.* There must be no serial correlation in the data for a return period analysis to be valid. This explains the focus on annual data as opposed to daily or monthly data. The return period approach provides a framework for design storm specification and professional accountability. Current and future studies will focus on farm- and field-scale trends plus regional and global trends. Hydrologic modeling software discussed in later chapters includes design precipitation event generation.

Problems and Questions

1. Determine the total precipitation to be expected once in 5, 25, and 100 years in Knoxville, Tennessee, and Tifton, Georgia, for a 3 hr storm using the Weiss method and the NRCS method. Compare the methods.
2. From precipitation given in Table 2.7, determine the maximum precipitation intensity for any 5, 30, and 240 min periods (from Schwab et al. 1993). *Hint:* Plot cumulative precipitation versus cumulative time to begin the analyses.
3. During a 60 min storm the following amounts of rain fell within each 15 min interval: 33 mm, 23 mm, 15 mm, 5 mm. What are (a) the maximum 15 min intensity, (b) maximum 30 min intensity, and (c) maximum 60 min intensity for this precipitation event?
4. Compute the mean, standard deviation, and coefficient of variation (C_V) for the following annual maximum precipitation intensities for a 5 yr period: 178, 127, 126, and 102 mm/hr.
5. For the total precipitation data shown in Figure 2.8 (or use data given by your instructor), determine the 10 yr annual precipitation amount using the following procedure: Plot probability versus precipitation points based on probability analyses. Next, show the theoretically estimated distribution, plotted as a straight line superimposed on the probability plot. Finally, read the 10 yr storm (or precipitation amount, etc.) from the plot as predicted by the theoretical and by the probability analyses.

Table 2.7. Data for Problem 2.

Time	Time interval (min)	Cumulative time (min)	Rainfall during interval (mm)	Cumulative rainfall (mm)	Intensity for interval (mm/hr)
6.50	—	—			
7.00	10	10	1	1	6
7.10	10	20	10	11	60
7.15	5	25	11	22	132
7.35	20	45	46	68	138
7.45	10	55	19	87	114
8.25	40	95	31	118	47
9.10	45	140	6	124	8
10.50	100	240	6	130	4

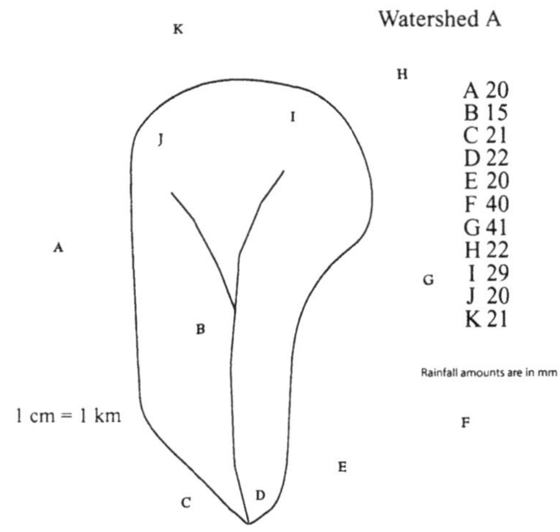

Figure 2.22. Watershed with rain gauges as noted. Amounts of rainfall received are shown on the right.

6. For the watershed shown in Figure 2.22, compute the weighted precipitation for the gauging networks shown with both the simple average and the Thiessen polygon method.
7. Plot the 50 yr–3 hr hyetograph for Clemson, South Carolina, or for a location specified by your instructor by using the NRCS method for generating synthetic storms.
8. For a precipitation data set given by the instructor, compute the minimum years of record needed using the Mockus procedure. What are the pitfalls of using the Mockus procedure to estimate adequacy of record length?
9. Compare the mean precipitation predicted from the isohyetal plot in Figure 2.22 to the prediction using the Thiessen method.
10. Define *rain, sleet, drizzle, precipitation intensity, precipitation hyetograph,* and *30 min intensity.*
11. Argue from practical considerations why we should use the lognormal distribution for precipitation analyses rather than the standard normal distribution.
12. What is the difference between the simple average and the Thiessen polygon approach for aerial average precipitation values?
13. Describe types of rain gauges used to collect precipitation data.
14. What is the probability associated with a 25 yr storm?
15. What does a *25 yr storm probability from a series of annual events* mean in simple English? What are the important assumptions?
16. Compute probability of being equaled or exceeded for any quantity with data supplied by your instructor. State the assumptions involved.
17. What steps do you use to extract the 30 min maximum intensity from a 24 hr precipitation event?
18. A developing country, which has the Atlantic Ocean as an eastern border, lies in a zone of east-west prevailing winds. The Pacific Ocean lies at the western boundary. Which coast would likely receive the most moisture? Which precipitation type (e.g., type I, Ia, II, or III) would you expect to be dominant based on your understanding of climates in the United States?

Notes

1. According to Biswas (1970) the first weighing gauge was built by Sir Christopher Wren (c. 1700). The tipping-bucket rain gauge was apparently in common use prior to A.D. 1364.
2. We can use each event for a partial duration series; the procedure outlined above is for an annual duration series. The following table (from Hershfield 1961) shows the relationship between the partial and annual series for regional precipitation analyses. For the 10 yr and larger storm, there is essentially no difference between the partial and annual duration series storms.

Return period (years)	Conversion factor applied to results of partial duration analysis (every storm used) to compute the equivalent storm value if the annual duration (using only the maximum storm for each year) had been used
2	0.88
5	0.96
10	0.99
>10	1

3. Earth Info, 5541 Central Ave., Boulder CO 80301. Mention of company names is not to be construed as a product endorsement. The name and address are given as a convenience to the reader.
4. Names of commercially available software packages are provided as a convenience to the reader and are not meant to imply an endorsement of the products.

References

Biswas, A. K. 1970. *History of hydrology.* North-Holland Publishing Co., Amsterdam, Holland.

Bouwer, H. 1978. *Groundwater hydrology.* McGraw-Hill, New York, NY.

Frederick, R. H., V. A. Myers, and E. P. Auciello. 1977. *Five- to 60-minute precipitation frequency for the eastern and central US.* NOAA Tech. Memo. HYDRO-35. Silver Spring, MD.

Haan, C. T., B. J. Barfield, and J. C. Hayes. 1994. *Hydrology and sedimentology of small catchments.* Academic Press, New York, NY.

Hershfield, D. N. 1961. *Rainfall frequency atlas of the United States.* U.S. Weather Bureau Technical Paper 40. Washington, DC.

Hintze, J. 2001. *Number cruncher statistical system (NCSS): Version 6.03 users guide.* Kaysville, UT. (Student version available; Email: ncss@ix.netcom.com.)

Holtan, H. N., N. E. Minshall, and L. L. Harrold. 1962. *Field manual for research in agricultural hydrology.* USDA-ARS Agricultural Handbook no. 224. U.S. Department of Agriculture, Washington, DC.

Kinard, D. J., and H. H. Tippins. 1982. *Weather summary for the Georgia Experiment Station, 1926–1981.* Research Report 406. University of Georgia College of Agriculture and Environmental Sciences, Athens.

Mockus, V. 1960. Selecting a flood frequency method. *Transactions of the ASAE* 3:48–51.

Novotny, V. 1995. *Nonpoint pollution and urban stormwater management.* Technomic Publishing Co., Lancaster, PA.

NRCS (US Soil Conservation Service). 1986. *Urban hydrology for small watersheds.* 2d ed. Technical Release TR-55. U.S. Department of Agriculture, Washington, DC.

Schwab, G. O., D. D. Fangmeier, W. J. Elliot, and R. K. Frevert. 1993. *Soil and water conservation engineering.* 4th ed. John Wiley & Sons, New York, NY.

Smith, J. A. 1993. Precipitation. In *Handbook of Hydrology,* ed. D. R. Maidment. McGraw-Hill, New York, NY.

Stedinger, J. R., R. M. Vogel, and E. Faofoula-Georgiou. 1993. Frequency analyses of extreme events. In *Handbook of Hydrology,* ed. D. R. Maidment. McGraw-Hill, New York, NY.

Steel, R. D. G., and J. H. Torrie. 1960. *Principles and procedures of statistics.* McGraw-Hill, New York, NY.

Surfer for Windows: Users guide. 1994. Golden Software, Golden, CO.

———. 1960. *Generalized estimates of probable maximum precipitation for the United States west of the 105th meridian for areas to 400 square miles and durations to 24 hours.* Technical Paper no. 38. U.S. National Weather Service, Washington, DC.

Tchobanoglous, G., and E. D. Schroeder. 1985. *Water quality: Characteristics, modeling, and modification.* McGraw-Hill, New York, NY.

U.S. Weather Bureau. 1947. *Generalized estimates of probable maximum precipitation for the United States east of the 105th meridian for areas to 400 square miles and durations to 24 hours.* Hydrometeorological Report no. 23. U.S. National Weather Service, Washington, DC.

Wanielista, M., R. Kersten, and R. Eaglin. 1997. *Hydrology: Water quantity and quality control.* John Wiley & Sons, New York, NY.

Weiss, L. L. 1962. A general relation between frequency and duration of precipitation. *Monthly Weather Review* 90:87–89.

Yoo, K. H., and C. E. Boyd. 1994. *Hydrology and water supply for pond aquaculture.* Chapman & Hall, New York, NY.

3

Infiltration

Infiltration is the movement of water through the soil surface into the soil. It is an integral part of the hydrologic cycle, affecting the amount and distribution of runoff. Infiltration is important for understanding water application dynamics in irrigation studies. Predicting infiltration is important in water quality assessments. Infiltration influences direct runoff and contributes to base flow in streams. The lack of infiltration is important in ponds, roads, and similar structures where one desires soil sealing.

Goals

- Identify the factors, such as soil type, moisture content, vegetative and surface cover, and time, that affect cumulative infiltration and infiltration rates.
- Apply Darcy's law to simple saturated flow scenarios.
- Empirically and theoretically model cumulative infiltration and infiltration rates via the Horton, Holtan, Phillip, and Green-Ampt equations.
- Understand pollutant transport processes.

Infiltration, Percolation, and Subsurface Flow Defined

Infiltration is the entry of water into the soil surface. *Percolation* is the downward movement of water through the soil profile. Infiltration and percolation are special cases (i.e., vertical flow) of generalized *subsurface flow,* which occurs in response to gradients in any direction.

Factors Affecting Infiltration

Soil Factors

The primary soil factor controlling the rate of infiltration is the number, size, and continuity of soil pores. Major factors influencing pore size include soil particle size, arrangement, and degree of aggregation. Raindrop impact and water flow over the surface cause surface sealing. Reductions in pore size at the surface occur during rainfall events and result in surface sealing. Mulches and crop residues reduce surface sealing, thereby maintaining higher infiltration rates. Large droplet size aggravates surface sealing, notably with bare soils.

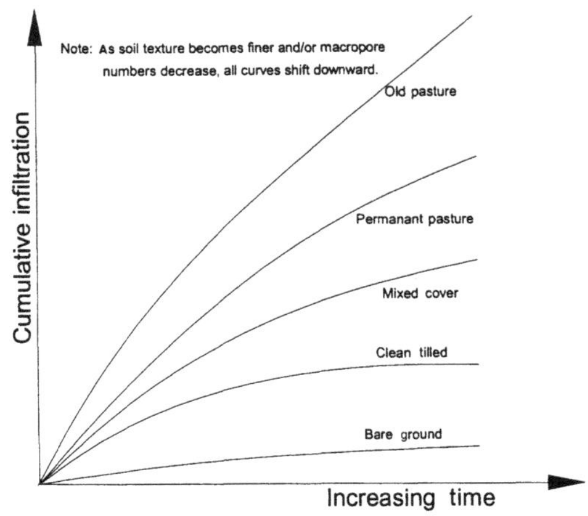

Figure 3.1. Infiltration rate as a function of time for different soil cover conditions.

Vegetation and Soil Organic Matter

Vegetation provides protection and reduces surface sealing. Infiltration rates in bare, crusted ground are as much as seven times lower than rates in old permanent pasture or heavy mulch (Holtan and Kirkpatrick 1950). Forage or grass results in higher infiltration rates, and the effect is proportional to the time under grass.

Other Factors

Soil water content can influence infiltration rates. Infiltration is highest in a dry soil. Matric potential, defined below, "pulls" water into a dry soil. Matric potential contributions to water infiltration decrease as initial soil moisture content increases. Some clays and colloids within the soil swell as they get wet. The volume change reduces the numbers and sizes of soil pores, reducing infiltration.

If an impeding layer such as a hardpan limits percolation, infiltration may approach zero. Vegetation and soil types can affect infiltration as shown in Figure 3.1.

Darcy's Law

Darcy's law is the fundamental physical relationship involved in infiltration analyses. Darcy's law describes laminar flow of liquids and gases[1] through porous media and appears as shown in equation 3.1:

$$q = fA = -AK_c\frac{\Delta H}{\Delta L_e} = -\frac{AE}{r} \tag{3.1}$$

where

q is flow rate (L^3/T);

f is the Darcian mass flux (L/T) (note that flux is the same as velocity of material through the cross-sectional area assuming no solids are present);

A is area normal to flow path (L^2);

K_c is hydraulic conductivity, proportional to soil pore size and number of continuous soil pores (L/T);

$\Delta H = H_2 - H_1$ is change in total hydraulic potential (L) over distance ΔL_e; see discussion below (*sign convention:* the flow rate q and flux f are positive when H_2 is located such that water reaches the H_2 measurement location *after* passing the H_1 measurement location);
E is analogous to electrical potential, thus analogous to $\Delta H/\Delta L$;
r is analogous to electrical resistance, thus analogous to $1/K_c$.

Compare the similarity of the Darcy equation with Fourier's law of heat conduction, Fick's law of mass diffusion, and Ohm's law of electrical current (see Hillel 1998).

The definition of hydraulic conductivity depends on both the fluid and the matrix. One may express the hydraulic conductivity as shown in equation 3.2a:

$$K_c = k\frac{\gamma}{\mu} \qquad \textbf{(3.2a)}$$

where

k is the intrinsic permeability (L^2/T);
γ is the fluid density (F/L^3);
μ is the absolute viscosity (F/LT).

The intrinsic permeability depends on the porous-media porosity, particle shape, and pore tortuosity (Bear 1972). Fluid properties are not included in intrinsic permeability. Equation 3.2a is useful for relating gas conductivity to water conductivity.

The total hydraulic head is a notion requiring some thought. Potential is the *capacity to do work* as measured from a suitable reference. The units of potential are often stated in units of length, which is the result of expressing potential in terms of energy per unit weight (e.g., $F - L$ per F). Potential is sometimes expressed in units of pressure. Potential at a point is pressure head, with reference of zero pressure, or distance above a datum. The datum is the elevation potential reference. The work done in Darcy flow problems is that of overcoming flow resistance between two points, 1 and 2. For many purposes, the total hydraulic potential may be viewed as three components: (1) the elevation potential, referenced from a datum; (2) the pressure head, referenced from zero pressure; and (3) the matric or capillary potential, referenced from zero. Texts such as Hillel 1998 discuss potential flow situations along with flow situations involving large differences in salt concentrations. In saline soils, additional potentials associated with salinity (see Chapter 14) may also contribute to flow.

One must first establish an elevation datum for the problem situation. Elevation potential is simply the distance above (positive) or below (negative) the established datum for the situation. Pressure head or potential is just that—the pressure of the fluid at the location. Principles of fluid statics are useful for assessing pressure in saturated soils and at boundaries. Another major component of total potential contributing to flow in unsaturated soils—matric potential—is discussed below. Fluid friction losses in the tubes leading to and away from the soil column are neglected due to the extremely low fluid velocities.

Figure 3.2 shows a scheme for measuring hydraulic conductivity. The scheme contains a water reservoir supplying water to a soil core that, in turn, discharges into the water catchment. First, one computes flow potentials. Suppose we wish to compute pressure and elevation potentials for points 1 and 2 indicated on Figure 3.2. Points 1 and 2 are both 25.8 cm above the indicated datum. Pressure at point 1 is computed using fluids statics approaches. The pressure potential at point 1 is the height of the water supply surface above

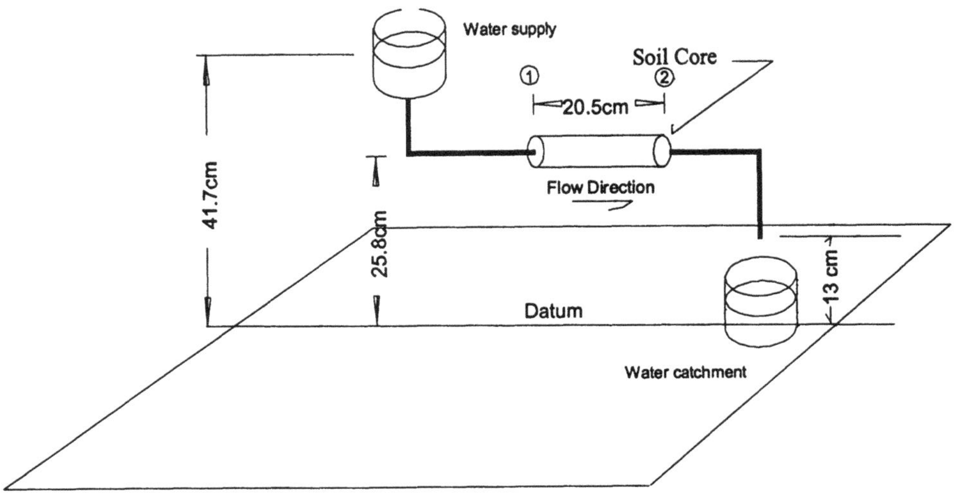

Figure 3.2. Potentials involved in flow through porous media. Points 1 and 2 indicate, respectively, head H_1 and H_2 measurement positions.

point 1. Since the supply is higher than point 1, the pressure is positive. One computes the pressure potential at point 2 from the difference in the capillary exit height and the height of point 2. The pressure potential is negative since the capillary tube exit is below point 2. *If a large noncapillary tube was used and the free water surface was at the point 2 exit level, the pressure potential would be zero. If the exit was above point 2, the pressure would be positive.* One could also tilt the soil core with respect to the datum, placing points 1 and 2 at different elevations. Problem exercises further explore tilted soil columns.

Matric potential is important when describing water flow in an unsaturated medium. The matric potential arises from the surface tension of the fluid. Suppose one idealized the soil pores by small tubes of radius r. Matric potential is given by $\psi = 2\sigma_{st}/r_p$ where σ_{st} is the surface tension of water ($\sigma_{st20°C} = 72.75$ dynes/cm). For example, suppose distilled water is at equilibrium in a tube having a 0.1 cm radius as shown schematically in Figure 3.3. Then $\psi = 2(72.75 \text{ dynes/cm})/0.1 = 1455 \text{ dynes/cm}^2$. One cm H_2O head

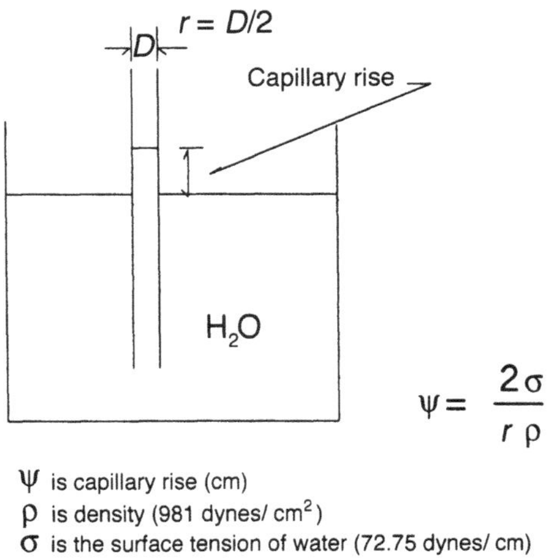

Figure 3.3. Schematic illustration of capillary, or matric, forces arising from surface tension of water interacting with a small tube.

is 981 dynes/cm^2 pressure. The capillary rise, or matric potential, is then 1.48 cm. The presence of contaminants lowers the surface tension. Thus, the observed level of capillary rise with ordinary tap water in tubes is not usually as high. Matric potential is zero at the soil-air interface (i.e., here $r \approx \infty$ and therefore $\psi \approx 0$). Hillel (1998) discusses the measurement of matric potential further. The primary application of matric potential in this text is with the development of the Green-Ampt infiltration equation.

Total head at points 1 and 2 is the algebraic sum of these three heads, respectively. *The sign convention noted in the definition of Darcy's law must be observed.* To get the difference in total potential, first establish the datum. Second, identify points 1 and 2 and get the distance between them for ΔL_e. Third, compute the total potential at each point. Distances above the datum are positive. Pressure is positive for pressure and negative for vacuum (e.g., a hanging water column).

Example 3.1: Compute the mass flux through the soil core in Figure 3.3.

Total potential at entrance of soil core: elevation = 25.8 cm; pressure = 41.7 − 25.8 = 15.9; matric potential = 0; therefore H_1 = 25.8 cm + 15.9 cm + 0 = 41.7 cm.

Total potential at exit of core: elevation = 25.8 cm; pressure = −25.8 cm + 13 cm = −12.8 cm; matric potential = 0; therefore H_2 = 25.8 cm −12.8 cm = 13 cm.

The flux $f = -K_c(H_2 - H_1)/L_e = -K_c[(+13\text{ cm}) - 41.7\text{ cm}]/20.5\text{ cm} = K_c(1.40)$.

One can measure hydraulic conductivity using the apparatus shown in Figure 3.3.

In the above example, suppose 100 cm^3 water moved through the soil core in 60 s. Further, suppose the cross-sectional area of the soil core in Figure 3.3 was 5 cm^2. Then one can compute the conductivity in the example as follows:

$$q = 100\text{ cm}^3/60\text{ s} = 5\text{ cm}^2\ (1.40)\ K_c$$
$$K_c = 0.238\text{ cm/s}$$

K_c usually represents saturated conditions (e.g., K_{sat}) unless otherwise indicated.

Drawing on analogies from electrical resistance enables direct calculation of the hydraulic conductivity of a distinctly layered soil. One may determine the resistance of each layer by dividing the layer conductivity into the layer thickness, giving a relative travel time. Sum the relevant travel times for the soil mass. Finally, divide the soil mass length by the relative travel time. This leads to equation 3.2b for composite conductivity:

$$K_{comp} = \frac{L_T}{\sum_{n=1}^{N} \frac{L_i}{K_{sat\,i}}} \qquad \textbf{(3.2b)}$$

where

K_{comp} is the composite conductivity (L/T);
L_T (= L_i) is the total length (L) across the composite section;
L_i is the length (L) of section i;
$K_{sat\,i}$ is the conductivity (L/T) of section i.

Equation 3.2b is similar to the physics equations describing current flow through a series resistance and heat flow through a composite material.

Example 3.2: What is the composite conductivity of a vertically layered soil composed of a 0.15 m sand layer with a K of 1 cm/s, a 0.25 m sandy loam layer with a K of 0.09 cm/s, and a 0.2 m compact clay with a K of 0.001 cm/s? The total length is 0.15 m + 0.25 m + 0.2 m = 0.6 m. From equation 3.2b we have

$$K_{comp} = 0.6\text{ m}/[(0.15\text{ m}/1\text{ cm/s}) + (0.25\text{ m}/0.09\text{ cm/s}) + (0.2\text{ m}/0.001\text{ cm/s})]$$
$$= 0.0029\text{ cm/s}$$

The clay layer effectively controls the conductivity in the above situation. Consult a soil physics text such as Hillel 1998 for further elaboration on factors affecting soil conductivity.

Example 3.3: A pond is to be sealed using a clay blanket. The clay blanket is effectively 0.75 cm thick and the compacted soil is 29 cm thick. The hydraulic conductivity of the clay blanket is 5×10^{-9} cm/s, and the hydraulic conductivity of the compacted soil is 1×10^{-6} cm/s.

First, compute the composite conductivity. Using equation 3.2b, we have

$$(29\text{ cm} + 0.75\text{ cm})/(29\text{ cm}/10^{-6}\text{ cm/s} + 0.75\text{ cm}/5 \times 10^{-9}\text{ cm/s})$$
$$= 1.662 \times 10^{-7}\text{ cm/s}$$

One may compute the flux from Darcy's law using potentials defined in Figure 3.4. The datum is chosen to be the bottom of the liner. H_1 is the pressure created by the overlying water (245 cm) plus the elevation above the datum (29.75 cm), or 274.75 cm. H_2 is assumed to be at zero pressure potential and is at the datum, or 0 cm. All matric potentials are neglected. The gradient is $H_2 - H_1 = -274.75$ cm/29.75 cm, or -9.235. Applying Darcy's law, the flux $f = -1.662 \times 10^{-7}$ cm/s(-9.235) = 1.53×10^{-6} cm/s, or 0.132 cm/day. Typical allowable seepage rates are 0.125 in per day, or 0.318 cm/day. Thus, the pond is within the allowable limit. Ideally, one would prefer to be an order of magnitude or more lower than the allowable seepage rate.

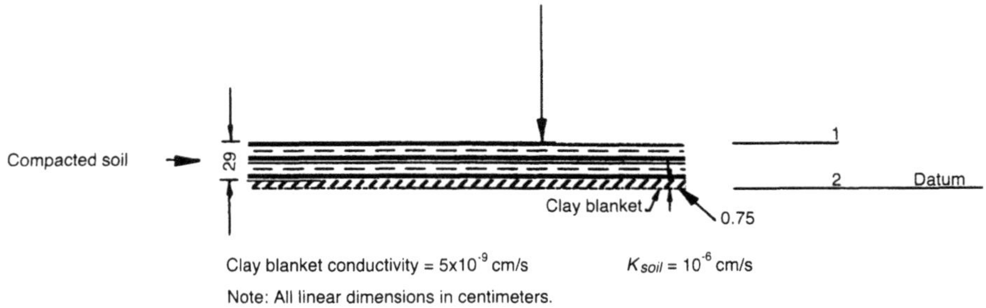

Figure 3.4. Infiltration scenario for lined pond of Example 3.3.

Table 3.1. Approximate Green-Ampt infiltration parameters for selected soil textures.

Soil texture	Porosity n_p	Equilibrium saturated water content (θ_s)	ψ (cm)	K_{sat} (cm/hr)
Sand	0.437	0.417	4.95	11.78
Loamy sand	0.437	0.401	6.13	2.99
Sandy loam	0.453	0.412	11.01	1.09
Loam	0.463	0.434	80.89	0.34
Silt loam	0.501	0.486	16.68	0.65
Sandy clay loam	0.398	0.330	21.85	0.15
Clay loam	0.464	0.309	20.88	0.1
Silty clay loam	0.471	0.432	27.3	0.1
Sandy clay	0.430	0.321	23.9	0.06
Silty clay	0.479	0.423	29.22	0.05
Clay	0.475	0.385	31.63	0.03

Source: Adapted from Rawls et al. 1983.

Notes: The negative sign for matric potential (ψ) is not shown; the sign is incorporated within the Green-Ampt model. Values shown assume dry soil. Values decrease to zero as soils approach saturation.

The variation about all variables should be estimated to be as much as 100% of the value due to the multitude of factors such as vegetation, vertical layering, etc.

In Example 3.3, the matric potential was assumed zero at the underside of the clay blanket. Because this boundary is not a soil-air or soil-water interface, this assumption may be incorrect. Depending on the texture and moisture level of the underlying soil, the matric potential may range from 0 to -30 cm H_2O or so. Table 3.1 provides values for matric potential (ψ) for a range of soil textures at dry conditions. The table does not show the negative sign. Infiltration equations have been derived to include the negative sign. Values of ψ for saturated soils approach zero, the usual case under a pond, analyzed in Example 3.3. If the matric potential in Example 3.3 was, say, -20 cm, one would have $H_2 = 0 + 0 + (-10)$, or $H_2 = -10$ cm. The flux is computed in the normal manner. Guidelines for estimating the matric potential for given soil textures are provided by the Green-Ampt infiltration model, discussed below.

Infiltration and Percolation Prediction

Infiltration models are sometimes used for relating measured infiltration values to soil properties. One may model infiltration by a variety of approaches. Many researchers have proposed empirical models. Two widely used empirical models, the Horton equation and the Holtan equation, are presented below. Researchers have also proposed theoretical models combining mass conservation and Darcy's law. A simple one-dimensional theoretical approach is exemplified by the Green-Ampt relationship given below.

Horton Equation

Horton (1939) expressed the infiltration curve as equation 3.3:

$$f_H = f_c + (f_0 - f_c)\exp(-k_{Hort}t) \qquad \textbf{(3.3)}$$

where

f_H is the Horton infiltration rate (L/T);
f_c is the steady-state infiltration rate (L/T);
f_0 is the initial infiltration rate (L/T);
k_{Hort} is an empirical constant ($1/T$);
t is time (T).

The Horton equation requires experimentally determined constants. A deficiency in the Horton equation is its lack of a physical basis. The Horton equation provides a simple rate-time empirical relationship.

Holtan Equation

Holtan and Kirkpatrick (1950) defined a root zone thickness and then further defined the water-holding capacity of this root zone based on the initial water content and the equilibrium saturated water content. Once the soil zone approaches the equilibrium saturated water content, the infiltration rate approaches a steady-state value. They used the following empirical relationship to describe the infiltration process. The Holtan model is

$$f_{Hol} = (GI)A_H F_{pH}^{n_H} + f_c \tag{3.4}$$

where

f_{Hol} is the Holtan infiltration rate (L/T);
F_{pH} is the unfilled portion of the soil water-holding capacity of the surface zone, or A horizon;
GI is the crop growth index, which varies from 0.1 at seeding to 1.0 at maturity;
A_H and n_H are empirical constants;
other nomenclature is as defined previously.

The Holtan model is similar to the Horton model in equation form. The Holtan model has a more physical basis and is used in some runoff models. Rawls et al. (1993) present estimates of constants for the Horton and Holtan infiltration models for various soil types and conditions. One may use plotting and regression techniques to estimate these constants in specific situations.

Phillip Equation

The Phillip (also known as the Richards) equation models the change in water content and potential in one, two, or three dimensions. One may write an equation analogous to the continuity equation for the constituent to describe the transport, starting with a shell balance:

(rate of mass incoming) − (rate of mass out)
− (mass lost (+) by uptake; gained (−) by injection) = (mass accumulation) **(3.5)**

The fundamental differential equation in one dimension (Klute 1952) appears as equation 3.6:

$$\frac{\partial\theta(h)}{\partial t} = \frac{\partial[K_c(h)g]}{\partial z} + \frac{\partial}{\partial z}\left[K_c(h)\frac{\partial h}{\partial z}\right] - R(z,t) \tag{3.6}$$

where

g is gravity (L/T^2);
h is soil suction head (L);
z is depth (L);
t is time (T);
θ is the volume of water per volume of soil, and it depends on the pressure head;
K_c is the hydraulic conductivity (L/T), which depends on h (in turn a function of soil saturation);
$R(z, t)$ represents a source/sink term such as root uptake (distributed sink) or drip irrigation (point source).

The flux $f(z, t)$ is found by solving Darcy's law.

Infiltration is usually a three-dimensional process. Finite difference or finite element methods are useful for solving the Phillip equation. One must develop the $h(\theta)$ relationship using ideas from soil physics. Solutions to the Phillip equation are computationally intensive but find a place in some specialty models. In theory, the Phillip equation models infiltration and percolation. In practice, infiltration is difficult to simulate numerically because the wetting front represents an infinite moisture gradient. The Phillip equation works well for a poorly defined wetting front, which is usually the case when soils are initially moist and after wetting has progressed to deeper soil layers.

Green-Ampt Equation

Green and Ampt (1911) proposed an infiltration equation based upon Darcy's law. This equation assumes a variable control volume that increases in volume as water infiltrates into the soil. The Green-Ampt model assumes soil properties are constant with depth. It further assumes the soil has one pore diameter instead of a distribution of pore diameters. The derivation of Green-Ampt starts with Darcy's law:

$$f_{GA} = -K_c \frac{\Delta H}{\Delta L} \tag{3.7}$$

where f_{GA} is the Green-Ampt flux (L/T).

Referring to Figure 3.5, one may write

$$-f_{GA}(t) = -K_{sat} \frac{L_{wf}(t) + Dp + \Psi}{L_{wf}(t)} \tag{3.8}$$

where

Dp is the ponded depth on the soil surface (L);
K_{sat} is the saturated hydraulic conductivity, replacing K_c in Darcy's law;
L_{wf} is the depth of the wetting front below the surface (L);
ψ is the soil matric potential (L).

One can evaluate ΔH and ΔL by first setting the datum at the lower $L_{wf}(t)$ boundary. Then, referring to Figure 3.5, H_1 is $L_{wf}(t) + Dp$ with no matric potential (because filled pores exert no capillarity). The pressure on the soil surface is equivalent to d. H_2 is $-\psi$. Therefore $H_1 - H_2 = \Delta H$ is $L_{wf}(t) + Dp + \psi$. The ΔL is the same as $L_{wf}(t)$.

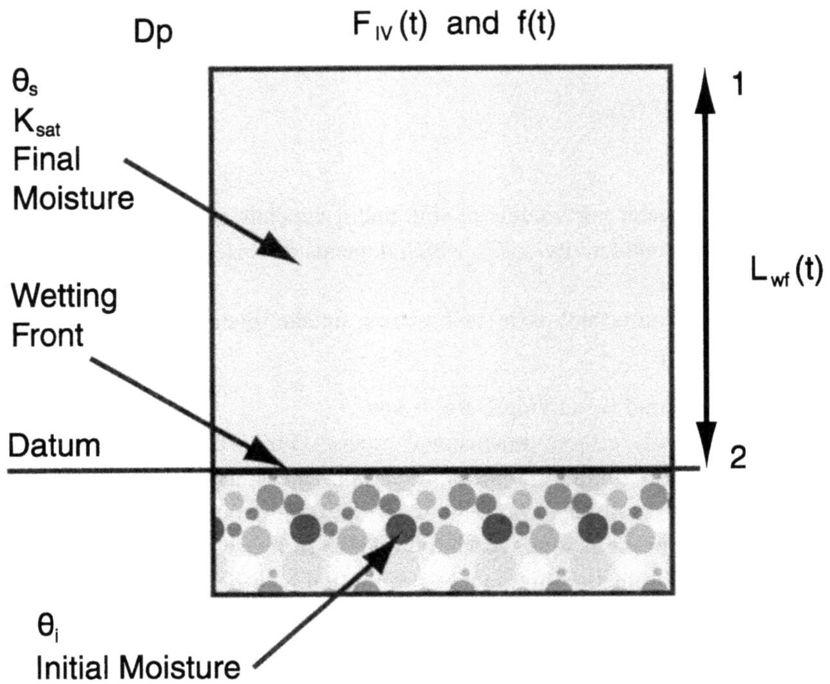

Figure 3.5. Definition sketch for the Green-Ampt equation solution.

The length L is the distance from the soil surface to the depth of infiltration, depending on time t. Refer to Figure 3.5. *The major insight of Green and Ampt is that the momentum effects of the moving wetting front may be neglected; a quasi-static condition may be used to set the datum and analyze the problem.* Function $f(t)$ is negative because ΔH was computed using $H_1 - H_2$ (see Figure 3.5 for placement of points 1 and 2). One can write the length to wetting front $L_{wf}(t)$ as

$$L_{wf}(t) = \frac{F_{IV}(t)}{\theta_s - \theta_i} \tag{3.9}$$

where

$F_{IV}(t)$ is the total depth of water infiltrated at time $t (L)$;
θ_i is the initial volumetric moisture content (L/L);
θ_s is the (near) saturated volumetric moisture content (L/L).

Putting equation 3.9 into equation 3.8 and assuming $Dp = 0$ results in the common form of the Green-Ampt equation for flux:

$$\frac{dF_{IV}(t)}{dt} = f_{GA}(t) = K_{sat}\left[\frac{1 + (\theta_s - \theta_i)\psi}{F_{IV}(t)}\right] \tag{3.10}$$

where

f_{GA} is the Green-Ampt infiltration rate (L/T);
K_{sat} is the saturated hydraulic conductivity (L/T);
other nomenclature is as defined above.

Separating the variables, integrating, and solving for time t results in the Green-Ampt equation for cumulative time:

$$t = \frac{F_{IV}(t) - \Psi(\theta_s - \theta_i)\ln\left[1 + \frac{F_{IV}(t)}{\Psi(\theta_s - \theta_i)}\right]}{K_{sat}} \qquad \textbf{(3.11)}$$

where t is the time required to infiltrate F_{IV} in the indicated soil conditions.

To use equation 3.11, assume values for cumulative infiltration and solve for time t, knowing values of other physical parameters. Table 3.1 gives estimates of soil parameters needed for applying the Green-Ampt infiltration method.

Given an opportunity time t and soil condition parameters, one may iterate for the infiltration volume. Rawls et al. (1993) present mathematical relations and extensive tables and graphs relating conductivity, porosity, density, and the suction parameter to soil texture. Use measured values for specific soils if they are available. After solving for time t and knowing cumulative infiltration, solve the differential form of the Green-Ampt relation to get infiltration rates versus time.

The easiest way to develop an infiltration relation is to do the following:

1. Assume several amounts of cumulative infiltration.
2. Compute the time to reach this amount of infiltration using equation 3.11 and soil factors in Table 3.1.
3. Compute the infiltration rate using equation 3.10.

The following example from Haan et al. 1994 shows the process.

Example 3.4: Calculate the infiltration curve for a silty clay loam at 30% effective saturation (S_e). Values for θ, ψ, and K_{sat} from Table 3.1 are 0.432, 27.3 cm, and 0.1 cm/hr, respectively. $\Delta L = (1 - S_e)\theta = (1 - 0.3)0.432 = 0.302$.

A vector or list of cumulative infiltration values is input into a spreadsheet or other equation-processing package. One can then solve equations 3.11 and 3.10. Plots for cumulative infiltration and infiltration rate for a generic silty clay loam soil (equilibrium moisture = 0.05, K_c = 0.1 cm/hr, Θ = 27 cm) appear in Figure 3.6.

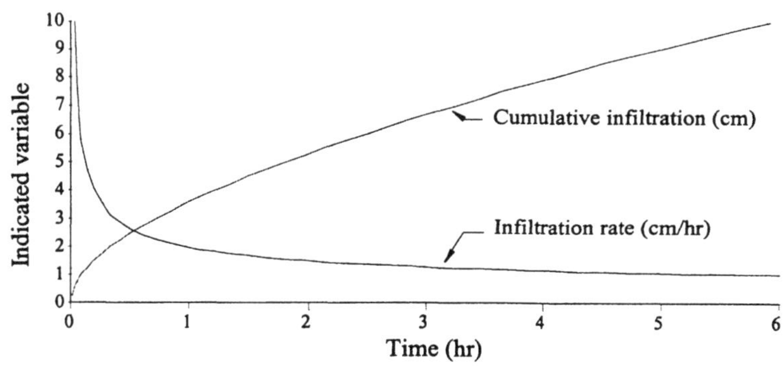

Figure 3.6. Cumulative infiltration and infiltration rate versus elapsed time in a silty clay loam soil.

Example 3.5: Given the soil conditions of Example 3.4, compute the infiltration volume and rate at a time of 1.05 hr.

The procedure of Example 3.4 was used to generate cumulative infiltration and infiltration rate as shown in Figure 3.6. Simply go to the time axis on the respective lines and estimate the corresponding values.

The Green-Ampt equation has several advantages over other equations. It is a relatively simple, physically based model, and the parameters used are based upon measurable physical properties of the soil. Problems with the Green-Ampt equation include (1) the model is one-dimensional; (2) soil properties usually vary widely over a field and with depth; (3) the solution may require iteration; (4) surface sealing dynamics do not enter into the Green-Ampt model (although other investigators have addressed this issue; see Hillel 1998); and (5) there is one effective pore size instead of a pore distribution (pore size distribution causes the wetting front to diffuse at deeper depths in well-graded soils). The Phillip equation, discussed above, addresses most of these concerns.

Conductivity is the most influential variable in the Green-Ampt model. Matric potential, initial water content, and cumulative infiltration effects are comparatively small.

Infiltration Measurement

The many factors discussed above (e.g., soil properties, vegetative cover, initial soil water content, rainfall intensity) determine infiltration rate. Infiltration is directly quantified in the field using the equipment shown in Figure 3.7. Variation in the above factors requires multiple measurements. Variations approaching 100% of the mean measured infiltration are common.

The infiltrometers shown in Figure 3.7 have inner and outer chambers, each supplied separately. The purpose of the outer chamber is to cause all flow from the inner chamber to be vertical only. Adjusting the float valve enables level control of the inner chamber.

Figure 3.7. Double-ring infiltrometers installed in a killed mulch for characterizing infiltration rate in a conservation tillage system.

Inner and outer chambers should be maintained at equal levels. Attach a flow meter to the line supplying the inner chamber to measure flow. In Figure 3.7, the inner chambers of three infiltrometers are measured with a single flow meter (not shown). Knowing the area of the inner chambers, one may convert flow to depth in millimeters per hour. Infiltration with rainfall impact can be simulated with nozzles that spray a droplet size distribution similar to "rain" on a location prepared for capturing excess rainfall. The infiltration rate is the difference between the water application rate and the excess runoff rate. The photograph in Figure 3.8 shows a rainfall-runoff simulator.

The Green and Ampt (1911) approach for modeling infiltration is one of several available models. Zhang (1997) presents a straightforward methodology for easily deriving the steady-state conductivity from experimental cumulative infiltration data curves generated with the apparatus shown in Figures 3.7 and 3.8. He summarizes approaches for estimating conductivity based on soil water-holding parameters as well (more fully discussed in Chapter 14). Carsel and Parrish (1988) summarize the parameters needed for the Zhang (1997) model. One may find additional details for infiltration measurement, hydraulic conductivity measurement approaches, and data interpretation in Klute 1986.

Infiltration and water movement through porous media such as soils are critical for understanding drainage and pollutant movement through soils. Infiltration is usually an indirect "force." It affects the relationship between the precipitation and runoff forces. It affects irrigation and bioremediation efforts. Quantification of infiltration has much practical value.

Figure 3.8. Instrumentation for studying infiltration rate in the presence of simulated rainfall. (Courtesy of Dr. D. E. Radcliffe, Department of Crop and Soil Science, University of Georgia.)

Pollutant Transport by Subsurface Flow

Flux has units of velocity. The travel time of a pollutant would seem to be the distance traveled divided by the flux. However, flux by itself does not account for the solid matrix. Flux assumes the entire volume is available for the flow. To account for the solids, one must divide the flux by the matrix porosity. Thus, the travel velocity is given as

$$V_{seep} = \frac{f}{n} \tag{3.12}$$

Water is a primary transporter of organic pollution (Figure 3.9). The organics may be spilled fuel, pesticides, fertilizer residues, etc. To describe organic pollution transport we need to add some additional notions to our knowledge of water transport. The presence of dissolved materials in water alters (generally lowers) the surface tension of the water and therefore the matric potential. The dissolved material will likely interact with the soil. The concentration of the dissolved phase will usually change with flow distance.

Transport of pollutants in soils differs from transport of water in one important respect. Pollutants often interact with the soil matrix. They may be absorbed by the matrix, which slows the effective travel velocity. As an approximation, one may divide the computed travel velocity by the retardance factors shown in Table 3.2 to adjust the computed travel times.

Both convection and diffusion contribute to constituent transport in a porous medium. Fick's law describes diffusion transport of dilute constituents (i.e., less than 1% concentrations):

$$F_{diff} = -D_{disp}\frac{\partial C}{\partial Z} \tag{3.13}$$

where
F_{diff} is flux (moles/m^2-s);
C is constituent concentration in moles/m^3;
D_{disp} is the diffusivity (m^2/s);
Z is distance differential x, y, or z (m).

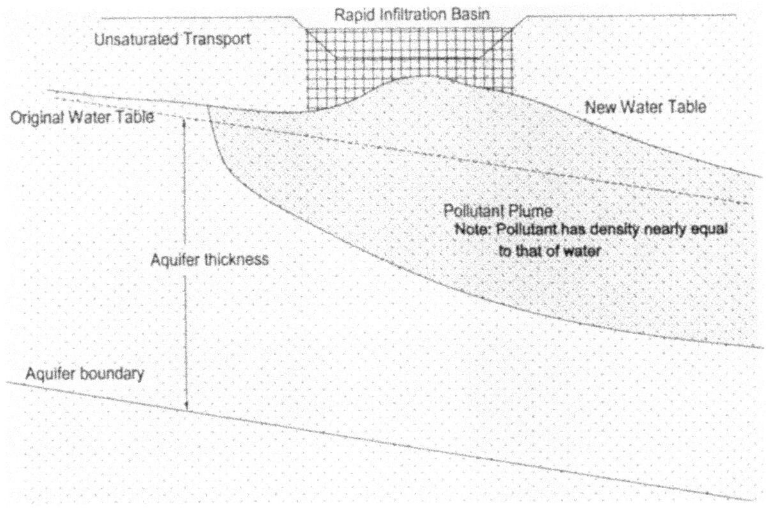

Figure 3.9. Schematic of a pollution transport problem.

Table 3.2. Retardation factors for selected compounds.

Material	Retardation factor R_d
Chloride, nitrate	1
Chloroform	3
Tetrachloroethylene	9
Toluene	3
Dichlorobenzene	14
Styrene	31
Chlorobenzene	25

Source: Adopted from Reed et al. 1995.

One may write the differential in Δ form as was done with Darcy's law. The flux is positive as the flux moves from point 1 to point 2. For low constituent concentrations, the convective transport may be expressed as

$$F_{conv} = qc \tag{3.14}$$

where

F_{conv} is the convective flux (moles/m^2-s);
q is the water flux f (m/s) times 1 unit area (m^2);
c is the constituent concentration (moles/m^3).

Total flux for concentrations above, say, 1% requires additional terms. Writing a shell balance leads to the convection-diffusion equation:

$$\frac{\partial(\rho\overline{V}c)}{\partial t} = \frac{\partial(qc)}{\partial z} - \frac{\partial}{\partial z}\left(D_{disp}\frac{\partial C}{\partial z}\right) + S(z, t) \tag{3.15}$$

Refer to Bird et al. 1960 or other texts on transport phenomena for these details.

Pollutant transport modeling is often computationally intensive. Density gradients must often be modeled. This problem has been introduced here only to show some of the ideas involved. Excellent software is available from commercial vendors and from the U.S. Environmental Protection Agency for modeling constituent transport. Pesticides, fertilizers, and industrial contaminants have been extensively studied. Serrano (1997) presents a student/professional-version software package (KYSPILL) for groundwater contamination analyses. Charbeneau and Daniel (1993) and Mercer and Waddell (1993) review software and give measurement approaches. They provide succinct summaries of the theory and practice of unsaturated and saturated contaminant transport.

Outlook

Infiltration is the movement of water into the soil, occurring at the time of natural rainfall or water application. Percolation is the subsequent downward movement of water within the soil in response to gravity. Infiltration and subsequent transport (in response to gravity or other hydraulic gradients) impact runoff, water quality, structure and surface sealing, and irrigation. The Green-Ampt model is the most physically based of the "simple" infiltration models. It enables physically based order-of-magnitude estimates.

ASAE 1983 is an excellent compilation of papers relating to infiltration physics. Computation of hydraulic heads is fundamental to subsurface drainage and pollutant transport through soils. Infiltration models find application across the spectrum of farm- and field-scale problems. Texts such as Bird et al. 1960 give an excellent discussion of transport. Soil physics texts (e.g., Hillel 1998) and unsaturated-zone hydrology texts (e.g., Tindall et al. 1999) give thorough discussions of contaminant transport processes in soil.

The U.S. Geological Survey and the U.S. Environmental Protection Agency (see links within http://www.usgs.gov and http://www.epa.gov) have downloadable software for modeling contaminant transport in porous media. A search on the Internet (use key words: e.g., porous media transport, infiltration modeling) will reveal several commercially available software packages for modeling porous media transport in one, two, or three dimensions based on the general convection-dispersion equation and Darcy's law.

Problems and Questions

1. Given a sandy loam soil initially at 5% of the equilibrium saturation value (e.g., rather dry), show the infiltration rate and cumulative infiltration curves for a total infiltrated amount of 100 mm. Follow Example 3.4. Rework for the sandy loam assuming an initial moisture content at 75% equilibrium saturation. Rework again, using a clay soil, 5% saturated. Summarize in a sentence or two the impact of soil type and of increasing the initial soil moisture level on infiltration rate.
2. Calculate the pore velocity of the sandy loam and clay soil flows in Problem 1 at the time 50% of the water is infiltrated, assuming an initial 5% saturation.
3. Compute the difference in total potentials and potential gradient in Figure 3.10: X = 10 cm; Y = 25 cm; Z = 8 cm; L = 100 cm. If K_{sat} = 1.5 cm/hr, compute the Darcian flux given the computed potential gradient. If the tube diameter D is 1.5 cm, compute the mass flux.
4. (a) Compute the hydraulic conductivity, given a flow of 1 cm^3/hr under the conditions shown in Figure 3.10, when X = 20, Y = 40, and Z = 10. The soil tube diameter is 2 cm. (b) Compute the flow for conditions shown in Figure 3.11, assuming the K to be 2 cm/hr. Is steady-state flow possible in Figure 3.11?

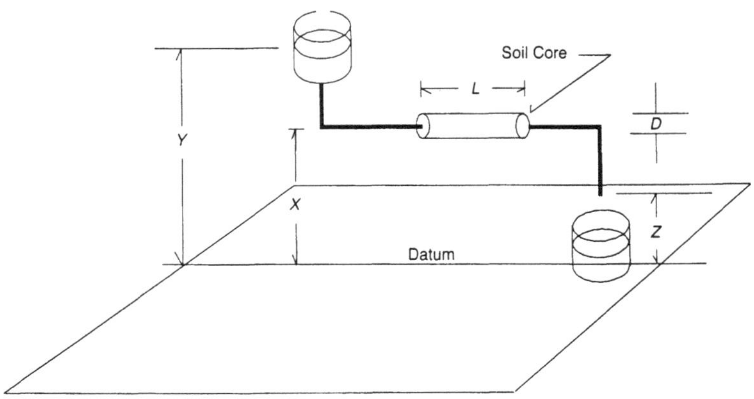

Figure 3.10. An infiltration problem situation.

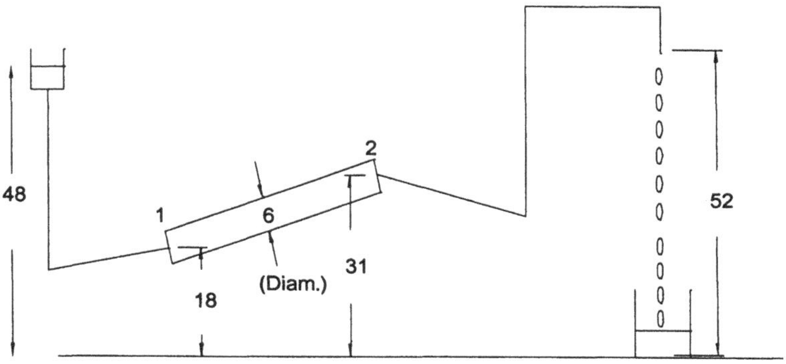

Figure 3.11. Darcian flow scenario.

5. Given a tube of 0.02 mm diameter, compute the capillary rise expected if one placed the tube in distilled water at 20°C.
6. Define the difference between infiltration and percolation.
7. Define the difference between Darcian flux (units of *L*/*T*) and Darcian flow (units of L^3/T).
8. Fit the generated infiltration data of Problem 1 to the Horton model using appropriate plotting techniques. Which model fits the data better? *Hint:* Estimate the long-term infiltration rate by determining the rate at the longest time analyzed in Problem 1. This is f_c; estimate f_0 by using the first computed infiltration rate from Problem 1; plot $f - f_c$ versus $(f_0 - f_c)e^{-kt}$ on semilog paper; determine best fit by manually minimizing the deviation of points from the best-fit line by eye; estimate *k* by computing the slope of the curve.
9. Suppose the clay blanket in Example 3.3 had an effective thickness of 1.5 cm and a hydraulic conductivity of 5×10^{-10} cm/s. Assume this problem to be identical to Example 3.3 otherwise. Compute the mass flux (i.e., flow rate) given these modifications.
10. Reconsider Example 3.3 wherein a matric potential of −25 cm exists at the bottom of the clay blanket. How sensitive (compared to Example 3.3) is the analysis to the assumption of a matric potential of zero at the lower clay blanket–soil interface?
11. Compare and contrast the time required to infiltrate 50 mm water into a sand, silt loam, and clay loam soil. *Hint:* Pick an arbitrary degree of saturation and representative values for the above-mentioned soil types and do the Green-Ampt analyses. Base your discussion on these results.
12. What is infiltration? How can it be quantified?
13. What is hydraulic conductivity? Describe an experiment for measuring it.
14. Define *potential,* giving units. What are elevation potential, matric potential, and pressure potential relative to an arbitrary datum?
15. Define *Darcian flux*. What is the difference between Darcian flux and seepage velocity?
16. How can you compute infiltration assuming ponded water on the surface? *Hint:* Review the infiltration models discussed, focusing on how to use each of them.
17. Using an Internet search, identify some commercially available software packages for modeling subsurface flow. Specify the key words and the name of the search engine used in your search.

Note

1. Convective laminar and turbulent flow of liquids and gases through soil (or packed bed) is described with an equation similar to but not identical to Darcy's law. Texts such as Bird et al. (1960) describe laminar and turbulent flow of liquids and gases through packed media. Turbulent flow is an issue particularly with gases.

References

ASAE. 1983. *Advances in infiltration*. Pub. 11-83. American Society of Agricultural Engineers, St. Joseph, MI.

Bear, J. 1972. *Dynamics of fluids in porous media*. Dover Publications, New York, NY.

Bird, R. B., W. E. Stewart, and E. N. Lightfoot. 1960. *Transport phenomena*. John Wiley & Sons, New York, NY.

Carsel, R. F., and R. S. Parrish. 1988. Developing joint probability distributions of soil water retention characteristics. *Water Resources Research* 34(5):755–69.

Charbeneau, R. J., and D. E. Daniel. 1993. Contaminant transport in unsaturated flow. In *Handbook of hydrology*, ed. D. R. Maidment. McGraw-Hill, New York, NY.

Green, W. H., and G. A. Ampt. 1911. Studies in soil physics I. The flow of air and water through soils. *Journal of Agricultural Science* 4:1–24.

Haan, C. T., B. J. Barfield, and J. C. Hayes. 1994. *Hydrology and sedimentology of small catchments*. Academic Press, New York, NY.

Hillel, D. 1998. *Environmental soil physics*. Academic Press, New York, NY.

Holtan, H. N., and M. H. Kirkpatrick. 1950. Rainfall, infiltration and hydraulics of flow in runoff computation. *Transactions of the American Geophysical Union* 31:771–79.

Horton, R. E. 1939. Analysis of runoff plot experiments with varying infiltration capacity. *Transactions of the American Geophysical Union* 20:693–711.

Klute, A. 1952. A numerical method for solving the flow equation for water in unsaturated materials. *Soil Science* 73:105–16.

———, ed. 1986. *Methods of soil analyses. Part 1, Physical and mineralogical methods*. Agronomy Monograph no. 9 (pt. 1). American Society of Agronomy, Madison, WI.

Mercer, J. W., and R. K. Waddell. 1993. Contaminant transport in groundwater. In *Handbook of hydrology*, ed. D. R. Maidment. McGraw-Hill, New York, NY.

Rawls, W. J., L. R. Ahuja, D. L. Brakensiek, and A. Shirmohammadi. 1993. Infiltration and soil water movement. In *Handbook of hydrology*, ed. D. R. Maidment. McGraw-Hill, New York, NY.

Rawls, W. J., D. L. Brakensiek, and N. Miller. 1983. Green-Ampt infiltration parameters from soils data. *Journal of Hydraulic Engineering, Proceedings of the American Society of Civil Engineering* 109(1):62–70.

Reed, S. C., R. W. Crites, and E. J. Middlebrooks. 1995. *Natural systems for waste management and treatment*, 2d ed. McGraw-Hill, New York, NY.

Serrano, S. E. 1997. *Hydrology for engineers, geologists, and environmental professionals*. HydroScience, Lexington, KY.

Tindall, J. A., J. R. Kunkel, and D. E. Anderson. 1999. *Unsaturated zone hydrology for scientists and engineers*. Prentice Hall, Upper Saddle River, NJ.

Zhang, R. 1997. Determination of soil sorptivity and hydraulic conductivity from the disk infiltrometer. *Soil Science Society of America Journal* 61:1024–30.

4

Evapotranspiration

Evaporation is the transfer of liquid water into the atmosphere. Water molecules in air and water are in rapid motion. Evaporation occurs when the number of moving molecules breaking from the water surface exceeds the reentering number. Transpiration is the transfer of water vapor into the atmosphere from the plant. Scientists and engineers often combine evaporation and transpiration to arrive at the term *evapotranspiration* (ET). In plant production, evapotranspiration occurs from wet soil surfaces between plant rows and from plant leaves. Knowledge of both evaporation and transpiration is required for predicting crop water needs. Evapotranspiration science has its roots in the British Commonwealth and the United States. Worldwide, the United Nations Food and Agriculture Organization (FAO) has contributed significantly to evapotranspiration prediction science.

Agriculture has historically been a major consumer of water. The biomass yield of a healthy, well-watered crop is well correlated to water transpired during growth. The *transpiration ratio* of a plant is a measure of its biomass conversion efficiency. It is the ratio of the mass of water transpired to the plant dry-matter weight. The transpiration ratio is an important plant selection consideration in arid regions. Transpiration ratio examples include 250 kg of transpired water per 1 biomass kg for sorghum, 350 for corn, 500 for wheat, 640 for potatoes, and 900 for alfalfa. Some investigators call the transpiration per unit mass "water use efficiency." The higher the transpiration ratio, the more water used per gram of plant dry weight. Plants with low ratios are usually less susceptible to water stress than plants with high ratios. Stress at unfavorable times (e.g., the time of blooming, harvestable yield development) can be devastating for any crop. This explains the interest of agriculture in protecting the extensive historical claim to water resources.

Goals

- Develop several empirical and physically based models (Myers aerodynamic equations; Bowen ratio; Penman-Monteith combination equation; Priestly-Taylor, Jensen-Haise, Blaney-Criddle, and Thornthwaite equations; and evaporation pan relationships) for predicting evaporation from climatological data and small evaporation pans.
- Predict evapotranspiration of crops by calculating evaporation from nearby water bodies and transpiration from well-watered crops.

Factors Affecting Evaporation

Evapotranspiration is driven by solar energy and is dependent on mass transfer relationships. The energy level depends upon factors such as hours of sunshine and solar intensity, cloud cover, reflectivity of surrounding surfaces, surface slope, and surface elevation. Sunshine hours and intensity, cloud cover, surface reflectivity, and surface slope affect the energy available for driving the evaporation process. Elevation affects energy received.

Wind speed, crop canopy characteristics, salinity of evaporating water, and relative humidity each affect mass transfer. Elevation also affects vapor pressure and vapor carrying capacity. Salinity affects the vapor pressure of the evaporating water, and relative humidity affects the vapor carrying capacity of the atmosphere.

Evaporation Prediction Approaches

The energy and mass factors are used in various ways, depending on modeling simplicity or thoroughness, in models for predicting ET. Methods of predicting evaporation are based on the following approaches:

- Mass transfer: Requires measurements of humidity and wind velocity at two or more locations and is based upon variations of Fick's law. Fick's law is a vapor concentration transport model analogous to Ohm's, Darcy's, and Fourier's law for electricity, water, and heat, respectively.
- Energy balance: Based on a heat balance; energy components are difficult to measure accurately. The Bowen ratio method, discussed by James (1988), involves measurement of temperature and humidity at two elevations. This method, originally proposed in 1926, is not widely used due to the expense of the measurements. The availability of modern instrumentation may be reversing this trend.
- Combination approaches: Address both energy and mass transfer aspects. The Penman-Monteith equation is the best combination approach.
- Pan evaporation: A physical analog based on use of a standard evaporation pan. One makes daily measurements of evaporated water and relates these measurements to evapotranspiration from a well-watered crop by applying a coefficient.
- Empirical methods: Developed from practical experience and field research; based primarily on the assumption that available energy for evaporation is proportional to temperature or radiation. Jensen-Haise and Blaney-Criddle are popular empirical approaches. The Priestly-Taylor method is used in crop modeling. Hatfield (1985) gives other examples.

Each method attempts to predict evaporation from a well-watered surface. Techniques such as the simple aerodynamic equation (E_{pond}), the Penman-Monteith (E_p), the modified Blaney-Criddle (E_{BC}), the Priestly-Taylor (E_{PT}), the Jensen-Haise (E_{JH}), and the pan evaporation method (E_{pan}) approximate the evaporation from a well-watered surface E_p. Multiplying the respective E_p value by a crop coefficient gives the evapotranspiration. This coefficient depends mainly on the physiological development of the crop.

Evaporation from Water Surfaces Based Solely on Aerodynamic Effects or Mass Balance

Dalton's law, shown in equation 4.1a, is the aerodynamic equation for predicting evaporation from free water surfaces:

$$E_{Pond} = C_{con}(pv_s - pv_a) \quad \textbf{(4.1a)}$$

where the nomenclature depends on the empirical proportionality constant C_{con}. We will discuss this constant further below.

Dalton's law is a mass transfer–based approach. It assumes that resistance to transport occurs in a small layer immediately next to the evaporating surface and that the gradient in vapor pressure is constant with position and time. These assumptions are usually meaningful only on the field and larger scales.

One must compute C_{con} at a prescribed height. The actual vapor pressure e_d is measured at the average of the daily maximum and minimum (1) temperatures or (2) relative humidity. Relations for estimating C_{con} include the Rohwer (1931) and the Meyer (1942) equations.

The Rohwer (1931) equation for estimating C_{con} is

$$C_{con} = (0.44 + 0.073v_{w0.15})(1.465 - 0.00073p_{atm}) \quad \textbf{(4.1b)}$$

where, for equations 4.1a and 4.1b,

C_{con} is the Dalton constant in mm/(day · mm Hg);
$v_{w0.15}$ is the average wind velocity at the 0.15 m height (km/hr);
pv_a is atmospheric pressure at 0°C (mm Hg, where 1 mm Hg = 133.322 Pa; 1 atm = 760 mm Hg = 101.325 kPa);
pv_s is the saturation vapor pressure (mm Hg) at an average of the maximum and minimum temperatures, at 7.6 m height;
pa_{ve} is the actual vapor pressure (mm Hg) at 7.6 m, based on the mean humidity;
E_{pond} is the evaporation from a water body computed using the appropriate C_{con} (mm/day).

One may correct the calculated free water surface evaporation E_{pond} when using the Rohwer relationship to obtain E for reservoirs by multiplying the free surface E_{pond} by a correction factor ranging from 0.6 to 0.8 (0.77 is suggested). The correction is needed because surrounding land forms impede air movement over the evaporating surface.

Meyer (1942) suggested an alternative relationship for the C_{con} parameter in the Dalton equation. For evaporation from small ponds and shallow ponds, he suggested equation 4.1b. He suggested using equation 4.1c for large water bodies.

$$C_{con} = 15 + 0.93v_{w7.6} \quad \textbf{(4.1c)}$$

$$C_{con} = 11 + 0.68v_{w7.6} \quad \textbf{(4.1d)}$$

where

C_{con} is the Dalton constant in mm/(month · mm Hg);
$v_{w7.6}$ is the average wind velocity in km/hr at 7.6 m above the evaporating surface.

Wind speed measurements at one height must occasionally be related to measurements at another height. Equation 4.2 relates wind speed to equivalent wind speed at an alternative height:

$$v_{w\,new} = v_{w\,known}(\text{new } v \text{ height/known } v \text{ height})^{0.2} \quad \textbf{(4.2)}$$

Due to the "oasis" effect and elevated water temperatures, evaporation depth is usually greater in small ponds and pans than in large lakes. Salinity depresses vapor pressure. Freshwater evaporation studies usually neglect salinity effects. Besides the mass transfer resistance, the constant C_{con} also embodies the energy driving the evaporation. One may convert evaporation rate to conventional units. Note that 1 mm depth of water over 1 square meter is equivalent to 1 kg water. Square meters of evaporating surface times millimeters per day evaporation gives kilograms of water evaporated per day.

Example 4.1: Compute the evaporation rate from a small reservoir in millimeters per day using the Dalton equation and estimating the Dalton coefficient from the Rohwer and Meyer equations. The wind velocity is 5 km/hr at 0.15 m height. Assume a standard atmospheric pressure of 760 mm Hg. The water temperature is 20°C, and the mean relative humidity is 50% (at a height of 7.6 m and near the water surface). Assume the air and water are at equilibrium temperatures.

First, we compute the C_{con} for the Rohwer and the Meyer equations. For Rohwer (equation 4.1b), C_{con} = [0.44 + 0.073(5 km/hr)][1.465 − 0.00073(760 mm Hg)] = 0.73 mm/(day · mm Hg). For Meyer (equation 4.1d), we first convert the 0.15 m (height) wind velocity to the equivalent 7.6 m height velocity. This is accomplished using equation 4.2. For the conditions of this problem we have $v_{w7.6}$ = 5 km/hr (7.6 m/0.15 m)$^{0.2}$. This leads to an estimate of $v_{w7.6}$ = 10.96 km/hr. Substituting into the Meyer equation for a small reservoir we have C_{con} = 11 + 0.68(10.96) = 18.4 mm/(mo · mm Hg). We can divide by 30 (days per month) to get C_{con} = 0.61 mm/(day · mm Hg) for a daily rate.

To apply the Dalton equation (equation 4.1a) one must compute the saturated and actual air vapor pressures. The relationship between vapor pressure and temperature (*Te*, in Celsius) and relative humidity (*rh*, as a decimal fraction) is

$$pv_d = \{3.38639[(0.00738Te_{7.6} + 0.8072)^8 - 0.00019(|1.8Te_{7.6} + 48|) + 0.0001316]\}7.502rh \tag{4.3}$$

where

$Te_{7.6}$ is temperature (°C) of air at the 7.6 m height;
rh is relative humidity, expressed as a decimal fraction; when computing the saturated vapor pressure, *rh* = 1.0 in equation 4.3;
pv_d is the vapor pressure in mm Hg; the factor 7.502 converts kPa to mm Hg.

Assuming equality between the air and water temperatures, from steam tables (or using relationships given in equation 4.3), the saturated pressure value is 17.15 mm Hg (e.g., 100% *rh* or 1.0 in equation 4.3). The actual air vapor pressure at 50% *rh* (use 0.5 in equation 4.3) is then 8.57 mm Hg.

Dalton's equation with the Rohwer constant is E_{pond} = 0.56 mm/(day-mm Hg)(17.15 mm Hg − 8.57 mm Hg) = 4.8 mm/day. Dalton's equation with the Meyer constant is E_{pond} = 0.61 mm/(day-mm Hg)(17.15 mm Hg − 8.57 mm Hg) = 5.23 mm/day. The near 8% difference in the evaporation predicted by the two methods is typical for evaporation predictions.

Evaporation Prediction from the Energy Balance: The Bowen Ratio Method

One may use the energy balance for predicting evaporation from a wet surface. One may write the energy balance as

$$\Delta Q = Q_s - Q_r + Q_a - Q_{ar} + Q_v - Q_{bs} - Q_e - Q_h - Q_w - Q_{con} \quad \textbf{(4.4)}$$

where

ΔQ is the increase in stored energy by the water body;
Q_s is the shortwave radiation incident at the water surface;
Q_r is the reflected shortwave radiation;
Q_a is the absorbed longwave radiation;
Q_{ar} is the reflected longwave radiation;
$Q_v = Q_{vi} - Q_{vo}$ is the net advected energy (bulk energy flow into and out of the body).
Q_{bs} is the backscattered longwave radiation emitted by the water body;
Q_e is the energy used in evaporation;
Q_h is the energy conducted from water body to air as sensible heat;
Q_w is the energy advected by the evaporating water;
Q_{con} is the heat conducted to the earth from the water body.

Units are in cal/(cm^2 · day), watts/m^2, or other consistent energy flux units.

Figure 4.1 shows in schematic form the energy balance of equation 4.4.

The ratio of sensible to latent heat as $B = Q_h/Q_e$ is customarily called the Bowen (1926) ratio. Energy advected by the evaporating water is calculated as $Q_w = C_p Q_e (T_e - T_b)/\lambda$ where c_p is the specific heat of water (kJ/kg − °C), T_e is the temperature (°C) of the evaporating water, T_b is the temperature (°C) of an arbitrary datum, and λ is the latent heat of water (mJ/kg). One may express evaporation E_{Br} as $Q_e/\lambda\rho$, where ρ is the mass density (Mg/m^3). Introducing these expressions into equation 4.4 and solving for E_{Br} results in

$$E_{Br} = \frac{Q_s - Q_r + Q_a - Q_{ar} - Q_{bs} - \Delta Q + Q_v - Q_{con}}{\rho[\lambda(1 + B) + c_p(T_e - T_b)]} \quad \textbf{(4.5a)}$$

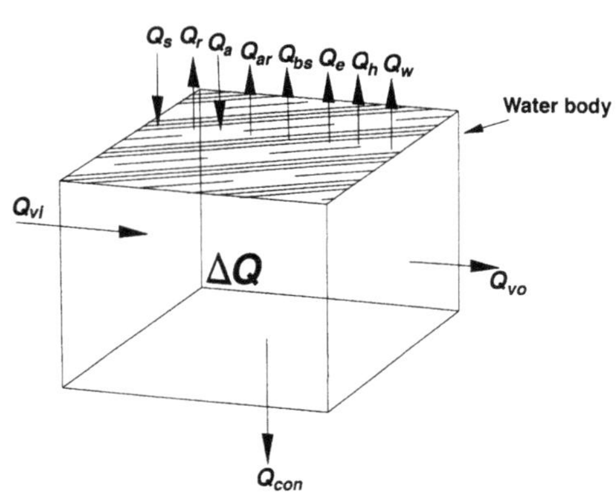

Figure 4.1. Heat balance on water body with top surface open to the atmosphere and sides, bottom, and incoming and outgoing streams adjacent to earth. In equation 4.4 Q_v is the net of Q_{vi} and Q_{vo}.

The Bowen ratio is required to predict evaporation with equation 4.5a. Going back to the definition of the Bowen ratio (Q_h/Q_e), one may write $Q_h = k\rho c_p A(T_o - T_a)$ and $Q_e + \lambda k_v A(e_o - e_a)$, where A is the surface area, k is the film resistance coefficient (watts/m^2 – °C), k_v is the vapor film conductivity (kg/s-mbar-m^2), and e_o and e_a are the vapor pressures at the surface and air (mbar), respectively. One may also use turbulent transfer coefficients for the sensible heat and vapor. Thus, the Bowen ratio (B) may be determined as shown in equation 4.5b, where Z represents height:

$$B = \frac{k\rho c_p T_o - T_a}{k_v\lambda\, e_o - e_a} = \frac{k\rho c_p\, \Delta T}{k_v\lambda\, \Delta Z} \tag{4.5b}$$

At standard atmospheric pressure (760 mbar, 101.325 kPa) the product of thermal conductivity, density, and specific heat divided by the product of vapor conductivity and heat of vaporization is approximately 0.46. Bowen (1926) gives further details of the derivation.

Measured or estimated values of the radiation heat transfer components of equation 4.5a are required. Incident longwave and shortwave radiation may be measured. One can estimate reflected longwave and shortwave radiation by applying appropriate reflection coefficients. Radiation terms are usually expressed as net radiation, discussed more fully below. The energy advection and storage terms require estimates of water temperatures and water budgets for the water body. Net advection and sensible heat accumulation are often neglected, as is conductive loss to soil. The Bowen ratio method as an approximate method for estimating evapotranspiration is discussed further below.

Evaporation (E_p) Prediction Accounting for Aerodynamic Effects and Energy Balance: The Penman-Monteith Combination Equation Method

The primary use of E_p estimates is to predict irrigation water requirements for growing plants.

The Penman-Monteith combination equation, sometimes called simply the Penman method (Campbell 1977), is one of the more theoretically correct evaporation equations. It builds on an energy balance coupled to an aerodynamic component. One can write the energy balance as

$$\lambda E_p = R_n - (G + H) \tag{4.6}$$

Figure 4.2a shows the energy balance pictorially. In Figure 4.2a,

R_n is the net radiation, equal to the incoming shortwave radiation to the surface minus reflected shortwave radiation minus the net longwave radiation, in W/m^2 or MJ/(m$^2 \cdot$ day); reflection coefficients appear in Table 4.1 and solar radiation available at the top of the atmosphere (R_{so}) appears in Table 4.2;
G is the sensible heat energy lost to the ground by conduction, in W/m^2 or MJ/(m$^2 \cdot$ day);
H is the sensible heat lost to the air, in W/m^2 or MJ/(m$^2 \cdot$ day);
$H + G$ represents the heat not available for evaporating water, in W/m^2 or MJ/(m$^2 \cdot$ day);
E_P = evaporation, in mm/day or kg/(m$^2 \cdot$ day), from a well-watered surface, from Penman-Monteith equation;
λ = heat of evaporation (MJ/kg).

In other words, energy for evaporation is the net radiation minus energy dissipated to surroundings. The G term is usually set to zero, although it is usually slightly positive in

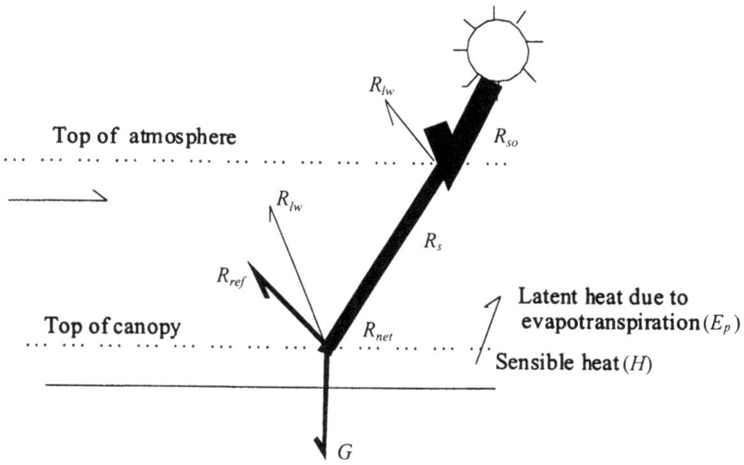

Figure 4.2a. Simplified evapotranspiration energy balance schematic.

spring and early summer (ground is warming) and slightly negative in late summer and early fall (ground is cooling).

The Penman-Monteith equation presumes a one-dimensional vapor balance with net movement in the vertical direction. Because vertical vapor movement is assumed, the method is most accurate away from field edges. The control volume top must be under the momentum, heat, and moisture boundary layer as shown in Figure 4.2b. Let us first recall some fundamental relationships for heat and mass transfer by convection (this is the aerodynamic component). Campbell (1977) describes sensible heat as

$$H = h\rho c_p(Te_{LW} - Te_a) = \rho c_p \frac{Te_{LW} - Te_a}{r_h} \tag{4.7}$$

where

H is the sensible heat flux, MJ/(m^2 · day)
h is the convection coefficient, m/day;

Table 4.1. Reflectivity (albedo) coefficients.

Condition	Coefficient
Moist black soil	0.08
Dry black soil	0.14
Moist gray soil	0.1
Dry gray soil	0.1 to 0.12
Spring wheat	0.01 to 0.25
Dense high grass	0.26
Tops of fir, pine, oak trees	0.1 to 0.18
Cotton	0.20 to 0.22
Alfalfa	0.23 to 0.33
Commercial irrigated crops	0.20 to 0.25
Clean snow	0.95
Water surface	0.05

Source: Excerpted from Merva 1995.

Table 4.2. Mean solar radiation at the top of the atmosphere, R_{so}, in units of MJ/(m^2·day).

	North latitude				
Month	0°	10°	20°	30°	40°
Jan.	28.18	25.25	21.65	17.46	12.27
Feb.	29.18	26.63	25	21.65	17.04
Mar.	30.02	29.43	28.18	25.96	22.9
Apr.	28.47	29.6	30.14	29.85	28.34
May	26.92	29.6	31.4	32.11	32.11
June	26.25	29.31	31.82	33.2	33.49
July	26.67	29.43	31.53	32.66	32.66
Aug.	27.76	28.76	30.14	30.44	29.18
Sep.	29.6	29.6	28.47	26.67	23.73
Oct.	29.6	28.05	25.83	22.48	18.42
Nov.	28.47	25.83	22.48	18.3	13.52
Dec.	26.8	24.41	20.5	16.04	10.76

Source: Adopted from Jensen et al. 1990.

c_p is the specific heat, MJ/(kg-°C);
ρ is the density, kg/m^3;
Te_{LW} is the temperature of the surface, °C;
Te_a is the ambient temperature, °C;
r_h is the reciprocal of the convection coefficient, day/m.

For mass transfer (e.g., the vapor transfer component) one can write

$$E_P = D_v(pv_{LW} - pv_a) = \frac{pv_{LW} - pv_a}{r_v} \tag{4.8}$$

where

E_P is the vapor flux rate, kg/(m^2·day) or mm/day;

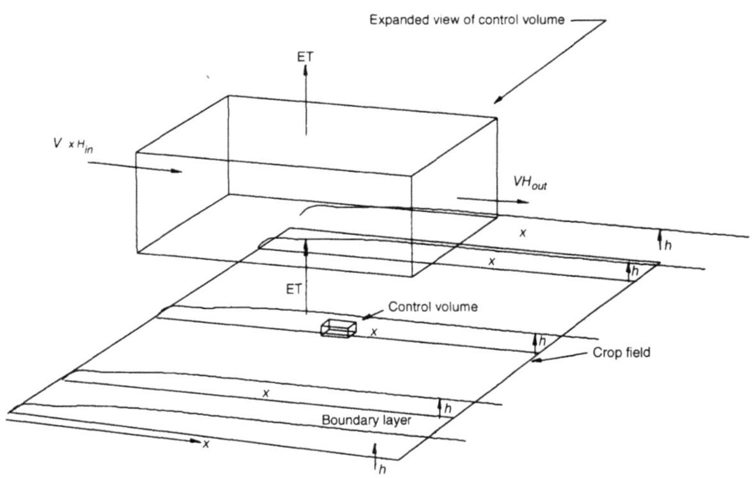

Figure 4.2b. Water vapor balance for the Penman-Monteith equation showing an expanded view of the control volume. There is also no net energy through the side walls of the control volume. Net sensible and latent energy moves vertically.

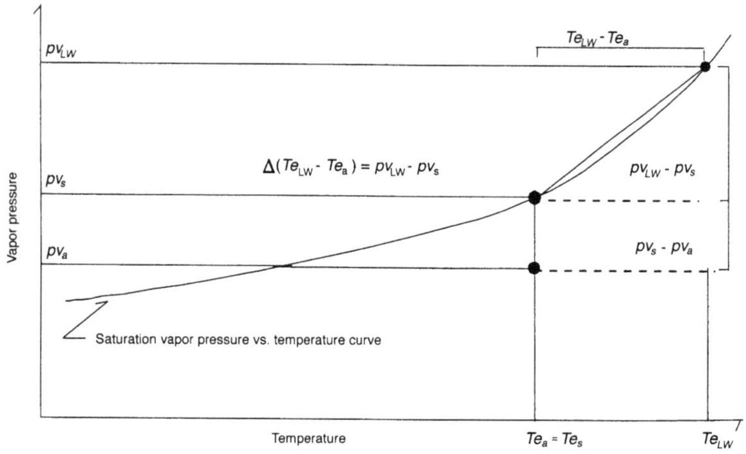

Figure 4.3. Vapor pressure density plot showing the Penman transformation where we assume the transpiring surface is a leaf or water surface. The plot indicates an evaporating surface that is hotter than the surrounding air. If the evaporating surface is cooler than the ambient air, the effect on the Penman transformation is a negative sign in front of the term containingΔ.

D_v is the vapor convection coefficient, m/day;
pv_{LW} is the vapor concentration at the leaf surface, kg/m^3;
pv_a is the vapor concentration in the air, kg/m^3;
r_v is the reciprocal of the vapor convection coefficient, day/m.

One may also write the units of the above equation using vapor pressure instead of vapor concentration, leading to units of D_v of kg/(m$^2\cdot$kPa$\cdot$day). Also, note that 1 mm depth of water over 1 m^2 is equivalent to 1 kg of water.

How can one relate vapor concentration at the leaf surface, which is driven by the surface temperature (hard to measure), to the saturated vapor pressure of the air (easy to measure)? The Penman transformation shown in Figure 4.3 accomplishes the relation.

One can write equation 4.9 based on Figure 4.3, where Δ represents the slope of the saturation vapor pressure curve:

$$pv_{LW} - pv_a = (pv_s - pv_a) + \Delta(Te_{LW} - Te_s) \tag{4.9}$$

where

pv_{LW} is the vapor pressure at the leaf or water body transpiring/evaporating surface, kPa;
pv_s is the air saturation vapor pressure at Te_a with 100% humidity, kPa;
pv_a is the actual air vapor pressure at Te_a at less than 100% relative humidity, kPa;
Δ is the slope of the saturation vapor curve, kPa°C;
Te_a is the air temperature, °C;
Te_{LW} is the temperature of the evaporating surface, °C.

One can write equation 4.10 due to the equality of air and saturated air temperature:

$$pv_{LW} - pv_a = (pv_s - pv_a) + \Delta(Te_{LW} - Te_a) \tag{4.10}$$

The left-hand side of this equation is similar to the moisture loss relation above. The right-hand term is similar to sensible heat loss. One can make some rearrangements as follows. First, multiply the above equation through by λ/r_v, giving a term identical to λE on the left-hand side. Second, rewrite the second term on the right-hand side of equation 4.10 as the sensible heat equation (equation 4.7) as shown in equation 4.11:

$$Te_{LW} - Te_a = Hr_h/c_p \tag{4.11}$$

The Penman transformation (equation 4.10) may then be rewritten

$$\lambda E_P = \frac{\lambda(pv_{LW} - pv_a)}{r_v} = \frac{\lambda(pv_s - pv_a)}{r_v} + \frac{\lambda \Delta H r_h}{\rho c_p r_v} \tag{4.12}$$

Noting $H = R_n - G - \lambda E$ and substituting into the second term on the right-hand side of equation 4.12 yields

$$\lambda E_P = \frac{\lambda(pv_s - pv_a)}{r_v} + \frac{\lambda \Delta (R_n - G) r_h}{\rho c_p r_v} - \frac{\lambda \Delta (\lambda E_P) r_h}{\rho c_p r_v} \tag{4.13}$$

Solving equation 4.13 for λE_P yields

$$\lambda E_P = \frac{\dfrac{\lambda(pv_s - pv_a)}{r_v} + \dfrac{\lambda \Delta r_h (R_n - G)}{\rho c_p r_v}}{1 + \dfrac{\lambda \Delta r_h}{\rho c_p r_v}} \tag{4.14}$$

Letting $\gamma_{Pen} = \rho c_p/\lambda$ and assuming $r_h/r_v \approx 1$ yields

$$\lambda E_P = \frac{\dfrac{\lambda(pv_s - pv_a)}{r_v} + \dfrac{\Delta(R_n - G)}{\gamma_{Pen}}}{1 + \dfrac{\Delta}{\gamma_{Pen}}} \tag{4.15}$$

Why does one use $r_h/r_v \approx 1$? Turbulent eddies drive heat transport and vapor transport in most field applications. The same turbulence structure carries both heat and mass. Therefore, the transport parameters (i.e., resistances) are similar in many situations—thus the assumption $r_v = r_h$. Rearranging the denominator of equation 4.15 yields

$$\lambda E_P = \frac{[\gamma_{Pen}\lambda(pv_s - pv_a)/\gamma_{Pen} r_v] + \Delta(R_n - G)/\gamma_{Pen}}{(\gamma_{Pen} + \Delta)\gamma_{Pen}} \tag{4.16}$$

One may simplify equation 4.16:

$$\lambda E_P = \frac{\Delta(R_n - G) + [\gamma_{Pen}\lambda(pv_s - pv_a)]/r_v}{\Delta + \gamma_{Pen}} \tag{4.17}$$

Next, an empirical function $Wind_f$ may be used to describe the inverse of vapor transport resistances:

$$\lambda E_P = \frac{\Delta(R_n - G)}{\gamma_{Pen} + \Delta} + \frac{\gamma_{Pen}}{\Delta + \gamma_{Pen}} Wind_f(pv_s - pv_a) \tag{4.18}$$

The wind function $Wind_f$ has several variants. In the following equation the Jensen et al. (1990) wind function, $Wind_f = 6.43(1 + 0.53v_{2m})$, is included. Solving for E_p yields

$$E_P = \left[\frac{\Delta}{\Delta + \gamma_{Pen}}(R_n - G_n) + \frac{\gamma_{Pen}}{\Delta + \gamma_{Pen}}[6.43(1 + 0.53v_{2m})](pv_s - pv_a)\right]\left[\frac{1}{\lambda}\right] \tag{4.19}$$

Table 4.3. Approximate crop coefficients K_c for a well-watered grass reference.

Date		Corn (grain), Midwest	Cotton, Southeast	Potatoes, Northwest	Soybeans, Midwest
Apr.	1–15	—	—	—	—
	16–30	—	0.1	—	—
May	1–15	0.2	0.2	0.1	0.1
	16–31	0.4	0.3	0.4	0.3
June	1–15	0.7	0.6	0.7	0.6
	15–30	0.9	0.9	0.9	0.8
July	1–15	1	1.2	1.1	1
	16–31	1	1.2	1.1	1
Aug.	1–15	1	1.2	1.1	1
	16–31	0.9	1.2	1	1
Sep.	1–15	0.6	1.1	0.8	0.8
	16–30	—	1.1	0.8	0.8
Oct.	1–15	—	0.5	—	—
	16–31	—	—	—	—

Source: Adapted from Jensen et al. 1990.

Note: Local irrigation guides typically contain detailed charts of crop coefficients for regionally significant crops.

One may now compute the predicted evapotranspiration ET_p by multiplying E_p by the crop coefficient K_c found in Table 4.3 (ET_p refers to the predicted evapotranspiration for a crop at a given stage of growth). The coefficients in Table 4.3 were empirically determined. *Equation 4.19 is the Penman-Monteith combination equation.*

Equations 4.20 through 4.32 provide inputs to equation 4.19. A general expression for vapor pressure is

$$pv_x = f(Te_y) \tag{4.20}$$

where pv_x is the vapor pressure (kPa) at condition x as a function of temperature Te_y (°C); Te_y may correspond to saturated or nonsaturated conditions.

Many calculation formulas exist for the psychrometric parameters (Jensen et al. 1990). Vapor concentration correlates to vapor pressure as with Dalton's law. One may use empirical relations such as equation 4.21 to compute the vapor pressure:

$$pv_x = f(Te_y) = 3.38639[(0.00738Te_y + 0.8072)^8 - 0.00019(|1.8Te_y + 48|) + 0.0001316] \tag{4.21}$$

Equation 4.21 is identical to equation 4.3, introduced in the discussion of the Dalton equation, except that pv_x in equation 4.21 is in units of kilopascals (kPa).

To compute the evaporating surface vapor pressure one can use equation 4.20 to write

$$pv_s = f(Te_{av}) \tag{4.22}$$

where Te_{av} is the daily average temperature, and the function f is defined in equation 4.21. To compute the vapor pressure of the air (p_a) one can use the dew point temperature in equation 4.20 to result in

$$pv_a = f(Te_{dp}) \tag{4.23}$$

where Te_{dp} is the dew point temperature.

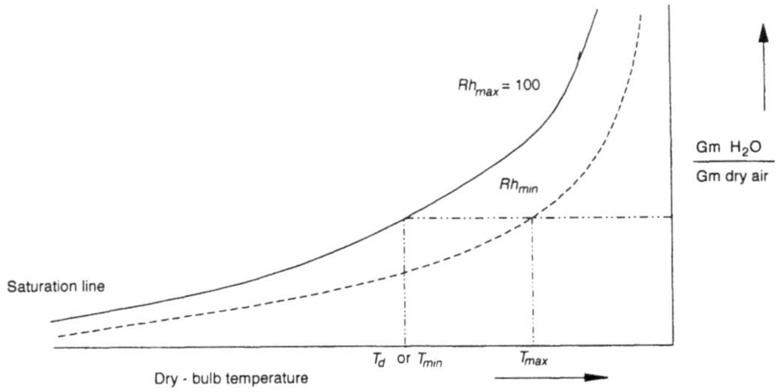

Figure 4.4. Schematic of psychrometric chart showing relationships between moisture content, saturation temperature, and dry-bulb temperature.

One may estimate the dew point temperature using the saturated moisture content at the evening low temperature. Parameter pv_s is the water vapor pressure, and p_s is the vapor pressure of the saturated ambient air. The 100% relative humidity defines saturation at the minimum or dew point temperature.

If one knows relative humidity but not a minimum temperature or dew point, pv_a may be computed using

$$pv_a = pv_s\, RH_{min} \tag{4.24}$$

The relative humidity is the mean relative humidity value occurring at the maximum temperature usually. The moisture level of the air is bounded by the saturation level at the evening minimum temperature, as evidenced by frequent dew events. As the air warms through the day, the moisture stays somewhat constant, barring passage of fronts or other incoming precipitation. Figure 4.4 shows the psychrometric relations schematically.

One may compute values for other parameters using the following equations. This text uses degrees Celsius for all temperatures and kilopascals for all pressures. The slope (Δ) of the saturation vapor pressure versus temperature Δ (kPa/°C) may be computed using the empirical relationship given in

$$\Delta = 0.2(0.00738Te_{av} + 0.8072)^7 - 0.000116 \tag{4.25}$$

The latent heat of vaporization λ (MJ/kg) may be computed using

$$\lambda = 2.501 - 0.002361Te_{av} \tag{4.26}$$

The atmospheric pressure p_{atm} (kPa) and the psychrometric constant γ_{Pen} (kPa/°C) may be computed using

$$p_{atm} = 101.325 - 0.01055E_l \tag{4.27}$$

$$\gamma_{Pen} = 0.00163\frac{p_{atm}}{\lambda} \tag{4.28}$$

where E_l is the elevation above sea level in meters.

The net radiation R_n, in MJ/(m^2-day), is given by

$$R_n = (1 - \alpha)R_s - \sigma(Te_{av} + 273)^4(0.34 - 0.139pv_a^{0.5})\left(0.1 + 0.9\frac{n}{N}\right) \tag{4.29}$$

$$R_s = \left(0.35 + 0.61\frac{n}{N}\right)R_{so} \tag{4.30}$$

where

R_{so} is tabulated in Table 4.2 and is in MJ/(m^2-day);
n/N is the fraction of the daytime hours not obscured by clouds;
Te_{av} is the average of the mean maximum and mean minimum temperature in °C;
pv_a is the vapor pressure computed from equation 4.23 using Te_{dp} for the mean dew point temperature (or the minimum temperature if the dew point is unknown);
σ is the Stefan-Boltzmann coefficient, 4.903×10^{-9} MJ/(m^2-day K^{-4});
α is the shortwave reflectivity coefficient (Table 4.1).

Relationships given for net radiation are largely empirical.

The net radiation R_n is the difference between the incoming shortwave radiation and the outgoing longwave radiation. The equation given above is one of several. Burman et al. (1980) and Shuttleworth (1993) discuss other variations. If presented with wind speed measured at a height other than 2 m, one can adjust velocity with the following adaptation of equation 4.4:

$$v_2 = v_x\left(\frac{2}{X}\right)^{0.2} \qquad \textbf{(4.31)}$$

where v_x is the measured wind velocity at height X above the surface (L/T). This relationship is identical to the velocity correction relationship presented in the Rohwer and Meyer (Example 4.1) calculation discussion.

Units for the above relationships are as follows. Elevation is in meters. Pressure is in kPa. Specific heat λ is in MJ/kg. The psychrometric constant γ and the saturation vapor pressure slope Δ are in units of kPa/°C. The wind velocity v_2 is the wind speed (m/s) at a 2 m height. Elevation E_l is in meters. Attention to detail regarding units is necessary with the Penman-Monteith equation. The above relationships lead to radiation units of MJ/(m^2-day). References such as Burman et al. (1980) give radiation in langleys per day, the same as cal/(cm$^2 \cdot$ day). One MJ/m^2-day is 23.92 cal/(cm$^2 \cdot$ day). This text uses pressure in units of kPa while Burman et al. (1980) use mbars. Ten mbars = 1 kPa. This text uses latent heat in MJ/kg while Burman uses cal/gram (239 cal/g = 1 MJ/kg). Burman et al. (1980) and Shuttleworth (1993) show additional methods for calculating radiation values and wind functions. One must select a system of equations in consistent units when making evaporation computations. *Don't mix results representing two or more unit systems!* NASA determined that the *Mars Challenger,* a $25 million satellite designed to map the surface of Mars, was lost due to *confusion in units.*

Application of the Penman-Monteith method requires a significant number of measured data. One may calculate expected E_P from a body of water or well-watered grass using the Penman-Monteith equation or its variants. Alfalfa is a standard reference grass. A crop coefficient (Table 4.3) is applied to obtain actual crop evapotranspiration. The crop coefficient accounts for the degree of canopy cover and for the residual evaporation from the dry soil surface. As the canopy grows to cover the soil surface, the crop coefficient approaches 1. Occasionally the crop factor may slightly exceed 1 since the crop canopy is multilayered.

Due to all the theoretical inputs, the Penman-Monteith combination equation is not perfect. As one can discern from the derivation, the model involves many assumptions. All terms require proper measurement. The selection of the reflectivity (α), the soil conductivity (G), and the fraction of daytime that the sky is cloudless (n/m) influence the predicted evapotranspiration result. All these terms relate to incoming radiation energy available for evaporation.

Example 4.2: Compute the estimated E_{crp} for cotton in July near Tifton, Georgia, with the combination equation. The site elevation is 100 m. Mean maximum temperature is 35°C; mean minimum temperature is 22°C. Skies are 80% cloudless. Mean dew point is 15°C. Wind at 6 m is 5 m/s. *Note:* If not given a dew point, assume the minimum temperature is the dew point. Assume zero net daily heat gain to soil (i.e., $G = 0$).

$Te_{av} = (35°C + 22°C)/2 = 28.5°C$

$\Delta = 0.2[0.00738(28.5°C) + 0.8072]^7 - 0.000116 = 0.2258$ kPa/°C (equation 4.25)

$P_{atm} = 101.3$ kPa $- 0.01055$ kPa/m (100 m) $= 100.245$ kPa (equation 4.27)

$\lambda = \{2.501$ MJ/kg $-$ [0.002361 MJ/(kg-°C)](28.5°C)$\} = 2.4337$ MJ/kg (equation 4.26)

$\gamma_{Pen} =$ [0.00163 MJ/(kg-°C)](100.245 kPa)/2.4337 MJ/kg $= 0.0671$ kPa/°C (equation 4.28)

$p_s = 3.38639\{[0.00738(28.5°C) + 0.8072]^8 - 0.00019|1.8(28.5°C) + 48| + 0.001316\} = 3.832$ kPa

$p_a = 3.38639\{[0.00738(15°C) + 0.8072]^8 - 0.000019|1.8(15°C) + 48| + 0.001316\} = 1.6629$ kPa (equations 4.21–4.23)

$\Delta/(\Delta + \gamma_{Pen}) =$ (0.2258 kPa/°C)/(0.2258 kPa/°C + 0.0671 kPa/°C) $= 0.7709$

$\gamma_{Pen}/(\Delta + \gamma_{Pen}) =$ (0.0671 kPa/°C)/(0.2258 kPa/°C + 0.0671 kPa/°C) $= 0.2291$

$R_{so} = 32.66$ MJ $\cdot$ m^{-2} $\cdot$ day^{-1} (Table 4.2)

$R_s = [0.35 + 0.61(0.8)]32.66$ MJ $\cdot$ m^{-2} $\cdot$ day^{-1} $= 27.37$ MJ $\cdot$ m^{-2} $\cdot$ day^{-1} (equation 4.30)

$R_n = (1 - 0.22)(27.37$ MJ $\cdot$ m^{-2} $\cdot$ day$^{-1}) - 4.903 \times 10^{-9}$ (28.5°C + 273°C)^4[0.34 $-$ 0.139Sqrt(1.6629 kPa)][0.1 + 0.9(0.8)] $= 16.01$ MJ $\cdot$ m^{-2} $\cdot$ day^{-1} (use $\alpha = 0.22$, Table 4.1; equation 4.29)

$w_2 = 5$ m/s(2 m/6 m)$^{0.2}$ $= 4.012$ m/s (equation 4.31)

$\lambda E_P = 0.7709(16.01$ MJ $\cdot$ m^{-2} $\cdot$ day$^{-1}) + [0.2291(6.43(1 + 0.53(4.012$ m/s$))$ (3.832 kPa $-$ 1.6629 kPa)] $= 22.320$ MJ $\cdot$ m^{-2} $\cdot$ day^{-1}

$E_p =$ (22.32 MJ m^{-2}day^{-1})/(2.4337 MJ/kg) $= 9.17$ kg/(m^2-day) $= 9.17$ mm/day (equation 4.19)

Applying a crop coefficient $K_c = 1.2$ from Table 4.3 yields $ET_P = 11.01$ mm/day water use. Multiply by days in a month to get monthly ET_P.

Empirical and Approximate Methods for *E* Prediction

Due to the complexity of the Penman-Monteith equation, researchers have investigated methods requiring fewer data. They include the Jensen-Haise, Blaney-Criddle, and Priestly-Taylor methods.

The Jensen-Haise Method

The Jensen-Haise method (Jensen 1966) is an energy-based empirical method for evaporation prediction. It is an empirical relationship developed mainly from data collected in the intermountain region of the western United States. The method is primarily radiation driven, with some elevation, temperature, and humidity inputs. Burman et al. (1980) elaborate on this method. The modified Jensen-Haise method is

$$\lambda E_{JH} = \left(\frac{1}{\left[38 - \frac{2E_l}{305}\right] + 7.3\left[\frac{5}{pv_{Temin} - pv_{Temax}}\right]} \right)\left(Te_{av} - \left[-2.5 - 1.4(pv_{Temin} - pv_{Temax}) - \frac{E_l}{550}\right]\right)R_s \tag{4.32}$$

where

E_{JH} is the evaporation (mm/day) computed with the Jensen-Haise equation;
pv_{Temin}, pv_{Temax} are vapor pressures in kPa (calculate using equations 4.20 and 4.21);
Te_{av} is temperature in °C;
R_s is radiation in MJ/m^2 · day (Table 4.2 and equation 4.30);
E_l is elevation in meters.

The Jensen-Haise equation was developed for monthly periods. Convert to the period total by multiplying the E_{JH} value by the appropriate number of days. One can apply the crop coefficients given in Table 4.3 to compute monthly ET_{JH}. To get seasonal ET_{JH}, repeat the above computation for each month or fraction of month encompassing the entire growing season. Add the monthly totals for the seasonal water requirement.

Example 4.3: Compute the estimated ET_{JH} for cotton in July near Tifton, Georgia. Mean maximum temperature is 35°C (ρ = 5.558 kPa); mean minimum temperature is 22°C (ρ = 2.5929 kPa), with a mean dew point of 15°C. The site elevation is 100 m. Mean temperature is 28.5°C. Skies are 80% cloudless. Wind velocity at the 6 m height is 5 m/s.

Substituting into the Jensen-Haise equation (equation 4.32) yields E_{JH} = 19.44 MJ/(m^2 · day) or 19.44 MJ/m^2·day^{-1}/2.4337 MJ/kg = 7.98 mm/day where 1 kg/m^2 = 1 mm. Apply the Table 4.3 crop coefficient K_c = 1.2 to get the ET_{JH} = 9.58 mm/day. Multiply by days in a month to get monthly ET_{JH}.

The FAO Modified Blaney-Criddle Method

The modified Blaney-Criddle method is a commonly used empirical method for estimating annual or seasonal crop water requirements. It is primarily a temperature-driven model with an indirect radiation component. Blaney and Criddle originally developed the empirical model for the United States. In subsequent revisions, the method was generalized using data from all over the world, covering all major agriculturally significant physiographic regions. The NRCS, United Nations FAO, and agricultural agencies of other countries contributed data to the model revisions. This method computes daily E_P using

$$E_{BC} = a + b[P_{BC}(0.46T_{av} + 8.13)] = a + bf \tag{4.33}$$

$$a = 0.0043rh_{min} - \frac{n}{N} - 1.41 \tag{4.34}$$

Table 4.4. Mean daily percentages (P_{BC}, expressed as percentages) of annual daytime hours for different latitudes on a daily basis, representing the 15th day of respective months.

Lat. (N or S)	Jan. July (%)	Feb. Aug. (%)	Mar. Sep. (%)	Apr. Oct. (%)	May Nov. (%)	June Dec. (%)	July Jan. (%)	Aug. Feb. (%)	Sep. Mar. (%)	Oct. Apr. (%)	Nov. May (%)	Dec. June (%)
60°	.15	.20	.26	.32	.38	.41	.4	.34	.28	.22	.17	.13
55°	.175	.215	.26	.15	.36	.385	.375	.33	.28	.23	.185	.165
50°	.19	.23	.27	.31	.34	.36	.35	.32	.28	.24	.2	.18
45°	.21	.235	.27	.3	.335	.355	.34	.315	.28	.24	.215	.2
40°	.22	.24	.27	.3	.32	.34	.33	.31	.28	.25	.22	.21
35°	.23	.25	.27	.29	.31	.32	.32	.3	.28	.25	.23	.22
30°	.24	.25	.27	.29	.31	.32	.31	.3	.28	.26	.24	.23
25°	.24	.26	.27	.29	.3	.31	.31	.29	.28	.26	.25	.24
20°	.25	.26	.27	.28	.29	.3	.3	.29	.28	.26	.25	.25
15°	.26	.26	.27	.28	.29	.29	.29	.28	.28	.27	.26	.25
10°	.26	.27	.27	.28	.28	.29	.29	.28	.28	.27	.26	.26
5°	.27	.27	.27	.28	.28	.28	.28	.28	.28	.27	.27	.27
0°	.27	.27	.27	.27	.27	.27	.27	.27	.27	.27	.27	.27

Source: Excerpted from Cuenca 1989.

Note: When multiplied by the number of days in the month, these entries give the percentage of annual daytime hours occurring for the indicated month.

$$b = 0.82 - 0.0041Rh_{\min} + 1.07\frac{n}{N} + 0.06v_{2mavg} - 0.006Rh_{min}\frac{n}{N} - 0.0006Rh_{min}v_{2mavg} \quad \textbf{(4.35)}$$

where

E_{BC} is daily evaporation (E_P) as estimated by Blaney-Criddle (mm);
Te_{av} is the mean daily temperature $[(Te_{max} + Te_{min})/2]$°C;
P_{BC} is the percentage of total daytime hours in the year on a daily basis; values are given in Table 4.4;
rh_{min} is the minimum relative humidity as a percentage; *relative humidity* rh_{min} *usually occurs at or near* Te_{max};
n/N is the ratio of cloudless to total sunshine hours in a day as a fraction;
v_{2mavg} is the average daily wind speed (m/s) at a 2 m height;
f is the original Blaney-Criddle term, $f = P_{BC}(0.46Te_{av} + 8.13)$.

Note: The quantity xx as percent $= xx\%$; the quantity as fraction $= 0.xx$.

The state-of-the-art version of Blaney-Criddle is the FAO modification by Doorenbos and Pruitt (1977), designated as the FAO Blaney-Criddle method. The sunshine-temperature factor is modified to include wind and humidity variations here.

To use the FAO modified Blaney-Criddle method:

- Compute the Blaney-Criddle $f = P_{BC}(0.46Te_{av} + 8.13)$ factors without the crop coefficient using the mean monthly temperature. The f term is a first estimate of E_{BC}.
- Determine the minimum relative humidity value and the representative daytime wind speed in units of m/s. *Note:* It is common for wind speed to be reported in wind run per day (i.e., km/day). When this happens, one must estimate the ratio of daytime wind run to nighttime wind run in order to calculate the daytime wind run. See Burman et al. 1980 for additional discussion of this situation.
- Estimate the approximate sunshine fraction and calculate E_{BC} using equations 4.33–4.35.
- Apply the crop constant K_c (Table 4.3) to get ET_{BC}.
- For seasonal computations, compute the ET_{BC} value for each month, multiply by days in the respective months, and sum the monthly results.

Example 4.4: Compute the estimated ET_{BC} for cotton in July near Tifton, Georgia. The site elevation is 100 m. Mean maximum temperature is 35°C, and the mean minimum temperature is 22°C; skies are 80% cloudless. Mean dew point is 15°C. Wind at 6 m is 5 m/s.

The wind at 2 m $= 5(2/6)^{0.2} = 4$ m/s (equation 4.30). For an annual percentage of solar radiation in July of 0.31 (Table 4.4) and a minimum relative humidity ranging from 20 to 50%, the f factor in equation 4.33 $= 0.31\{0.46[(35 + 22)/2] + 8.13\} = 6.58$ mm/day. Using the calculation formulas (equations 4.33 and 4.34) for a and b, first estimate the rh_{min} to be about 40% from a psychrometric chart where one expects the minimum temperature of 22°C to occur at dew point. The maximum temperature of 35°C on a line of constant moisture ratio then gives a rh_{min} of 40%. Then $a = -2.04$ using rh_{min} as 40%; $b = 1.49$. This results in a calculated value of 7.77 mm/day, lower than the Penman-Monteith value. Apply the Table 4.3

crop coefficient of 1.2 to get the daily ET_{BC} of 9.32 mm/day. Multiply by days in a month to get monthly ET_{BC}.

To get seasonal E_{BC}, repeat the above computation for each month or fraction of month in the entire growing season. Add the monthly totals for the seasonal water requirement.

The Priestly-Taylor Method

The Priestly-Taylor method (Priestly and Taylor 1972) is an energy-based method for calculating E_P:

$$\lambda E_{PT} = \alpha_{PT}\left(\frac{\Delta}{\Delta + \gamma_{Pen}}\right)R_n \tag{4.36}$$

where E_{PT} is E_P as estimated using the Priestly-Taylor method. The parameter α_{PT} is an empirical constant usually taken as 1.26. Kustas et al. (1996) have researched specific factors affecting the α_{PT} value. The Priestly-Taylor equation is simply the energy portion of the Penman-Monteith equation, modified with an empirical constant accounting for the wind-humidity component. The above equation implies that wind and humidity are correlated with direct radiation. Unfortunately, the correlation is not high. Good correlations usually exist between radiation and temperature. Thus, given temperature, one can calculate all the parameters of the Priestly-Taylor equation. The Priestly-Taylor equation is more appropriate in the humid eastern United States, where humidity is generally higher and wind run lower. This method is convenient for computer-modeling applications.

Example 4.5: Compute the estimated ET_{PT} for cotton in July near Tifton, Georgia. The site elevation is 100 m. Mean maximum temperature is 35°C, and the mean minimum temperature is 22°C; skies are 80% cloudless. Mean dew point is 15°C. Wind at 6 m is 5 m/s.

We computed components of the Priestly-Taylor equation for this example using data for the Penman-Monteith equation example. From equation 4.36,

$$E_{PT} = [1.26(0.7709)(16.01 \text{ MJ/m}^2\cdot\text{day}^{-1})]/2.4337 \text{ MJ/kg} = 6.39 \text{ mm/day}$$
$$ET_{PT} = 6.39(1.2) = 7.7 \text{ mm/day}$$

The Priestly-Taylor values are somewhat low. The constant α_{PT} could be calibrated if the Priestly-Taylor method were used routinely.

The Bowen Ratio ET Method

The Bowen ratio approach is fundamentally an energy-based approach and was discussed above, at equations 4.4 and 4.5. For *ET* predictions, one may substitute a crop canopy for the water body in Figure 4.1. For a surface with no advection, a steady temperature, and a water density of 1 Mg/m^3, equation 4.5a appears as equation 4.37:

$$ET_B = \frac{R_n - G}{\lambda(1 + B)} \tag{4.37}$$

where R_n is the net incident longwave and shortwave radiation as discussed with equation 4.19, and G refers to all heat conducted to the ground as in equation 4.19. One may estimate the Bowen ratio (B) from equation 4.5b using the differentials on the right-hand side. The differential form allows us to determine temperature and humidity gradients at points other than the water or crop surface.

The Bowen ratio method is undergoing renewed interest as instrumentation improves. The Bowen ratio method does not require crop coefficients and thus provides a direct measure of *ET*. Hatfield et al. (1996), along with several others (see Camp et al. 1996), further discuss the Bowen ratio method for *ET* prediction. A spin-off of the Bowen ratio approach known as the *eddy correlation method* is based on direct measurements of turbulent heat and vapor flux over the crop canopy. Eddy correlation enables us to directly measure the respective fluxes instead of making gradient calculations. This instrumentation has become commercially available in the past few years.

The Thornthwaite Method

The Thornthwaite method presented by Gray (1970) is similar to the Blaney-Criddle method. It factors in temperature, length of day, and a fixed proportion of the thermal energy budget. The method was put forward to give evaporation estimates over large physiographic regions. Unlike Blaney-Criddle, the Thornthwaite method does not use a crop coefficient. The method is not as reliable as Blaney-Criddle and thus will not be discussed in detail here. Older texts such as Gray 1970 give procedures and example applications of the Thornthwaite method, and it still appears in some agency regulatory documents.

Comprehensive Soil-Canopy Simulation

Computer simulation is a viable approach for estimating moisture flux from crop canopies. Lascano and van Bavel (1986) developed an energy and water balance model (ENWATBAL) simulating moisture and heat flux from various bare soil and crop canopy configurations. A mulch component was recently added, enabling simulation of many conservation tillage and water management practices (van Donk et al. forthcoming). The model ENWATBAL requires numerous soil, climatological, and crop inputs. The model provides temperature and moisture predictions enabling evaluations of temperature and moisture effects on phenomena such as nutrient cycling and root crop development.

A variety of other techniques of varying levels of empiricism exist. ASAE 1985 discusses a wide range of theoretical and empirical irrigation scheduling techniques.

Physical Analog: The Standard Evaporation Pan

Climatologists and growers often estimate E_p from measurements recorded in a class A evaporation pan. Any grower can set a class A pan. The initial investment is not large; however, one must consistently resupply the pan with water. Thus, maintenance of the pan requires constant attention. To calculate E from the pan measurement one applies a correction factor. An average correction factor of approximately 0.7–0.75 is used to correlate the pan to a body of water, or theoretical E. A crop coefficient, depending on the crop and its stage of development, is then used to estimate crop *ET*.

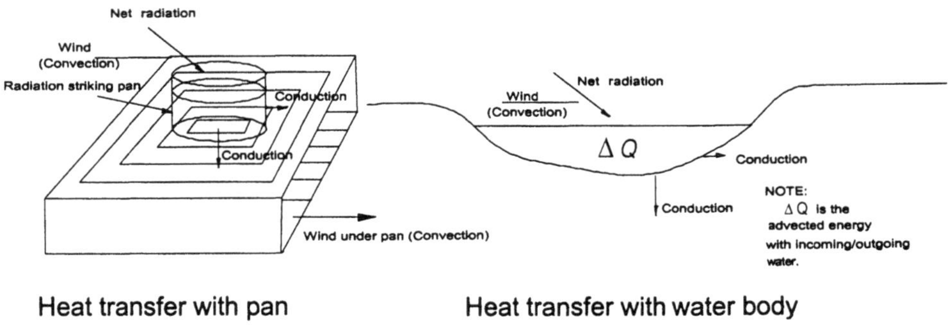

Figure 4.5. Schematic showing energy movement paths for an evaporation pan and for a small reservoir.

The class A evaporation pan has specified dimensions (galvanized iron, 4 ft [1.22 m] in diameter, 10 in [254 mm] deep, mounted 12 in [305 mm] above ground). The pan is filled with 8 in (203 mm) of water. One must refill the pan when the level drops to 7 in (178 mm). One may measure pan water level electronically or manually at a consistent time, usually early morning. Manual readings involving a hook or point gauge with a vernier scale accurate to 0.01 in or 0.1 mm are accurate. Evaporation is computed from differences in observed levels after correcting for precipitation. The class A pan is mounted on an open wooden frame, and the pan's bottom should be about 15 cm above the soil surface. To qualify as a standard class A evaporation pan there must be a well-watered, green grass crop upwind from the pan. Figure 4.5 shows energy relations for evaporation from a pan and a small reservoir. Compare the energy balance in Figure 4.5 with that of Figure 4.1.

Heat and mass transfer to the pan is different from that to the water body. Water will usually evaporate faster from the evaporation pan than it would from an equivalent water body because sensible heat transfer to the pan is usually greater. Qualitative differences in heat transfer are usually expressed as a single coefficient, K_p. Equation 4.38 is used to predict E_{pan}:

$$E_{pan} = K_p E_{panmeas} \quad \textbf{(4.38)}$$

where

E_{pan} is the evaporation from a well-watered surface as estimated by the pan method;
K_p is the pan coefficient;
$E_{panmeas}$ is the actual measured daily evaporation from the pan.
E_{pan} is directly comparable to other computed measures of evaporation such as E_{pond}, E_{PT}, E_P, E_{BC}, and E_{JH}.

The pan coefficient K_p usually ranges from 0.6 to 0.8 and is often set to 0.7. Our inability to precisely model the pan coefficient is the largest problem with the use of the evaporation pan. *Accepted practice is to use $K_p = 0.7$.* Experience has shown that one should measure the pan level at a consistent time of day. The only way to bypass the need to use a crop coefficient is to install a field lysimeter. Lysimeter systems are extremely expensive initially and require much management effort. Lysimeters are mainly limited to specialized research installations.

To calculate consumptive use by a crop, equation 4.38 may be expanded:

$$ET_{pan} = K_p E_{panmeas} K_c = E_{pan} K_c \quad \textbf{(4.39)}$$

where K_c is the crop coefficient tabulated in Table 4.3.

Example 4.6: Compute the estimated ET_{pan} for cotton in July near Tifton, Georgia. The site elevation is 100 m. Mean maximum temperature is 35°C, and mean minimum temperature is 22°C; skies are 80% cloudless. Mean dew point is 15°C. Wind at 6 m is 5 m/s. The evaporation pan was well maintained and was 1000 m within well-watered grass. The pan evaporation was 12 mm.

The wind at 2 m = 5 m/s(2 m/6 m)$^{0.2}$ = 4 m/s = 345.6 km/day from equation 4.30. From a psychrometric chart, one can estimate the maximum humidity to be 60% and the minimum humidity to be 30% for a mean humidity of 45%. From equation 4.39,

$$ET_{pan} = K_p E_{panmeas} K_c = 0.7(12\text{ mm})1.2 = 10.1\text{ mm}.$$

Evaporation and Evapotranspiration Measurement

Figure 4.6 shows a typical micrometeorological observation station containing an evaporation pan. Like other meteorological instruments, the pan should not be shaded and obstructions should not be taller than twice the distance to the pan. The pan should be level with the horizontal and should be regularly serviced to ensure that rodents and insects are not causing aberrations.

Irrigation engineers often combine evaporation (E) and transpiration (T) and denote the combination as evapotranspiration (ET). Computations for evaporation assume a wet soil surface and/or a well-watered crop (e.g., a crop that has received adequate rainfall or irrigation). One can measure ET of growing crops with lysimeters, used mainly for research. Figure 4.7 shows a schematic of a lysimeter. The lysimeter is the basic tool for validating ET models.

Lysimeters provide valid results only if the crop conditions in the lysimeter are the same as with the surrounding crop. This condition is difficult to establish in production systems. Lysimeters are excellent research tools for studying crop moisture loss in both the well-watered and the stressed conditions. Evaporation from a dry soil is near zero. Applying

Figure 4.6. Class A evaporation pan and anemometer as part of a weather station.

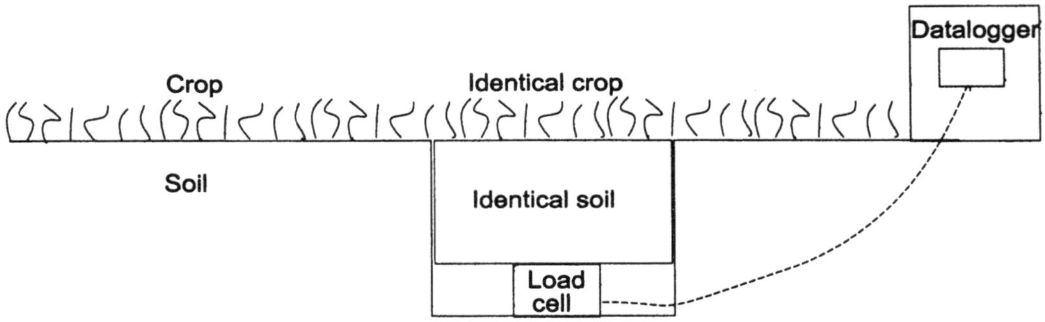

Figure 4.7. Schematic of a lysimeter.

a crop factor to the evaporation value enables estimation of *ET* from a well-watered crop partially covering a dry soil by computing for the well-watered surface. *ET* from crops under water stress requires additional plant physiology concepts beyond the scope of this text.

Pollutants and Evaporation

Pollutants affect the evaporation process mainly through the vapor pressure. Inorganic salts result in suppression of vapor pressure compared with distilled water in comparable temperature conditions. Materials such as ethylene glycol (antifreeze) suppress vapor pressure. Light nonaqueous materials float on water. Their evaporation depends upon their vapor pressure. Thin films of some organics are used in rain gauges to suppress evaporation. Dense nonaqueous materials sink and thus persist. Advanced courses on transport phenomena treat other aspects of pollution transport.

Outlook

Evaporation, the transformation of water from the liquid phase to the gaseous phase, is a hydrologic process occurring along the path of precipitation, surface runoff, and surface storage. Annual evapotranspiration is well correlated to annual yield for many agronomic crops. Multiple methods exist for predicting evapotranspiration. The Meyer aerodynamic equations, the Bowen ratio, the Penman-Monteith combination equation, and the Priestly-Taylor, Jensen-Haise, Thornthwaite, and Blaney-Criddle models were evaluated. Each model has its strengths and weaknesses.

The Penman-Monteith method is readily applicable to other farm/field-scale problems, assuming adequate data are available. It most closely models the parameters affecting evaporation. This model and its variants imply a reasonably steady wind from a constant direction. One needs the wind to create the boundary conditions assumed with the mass and energy transport relations in the derivation. The fraction of cloudiness corrects for portions of the day with a solid cloud cover. Cumulus clouds are not factored into the *percent cloudiness* term. Soil heating and cooling are usually neglected. Soil heating is usually less than 2% of the daily radiation. The Penman-Monteith value is the most trustworthy, provided one meets the assumptions. Other methods are much more empirical, more site specific, or more discipline specific.

Evapotranspiration from crops under water stress or from less-than-saturated soil surfaces is complicated by the presence of stomatal, soil, and root resistance terms. These are beyond the scope of this discussion. If soil is saturated, it evaporates water similarly to a water surface in comparable climatic conditions. As water content decreases, the rate of evaporation becomes limited by soil resistance. Mulches help reduce evaporation by providing resistance to vapor transport. References cited herein give additional details concerning water movement from nonwet surfaces, including crops under water stress. Evaporation and transpiration affect infiltration because the soil infiltration capacity depends on unfilled pores. Evapotranspiration is the driving force behind the need to irrigate growing crops. Doorenbos and Pruitt (1977) give an excellent discussion relating to evapotranspiration estimation from a worldwide perspective. The computer model ENWATBAL is a promising approach based on a comprehensive energy and mass balance at the canopy top. An Internet search (use the search term *irrigation scheduling*) will reveal several software packages, some of which are immediately downloadable, for computing evapotranspiration in several different ways.

Problems and Questions

1. Corn is being grown in July, at latitude 40° north, elevation of 500 m, average temperature of 18°C. The daily average wind speed is 3 m/s at 2 m height, the sunshine is medium (i.e., n/N is about 0.7), and rh_{min} is 30%. The minimum temperature is 12°C, and the maximum temperature is 24°C. Estimate *ET* using the FAO modified Blaney-Criddle method. Next, estimate the *ET* with the Penman-Monteith equation. State all assumed parameter values. What is the percentage difference between the two estimates?
2. Cotton is being grown in June, at latitude 30° north. The average minimum temperature is 15°C, and the maximum temperature is 25°C. The site elevation is 100 m, and the cloudless ratio is 0.8. Wind speed is 3 m/s at a height of 5 m. State all assumptions. Estimate the *ET* using the FAO modified Blaney-Criddle method. (*Hint:* Use a psychrometric chart to estimate rh_{min}.) Next, estimate the *ET* using the Penman-Monteith equation. What is the percentage difference between the two estimates?
3. Estimate the average daily *ET* under the conditions of Problem 1 using the Jensen-Haise method. How does the Jensen-Haise result compare with the Penman-Monteith? With the FAO modified Blaney-Criddle?
4. Given the conditions of Problem 1 using the Blaney-Criddle method, what would be the equivalent evaporation in a standard evaporation pan with a pan coefficient of 0.7? Compare this result with the value expected when using the appropriate Meyer equation.
5. Plot the effect of *ET* versus wind speed using the conditions of Problem 1. Vary the wind from 0 to 10 m/s and plot at increments of 1 m/s over a range of 0–5 m/s. Use the Penman-Monteith method.
6. Plot the effect of *ET* versus radiation, varying the radiation from 20 to 40 MJ/m^2, using other conditions as expressed in Problem 1. Use the Penman-Monteith method, plotting at least five points.
7. A standard evaporation pan is placed near a pond to enable partitioning between infiltration losses and evaporation losses. The pan coefficient is 0.8. If 10 mm losses were measured in the pan and the pond level loss based on measurements was 20 mm, what percentage of the pond losses was due to infiltration? No rainfall events or other inflow/outflow occurred during the measurement period. *Note:* One may also use

Penman-Monteith, Blaney-Criddle, or other evaporation models to estimate evaporation losses in pond sealing studies. Others have tried using sunken pans (e.g., pan in the pond submerged to within 1 in of the pan top, no waves) with pan coefficients approaching 1.0 to estimate evaporative losses.

8. Define the difference between *evaporation* and *transpiration.*
9. Compare and contrast the methods for predicting evaporation/transpiration that were discussed in this chapter with respect to assumptions made.
10. Review the various approaches to predicting the wind function listed in Burman et al. 1980. What is the maximum percentage of variation in evaporation based on the Penman-Monteith model that one may expect at your location with the various wind functions?
11. Compare the pros and cons of various methods discussed in class for predicting *ET*.
12. Derive the energy balance approach for the Bowen ratio method in equation 4.37 from equation 4.4 and Figure 4.1 by lumping appropriate terms and developing a rationale for setting other terms in equation 4.4 to zero.
13. Suggest two ways to compute *ET* and specify the conditions under which each should be considered (e.g., base your discussion on data requirements, interval of interest).
14. Suggest a practical, most direct way to estimate cumulative *ET* in a given location.
15. What is the difference between the pan coefficient and the crop coefficient? When are these coefficients used?
16. Using the Penman-Monteith equation, determine the *ET* for corn in July, at latitude 40° north, and average temperature of 18.3°C. The daily average wind speed is 3 m/s at 2 m height, the sunshine is medium (i.e., $n/N = 0.8$), and rh_{min} is 30%. What are the values of r_h, r_v, wind function, sensible heat, and latent heat? Estimate leaf temperature. Identify assumptions of the combination equation. *Hint:* Review the Penman transformation and the remainder of the derivation closely.
17. Identify at least one software package for computing evapotranspiration using an Internet search. Specify the key words and the search engine used in your search.

References

ASAE. 1985. *Advances in evapotranspiration: Proceedings of the national conference on advances in evapotranspiration.* December 16–17, 1985. Chicago. ASAE Publication 14-85. American Society of Agricultural Engineers, St. Joseph, MI.

Bowen, I. S. 1926. The ratio of heat loss by conduction and by evaporation from any water surface. *Physics Review* 27:779–87.

Burman, R. D., P. R. Nixon, J. L. Wright, and W. O. Pruitt. 1980. Water requirements. In *Design and operation of farm irrigation systems.* ASAE Monograph no. 3. American Society of Agricultural Engineers, St. Joseph, MI.

Camp, C. R., E. J. Sadler, and R. E. Yoder. 1996. *Evapotranspiration and irrigation scheduling: Proceedings of the international conference.* American Society of Agricultural Engineers, St. Joseph, MI.

Campbell, G. S. 1977. *An introduction to environmental biophysics.* Springer-Verlag, New York, NY.

Cuenca, R. H. 1989. *Irrigation system design: An engineering approach.* Prentice Hall, Englewood Cliffs, NJ.

Doorenbos, J., and W. O. Pruitt. 1977. *Guidelines for predicting crop water requirements.* 2d ed. FAO Irrigation and Drainage Paper no. 24. FAO, Rome, Italy.

Gray, D. M. 1970. *Handbook on the principles of hydrology.* Water Information Center, Port Washington, NY.

Hatfield, J. L. 1985. Estimation of regional evapotranspiration. In *Advances in evapotranspiration: Proceedings of the national conference on advances in evapotranspiration,* ASAE Publication 14-85. American Society of Agricultural Engineers, St. Joseph, MI.

Hatfield, J. L., J. H. Prueger, and T. J. Sauer. 1996. Comparison of evapotranspiration over different surfaces. In *Evapotranspiration and irrigation scheduling: Proceedings of the international conference,* ed. C. R. Camp, E. J. Sadler, and R. E. Yoder. American Society of Agricultural Engineers, St. Joseph, MI.

James, L. G. 1988. *Principles of farm irrigation system design.* John Wiley & Sons, New York, NY.

Jensen, M. E. 1966. Empirical methods of estimating or predicting evapotranspiration and irrigation water requirements. In *ASAE conference proceedings: Evapotranspiration, its role in water resource management.* American Society of Agricultural Engineers, St. Joseph, MI.

Jensen, M. E., R. D. Burman, and R. G. Allen, eds. 1990. *Evapotranspiration and irrigation water requirements.* ASCE, New York, NY.

Kustas, W. P., D. I. Stannard, and K. J. Allwine. 1996. Variability in the surface energy flux partitioning during Wishita 1992: Resulting effects on the Penman-Monteith and Priestly-Taylor parameters. *Agricultural and Forest Meteorology* 82:171–93.

Lascano, R. J., and C. H. M. van Bavel. 1986. Simulation and measurement of evaporation from a bare soil. *Soil Science Society of America Journal* 50:1127–32.

Merva, G. E. 1995. *Physical principles of the plant biosystem.* American Society of Agricultural Engineers, St. Joseph, MI.

Meyer, A. F. 1942. *Evaporation from lakes and reservoirs.* Minnesota Resources Commission, St. Paul, MN.

Priestly, C. H. B., and R. J. Taylor. 1972. On the assessment of surface heat flux and evaporation using large scale parameters. *Monthly Weather Review* 100:81–92.

Rohwer, C. 1931. *Evaporation from free water surfaces.* USDA Technical Bulletin 271. Government Printing Office, Washington, DC.

Shuttleworth, W. J. 1993. Evaporation. In *Handbook of hydrology,* ed. D. R. Maidment. McGraw-Hill, New York, NY.

van Donk, S. J., E. W. Tollner, J. L. Steiner, and S. R. Evett. Forthcoming. A mulch submodel for ENWATBAL. *Agronomy Journal.*

5

Runoff

Runoff is that portion of rainfall not infiltrating or evaporating. Runoff prediction techniques integrate hydrologic cycle components into volume prediction techniques, peak flow predictions, and composite storm hydrograph development. Peak runoff is the key determinant in downstream channel and structure selection and sizing. Runoff transports sediment, chemicals adsorbed to the sediment, and dissolved chemicals. Runoff from rainfall is our primary interest here. Runoff volume and flow rate depend on factors such as surface storage, evapotranspiration, infiltration, rainfall amount, and rainfall distribution.

Goals

- Identify factors such as precipitation intensity, watershed slope, watershed roughness, cover, soil type, and preexisting water content affecting runoff volume and peak flow predictions.
- Compute runoff volume, peak flow, and hydrograph shape resulting from design or actual storms over compact watersheds drained by natural channels and constructed diversions.

Background

Soil has a limited capacity to absorb rainfall. The wetter the soil, the lower the infiltration. Soils with high clay contents tend to have low infiltration capacities, whereas sandy soils may have high infiltration capacities. In many situations, the infiltration rate at the beginning of a rainfall event is very high. The infiltration rate then decreases rapidly to a near steady-state rate. Figure 5.1 shows how a uniform 2 cm/hr rainfall, runoff, and infiltration are typically related for a wet and a dry sandy loam soil. A more moist soil, such as the sandy loam that is initially 70% saturated in Figure 5.1, results in more runoff than from the same soil in a dry condition. Elapsed time from onset of rain to runoff is quicker in the wet soil. Figure 5.2 contrasts infiltration in two soils with different covers.

Evaporation may reduce the actual runoff volume. Evaporation affects runoff volume by influencing the initial water content. Increasing the rainfall rate gives increased runoff.

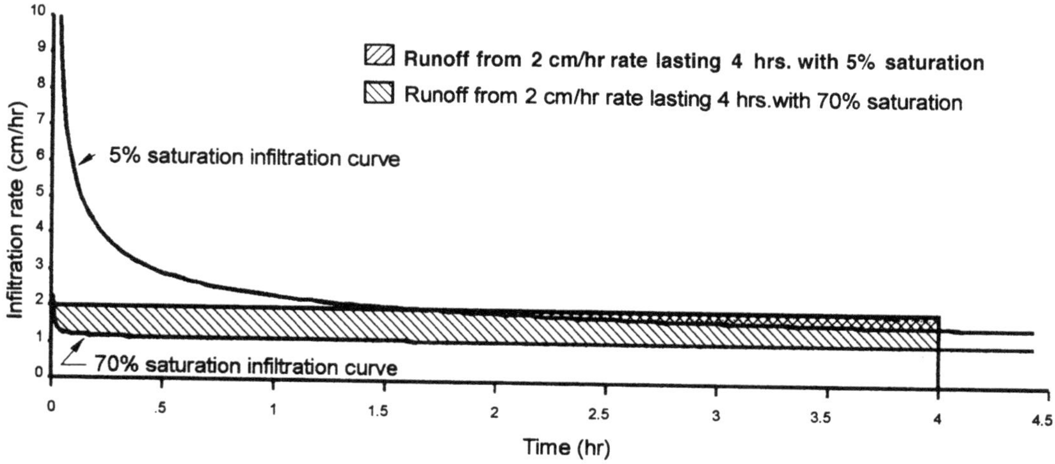

Figure 5.1. Infiltration rate versus elapsed time with a 2 cm/hr uniform rainfall event superimposed to show runoff volume. Runoff begins within 20 min on the saturated soil versus 1.7 hr on the dry soil.

Runoff is the portion of rainfall that does not infiltrate (excess rainfall). One approach to modeling excess rainfall is to predict infiltration and subtract it from the precipitation. The Green-Ampt approach, discussed in Chapter 3 and at the end of this chapter, exemplifies the use of excess rainfall for runoff hydrograph generation.

Definitions

A discussion of runoff requires familiarity with terms used in water science. Some definitions follow:

Compact watershed: A drainage catchment that is roughly semicircular and has one main drainage channel, as opposed to being long and narrow and/or having multiple main channels.

Figure 5.2. A field with bare soil (right) and a vegetatively covered field (left) after a rainfall. More standing water is apparent on the bare soil.

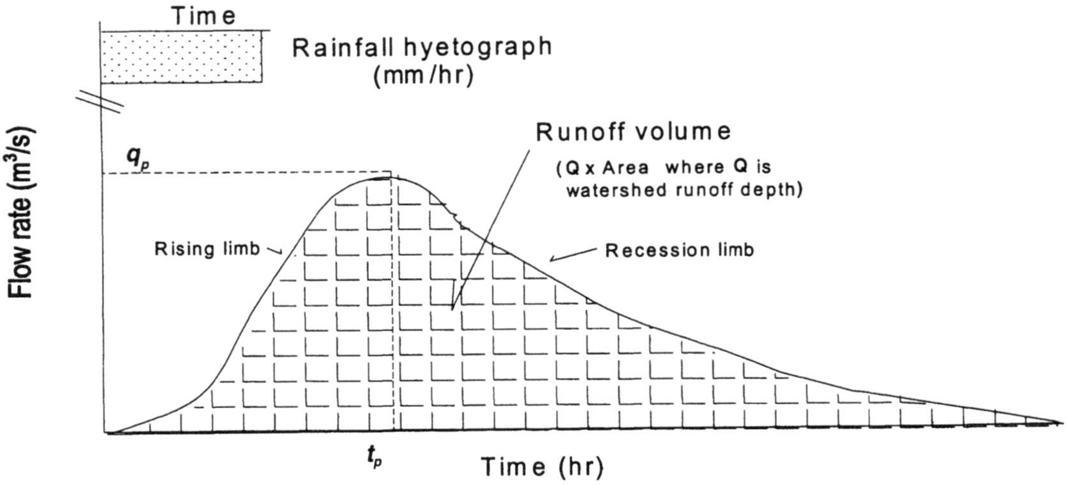

Figure 5.3. Schematic of hypothetical hyetograph and hydrograph resulting from rainfall over a compact watershed (not to scale).

Complex watershed: A watershed consisting of networked compact watersheds. Superimposing and adding hydrographs of subsections of compact watersheds will give the hydrograph for the complex watershed. Detailed complex watershed analysis is beyond the scope of this text.

Hydrograph: The relationship between flow rate and time that results from a precipitation event. Figure 5.3 shows an example.

Unit hydrograph: The hydrograph resulting from a 1 in (25.4 mm), 1 cm, or 1 mm runoff volume (the volume being the product of 1 in [or 1 cm or 1 mm] times the area of the watershed). The unit hydrograph is by definition produced by a uniform hyetograph. The runoff volume is the area under the hydrograph. The unit hydrograph has an area of 1 equivalent depth unit (inches, centimeters, or millimeters) over the watershed area. The shape of the unit hydrograph models the shape of the hydrograph produced by the compact watershed.

Peak flow: The maximum flow on the hydrograph.

Rising limb: The portion of the hydrograph in which flow is increasing with time. The duration of the rising limb is the time to peak t_p.

Recession limb: The portion of the hydrograph in which flow is decreasing with time. The duration of the recession limb is from t_p to the ceasing of runoff.

Runoff volume: The portion of rainfall leaving the watershed from a specified rainfall event. One may express runoff volume in watershed millimeters or inches. If so, one multiplies the effective precipitation depth by the watershed area to obtain true volume units.

Watershed or Catchment Delineation

We define a water catchment or watershed with respect to the stream gauging station location, where one measures runoff.[1] A watershed, or drainage basin, is a topographically defined area drained by a system of connecting rivers/streams such that all outflows discharge at a single outlet.

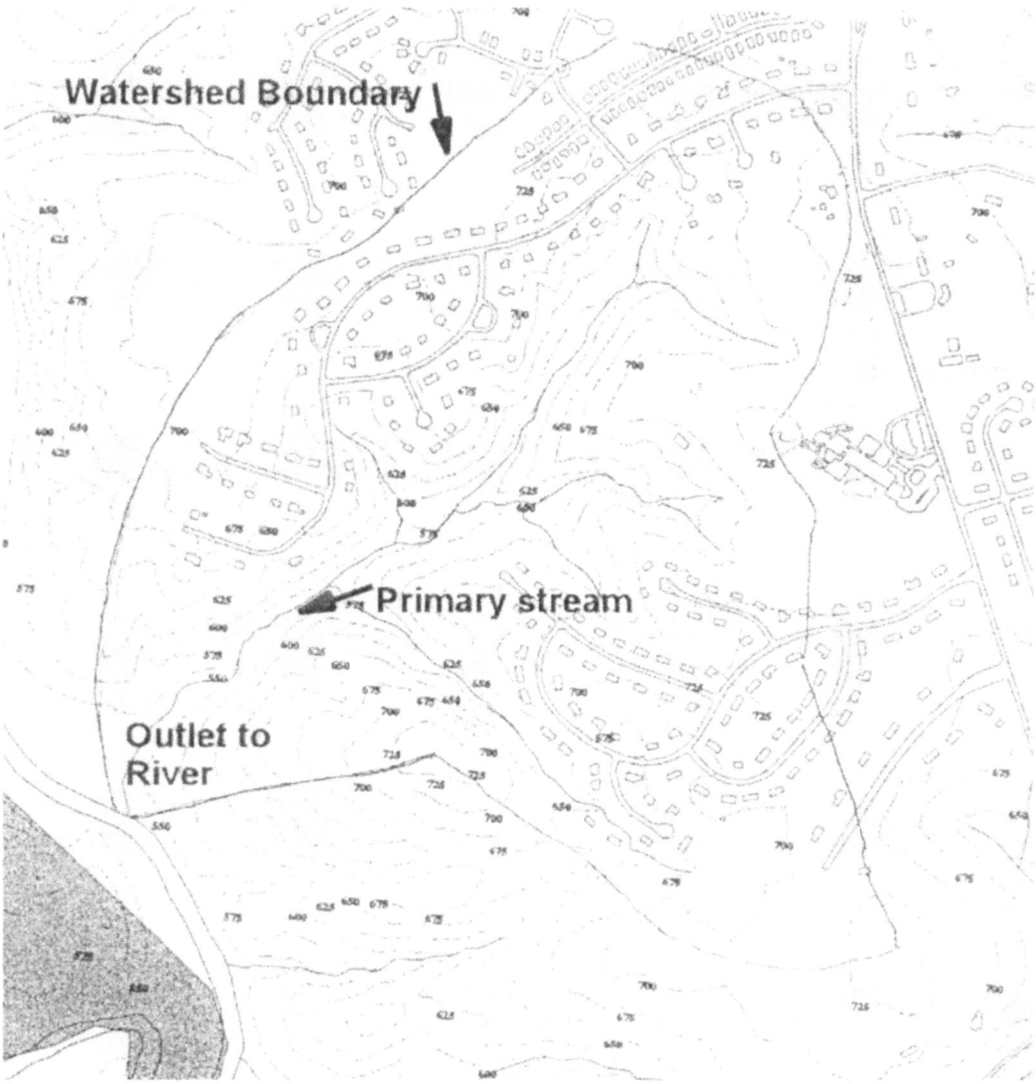

Figure 5.4. Delineation of a watershed in Clarke County, Georgia, on a Geographic Information System (GIS)–produced topographic map. Possible alteration of drainage paths by the indicated development is not considered, though such alterations may be considerable. Site visits are usually required to determine effects of development on flow paths.

The catchment area includes all points above the elevation of the gauging station and within the boundaries of the *topographic divide* separating adjacent watersheds. Figure 5.4 shows an example of a delineated watershed. The watershed divide, or watershed perimeter, follows the ridgeline between hydrologic units. The watershed divide is a closed traverse identifying a boundary that runoff water does not cross. The following steps may be useful in delineating a watershed:

1. Start at the outlet. We must specify the outlet and possibly also a dam, bridge, or other particular point of interest for studying flow rate or water quality.
2. If obvious stream channels and tributaries flow to the outlet, highlight them with a marker or trace them with your finger.
3. Extend the channel network by looking for contours that point upstream.

4. Start at the outlet, and with a pencil, draw a line away from the outlet that is perpendicular to the closest contour line. Keep drawing perpendicular to the contour line. Try to find ridges and split them in half. Eventually the boundary line should close back at the outlet point. Start on the other side of the outlet if problems arise.
5. Test your boundary at several points by making a dot on either side of the preliminary line. Ask, "If rain fell on this spot where would it run?" Obviously, downhill—so determine which direction is downhill. If points inside the trial boundary *do not* drain to the outlet, or if points outside the trial boundary *do* drain to the outlet, change the boundary accordingly. Remember, in running downhill, water will take the steepest (easiest) path down. In other words, flow paths are always perpendicular to the contours.
6. There are some tricky situations. Flat areas can be difficult. In general, split flat areas in half. In practice, one should make a site visit to measure elevations when in doubt. Be careful about installed structures that may divert flow from the original path. Highway embankments, roadside ditches, etc. may appear on small-scale maps. When in doubt, make a site visit.

Runoff Volume for a Compact Watershed

Runoff volume estimates are important for designing terraces with tile outlets, designing flood control ponds and spillways, and in estimating storage volumes required for runoff impoundment. The NRCS curve number method is applicable for estimating impoundment water volume for compact watersheds that are less than 64 km^2 (25 mi^2). The NRCS approach is embedded in water quality computer models such as AGNPS (USDA-Agricultural Research Service). These models extend compact watershed techniques to complex watersheds.

NRCS Runoff Volume (Curve Number Equation)

We may derive the curve number equation by first defining a simple proportion:

$$\frac{runoff}{precip} = \frac{Q}{P} = \frac{actual\ infiltration}{potential\ \mathrm{max}\ infiltration} = \frac{F}{S} \tag{5.1}$$

where

Q is runoff volume in watershed mm or inches;
P is the rainfall depth (inches or mm);
F is the actual infiltration (inches or mm);
S is the potential maximum infiltration (inches or mm) for the soil and cover condition.

Thus, we may write

$$F = P - Q \tag{5.2}$$

One may ignore evaporation in this balance because runoff occurs in a very short time. Solving equation 5.1 for Q after substituting equation 5.2 yields

$$Q = \frac{P^2}{p + S} \tag{5.3}$$

Not all precipitation is available for possible infiltration. Surface cover intercepts some precipitation and some evaporates before infiltrating. An initial abstraction, I_a, is removed from the precipitation by vegetation interception and by surface depressions. This enables evaporation. After adjusting precipitation for initial abstraction, we have

$$Q = \frac{(P - I_a)^2}{(P - I_a) + S} \tag{5.4}$$

The potential maximum retention S is

$$\text{S} = (25{,}400/\text{CN}) - 254 \quad (\text{S in mm}) \tag{5.5a}$$

$$\text{S} = (1000/\text{CN}) - 10 \quad (\text{S in inches}) \tag{5.5b}$$

where CN is the NRCS curve number, developed from many instrumented watersheds representing many soil-cover conditions throughout the United States. Tables 5.1 and 5.2 summarize these values. Table 5.3 defines hydrologic soil groups.

Table 5.1. Runoff curve numbers (*CN*) for agricultural soil-cover complexes for antecedent moisture condition II.

			Hydrologic soil group			
Land use	Treatment or practice	Hydrologic condition	A	B	C	D
Fallow	Straight row		77	86	91	94
Row crops	Straight row	Poor	72	81	88	91
	Straight row	Good	67	78	85	89
	Contoured	Poor	70	79	84	88
	Contoured	Good	65	75	82	86
	Terraced	Poor	66	74	80	82
	Terraced	Good	62	71	78	81
Small grain	Straight row	Poor	65	76	84	88
	Straight row	Good	63	75	83	87
	Contoured	Poor	63	74	82	85
	Contoured	Good	61	73	81	84
	Terraced	Poor	61	72	79	82
	Terraced	Good	59	70	78	81
Close-seeded legumes or rotation meadow	Straight row	Poor	66	77	85	89
	Straight row	Good	58	72	81	85
	Contoured	Poor	64	75	83	85
	Contoured	Good	55	69	78	83
	Terraced	Poor	63	73	80	83
	Terraced	Good	51	67	76	80
Pasture or range		Poor	68	79	86	89
		Fair	49	69	79	84
		Good	39	61	74	80
	Contoured	Poor	47	67	81	88
	Contoured	Fair	25	59	75	83
	Contoured	Good	6	35	70	79

Source: Excerpted from U.S. Soil Conservation Service 1986.

Note: See Table 5.4 for antecedent moisture condition classifications.

Table 5.2. Runoff curve numbers (CN) for antecedent moisture condition II, for mainly nonagricultural applications.

	Hydrologic soil group			
Land use description	A	B	C	D
Cultivated land:				
Without conservation treatment	72	81	88	91
With conservation treatment	62	71	78	81
Pasture or rangeland:				
Poor condition	68	79	86	89
Good condition	39	61	74	80
Meadow: Good condition	30	58	71	78
Wood or forest:				
Thin stand, poor cover, no mulch	45	66	77	83
Good cover, no grazing	25	55	70	77
Open urban spaces:				
Grass on 75% or more	39	61	74	80
Grass on 50–75% of area	49	69	79	84
Commercial and business (85% impervious)	89	92	94	95
Industrial districts (72% impervious)	81	88	91	93
Residential:				
1/8 acre lot, 65% impervious	77	85	90	92
1/4 acre lot, 38% impervious	61	75	83	87
1/3acre lot, 30% impervious	57	72	81	86
1/2 acre lot, 25% impervious	54	70	80	85
1 acre lot, 20% impervious	51	68	79	84
Parking lots, roofs, driveways	98	98	98	98
Streets and roads:				
Paved with curbs and storm sewers	98	98	98	98
Gravel	76	85	89	91
Dirt	72	82	87	89

Source: Excerpted from U.S. Soil Conservation Service 1986.
Note: See Table 5.4 for antecedent moisture condition classifications.

The initial abstraction I_a is approximated by $I_a = 0.2S$. This brings us to equations 5.6a and b, the standard form of the NRCS runoff volume relationship. Equations 5.6a and b are called the curve number equations.

$$Q = \frac{(P - 0.2S)^2}{P + 0.8S} \quad \text{if} \quad P - 0.2S \geq 0 \qquad \textbf{(5.6a)}$$

$$Q = 0 \quad \text{if} \quad P - 0.2S \leq 0 \qquad \textbf{(5.6b)}$$

The above condition states that infiltration cannot exceed total precipitation. A rainfall not exceeding initial abstraction cannot produce runoff. Figure 5.5 shows runoff for precipitation values up to 250 mm for a range of curve numbers. Figure 5.6 presents the nomenclature.

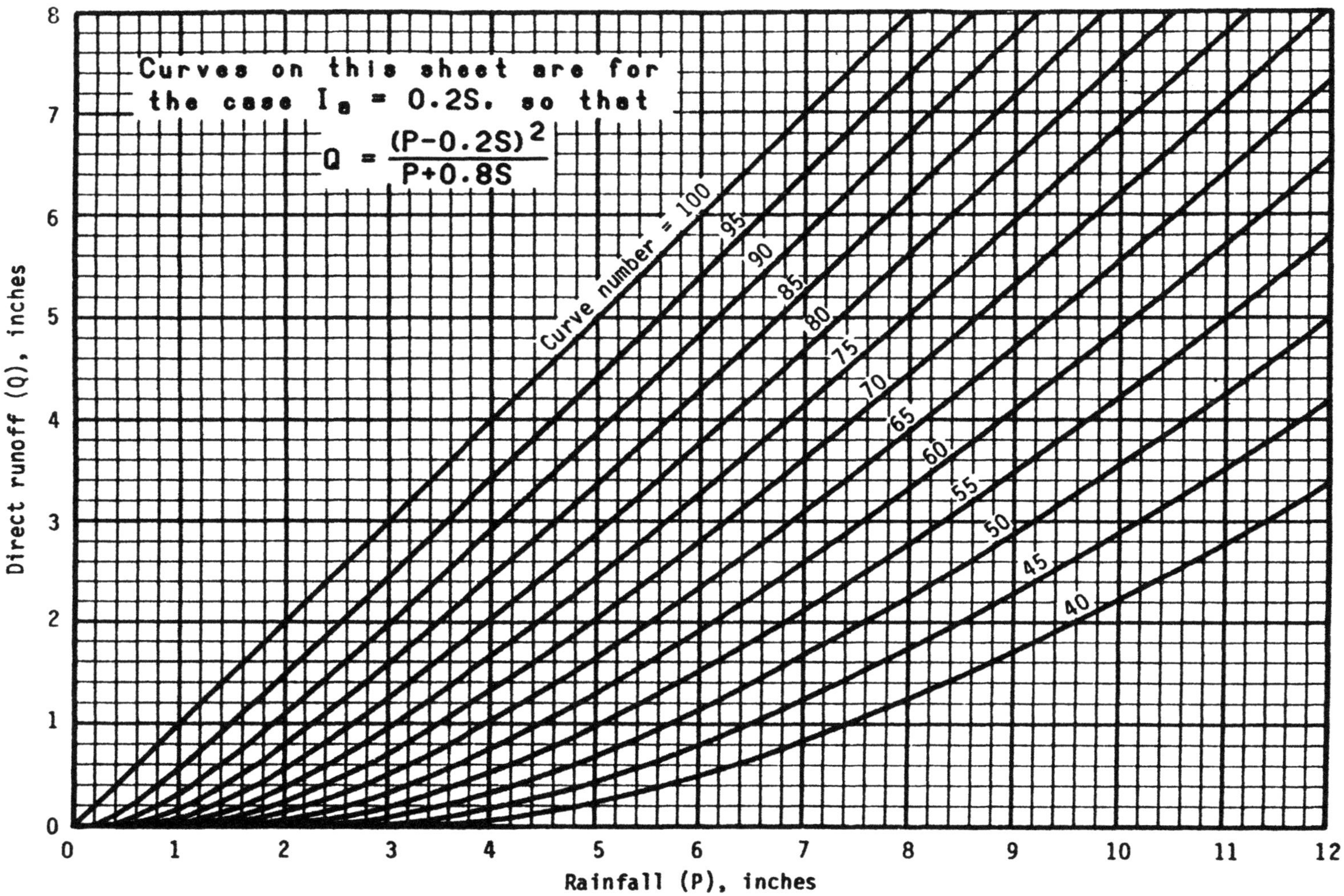

Figure 5.5. Plot of the NRCS curve number equation. (From NRCS 1986.)

Table 5.3. Hydrologic soil group and hydrologic condition definitions.

Soil group	Description	Final infiltration rate (mm/hr)
A	Lowest runoff potential: deep sands with very little silt and clay; deep, rapidly permeable loess.	8–12
B	Moderately low runoff potential: Mostly sandy soils less deep than A, and loess less deep or less aggregated; above-average infiltration after thorough wetting.	4–8
C	Moderately high runoff potential: shallow soils and soils containing considerable clay and colloids, though less than D; below-average infiltration after saturation.	1–4
D	Highest runoff potential: mostly clays of high swelling percentage but also some shallow soils with nearly impermeable subhorizons near the surface.	0–1
Good hydrologic condition	Combinations of newly installed practices, top of sloping soil, few rocks in soils.	
Fair hydrologic condition	Between good and poor.	
Poor hydrologic condition	Combinations of old practices, steep slopes, rocky soils, weathered outcrop zones, highly eroded slopes.	

Source: Adapted from Schwab et al. 1993.

The NRCS curve number method is approximate. Both the coefficient (0.2) and the value for S are influenced by the storm time or intensity. The coefficient is probably greater than 0.2 for short, low-intensity storms. The S value probably decreases for the low-intensity storms due to reduced soil sealing. The sealing effect is likely greatest for soils with minimum vegetation and poor structure. One may modify the CN base value depending on soil hydrological group, hydrologic condition, and antecedent moisture condition. Antecedent rainfall indirectly accounts for evapotranspiration. Table 5.4 defines common antecedent

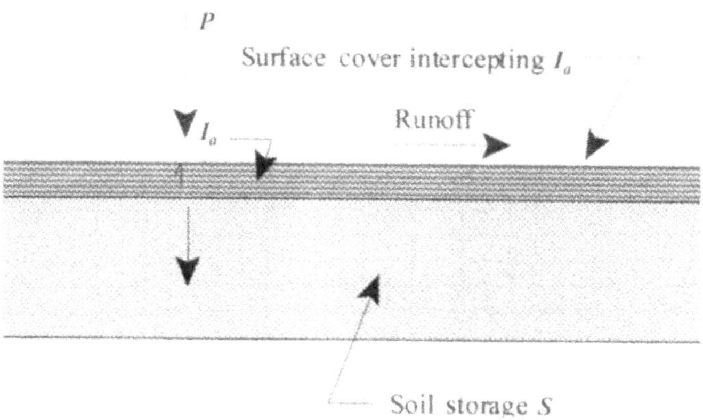

Figure 5.6. Nomenclature for the NRCS curve number equation.

Table 5.4. Antecedent moisture conditions.

Condition	Description	5-day antecedent rainfall (mm) during: Dormant season	Growing season
I	Optimum condition	<13	<36
II	Average value	13–28	36–58
III	Recent heavy rain or cool temperature	>28	>58

Source: US NRCS 1972.

moisture conditions. To convert the condition II factors to condition I or condition III values, multiply by the factor in Table 5.5. Antecedent moisture conditions I and III enable semiquantitative (the conversions in Table 5.5 are at best approximate) runoff comparisons between very wet and very dry watersheds. Conversion to condition I results in a lower *CN* (lower runoff), and conversion to condition III results in a higher *CN* (greater runoff).

Many believe that because of the imprecise nature of the curve number data, the precision suggested by making the antecedent moisture adjustments is not warranted. Thus, many assume antecedent condition II for all design purposes. The curve number adjustment for antecedent rainfall is followed in this text, however.

The next issue concerns hydrologic soil grouping and hydrologic condition. Soil group A has the lowest runoff potential and includes deep sands with little silt and clay. Rapidly permeable loess soils of the Midwest may also fall into this category. Group D, at the other extreme, includes mostly clays of high swelling percent. Group D can include some shallow soils with near impermeable horizons near the surface. Groups B and C are variations between these extremes.

Table 5.5. Curve number (*CN*) conversion factors related to antecedent conditions.

CN for condition II	Factors for converting *CN* for condition II to: Condition I	Condition III
10	0.40	2.22
20	0.45	1.85
30	0.50	1.67
40	0.55	1.50
50	0.62	1.40
60	0.67	1.30
70	0.73	1.21
80	0.79	1.14
90	0.87	1.07
100	1	1

Source: US NRCS 1972.

The hydrological soil group for soils is dependent on soil mineralogy. The grouping for all U.S. soils is available from soil surveys. The hydrologic conditions account for factors such as age of treatment or practice and position on slope. Antecedent rainfall modifies the base values depending on the recent rainfall. Wet soils do not absorb as much (for obvious reasons!) as do dry soils. The effect becomes progressively smaller as the base *CN* approaches 100 (the impervious condition).

The NRCS curve number method computes runoff volume. The runoff volume is the area under the runoff hydrograph. Only watershed cover, soil grouping, cultural practice, and antecedent precipitation influence the runoff volume calculation. Slope indirectly influences runoff volume in that steeper slopes are usually in a poorer hydrologic condition. One usually encounters different soil groups in a watershed. This necessitates an area-weighted calculation.

Example 5.1: Suppose a watershed with 25 acres (10.12 ha) of soil group A in pasture and with 50 acres (20.24 ha) of soil group C in forest, antecedent moisture condition III, received a 1.95 in (49.53 mm) rainfall. What is the runoff volume, assuming the watershed is in hydrologically good condition?

We may choose curve numbers from Tables 5.1 and 5.2. Judgment is required in using these tables. The *CN* for soil group C, good cover, is 70 (Table 5.2); the *CN* for soil group A, good pasture, is 39 (Tables 5.1 and 5.2). The weighted *CN* is [25 ac(39) + 50 ac(70)]/(25 ac + 50 ac) = 59.67. The antecedent condition III (see Table 5.4 if prior rainfall amount is given) multiplier is 1.3 (Table 5.5), resulting in $CN = 77.57$. Using equations 5.5 and 5.6,

$$S = 25{,}400/77.57 - 254 = 73.44 \text{ (mm)}$$

$$S = 1000/77.57 - 10 = 2.89 \text{ (in)}$$

$$Q = [1.95 \text{ in} - 0.2(2.89 \text{ in})]^2/[1.95 \text{ in} + 0.8(2.89 \text{ in})] = 0.44 \text{ in}$$

$$= [49.53 \text{ mm} - 0.2(73.44 \text{ mm})]^2/[49.53 \text{ mm} + 0.8(73.44 \text{ mm})] = 11.21 \text{ mm}$$

$$\text{volume} = (75 \text{ ac})(43{,}560 \text{ ft}^2/\text{ac})[0.44\text{in}/(12 \text{ in/ft})] = 119{,}790 \text{ ft}^3$$

$$= (30.36 \text{ ha})(10{,}000 \text{ m}^2/\text{ha})[11.21 \text{ mm}/(0.001 \text{ m/mm})] = 3403.356 \text{ m}^3$$

Small changes in both precipitation and curve number cause major changes in predicted runoff. Runoff prediction is inherently imprecise.

The NRCS curve number method accounts for soil hydraulic conductivity and for differences in matric potential via the hydrologic soil grouping. It accounts for surface cover via the individual curve numbers associated with various practices (and the designation of good or poor hydrologic condition). It accounts for soil level of saturation via the antecedent moisture index. These associations are somewhat weak but seem to work fairly well for general purposes.

φ Index

The older literature sometimes mentions a φ index. The φ index is a ratio between actual runoff volume and measured rainfall volume. The NRCS curve number equation modernizes the φ index by accounting for topography and cover variation. For small (i.e., less than 2 ha), steep, and impervious watersheds, the φ index is sometimes useful.

PEAK RUNOFF RATE WITH COMPACT WATERSHEDS

Peak runoff rate is used for designing culverts, various channels, and pond spillways. Peak runoff rate is the design flow rate for these structures. The rational method and the NRCS TR55 method, among others, provide peak runoff estimates for compact watersheds. The rational and NRCS TR55 methods are applicable only if the watershed gradient is greater than 0.5%. If the gradient is 0.5% or less, the American Society of Agricultural Engineers drainage ditch formula is applicable. Other triangular hydrograph-based methods are also available. Other specialized approaches are application specific.

Rational Method

Civil engineers originally developed the rational runoff equation for design of parking lot drainage systems and for highway culvert sizing problems. The watershed was typically less than 25 ac (10 ha). The rational method has been adapted for rural watershed peak flow prediction by the development of appropriate runoff coefficients (see Figure 5.7).

The primary assumptions for the rational method are (1) a uniform excess rainfall over the watershed and (2) equal rising and recession hydrograph limbs, t_c. The rational method in metric units appears as

$$q_{prat} = 0.0028 C_{rat} i A \tag{5.7a}$$

where

q_{prat} is the rational method peak runoff, m^3/s.
C_{rat} is the rational runoff coefficient, dependent on soils, topography, and rainfall intensity.
i is the rainfall intensity in mm/hr for the design storm return period; storm duration is taken as the time of concentration t_c of the watershed and is found by $P_{xx\ yr\text{-}tc} / t_c$ where $P_{xx\ yr\text{-}tc}$ is the precipitation from the xx yr storm with a duration of t_c. Time of concen-

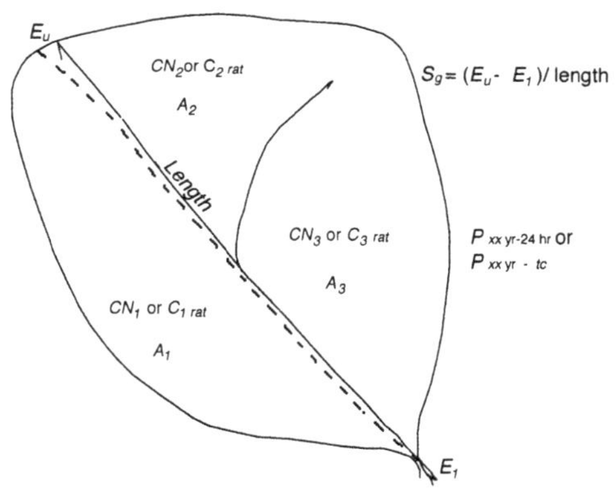

Note: E_u and E_1 refer to upper and lower elevations at the highest and lowest points of the watershed.

Figure 5.7. A hypothetical compact watershed with a watershed gradient (S_g) and composed of three subareas (A_i) with different curve numbers (CN_i) and runoff coefficients ($C_{i\ rat}$).

tration is the time required for water to travel from the farthest point of a watershed to the outlet. Time of concentration computation is discussed below.
A is the watershed area in ha.

In English units, the rational equation appears as

$$q_{prat} = C_{rat} iA \tag{5.7b}$$

where

q_{prat} is in cubic feet per second (cfs);
i is in in/hr;
A is in ac.

(Units exercise: Verify that ac-in/hr is nearly the same as cfs, therefore the units constant of 1.0 above. Also, verify the units constant of 0.0028 for SI units.)

The rational hydrograph describes the extreme case where the storm duration happens to be the same as the time of concentration. At the t_c storm duration the entire watershed contributes to runoff. Although not apparent in Figure 5.8 due to scaling, the excess rain is equal to the runoff volume.

Time of Concentration (Compact Watersheds)

For compact watersheds not dominated by a single ditch or channel, t_c computed using the Kirpich (1940) formula is appropriate. The Kirpich formula is

$$t_{c\ Kirpich} = 0.0195 L^{0.77} S_g^{-0.385} \tag{5.8a}$$

where

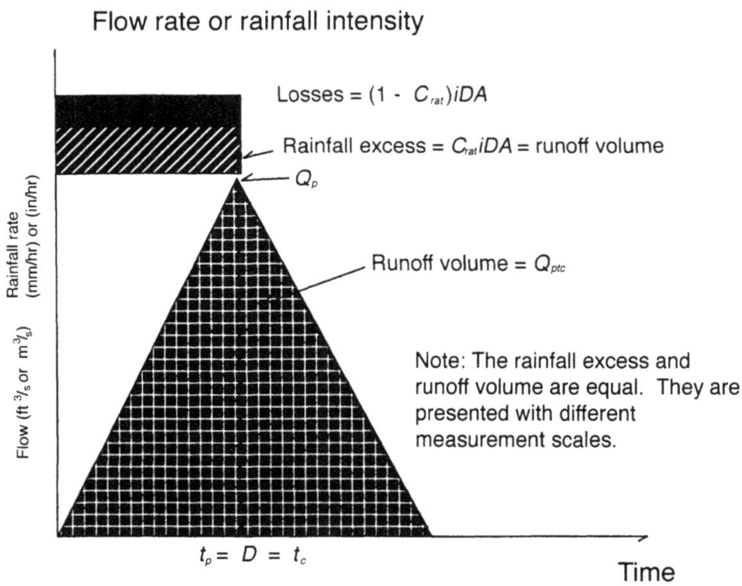

Figure 5.8. Schematic of hydrograph with peak flow predicted by the rational method.

L is length in meters;
$t_{c\ Kirpich}$ is time of concentration using the Kirpich method (in min);
S_g is watershed gradient (or elevation difference between most remote location and outlet divided by L) expressed as a fraction.

In English units the Kirpich formula appears as

$$t_{c\ Kirpich} = 0.0078L^{0.77}S_g^{-0.385} \quad \textbf{(5.8b)}$$

where

L is length in feet;
$t_{c\ Kirpich}$ is in minutes;
S_g is a fraction (e.g., L/L).

Equations 5.8a and 5.8b do not account for the effect of cover on watershed response.

Time of Concentration via Combination of Overland Flow and Obvious Channel Flow Paths

When the flow is overland (e.g., traversing a forested area, field, or other space in the absence of a defined waterway, then moving to a defined waterway), an alternative approach is to use equations 5.9a–c (see also Figure 5.9):

$$t_{c\ ovlf} = \frac{L}{a_{of}\sqrt{s_g}} \quad \textbf{(5.9a)}$$

$$t_{c\ ditch} = \frac{L_c}{v_p} \quad \textbf{(5.9b)}$$

$$t_{c\ ovl\ dch} = t_{c\ ovlf} + t_{c\ ditch} \quad \textbf{(5.9c)}$$

where

$t_{c\ ovlf}$ is the time of travel while in overland flow in s;
$t_{c\ ditch}$ is the time of travel while in the ditch;
$t_{c\ ovl\ dch}$ is the time of concentration using the overland flow method;
L is the maximum length of flow in feet or meters, or length of overland flow or length of channel reach;

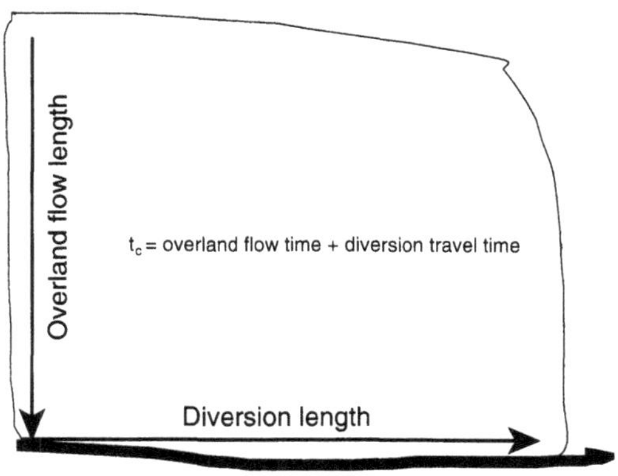

Figure 5.9. Schematic of watershed showing overland flow and ditch travel contributions to total travel time, equaling time of concentration.

Table 5.6. Values of overland flow time-of-concentration coefficient a_{of} for use in equation 5.9a.

	a_{of}	
	(ft/s)	(m/s)
Overland flow surface type		
Forest with heavy ground cover	2.5	0.76
Hay, meadow	2.5	0.76
Trash fallow; minimum tillage	5.1	1.55
Contour; strip cropped	5.1	1.55
Woodland	5.1	1.55
Short grass	7	2.13
Straight row cultivation	8.6	2.62
Bare; untilled	10.1	3.08
Paved	20.3	6.19
Shallow concentrated flow		
Alluvial fans	10.1	3.08
Grassed waterways	16.1	4.91
Small upland gullies	20.3	6.19

Source: Adapted from Haan et al. 1994.

L_c is the length of the channel (feet or meters) from the farthest overland flow entry point to the channel discharge;

v_p is the permissible velocity in a channel: 6 ft/s, or 1.8 m/s, in a vegetated channel or natural channel and 2.5 ft/s, or 0.61 m/s, in an earthen channel; for now, use 6 ft/s or 2 m/s; more specific values for design work are given in Table 8.2, discussed later;

S_g is the watershed gradient expressed as a fraction;

a_{of} is the runoff coefficient given in Table 5.6, selecting appropriate units.

If a diversion ditch is prominent, as shown in Figure 5.9, we may use the combination overland flow–ditch travel time approach. Otherwise, Kirpich or related formulas are used. Haan et al. (1994) and Wanielista et al. (1997) discuss additional approaches for computing t_c values. Warner (1998) uses this approach for computing t_c values for watersheds in the SedCad[2] surface hydrology modeling program.

The design storm duration is equal to the time of concentration. Compute the precipitation, dividing by the duration to get the intensity, i, in the rational equation.

Selection of the C_{rat} coefficient is the remaining step, and proper selection of the C_{rat} coefficient is crucial. Haan et al. (1994) discuss C_{rat} coefficient selection. The coefficient C_{rat} is lower for open-textured soils (e.g., sands, loess soils) and higher for tight clays. Impervious surfaces such as paved roads have C_{rat} values approaching 1.0. Again, the watershed or subwatershed should be smaller than 100 ac.

Horn and Schwab (1963) found some suggested C_{rat} coefficients for soil group B, shown in the Table 5.7a. We then convert these factors to group A, C, or D by applying the appropriate factors in Table 5.8.

One can calculate area-weighted C_{rat} values to describe watersheds with different land uses. Most watersheds do vary, and weighted C_{rat} values are the typical way to account for such variation over the watershed. The watershed should be reasonably compact (e.g., not long and narrow but semicircular) for the weighted average to be reasonable. We sum-

Table 5.7a. Rational equation runoff coefficients (C_{rat}) for specific agriculture conditions.

Crop and hydrologic condition	Coefficient C_{rat} for rainfall rates, independent of soils		
	25 mm/hr	100 mm/hr	200 mm/hr
Row crop, poor practice	0.63	0.65	0.66
Row crop, good practice	0.47	0.56	0.62
Small grain, poor practice	0.38	0.38	0.38
Small grain, good practice	0.18	0.21	0.22
Meadow, rotation, good	0.29	0.36	0.39
Pasture, permanent, good	0.02	0.17	0.23
Woodland, mature, good	0.02	0.1	0.15

Sources: Adapted from Schwab et al. 1993 and NRCS 1972.

Note: The values apply to hydrologic group B soils. Use factors in Table 5.8 to correct to other hydrologic groups.

Table 5.7b. Rational equation runoff coefficients (C_{rat}) for general watershed cover conditions.

	Coefficient C_{rat} for indicated soils		
	Sandy	Silt/clay	Tight clay
Flat woodland, less than 5% slope	0.10	0.3	0.4
Rolling woodland, 5–10% slope	0.25	0.35	0.5
Hilly woodland, 10–30% slope	0.3	0.5	0.6
Flat pasture	0.1	0.3	0.4
Rolling pasture	0.16	0.36	0.55
Hilly pasture	0.22	0.42	0.6
Flat cultivated	0.30	0.5	0.6
Rolling cultivated	0.4	0.6	0.7
Hilly cultivated	0.52	0.72	0.82

Source: Adapted from NRCS 1973.

Note: Values are independent of rainfall intensity. Do not apply the factors in Table 5.8.

Table 5.7c. Rational equation runoff coefficients (C_{rat}) for urban conditions.

Asphalt and concrete	0.7–0.9
Brick	0.7–0.85
Roofs	0.75–0.95
Sandy, flat lawns	0.05–0.10
Sandy, average (2–7% slope lawns)	0.10–0.15
Sandy, steep lawns	0.15–0.20
Clayey flat lawns	0.13–0.17
Clayey average lawns	0.18–0.22
Clayey steep lawns	0.25–0.35
Playgrounds, railroad yard areas	0.2–0.35
Parks, cemeteries	0.1–0.25
Unimproved areas	0.1–0.35

Source: Adapted from Haan et al. 1994.

Note: Use judgment to select within the indicated ranges depending on the degree of expected traffic or use. Higher values should also be used when critical conditions occur with storms of high intensity. Do not apply the factors in Table 5.8.

Table 5.8. Factors to modify rational coefficients for group B to other soil groups.

	Factors for converting the runoff coefficient C_{rat} for group B soils to:		
Crop and Hydrologic condition	Group A	Group C	Group D
Row crop, poor practice	0.89	1.09	1.12
Row crop, good practice	0.86	1.09	1.14
Small grain, poor practice	0.86	1.11	1.16
Small grain, good practice	0.84	1.11	1.16
Meadow, rotation, good	0.81	1.13	1.18
Pasture, permanent, good	0.64	1.21	1.31
Woodland, mature, good	0.45	1.27	1.4

Source: Excerpted from NRCS 1972.

marize the method as follows:

1. Total area should not exceed 25 ac (10 ha).
2. Obtain time of concentration using Kirpich equations (equations 5.8a–b) or, in the case of overland flow or ditch or channel flow, using equations 5.9a–c. See Table 5.6 for overland flow coefficients.
3. Obtain the precipitation knowing the duration = time of concentration and the return period using the Weiss method (equation 2.5), available map, or other procedure.
4. Find the intensity i by dividing precipitation by the relevant time of concentration.
5. Evaluate a weighted C_{rat} (see Tables 5.7 and 5.8) value for the watershed based on soils and land use. Focus on Table 5.7b for most natural land uses or Table 5.7c for urban conditions. Use Table 5.7a and Table 5.8 for specific agricultural land uses.
6. Substitute values into the rational formula (equations 5.7a–b) to get the peak flow q_{prat}.

Example 5.2: Suppose a watershed with 25 acres (10.12 ha) of soil group C (silty clay) in rolling pasture, good condition, and with 50 acres (20.24 ha) of soil group C (silty clay) in forest, poor condition, had a watershed gradient of 2% and was 2624 ft (800 m) from outlet to farthest end. Find the 50 yr runoff assuming the watershed is in Athens, Georgia.

The time of concentration is then

$$t_c = 0.0078\ (2624\ \text{ft})^{0.77}\ (0.02)^{-0.385} = 15.12\ \text{min}$$
$$= 0.0195(800\ \text{m}^{0.77})(0.02)^{-0.385} = 15\ \text{min}$$

(equations 5.8a–b). Applying the Weiss method for precipitation prediction, the 50 yr–15 min storm intensity is 1.95 in/0.25 hr = 7.8 in/hr (198 mm/hr). The C_{rat} for the group C rolling pasture is 0.36 (Table 5.7). The C_{rat} for the group C poor-condition forest is 0.3 (Table 5.7). The group C soil adjustments are 1.21 and 1.27, respectively (Table 5.8). The weighted C_{rat} factor (considering the soil groups; we may also make the C_{rat} selection based on rainfall intensity) is

$$C_{rat} = [25\ \text{ac}(0.36)(1.21) + 50\ \text{ac}\ (0.3)(1.27)]/75\ \text{ac} = 0.399$$

(see Tables 5.7 and 5.8). As with *CN* selection, C_{rat} selection requires judgment.

$$q_p = 0.399(7.8 \text{ in/hr})(75 \text{ ac}) = 233.5 \text{ ac-in/hr} \sim 233.5 \text{ cfs}$$
$$= 0.0028\ (0.399)\ (198 \text{ mm/hr})(30.36 \text{ ha}) = 6.71 \text{ m}^3\text{/s}$$

(equations 5.7). The large size of the watershed causes the estimates to be somewhat suspect.

Suppose the watershed had a 551 m diversion ditch plus 551 m overland flow through forest sloped at 2%. The area is $(551 \text{ m} \times 3.28 \text{ ft/m})^2 \times 1 \text{ ac}/43{,}560 \text{ ft}^2 = 75$ ac (same as above). The diversion ditch has a permissible velocity of 6 ft/s. Then

$$t_c = \{551 \text{ m}(3.28 \text{ ft/m})/[2.5\ (0.02)] + 551 \text{ m}(3.28 \text{ ft/m})/(6 \text{ ft/s})\}/3600 \text{ s/hr}$$
$$= 1.5 \text{ hr}$$

(equations 5.9a–c, Table 5.6). The 50 yr–1.5 hr rain is 3.58 inches in Athens, Georgia (computed using the Weiss equation). The rational coefficient for a forest with slope less than 5% and silty clay soil, modified for hydrology soil group, is 0.36×1.21, or 0.44 (Tables 5.7 and 5.8)

$$q_p = 0.44\ (3.58 \text{ in}/1.5 \text{ hr})(75 \text{ ac}) = 78.76 \text{ cfs}$$

Watershed configuration and composition can make a great difference!

The predicted runoff rate is sensitive to the rational coefficient and precipitation used. The analysis does not show the influence of the method of predicting time of concentration. This effect may also be large.

Peak Flow Based on Excess Rain and Hydrograph Shape Estimates

Peak flow for a given storm is related to the storm runoff volume. Topography and cover affect the relationship. The relationship between peak flow and runoff volume stems from the hydrograph itself, with the peak flow being the maximum value of the hydrograph and the volume being the area under the hydrograph. The USDA-ARS, along with state and privately supported researchers around the country, has instrumented watersheds that have a wide variety of topography and surface cover. This runoff monitoring equipment has enabled evaluation of peak flows, time to peak, time of recession, and runoff volume. Data from these watersheds have shown a variety of hydrograph shapes in terms of the time to peak versus the time of recession. The following discussion explores some of the commonalities in the findings. Most important, weaknesses in the rational method for peak flow prediction should become apparent.

NRCS Triangular Hydrograph with Design Storm Duration Equal to t_c (Version 1)

The rational method may be generalized by including the effect of topography and cover on hydrograph shape. We again consider the triangular hydrograph. The triangular hydrograph method idealizes storms as uniform storms resulting in triangular-shaped runoff events (see Figure 5.10). *The rational method presumes time to peak* (t_p) *is equal to time of recession* (t_{recess})*; thus* $t_{recess}/t_p = 1$. The NRCS hydrograph modifies the hydrograph by stretching the recession limb. The NRCS modification has a longer recession limb than the rational storm based on empirical observation of longer recession times in typical agricultural watersheds.

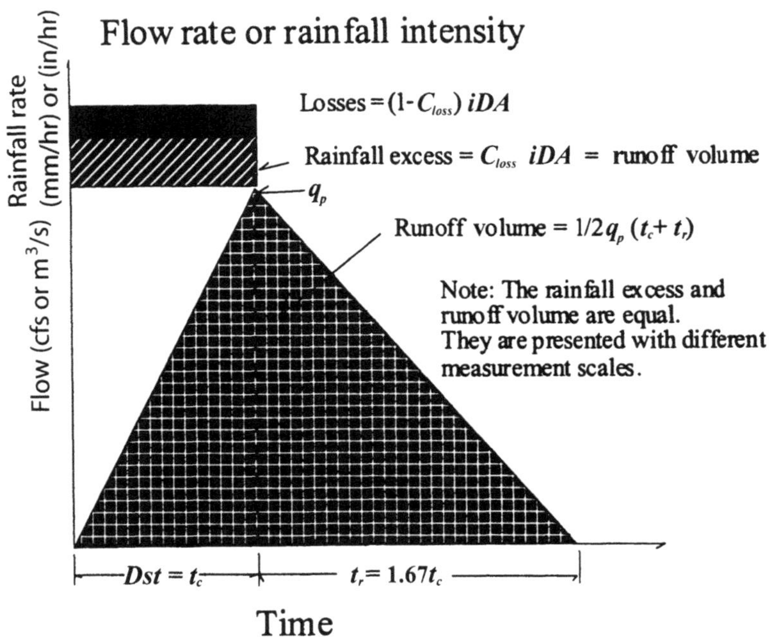

Figure 5.10. Runoff hydrograph for the NRCS hydrograph method assuming a design storm equal to the time of concentration.

Knowing the storm volume (from the NRCS curve number formula), storm duration (given here), and the time of concentration (computed using methods discussed above), we can estimate the peak flow from

$$q_p = 0.75C_{excess}iA \quad \text{(English)} \tag{5.10a}$$

where C_{excess} represents a modified runoff coefficient. Area is in acres (recall 1 cfs ~ 1 ac-in/hr). Alternatively, one may use

$$q_p = \frac{0.46C_{excess}iDstA}{(Dst/2) + 0.6t_c} \tag{5.10b}$$

Other nomenclature is as defined above. We may estimate C_{excess} by calculating the ratio Q/P for the design storm. Alternatively, we can calculate the runoff volume (QA) from the NRCS curve number equation and find $q_p = 0.75$ volume/t_c (English units). Insert the factor 0.0028 for metric units. Other values of t_{recess}/t_{rising} could be used to make the peak flow computations. Wanielista et al. (1997) summarize the variation of the 0.75 factor with watershed topography. The value 0.75 represents the typical NRCS watershed with area expressed in acres. Values range from 1 (paved slopes) to 0.16 (very flat rural watersheds).

NRCS Triangular Hydrograph with Actual Storm Duration Less than t_c and/or the Duration Less than t_c (Version 2)

The NRCS developed another triangular hydrograph useful for storm durations of 2 hr or less. We introduce two new ideas: (1) a formulation for the time to peak t_p that is something other than storm duration and (2) a larger ratio of recession limb to rising limb. *The new formulation for t_p attempts to correct for the reality that excess rainfall is not uniform.*

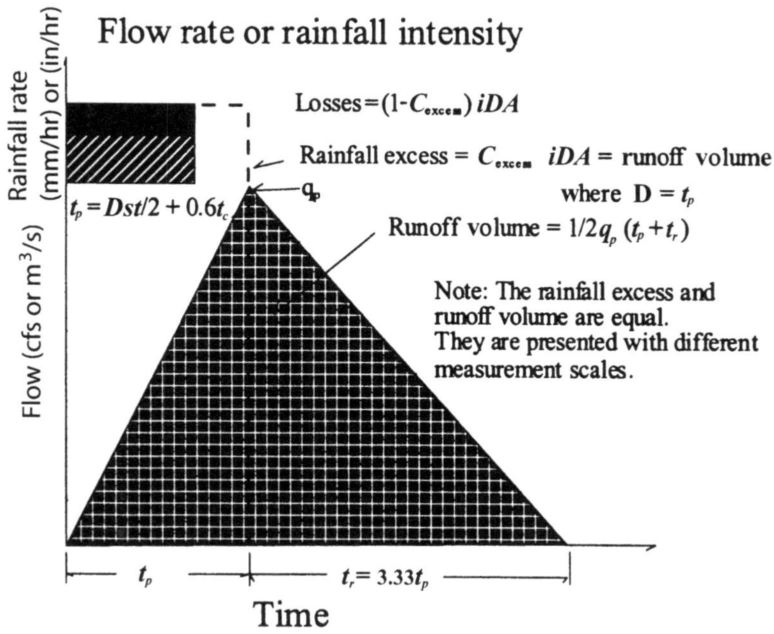

Figure 5.11. Triangular unit hydrograph where the time to peak is a common NRCS function of storm duration and time of concentration and the recession limb to rising limb duration is as indicated.

Increasing the ratio of recession to rising limb is an attempt to account for watershed slope and cover. Figure 5.11 shows the details.

Relating the rainfall excess and the runoff volume and solving as before yields

$$q_p = \frac{0.46C_{excess}iADst}{Dst/2 + 0.6t_c} \tag{5.11}$$

where C_{excess} is a modified runoff coefficient estimated by taking Q/P as above. Had we taken t_p as t_c instead of Dst/2 + 0.6 t_c, the function would have been identical to the version 1 function, except for the coefficient. The coefficient decreased to 0.46 as time of recession to time of rising increased.

Alternatively, using the curve number formula to obtain the volume (QA = Vol) results in

$$q_p = \frac{QA}{1.08Dst + 1.3t_c} \tag{5.12a}$$

If $t_p = t_c =$ storm duration (Dst), then we can easily show q_p to be

$$q_p = \frac{0.46QA}{t_c} \tag{5.12b}$$

The NRCS developed the relationship $t_p = Dst/2 + 0.6\ t_c$ for approximate unit hydrographs for small watersheds and short storm durations (2 hr or less). This formulation for t_p is useful for developing synthetic hydrographs, as discussed below.

Wanielista et al. (1997) suggest that the ratio of recession limb to rising limb increases as the slope decreases. The recession limb to rising limb ratio of 1 (the rational formula) describes very steep slopes and fast-responding watersheds. The common NRCS value of 1.67 describes typical watersheds. The value of 3.33 describes rural rolling hills. The ratio

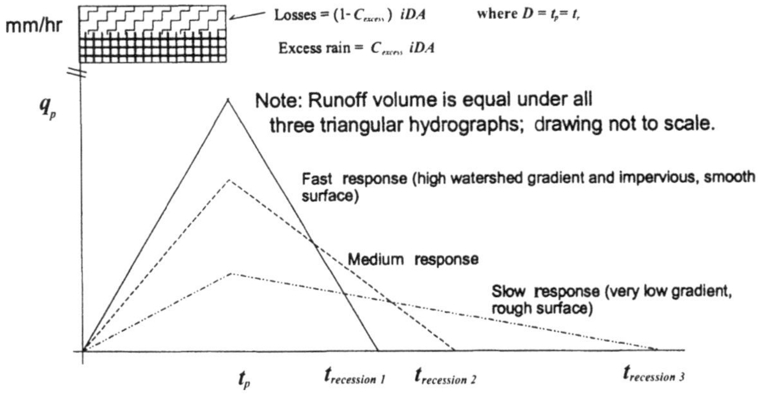

Figure 5.12. Impact of decreasing watershed slope and increasing surface roughness on expected peak flow given a storm of equal duration and given the excess rain or runoff volume is the same.

may approach 12 for extremely flat lands. This corresponds to a factor of 0.16 instead of 0.75 in equation 5.10a. Both the excess rain coefficient and the recession to rising limb ratio account for watershed topography. The rational coefficients should not be used with other triangular hydrograph methods. Figure 5.12 shows the effect of watershed topography on hydrograph relations.

One may estimate excess rain coefficients using the NRCS curve number method and then estimate hydrograph shape based on watershed slope and cover. This discussion shows the limitation of the classical rational method because the ratio of recession to rising limb is usually not 1 unless topography is very steep and there is an impervious cover. The NRCS suggests this empirical relation for t_p ($= Dst/2 + 0.6t_c$) with typical rainfall patterns and small watersheds (e.g., less than 100 ac, or 40 ha).

Erosion modeling software such as SedCad (Warner 1998) invokes different hydrograph shapes by inputting fast, medium, or slow watershed responses. A fast response corresponds to a recession limb to rising limb ratio of approximately 1. A medium response corresponds to a value of 3–6. A slow response corresponds to values of 7–12 or more. Figure 5.12 shows typical watershed responses.

Example 5.3: Suppose a watershed with 25 acres (10.12 ha) of soil group C (silty clay) in pasture, good condition (CN = 74, Table 5.1), and with 50 acres (20.24 ha) of soil group C in forest, poor condition (CN = 77, Table 5.2), had a watershed gradient of 2% and was 2624 ft (800 m) from outlet to farthest end. Find the 50 yr runoff assuming the watershed is in Athens, Georgia, using the recession limb to rising limb ratios of 1, 1.67, and 3.33. Compare the results with runoff volume equations and the excess rainfall coefficient equations.

For the above watershed, the weighted CN is equal to 75 and the total acreage is 75 ac (30.36 ha).

$$t_c = 15 \text{ min} = 0.25 \text{ hr}$$
$$t_p = 0.25 \text{ hr}/2 + 0.6(0.25 \text{ hr}) = 0.275 \text{ hr}$$

The 50 yr–15 min storm is 1.95 in (from the Weiss equation). Runoff volume Q = 0.35 in (8.89 mm) from the NRCS curve number formula. Therefore, for the recession limb to rising limb ratio of 1 (t_{recess}/t_p, see Figure 5.8) we have

$$q_p = (QA)/t_c = 0.35 \text{ in } (75 \text{ ac})/0.25 \text{ hr} = 105 \text{ cfs}$$

(The student is encouraged to derive this from the data given in the rational hydrograph definition sketch above.)

For the t_{recess}/t_p ratio of 1.67 we have, using equation 5.12a,

$$q_p = [0.75(0.35 \text{ in})75 \text{ ac}]/0.275 = 71.6 \text{ cfs}$$

where $t_p = 0.275$. If $t_p = t_c = 0.25$ hr, using equation 5.12b, then

$$q_p = [0.75(0.35 \text{ in})75 \text{ ac}]/0.25 \text{ hr} = 78.75 \text{ cfs}$$

For the t_{recess}/t_p ratio of 3.33 we have (equation 5.12a)

$$q_p = (QA)/(1.08Dst + 1.3t_c)$$
$$= 0.35 \text{ in}(75 \text{ ac})/[1.08(0.25 \text{ hr}) + 1.3(0.25 \text{ hr})] = 44 \text{ cfs}$$

If $t_p = t_c$, then

$$q_p = [0.46(0.35 \text{ in})75 \text{ ac}]/0.25 \text{ hr} = 48.5 \text{ cfs}$$

Using the excess rainfall coefficients and a ratio of 1 (see Figure 5.8), we calculate C_{excess} using the Q/P ratio:

$$C_{excess} = 0.35 \text{ in}/1.95 \text{ in} = 0.18$$
$$i = 1.95 \text{ in}/0.25 \text{ hr} = 7.8 \text{ in/hr}$$

In the rational method example $C_{rat} = 0.375$. The rational method is likely an overdesign here. The rational method is more realistic as C_{rat} approaches 1.

$$q_p = C_{excess}iA = 0.18(7.8 \text{ in/hr})75 \text{ ac} = 105.3 \text{ cfs}$$

If the t_{recess}/t_p ratio is 1.67, then

$$q_p = 0.75(0.18)(7.8 \text{ in/hr})(75 \text{ ac}) = 78.9 \text{ cfs}$$

(see equation 5.10a). If the t_{recess}/t_p ratio is 3.33, then

$$q_p = 0.46(0.18)(7.8 \text{ in/hr})(75 \text{ ac}) = 48.4 \text{ cfs}$$

(see equation 5.12b).

Note: We invoke the numerical equivalency of flows in ac-in/hr and cfs units here.

These values agree with those computed using the runoff formulations when $t_p = t_c$. The student is encouraged to check this using metric units.

This example shows the effect of the recession limb to rising limb ratio, or watershed response, on the predicted flow rate with identical excess rain coefficients. Again, the rational coefficient approximates the C_{excess} value. Perhaps a better measure of the excess rain coefficient would be to estimate *Q/P* for the storm.

The other triangular hydrographs were presented to show how hydrograph shape, dependent upon watershed gradient and surface roughness, might affect peak flow. They also provide a foundation for the unit hydrograph discussion. Even though excess rain variation, watershed topography, and watershed cover figure into the runoff prediction, one should observe the compactness assumption and the size limit of 100 ac (40 ha) for best results.

The discussion has shown that (1) the rational method presumes time to peak is equivalent to time of recession; (2) the time of recession may be up to 12 or more times the time to peak, depending on watershed characteristics; and (3) it is necessary to compensate for nonuniform rainfall distribution over the watershed. *For design purposes, the smaller the rational coefficient, the less confidence one should place in the rational method.* The purpose of this discussion was to exemplify the approximate nature of peak flow specification.

ASAE (1997) Drainage Ditch Design Flow Rate

The ASAE (1997) drainage ditch formula is a published standard for drainage ditch design. This is a specialized approach specifically for drainage ditches in areas where the land slope is less than 0.5% and in watersheds less than 4000 ha (10,000 ac). It is an adaptation of the rational method with a hydrograph shape correction. The empirical formula is given by

$$q_{pASAE} = 0.013 C_{ASAE} A^{0.833} \quad \textbf{(5.13a)}$$

where

q_{pASAE} is the runoff rate in m^3/s;
C_{ASAE} is a constant;
A is watershed area in km^2.

The slope is no more than 0.5% for this method! In English units,

$$q_{pASAE} = C_{ASAE} A^{0.833} \quad \textbf{(5.13b)}$$

where

q_{pASAE} is in cfs;
A is in (statute) mi^2.

Table 5.9 gives values for C_{ASAE}. One may rearrange the ASAE drainage formula to fit the triangular hydrograph form as follows:

$$q_{pASAE} = C_{dim} A^{-0.17} C_{excess} i_{2\,yr-48\,hr} A = C_{dim} F(t_{recession}/t_{rising}) C_{excess} i_{2\,yr-48\,hr} A \quad \textbf{(5.13c)}$$

Table 5.9. Runoff coefficients for the ASAE 1997 land drainage equation.

Soil condition	Drainage runoff coefficient C_{ASAE}
Coastal plain, cultivated	45
Delta, cultivated	40
Cornbelt and Northeast, cultivated	37
Coastal plain, pasture	30
Corn Belt and Northeast, pasture	25
Delta and coastal, riceland	22.5
Red River Valley (Minnesota and North Dakota)	20
Southwestern range	15
Coastal plain woodland	10

Source: ASAE 1997.

where

C_{dim} is a dimension system constant;
C_{excess} is the excess rain coefficient estimated by $Q_{2\ yr\text{-}48\ hr}/P_{2\ yr\text{-}48\ hr}$;
$i_{2\ yr\text{-}48\ hr}$ is the 2 yr–48 hr storm intensity;
$F(t_{recession}/t_{rising})$ is the function of the ratio of recession to rising limb time durations = $A^{-0.17}$ (dimensionless);
Q is the runoff in watershed inches or other length;
P is the storm precipitation in the same units as Q;
other nomenclature is as previously defined.

Interestingly, the ratio of recession limb to rising limb duration decreases as A increases, consistent with Wanielista et al. 1997. We may then write

$$C_{ASAE} = C_{dim}(Q_{2\ yr\text{-}48\ hr}/P_{2\ yr\text{-}48\ hr})i_{2\ yr\text{-}48hr} = 640(Q_{2\ yr\text{-}48\ hr}/48\ \text{hr})$$

in English units, thus suggesting a method for estimating C_{ASAE} values for specific locations. Table 5.9 presents published factors for regions of major surface drainage construction in the United States.

Example 5.4: Suppose a client wants to drain a 200 km² region in the coastal plain (woodland). Estimate the capacity for a drainage ditch. Slope is essentially flat.

Since the slope is flat, assume the slope is less than 0.5%, making the ASAE drainage formula applicable. The C_{ASAE} factor for the drainage equation (equations 5.13a–c) is 10 (Table 5.9). Using the ditch drainage formula gives

$$q_{pASAE} = 0.013(10)(200\ \text{km}^2)^{0.833} = 10.56\ \text{m}^3/\text{s}$$

This concludes our discussion of hydrograph shape. In summary, the rational method hydrograph directly applies to small, impervious watersheds with a significant watershed gradient. In other words, watersheds meeting these conditions will produce an isosceles triangular hydrograph. The rising limb mirrors the falling limb. (This enables a prediction of runoff volume with knowledge of the rational coefficient.) As watersheds become larger in size and more diverse in cover, complex flow-cover relationships plus infiltration and reemergence as springs and seeps explain why the recession limb may lengthen as watersheds are larger, flatter, and more pervious. Wanielista et al. (1997) provide limited guidance for choosing the appropriate recession to rising limb ratio value.

NRCS TR55 Graphical Method

The example in the triangular hydrograph discussion shows the difficulty of using a hydrograph shape to relate peak flow and volume. Clearly defined criteria for selecting a recession limb to rising limb ratio value are usually unavailable. Due in part to this difficulty, the NRCS considered a different approach, which factors in runoff volume and watershed characteristics. They ran a large number of simulations and developed a graphical procedure based on a unit discharge depending on precipitation, time of concentration, storm type, and soil characteristics. This analysis culminated in the NRCS TR55

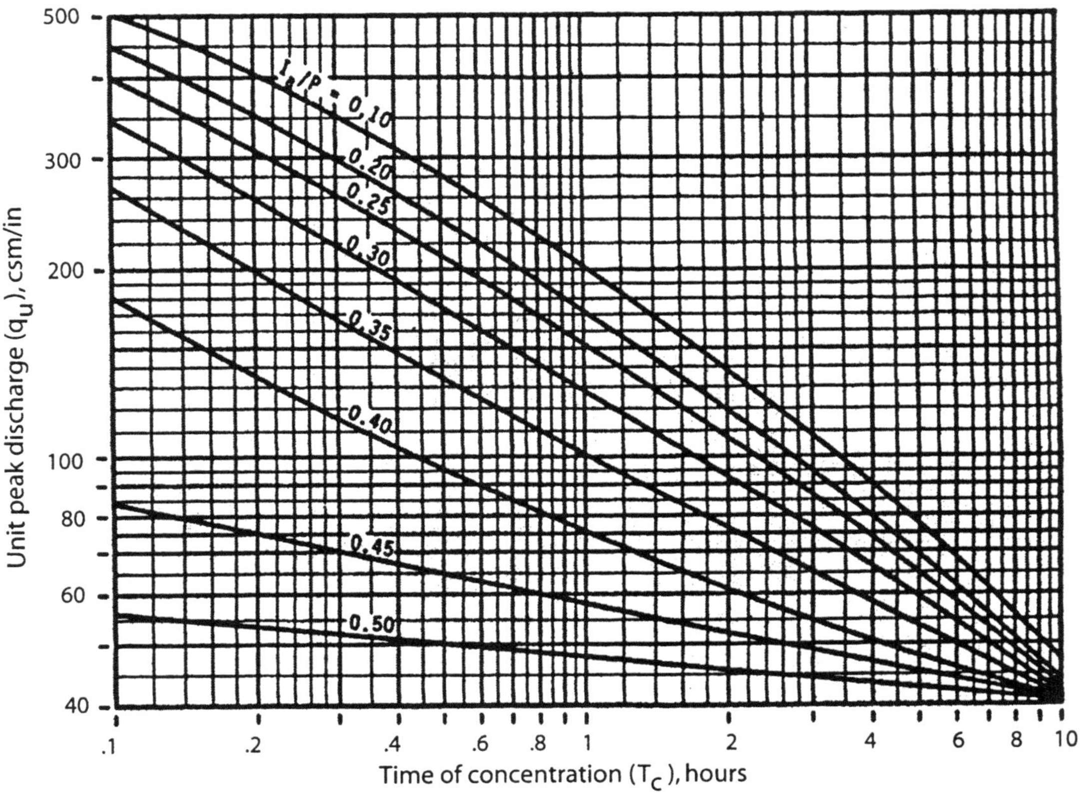

Figure 5.13. NRCS runoff curves for the type I storm. Multiply csm/in by 4.3055 × 10^{-6} to get value in units of $m^3/s \cdot mm \cdot ha$. (From NRCS 1986.)

graphical method. One should use the NRCS TR55 method when the watershed area exceeds 40 ha (100 ac). The area should not exceed 5200 ha (20 mi²). The basic equation for this method is (Haan et al. 1994)

$$q_{pTR55} = q_u A Q_{x\,yr-24hr} F_p \tag{5.14}$$

where

q_u is the unit peak flow in units of $ft^3/(mi^2 \cdot in \cdot s)$ from Figures 5.13–5.16); these units are abbreviated as csm/in in Figures 5.13–5.16. The csm is defined as cubic feet per second per square mile. One may convert these units to $m^3/(s \cdot mm \cdot ha)$ by multiplying the English value by 4.3055 × 10^{-6} (check this!). Units of miles are statute miles.
q_{pTR55} is the NRCS TR55 peak flow (cfs, m^3/s);
A is area in square miles (640 ac/mi²; 259 ha/mi²);
$Q_{x\ yr\text{-}24\ hr}$ is runoff volume for the 24 hr storm of the desired (x yr) return period (in);
F_p is the swamp adjustment factor tabulated in Table 5.10.

The time of concentration $t_{c\ NRCS}$ is the length of time required for water to travel from the farthest location in the watershed to the outlet:

$$t_{c\,NRCS} = \frac{L^{0.8}\left(\frac{1000}{CN} - 9\right)^{0.7}}{11{,}400\sqrt{S_g}} \tag{5.15}$$

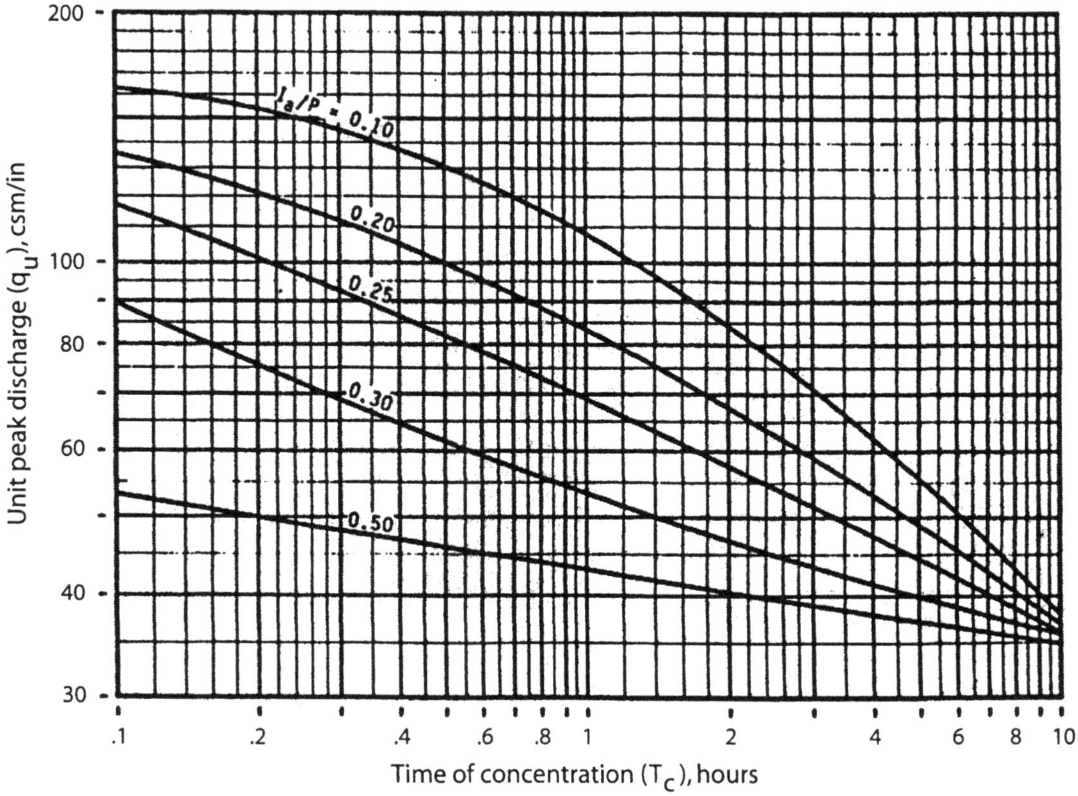

Figure 5.14. NRCS runoff curves for the type Ia storm. Multiply csm/in by 4.3055×10^{-6} to get value in units of $m^3/s \cdot mm \cdot ha$. (From NRCS 1986.)

where

$t_{c\ NRCS}$ is in hours;
L is in feet;
S_g is expressed as a fraction;
CN is the NRCS curve number, adjusted for antecedent moisture as above.

Equation 5.15 brings in the curve number, which factors in watershed cover effects on watershed response.

Table 5.10. Swamp factor F_p for the graphical NRCS TR55 runoff equation.

Percentage of land in swamp and wetland areas	F_p
0	1.00
0.2	0.97
1.2	0.87
3.0	0.75
5.0	0.72

Source: NRCS 1986.

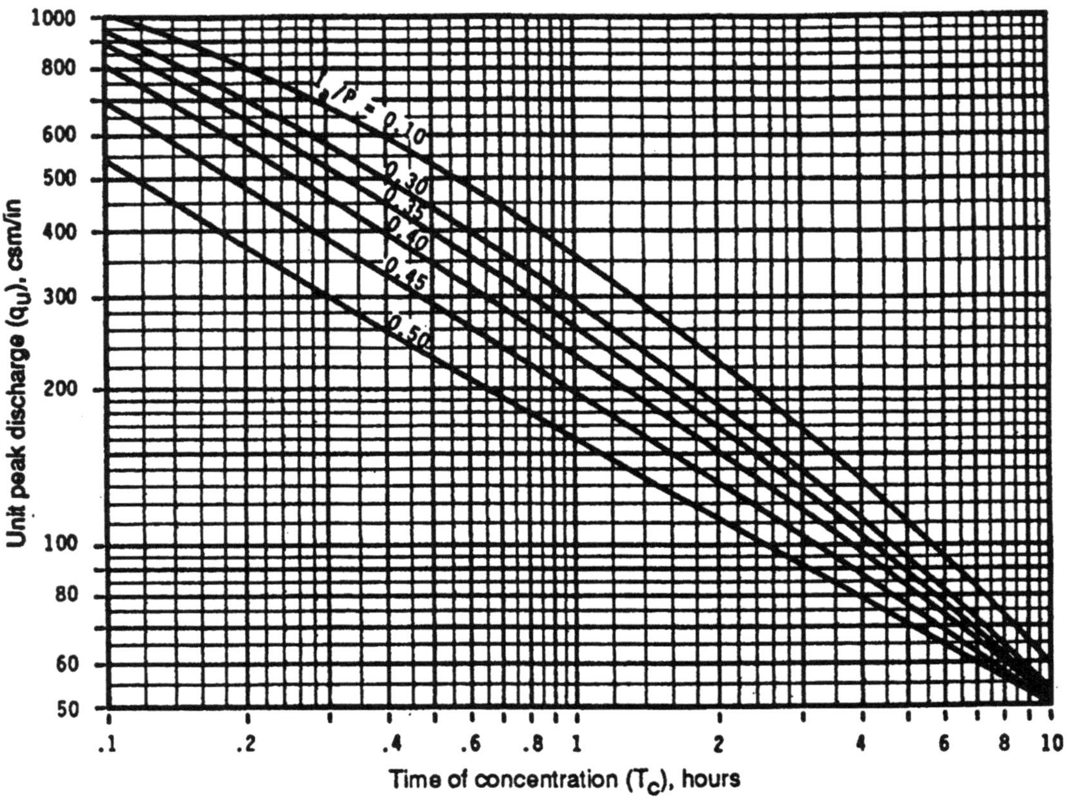

Figure 5.15. NRCS runoff curves for the type II storm. Multiply csm/in by 4.3055×10^{-6} to get value in units of $m^3/s \cdot mm \cdot ha$. (From NRCS 1986.)

In SI units equation 5.15 appears as

$$t_{c\,NRCS} = \frac{L^{0.8}\left(\frac{1000}{CN} - 9\right)^{0.7}}{4407\sqrt{S_g}} \tag{5.16}$$

where

$t_{c\,NRCS}$ is in hours;
L is in meters;
S_g is expressed as a fraction.

To solve, read the unit peak flow from the appropriate unit peak flow plot depending on rainfall type. See the map in Figure 2.19 for the appropriate rainfall type. Watershed size should not exceed 800 ha. The slope should exceed 0.5%. The watershed should be compact. Usually use an initial abstraction $I_a = 0.2S$ where S is found using equations 5.5a–b.

Unit peak flow plots for all storm types appear in Figures 5.13–5.16. Units here are English, csm/in or $ft^3/(mi^2 \cdot in \cdot s)$. Notice the variation in unit peak runoff values at a given t_c value of, say, 1 hr for the various storm types.

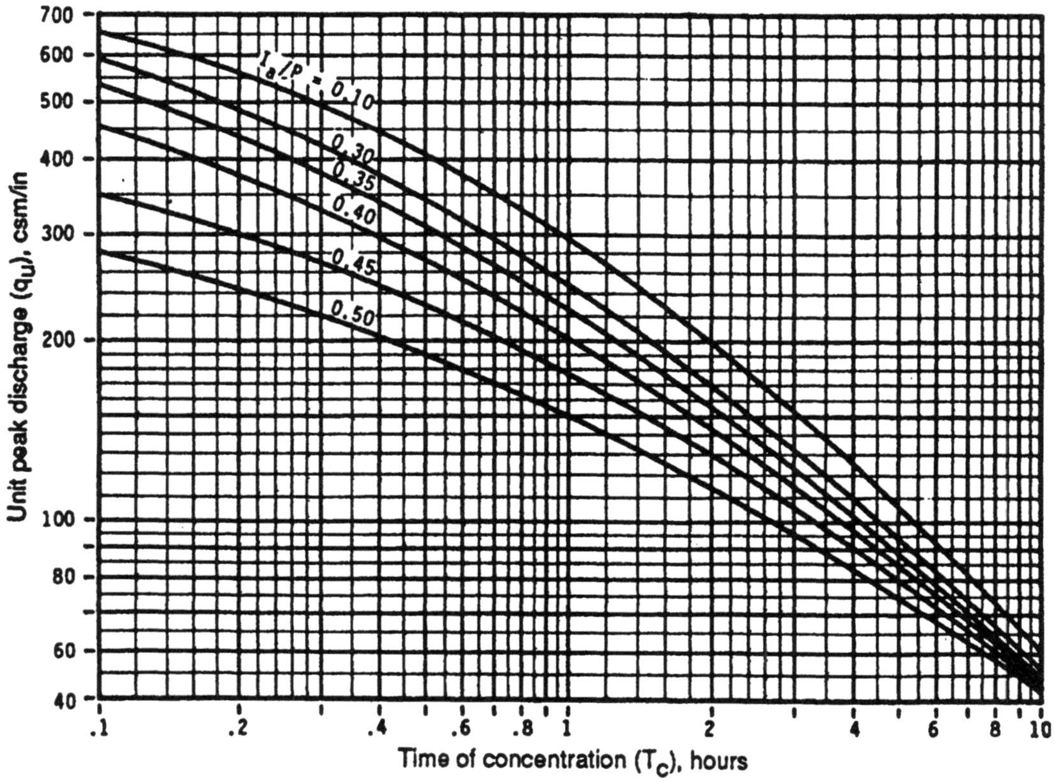

Figure 5.16. NRCS runoff curves for the type III storm. Multiply csm/in by 4.3055×10^{-6} to get value in units of $m^3/s \cdot mm \cdot ha$. (From NRCS 1986.)

NRCS TR55 Method Summary

1. Determine the design storm return period.
2. Obtain the area-weighted curve number (Tables 5.1 through 5.5).
3. Obtain $P_{x\ yr\text{-}24\ hr}$ from precipitation maps or methods such as the Weiss calculation.
4. Obtain the S value from the S equation of appropriate units (equations 5.5a–b).
5. Calculate the runoff $Q_{x\ yr\text{-}24\ hr}$ from the curve number equation (equations 5.6a–b).
6. Calculate the time of concentration from the appropriate $t_{c\ NRCS}$ equation (equation 5.15).
7. Determine the storm type based on region (Figure 2.19).
8. Compute $0.2S/P_{x\ yr\text{-}24\ hr}$ to determine the appropriate line in the unit discharge curves.
9. Read the base q_u from the appropriate graph (i.e., type of storm) (Figures 5.13–5.16).
10. Estimate the swamp percentage from the map and estimate the F_p factor (Table 5.10).
11. Substitute the values into the TR55 runoff formula to arrive at the peak flow (equation 5.14).

Example 5.5: Suppose a watershed with 25 acres (10.12 ha) of soil group C in pasture, good condition ($CN = 77$), and with 50 acres (20.24 ha) of soil group C in forest, thin stand, poor cover ($CN = 74$), has a watershed gradient of 2% and is 800 m

(2624 ft) from outlet to farthest end. Find the 50 yr runoff assuming the watershed is in Athens, Georgia. There are no swamps present.

The weighted curve number is [25 ac(77) + 50 ac(74)]/75 ac = 75. For the above watershed,

$$t_{c\ NRCS} = (800\ \text{m})^{0.8}(1000/75 - 9)^{0.7}/[4407\ (0.02^{0.5})] = 0.94\ \text{hr}$$

(equation 5.15).

$$S = 25400/75 - 254 = 84.67\ \text{mm} = 3.33\ \text{in}$$

(equation 5.6). The 50 yr–24 hr precipitation is 185.16 mm (7.3 in) by the Weiss formula. The corresponding $Q_{50\ yr\text{-}24\ hr}$ is 111.76 mm (4.4 in) by the NRCS runoff volume formula (equations 5.6a–b). The factor $0.2S/P_{50\ yr\text{-}24\ hr} = 0.2(84.67\ \text{mm})/185.16\ \text{mm} = 0.09$. The absence of swamps leads to a swamp factor $F_p = 1.0$ from Table 5.10. Read q_u of 375 $\text{ft}^3/(\text{mi}^2 \cdot \text{in} \cdot \text{s})$ from the type II graph (Figure 5.15), entering the $t_{c\ NRCS}$ on the x axis, reading up to $I_a/P = 0.09$, then left to the y axis. From equation 5.14, peak flow is then

$$\begin{aligned} q_{pTR55} &= 375\ \text{ft}^3/(\text{mi}^2 \cdot \text{in} \cdot \text{s})\ [75\ \text{ac}/(640\ \text{ac/mi}^2)](4.4\ \text{in})(1) \\ &= 193\ \text{cfs}(1\ \text{m}^3/\text{s})/35.28\ \text{cfs} = 5.5\ \text{m}^3/\text{s} \end{aligned}$$

Based on a sensitivity analyses, a 5% change in curve number very strongly affects the predicted runoff flow rate. Precipitation, followed by the choice of swamp factor is also influential. The choice of curve number is crucial. The TR55 method effectively shows the impact of watershed land use changes.

The Design Storm

The design storm for peak flow depends on the return interval and watershed characteristics. Agency or regulatory guidelines usually specify the return interval. As a rule of thumb, a 2 yr storm may be used for temporary structures, a 10 yr storm is used when property damage on project failure is minor, 25 or 50 yr is used when property damage on failure may be substantial, and 100 years or greater is used when loss of life is likely on project failure. Storm duration for peak flow is determined as follows. From the discussion of intensity-duration-frequency (see Figure 2.14), recall that intensity at a given frequency increases as the duration (time) decreases. *The time of concentration is the time required for runoff to reach the outlet from the watershed's farthest point. Since the entire watershed must contribute for maximum flow, it follows that storm duration must be at least equal to the time of concentration.* Based on the intensity-duration-frequency curve, the storm should not exceed the time of concentration for a given frequency. Return interval and duration specification complete the design storm specification.

Similar considerations enter into determining the interval for runoff volume calculations. With runoff volume, agency policy usually sets the storm duration as well as the storm return period. The 24 or 48 hr storm duration is often used.

Runoff and Peak Flow Using Frequency Analyses

Obviously, we may assess peak flow and possibly runoff volume for a given design storm duration if a gauging station is present at the site. Peak flows may be determined using frequency analyses if an adequate period of record of peak stage exists. Likewise, we may analyze flow volumes if the gauge was a recording model. We would integrate the stage relationship, apply the stage-flow curve for the gauge, and perform a frequency analysis on the flow volumes. Steps outlined in the precipitation discussion are applicable. *As before, data must be independent, with no serial correlation; in general monthly, weekly, or daily data cannot be analyzed using this technique.*

Consult a hydrologist in the following cases: (1) if a gauge is present nearby on the stream in question, (2) the record is very short, or (3) if gauges are present on other streams in the region. One should use gauge records or supplement computed runoff predictions with gauge records if gauge data are available. If no gauge is present, we must use calculation methods. This is the usual case for farm- and field-scale problems. Therefore, prediction of runoff volume, peak runoff rate, and runoff hydrograph shape is of interest.

Runoff Hydrograph Development

The analysis of complex watersheds requires simulation of the entire runoff hydrograph. Hydrograph development begins with a rainfall hyetograph simulation as discussed in Example 2.7. We divide the hyetograph into increments. Each increment is processed using a synthetic unit hydrograph. Hydrographs are lagged and summed to give the composite storm hydrograph. One may route hydrographs for subwatersheds downstream to the complex watershed outlet. This discussion will be concerned only with hydrograph generation for a compact watershed or one subwatershed of a complex watershed. Hydrograph simulation is computationally intensive. Complete hydrographs are also useful for routing storms down streams and through structures, discussed in Chapters 11 and 12.

The Unit Hydrograph

We have thus far considered hydrograph peak flow and runoff volume (area under hydrograph) prediction for extreme-case storms. We need the peak excess rainfall, computed as part of simulating a runoff hydrograph, for predicting the sediment delivery ratio, discussed in Chapter 6. Sometimes we may want to predict the hydrograph from a storm duration different (shorter or longer) from the time of concentration. We may have an actual hyetograph from a recording rain gauge. Alternatively, we may want to use a synthetic storm.

In this section, we use a synthetic storm to generate a runoff hydrograph storm duration exceeding the time of concentration. The procedure is to (1) develop a unit hydrograph; (2) develop the synthetic storm (or apply recorded rainfall); (3) determine the excess rainfall; and finally (4) apply the unit hydrograph to each increment of excess rainfall and sum the components. The starting point for hydrograph development is the unit hydrograph.

Unit Hydrograph Definition

The unit hydrograph is the hydrograph resulting in 1 in, or 25.4 mm, runoff from a watershed, given a rainfall of uniform intensity over the entire watershed area. Rainfall begins and ends simultaneously over the entire watershed.

One may scale the shape of the unit hydrograph depending on the actual runoff volume. For many situations, we can apply the unit hydrograph to the gross runoff volume for sufficient design precision. For other situations, the unit hydrograph is applied to increments of effective runoff[2] (e.g., instances in which rainfall exceeds infiltration), and we sum the results. We first develop the simple hydrograph.

Basic Runoff Hydrograph for Design Runoff Events

We now refine the hydrograph shape from the triangular to a more realistic shape. Haan et al. (1994), building upon the NRCS unit hydrograph, described the basic runoff hydrograph as

$$\frac{q_u(t)}{q_{pu}} = \left[\frac{t}{t_p}\exp(1 - t/t_p)\right]^{K_{Hahn}} \tag{5.17}$$

where

t_p is approximated by $Dst/2 + 0.6t_c$;
q_{pu} is the peak flow of the unit hydrograph;
$q_u(t)$ is the flow at time t;
K_{Hahn} is 3.77 for watersheds with a recession to rising limb ratio of 1.67. See Haan et al. 1994 for values relating to other watershed response types.

The above equation shape is a modification of the triangular hydrograph presented above. Area under the two curves (runoff volume) is the same in each case. The triangular relationship between runoff volume and peak flow is used with the above equation also.

To apply the latter hydrograph shape we need the following. First, we need the time to peak as calculated above. Second, calculate q_{pu} from

$$q_{pu} = \frac{484A}{t_p} \tag{5.18}$$

where

A is the area in square (statute) miles; volume is 1 watershed inch (if A is in acres, the constant is 0.75; if A is in hectares, runoff volume is 25.4 mm, and flow is in m^3/s, then the constant is 0.0021);
q_{pu} is the unit peak flow, ft^3/s or m^3/s;
t_p is the time to peak in hours;
the volume is 1 watershed inch.

The above peak flow equation comes straight from a combination of versions 1 and 2 of the NRCS triangular hydrograph. One may use version 1 with the version 2 definition of t_p. By letting time t range up to 5 times t_p and solving for $q_u(t)/p_{pu}$ and then for $q_u(t)$, we define the unit hydrograph. This hydrograph shape is more consistent with field observations. Figure 5.17 shows an example of a unit hydrograph.

Next, scale the unit hydrograph to get the actual hydrograph for a storm with effective runoff duration of t_c. Take a ratio of q_p over q_{pu}. Multiply each value of $q_u(t)$ by this ratio to get $q(t)$. If the duration of the effective runoff (the time the rainfall rate exceeded the infiltration and evaporation rate) was equal to the time of concentration of the watershed, then one would simply scale the curve by proportionally scaling $q(t)$ by the ratio. Figure 5.17 also shows a unit hydrograph scaled for a 2.5 mm excess rain. This method is reasonable for storm durations of 2 hr or less, assuming the t_c of the watershed is greater than the storm duration.

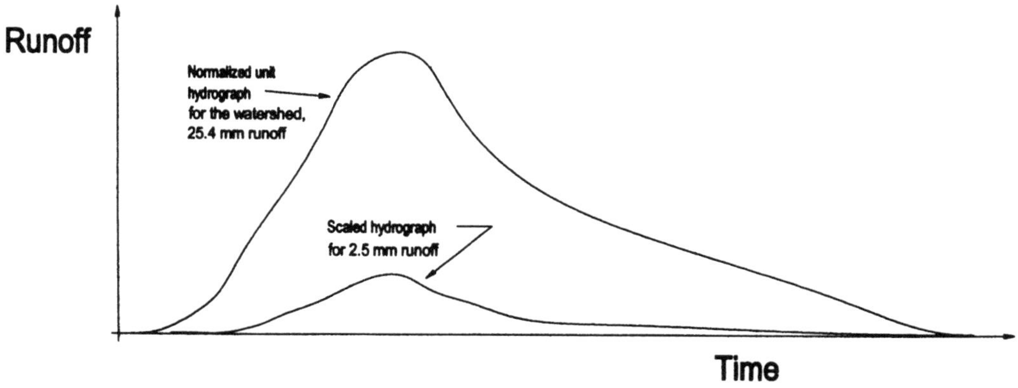

Figure 5.17. Normalized unit hydrograph obtained using equations 5.17 and 5.18. The scaled hydrograph was computed on the indicated smaller volume of effective runoff.

Generalized Approach to Hydrograph Development

The following generalized runoff approach enables us to generate hydrographs from actual or synthetic storms whose duration may exceed the time of concentration of the watershed. The generalized hydrograph starts from the unit hydrograph and the effective runoff analyses. Here we will select rainfall increments of small duration—regard the storm as a series of pulses. The t_p will be approximately $D_g/2 + 0.6t_{c\ NRCS}$. The value D_g is the increment consistently used throughout the problem. The t_p should be 3–5 times D_g. For example, if $0.6t_c$ is about 15 min (round to the nearest 5 min typically) then a 5 min duration would be sufficient. This would lead to a t_c of 20 min. Wanielista et al. (1997) give additional discussion on hydrograph development.

Two approaches for effective rainfall are suggested. One may correct the precipitation for infiltration using the Green-Ampt method (see Chapter 3). One may also use the NRCS methods. After computing effective runoff, we then compute a unit hydrograph using, for time to peak, one-half the duration of an increment plus the t_c of the watershed. Generally, t_c will be much greater than half the duration. One should define rainfall increments one-third to one-fifth the time to peak. One may then apply the unit graph to each rainfall increment, lagging successive increment results by the number of increments elapsed. The composite hydrograph is obtained using superposition.

Green-Ampt Effective Runoff Calculation

The Green-Ampt runoff calculation in effect uses the Green-Ampt infiltration model to determine infiltrated volume versus elapsed storm time. It brings to bear the concepts shown in Figure 3.6. The rainfall may vary with time. If the rainfall at time t is less than the infiltrated volume, we take infiltration as the rainfall. If the rainfall exceeds the infiltrated amount, the excess rain is treated as runoff. Table 5.11 gives an example of the effective rainfall calculation based on the Green-Ampt infiltration model.

The selection of the time interval in the above table is important. It should range from 5 to 15 min normally. The need to have 3 or 4 increments pass before reaching the peak of the hydrograph governs the selection of the time interval. This method is very useful for simulating irrigation machine passage over soils. This method is not commonly used in watershed-scale runoff studies, because of the wide variation in soil properties over a watershed.

Table 5.11. Example effective rainfall computation based on the Green-Ampt infiltration method.

	Time (hr)	Incrm rain (cm)	f_p (cm/hr)	Δf_p (mm in interval)	Actual infil. (cm)	F_p (cum. Infil.)	Surface storage	Eff. rain
Element	1	2	3	4	5	6	7	8
1	0	0.01					0	0
2	0.05	0.481	37300	1870	0.481	0.481	0	0
3	0.1	0.481	783	39.2	0.481	0.961	0	0
4	0.15	0.481	395	19.7	0.481	1.44	0	0
5	0.2	0.481	265	13.3	0.481	1.92	0	0
6	0.25	0.481	201	10	0.481	2.4	0	0
7	0.3	0.481	162	8.09	0.481	2.88	0	0
8	0.35	0.481	136	6.8	0.481	3.36	0	0
9	0.4	0.481	117	5.87	0.481	3.84	0	0
10	0.45	0.481	104	5.18	0.481	4.33	0	0
11	0.5	0.481	92.8	4.64	0.481	4.81	0	0
12	0.55	3.8	84.2	4.21	3.8	8.61	0	0
13	0.6	3.8	49.8	2.49	2.49	11.1	0	1.31
14	0.65	3.8	40.1	2.01	2.01	13.1	0	1.8
15	0.7	3.8	35	1.75	1.75	14.9	0	2.06
16	0.75	3.8	31.6	1.58	1.58	16.4	0	2.22
17	0.8	3.8	29.2	1.46	1.46	17.9	0	2.34
18	0.85	3.8	27.4	1.37	1.37	19.3	0	2.44
19	0.9	3.8	25.9	1.29	1.29	20.6	0	2.51
20	0.95	3.8	24.7	1.23	1.23	21.8	0	2.57
21	1	3.8	23.6	1.18	1.18	23	0	2.62
22	1.05	0.721	22.7	1.14	0.721	23.7	0	0
23	1.1	0.721	22.3	1.11	0.721	24.4	0	0
24	1.15	0.721	21.8	1.09	0.721	25.1	0	0
25	1.2	0.721	21.3	1.07	0.721	25.9	0	0
26	1.25	0.721	20.9	1.05	0.721	26.6	0	0
27	1.3	0.721	20.5	1.03	0.721	27.3	0	0
28	1.35	0.721	20.2	1.01	0.721	28	0	0
29	1.4	0.721	19.8	0.991	0.721	28.7	0	0
30	1.45	0.721	19.5	0.974	0.721	29.5	0	0
31	1.5	0.721	19.2	0.958	0.721	30.2	0	0
32	1.55	0.37	18.9	0.943	0.37	30.6	0	0
33	1.6	0.37	18.7	0.936	0.37	30.9	0	0
34	1.65	0.37	18.6	0.929	0.37	31.3	0	0
35	1.7	0.37	18.4	0.921	0.37	31.7	0	0
36	1.75	0.37	18.3	0.914	0.37	32	0	0
37	1.8	0.37	18.2	0.908	0.37	32.4	0	0
38	1.85	0.37	18	0.901	0.37	32.8	0	0
39	1.9	0.37	17.9	0.894	0.37	33.1	0	0
40	1.95	0.37	17.8	0.888	0.37	33.5	0	0

Notes:

Col 1. Choose a time interval one-third to one-fifth the time to peak of the unit hydrograph. Rounding to the nearest five minutes is acceptable. Usually 5, 10 or 15 minutes work well. If one is using a hyetograph, and one does not intend to develop a hydrograph, use the time intervals of the hyetograph.

Col 2. Rainfall hyetograph (synthetic or actual).

Col 3. Green-Ampt infiltration (equation 3.10) or other function.

Col 4. Infiltration volume, $f_p t$.

Col 5. Actual infiltration volume = Minimum (rain hyetograph, infil volume).

Col 6. Cumulative infiltration (the sum of actual infiltration to this point). For the first entry use 0.01 to keep from dividing by zero in column 3.

Col 7. Assume surface storage to be zero. Haan et al. (1994) further discuss surface storage.

Col 8. Effective runoff is rain increment minus actual infiltration (not less than zero).

NRCS Effective Runoff Calculation

This method is based on the NRCS curve number method for computing runoff. Table 5.12 summarizes the steps in applying the NRCS method. A 3 hr storm is used; however, we can easily expand or contract the table for longer or shorter storm durations. See Haan et al. 1994 for further discussion of storm simulations. We can use any infiltration function for which coefficients are available.

The next step is to convolve the unit hydrograph with the effective precipitation at each effective precipitation increment. One can do this numerically by forming a table with

Table 5.12. Calculation table for NRCS curve number method for a 3hr storm.

Element	Storm time T	Depth at T	Effective rain	Incremental effective
1	10.5	0.809	0	0
2	10.6	0.829	0	0
3	10.7	0.85	0	0
4	10.7	0.87	0	0
5	10.8	0.891	0	0
6	10.9	0.911	0	0
7	11	0.931	0	0
8	11.1	0.963	0	0
9	11.2	0.995	0	0
10	11.2	1.03	0	0
11	11.3	1.06	0	0
12	11.4	1.09	0	0
13	11.5	1.12	0	0
14	11.6	1.37	0.00483	0.00483
15	11.7	1.62	0.028	0.0231
16	11.7	1.87	0.0682	0.0403
17	11.8	2.13	0.124	0.0555
18	11.9	2.38	0.193	0.0692
19	12	2.63	0.274	0.0815
20	12.1	2.68	0.291	0.0167
21	12.2	2.72	0.308	0.0171
22	12.2	2.77	0.326	0.0175
23	12.3	2.82	0.344	0.0179
24	12.4	2.87	0.362	0.0182
25	12.5	2.91	0.381	0.0186
26	12.6	2.94	0.39	0.0097
27	12.7	2.96	0.4	0.00979
28	12.7	2.99	0.41	0.00988
29	12.8	3.01	0.42	0.00998
30	12.9	3.04	0.43	0.0101
31	13	3.06	0.44	0.0102
32	13.1	3.08	0.448	0.00747
33	13.2	3.1	0.455	0.00751
34	13.2	3.11	0.463	0.00756
35	13.3	3.13	0.47	0.00761
36	13.4	3.15	0.478	0.00765
37	13.5	3.17	0.486	0.0077

time increments across the top axis and with effective runoff along the left side. Apply the unit graph to the first increment of rainfall, putting the entries in the row corresponding to that rainfall increment. Do the same with the second, shifting one column to the right and one row down corresponding to the current increment. Continue until all increments have been processed. Now, sum the rows corresponding to each time increment. Plotting the row sums versus time defines the storm hydrograph. Wanielista et al. (1997) present matrix methods of hydrograph development.

Example 5.6: Calculate the runoff from a 25 yr–3 hr storm on a watershed near Athens, Georgia, with the following characteristics: $CN = 70$, antecedent condition II; area = 100 ac; watershed is 5000 ft long; and the watershed gradient is 5%.

First, compute the unit hydrograph for the watershed: Duration here is the span of a rainfall increment, which we do not yet know. Compute $0.6t_c$ and use a duration from one-third to one-fifth of $0.6t_c$. We suggest a duration of 5, 10, 15, or 20 min, whichever meets the criteria. Set t_p equal to 3–5 times the selected duration.

Suppose $0.6t_c$ was 0.7 hr, or 42 min. Then a duration of 15 min would be reasonable. A value of t_p of $15/2 + 42 = 47.5 \sim 45$ min. The time 45(0.75 hr) is an even multiple of 15 and is 3×15, meeting the $1/3$ to $1/5$ criterion.

Use equation 5.17 with $K_{Hahn} = 3.77$ to approximate the unit hydrograph. We can compute time $t_{c\ NRCS}$ from equation 5.15. Calculate q_p from equation 5.18. Taking t_p as 0.75 and the increment as 0.25 (this is three times smaller than the t_p), one obtains the unit hydrograph tabulated in Table 5.13.

Second, obtain or simulate the rainfall mass curve: We can measure this curve (representing an actual event) or we can simulate the event from the NRCS ordinates. Compute the 24 hr–25 yr storm using the storm ordinates. Remember to take the duration of the storm over 2 and center the event on the 12 hr point in the table. Using the increment of 0.25 hr (15 min) determined above, one obtains Table 5.14. Again, we can simulate the array depthM in Table 5.14 as before; or we can input a measured mass curve. The interval must be the same one used for the unit hydrograph.

Third, generate the effective runoff curve: Apply the Green-Ampt or other technique to get the effective runoff curve. Haan et al. (1994) discuss the development of a Green-Ampt and NRCS effective runoff. We apply the runoff equation to each value

Table 5.13. Computed unit hydrograph for Example 5.6.

Ele.	Time (hr)	q_{pu} (cfs)	Ele.	Time (hr)	q_{pu} (cfs)
1	0	0	11	2.5	1.46859174
2	.25	20.3597139	12	2.75	0.59867084
3	.5	79.0480855	13	3	0.236530735
4	.75	103.750931	14	3.5	0.091028487
5	1	87.3469293	15	3.75	0.034257067
6	1.25	57.6547366	16	4	0.012645853
7	1.5	32.6275821	17	4.25	0.004590426
8	1.75	16.6039298	18	4.5	0.001641911
9	2	7.81766633	19	4.75	0.000579657
10	2.25	3.46863611	20	5	0.00020227

Table 5.14. Simulated storm for Example 5.6.

	Stormtime	Stoord	DepthT	Plottime	IncDepth	DepthM (in)
El	A	B	C	D	E (in/hr)	F
1	10.5	.204	1.34450688	.25	0	0
2	10.75	.2195	1.44666304	.5	.40862464	.10215616
3	11	.235	1.5488192	.75	.40862464	.20431232
4	11.25	.259	1.70699648	1	.63270912	.3624896
5	11.5	.283	1.86517376	1.25	.63270912	.52066688
6	11.75	.473	3.11741056	1.5	5.0089472	1.77290368
7	12	.663	4.36964736	1.75	5.0089472	3.02514048
8	12.25	.699	4.60691328	2	.94906368	3.2624064
9	12.5	.735	4.8441792	2.25	.94906368	3.49967232
10	12.75	.7535	4.96610752	2.5	.48771328	3.62160064
11	13	.772	5.08803584	2.75	.48771328	3.74352896
12	13.25	.7855	5.17701056	3	.35589888	3.83250368
13	13.5	.799	5.26598528	3.25	.35589888	3.9214784

Note: A is ordinate time, B is ordinate value; C is B times 24 hour value; D is plottime (starts at 12-3/2 = 10.5); E = [col C(t) − col C(t − 1)]/Δt; F = col F(t − 1) + [col E(t)]Δt.

of the mass curve depthM to get rneff. Next, determine the increments by subtracting the rneff value just above from the current rneff value and enter inceffr. Table 5.15 shows the result. The incremental effective runoff is the same as one would have in the right-hand column of the NRCS method discussed previously. We could also use the Δrc column on the right hand of the Green-Ampt table (Table 5.11).

Fourth, compute the component hydrographs for each increment of effective runoff and lag by time increments used above: Compute a hydrograph for each element of

Table 5.15. Incremental rainfall for Example 5.6.

	Time	Rneff	Inceffr
El	A	B	C
1	0	0	0
2	0.25	0	0
3	0.5	0	0
4	0.75	0	0
5	1	0	0
6	1.25	0.161226934	0.161226934
7	1.5	0.728296174	0.56706924
8	1.75	0.864640843	0.136344669
9	2	1.00789785	0.143257005
10	2.25	1.08397735	0.076079502
11	2.5	1.16161574	0.077638395
12	2.75	1.21920954	0.057593793
13	3	1.27756308	0.058353545

Note: A is stormtime, B is Q from the NRCS runoff volume equation, and C = Col $B(t)$ − Col $B(t-1)$.

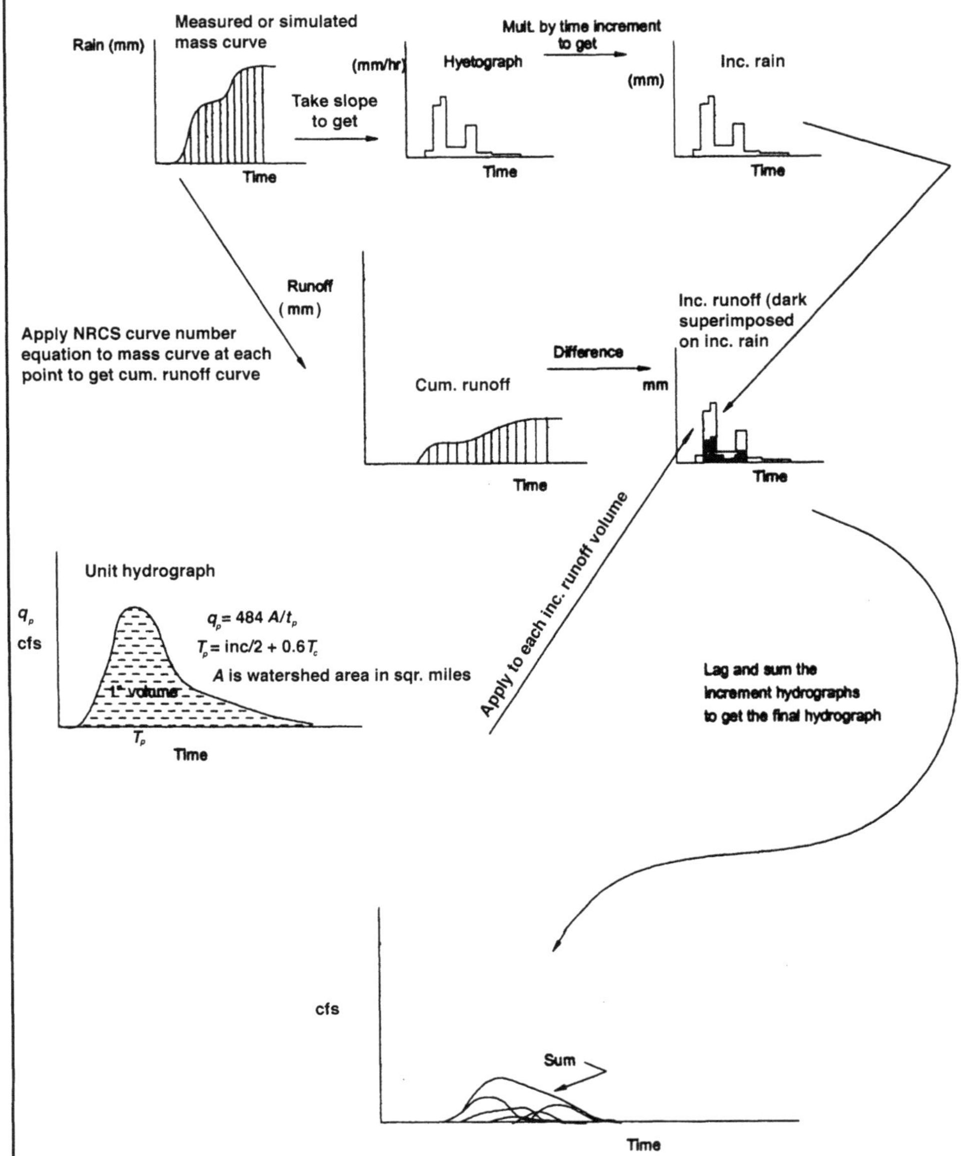

Figure 5.18. Process schematic for computing a hydrograph for a specified storm duration. We may measure the rainfall or simulate it using the NRCS distribution ordinates.

rainfall above, starting with the first nonzero rainfall. We compute the unit hydrograph with the same time increment as used in prior steps. This is important! Compute for successive times until the magnitude gets very low, as shown below. Table 5.16 shows partial results. Now, sum the columns along each row of Table 5.16 to the right. The result is the desired hydrograph as shown in Table 5.17. Figure 5.18 summarizes the hydrograph generation process.

Table 5.16. Partial results of applying the unit hydrograph to each element of incremental rainfall in Example 5.6.

Element	uh3l	uh4l	uh5l	uh6l	uh7l	uh8l	uh9l	uh10
1	3.28253426	0	0	0	0	0	0	0
2	12.7446805	11.5453675	0	0	0	0	0	0
3	16.7274446	44.8257378	2.77593845	0	0	0	0	0
4	14.0826776	58.8339618	10.777785	2.91667163	0	0	0	0
5	9.29549642	49.5317568	14.1458864	11.324192	1.5489569	0	0	0
6	5.26044503	32.6942277	11.9092882	14.8630476	6.01393898	1.58069552	0	0
7	2.6770007	18.5020982	7.86091596	12.5130595	7.89331919	6.13716651	1.173	0
8	1.26041837	9.41557786	4.44859687	8.25944486	6.64531088	8.05505581	4.553	1.188061489586264
9	.559237565	4.4331581	2.26385731	4.67412968	4.38634365	6.78147542	5.975	4.61273
10	.236776544	1.96695684	1.06589713	2.37862925	2.4822902	4.47622122	5.031	6.054234
11	.096521864	.832793202	.472930041	1.11993546	1.26321871	2.53315311	3.321	5.097
12	.038135125	.339487819	.200234654	.496906419	.594764161	1.28910246	1.879	3.364
13	.014676244	.134129304	.081625577	.210386054	.263892108	.606951068	.9563	1.9039
14	.005523162	.051619455	.032249705	.085763791	.111729728	.269299341	.4502	.968898170758942
15	.002038852	.019426129	.012411249	.033884685	.045546579	.114019106	.1998	.4561885463806415
16	.0007401	.007171074	.004670768	.013040468	.017995141	.046479843	.0846	.2024072143335115
17	.00026472	.00260309	.001724195	.004907565	.006925402	.018363867	.0345	.08569753465882209
18	9.34564E-5	.000931077	.00062588	.001811607	.002606261	.007067306	.0136	.03493456601472501
19	3.26114E-5	.000328706	.000223866	.000657611	.00096209	.002659664	.0052	.01380240695332032
20	1.12613E-5	.000114701	7.90332E-5	.000235215	.000349237	.000981804	.002	.005311834921275506
21	3.85219E-6	3.96083E-5	2.75784E-5	.00008304	.000124916	.000356393	.0007	.00199902132536326
22	1.30651E-6	.000013549	9.52332E-6	2.89766E-5	.0000441	.000127475	.0003	.0007379303631686112
23	4.39672E-7	4.59526E-6	3.25768E-6	1.00061E-5	1.53886E-5	4.50037E-5	9E-5	.0002678676510912217
24	1.46909E-7	1.54642E-6	1.10487E-6	3.42284E-6	5.31395E-6	1.57039E-5	3E-5	.0000958113521989478

Table 5.17. Hydrograph for Example 5.6.

El	Time (hr)	Sum (cfs)	El	Time (hr)	Sum (cfs)
1	0	3.28253426	12	2.75	8.20213513
2	0.25	24.290048	13	3	4.17187874
3	0.5	64.3291208	14	3.25	1.97533241
4	0.75	86.6110961	15	3.5	0.883287056
5	1	85.8462885	16	3.75	0.377086378
6	1.25	72.321643	17	4	0.154966097
7	1.5	56.7561531	18	4.25	0.061692855
8	1.75	43.8251452	19	4.5	0.02390963
9	2	33.6863474	20	4.75	0.009056081
10	2.25	23.6916468	21	5	0.003362732
11	2.5	14.7361103			

Impact of Errors on Runoff Assessment

We have purposefully not yet discussed the techniques of measuring runoff rate and volume. One may measure runoff rate using flumes or weirs. Direct integration of velocity over cross sections is also used. We discuss flumes and weirs for runoff measurement in Chapter 11. Runoff measurement in natural channels is discussed in Chapters 11 and 15. Integrating the hydrograph with respect to time gives runoff volume.

Error analysis is important in the practical application and analysis of hydrological data. Suppose you are presented with the following problem: Given infiltration and change in soil water storage along with runoff data and evapotranspiration, can we estimate rainfall via a backdoor approach? Consider the diagram in Figure 5.18. Based on knowing four of the five variables we can *theoretically* solve for the other variable, here rainfall. From an engineering practicality point of view, we must consider the errors in Figure 5.19.

The short confidence interval spread reflects reasonably precise measurements for each parameter. The high sides of the indicated ranges reflect a typical field situation with crude parameter estimates. The relative cost figures show that precipitation measurement is the most economical of the measurements. Costs plus the broad indicated ranges both suggest focusing on direct rainfall.

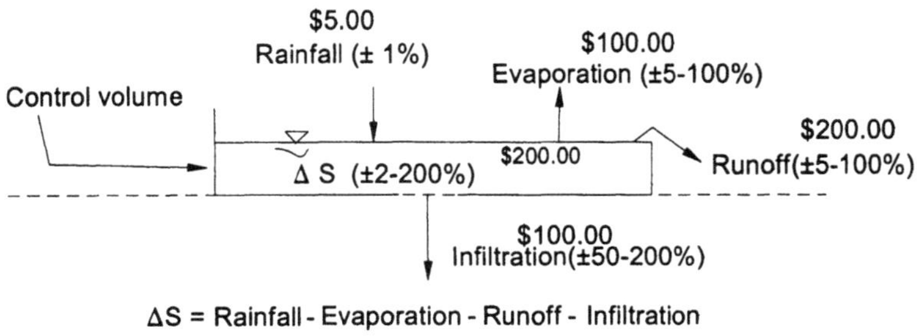

Figure 5.19. Schematic showing the water balance with estimated costs and estimated cost ranges associated with each measurement.

Outlook

Runoff is a complex hydrologic process. Design models for runoff volume and peak flow estimation all contain considerable empiricism. Selected NRCS approaches and civil engineering approaches were evaluated. With the NRCS curve number model, we have now taken a precipitation event, empirically partitioned it into runoff and infiltration events, and indirectly accounted for evaporative loss via the curve number antecedent rainfall correction. The NRCS curve number approach, even if crude, works well for many routine engineering problems. Advanced techniques are available for peak, volume, and hydrograph determination for compact and complex watersheds. Instrumented watersheds are the best approach for collecting curve number equation parameters.

One is often interested in the peak flow for watersheds. If there is a gauge with many years of record near the point of interest, a frequency analysis can provide the design flows. Otherwise, one must calculate peak flows from design precipitation events. With a nearly impervious surface, steep watershed gradient, and small runoff area, the rational method (rising limb $\approx$ recession limb) may provide a reasonable estimate of peak flow. The rational method, originally developed for paved surfaces such as parking lots, is applicable for small watersheds less than 10 ha. The NRCS curve number method is a generalized approach toward runoff volume prediction. The NRCS method works well for watersheds less than 25 mi^2. Areas exceeding 25 mi^2 should probably be modeled using complex modeling techniques. Techniques such as the ASAE drainage formula are published for specialized applications such as drainage ditch design.

Most watersheds do not conform to the definition of the compact watershed. Thus, in practice, computer modeling approaches are used to describe complex watershed hydrology. We divide the complex watershed into several compact watersheds. Runoff hydrographs from each subwatershed are routed and appropriately combined. Models such as SedCad[3] and others described in DeVries and Hromadka (1993) use basic approaches in this chapter coupled with streamflow routing techniques discussed in watershed hydrology texts to model complex watersheds. The NRCS Internet page (http://www.nrcs.usda.gov) presents downloadable software implementing the TR55 approach. The USEPA internet page (http://www.epa.gov) also provides software (storm water management model, SWMM) for modeling runoff from single precipitation events on watersheds with natural and manmade drainage patterns. An Internet search (e.g., use *hydrologic model* and *hydrology* as key words) will provide links to other commercially available software packages for hydrologic modeling. Haan et al. (1982) give an excellent theoretical background and historical perspective on watershed runoff modeling.

Problems and Questions

1. Analyze the data for the annual maximum stages and corresponding discharges for each of the 39 years given in Table 5.18 using the following approach.

   Using a spreadsheet, enter the stage in the first column and the corresponding discharge in the second column, then analyze the discharge as was done with precipitation in Chapter 2. Plot the calculated probability versus the ranked discharge on log-probability paper. Compute two or three probabilities using the theoretical lognormal distribution and show these on the plot. Finally, develop a scatter point plot of stage versus discharge (column a vs. column b) on a log-log plot. The stage discharge curve

Table 5.18. Stage-discharge relationship for Problem 1.

Stage (ft)	Discharge (cfs)	Stage (ft)	Discharge (cfs)
23	200,000	18.12	97,200
11.8	42,000	22.82	179,000
6.4	11,300	19.55	124,000
10.4	32,400	19.48	110,000
18.7	108,000	18.5	114,000
15	73,000	14.93	70,200
15.3	76,500	15.3	70,700
12.1	47,800	17.6	92,800
9.5	28,200	21.45	135,000
10.6	33,700	10.48	25,800
9.3	25,700	8.8	17,500
6.4	11,700	9.07	18,700
16	77,800	12.71	36,300
9.9	26,600	14.64	49,200
13	47,500	21.41	120,000
16.44	75,600	14.86	56,800
8.48	19,200	14.65	54,800
10.26	27,800	21.62	158,000
13.59	51,000	21.22	165,000
18.54	94,000		

is in effect a calibration curve for the flow-measuring device. We will further develop flow calibration of these devices in Chapter 11. What is the difference between the 20 yr discharge as determined from the plot and the modeled 20 yr discharge?

2. Using the rational method, calculate the 10 yr return period peak runoff for a 10 ha watershed having a runoff coefficient of 0.4. The maximum length of flow of water is 610 m and the fall along this path is 6 m. Assume the watershed is in central Tennessee with a group C soil. Repeat the solution using the NRCS TR55 method. Assume the watershed is predominantly pasture, in good hydrologic condition, and in antecedent moisture condition II.
3. By the rational method, determine the design peak runoff for a 50 yr storm. Assume the watershed is in central South Carolina and is one-third pasture, one-third rotation meadow, and one-third row crops on the contour. The soil is hydrologic group D. The watershed is roughly rectangular, having a main channel of 10,000 ft and overland flow path through a pasture of 1000 ft. The total area is 10 ac. The elevation change over the overland flow portion is 20 ft. The elevation change over the main channel is 50 ft. What is the time of concentration? What is the peak flow?
4. Determine the runoff volume in millimeters of depth from a 50 yr return period storm in eastern Ohio assuming antecedent moisture condition III, $CN = 70$, and 3 hr duration.
5. Given a compact watershed in the Atlanta, Georgia, area that is two-thirds row crop and one-third forest, is 2500 ft long, is 210 ac, has a slope of 0.05, is in antecedent moisture condition III, has no ponds or wetlands, is soil group C, is in good hydrologic condition, and is terraced, compute the NRCS TR55 peak runoff for a 10 yr storm. Compare the Atlanta peak flow with that at Savannah, Georgia, and that at Mobile, Alabama.
6. Calculate the 50 yr–3 hr synthetic hydrograph for a watershed in Columbia, Missouri. The watershed is forested. The average slope is 5%, length is 6000 ft, and area is 3 mi^2.

7. Derive the rational equation assuming a triangular unit hydrograph wherein the recession limb to rising limb ratio is 1.
8. Derive a runoff equation assuming the rising limb is storm duration $Dst/2 + 0.6t_c$ and the recession limb to rising limb ratio is 12.
9. What is the difference between a hyetograph and a hydrograph?
10. Prove that 1 ac-in/hr $\approx$ 1 cfs. Prove that $q_p = 0.0028CiA$ where intensity i is in mm/hr and area A is in ha.
11. Why is the rational equation a reasonable approach for predicting runoff volume with small, compact, impervious, steep watersheds?
12. From your library, obtain a 7.5′ quadrangle for the region of your (or your instructor's) choice. In consultation with your instructor, select a watershed and delineate the boundary.
13. How are the following terms related: runoff volume Q in watershed inches (or mm), flow rate q and runoff volume in L^3/T?
14. Compare and contrast the various ways of estimating the time of concentration discussed in this chapter. Focus your discussion on the type and number of parameters used by each method.
15. Compare and contrast the NRCS TR55 method, the rational method, and the triangular hydrograph method for estimating peak flow.
16. Why does one use the time of concentration as the design storm duration with the rational method for peak flow prediction?
17. What parameter in the rational equation poses the most difficulty in terms of consistent predictions among professionals?
18. Define and differentiate between the rainfall hyetograph and the runoff hydrograph.
19. What are the steps in performing a frequency analysis and what is the major attribute of data valid for a frequency analysis?
20. What is the major problem with computing a return period for a series of runoffs using daily data?
21. What is *time of concentration*?
22. What are three basic ways to compute peak flow? Compare the strengths and weaknesses of each method.
23. For the rational method, we gave two ways to compute time of concentration. What were they and under what conditions is each applicable?
24. What is a unit hydrograph?
25. Define the difference between a unit hydrograph and a runoff hydrograph.
26. What are the similarities and differences among a rainfall mass curve, rainfall hyetograph, effective rain and runoff hydrograph?
27. Using an Internet search identify a package (other than those mentioned) for hydrologic modeling. Specify the key words and search engine used in your search.

Notes

1. I am indebted to Dr. Sharyl Walker, Department of Agricultural Engineering, University of Illinois, Urbana, for significant contributions to this section.
2. *Effective runoff* is the portion of the rainfall not initially abstracted: not infiltrating or evaporating. Some authors (e.g., Haan et al. 1994) designate this as *effective rainfall*.
3. Available from Civil Software Design, P.O. Box 706, Ames, IA 50050. Mention of this product is solely for the convenience of the reader and is not to be construed as an endorsement of the product.

References

ASAE. 1997. *Agricultural drainage outlets, B: Open channels.* ASAE EP407.1 DEC96. American Society of Agricultural Engineers, St. Joseph, MI.

DeVries, J. J., and T. V. Hromadka. 1993. Computer models for surface water. *In Handbook of hydrology,* ed. D. R. Maidment. McGraw-Hill, New York, NY.

Haan, C. T., H. P. Johnson, and D. L. Brakensiek. 1982. Hydrologic modeling of small watersheds. ASAE Monograph no. 5. American Society of Agricultural Engineers, St. Joseph, MI.

Haan, C. T., B. J. Barfield, and J. C. Hayes. 1994. *Design hydrology and sedimentology for small catchments.* Academic Press, New York, NY.

Horn, D. L., and G. O. Schwab. 1963. Evaluation of rational runoff coefficients for small agricultural watersheds. *Transactions of the ASAE* 6(3):195–98, 202.

Kirpich, P. Z. 1940. Time of concentration of small agricultural watersheds. *Civil Engineering* 10:362.

NRCS (U.S. Soil Conservation Service). 1972. *National engineering handbook.* Section 4, Hydrology. Washington, DC.

———. 1986. *Urban hydrology for small watersheds.* 2d ed. Technical Release TR55. Soil Conservation Service, U.S. Department of Agriculture. Washington, DC.

Schwab, G. O., D. D. Fangmeier, W. J. Elliot, and R. K. Frevert. 1993. *Soil and water conservation engineering.* 4th ed. John Wiley & Sons, New York, NY.

Wanielista, M., R. Kersten, and R. Eaglin. 1997. *Hydrology: Water quantity and quality control.* John Wiley & Sons, New York, NY.

Warner, R. C. 1998. SedCad[7] 4 for Windows95/NT: Notes for 5–7 Aug. 1998 short course. Department of Biosystems and Agricultural Engineering, University of Kentucky, Lexington.

6

Erosion by Water

World historians have attributed the downfall of many civilizations to soil erosion. Awareness of the problem developed slowly. The Bible contains a few admonitions concerning land stewardship. The Greek writers Homer and Plato began to associate floods and erosion with destruction of forests. The Roman writers Virgil and Pliny advocated what we today would call a form of conservation farming. Their voices were largely unheard. Up to the end of the nineteenth century there continued to be a few lone voices for conservation but no general awareness (Hudson 1971). With the dust bowl as a backdrop, H. H. Bennett, a leading conservationist of the 1930s and 1940s, played a major role in establishing public awareness of the erosion problem. Erosion by wind can cause significant problems with crop stand establishment even in humid upland areas. A detailed discussion of erosion by wind, however, is beyond the scope of this text.

Sediment is the greatest pollutant when expressed as pollutant mass. Erosion control practices have reduced cropland erosion in the United States from 8 tons per acre in 1982 to 5.2 tons per acre in 1997. Worldwide, however, erosion is out of control. Many countries have the equivalent of the NRCS; nevertheless, their efforts have only slightly reduced soil loss (Hudson 1971). Erosion from urban areas is an increasing problem. Between 1995 and 1997, 1.8 million acres of cropland were lost, due mainly to urbanization (NRCS 1998). Most sediment results from water erosion processes in humid upland areas. Erosion represents a real economic loss to plant and animal producers. Sediment represents a hidden cost to many downstream water users due to the need to remove it.

Water erosion is the detachment and transport of soil from the land by water, including runoff from melting snow and ice. Types of erosion by water include interrill, rill, gully, and stream channel. Erosion leads to sediments that carry a variety of plant nutrients and pesticides. Soil erosion and sediment yields are similar but different concepts. *Soil erosion* represents detached soil mass. *Sediment yield* represents sediment that is deposited at a location. The difference between erosion and sediment yield is that soil eroded is not necessarily transported to the point of yield measurement. Eroded soil in a flow may be deposited anywhere along the way. Runoff may cause additional stream or gully erosion. Figure 6.1 shows a case of serious gully erosion.

Goals

- Identify factors such as climate, vegetative cover, and topographical parameters relating to soil detachment at the field and smaller scales.

Figure 6.1. Extreme example of gully erosion in the southeastern United States.

- Apply forms of the universal soil loss equation to predict average annual soil loss from agricultural and other land-disturbing activities and soil loss from storms.
- Determine sediment yield at selected locations.

FACTORS AFFECTING EROSION BY WATER

Climate

Erosion is a natural geologic process. Precipitation, temperature, wind, humidity, and solar radiation are the predominant climate factors affecting erosion. Precipitation is influential and is expressed in terms of falling-raindrop energy. Hudson (1971) relates erosion to mean annual rainfall, shown in Figure 6.2. Temperature and other climate factors have an indirect impact, influencing soil organic matter and structure over the long term.

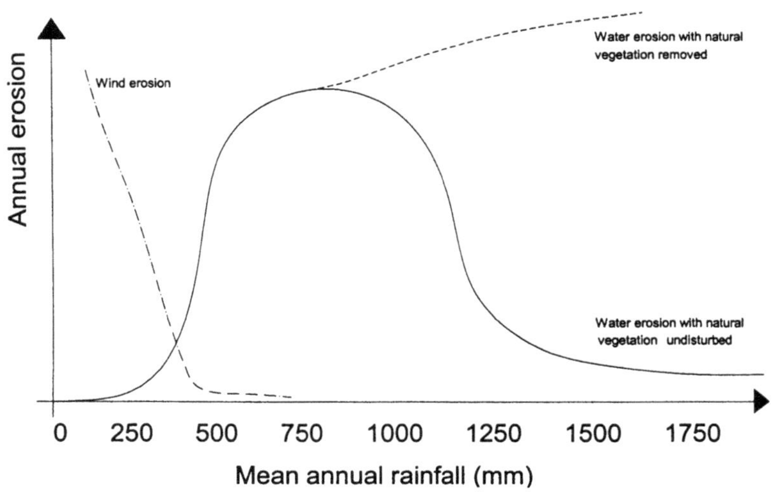

Figure 6.2. Relationships between mean annual rainfall and annual erosion. (Data provided by Hudson 1971.)

Soil

Soil structure, texture, organic matter, water content, clay mineralogy, and density impact soil erodibility. Soil detachability and soil transportability vary with organic content, texture, and mineralogy. Small particles are harder to detach but easier to transport than large particles.

Vegetation

Vegetation can reduce erosion by intercepting rainfall, decreasing runoff velocities, physically restraining soil particles, improving structure with residues and roots, increasing biological activity, and increasing evapotranspiration and thus the potential for infiltration. Interception by vegetation effectively reduces rainfall energy absorbed by the soil.

Water erosion is low with low rainfall for obvious reasons. However, with 250–500 mm mean annual rainfalls, erosion increases because the annual rainfall does not support extensive vegetative cover. When rainfall exceeds 1000 mm annually and vegetation is maintained, erosion is low because the vegetation protects soil from direct rainfall impact.

Agricultural and urbanization activities exacerbate erosion because of the removal of vegetative cover. Hudson (1971) estimated that 80% of the potential arable land of the world (1600 million hectares) was under agricultural production. Much of the remaining 20% is steep and/or has low productivity. Urban encroachment is a factor on much of this land. Urban encroachment adds to erosion likelihood owing to cover removal. Lal (1994) summarized data suggesting that, outside North America, the problem is getting worse due to extreme population pressures.

Topography

Steepness, slope length, and shape define the effect of topographical variables on potential erosion. Overland flow is faster on steep slopes. Long slopes allow for flow accumulation, and thus flows attain faster velocities than on short slopes under comparable rainfall and soil conditions. One may encounter three characteristic slope shapes: convex, uniform, and concave. Figure 6.3 shows these characteristic shapes.

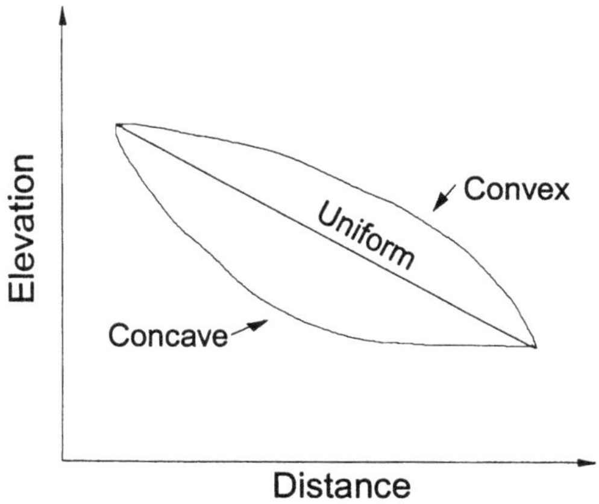

Figure 6.3. General slope shape categories.

The uniform slope is less erodible than the convex but more erodible than the concave under similar conditions. This text will confine the discussion to uniform slopes.

Soil Erosion versus Sediment Yield

Soil erosion loss is a measure of soil detached from an area. Sediment yield is a measure of eroded soil transported to a given location. Soil erosion losses and sediment yield do not necessarily balance because of deposition opportunities between the detachment location and the sediment yield measurement location.

Soil Loss Tolerance

Soil formation is a part of the natural geologic weathering process. The soil formation, or accretion, rate in the temperate zone is 25 mm every 150–200 years. This translates roughly to 1–3 tons/ac-yr (2.24–6.7 Mg/ha-yr where 1 ton/ac = 2.24 Mg/ha; 1.1–3.3 metric tons/ha-yr where 1 ton/ac = 1.1 metric ton/ha). Soil organic-matter additions often accelerate soil formation. Soil scientists make informed judgments on the sustainable rate of erosion based on anticipated soil accretion rates, amount of topsoil present, desired productivity, expected cultural practices, and past erosion (Schmidt et al. 1982). Soil loss is sustainable if the erosion does not exceed, over the long term, the soil formed. Soil surveys tabulate sustainable soil loss tolerances for each soil mapping unit. Soil loss tolerances in the temperate zone commonly range from 3 to 5 tons per acre per year, or roughly 7–11 Mg per hectare per year. The notion of soil loss tolerance remains valuable for erosion planning purposes.

Types of Water Erosion

Interrill Erosion

Interrill erosion includes erosion by raindrop impact and by sheet flow across the soil surface. Raindrops detach soil particles as they strike the soil surface. Raindrops may throw soil particles 0.6 m high and 1.5 m laterally. Raindrop erosion increases with the energy in the raindrops. Generally, the higher the rainfall intensity, the greater the erosive energy. Raindrops break up soil aggregates.

The concept of "interrill" may be outdated though the term remains in widespread use. Current thought is that microscopic rills form immediately on runoff initiation. Water flows through the microscopic rills and not "uniformly" across the surface. The erosive and transport capabilities of interrill flow are functions of rainfall intensity, infiltration rate, slope steepness, surface condition, and particle densities. Figure 6.4 shows a simple qualitative technique for measuring interrill erosion. Erosion to a depth equal to the thickness of a dime is 4 tons/ha.

Rill Erosion

Rill erosion is the detachment and transport of soil by a concentrated flow. It is the predominant form of erosion. Rill erosion is a concentrated flow path resulting in parallel channels. Rills may relocate in slightly different locations if erased by tillage.

Figure 6.4. Interrill erosion demonstrated near Griffin, Georgia, by placing washer over stake before a rainfall event and noting the difference in soil level after the rainfall event.

Tillage usually destroys rills. Figure 6.5 shows ongoing rill erosion in an erosion measurement plot.

Gully Erosion

Gully erosion produces channels larger than rills. After being erased, gullies will form again in the same location. Field tillage tools cannot routinely destroy gullies. Gully erosion processes include head cutting and associated mass failures and erosion by flowing

Figure 6.5. Rill erosion within a bounded soil loss measurement test area.

water and freezing/thawing. These contribute to soil mass sloughing. Most erosion models do not include gully erosion contributions. Advanced modeling approaches such as WEPP (see below) partially model the gully erosion process. Figure 6.1 shows extreme gully erosion.

Stream Channel Erosion

Stream channel erosion occurs in lower headwater regions and in permanent streams. Stream bank erosion occurs from water running over the bank into the stream and from water undercutting the bank. Water velocity in the channel contributes to erosion. Factors operative in gully erosion work on a larger scale in stream channel erosion.

Erosion Mechanics

Raindrop erosion is a product of two related phenomena: raindrop dislodgment and flow-induced erosion.

Raindrop Dislodgement

Raindrops falling at terminal velocity downward plus a horizontal (due to wind) velocity component strike the soil surface, dislodge soil particles, and compact soil. This kinetic energy due to rainfall, proportional to the square of the falling-drop velocity, accounts for interrill erosion and some rill erosion. Figure 6.6 shows schematically the processes of raindrop impact.

The terminal velocity of a 2 mm diameter drop ranges from 5 to 7 m/s. As the drop size approaches a diameter of 6 mm, the velocity increases to 9–12 m/s. Drop size in rainfall events ranges from extremely small diameters (mist) to the 5 or 6 mm diameter. High-intensity storms usually have drop size distributions skewed toward the larger drop

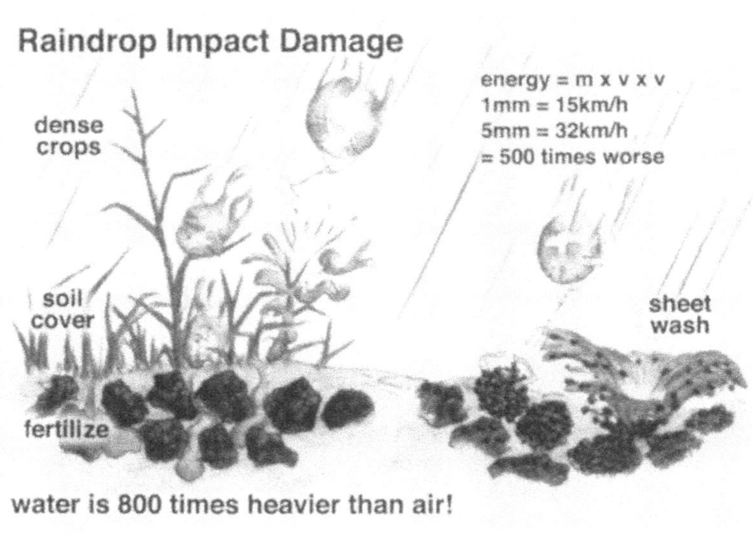

Figure 6.6. The explosive impact of raindrops on soil leads to detachment and transport. (Courtesy of Seafriends Foundation, New Zealand.)

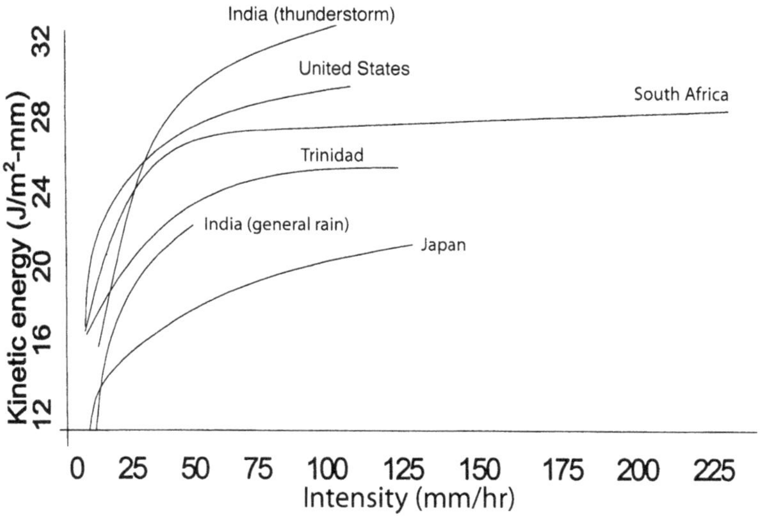

Figure 6.7. Relationship between rainfall kinetic energy and rainfall intensity for a variety of sites around the world. The units on the vertical axis are joules per meter soil area-millimeter rainfall depth. Kinetic energy is computed using $V^2/2g$ where V is velocity (m/s) and g is gravity (m/s^2). Terminal velocity V_{term} may be approximated using $V_{term} = 3.86d_p^{0.67}$ where d_p is the droplet diameter in mm. (Redrawn following Hudson 1971.)

diameters. Figure 6.7 shows kinetic energy versus mean rainfall intensity at several sites. The high-kinetic-energy events correlate directly to large mean drop sizes. One may approximate terminal velocity V_{term} using $V_{term} = 3.86{d_p}^{0.67}$ (Smith 1993) where V_{term} is in meters per second and d_p is drop diameter (mm).

The immediate message of Figure 6.7 is that the same rainfall intensity can result in different energies for eroding soil. The energy depends on drop size distributions. Precipitation due to orographic lifting has many small drops and low erosion potential. High-energy rainfalls, which are often associated with convective thunderstorms, have high erosion potential.

Flow-Induced Erosion

Tractive force induced by sheet flow or channelized flow is the second major mechanism for erosion. Water begins to flow over the surface when infiltration exceeds soil intake. As water flows, shear stress induced by the flowing water dislodges more particles and transports particles. Rills form essentially at random over the surface.

Some soils are more prone than others to form rills. As water flows accumulate due to topography, gullies can result from the large tractive forces that develop as flows deepen and increase in velocity. Gullies, unlike rills, are not removable with tillage equipment. Streams are gullies grown large. Figure 6.8 shows these mechanisms in a field context. Strategies are discussed in the following sections for protecting soils from the tractive forces potentially present in gullies and streams. As can be seen from the variety of forces evident on the third amplification in Figure 6.8, erosion on the field scale is a complicated process, widely varying spatially and temporally. The complexity of the various erosion and transport mechanisms coupled with the many factors known to affect detachability and transportability explains the empirical nature of much erosion literature.

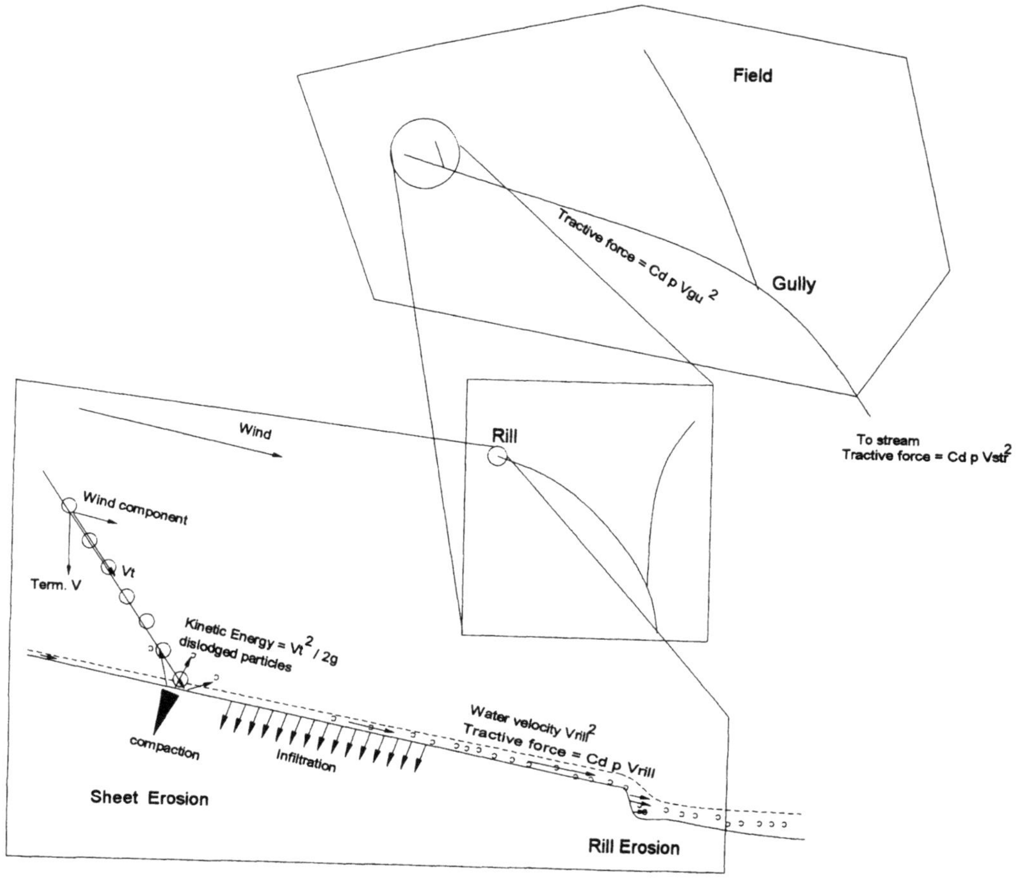

Figure 6.8. Amplified view of a point in a field showing forces due to raindrop impact and tractive force erosion mechanisms and how they relate to interrill, rill, gully, and stream channel erosion. In reality all the processes occur over the entire field, making process modeling very difficult.

Predicting Soil Detachment and Upland Erosion

The original standard for soil loss prediction from various management practices is the universal soil loss equation (USLE; Wischmeier and Smith 1978). This equation also applies to disturbed urban areas. This equation has gone through several revisions. The USLE and its variants are in use around the world for erosion prediction (Bagarello and Ferro 1999).

An interim revision is the modified universal soil loss equation (MUSLE; Williams 1974). It uses a different approach for the rainfall factor computation. It incorporates peak runoff rate and runoff volume and thus predicts soil loss according to the area contributing to the point of measurement. The Williams (1974) energy form embodies the idea that work is required to dislodge and suspend surface sediments. The modification is that one computes the R coefficient based on runoff instead of the 30 min storm. The Williams (1974) sediment delivery ratio approach is discussed at the end of the chapter. Haan et al. (1994) summarize the use of this approach.

The latest revision is the revised universal soil loss equation (RUSLE). Renard et al. (1991, 1997) document RUSLE, available as a computer-based erosion prediction tool. The revisions were made, not to the basic structure of the USLE equation, but to details in computing the USLE coefficients. RUSLE reflects many current cultural practices such as no-tillage, minimum tillage, and other conservation approaches and urban construction applications. The earlier USLE and MUSLE reports reflected extensive moldboard primary tillage agriculture. RUSLE is still not a rigorous process-based model. It still does not address stream and gully effects. The RUSLE is similar to the USLE, but the evaluation of factors is more involved. This text presents a "hybrid form" emphasizing basic principles. The soil loss equation appears as

$$A_{Ero} = R_{rf} K_{sl} L_f S_f C_{prac} P_{prac} \tag{6.1}$$

where

A_{ero} (Mg/ha) is the soil loss due to surface transport and overland flow (not including gully and stream erosion; to convert to tons/acre divide by 2.24);
R_{rf} is the rainfall factor, in MJ · mm/(ha · hr · storm) or MJ · mm/(ha · hr · yr);
K_{sl} is the soil factor, in metric tons · ha · hr/(ha · MJ · mm);
L_f is the length factor (−);
S_f is the slope factor (−);
C_{prac} is the cover factor (−);
P_{prac} is the practice factor (−).

A more complete discussion of the components of the USLE follows.

R_{rf}

The factor R_{rf} is the rainfall and runoff erosivity index for geographic locations. The R_{rf} factor quantifies the available energy for erosion, on the average annual, single-storm, or recurrence interval basis. Figure 6.9 shows the average annual R_{rf} mapped over the eastern United States. For a single storm with recording rain gauge output available, one may compute single storm R_{rf} values using equation 6.2 (from Renard et al. 1997, converted to SI units):

$$R_{rfstorm} = \frac{\Sigma(EI_{30})}{100} = 0.29(i_{30\,\max}) \sum_{i=1}^{N} P_i(1 - e^{-0.05 i_i}) \tag{6.2}$$

where

$R_{rfstorm}$ is the single-storm erosivity, in MJ · mm/(ha · hr · storm);
P_i is the precipitation (mm) occurring during the ith storm segment;
I_i is the intensity in mm/hr during the ith storm segment;
N is the number of storm segments;
i_{30max} is the 30 min storm portion yielding the highest rainfall intensity; this 30 min zone may extend over more than one segment.

Figure 6.10 demonstrates calculation of R_{rf} for a single storm using a rainfall mass curve. The analyses began by subdividing the storm into three zones, each having an approximately uniform rainfall intensity. One may compute the rain falling within each interval and divide the intensity for the interval by the time of the interval. The maximum 30 min intensity is likewise computed. Then, substitute values into equation 6.2.

The relationship given in equation 6.2 is one method for computing single-storm R_{rf} values from historical records. The annual R_{rf} value is the sum of all storm events over a year. The average annual R_{rf} is the average of annual R_{rf} values over many years. One

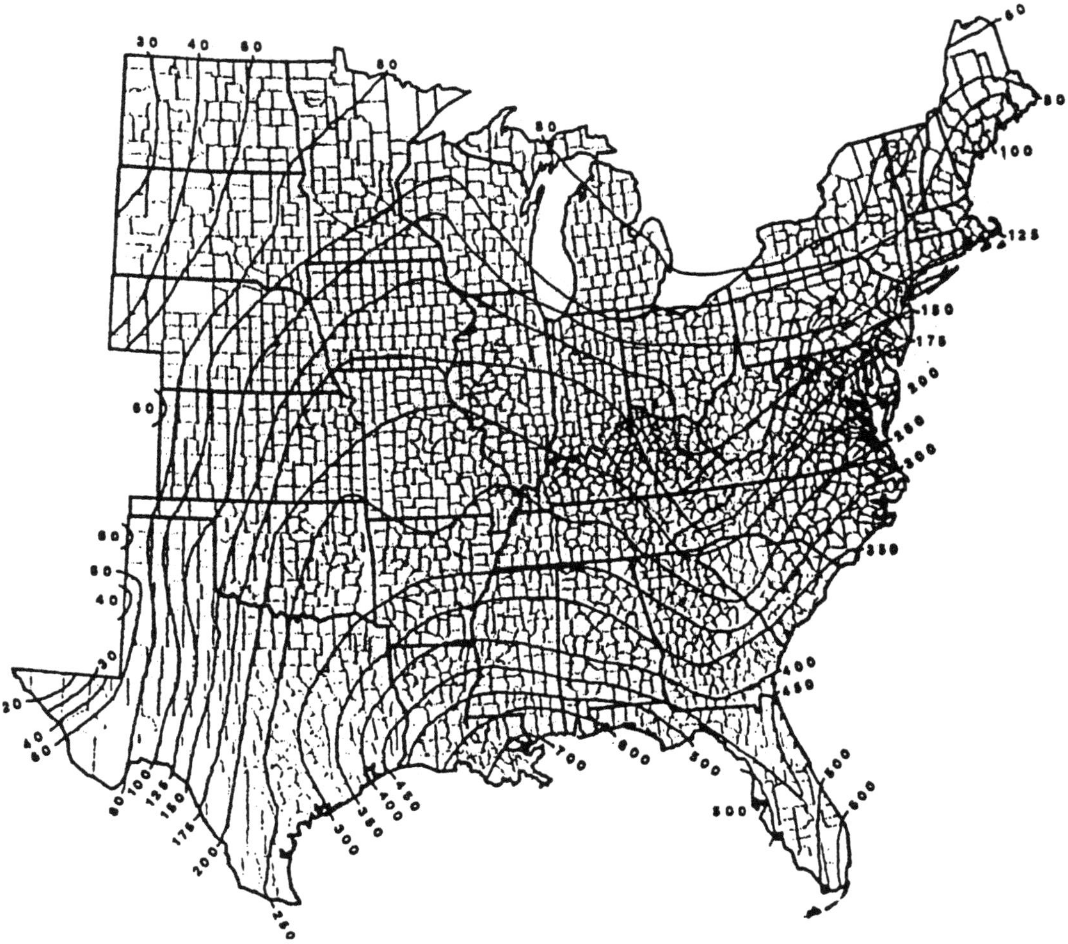

Figure 6.9. Isoerodent map showing average annual erosivity R_{rf} for the USLE in English units for the humid eastern United States. Units are in 100s ft tons in/(ac-in-yr). To convert to SI units (MJ mm/(ha-yr), multiply indicated values by 17.02. (From Renard et al. 1997, who also provides maps for other parts of the United States.)

may perform frequency analyses on R_{rf} as for precipitation or runoff. For conservation planning, the average annual R_{rf} is often used. Figure 6.9 shows values for average annual R_{rf} over the 48 contiguous states above. Single-storm and annual R_{rf} values are tabulated by return period in Table 6.1 for selected locations. Onstad and Foster (1975) suggest an alternative relationship for predicting R_{storm} from peak flow and runoff volume. Haan et al. (1994) and Renard et al. (1997) discuss other approaches for computing rainfall energy for specific regions. Renard et al. (1997) present isoerodent maps for the 10 yr R_{rf} factor for the United States.

K_{sl}

The factor K_{sl} is the soil erodibility factor. K_{sl} represents the eroded soil in metric units, metric ton · ha · hr/(ha · MJ · mm). K_{sl} estimates the ability of a soil to withstand rainfall energy. To convert to English units of ton · ac · hr/(hundreds ac · ft · tonf · in) multiply by 7.59.

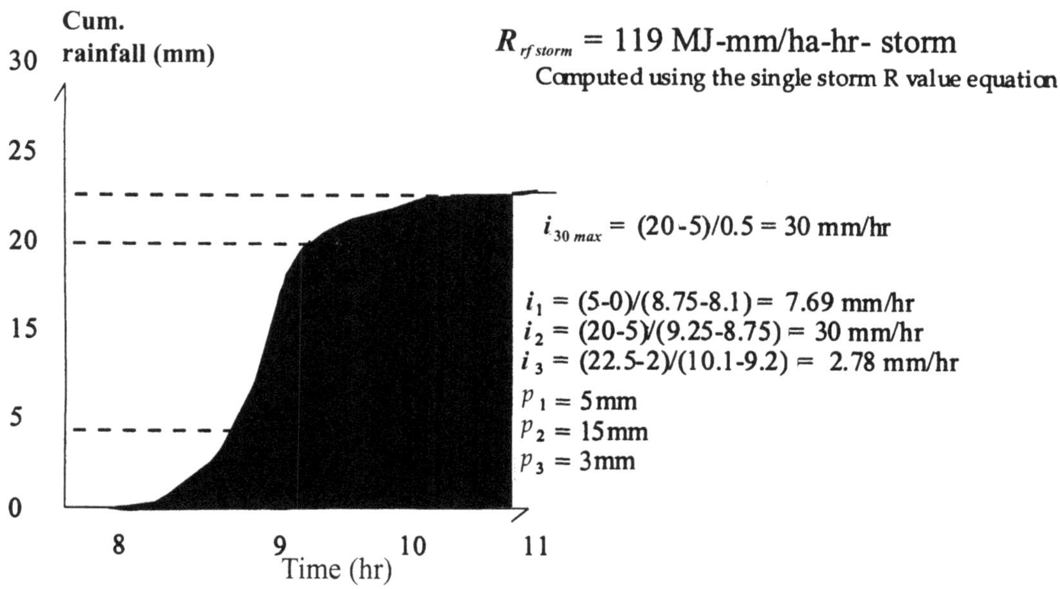

Figure 6.10. Example calculation for single-storm R_{rf} given a rainfall mass curve.

The K_{sl} factor represents soil erodibility under unit plot conditions. The unit plot is 22 m long, with a uniform lengthwise slope of 9% in continuous fallow, tilled up and down the slope.

In early versions of the USLE, researchers developed the K_{sl} factor for each soil type. From these data, a nomograph was developed relating K_{sl} to organic matter, texture, structural class, and permeability class. Equation 6.3 relates these parameters in most agricultural soils (Renard et al. 1997):

$$K_{sl} = 2.8 \times 10^{-7} M p^{1.14}(12 - a) \times 4.3 x\, 10^{-3}(b - 2) + 3.3 \times 10^{-3}(c - 3) \quad \textbf{(6.3)}$$

Table 6.1. Rainfall erosivity values for selected locations.

	Single-storm R by return period (MJ-mm/ha-hr-yr)[a]		
Location	2	5	20
Little Rock, AR	1174	1957	3591
Indianapolis, IN	698	1021	1532
Devils Lake, ND	460	664	1004
Watkinsville, GA	1208	1668	2417
	Average annual R by return period		
Little Rock, AR	5242	7182	9684
Indianapolis, IN	2825	3830	5140
Devils Lake, ND	953	1532	2417

Source: Excerpted from Wischmeier and Smith 1978.

[a]The English units are ft-tons-in/acre-hr-yr. To convert from the above metric units to English units, divide by 17.02.

where

K_{sl} is the RUSLE soil factor, measured erosion per unit area over ΣEI_{30} in some 55 benchmark soils under standard test conditions; units are metric ton · ha · hr/(ha · MJ · mm) or ton · ac · hr/(hundreds ac · ft · tonf · in);

Mp is the particle size parameter (%silt + %very fine sand) × (100 − %clay);

a is the percent organic matter (om);

b is the soil structure code (1 = very fine granular; 2 = fine granular; 3 = medium to coarse granular; 4 = blocky, platy, or massive);

c is the soil permeability code (1 = rapid; 2 = moderate to rapid; 3 = moderate; 4 = slow to moderate; 5 = slow; 6 = very slow).

One should not use equation 6.3 with artificial soils (e.g., beach sand or horticultural potting mixes). Renard et al. (1997) give additional relationships between K_{sl} and soil physical, chemical, and biological properties.

Rainfall simulators such as shown in Figure 6.11 enable evaluation of K_{sl} factors for many soil types under field conditions. Figure 6.12 shows a technique for measuring flow while sampling for erosion using a Coshocton wheel. Small simulators are also useful for studying soil erodibility.

Interestingly, the product of $R_{rf} \times K_{sl}$ gives ton/unit area-storm where storm may represent one erosion event. This product may also represent an event equivalent to the average annual erosion under unit plot conditions. The remaining USLE factors are dimensionless. They adjust the unit plot condition to other slope, slope length, and crop conditions.

L_f

The factor L_f is the slope length factor. Terracing, discussed in Chapter 10, alters this factor. One may also alter slope length by choosing appropriate field layouts. The slope length

Figure 6.11. Rainfall simulator and runoff sampling apparatus for quantifying soil erosion losses under field conditions.

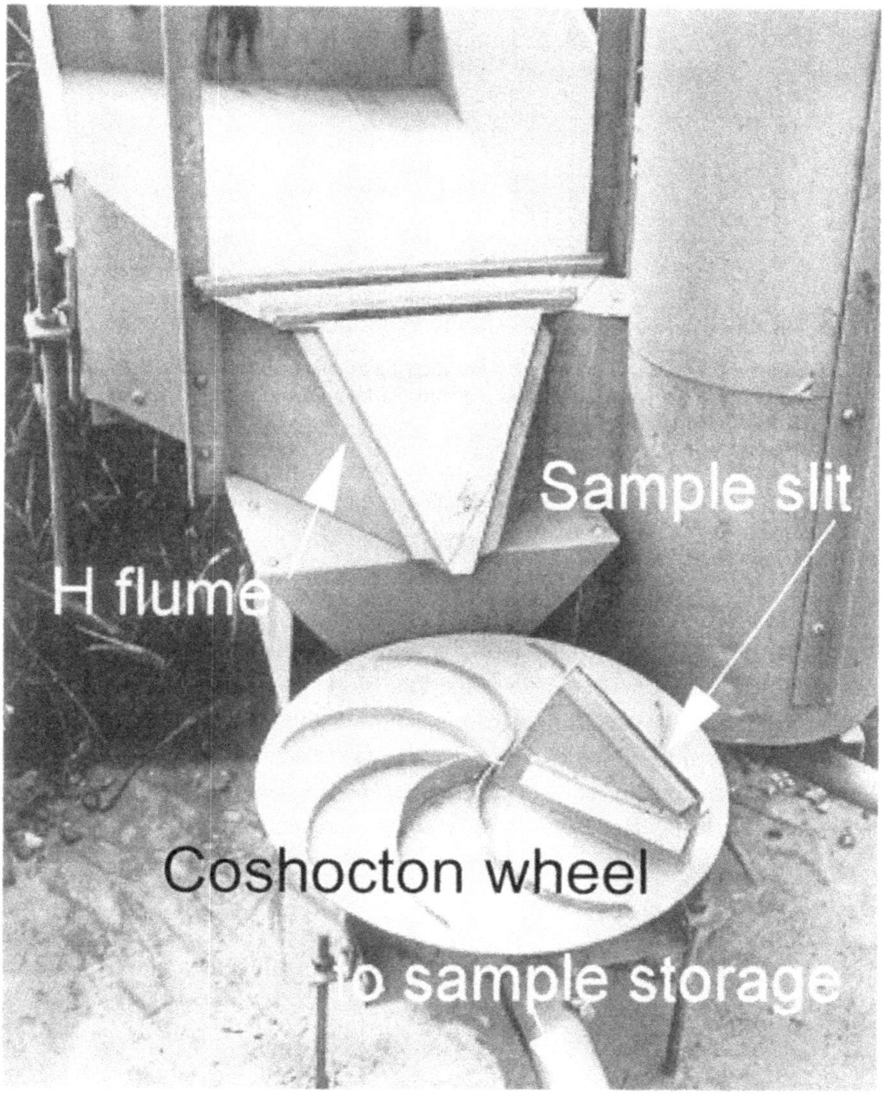

Figure 6.12. Close-up of apparatus for quantifying runoff flow rate and sediment concentration from research plots.

factor is dimensionless. Slope length is computed using

$$L_f = \left(\frac{l'}{22}\right)^{m_{er}} \tag{6.4}$$

where

l' = length along the slope face (m); change the denominator in equation 6.4 to 72.16 if length is given in feet;

m_{er} is a factor relating angle to slope length erosion severity.

Find m_{er} using equations 6.5 and 6.6:[1]

$$m_{er} = \frac{\sin(\theta)}{\sin(\theta) + 0.269[\sin(\theta)]^{0.8} + 0.05} \tag{6.5}$$

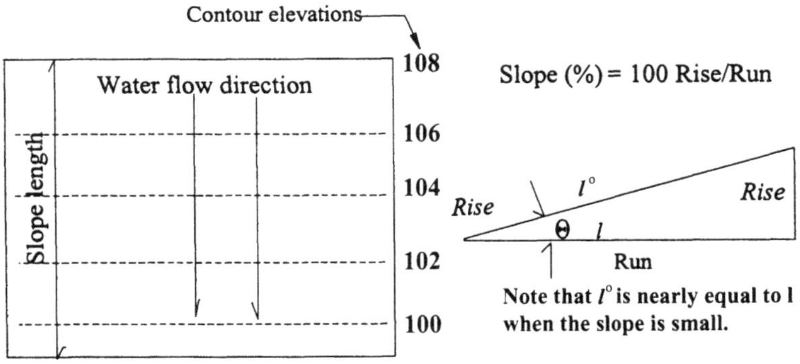

Figure 6.13. Sketch of a simple field showing practical slope length l' along the field slope in the direction of flow, combined with a definition sketch showing the rise/run calculation.

The parameter θ is the slope angle and relates to field slope s (%) using

$$\theta = \tan^{-1}\left(\frac{s}{100}\right) \quad \textbf{(6.6)}$$

The definition sketch in Figure 6.13 shows how the parameters relate to the measurements made on the field surface. The term s is the field slope rise over run in percent.

The original USLE test protocol was based on plots with a standard 22 m length and a 9% slope. This explains the division by 22 m. The m_{er} factor adjusts for slopes. At the standard slope of 9%, one can readily show the exponent to be 0.5. The exponent increases as slope increases and vice versa. The length factor becomes ever more critical as slope increases.

S_f

The factor S_f is the slope steepness factor. This factor is dimensionless. One computes S_f using

$$S_f = 3.0[\sin(\theta)]^{0.8} + 0.56 \quad \text{if the slope length 1 is shorter than 4 m} \quad \textbf{(6.7a)}$$

$$S_f = 10.8\sin(\theta) + 0.03 \quad \text{if the slope length 1 is longer than 4 m and } s < 9\% \quad \textbf{(6.7b)}$$

$$S_f = 16.8\sin(\theta) - 0.5 \quad \text{if the slope length 1 is longer than 4 m and } s \geq 9\% \quad \textbf{(6.7c)}$$

Slope shape affects erosion. A convex slope is more erodible than a concave slope (see Figure 6.3). The physical erosion process is a highly localized balance between detachment and deposition. The detachment/deposition process clearly changes as slope changes. Convex slopes (steep at the end) favor detachment as the transporting flow leaves the segment, whereas concave slopes favor deposition as the transporting flow leaves. The concave slope enables deposition of soil detached on the upper end of the slope, whereas the convex slope does not. Renard et al. (1997) present a series of subfactors to account for concave and convex slopes. The RUSLE software can handle nonuniform slopes. This discussion will be concerned with uniform slopes.

C_{prac}

The factor C_{prac} is the cover management factor and is a product of five subfactors in agricultural settings: prior use, canopy cover, surface cover, surface roughness, and soil moisture (Renard and Foster 1995). Haan et al. (1994) discuss an implementation of USLE for urban construction-related purposes, where the C_{prac} factor is composed of seven subfactors: time, density, prior use, canopy cover, surface cover, surface roughness, and soil moisture. One should select these factors in consultation with an experienced soil scientist.

The C_{prac} factor is dynamic, reflecting crop growth and field operations or construction site conditions. For purposes of surveying the method, C_{prac} factors from Table 6.2 will be used. This table shows data for several crops at various growth stages including the fallow stage. The tabulated C_{prac} factor embodies variation in canopy cover, surface roughness, and, perhaps indirectly, soil moisture. The RUSLE software in effect computerizes the Wischmeier and Smith (1978) C_{prac} factor table, using the five subfactors.

P_{prac}

The factor P_{prac} is the conservation practice factor. This factor is dimensionless and has a maximum value of 1 for no conservation practices (e.g., plowing up and down slope). P_{prac} is subdivided into three components.

P_c is the contouring-practice subfactor, which depends on land slope. P_s is the strip-crop practice subfactor. Table 6.3 gives values of these subfactors for agricultural or non-construction applications of our hybrid model of USLE. Haan et al. (1994) give a more exhaustive table for the USLE. If the slope length in Table 6.3 is exceeded for the given slope, then use $P_c = 1$.

The strip-crop factor depends on the crop rotation. The maximum allowable strip for the factor to be valid depends on the slope. If strip-crop widths are equal to or less than the strip-crop widths given in Table 6.3, P_s is 1 for contouring only; it is 0.75 for 4 yr rotation with 2 yr row crop and 2 yr meadow or small grain; and 0.5 with 1 yr row crop with 3 yr meadow or small grain. Assume P_s is 1 with shorter rotations and when the strip width for a given slope is exceeded.

The third subfactor, P_t, the terracing factor, is 1 for no terraces, 0.2 for terraces with graded sod outlets, and 0.1 with tile outlet terraces. P_{prac} is 1 for construction, mining, and usually for clear-cut forest harvest areas. If one exceeds slope length for given land slopes in Table 6.3, then $P_c = 1$ also.

Compute P_{prac} using

$$P_{prac} = P_c P_s P_t \tag{6.8}$$

Renard et al. (1997) extend the P_{prac} factor to include P_{prac} values for filter strips and subsurface drainage.

The above discussion briefly introduces soil erosion by water. More detailed summaries of soil erosion research state-of-the-art are available in Follett and Stewart 1985, and Agassi 1996.

Computer software is available for RUSLE (Renard and Foster 1995; Renard et al. 1997). When computing average annual erosion, one must calculate a composite $R_{rf}C_{prac}$ factor since the rainfall and the C_{prac} factor are not uniformly distributed throughout the year.

Table 6.2. Abbreviated soil loss ratio table giving C_{prac} factors related to crop growth stage and C_{prac} factors for selected nonagricultural practices.

	Spring planting		Soil loss ratio (%)[d] for crop stage period[e] and canopy covers (Convert C_{prac} factor to decimal fraction before inserton in the RUSLE)							
Cover, sequence, and management[a]	Residue (kg)[b]	Cover (%)[c]	F	SB	1	2	3:80	3:90	3:96	4L
Continuous:										
Corn, RdL, sprg, TP	5000		36	60	52	41		24	20	30
Small grain	5000	60		16	14	12	7	4	2	
Meadow							1			
Rotation:										
Row crop after meadow			12	27	23	20		14	12	21
Corn after beans			47	78	65	51		30	25	37
Beans after corn, sprg, TP			39	64	56	41		21	18	
Conserv.										
Tillage:										
Beans or corn after corn	5000	60		13	11	10		10	8	20
Corn after beans		30		33	29	25	22	18	14	33
Small grain after corn	5000	60		16	14	13	7	4	2	

Material condition C_{prac}
(Convert C_{prac} factors to decimals before insertion in the RUSLE)

Mulch	0.5t/ac	30	mech. prepared	good soil	24	bare soil	compact and root racked	90–120
	4	2		poor soil	36	disk tillage	Fresh	100
Woodchips	0.5 t/ac	90	burned	good soil	26		after one rain	89
	4	22		poor soil	45			
Chemical binders	asphalt	14–52	bare soil	scraped	100–130	open[f]-weave seeded mats	initial	8
	terratac	67		compact smooth	100–140		after emergence	2

Sources: Adapted from Schwab et al. 1993 and Haan et al. 1994.

[a]Rdl = crop residue left in field, sprg = spring tillage; TP = plowed with moldboard.

[b]Dry mass per hectare after winter loss and reductions by grazing or partial removal; 5000 kg/ha represents a yield of 6–8 Mg/ha.

[c]Percentage of soil surface covered by plant residue mulch after crop seeding. They reflects the difference between spring residue and surface residue after crop seeding in the soil loss ratios as residues mixed with the topsoil.

[d]The soil loss ratios assume the indicated crop sequence and consistent performance. One-year deviations from normal practices do not have the effect of a permanent change.

[e]F = rough fallow; SB = seedbed until 10% canopy cover; 1 = establishment until 50% canopy cover; 2 = development until 75% canopy cover; 3:xx = maturing until harvest for the indicated degrees of canopy cover; 4L = residue or stubble.

[f]Estimated.

Table 6.3. Values for the contouring practice subfactor, P_c, and the strip-crop practice subfactor, *Ps*, provided maximum slope length and maximum strip-crop width are maximum values for the corresponding land slope.

Land slope (%)	P_c	P_s (depends on rotation)	Maximum slope length (m)	Maximum strip-crop width (m)
1–2	0.6	1 = contouring only	≤ 120	≤ 40
3–5	0.5	0.75 = 4 yr	≤ 90	≤ 30
6–8	0.5	rotation with 2 yr	≤ 60	≤ 30
9–12	0.6	meadow with 2 yr row	≤ 36	≤ 24
13–16	0.7	crop; 0.5 = 4 yr	≤ 24	≤ 24
17–20	0.8	rotation with 3 yr meadow	≤ 18	≤ 18
21–25	0.9	and 1 yr row crop	≤ 15	≤ 15

Source: Adapted from Schwab et al. 1993 and Wischmeier and Smith 1978.

Note: If one exceeds maximum slope length for the appropriate tabulated land slope, use $P_c = 1$. If one exceeds maximum strip-crop width for the appropriate tabulated slope, use $P_s = 1$. Haan et al. (1994) provides a more exhaustive table of P_c factors when the critical length is exceeded.

The Weighted $R_{rf}C_{prac}$ Factor for Average Annual Erosion Computation

The weighted $R_{rf}C_{prac}$ factor is necessary because (1) rainfall energy is not distributed uniformly over the year and (2) the C_{prac} factor changes over the year as the crop develops or other practices are imposed. Plots in Figure 6.14 show rainfall energy distribution for several regions. Table 6.2 shows how C_{prac} varies with the crop growth stage.

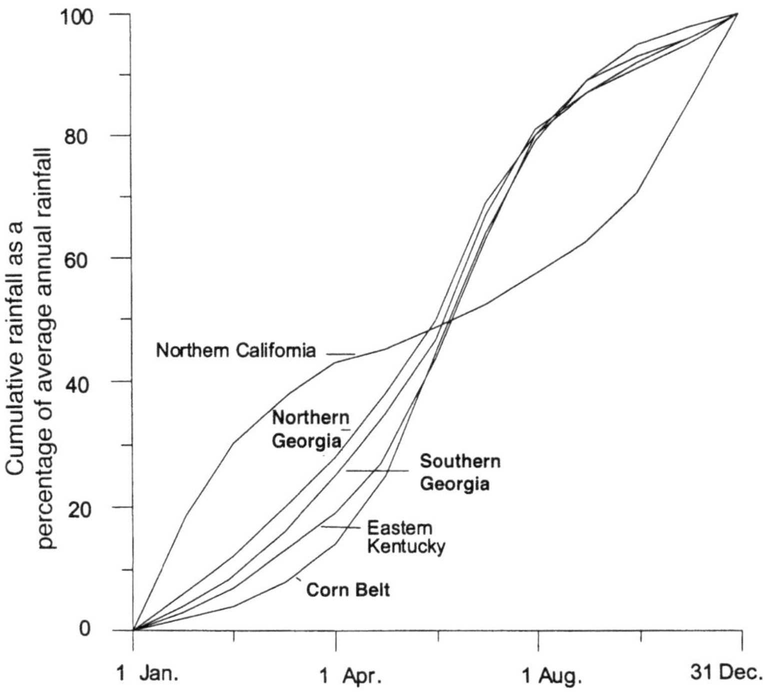

Figure 6.14. Examples of rainfall distribution curves for several regions of the United States.

Calculation of the weighted $R_{rf}C_{prac}$ factor is crucial for properly applying the USLE or RUSLE on an average annual basis.

The weighted $R_{rf}C_{prac}$ factor calculation goes as follows. First, proportionalize the average annual R_{rf} rainfall energy factor. This factor is region specific. Second, determine the appropriate C_{prac} factor for the crop-rotation-tillage practice and time of development. Proper selection of the C_{prac} factor requires practical knowledge of cropping practices in a region. Some typical rainfall distributions and an abbreviated C_{prac} factor table sufficient for this introduction are shown in Figure 6.14 and Table 6.2. Table 6.4 exemplifies the process for a Georgia crop rotation.

Erosion problems lend themselves nicely to the use of a spreadsheet. First, enter the cumulative rainfall distribution for the year by selecting points corresponding to months of the year (sometimes one may want to use 2 week intervals). Next, compute the incremental rainfall for the period. Then enter the appropriate C_{prac} factor for the period. Take the product of the rainfall increment and C_{prac} factor, expressing both as decimals before multiplication. Sum these products for the year. Add the yearly sums that make up the complete rotation. Divide the sum by the number of years to get the weighted C_{prac} factor. Haan et al. (1994) and Renard et al. (1997) tabulate the cumulative rainfall distributions for 140 areas that make up the 50 states.

Does the C_{prac} factor along with other factors of the USLE result in an average annual soil loss falling within the soil loss tolerance? Use creativity along with practical judgment to arrive at a workable answer. For the client, the workable answer must be economically feasible. This means one must strive to be consistent with crops, equipment, etc.

Several strategies are useful for reducing soil loss in agriculture. The C_{prac} factor and P_{prac} factor are the most controllable. Inclusion of meadows to lengthen rotations (C_{prac} factor manipulation) and the use of terraces, tile outlet terraces (P_{prac} factor and L_f factor manipulation), and conservation tillage (C_{prac} factor manipulation) all can reduce soil losses. One can indirectly influence the K_{sl} factor by long-term organic matter buildup and soil structure management. However, control of K_{sl} is indirect. Meadows are uneconomical in the short run. Terraces were popular in the 1940s through 1960s and then were abandoned due to the constraints they imposed on large machinery. Conservation tillage is now the method of choice for row crop production. Terraces may have a place for high-value crops even now. Growers are accepting the economics of multiple crops and meadows because of pest pressures associated with monoculture.

Table 6.4. Weighted C factor calculation for a northern Georgia crop rotation (peanut-corn-meadow—meadow).

Example Problem—Erosion				
C factor calculator		Peanut-corn-meadow-meadow		
Month	Cum. rain (%)	Inc. rain (%)/100	Tabulated C_{prac}/100	Weighted C_{prac}
	(NORTH GEORGIA)			
Jan.	6	0.06	0.01	0.0006
Feb.	12	0.06	0.01	0.0006
Mar.	20	0.08	0.01	0.0008
Apr.	28	0.08	0.12	0.0096
May	38	0.10	0.27	0.0270
June	50	0.12	0.23	0.0276
July	69	0.19	0.20	0.0380

(Continued)

Table 6.4. (*Continued*)

Example Problem—Erosion					
C factor calculator		Peanut-corn-meadow-meadow			
Month	Cum. rain (%)	Inc. rain (%)/100	Tabulated C_{prac}/100	Weighted C_{prac}	
	(NORTH GEORGIA)				
Aug.	80	0.11	0.14	0.0154	
Sep.	87	0.07	0.12	0.0084	
Oct.	92	0.05	0.21	0.0105	
Nov.	96	0.04	0.21	0.0084	
Dec.	100	0.04	0.21	0.0084	
	YR 1 SUMS			0.1553	
Jan.	6	0.06	0.21	0.0126	
Feb.	12	0.06	0.21	0.0126	
Mar.	20	0.08	0.21	0.0168	
Apr.	28	0.08	0.21	0.0168	
May	38	0.10	0.47	0.0470	
June	50	0.12	0.78	0.0936	
July	69	0.19	0.65	0.1235	
Aug.	80	0.11	0.30	0.0330	
Sep.	87	0.07	0.37	0.0259	
Oct.	92	0.05	0.16	0.0080	
Nov.	96	0.04	0.14	0.0056	
Dec.	100	0.04	0.12	0.0048	
		YR 2 SUMS			0.4002
Jan.	6	0.06	0.07	0.0042	
Feb.	12	0.06	0.04	0.0024	
Mar.	20	0.08	0.02	0.0016	
Apr.	28	0.08	0.01	0.0008	
May	38	0.10	0.01	0.0010	
June	50	0.12	0.01	0.0012	
July	69	0.19	0.01	0.0019	
Aug.	80	0.11	0.01	0.0011	
Sep.	87	0.07	0.01	0.0007	
Oct.	92	0.05	0.01	0.0005	
Nov.	96	0.04	0.01	0.0004	
Dec.	100	0.04	0.01	0.0004	
		YR 3 SUMS			0.0162
Jan.	6	0.06	0.01	0.0006	
Feb.	12	0.06	0.01	0.0006	
Mar.	20	0.08	0.01	0.0008	
Apr.	28	0.08	0.01	0.0008	
May	38	0.10	0.01	0.0010	
June	50	0.12	0.01	0.0012	
July	69	0.19	0.01	0.0019	
Aug.	80	0.11	0.01	0.0011	
Sep.	87	0.07	0.01	0.0007	
Oct.	92	0.05	0.01	0.0005	
Nov.	96	0.04	0.01	0.0004	
Dec.	100	0.04	0.01	0.0004	
		YR 4 SUM			0.01

Four year rotation $C_{prac}=(0.1553+0.4002+0.0162+0.01)/4=0.1454$
Three year rotation (one meadow instead of two) $C_{prac}=(0.1553+0.4002+0.0162)/3=0.1906$

Consider a construction-related erosion problem. One can easily show the impact of construction timing using the weighted $R_{rf}C_{prac}$ factor. The C_{prac} factor is 1 with exposed soil. Further, assume a construction duration of 3 months. After construction, assume establishment of an appropriate grass, estimating coefficients for a meadow. The student should calculate the sums where we show xxx in Table 6.5. Substantial differences in the soil lost versus construction initiation season then become obvious. Inserting these

Table 6.5. Weighted C factor calculation: Construction-induced erosion in northern Georgia.

Month	Cum. rain	Inc. rain (%)/100	Spr. cons C_{prac}/100	Weighted C_{prac}-Spr.	Sum cons C_{prac}/100	Weighted C_{prac}-Sum	Fall cons C_{prac}/100	Weighted C_{prac}-Fall
	(NORTH GEORGIA)							
Jan	6	0.06	0.01	0.0006	0.01	0.0006	0.01	0.0006
Feb	12	0.06	0.01	0.0006	0.01	0.0006	0.01	0.0006
Mar	20	0.08	1.00	0.0800	0.01	0.0008	0.01	0.0008
Apr	28	0.08	1.00	0.0800	0.01	0.0008	0.01	0.0008
May	38	0.10	1.00	0.1000	0.01	0.0010	0.01	0.0010
Jun	50	0.12	0.16	0.0192	1.00	0.1200	0.01	0.0012
Jul	69	0.19	0.14	0.0266	1.00	0.1900	0.01	0.0019
Aug	80	0.11	0.12	0.0132	1.00	0.1100	0.01	0.0011
Sep	87	0.07	0.10	0.0070	0.16	0.0112	1.00	0.0700
Oct	92	0.05	0.10	0.0050	0.14	0.0070	1.00	0.0500
Nov	96	0.04	0.10	0.0040	0.12	0.0048	1.00	0.0400
Dec	100	0.04	0.10	0.0040	0.10	0.0040	0.16	0.0064
			YR 1 SUMS	0.3402		xxx		xxx
Jan	6	0.06	0.10	0.0060	0.10	0.0060	0.16	0.0096
Feb	12	0.06	0.10	0.0060	0.10	0.0060	0.16	0.0096
Mar	20	0.08	0.10	0.0080	0.10	0.0080	0.14	0.0112
Apr	28	0.08	0.05	0.0040	0.10	0.0080	0.12	0.0096
May	38	0.10	0.02	0.0020	0.10	0.0100	0.10	0.0100
Jun	50	0.12	0.01	0.0012	0.05	0.0060	0.05	0.0060
Jul	69	0.19	0.01	0.0019	0.02	0.0038	0.05	0.0095
Aug	80	0.11	0.01	0.0011	0.01	0.0011	0.01	0.0011
Sep	87	0.07	0.01	0.0007	0.01	0.0007	0.01	0.0007
Oct	92	0.05	0.01	0.0005	0.01	0.0005	0.01	0.0005
Nov	96	0.04	0.01	0.0004	0.01	0.0004	0.01	0.0004
Dec	100	0.04	0.01	0.0004	0.01	0.0004	0.01	0.0004
			YR 2 SUMS	0.0322		xxx		xxx
Jan	6	0.06	0.01	0.0006	0.01	0.0006	0.01	0.0006
Feb	12	0.06	0.01	0.0006	0.01	0.0006	0.01	0.0006
Mar	20	0.08	0.01	0.0008	0.01	0.0008	0.01	0.0008
Apr	28	0.08	0.01	0.0008	0.01	0.0008	0.01	0.0008
May	38	0.10	0.01	0.0010	0.01	0.0010	0.01	0.0010
Jun	50	0.12	0.01	0.0012	0.01	0.0012	0.01	0.0012
Jul	69	0.19	0.01	0.0019	0.01	0.0019	0.01	0.0019
Aug	80	0.11	0.01	0.0011	0.01	0.0011	0.01	0.0011
Sep	87	0.07	0.01	0.0007	0.01	0.0007	0.01	0.0007
Oct	92	0.05	0.01	0.0005	0.01	0.0005	0.01	0.0005
Nov	96	0.04	0.01	0.0004	0.01	0.0004	0.01	0.0004
Dec	100	0.04	0.01	0.0004	0.01	0.0004	0.01	0.0004
			YR 3 UMS	0.01		xxx		xxx

Note: Students are asked to arrive at the values indicated xxx in homework problem.

weighted C_{prac} factors along with the R_{rf}, K_{sl}, L_f, S_f, and P_{prac} factors into the USLE/RUSLE leads to the average annual erosion in both cases.

One may reduce construction site soil loss by exposing soil at the time of the year with the least rainfall energy. Diversions are also used to reduce slope lengths.

Example 6.1: A soil in Athens, Georgia, has 50% silt and fine sand and 50% clay. The soil has 1% organic matter. The soil is fine granular (b = 2) and is of moderate permeability (c = 3). With a field slope of 5% and a 100 m slope length, compute the average annual soil loss with a peanut-corn-meadow rotation on the contour with no strips or terraces.

$$A_{ero} = R_{rf}K_{sl}L_fS_fC_{prac}P_{prac}$$

(equation 6.1).

$$R_{rf} \text{ for Athens} = 5000 \text{ MJ} \cdot \text{mm/(ha} \cdot \text{hr} \cdot \text{average annual storm)}$$
$$[294 \text{ ft} \cdot \text{tons} \cdot \text{in/(ac} \cdot \text{in} \cdot \text{yr)}] \times 17.02$$

(Figure 6.9).

$$K_{sl} = 2.8 \times 10^{-7}(50 \times 50)^{1.14}(12 - 1) + 4.3 \times 10^{-3}(2 - 2) + 3.3 \times 10^{-3}(3 - 3)$$
$$= 0.023 \text{ metric ton} \cdot \text{ha} \cdot \text{hr/(ha} \cdot \text{MJ} \cdot \text{mm)}$$

(equation 6.3).

$$\theta = \tan^{-1}(5/100) = 2.8624 \text{ degrees}$$

(equation 6.6).

$$m_{er} = \sin(2.8604)/[\sin(2.8604) + 0.269\sin(2.8604)^{0.8} + 0.05] = 0.4015$$

(equation 6.5).

$$L_f = (100/22)^{0.4015} = 1.8366$$

(equation 6.4).

$$S_f = 10.8\sin(2.8604) + 0.03 = 0.5689$$

(equations 6.7a–c).

$$C_{prac} = 0.1906$$

(from weighted RC calculation; use a spreadsheet as in Tables 6.4 and 6.5)

$$P_{prac} = 1(1)(1) = 1$$

(equation 6.8, Table 6.3, noting the slope length exceeds value for P_c at the 5% slope; thus P_c defaults to 1 instead of the Table 6.3 value).

A_{ero} = 5000 MJ · mm/(ha · hr · average annual storm) [0.023 Mg · ha · hr/(ha · MJ · mm)] (1.8366)(0.5689)(0.1906)(1) = 22.9 Mg/ha · yr

This soil loss level is on the high side. What steps might one take to reduce the soil loss? One could put in a terrace (reduce P_{prac}, L_f, rarely S_f with massive excavation), use conservation tillage (reduce C_{prac}), or add meadow (reduce C_{prac}). The K_{sl} factor is for practical purposes constant. The slope factor is constant (unless one purposely modifies the contour; not usually practical). Location sets the rainfall R_{rf}. Tillage or construction timing may affect the influence of the rainfall distribution on the C_{prac} factor. *Challenge:* Analyze each factor of the USLE from a construction viewpoint.

NOTE ON UNITS

SI units are emphasized because of the current societal push towards SI units. We show SI units here. Renard et al. maintained English units in their latest erosion handbook (1997). One should be familiar with both systems because the literature abounds in English units.

Foster et al. (1981) provides a thorough analysis of both SI and English units with respect to erosion prediction. NASA determined that the *Mars Challenger,* a satellite designed to map the surface of Mars, was lost due to *confusion in units of measurement.*

Prediction of Sediment Yield

The USLE/RUSLE predicts detached soil, not sediment yields. The prediction of sediment yield is an ongoing research problem. The sediment delivery ratio approach is sometimes used to estimate sediment yield. The sediment delivery ratio (SDR) is the ratio of the sediment delivered at a particular measurement point to the gross erosion (detached soil) from the drainage area. SDRs are necessary because the USLE/RUSLE equations cannot predict deposition of sediment. One may make simple estimations of delivery ratios based on watershed area or the watershed gradient (Haan et al. 1994). Figures 6.15 and 6.16 show plots for delivery ratio estimation.

A more thorough approach for computing sediment delivery ratio is based on the U.S. Forest Service (1980) method. This method considers delivery distance, surface roughness, slope gradient, surface runoff per foot of slope width, texture of eroded material (percent silt and smaller), and percent ground cover. One may arrange these parameters in a Stiff diagram, shown in Figure 6.17.

Compute values for each parameter in Figure 6.17, connect the points with a line resulting in a polygon, then determine the area enclosed by the polygon. Then form a ratio of area enclosed by the polygon to the bounding rectangle. One enters this ratio value into the SDR prediction curve.

The factors on the Stiff diagram are obvious influences on the ability of runoff to transport sediment. The slope is that of the zone (e.g., a riparian zone) between the sediment producing zone and the receiving channel. The slope shape of 0 represents a convex slope, while at the other extreme a factor of 4 represents a concave slope. A factor of 2 represents a uniform slope.

Surface roughness is a judgment call. A factor of 0 represents a smooth (e.g., prepared by tillage plus dragging to smooth) surface, and a factor of 4 represents a rough (e.g., many clods, trenches, holes) surface. As with the slope, the zone in question is that between the sediment-producing zone (e.g., a riparian zone) and the receiving channel.

One can compute the surface runoff factor using

$$q_{sp} = 2.31 \times 10^{-5}[\text{peak excess rain (in/hr)}][\text{disturbed area slope length(ft)}] \quad \textbf{(6.9)}$$

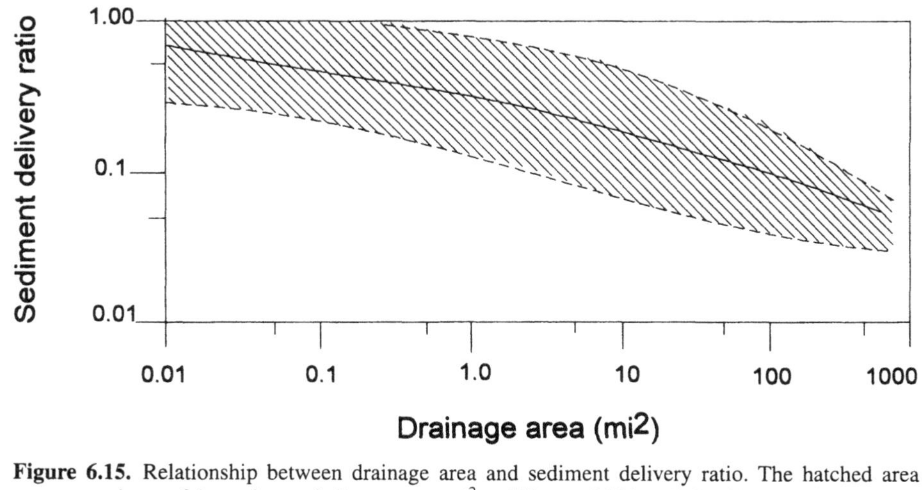

Figure 6.15. Relationship between drainage area and sediment delivery ratio. The hatched area represents the confidence interval. One statute mi^2 = 640 ac = 259 ha. (After Haan et al. 1994.)

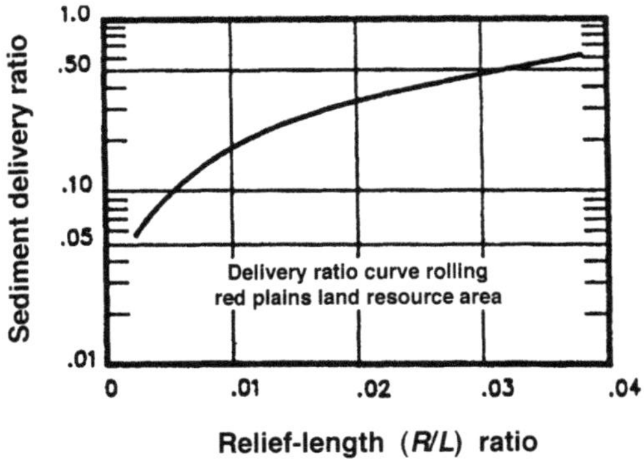

Figure 6.16. Relationship between the relief ratio, or watershed gradient, to sediment delivery ratio. (After Renfro 1975; courtesy of USDA-ARS.)

where q_{sp} is the flow in cfs per foot of eroded area. One in/hr equals 25.4 mm/hr, and cfs/ft equals 0.093 m^3/s-m. *We must compute or simulate the excess rain hyetograph to find the peak excess rain required by this method.* The sigmoidal curve shown in Figure 6.18 relates the SDR to the area bounded within the Stiff diagram.

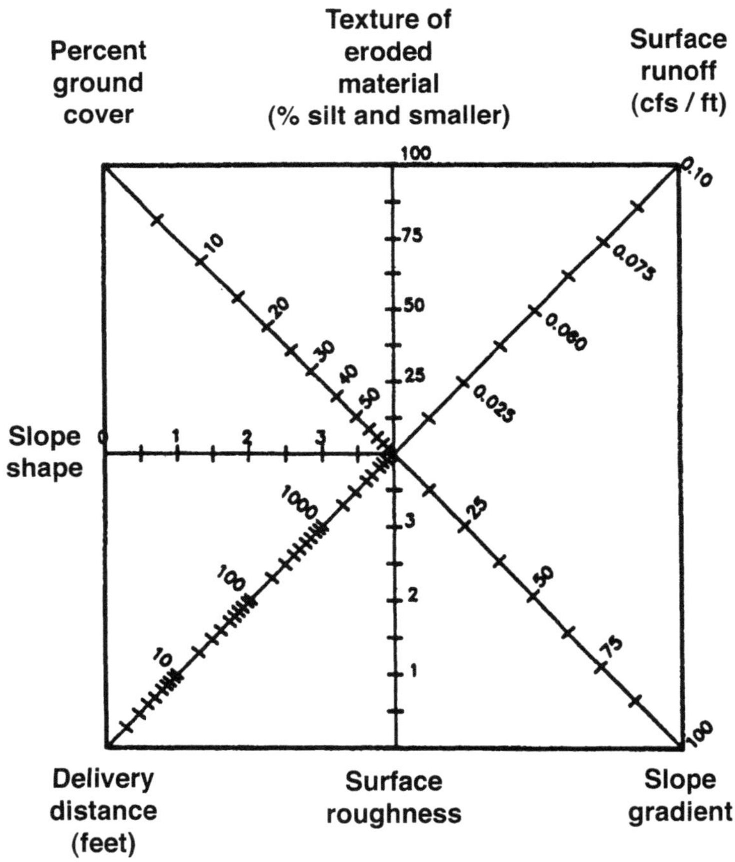

Figure 6.17. Stiff diagram for determining relative contribution of the indicated factors to the reduction of sediment; 1 m = 3.28 ft; 1 cfs/ft = 0.093 m^3/s-m. (From U.S. Forest Service 1980.)

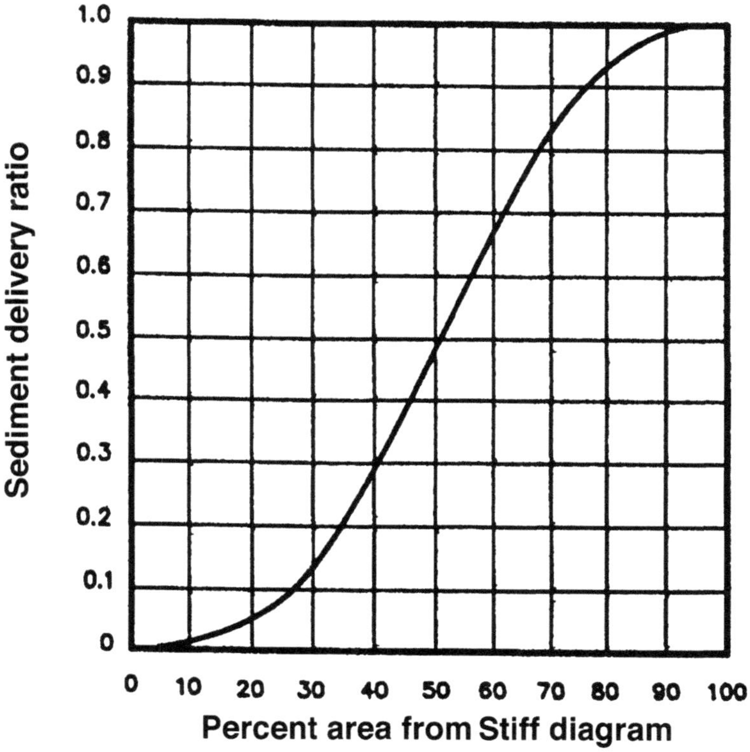

Figure 6.18. Relationship between area within the U.S. Forest Service (1980) Stiff diagram and the sediment delivery ratio.

These predicted values are useful in design of sediment basins and related structures. They give estimates. More important, the erosion prediction equations and the delivery ratio prediction give valuable insights into steps for reducing sediment production and delivery compared with prior or current conditions.

Example 6.2: A construction operation is disturbing a large area in a forest watershed upslope of a sensitive stream. The regulatory authority requires an undisturbed riparian zone on either side of the stream. During a storm, we estimate the sediment yield to the edge of the forest from the RUSLE to be 30 tons/ac. The eroded soil distribution is 50% sand and 50% silt plus clay. The disturbed area has a slope length of approximately 800 ft. The width is 2000 ft, leading to an area of 1,600,000 ft^2, or 34.731 ac (43,560 ft^2 per acre). Based on storm simulation, the peak rainfall excess is 4.5 in/hr. The riparian zone has the following characteristics:

Slope shape: moderately convex
Change in elevation across the riparian zone: 50 ft
Average slope: 17%
Ground cover: 100%
Surface roughness: rough
Flow delivery distance: 300 ft

Compute the SDR and the expected gross erosion.

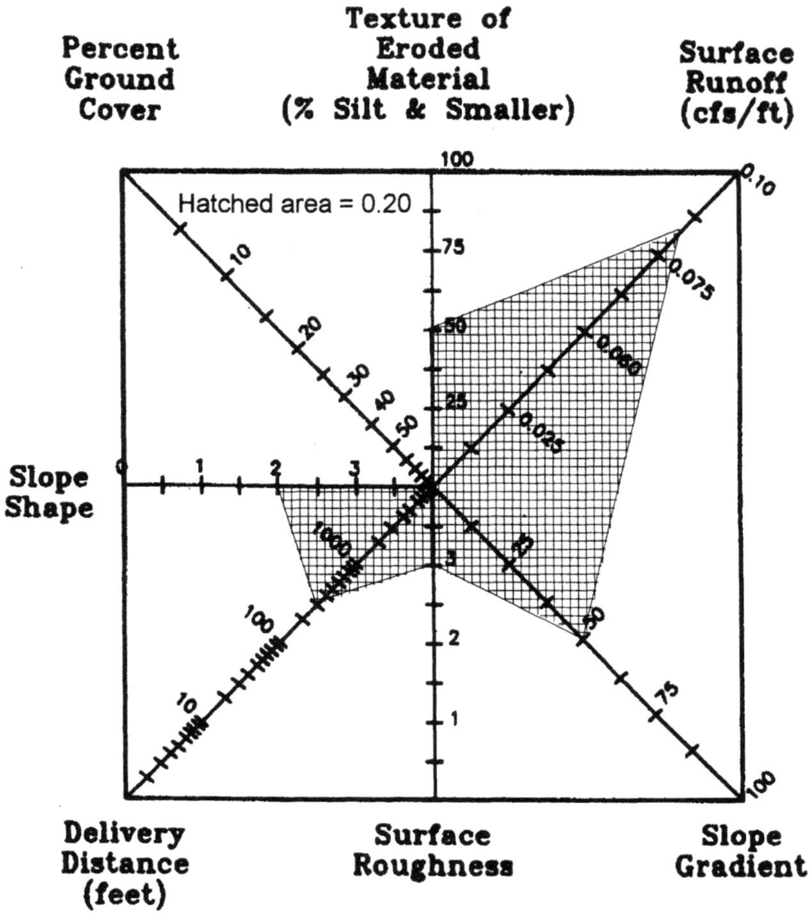

Figure 6.19. Stiff diagram for conditions of Example 6.2.

From equation 6.9, we can compute the runoff parameter to be

$$q_{sp} = 2.31 \times 10^{-5}\ (4.5\ \text{in})(800\ \text{ft}) = 0.083\ \text{cfs/ft}$$

One may plot relevant information on the Stiff diagram as shown in Figure 6.19. The percentage of the Stiff diagram covered by the cross-hatching is 13.5%, leading to an estimated SDR of 0.04 from Figure 6.18. The gross erosion is 30 tons/ac × 34.73 ac × 0.04 = 44.1 tons = 40.07 Mg.

Modeling Approaches for Sediment Yield Prediction

More numerically amenable approaches have begun to appear for predicting sediment yields. The approach outlined herein, called the Kentucky reclamation approach, is discussed in detail in Haan et al. 1994. This approach, drawing from agricultural and reclamation experiences around the United States, is applicable to erosion control engineering on the farm and field scale. It is also amenable to mining and construction site planning. Models such as SedCad[2] (Warner 1998) have used the Kentucky mining reclamation

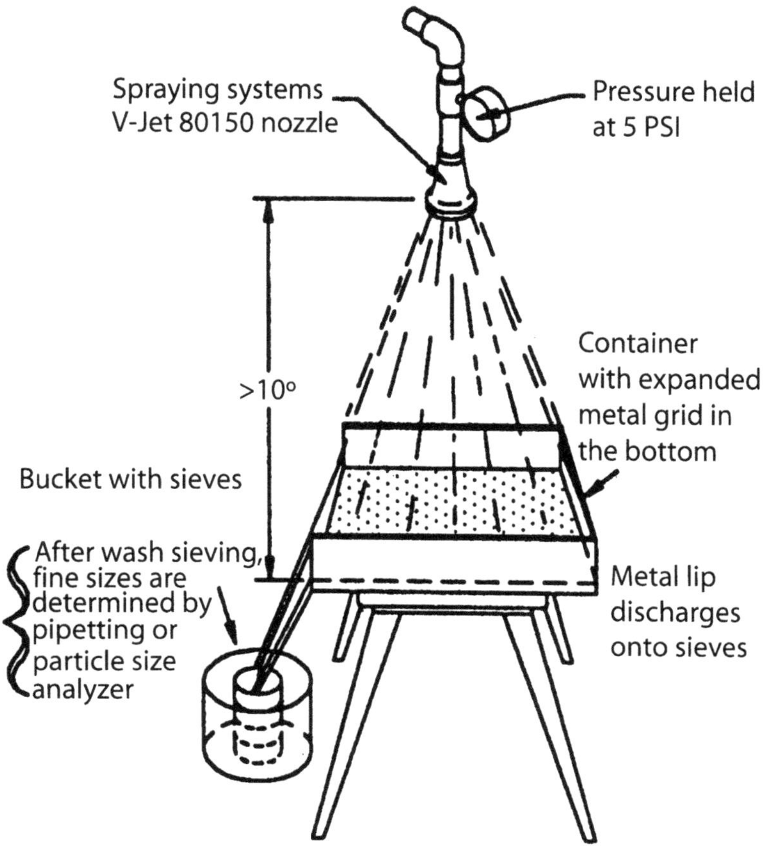

Figure 6.20. Soil erodibility test apparatus using a single nozzle rainfall simulator. (From Haan et al. 1994; used with permission of Academic Press.)

approach. This method models sediment yield beginning with a measured eroded particle size distribution. Particle size distribution is measured using the apparatus shown in Figure 6.20.

The V-jet nozzle shown simulates drops and kinetic energy approximating those of 2–13 mm/hr rainfall intensities. The average (D_{50}) particle size may be determined for each of n homogenous subwatersheds composing the entire watershed.

Williams (1975) proposed the following empirical procedure for determining sediment yields. For each subwatershed write

$$Ayld_i = Aero_i \exp(-BT_{ti}\sqrt{D_{50}}) \quad \textbf{(6.10)}$$

where

$Ayld_i$ is the contribution of subwatershed i to the sediment yield at the point of measurement;
$Aero_i$ is the predicted erosion at the exit of subwatershed i as predicted by versions of the USLE;
T_{ti} is the travel time from the subwatershed to the point of measurement;
D_{50} is the measured average particle size of eroded sediment;
B is a routing coefficient.

Summing the contributions of all subwatersheds gives

$$AyldT = \sum_{i=1}^{n} A_{ero\,i} \exp(-BT_{Ti}\sqrt{D_{50}}) \tag{6.11}$$

where $Ayld_i$ is the cumulative sediment yield from n subwatersheds.

One may find the routing coefficient B by iterating for a best fit using

$$(Qq_p)^{0.56} = \sum_{i=1}^{n} (Q_i q_{pi})^{0.56} \exp(-BT_{Ti}\sqrt{D_{50}}) \tag{6.12}$$

where

Q is the runoff volume (ac-ft) of the entire watershed;
q_p is the peak runoff flow rate (cfs) of the entire watershed;
Q_i is the runoff volume (ac-ft) of subwatershed i;
q_{pi} is the peak runoff flow rate (cfs) of subwatershed i.

The $(Qq_p)^{0.56}$ term is proportional to the energy available for erosion. It follows from the Williams MUSLE.

The size distribution of the sediment from each subwatershed may be computed using

$$PF_{2i} = \frac{PF_{1i}}{D1_i} \quad PF_{2i} \leq 100 \tag{6.13}$$

where PF_{2i} is the particle size distribution resulting from routing the original measured distribution (PF_{1i}) at the exit of subwatershed i to the measurement location. The delivery ratio D_{1i} is an energy ratio given by

$$D1_i = \frac{95(Q_i q_{pi})^{0.56}}{R_{rfstorm} A_i} \tag{6.14}$$

The nomenclature was previously defined. Equation 6.13 describes the sedimentation because the largest particles settle first. Equations 6.13 and 6.14 may be applied in successive reaches by recomputing D_{2i}, D_{3i} . . . using appropriate Q and q_p for the area above the stream reaches. Figure 6.21 shows the effect of two successive routings.

After routing each subwatershed down the successive reaches to the outlet, one may accumulate the mass of material within each size category and replot the size distribution.

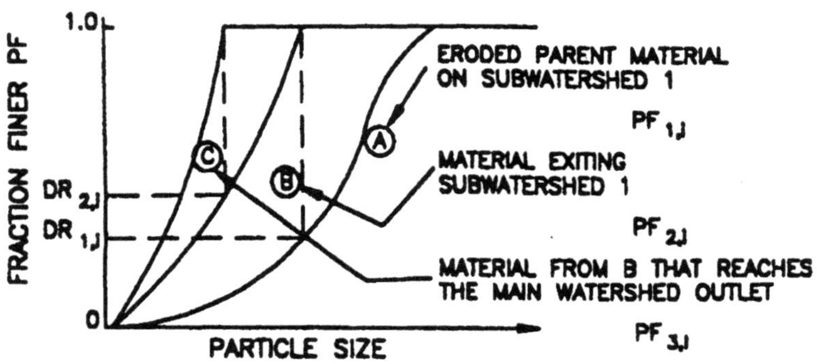

Figure 6.21. Effect of travel downstream on sediment particle size distribution. (From Haan et al. 1994; used with permission of Academic Press.)

One can construct a relationship of sediment concentration versus time using the procedure outlined by Haan et al. (1994). The relation between sediment concentration versus time is called a *sedigraph.* They began by writing

$$c = kq^a \tag{6.15}$$

where

k and a are fitted parameters;
c is the sediment concentration (kg solids/m^3);
q is the flow rate (m^3/s).

The mass flow rate of sediment is given by

$$q_s = cq \tag{6.16}$$

where q_s is the mass flow rate of sediment (e.g., kg/s). One may write

$$q_s = kq^{a+1} \tag{6.17}$$

To obtain the total sediment yield A_{ero} one must integrate the entire storm hydrograph, leading to

$$Aero = \int_0^{Dst} kq^{a+1} \tag{6.18}$$

where Dst is the storm duration. Since k is constant, one may write

$$k = \frac{Aero}{\int_0^{Dst} q^{a+1}dt} = \frac{Aero}{\sum_0^{Dst} q^{a+1}\Delta t} \tag{6.19}$$

One may optimize values for k and a using nonlinear optimization techniques if observed data are available. Haan et al. (1994) suggest values ranging from 0.5 to 1. The value of k may then be found from equation 6.19.

Three core processes underlie SDR modeling in the Kentucky mining approach: (1) largest particles settle first; (2) deposition rates are proportional to the product of travel time and the square root of particle size; and (3) the delivery ratio is proportional to the Williams modified soil loss energy term and the USLE rainfall energy term. Sediment yields and size distributions from watersheds built up from a series of homogenous subwatersheds are, at best, estimates. The data are difficult to collect, and validation of models is sparse. Haan et al. (1994) further discuss sediment yield modeling. We will discuss the effect of structures on sediment yield in Chapter 11.

Erosion Control Practices in Agriculture

Erosion control practices strive to reduce soil detachment by manipulating RUSLE factors in ways that make economic sense. We discuss several below.

Contouring

Contouring manipulates P_{prac}. Contouring is the performing of field operations such as tillage, planting, harvesting, etc., on lines of constant elevation. Contouring creates small depressions that temporarily impound water and provide infiltration opportunity. It also decreases rill formation. Many U.S. growers practice contouring. Contouring on steep slopes requires care. Surface sealing and soil compaction result in decreased effectiveness of contouring over time. Runoff may be reduced by 75–80% initially and then may drop to 20% by year's end. One cannot always follow true contours without creating many short rows. Contour farming systems also have the advantages of fuel savings, reduced equipment wear, and improved equipment operation.

Strip-Cropping

Strip-cropping is an option for reducing slope length by interspersing strips of close-growing crops such as grasses with row crops. The close-growing crop serves to trap sediment arising from the row-cropped area. Strip-cropping in effect reduces C_{prac} by always keeping some grass on the field. It also reduces P_{prac}. Three types of strip-cropping are:

- Growing alternate strips of different crops in the same field. For water erosion control, strips are placed on contours (see Figure 6.22).
- Placing rows and strips of uniform width across the general slope of the field.
- Placing buffer areas in critical slope areas between uniform strips. Conservationists and engineers commonly call buffer zones *filter strips* when placed along streams. They call them *buffer strips* when nonrotated, grassed areas are placed along slopes between rotated areas. Buffer areas reduce erosion or cause deposition. Buffer areas result in erosion reductions similar to those of strip-cropping, the difference being that buffer strips are not rotated, whereas one rotates strip-cropping. Figure 6.23 shows a riparian zone with filter strips in an agricultural setting.

Rotations including strips of low-growing grasses and legumes alternating with grain and row crops are effective.

Insertion of pastures and long-term forests into erosion control plans gives more soil cover for longer increments of time, resulting in lower average annual soil losses. Cover crops reduce erosion by maintaining vegetation cover over the winter months.

Figure 6.22. Strip-cropping showing small grain (being harvested) and a row crop. (Courtesy of Georgia Soil and Water Conservation Commission.)

Figure 6.23. Riparian filter strips protecting streams around the boundary of the agricultural field. (Courtesy of Georgia Soil and Water Conservation Commission.)

Terracing

Terracing reduces slope length, slope (when major reshaping of the land surface is undertaken (e.g., see Figure 10.1), and the practice factor. Shorter slope lengths limit the velocity of water and therefore can reduce the transport capacity of the runoff. Terrace channels provide opportunity for deposition, especially when using tile outlets.

Conservation Tillage

Conservation tillage goes to the heart of erosion reduction by reducing the cropping factor C_{prac}. Conservation tillage schemes, such as shown in Figure 6.24, reduce exposure of bare soil by preserving surface residue. Conventional tillage is usually deleterious to soil conservation. Tillage is used primarily for preparation of the seedbed. Tillage is becoming less used for weed control. The impact of tillage operations on erosion depends on factors such as surface residue, aggregation, infiltration, surface

Figure 6.24. Cotton growing in a conservation tillage scheme. The crop residue forms the surface cover. (Courtesy of Georgia Soil and Water Conservation Commission.)

sealing, and resistance to movement by water and wind. Excessive tillage or tillage done under poor conditions destroys soil structure, thus increasing susceptibility to erosion. Some tillage practices, such as minimum tillage and furrow-diking, can significantly reduce erosion.

Sediment Control Practices in Construction

One goal of the architectural design phase of a development is to accomplish a match between the development and the terrain. Sediment control practices employed in construction-related activities involve manipulating factors contained within the RUSLE and SDR computations. One should employ the following principles (Goldman et al. 1984) when devising an appropriate mix of conservation practices for erosion and sediment control during construction:

Fit development to terrain: This reduces bare soil exposure.

Construction timing: Time construction when rainfall energy is low. Fall is the season with lowest rainfall energy in the U.S. regions receiving type II storms. Proper timing and sequencing of excavation operations may result in reduced soil exposure. Close coordination is required in projects involving multiple contractors.

Vegetation: Retain existing vegetation wherever possible. Apply mulching and temporary vegetation when possible when level of activity will permit. Hydroseeders are often used to apply seed to construction areas. Apply asphalt-based tackifiers and biodegradable oils to bare soil surfaces to reduce erosion by wind and rain. Wood chips and other biodegradable materials such as jute or coconut fiber mats have potential for mulching.

Runoff velocity control and sediment management: Keep runoff velocities low by reducing length and steepness of slopes when possible after disturbance. Surface roughness also aids in keeping velocities low. Prepare drainage ways and outlets to handle concentrated or increased runoff. Trap sediment on site by creating mini-impoundments. This is the basic idea behind the silt fences or compost berms used on many construction sites. Use diversion ditches to divert runoff away from denuded areas. One should have an inspection and maintenance program to ensure the continued functionality of the control structures. Figure 6.25 shows a straw bale barrier for excluding sediment from a culvert inlet. Additional discussion is given in Chapter 10.

Erosion and Pollution

Many metals and organics bind to the clay fraction of soils. Prediction of the clay fraction of soil loss is useful for qualitatively predicting the transport of these pollutants. Controlling erosion not only sustains the long-term productivity of soil but also reduces pesticide and nutrient movement into streams. For example, if one determines a pollutant concentration in mg pollutant/kg dry soil (ppm), then one can estimate the pollutant mass from lost-soil mass. More advanced erosion models such as CREAMS, AGNPS, and WEPP contain components for modeling the transport and runoff of agricultural chemicals (e.g., fertilizers and pesticides).

Figure 6.25. Straw bale barrier for excluding larger sediment size fractions from a culvert inlet.

Future Directions of Water Erosion Prediction

The ideas of USLE and its successors have been incorporated into larger, more comprehensive computer models. Knisel (1980) discusses CREAMS (a field-scale model for chemicals, runoff, and erosion from agricultural management systems). He incorporated basic precipitation, radiation, temperature, land use, cultural practices, plant nutrients, and pesticide components in his model. He modeled soil loss and sediment delivery. CREAMS models single storms and estimates annual soil loss. It also incorporates some structures for sediment control such as ponds. The CREAMS model attempts to estimate sediment yield using advanced sediment transport ideas beyond the scope of this text.

Software packages such as SedCad (Warner 1998) simplify the engineering design of sediment basins and other sediment control approaches. Sediment design packages model hydrograph and sedigraphs (relations between sediment concentration and time) produced by watersheds. They enable one to model complex watersheds by subdividing them into several simple watersheds, where previously discussed techniques apply. Engineering design packages continue to incorporate sediment control structures such as sediment basins, riparian buffers, and temporary structures such as straw bale check dams and silt fences.

The USDA is engaged in a massive effort to improve erosion prediction techniques: the result is WEPP (Water Erosion Prediction Project). This effort addresses interrill, rill, gully (scouring only), and stream erosion (scouring only) as separate processes. WEPP (Foster 1988) enables estimation of both the temporal and the spatial distribution of soil loss. WEPP requires extensive data.

Outlook

The USLE/RUSLE estimates the *gross* sheet and rill erosion occurring within the upper reaches of a watershed. It does not account for gully or stream channel erosion in downstream areas. The USLE and its successors are at best approximate. The text discussed a hybrid USLE model, using mainly the RUSLE rainfall factor, slope and slope length factors, and the USLE crop and practice factors. Pragmatically, software is required to implement the RUSLE model for agricultural (Renard and Foster 1995) and construction activities (Toy et al. 1998). RUSLE is an empirical model.

The model WEPP is a much more detailed physically based erosion model. WEPP, with all its intricacies, still only provides estimates. Soil erosion is a major contributor to water quality problems downstream. WEPP allows for soil deposition or erosion. The USLE/RUSLE describes soil erosion (detachment) and the U.S. Forest Service gives a comprehensive SDR computation. USLE-detached soil may be deposited elsewhere in a watershed. Computer models such as SedCad ease the design of erosion control plans for farm- and field-scale projects. They effectively break a complex watershed into small, compact subwatersheds, onto which hybrids of USLE/RUSLE are used.

Strategies for erosion control are constantly changing. The websites of the International Erosion Control Association (IECA, http://ieca.org/) and the Land Improvement Contractors Association (LICA, http://ourworld.compuserve.com/homepages/delmarva/), along with ASAE (http://www.asae.org) and American Society of Civil Engineers (ASCE, http://www.asce.org), provide links to modern approaches and application methods for water erosion control. Agencies such as the NRCS and USEPA provide software for modeling erosion on cropland and in urban environments. WEPP is available at links within http://www.nrcs.usda.gov. The USEPA provides the SWMM (storm water management model) at links within http://www.epa.gov.

Going from Engineering Science to Engineering Design: Why Some Approaches Are Empirical

One may conclude that erosion prediction is highly empirical. This conclusion is correct. One may wonder why, after having taken many courses in engineering science, mathematics, and physics, we are satisfied with such empiricism. To answer this we must consider the following practical reality: engineering science courses generally address reasonably small control volumes or small interconnected volumes. Many valid analyses can be done using this approach. However, some problems do not readily lend themselves to this approach. Consider the impact of a water droplet on a soil surface, discussed in the text. One cannot readily identify all the forces at work on one soil particle, let alone the uncountable totality of them.

To begin to understand the erosion problem, researchers thus took the following approach: (1) collect samples from each of many soils; (2) subject each to controlled rainfall energy; (3) determine the amount of detached sediment per unit of rainfall energy; (4) at selected conditions, study the impact of many crops at various stages of their growth; (5) at selected conditions, study the impact of slope length and slope steepness; (6) at selected conditions, study contouring, terracing, and strip-cropping factors. The USDA-ARS developed the USLE at the National Runoff and Soil Loss Data Center in 1954 in cooperation with Purdue University. Federal and state cooperative research projects at 49 locations (including Watkinsville and Tifton, Georgia) contributed more than 20,000 plot years of data to the ARS facility at Purdue University. Data from 16 states having rainfall simulators filled gaps after 1960. More details of the history of the USLE can be found in Wischmeier and Smith 1978. The alternative to a rigorous analysis is expensive and not recommended unless there is clearly no other alternative available. The Water Erosion Prediction Project (WEPP) and models such as CREAMS are attempts to add increasing rigor to erosion prediction.

Empirical approaches are only as good as the underlying data. That is the reason for the large number of locations used in the development of USLE. Such approaches are only as good as the crops evaluated, and they are only as good as the conservation practices evaluated. As cultural practices change, research must continue. This is a major disadvantage of the empirical approach. The engineer often must make judgment calls about what one can reasonably extrapolate to new and current conditions. Empirical approaches such as these provide order-of-magnitude answers. The more highly developed the empirical procedure, the better it is for anticipating the impacts of changes on key variables. Clients often seek help in anticipating the effects of change. Thus, the USLE and its children, along with the wind erosion equation and the various runoff equations, though empirical, are still highly valuable tools for the environmental conservation engineer.

Problems and Questions

In cases below where you are requested to suggest ways of reducing erosion rates, be sure to provide pragmatic, economically feasible alternatives as much as possible. Soils and rainfall (except in the case of construction timing) are largely invariant at a location; slope lengths, cropping, or cover factors and practice factors may be altered with modified farming or construction practices.

1. If the soil loss at Cincinnati, Ohio, for a given set of conditions is 11.2 Mg/ha, what is the expected loss in Athens, Georgia, assuming the same topography, field shape, soil conditions, cropping, and cultural practice conditions apply?
2. Two 60 m long slopes, one with a 2% slope and one with a 10% slope, are located on the same farm. Rainfall energy, soil, cropping rotations, and conservation practices are constant. What is the relative difference in water erosion on the slopes?
3. Determine the average annual soil loss for a field in Athens, Georgia, if $K_{sl} = 0.014$, $l = 91$ m, $S = 10\%$, and upslope-downslope farming of continuous corn is practiced. Suggest conservation practice(s) to reduce the average annual soil loss to less than 7 Mg/ha. Prove the efficacy of the suggested practice(s) by RUSLE-type calculations.
4. Calculate the average annual soil loss for a 500 ac housing development in Athens, Georgia, when
   (a) constructing in spring
   (b) constructing in summer
   (c) constructing in fall

   Assume the entire 500 ac will be exposed as bare soil once construction begins. The site is roughly square with one boundary formed by a stream. In each case, evaluate over a 3 yr period. The construction site was initially a permanent pasture. Sites are to be landscaped with fescue grass. The development sits on a 1% slope on a Cecil sandy loam soil with a $K_{sl} = 0.015$ (SI units). When is the best time to build in terms of minimizing soil loss?
5. Given a corn-soybean-meadow rotation in Lexington, Kentucky, compute the estimate of C_{prac} by computing the weighted $R_{rf}C_{prac}$ factor for this crop rotation, where corn and soybeans are grown using conventional tillage techniques. Next, compute the results when corn is grown using conservation tillage. What is the percentage reduction in C_{prac}?

6. Given a corn-peanut-meadow rotation on a 10% slope, 100 m slope length, on a Tifton sandy loam soil (sand 50%, silt 30%, clay 20%, om 1%, very permeable, granular structure), in Tifton, Georgia, suggest and prove by USLE calculation methods a practice/crop rotation combination that will bring average annual erosion within 4.7 Mg/ha.
7. Compute the average annual soil loss on a continuous corn rotation (conventional tillage, contour, terraced) with 10% slope, 120 m slope length, on a Tifton sandy loam (sand 50%, silt 30%, clay 20%, very permeable, granular structure), at Tifton, Georgia. Repeat using a no-till cultural practice on contour. What is the difference between conventional tillage-terraces and the no-till practice?
8. Given the storm hyetograph in Table 6.6, compute the $R_{rf\,storm}$ value.
9. Suppose a 100 m forested riparian strip between a stream and a 25 ha square field having an average annual soil loss of 14 Mg/ha and sloped at 18% from the strip inlet to strip outlet. Compute the sediment delivery ratio for the buffer using the U.S. Forest Service method. Assume the equivalent of average annual sediment is produced by a storm having a peak excess rainfall rate of 4 in/hr. The slope is convex. The surface of the producing field is moderately smooth. Silt plus clay percentage is 50%. Ground cover is 80%.
10. Given the cumulative rainfall distributions in Table 6.7, compare the average annual erosion of the rotation given in the text for north Georgia with an identical rotation in one (you select) of the other zones indicated above.
11. Compute the average annual erosion for a corn-corn-meadow rotation given a slope of 7%, sand = 30%, silt = 20%, clay = 50%, om = 2%, blocky structure, moderately impermeable. The location is central Ohio. The 10 ha field is square, with runoff moving uniformly off one edge. Conventional tillage on the contour is used. There are no terraces.
    (a) If the tolerance is 6 Mg/ha, how can the soil loss situation be brought under control?
    (b) Repeat the calculation, including terraces at 120 ft intervals. What is the effect of terracing on the average annual soil loss?
    (c) Assuming a riparian buffer of 30 m, estimate the SDR assuming no practices other than contouring are used. Assume a peak excess rain of 20 mm/hr; riparian buffer has medium roughness, uniform slope, and same slope as the eroding field.

Table 6.6. Storm hyetograph for Problem 8.

Hours elapsed	Rainfall(mm)
0	0
0.5	5
1	15
1.25	30
1.5	50
1.7	60
1.9	65
2.	66
3	66

Table 6.7. Cumulative rainfall distribution for problem 10.

Month	North Georgia	North California	South Georgia	East Kentucky	Corn Belt
0	0	0	0	0	0
1	6	18.5	4	3	2
2	12	30.2	8	7	4
3	20	38.9	16	13	8
4	28	42.9	25	19	14
5	38	45.2	35	27	25
6	50	50.8	47	44	45
7	69	52.5	67	63	64
8	80	57.4	81	80	79
9	87	60.1	87	89	89
10	92	69.6	91	93	95
11	96	85.4	95	96	98
12	100	100	100	100	100

12. Suggest a strategy for determining the 50 yr average annual USLE R_{rf} value for Athens, Georgia. Do the same for a 50 yr single-storm R_{rf} value.
13. Define all the components of the USLE.
14. How may one relate the rainfall factor R_{rf} in the USLE to energy in a given rainfall assuming a known mass curve?
15. What are five strategies for reducing average annual soil loss for agriculture? For construction?
16. Explain why one must compute a weighted $R_{rf}C_{prac}$ factor when computing average annual erosion.
17. What are four factors contributing to the fact that the sediment delivered at a point is less than total erosion in the drainage area?
18. Define *sediment delivery ratio.*
19. Given the following rainfall mass curve (Figure 6.26), (a) compute the single-storm R_{rf} value. (b) How many liters of diesel fuel are equivalent to the energy received

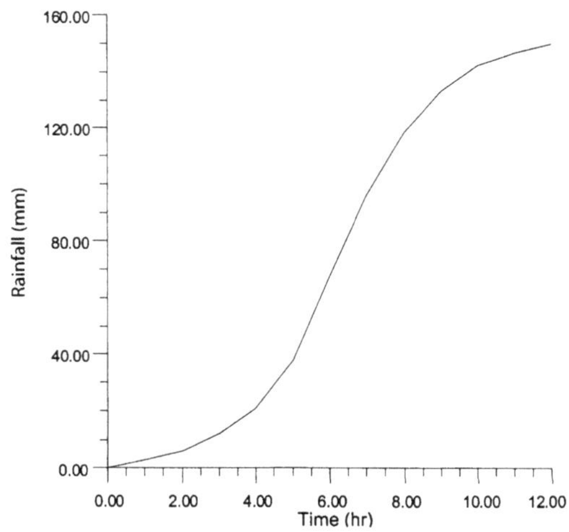

Figure 6.26. Mass curve for R determination.

from the rainfall onto 1 ha, assuming a fuel efficiency of 20%? Fuel efficiency is the energy converted into mechanical work divided by the chemical energy in the fuel.

20. Given the rainfall of Figure 6.26, compute the single-storm erosion from this storm on a field with a slope length of 100 m, gradient of 8%, contouring, no terraces, and no strip-cropping. The soil is a silty loam with 30% sand, 37% silt, 30% clay, and 3% om. It has a blocky structure and is moderately permeable. The site is freshly plowed at the time of the storm. What is the erosion from the same site with permanent pasture?
21. Compare the average annual soil loss with continuous beans and continuous corn in western Kentucky. The area is planted with a small-grain cover crop plowed under in the spring. Assume the land slope is 5%, the 5 ha field is square with one side along a creek, the soil is a moderately well structured soil with mineral fraction being 50% sand, 20% silt, and 30% clay. The organic fraction is 2%. There are no terraces or strip-crops. The land is planted and tilled on the contour. What if a rotation with corn in year 1, cotton in year 2, and meadow in year 3 were employed? Repeat assuming conservation tillage for both row crops is used instead of conventional tillage. What is the impact of conservation tillage?
22. A 1 ha site with a slope of 20% has been cleared for development. A thunderstorm with a single-storm R_{rf} of 3000 moves through. Estimate the single-storm soil loss given that the soil is poorly structured and its mineral fraction is 40% sand, 20% silt, and 40% clay. The organic fraction is 0.5%. Assume no conservation practices and a P_{prac} factor of 1.
23. Your client proposes a corn, cotton, and peanut rotation for his farm. Compute the average annual C_{prac} value. Meadow for 1 yr could be added if needed. Suggest a combination of conservation practices leading to a C_{prac} value of 0.25 or less.

Notes

1. The m_{er} factor relationship follows from algebraic manipulations in Renard et al. 1997. The m_{er} factor is related to the ratio of rill to interrill erosion predicted by slope steepness. Renard et al. (1997) present a table to adjust the equation 6.5 value upward if the soil is susceptible to rill erosion and downward if the soil is not susceptible to rill erosion. Runoff (e.g., slope length), cover, soil type and condition, and management impact m_{er}. One should probably increase the calculated m_{er} by 20% under construction conditions.
2. Available from Civil Software Design, P.O. Box 706, Ames, IA 50010. Mention of this product is solely for the convenience of the reader and is not to be construed as an endorsement.

References

Agassi, M. 1996. *Soil erosion, conservation, and rehabilitation.* Marcel Dekker, New York, NY.

Bagarello, V., and V. Ferro. 1999. Assessment of soil erosion. In *CIGR handbook of agricultural engineering: Land and water engineering,* ed. H. N. van Lier, L. S. Pereira, and F. R. Steiner. American Society of Agricultural Engineers, St. Joseph, MI.

Barfield, B. J., I. D. Moore, and R. G. Williams. 1979. Sediment yield in surface mined watersheds. In *Proceedings, Symposium on surface mine hydrology, sedimentology, and reclamation.* University of Kentucky, Lexington.

Follett, R. F., and B. A. Stewart. 1985. *Soil erosion and productivity*. American Society of Agronomy, Madison, WI.

Foster, G. R. 1988. User requirements, USDA water erosion prediction project (WEPP). USDA-ARS National Soil Erosion Laboratory, West Lafayette, IN.

Foster, G. R., D. K. McCool, K. G. Renard, and W. C. Moldenhauer. 1981. Conversion of the universal soil loss equation to SI metric units. *Journal of Soil and Water Conservation* 36:355–59.

Goldman, S. J., K. Jackson, and T. A. Brusztynsky. 1984. *Erosion and sediment control handbook*. McGraw-Hill, New York, NY.

Haan, C. T., J. C. Hayes, and B. J. Barfield. 1994. *Hydrology and sedimentology of small catchments*. Academic Press, New York, NY.

Hudson, N. 1971. *Soil conservation*. Cornell University Press, Ithaca, NY.

Knisel, W. G. 1980. CREAMS: A field scale model for chemicals, runoff, and erosion from agricultural management systems. Conservation Research Report 24. U.S. Department of Agriculture, Washington, DC.

Lal, R. 1994. Soil erosion by wind and water: Problems and prospects. In *Soil erosion research methods*, 2d ed., ed. R. Lal. Soil and Water Conservation Society, Ankeny, IA.

NRCS. 1998. *National Resource Conservation Service national resources inventory, 1997 update* (http://www.nrcs.usda.gov). Washington, DC.

Onstad, C. A., and G. R. Foster. 1975. Erosion modeling on a watershed. *Transactions of the ASAE* 18:288–92.

Renard, K. G., and G. R. Foster. 1995. *RUSLE users guide (version 1.04)*. Soil and Water Conservation Society, Ankeny, IA.

Renard, K. G., G. R. Foster, G. A. Weesies, and J. P. Porter. 1991. RUSLE revised universal soil loss equation. *Journal of Soil and Water Conservation* 46:30–33.

Renard, K. G., L. J. Lane, G. R. Foster, and J. M. Laflen. 1994. Soil loss estimation. In *Soil erosion, conservation, and rehabilitation*, ed. M. Agassi. Marcel Dekker, New York, NY.

Renard, K. G., G. R. Foster, G. A. Weesies, D. K. McCool, and D. C. Yoder. 1997. Predicting soil erosion by water: A guide to conservation planning with the revised universal soil loss equation (RUSLE). USDA-ARS Agriculture Handbook 703. U.S. Department of Agriculture, Washington, DC.

Renfro, G. W. 1975. Use of erosion equations and sediment delivery ratios for predicting sediment yield. In *Present and prospective technology for predicting sediment yields and sources*. USDA-ARS Publication ARS-S40. Agricultural Research Service, U.S. Department of Agriculture, Washington, DC.

Schiechtl, H. 1980. *Bioengineering for land reclamation and conservation*. University of Alberta Press, Edmonton.

Schmidt, B. L., R. R. Allmaras, J. V. Mannering, and R. I. Papendick. 1982. *Determinants of soil loss tolerance*. American Society of Agronomy Special Publication no. 45. American Society of Agronomy, Madison, WI.

Schwab, G. O., D. D. Fangmeier, W. J. Elliot, and R. K. Frevert. 1993. *Soil and water conservation engineering*. 4th ed. John Wiley & Sons, New York, NY.

Smith, J. H. 1993. Precipitation. In *Handbook of hydrology*, ed. D. R. Maidment. McGraw-Hill, New York, NY.

Toy, T. J., G. R. Foster, and J. R. Galetovic. 1998. *Guidelines for the use of the revised universal soil loss equation (RUSLE) on mined lands, construction sites, and reclaimed lands (version 1.06)*. Office of Surface Mining, Western Regional Coordinating Center, Denver, CO.

U.S. Forest Service. 1980. An approach to water resources evaluation of nonpoint silvicultural sources (a procedural handbook). EPA 600/8-80-012. Environmental Protection Agency, Washington, DC.

Warner, R. C. 1998. SedCad[7] 4 for Windows95/NT: Notes for 5–7 Aug. 1998 short course. Department of Biosystems and Agricultural Engineering, University of Kentucky, Lexington.

Williams, J. R. 1974. Sediment yield prediction with universal equation using runoff energy factor. In *Present and prospective technology for predicting sediment yields and sources*. USDA-ARS Publication ARS-S-40. Agricultural Research Service, U.S. Department of Agriculture, Washington, DC.

———. 1975. Sediment routing for agricultural watersheds. *Water Resources Bulletin* 11(5):965–74.

Wischmeier, W. H., and D. D. Smith. 1978. Predicting rainfall erosion losses: A guide to conservation planning. USDA Handbook 537. Government Printing Office, Washington, DC.

7

Water Quality and Upland Nonpoint Pollution

Water must be suitable for its intended use, which may be drinking, crop irrigation, bioremediation applications, support of aquatic life, or other uses. Surface water impairment increases the cost of treating water for municipal use and decreases the usefulness of water for agriculture and for recreation. In 1988, up to 65% of the streams in the United States fell short of water quality standards for fishing and recreation. Worldwide, water quality status is as bad as or worse than erosion status. The United States and the European Union are the leaders in raising public awareness of the importance of sustainable water quality. Water quality is problematic around the world (Greppi and Preti 1999).

The USEPA mandate for states to establish total maximum daily load (TMDL) values for streams continues to drive pollution reduction engineering activities. The TMDL must account for point and nonpoint pollution sources and include a safety factor. Comprehensive nutrient management plans (CNMPs) are becoming required for large agricultural operations as a strategy to reduce nonpoint source pollution, a major factor contributing to stream pollution. The USEPA is also requiring management of nonpoint source discharge in urban areas. The goal is to reduce nonpoint pollution from industries and municipalities. Upland nonpoint sources of agricultural and urban pollutants are major sources of downstream water quality problems. Management of upland, nonpoint pollution is the focus of this chapter. Best management practices (BMPs) and CNMPs are the primary strategies for improving upland and stream water quality. How to identify and select BMPs and understanding how BMPs preserve and enhance water quality are major objectives of this chapter. Chapter 15 further discusses nonpoint pollution as related to permanent streams.

Agricultural crop and animal production are major nonpoint pollution generators (see Figures 7.1 and 7.2). One may apply nutrients and handle waste incorrectly or appropriately, depending on the knowledge and care of the operator. Urbanizing activity also contributes to water quality degradation. Forestry operations cause nonpoint pollution during harvest and replanting episodes. Analyses of water quality problems require multidisciplinary expertise. Agricultural, civil, and biological engineers are in unique positions to lead teams addressing water quality problems.

Erosion reduction is a logical starting point for water quality improvement. The focus in this chapter is on water quality determination and management in uplands. The text focuses on uplands where streams are intermittent or ephemeral. Some pollutant concentrations attenuate naturally. Others become enriched. Mechanisms for natural attenuation

Figure 7.1. Poultry (broiler) litter application to a pasture using a spreader truck. Nutrient delivery trucks are currently being upgraded to enable variable-rate application controlled by field nutrient maps and satellite positioning systems. (Courtesy of Georgia Soil and Water Conservation Commission.)

and enrichment and strategies for minimizing pollutant transport are the focus. Chapter 15 carries the discussion forward to the notion of stream TMDLs, which necessarily involves permanently flowing streams.

Goals

- Identify various pollution indicators and learn definitions of various types of pollution.
- Recognize the processes that affect water quality on a large scale, mainly nonpoint pollution in upland areas.
- Visualize how overlapping scale effects, nonlinearities, and delayed responses create difficulty in determining precise cause and effects with nonpoint source pollution.

Figure 7.2. Cattle on bare ground can cause extensive nonpoint or point pollution.

- Know how BMPs work by increasing opportunity time for physical settling, ultraviolet breakdown of some organics, volatilization of some compounds, and cycling of carbon, nitrogen, and sulfur.
- Know how to utilize BMP options while avoiding such problems as transferring pollution from water to soil (e.g., by precipitating heavy-metal compounds, which will require expensive remediation in the future.

Important Concepts and Selected Pollution Measurement Techniques

Figure 7.3 summarizes the relationships between the water body and physical, chemical, biologic, and energy variables. The following terms appear often in the literature on water quality.

Concepts

Biogeochemical cycles: Natural progressions of solar-driven transformations of nutrients such as carbon, nitrogen, and phosphorus. Figures 7.4a–c depict cyclical constituent transformations among the carbon, nitrogen, and phosphorus nutrient pools. These transformations enable sustainable loading rates equaling rates of volatilization. Phosphorus and heavy metals, however, do not cycle and remain in the soil for long periods. Residues and other biological wastes commonly contain much carbon, nitrogen, sulfur, and phosphorus. They may or may not contain heavy metals or other industrial organics. Many industrial organics are slowly degradable. The rates of transformation depend on the makeup of the waste and environmental conditions. Ecology texts such as Odum 1983 discuss additional cycles.

BMP: A best management practice is a management strategy proven to reduce pollution impacts. BMPs can be managerial or structural and are site specific.

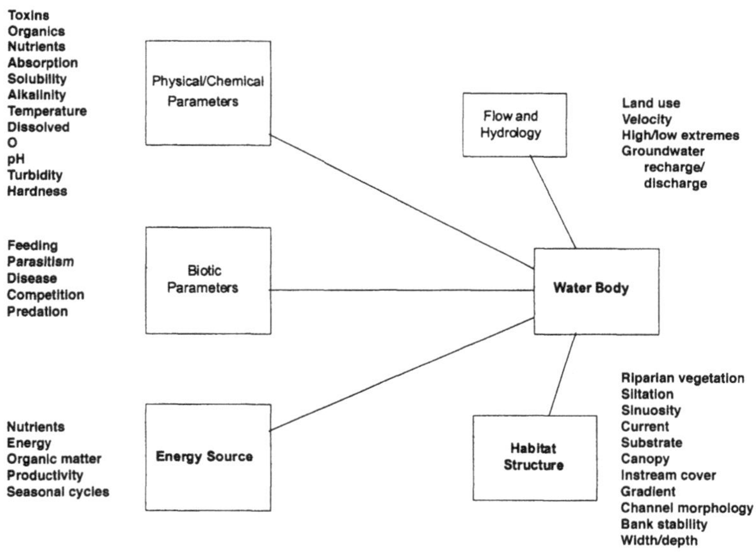

Figure 7.3. Relationships between the water body and energetic, biotic, physical, and chemical variables. This chart shows the all-encompassing scale associated with nonpoint source pollution management. (Adapted from Novotny 1995.)

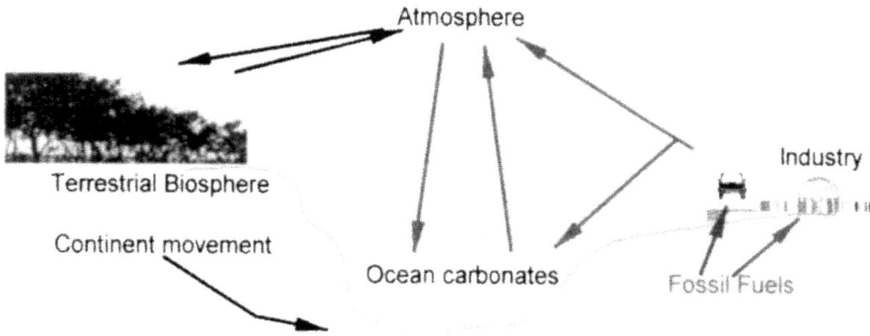

Figure 7.4a. The carbon cycle, showing natural cycling between the atmosphere and the flora and fauna on the left and cycling between the atmosphere and earth due to humankind's activities on the right. The ocean is the largest sink, followed by fossil fuels and plants. (After Odum 1983.)

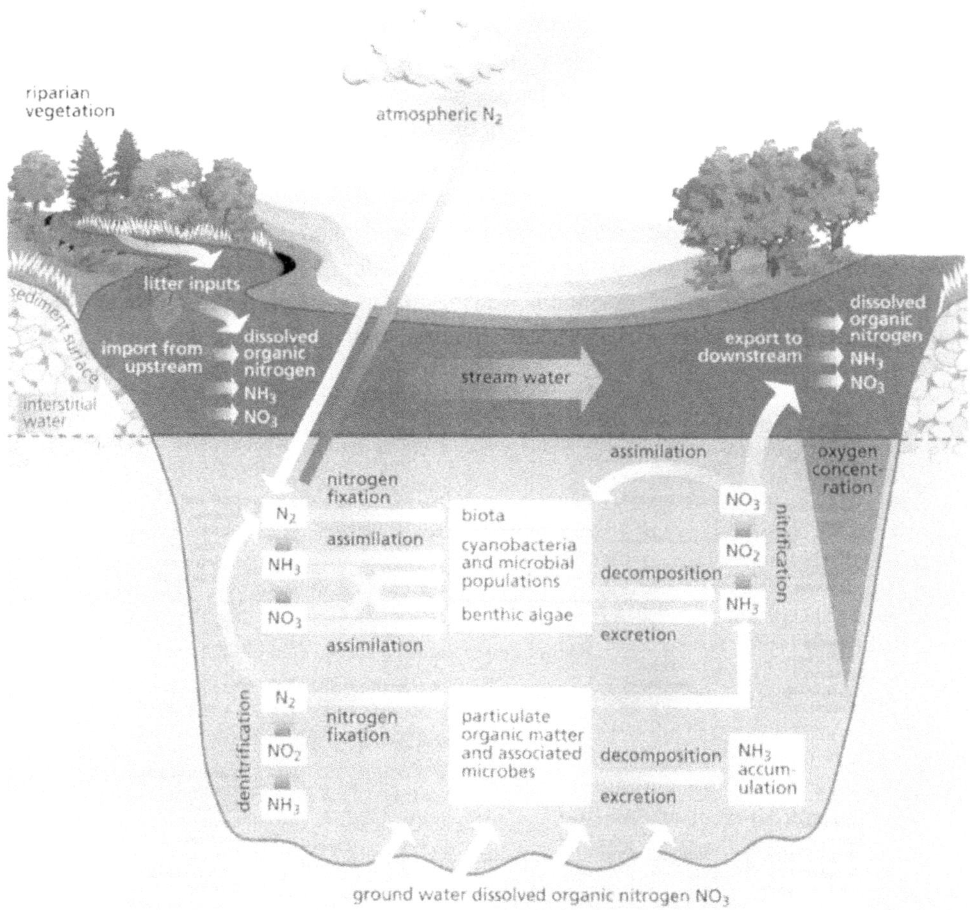

Figure 7.4b. Nitrogen cycle. (Courtesy of NRCS.)

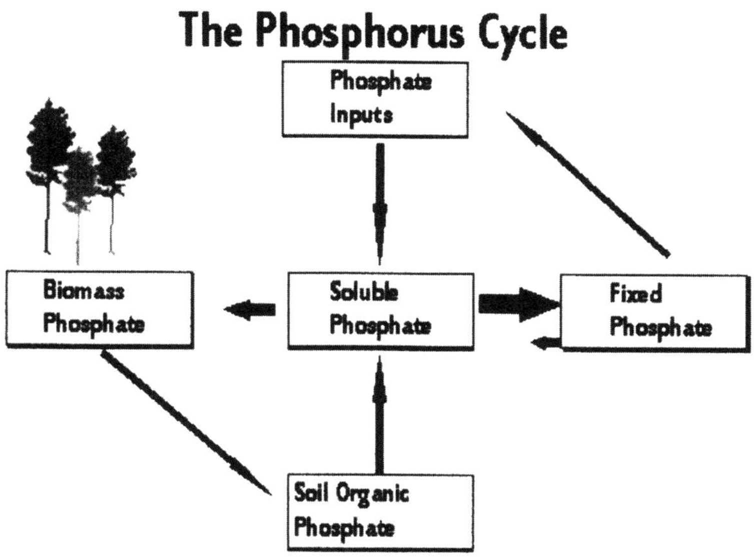

Figure 7.4c. Phosphorus cycle. (Courtesy of Richard Lewis, Brigham Young University.)

CNMP: A comprehensive nutrient management plan is required for larger agricultural operations (requirements vary with state and enterprise type). A joint task force of the USEPA and USDA-NRCS has mandated CNMPs for animal production operations with more than 300 animal units.

CWA: The Clean Water Act is administered by the USEPA and state counterparts.

Heavy metals: These are chemical constituents that remain absorbed in soils for long periods. They may or may not be trace nutrients.

NPDES: The National Pollutant Discharge Elimination System is a regulatory framework for mandating and enforcing point source water quality standards. It is managed by the USEPA and state counterparts.

Pollution: Pollution is a human-induced or naturally occurring alteration of the physical, chemical, biological, or radiological integrity of water. Sediment is an example of physical pollution. Salinity, an excess of dissolved monovalent and divalent cations (some of which may be nutrients), is an example of chemical pollution. Microorganisms are an example of biological pollution. Radionuclides lead to radiological pollution. In other words, *pollution is the presence of substances that interfere with beneficial uses of the water.*

Point source pollution: This type of pollution arises from discrete locations and thus allows easy characterization and correction; it causes acute and chronic problems. Municipal and industrial wastewater effluents are traditional examples. Large feedlots are point sources. Point source pollution is most readily treated by structures and processes beyond the scope of this text.

Nonpoint source pollution: This form of pollution has one or more of the following defining characteristics:

- Meteorological events cause pollution transport from diffuse areas.
- Generation occurs over extensive land areas before the pollution reaches surface waters or shallow aquifers.
- Use of management practices is more practical than treatment with various structures.
- An identifiable source is difficult to pinpoint.

- Suspended solids, nutrients, and toxic compounds are usually the greatest constituents.
- Compliance monitoring occurs on land rather than water.
- Chronic health symptoms as opposed to acute symptoms occur.

TMDL: The total maximum daily load is the maximum allowable point and nonpoint source loading within a designated stream reach, including a safety factor. It is the maximum amount of a pollutant that a water body can receive and still meet water quality standards. The standard is enforced by state and federal regulatory agencies.

Volatilization: This process is the change from the liquid to the gas phase.

Physical Measures of Pollution

Ash: Total solids remaining at 550°C; considered the inorganic portion of total solids; also called fixed solids; reported in ppm or mg/l.

Color: Visible appearance; an example is color caused by dissolved tannin compounds in water.

Filterable solids: Solids that can be filtered from a sample as opposed to dissolved solids; reported in mg/l. One may refer to filterable solids as settleable or suspended solids. Physical, as opposed to chemical, processes remove filterable solids.

Odor: Smells caused by volatile chemicals and small particulates.

TDS: Total dissolved solids; solids left after filterable solids have been removed; reported in ppm or mg/l.

TS: Total solids; residues left on evaporation at 105°C; a measure of the total solids in the sample. Total solids represent organic plus inorganic pollution; reported in ppm or mg/l.

Turbidity: Total solids as measured using Jackson turbidity units (JTU) or Secchi disks. Water treatment plants and limnologists report total solids in turbidity units. No strong correlation exists between total solids expressed in mass units or total solids expressed in turbidity units because optical properties of sediments at equivalent mass concentrations vary widely.

VS: Volatile solids; total solids volatilizing between 105°C and 550°C; the organic portion of TS. VS correlates with organic pollution; reported in mg/l.

Chemical Measures of Pollution

COD: Chemical oxygen demand; oxygen used to oxidize carbon and possibly other substrates under a simulated or actual hot reflux digestion. COD is a measure of easily degradable plus slowly degradable organic pollution and often correlates with VS; reported in mg/l.

DO: Dissolved oxygen in a sample measured chemically or electronically; reported in mg/l.

EC: Electrical conductivity; a measure of the dissolved anions and cations in solution. EC measures salinity; reported in decisiemens/m.

pH: The logarithm of the inverse of the hydrogen ion concentration; 7 is neutral, less than 7 indicates an acid condition, and greater than 7 indicates an alkaline condition; measured electronically with a glass electrode.

Salinity: The presence of divalent and/or monovalent cations and anions in solution; reported in meq/l.

TKN: Total Kjeldahl nitrogen; a measure of total organically bound and free ammonia-nitrogen (NH_4^+) contained in a sample; reported in mg/l.

Biological Measures of Pollution

BOD: Biochemical oxygen demand; oxygen used by microorganisms while metabolizing dissolved carbon and possibly other substrates in a controlled-temperature environment; measures easily degraded organic carbon or that treatable by microbial processes. The BOD test may extend for 5 or more days. The standard 5-day BOD is customarily reported as BOD_5 (mg/l). BOD is always less than or equal to COD. COD exceeds BOD since COD conditions usually oxidize nearly all the biodegradable carbon.

Coliforms: E. coli and related microorganisms as detected using specialized growth media. Coliforms indicate possible contamination by fecal material from warm-blooded animals. This indicates the possible presence of pathogenic organisms. Fecal streptococcus presence is an additional, somewhat more robust, indicator of animal waste presence.

Indicator species: Coliforms, streptococcus organisms, and various bacteriophages find application as indicators of the presence of human and animal waste. The presence of abundant daphnia (water fleas), mayflies, stone flies, caddis flies, minnows, and brine shrimp and other filter-feeding crustaceans indicates satisfactory biological health of streams while their absence may indicate ecological problems. Chapter 15 further discusses stream health.

Texts such as Tchobanoglous and Schroeder 1985 and Sawyer et al. 1994 give additional measurements of physical, chemical, and biological water status. Water quality sampling is similar to sampling for sediment, with two important differences. One must properly preserve chemical and biological samples, and one must ensure that the collection apparatus does not contaminate the sample. Stainless steel or glass is the preferred material for apparatus construction. Samples for microbial analyses (e.g., coliforms) require proper sterilization to avoid contamination. Clesceri et al. (1998) give many analytical methods for sampling and quantifying physical, chemical, and biological pollution.

WATER QUALITY BACKGROUND

Proficiency in solving water quality problems requires familiarity with basic inorganic and organic chemistry, assumed herein. An understanding of biology, microbiology, and ecology is also necessary. Many engineers will not have sufficient background to solve these problems; however, they may expect to work in teams that collectively possess the expertise. Thus, familiarity with nomenclature used by these professionals is helpful.

Ideally, all water in the environment would match standards and criteria for drinking water. Table 7.1 shows selected standards for drinking water as of 1999. Regulatory agencies developed drinking water and surface water standards in response to the Clean Water Act. Chapter 14 discusses standards for irrigation waters. Not all natural sources of water are in total compliance with standards in Table 7.1. Tchobanoglous and Schroeder (1985) give a much more detailed listing of pollutant maximums. The criteria in Table 7.1 are sometimes hard to satisfy. For example, widespread pesticide use in an area virtually ensures that eventually pesticide limits will be exceeded. Regions containing widespread

Table 7.1. Selected USEPA 1999 drinking water standards.

Type of contaminant	Name of contaminant	Maximum contaminant level
Inorganic chemicals	Arsenic	0.05 mg/l
	Fluoride	4.0 mg/l
	Lead	0.015 mg/l
	Copper	1.3 mg/l
Disinfection by-products	THM	0.10 mg/l
Turbidity	Turbidity	Not more than 1 NTU as monthly average or not more than 5 NTU as average of 2 consecutive days
Microorganisms	Total coliform	< 40 samples/month with no more than 1 positive ≥ 40 samples/month with no more than 5% positive
	Giardia lamblia	3 log (99.9%) removal
	Enteric viruses	4 log (99.99%) removal
Radionuclides		
Natural	Adjusted gross alpha emitters	15 pCi/l
	Combined Radium 226 & Radium 228	5 pCi/l
	Radon	300 pCi/l
	Uranium	20 ugm/l
Man-made	Beta particles and photon emitters	4 mrems/yr (applies only to systems determined by the government to be vulnerable and surface water systems serving >100,000)
Secondary elements	Aluminum	0.05–0.2 mg/l
	Chloride	250 mg/l
	Color	15 color units (CU)
	Flouride	2 mg/l
	Foaming agents	0.05 mg/l
	Iron	0.3 mg/l
	Manganese	0.05 mg/l
	Silver	0.1 mg/l
	Sulfate	250 mg/l
	Total dissolved solids (TDS)	500 mg/l
	Zinc	5 mg/l

animal or poultry production operations sometimes exceed nutrient limits. The diffuse, as opposed to concentrated, runoff from many farm-scale situations makes treatment difficult. Natural rainfall, geology, and humankind's activities all affect water quality. As discussed in Chapter 2, one may express water quality using measurable quantities and narrative statements.

The emphasis on water quality reflects the reality that freshwater supplies are limited. The regulatory environment for animal agriculture is becoming more stringent. This arises in part because many rivers and streams have little additional carrying capacity for pollutant loads. Each intended use has associated parameters expressing water quality. The overall goal is to maintain the integrity of water for maximum future usefulness.

Historical environmental approaches began evolving from positions approximated by "assume all naturally occurring water is suitable for any use." The current position of "assume water bodies renovate themselves" and "we can economically treat the water at the point of withdrawal" is also becoming obsolete. Water bodies do in general "renovate themselves"; however, the self-renovation requires time and depends on regional loading. Many water bodies have little additional capacity to carry more pollution and continued development is mandating pollution limits. The standard is becoming "return the discharge and surface runoff with the best possible quality or least pollutant load." With respect to diffuse runoff sources from farm-scale operations, the control of incoming pollution is a challenge.

The major federal legislation designed to protect the nation's waterways is the 1972 Clean Water Act (CWA), intended to address problems with point pollution (Segars 1997). Although the CWA directly addresses point pollution, the provisions are being indirectly applied to nonpoint source pollution via requirements for municipalities to demonstrate the lack of negative consequences of expanding existing point source permitted flows. Pollution problems in the Chesapeake Bay motivated Congress to amend the CWA in 1987 to include nonpoint pollution. Interpretation of the provisions of the act is becoming more stringent. Regulators are moving to establish water quality standards for all waters. Vigorous enforcement of stream water quality provisions on historically noncompliant streams is a regulatory goal.

The 1996 Farm Bill required development of a Certified Conservation Compliance Plan to protect water quality and prevent erosion on highly erodible lands. This program protects about 98% of all erodible lands. The Coastal Zone Management Act of 1990 requires farmers and others to carry out EPA-approved BMPs in all coastal zone areas of the 34 coastal states and territories. Agriculture has been the major contributor to nonpoint pollution. EPA regulations categorize animal production into Animal Feeding Operations and Concentrated Animal Feeding Operations. An Animal Feeding Operation by definition stables, confines, and feeds or otherwise maintains animals for a total of 45 days or more a year. A Concentrated Animal Feeding Operation (CAFO) is an Animal Feeding Operation that (1) has more than 1000 animal units or (2) has between 301 and 1000 animal units discharging wastes into streams or (3) has been designated a CAFO based on particular circumstances. Facilities discharging into streams must have a discharge permit under recent provisions of the CWA NPDES. One computes animal units by multiplying the number of animals of each species by an appropriate multiplier. Multipliers for animal unit computation are as follows: feeder cattle, 1.0; dairy cattle, 1.4; swine weighing over 25 kg, 0.4; sheep, 0.1; horses, 2.0; poultry, not specified, other factors usually dictate. One must be able to contain and reuse runoff from 25 yr–24 hr storms on site; otherwise, a point source discharge permit is required.

Point source discharge permits are harder to obtain for any animal operation. Implementation of specific regulations varies state by state. Federal EPA and various state environmental regulations are multiplying.

Crop production is also a major source of nonpoint pollution. The Federal Insecticide, Fungicide, and Rodenticide Act (FIFRA), passed by Congress in 1947 and amended in 1972, 1975, 1978, and 1988, requires registration by the EPA, mandates restricted-use labeling, and requires all applications be done with trained and certified applicators (see *Apply Pesticides Correctly* 1993). Meeting the FIFRA requirements does not always guarantee zero pollution. Congress amended the CWA in 1987 by adding section 319, which addresses nonpoint pollution. Section 319 mandates the development of BMPs for nutrient controls. Farm service agencies indirectly regulate nutrient application by requiring nutrient management and erosion control plans before federal assistance can be received.

Congress established the Comprehensive Environmental Response, Compensation, and Liability Act (CERCLA) in 1980, amended in 1986, commonly called the Superfund, to address these controversial problems. Accidents resulting in leaks may in effect create a nonpoint source by diffusing pollutants.

The USEPA and state counterpart agencies regulate pollution associated with industrial activity through the CWA NPDES. CWA provisions impacting nonpoint pollution have begun to be implemented by presidential mandate to clean up rivers and streams by 2008. Nonpoint pollution may be due to past practices or to inappropriate point source permit expansions. Storm water runoff associated with various industries may also contribute to nonpoint pollution by diffusing pollutants. The EPA has promulgated detailed regulations to attempt to minimize storm water impacts (Dodson 1999). Nyer et al. (1996) have evaluated several strategies for recovering some organic solvents and other water-soluble pollutants (see Figure 1.8).

Storm water regulations for new development now require that

- one cannot divert water from a watershed,
- one cannot increase the runoff from a site, and
- one cannot concentrate runoff from a site.

The USEPA and USDA-NRCS, following the presidential mandate to clean up U.S. rivers and streams, developed the CNMP approach for agricultural operations with more than 300 animal units. State agencies now require certification of farm operators for animal production. Requirements to have certified individuals develop CNMPs for each animal production farm (see Figure 1.9 for an example of information required in the planning process) now exist. The CNMP represents a detailed management plan for a farm unit. In many regions, these farms also must have land application system permits (LAS) for spray field operation. Others must now have NPDES permits for discharge. Storm water discharge from municipalities and industries is coming under the regulatory umbrella. Agricultural and urban nonpoint pollution regulation and minimization are integral parts of the strategy to enable municipal growth without impairing receiving waters.

The above survey barely skims the surface of environmental legislation. Bookchin and Farnsworth (1997) give an excellent summary of pollution control laws for air, water, and soil, with emphasis on urban activities. Environmental legislation and regulation are rapidly moving targets. One should consult with regulatory personnel to keep abreast of current regulations.

Scale Effects

Based on the diffuse character, intricate chemistry, and complex cycling of some nonpoint pollution constituents, pollution impact is not immediate. This means pollution is often recognized well after the fact. In other words, we are "in the soup" after the "soup is hot and getting hotter." Flow and hydrology, biotic forms, chemical properties, energy source, and habitat structure affect water quality. A suite of measurements documents each of these mechanism categories.

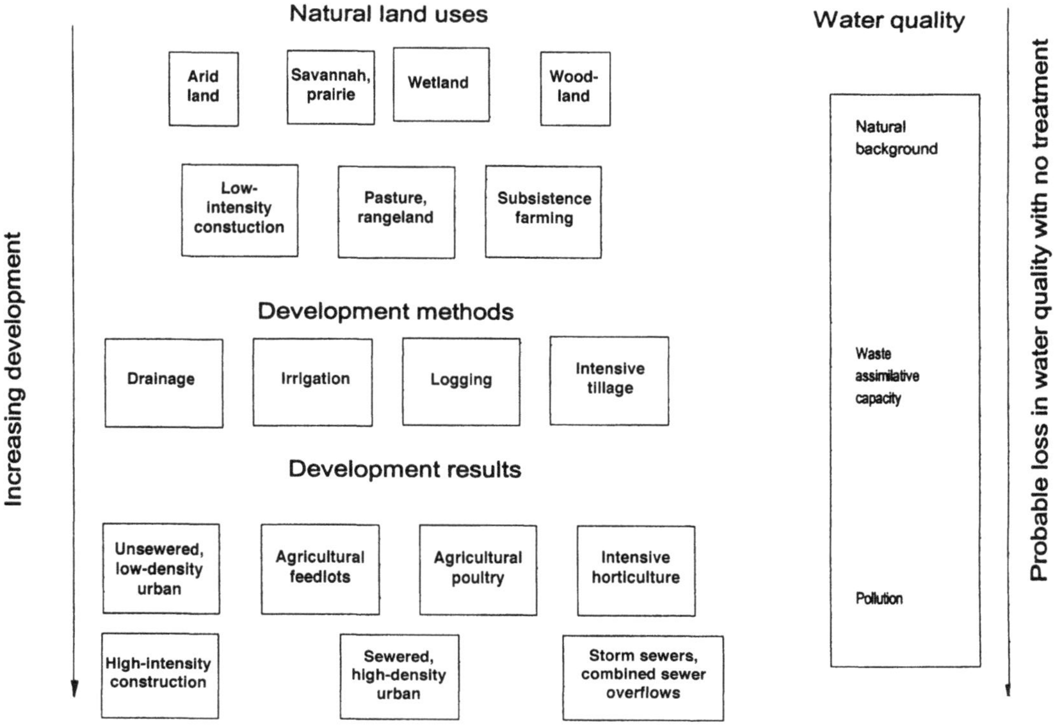

Figure 7.5. Developmental activities and land use transitions influencing water quality via nonpoint and point pollution. (Modified from Novotny 1995.)

Considering the various temporal and spatial scales at work in the environment, along with the multitude of processes associated with a given waste application, one can appreciate the difficulties of discovering cause-and-effect relationships. Various volatilizations decouple effects from their causes. Similarly, dilutions also have a decoupling effect. It can be difficult to identify the original pollutant and its source. Change in societal values over time and space further complicates analyses of cause and effect.

Figure 7.5 shows land transition impacts on pollution. The spatial and temporal characteristics and the variety of processes establish the nonpoint nature. The four blocks at the top of Figure 7.5 represent original, "natural" land uses. The other blocks represent land uses with varying degrees of managed activities. The multitude of nonpoint pollution sources increases with intensity of management. Development will exceed the natural assimilative capacity of water bodies, including even the oceans, if treatment is not adequate.

Development activity above a water body usually decreases travel time of surface and groundwater to the water body. This gives less time for biochemical cycles to operate. Development usually increases creation of sediment and reduces retention of sediment. Thus, the water body must assimilate more pollution. The assimilative capacity of water bodies important to urban centers is often the controversial crux of land use debates. Sometimes the assimilative capacity becomes overwhelmed, rendering the water body unusable for the intended purpose.

Nonpoint pollution assessment is complex, given the varying spatial and temporal scales and the variety of simultaneous processes. Figures 7.6 and 7.7 give approximate spatial

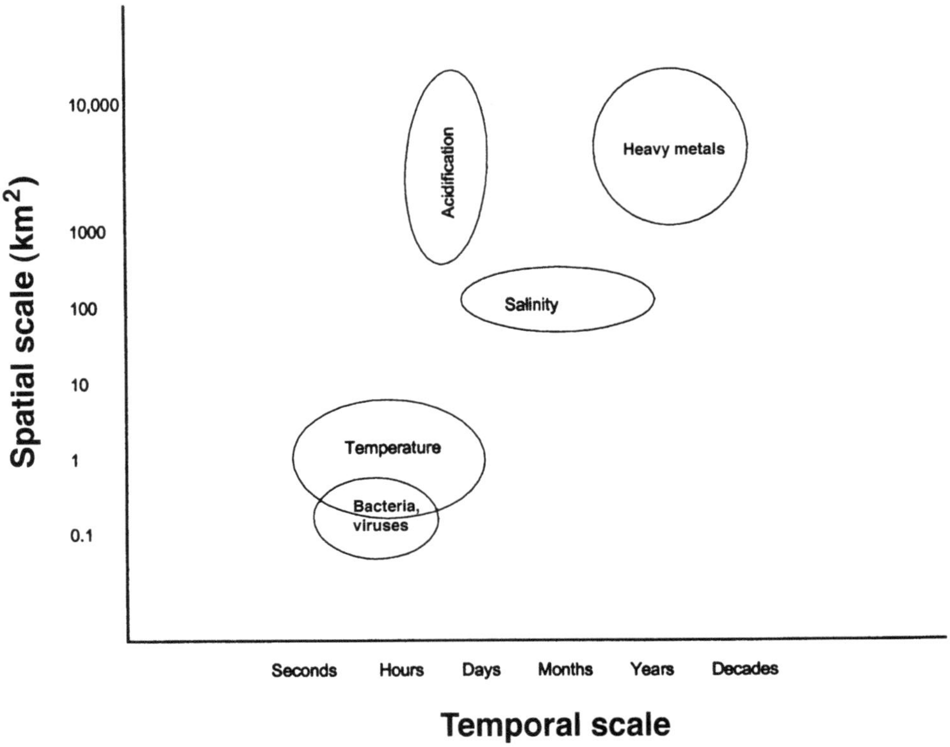

Figure 7.6. Approximate spatial and temporal scales of five water quality parameters affected by a forestry operation. Spatial and temporal scale effects confound the determination of cause-and-effect relations in the environment. (Redrawn from Perry and Vanderklein 1996.)

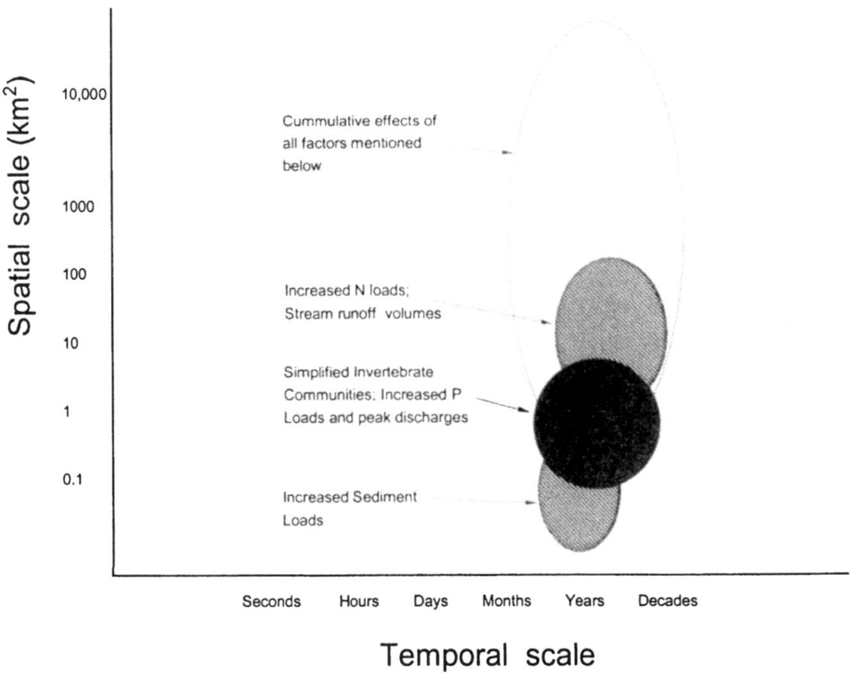

Figure 7.7. Approximate spatial and temporal scales applicable when a forested watershed is harvested. (Redrawn from Perry and Vanderklein 1996.)

and temporal scales for water quality parameters affected by a forestry operation. For example, bacteria and viruses usually cause local, short-duration effects, whereas heavy metals become distributed over large areas and persist for long times. Constituents with large spatial and temporal time scales are exceedingly difficult to remediate. See Appendix A for further discussion of spatial and temporal scales.

The essential message of Figures 7.5 through 7.7 is that scale effects can explain why some natural processes take decades to become manifest in some locations. For example, pesticide breakthroughs to groundwater may take decades in some areas but may become manifest much faster in other locations. *Process complexity with numerous linear components (e.g., Figure 7.5) and cyclical components (e.g., Figures 7.4a–c) coupled with continuously varying climate leads to diverse solutions. This complexity may manifest slowly over time (e.g., Figures 7.6 and 7.7).* Temporal and spatial scale continua are defining characteristics of nonpoint pollution.

Some Best Management Practices for Nonpoint Pollution Abatement

Nonpoint pollution abatement is largely concerned with identifying and implementing economical and efficacious BMPs for land areas. The goal is to maintain the long-term chemical, biological, and physical integrity of downstream water. Federal EPA and state officials identify a proposed management practice as a BMP following rigorous evaluation via agency and university research. Principles discussed earlier are useful for selecting an appropriate BMP for a specific situation. Conservationists design BMPs to implement fundamental approaches for controlling nonpoint pollution. These fundamentals include

- control measures to reduce emission rates of pollutants from the source area (e.g., a waste generation site or a polluted soil);
- hydrological modification of watersheds to reduce the hydrological activity of source areas and consequently the polluting runoff (e.g., create or capitalize on naturally occurring wetlands, create and maintain grass-covered areas in upland watersheds);
- delivery control measures that increase natural attenuation of pollutants between the source area and the receiving water body (e.g., impoundments that increase the total hydraulic residence time and the opportunity for biogeochemical cycling to remove components that cycle into the gaseous phase; control measures that facilitate volatilization, ultraviolet light interactions, and energy exchange);
- treatments and/or production practices that remove (or withhold from the process) pollutants from runoff before entering the receiving water body.

One designs the above approaches to limit pollutant entry into the system and/or increase residence time. This allows biochemical cycles and volatilization as much time as possible to cycle nutrients into volatile forms. When removing pollutants from incoming water, buildups of the element in the soil occur unless the pollutants are volatile or participate in biochemical cycles with volatile components. Likewise, clean water running over dirty soil will acquire some pollutant load. One must select BMPs to capitalize on these realities.

Selected BMP Nomenclature

The following definitions are used in discussions of BMPs. The ASAE (1999) standard ASAE S526.1 MAR95 contains additional definitions.

Buffer: Any zone such as a floodplain, beach, wetland, or other zone wherein access and development activities are limited for the sake of nutrient attenuation, sediment trapping, or hydrograph peak reduction purposes.

Composting: Bioconversion of organic compounds such as proteins, carbohydrates, and fats commonly found in animal wastes and plant residues to environmentally friendly soil conditioners.

Conservation tillage: Tillage practice that leaves plant residues on the soil surface. One must leave at least 30% of residue cover on the surface after the planting operation to qualify as conservation tillage.

Contour: Lines around a site with equal elevation.

Contour farming: System of farming used for erosion control where operations are performed on the land surface contour.

Cover crops: Close-growing crop providing soil protection, seeding protection, and soil improvement between periods of normal crop production.

Dike: Temporary nonengineered channel.

Gabions: Rectangular packages of rock enclosed by wire for stream bank erosion prevention.

Grass waterway: Natural or constructed channel covered with an erosion-resistant grass that transports surface runoff to a suitable discharge location at a nonerosive velocity.

Gully control structures: Combinations of spillway and chute/pipe structures used for renovating gullied areas.

Infiltration basin: Basin for water disposal or groundwater recharge in areas with porous soils.

Irrigation management: Applying proper amounts of water at proper times for optimal crop water use.

Pasture management: Regular mowing and fertilization, coupled with timing, to minimize fertilizer nutrient loss to nearby streams.

Pipe slope drain or chute: Structures that carry concentrated runoff down a steep slope.

Pond: Structure for recreation or agricultural water storage that also reduces peak flows.

Porous pavement: Pavement allowing infiltration, thereby reducing runoff from urban areas.

Riparian buffer or filter strip: Protected zone between stream channels and agricultural or urban development–related activities that allows sediment and nutrient attenuation.

Riprap: Large rocks or other materials for stream bank erosion protection or for culvert and bridge outlet protection.

Sediment retention basin: Basin enabling removal of sand and silt from runoff by mechanical settling.

Strip-cropping: Growing of different crops in alternating strips or bands along contours or across the prevailing wind direction to serve as vegetative barriers to wind and water erosion.

Swale: Diversion ditch, usually shallow and broad, cut into the soil to intercept sheet flow and divert it to a permanent channel, drain, or sediment retention structure.

Terrace: A broad channel, bench, or embankment constructed across the slope (across the contour) to intercept runoff and detain or channel it to protected outlets.

Wetlands: Areas of wet soil under normal conditions that support a prevalence of hydrophytes (e.g., water-loving) plants.

When selecting BMPs for nonpoint pollution control, one must first know the extent to which existing environmental regulation is applicable. Urban and/or industrial areas are more closely regulated than rural areas. Agencies regulate storm water and non–storm water discharge from municipalities through the NPDES permitting system. Mandated by the National Storm Water Program, the EPA regulates storm water discharge from industrial areas through a complex tiered system within the NPDES. They specify treatment approaches with almost no discretion. Industrial water discharges are regulated through NPDES permitting. The EPA also can issue general multisector permits emphasizing the discretionary use of BMPs for various industry sectors. Dodson (1999) gives details for each of approximately 30 industry sectors. The U.S. Forest Service (1980) suggests strategies for erosion control in forestry operations.

Storm water runoff from rural areas is regulated via the "carrot-and-stick approach," the "carrot" being the possibility of federal assistance if mandated practices are put in place. Section 319 of the CWA mandates that conservationists identify BMPs for agricultural areas. Structural and nonstructural practices are available. Structural processes are suited for channelized flows or for channelizing the flow. Nonstructural processes are more appropriate to fields and field management. Conservationists have also identified and categorized BMPs for cropping and animal operations. BMP refinements will occur as more CNMPs are implemented and evaluated.

Rural area BMP selection is emphasized in this text. Discretion becomes less to nonexistent in urban settings. Refer to texts such as Dodson 1999 for additional details. One should keep in mind scale effects, removal kinetics, and site factors when considering the various BMP options.

Structural versus Nonstructural (Managerial) Approaches

BMPs for accomplishing the above strategies have been separated into structural and managerial approaches (Novotny 1995) depending on activity intensity. The structural approach involves a physical structure (e.g., swale, channel, pond, wetland) for removing or treating the pollution problem, while the nonstructural, or managerial, approach minimizes pollution or treats it very near the source (e.g., land management, riparian zones). The efficacy of a given BMP depends on the nature of the pollutant. Table 7.2 identifies and rates nonstructural approaches. Nutrient management is one of the more effective nonstructural approaches. Management of activities on designated coastal zones, maintenance of environmental reserves, site planning, erosion and sediment control measures, and postdevelopment management demonstrate societal desires to control nonpoint pollution. Table 7.3 summarizes structural approaches. One should keep several principles in mind when analyzing Table 7.3. To remove suspended solids, filtration or settling must occur, requiring flow through a porous medium or residence time within a structure. Cycling is a strategy to enable entrapped biological materials to enter carbon and nitrogen cycling for further breakdown. Successful treatment of pathogens requires long retention times. Porous media–based approaches offer another alternative for pathogen control.

Table 7.2. Efficacy of selected nonstructural BMPs for general nonpoint pollution constituent control.

Selected BMPs	Nutrient control	Shellfish	Estuarine	Sedimentation	Sediment toxins	Sto. water controls	Cost feasibility	Maint. burden	Longevity	Acceptance	Secondary impacts	Developer cost	Govt. cost	Implementation	Site data reqd.	Water needed
Environ. Reserve Areas (Regional Scale)																
Intensely trafficked zones	2	2	2	2	2	2	4	1	1	2	2	2	3	2	3	–
Rural regions	2	34	3	3	3	3	4	1	1	2	3	3	3	2	3	–
Protected areas (e.g., park)	3	4	4	4	4	4	4	1	1	2	4	4	3	3	3	–
Detailed property zoning	3	3	3	3	3	4	4	1	1	2	3	3	3	3	2	–
Environ. Reserve Areas (Field Scale)																
Stream buffers	3	3	4	3	2	2	4	4	4	4	4	3	4	3	4	2
Wetland buffers	3	2	4	3	2	2	4	4	4	4	4	2	4	4	4	4
Coastal buffers	3	4	4	3	2	2	4	4	4	4	4	2	3	2	4	4
Floodplain buffers	3	2	4	2	2	3	4	4	4	3	4	3	4	3	3	2
Steep soil limits	2	2	3	4	2	3	4	4	4	3	4	3	4	4	4	3
Septic limits	3	4	2	2	2	2	4	4	4	3	4	2	3	2	3	3
Wetland protection	4	32	4	4	2	3	4	4	4	4	4	3	4	4	2	4
Forest protection	2	3	3	3	2	2	4	4	4	4	4	2	4	3	3	4
Habitat protection	2	2	4	3	2	2	4	4	4	4	4	3	4	3	2	4
Open space protection	2	2	3	2	20	2	4	4	4	4	4	2	4	3	3	4
Site Planning																
Cluster	2	3	2	2	2	2	4	4	4	3	4	4	4	2	3	4
Performance criteria	4	4	4	4	4	4	4	2	2	2	2	1	2	2	2	4
Minimum impervious area	2	2	2	2	2	3	4									
Erosion and Sediment Control																
Construction timing	3	2	4	3	2	2	4	4	4	4	4	4	3	4	3	4
Postdevelopment																
Urban area maintenance	2	3	2	3	3	2	4	3	2	3	3		4	4	4	4
Fertilizer application limits	3	2	1	1	2	2	3	2	2	3	4		3	3	3	3
Septic tank maintenance	3	4	2	1	2	2	4	2	3	3	3		3	3	2	3
Household hazardous waste disposal regs.	2	1	2	1	4	2	3	23	2	3	3		3	3	2	3

Source: Adapted from Novotny et al. 1995.

Note: 4 = most desirable, most effective or directly protecting, least burdensome, simplest, least costly; 1 = least desirable, least effective, most burdensome; – = not applicable.

Volatilization and/or surface adsorption with subsequent breakdown remove trace organics. Again, cycling is important and concentrations must not exceed toxicity limits for growing vegetation. Nutrient removal happens mainly by surface adsorption and, sometimes, by subsequent nutrient cycling into gaseous forms. Porous media with significant clay or organic fractions having high specific surface area absorb the most nutrients. Media containing mostly sand have low specific surface area and absorb the fewest nutrients. Cropping with harvest is an effective strategy for nutrient removal. Nitrogen and sulfur cycles can result in much volatilization of these nutrients. Without harvest, nutrients accumulate and can ultimately become pollution hazards.

Table 7.3. Structural BMPs for general nonpoint pollution reduction.

Structural BMPs	Suspended solids	Nitrogen	P	Path	Metals	Land req.	Drainage area	Soil conditions	Groundwater elevation
Flow inhibiting:									
Ponds	Med-high	Med	Med	Low	Med-high	Large	Med-large	Imperm.	Near surface
Constr. Wetlands	Med-high	Low	Low-Med	Low	Med-high	Large	Large	Imperm.	Near surface
Infiltration facilities:									
Infil. Basins	Med-high	Med-high	Med	High	Med-high	Large	Small-med	Permeable	Below
Trenches	Med-High	Med-high	Low-	Med-	Med-	Small	Small	Permeable	Below
Porous pavement	High	High	Med	High	High	n/a	Small-med	Permeable	Below
Vegetative practices:									
Grassed	Medium	Low-med	Low-	Low	Low-	Small	Small	Permeable	Below
Filter strips	Med-high	Med-high	Med-high	Low	Med	Varies	Small	Depends on type	Depends on type
Filtration methods:									
Filtration	Med-high	Low	Med-	Low	Med-	Large	Med-	Permeable	Below
Sand filters	High	Low-med	Low	Low	Med-high	Varies	Low-med	Depends on type	Depends on type
Others:									
Inlet Filter	Low-med	Low	Low	Low	Low	n/a	Small	n/a	n/a

Source: Adapted from Novotny et al. 1995.

Metals removal can be good because many soils and other surfaces have high affinities for metals. Metals do not volatize. Although harvestable crops do not usually uptake metals to a significant extent, crops grown in media with high metal content can contain enough of the metal to pose health threats. Extended application (e.g., 50 yr) of poultry litter can be harmful to land due to the presence of metal feed additives in the litter.

Crop versus Animal

BMPs for reducing nonpoint pollution by crop and animal enterprises are presented in Tables 7.4 and 7.5. From a land management viewpoint, farmers should eliminate excessive fertilizer applications and excessive pesticide applications. Applications of all nutrients and pesticides should be timed to need. Rapidly degrading pesticides are obviously desirable. Physical-mechanical (e.g., tillage) weed controls reduce chemical usage but are usually not economical. The tillage approach to weed control may lead to additional erosion hazards. Standard erosion control methods are useful for pollution reduction. For animal production,

Table 7.4. Managerial and structural BMP summary for agricultural crop production.

	Surface water			Groundwater	
BMP	Sediment	Soluble pollutants	Adsorbed pollutants	N loss	Pesticide loss
Soil test/plant analyses	0	++	+	++	0
Setting realistic yield goal	0	++	++	++	+
Site-specific management (precision farming)	0	+	+	+	+
Equipment maintenance	0	+	+	+	+
Application methods and timing	0	++	++	+	+
Water control structure	++	+	++	–	–
Subsurface drainage	?	–	?	++	++
Env. friendly integrated pest management	0	++	++	0	++
Pesticide selection and rotation	0	++	++	0	+
Conservation tillage	++	?	++	?	?
Contours and terraces	++	+	++	–	–
Strip-cropping	+	+	+	0	0
Cover crops	+	+	+	+	0
Crop rotation	++	+	++	0	+
Irrigation management	+	++	+	++	+
Pasture management	+	0	+	0	0
Waterbody protection	++	+	++	0	?
Grassed waterways	+	0	+	0	0
Critical area planning	++	+	+	0	0
Stream buffer	++	+	++	+	?
Gully control structures	+	0	0	0	0

Source: Adapted from Adams and Risse 1996.

Key: ++ medium to high effectiveness; + low to medium effectiveness; 0 little or no effectiveness; – could be detrimental; ? research needed. BMPs are site specific in nature.

Table 7.5. Managerial and structural BMP summary for agricultural crop and livestock production.

	Surface Water			Groundwater		
BMP	Sediment	Soluble solids	Adsorbed pollutents	Bacteria	N Loss	Pesticide loss
BMPs primarily used in feed production						
Conservation tillage	++	?	++	?	?	?
Contours and terraces	++	+	++	+	–	–
Strip-cropping	+	+	+	+	0	0
Cover crop	+	+	+	0	+	0
Crop rotation	++	+	++	0	0	+
BMPs for confined animal operations						
Waste management system	++	++	++	++	++	0
Diet manipulation	0	+	+	+	+	0
BMPs for both confined and unconfined operations (see Table 7.6 also)						
Nutrient management	0	++	++	+	++	0
Pest management	0	++	++	0	0	++
Pasture management	+	0	+	+	0	+
Waterbody protection	++	+	++	+	0	?
Grassed waterways	+	0	+	0	0	0
Critical-area planning	++	+	+	0	0	0
Stream buffers	++	+	++	++	+	?
Composting	0	++	++	+	++	0

Source: Adapted from Risse and Cheadle 1996.

Key: ++ medium to high effectiveness; + low to medium effectiveness; 0 little or no effectiveness; – could be detrimental; ? research needed. BMPs are site specific in nature.

proper diet is helpful in managing nutrients. Effective waste-handling facilities are essential. Stream buffers and critical-area planning are essential.

BMPs and Ecology

Principles of ecology may guide selection of BMPs for mitigating chemical and biological pollution. Jorgensen and Mitsch (1989) analyzed aspects of natural wetlands and determined several basic principles leading to operational success:

1. The best outcomes result when physical, biological, and chemical functions are in harmony. Self-selection of biological species should be practiced as much as possible, followed by introduction of species likely to thrive.
2. Harmony requires recognition of time and space scales. Allow adequate room for species needing room.
3. Systems are more resilient when adequate chemical and biological diversity exists.
4. Systems with pulsing patterns seem to have advantages. Allow for rest periods.

Tables 7.2–7.5 directly or indirectly embody these ecological principles. Tables 7.2 and 7.3 present general ratings for a variety of nonstructural and structural measures applicable to regional and field problems. Tables 7.4 and 7.5 present practices applicable to crop and animal production. One may combine suggested BMPs in Tables 7.2–7.5 as needed to arrive at a suitable mix. Producers are beginning to realize that nutrients

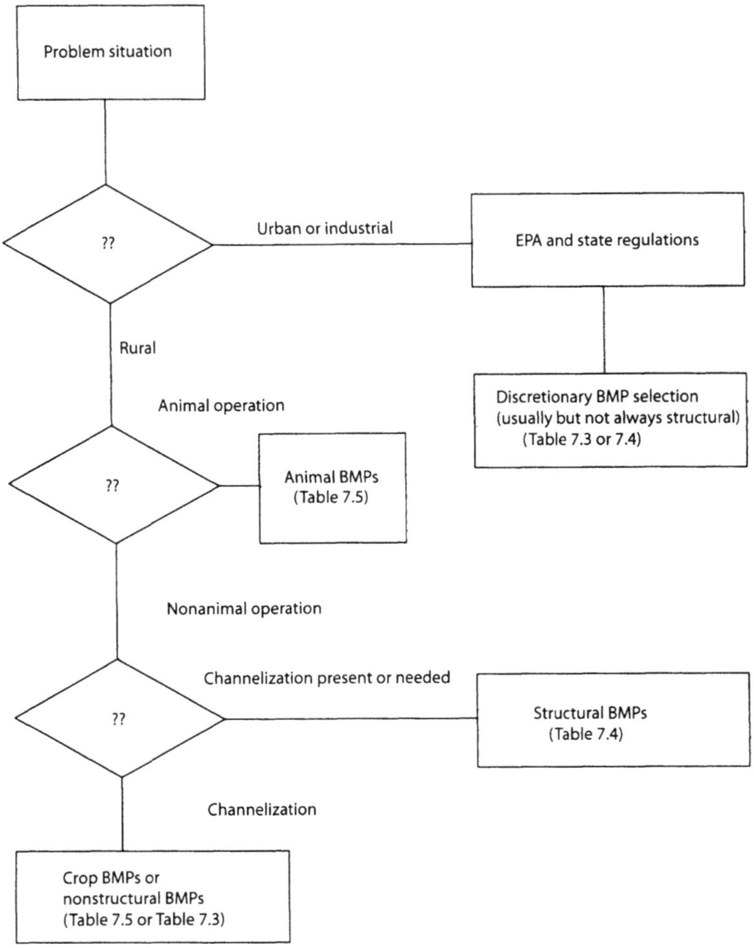

Figure 7.8. Decision aid for BMP selection.

causing pollution are in fact nutrients wasted. They represent process and system inefficiencies and thus a avoidable cost or forfeited profit. The BMPs are strategies for reducing the impact of waste on the environment. The International Standards Organization (ISO) has developed a series of standards (ISO 14000) to guide manufacturers in tracking environmental management. These standards provide guidelines for quality audits and risk assessment. The cost effectiveness of environmental stewardship motivates them (Puri 1996; Tibor and Feldman 1997). ISO 14000 outlines a philosophy of continuous improvement in environmental management. The paradigm wherein environmental stewardship is cost effective represents a major change in agricultural and industrial thinking. The impact of this ongoing change of approach is only now becoming realized. ISO 14000 will likely influence environmental approaches for years to come. Implementation of BMPs on a site does not result in instant improvement or compliance (Darling 1997). Examples in Florida and Vermont show that BMPs for phosphorus control may require decades before phosphorus reductions appear. Once a site becomes loaded with a pollution constituent, restoration to background conditions is problematic.

Background concentrations and degree of existing site loadings are both important to weight components of nonpoint pollution adequately. For example, one should not expect to reduce any constituent below rainfall concentrations in the region. Do not expect to reduce any constituent below background amounts such as those given in Young et al. 1994. Once a region is loaded with a pollution constituent, processes may require much time to equilibrate with the background condition.

Soils receiving acid rain renovate runoff by raising the pH as runoff occurs, provided the soil has the buffer capacity to supply the calcium and other nutrients. Acid rain lowers the pH of unbuffered soil. Acid rain has a serious impact in the southeast United States because many soils there have inherently low buffer capacity.

The wetland removes pollution because of the extreme diversity of the biota. The extreme diversity, however, greatly complicates accurate prediction of pollutant removal. Spatial and temporal scales vary with organism type. The wide variability in spatial and temporal scales greatly complicates verification of relationships describing chemical and microbial process kinetics. Thus, one should treat these relationships as rough approximations.

Selection of BMPs for given situations may seem daunting. Figure 7.8 may aid in structuring the BMP selection decision.

Quantitative Removal and Renovation of Selected Pollution Constituents

One may classify key nonpoint pollution categories as shown in Table 7.6. BOD represents carbonaceous biodegradable material. Nitrogen, sulfur, and phosphorus are the primary nutrient constituents. Sediments represents essentially inert solids, and organics represents a wide range of pesticides and other industrial wastes. A given waste

Table 7.6. Removal and renovation mechanisms for selected nonpoint pollutant classes.

Pollutant class	Predominant removal mechanisms	Predominant or possible renovation mechanisms
BOD	Absorption and sedimentation	Biochemical cycling, organic matter harvest
Nitrogen	Absorption, limited cation exchange, and sedimentation	Biochemical cycling, organic matter harvest
Sulfur	Absorption and sedimentation	Biochemical cycling
Phosphorus	Absorption, precipitation	Chemical treatment, enrichment, mass removal
Pesticides	Absorption	Degradation, photochemical effects, enrichment
Heavy metals	Precipitation	Chemical treatment, enrichment, mass removal
Suspended (inert) solids	Sedimentation and filtration	Mass removal
Organic solvents	Absorption	Volatilization, photochemical effects, selective solvents, mass removal

material usually contains multiple pollutant classes. One must first capture the offending material from the transport phase (e.g., air or water). The treatment system must possess a renovation mechanism if the system is to continue in effectiveness. Biochemical cycling, if gaseous forms exist and are common, is helpful. Volatilization is also a useful transformation. Scale effects are obvious when biochemical cycles are involved. Mass removal is the removal of physical mass from the treatment system to control offending materials. Mass removal and treatment are expensive. Photochemical effects refer to the action of ultraviolet rays on double-bonded carbon in some pesticides and organic solvents. Photochemical action requires exposure to sunlight. Enrichment is discussed further below.

Transport processes alter pollution delivery to the mouth of a watershed. As with sediment, pollution concentration usually peaks before the hydrograph peaks. Factors affecting sediment delivery also affect attached nutrients. Refer to the discussion of the sediment delivery ratio in Chapter 6. The volatilization process reduces concentrations of many compounds. Absorbed pollutants enrich transported sediments. Absorbed pollutants include phosphates, ammonium, metals, and pesticides. One does not apply the enrichment ratio to dissolved, nonabsorbing constituents such as nitrates.

Novotny (1995) suggests an exponentially decreasing delivery ratio with overland flow length and a logarithmic increase in the enrichment ratio with overland flow length. Selective absorption on small particles as opposed to large particles causes enrichment of absorbing/adsorbing pollutants on the transported materials. Large particles settle out of the flow and thus do not transport to the same extent as small particles. The large particles account for most sediment mass. The small particles contain the most surface area for adsorption. The small transported sediments have an overall higher concentration (mg pollutant/kg sediment) of pollutants than the original soil. Thus, the transported soils are "enriched" with pollutants. Clean water with an unused capacity to carry constituents may also "enrich" the transporting fluid. Figure 7.9 presents enrichment and delivery ratio curves. In summary, the delivery ratio operates when precipitation or sedimentation is removing incoming pollution. The enrichment ratio operates when incoming flow is removing a constituent by dissolution. Enrichment likewise occurs with adsorption on a moving clay particle. Novotny (1995) provides several calculation examples involving removal of pollution constituents by soil adsorption and removal of adsorbed constituents by flowing water.

Young et al. (1994) use an empirical filtering relationship approach to account for soluble N and P removal from overland flow (through grass). They do detailed sediment routing to account for sediment deposition and reentrainment from the eroding source to the watershed mouth. They compute an enrichment ratio for N and P based on an exponential model adjusted by soil texture.

Prediction of nutrients is a matter of determining the available portion of soluble nutrients in the soil and nutrients lost due to erosion processes. Modeling these dynamic, time-dependent processes requires computers. One such model is AGNPS (Young et al. 1994), which simulates uplands erosion and nutrient transport. The USEPA model SWMM (storm water management model) also provides water quality simulation for single events in natural and man-made drainage systems (USEPA 1994). They use empirical relationships accounting for background amounts in rainfall to predict N and P washoff, along with clay and organic-matter content. Models describe N and P dynamics, albeit crudely.

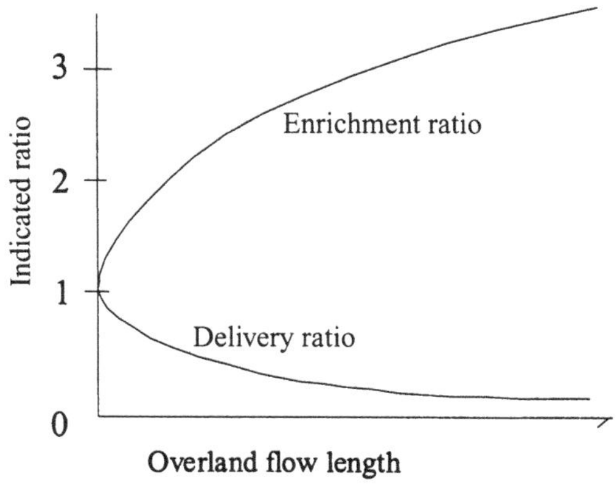

Figure 7.9. Enrichment ratio versus the delivery ratio as a function of overland flow length.

Sediment-attached N and P are modeled using variations of the USLE for soil loss calculations. Soil is then broken into clay, silt, small aggregates, large aggregates, and sand classes. Clay and silt fractions are mobile whereas other classes are less mobile. Sediment routing techniques can model deposition and entrainment processes. This topic is beyond the scope of this text.

Organics in runoff are difficult to model. Processes occurring in constructed wetlands, infiltration systems, and overland flow systems are similar to watershed-scale processes. In a BOD-rich effluent (e.g., from a feedlot), BOD attenuation occurs in overland flow processes. Mitsch and Gosselink (1993) and Reed et al. (1995) show that the BOD remaining in the effluent decreases linearly as travel distance down the slope increases. Cycling the application to allow for drying time is also important. Drying time gives dynamic nutrient/carbon biochemical cycles opportunity time. Other nutrients are removed as linear distance or as exponential distance increases. Coëfficient of variation values are usually large owing to the multitude of factors influencing these processes. AGNPS, for feedlot nutrient sources, bases soluble COD predictions on background data measured from a variety of land uses and locations. Soluble nutrients (N and P) attenuate following a logarithmic (N) or linear function (P) based on travel time in a buffer strip or grassed waterway. *Balancing the modeled point plus nonpoint contributions to pollutant loading with modeling of the natural attenuation process along with a safety factor is the essence of TMDL determination for a stream.* Chapter 15 discusses these processes further.

Outlook

The following societal trends aggravate the likelihood of nonpoint pollution:

- Population increase: increase of living standards throughout the developing world and increased urban concentrations
- Land use transition resulting from urbanization, deforestation, wetland alteration, and intensive agriculture
- Urbanization: construction activities, unsewered urban development, industrial contamination
- Surface mining activities
- Agricultural and silvicultural activities, especially as operations become concentrated

People are beginning to realize the cost of and forfeited profit from nutrient or sediment pollution. Design for nonpoint water quality management is a rapidly evolving subject in the urban and rural contexts. CNMPs and storm water management plans will mandate the use of BMPs.

Depending on current situation and history, soils may add pollution to or remove it from incoming waters. BMPs are available for both scenarios. One may help or hinder integrity of downstream water bodies depending on the practices followed. Ideally, *one should select economical BMPs that slow runoff to provide increased time for physical settling or filtering, chemical transformation, biogeochemical cycling, and ultimately beneficial use by natural flora and fauna.*

The following principles of erosion and sediment control (Goldman et al. 1986) and riparian zone design (Jorgensen and Mitsch 1989) lead to reduced nonpoint pollution:

- Proper planning and anticipation are the best strategies for avoiding nonpoint pollution.
- Implement control measures to reduce emission rates of pollutants from the source area.
- Hydrologically modify watersheds to reduce the hydrological activity of source areas and consequently the volume of polluting runoff.
- Take delivery control measures that increase natural attenuation of pollutants between the source area and the receiving water body.
- Impose treatments to remove or eliminate pollutants from the runoff before it enters the receiving water body.

An alternative way of stating these principles is to bring in the basic soil erosion control principles of Chapter 6, supplemented using the following ecological principles:

- Distribute pollutant over large areas with provision for pulsing in time and space to minimize continual application to one area.
- As much as practical with riparian zones and buffers, maintain high flora and fauna diversity and support naturally occurring cycles in flora and fauna, pulsing inputs instead of maintaining steady-state programs.
- Avoid using the soil as a "sink" for insoluble precipitates; this entails an expensive future cleanup.
- Use production approaches facilitating internal recycling and reuse (e.g., CNMPs) to minimize potential pollutant discharge.
- Use development approaches with BMPs to avoid diversion of runoff from watersheds, runoff increases, and runoff concentration.

Strategies for erosion control and water quality protection are constantly improving for nonagricultural applications. The NRCS (http://www.usda.nrcs.gov) provides excellent downloadable publications relating to BMPs and water quality. The *NRCS Waste Management Handbook,* an excellent reference, is available at links within the NRCS site. The USEPA provides the BASINS and SWMM models at links within its site (http://www.epa.gov). The NRCS and EPA sites are moving toward the use of JAVA scripts, which will run models with user input and selected input from their central server via the web.

The Internet water quality database browser (http://www.hermes.ecn.purdue.edu:8001/server/water/bib/browse.html) provides excellent information on water quality control strategies.

This discussion has focused upon uplands that have many intermittent or ephemeral streams. Chapter 15, an overview of streamflow processes, carries forward the discussion of CNMPs, BMPs, and storm water drainage management to the notion of TMDLs for watersheds, which necessarily involves permanently flowing streams.

Problems and Questions

1. One is converting a land parcel from agricultural zoning to commercial zoning. Analyze the potential pollution impacts considering Figures 7.4 and 7.5. Suppose the development is on the higher elevation of a mildly rolling topography. Suggest some BMPs abating downstream pollution.
2. Given basic principles of ecological engineering discussed in Appendix A, how may we implement these principles in a wetland receiving runoff from a dairy feedlot?
3. A farmer is evaluating ways to minimize nutrient movement off his farm. Lands are rolling. For a feedlot overlooking a stream, suggest some BMPs for minimizing pollution potential. Assume plenty of available land. Reconsider, assuming minimal land area between stream and feedlot.
4. Select one of the BMP summary tables (Tables 7.2–7.5). Compare and contrast the table recommendations with the suggested enhancement steps in the Outlook section.
5. Refer to the ecological engineering principles listed at the end of the BMP discussion. Which BMPs support these principles? Suggest some new BMPs based on these principles. (Your BMPs may be "hybrids," or combinations of any mentioned in this chapter, including Tables 7.2–7.5.)
6. What are the greatest barriers to taking a comprehensive systems view of a watershed?
7. Analyze the pros and cons of destroying a wetland for construction of a road, factoring in the ethics discussion of Appendix A. Assume the area is a rural, developing county. Assume wetlands exist in other parts of the county.
8. Suppose an endangered species is completely lost due to commercial development. Describe the potential impact of this loss.
9. Given a 1 ha square construction site that is uniformly sloped at 20% along one side to a stream, suppose a single storm with an R_{rf} equivalent to the average annual value for Knoxville, Tennessee, falls in the summer at peak construction activity. If the soil is a group C, moderately permeable, massive structure, 40% sand, 30% silt, 30% clay, and 0.5% organic, what is the average annual soil loss? Rework the problem assuming well-maintained silt fences were installed at 50 m intervals across the contour. What is the impact of the fences on average annual soil loss? Given the runoff volume expected and the sediment detached, predict the average sediment concentration with and without the silt fences. Suggest (but do not quantify) some BMPs for ameliorating the sediment problem.
10. Given a 1 ha square construction site in Fargo, North Dakota, that is uniformly sloped at 20% along one side to a stream, suppose a single storm with an R_{rf} equivalent to the average annual value for Atlanta, Georgia, falls in the summer at peak construction activity. If the soil is a group A, moderately permeable, massive structure, 20% sand, 50% silt, 27% clay, and 3% organic, what is the average annual soil loss and projected sediment concentration with silt fences spaced at 50 m, assuming the silt fences intercept and channel the flow similar to a terrace?

11. Four basic approaches to the development of BMPs were discussed in this chapter. Identify these principles and discuss how they apply to BMPs of your choice described in this chapter. Use information in Tables 7.2–7.5 in developing your answers.
12. Suppose you are helping a client in a mountainous region develop a CNMP for a poultry operation. Select one efficacious BMP from each of Tables 7.2–7.5. Given the four principles for BMP selection discussed in this chapter, why are your selected practices highly rated? Repeat for a feedlot operator in central Iowa.
13. Why are cause and effect so hard to quantify in environmental analyses of nonpoint pollution?
14. The EPA has declared that a long-term lead-acid battery recycling site is in need of cleanup. From the ethical positions outlined in Appendix A, make an argument for one or more of the following groups to pay for the cleanup: general public; individuals bringing batteries to the site; firms who paid individuals to bring batteries to the site; purchasers of the recycled products. Identify the values on which your argument is based.
15. Name and define three chemical methods for assessing aspects of water quality. Repeat for physical and biological methods.
16. What is the significance of the biochemical cycles for manures and other biological residues?
17. Name four physical characteristics/measurements of water quality.
18. Name five chemical characteristics/measurements of water quality.
19. What are five categories of agricultural nonpoint pollutants?
20. Discuss the difference(s) between nonstructural and structural best management practices.
21. List five strategies (e.g., BMPs) for minimizing pollution production from crop production in an area with a rolling topography.
22. Animal (livestock) production can result in detrimental environmental effects if not managed properly. Describe four specific types of pollution resulting from animal production.
23. List four BMPs for minimizing pollution from livestock production.
24. With what type of pollutants are riparian buffers most effective? Least effective?
25. Discuss the difference between enrichment ratio and delivery ratio, and explain how these ratios can change as overland flow length increases.
26. List five ways in which natural wetlands are "valuable."
27. List four ways in which constructed wetlands improve water quality.
28. Describe the major steps in designing a constructed wetland for the treatment of nonpoint pollution.
29. What are the major mechanisms at work in a natural or constructed wetland?
30. Summarize ISO 14000 from a water quality point of view, using references given in the literature cited above.
31. A local park has asked you to design a land use plan for optimal public benefit. Outline the steps required to achieve optimal land use.
32. A water sample has $BOD_5 = 20$ mg/l and COD $= 200$ mg/l. What can one say regarding the biodegradability of the sample?

33. Considering the BMPs in Tables 7.5 and 7.6, which are structural and which are managerial, based on the definitions given?

References

Adams, T. T., and L. M. Risse. 1996. *Pollution prevention in agricultural crop production.* Report submitted to the Pollution Prevention Assistance Division, Georgia Department of Natural Resources, Atlanta, GA.

Apply pesticides correctly. 1993. Special Bulletin 15. Georgia Cooperative Extension Service. University of Georgia College of Agricultural and Environmental Science, Athens, GA.

ASAE. 1999. Soil and water terminology. ASAE standard S526.1 MAR95. American Society of Agricultural Engineers, St. Joseph, MI.

Bookchin, D., and D. Farnsworth. 1997. Environmental laws and regulations. In *Environmental engineers' handbook,* 2d ed., ed. D. H. F. Liu, B. G. Liptak, and P. A. Bouis. Lewis Publishers, Boca Raton, FL.

Clesceri, L. S., A. E. Greenberg, and A. D. Eaton. 1998. *Standard methods for the examination of water and wastewater.* 20th ed. American Public Health Association, Washington, DC.

Darling, W. A. 1997. Dealing with the challenges of animal waste regulation in the southeast. In *Southeastern sustainable animal waste management workshop* proceedings, 11–13 Feb., Rural Development Center, Tifton, GA.

Dodson, R. D. 1999. *Storm water pollution control.* 2d ed. McGraw-Hill, New York, NY.

Goldman, S. J., K. Jackson, and T. A. Brusztynsky. 1986. *Erosion and sediment control handbook.* McGraw-Hill, New York, NY.

Greppi, M., and F. Preti. 1999. Water quality in agriculture. In *CIGR Handbook of agricultural engineering: Land and water engineering.* American Society of Agricultural Engineers, St. Joseph, MI.

Jorgensen, S. E., and W. J. Mitsch. 1989. Ecological engineering principles. In *Ecological engineering: An introduction to ecotechnology,* ed. W. J. Mitsch and S. E. Jorgensen. John Wiley & Sons, New York, NY.

Mitsch, W. J., and J. G. Gosselink. 1993. *Wetlands.* 2d ed. Van Nostrand Reinhold, New York, NY.

Novotny, V. 1995. *Nonpoint pollution and urban stormwater management.* Technomic Publishing Co., Lancaster, England.

Nyer, E. K., D. F. Kidd, P. L. Palmer, T. L. Crossman, S. Fam, F. J. Johns II, G. Boettcher, and S. S. Suthersan. 1996. *In situ treatment technology.* Lewis Publishers, Boca Raton, FL.

Odum, E. P. 1983. *Basic ecology.* Saunders College Publishing, New York, NY.

Perry, J., and E. Vanderklein. 1996. *Water quality: Management of a natural resource.* Blackwell Science, Cambridge, MA.

Puri, S. 1996. *Stepping up to ISO 14000: Integrating environmental quality with ISO 9000 and TQM.* Productivity Press, Portland, OR.

Reed, S. C., R. W. Crites, and E. J. Middlebrooks. 1995. *Natural systems for waste management and treatment.* McGraw-Hill, New York, NY.

Risse, L. M., and S. A. Cheadle. 1996. *Pollution prevention in agricultural livestock production.* Report submitted to the Pollution Prevention Assistance Division, Georgia Department of Natural Resources, Atlanta, GA.

Sawyer, C. N., P. L. McCarty, and G. F. Parkin. 1994. *Chemistry for environmental engineering.* 4th ed. McGraw-Hill, New York, NY.

Segars, W. I. 1997. Overview of federal and state regulations affecting livestock and poultry operations. *Southeastern sustainable animal waste management workshop* proceedings, 11–13 Feb., Rural Development Center, Tifton, GA.

Tchobanoglous, G., and E. D. Schroeder. 1985. *Water quality: Characteristics, modeling, and modification.* Addison-Wesley, Reading, MA.

Tibor, T., and I. Feldman. 1997. *Implementing ISO 14000.* McGraw-Hill, New York, NY.

———. 1994. *SWMM—Storm water management model.* EPA 823-C-94-001. USEPA, Washington, DC.

U.S. Forest Service. 1980. *An approach to water resources evaluation of nonpoint silvicultural sources (a procedural handbook).* EPA 600/8-80-012. USEPA, Washington, D.C.

Young, R. A., C. A. Onstad, and D. D. Bosch. 1994. *Agricultural nonpoint source pollution model (AGNPS), version 4.03 user's guide.* USDA-ARS North Central Soil Conservation Research Laboratory, North Iowa Ave., Moris, MN.

8

Open Channel Hydraulics

Natural and engineered channels are the primary means for conveying storm water runoff in humid regions. Worldwide, irrigation water delivery to farms and ranches via open channels is the norm. Increasing irrigation demands and increasing aquacultural production in developing countries and increasing urbanization and water quality improvement in developed countries create a continuing demand for stream channel engineering services (see FAO 1998 and http://www.nhq.nrcs.usda.gov). Engineers around the world use similar methods and approaches for channel design, with only minor variations in specific design equations (see Hudson 1971; Beltran 1999). This text emphasizes U.S. methods.

Open channels discussed in this and the following chapters are an integral part of BMPs for water quality improvement. One sometimes reconfigures or relines natural channels to handle additional flow, requiring design calculations. Figure 8.1 shows a trapezoidal unlined channel under construction for drainage.

Channel design requires knowledge of the most basic aspects of engineering science. Mass, energy, and momentum relationships provide the foundation for design.

Goals

- Identify commonly used terms in channel design literature.
- Review the continuity, energy, and momentum equations.
- Design stable, erodible, and optimal channels using the Manning and continuity equations.
- Identify when stable, erodible, and optimal channel design approaches are relevant to given design scenarios.
- Design channel transitions using the energy and Manning equations.

Fundamentals of Hydraulics

Channel Elements and Important Nomenclature

Natural channels are *ephemeral* if they do not flow constantly. They are *perennial* if they flow constantly or are below the water table. Designed channels that do not constantly flow are designated *intermittent.* Otherwise, they are designated *continuous.*

Figure 8.1. Trapezoidal channel under construction with a dragline in a high water table soil.

One can describe flow channels with the following elements (see Figure 8.2):

Flow depth (y): the actual depth in the channel (L).

Area (A): the portion of the channel cross section wetted by the fluid (L^2).

Wetted perimeter (P): the portion of the channel perimeter wetted by the fluid (L).

Hydraulic radius (R): flow area divided by wetted perimeter (L).

Top width (t): the length of the cross section exposed to air. Figure 8.3 shows top width is equal to the differential of area with respect to depth ($t = dA/dy$). This equality is useful in some manipulations of the energy and other equations (L).

Side slope (z): the slope of the side of a trapezoidal, triangular, or rectangular channel. Designers express side slope as $z:1$, run to rise, where z is the horizontal run and the vertical rise is unity. The rectangular cross section has a $z = 0$. Side slope z is dimensionless.

Channel gradient or slope (S_o): the longitudinal slope of the channel; normal to the cross-sectional area, along the direction of flow. Slope is dimensionless.

Hydraulic depth (D_h): defined as the cross-sectional area (A) divided by the top width (t); used in Froude number calculations (L). Refer to Figure 8.3.

Freeboard: the extra depth increment added to the final design to account for settling, wave action, possible sedimentation, capillary action in soil, and the design safety factor.

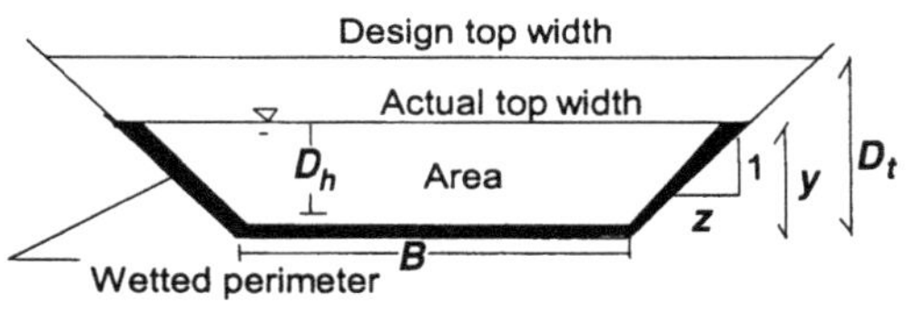

Figure 8.2. Channel elements of a trapezoidal channel. The hydraulic depth is a calculated quantity. The slope (S_o) is measured perpendicular to the cross-sectional area shown.

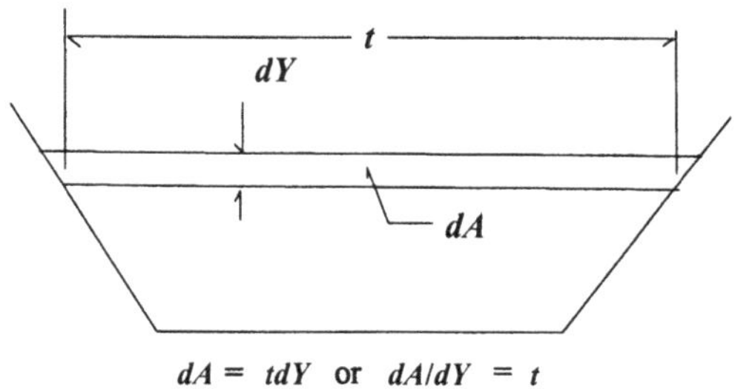

Figure 8.3. Definition sketch showing how top width relates to differentials of depth and area.

Freeboard may be a constant amount, fraction of design depth, fraction of the velocity-depth product, or other agency policy–determined amount.

Velocity (v): the channel mean velocity (L/T). We may approximate velocity by recording the distance moved by a float per unit time.

Flow rate (q): the channel flow rate (L^3/T). One may approximate flow rate by multiplying velocity by the cross-sectional area of the channel.

The trapezoidal, triangular, circular, and parabolic cross sections are most often used in agricultural and urban drainage applications. Table 8.1 gives relationships for computing hydraulic elements for common channel cross sections. Figure 8.4 gives hydraulic elements of a circle in graphical form. The hydraulic elements exhibit several noteworthy trends. For the triangular geometry, z is weakly dependent on hydraulic radius R. Similarly, for the parabolic geometry, top width T is weakly dependent on R. This is why an approximate relationship for hydraulic radius and depth for the triangular and parabolic geometries often suffices. As the trapezoidal channel becomes wide, depth more closely approximates the hydraulic radius. Graphical solution aids show element sensitivity to changes in other elements. Chow (1959) provides graphical aids for several common channel cross sections. Charts interrelating various hydraulic elements are helpful; however, more than one solution to the problem may exist.

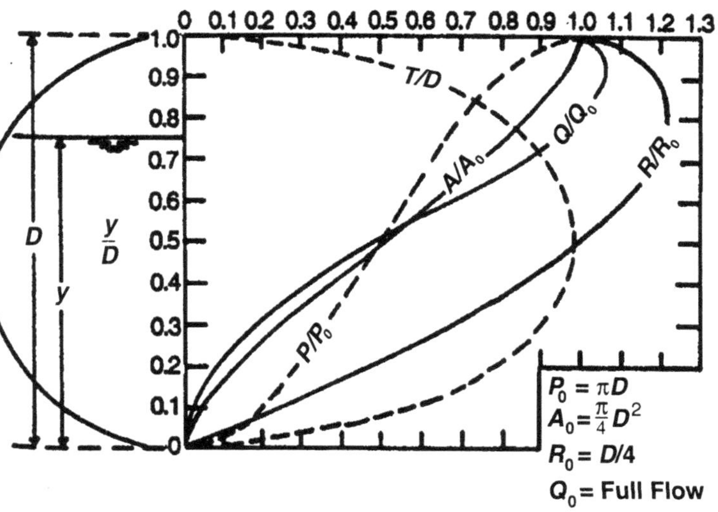

Figure 8.4. Hydraulic elements for a circular cross section as fraction of hydraulic elements at full flow.

Table 8.1. Hydraulic elements for common channel cross sections.

	Cross-sectional area A	Wetted perimeter P	Hydraulic depth D_h	Hydraulic radius R	Top width T
Trapezoid 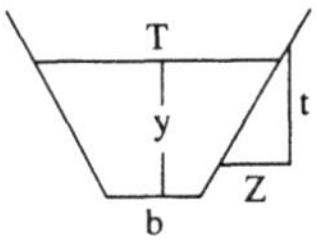	$by + Zy^2$	$b + 2y\sqrt{Z^2 + 1}$	A/t	A/P	$t = b + 2yz$ $T_f = b + 2y_f z$
Triangle 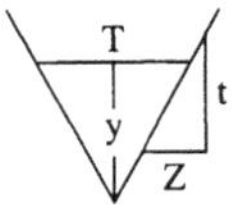	zy^2	$2y\sqrt{Z^2 + 1}$	A/t	A/P $\approx \frac{y}{2}$ *when* $Z \geq 3$	$t = 2yz$ $T_f = 2y_f z$
Rectangle 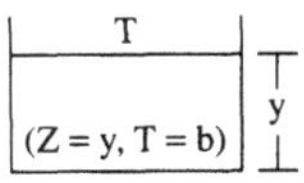	by^2	$b+2y$	y	A/P $\approx y\ (b/y \geq 25)$	b
Parabola 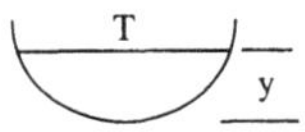	$\frac{2}{3}ty$	$t + \frac{8y^2}{3t}$ *where* $0 < 4y/t \geq 1$ (See Chow (1959) if this condition fails; failure is rare for broad, shallow channels)	A/t	A/P $\approx \frac{2}{3}y$	$t = \frac{A}{0.67y}$ $T_f = t\left(\frac{y_f}{y}\right)^{1/2}$
Circle 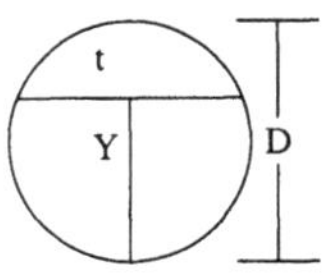	See Figure 8.4 (See Chow, 1959 for exact equations)	See Figure 8.4	A/t	See Figure 8.4	See Figure 8.4

Note: Subscript "f" refers to the freeboard condition.

Note: See Figure 8.3 for other hydraulic elements; Nomenclature: Y = design flow depth; D_f = depth with freeboard; Z = side slope ratio; t = top width before freeboard; A = cross-sectional area before freeboard; T_f = top width after freeboard; B = bottom width; D_h = hydraulic depth; D = circle diameter.

Fluid Statics Review

The most important relations of fluid mechanics with respect to design of open channels include fluid pressures and forces. Fluid pressure is given by $P = \rho g y = \gamma y$. The total force on a vertical wall, per unit length, is equal to $\gamma y^2/2$ acting at a point 2/3 down the wall. The unit weight of water $\gamma = \rho g$ the density of water times gravity. Figure 8.5 illustrates important fluid pressures. Multiplying the pressures by respective surface areas will convert the pressures shown in Figure 8.5 to forces.

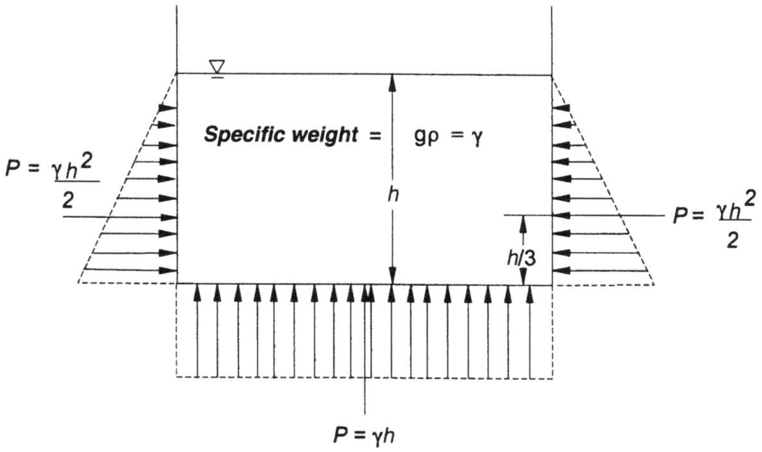

Figure 8.5. Hydrostatic pressures arising from the presence of a fluid. Pressures may be converted to forces by taking the area of the appropriate surface.

General and Uniform Flow Relationships

Continuity Equation

One can write the continuity equation (CE) in macroscopic and microscopic terms. This text is concerned with the macroscopic form. A practical form of the CE with steady flow is[1] given in equation 8.1:

$$A_1 v_1 = A_2 v_2 = q \qquad \textbf{(8.1)}$$

where

A_i is cross-sectional area (L^2) at point *I*;
v_i is the velocity (L/T) at point *i*;
q is the flow rate (L^3/T).

When flow is unsteady as with flow routing, one can write

$$q_i dt = q_o dt + \Delta Vol \qquad \textbf{(8.2)}$$

where

q_i is inflow rate (L^3/T);
q_o is the outflow rate (L^3/T);
ΔVol is the change in storage (L^3/T);
dt is time increment (T).

Froude Number and Critical Flow

The Froude number (Fr) is defined as

$$Fr = \frac{v}{\sqrt{gD_h}} = \frac{q}{A\sqrt{gD_h}} \qquad \textbf{(8.3)}$$

where g is gravity (L/T^2).

The formulation for Froude number is general. The hydraulic depth D_h is cross-sectional area divided by the top width as discussed above. The hydraulic depth is y for rectangular cross sections. Flows with a Froude number exceeding 1 are supercritical. Flows with Froude numbers less than 1 are deemed subcritical or tranquil.

Flows with Froude number = 1 are deemed critical. Critical flow is the basis of flow measuring devices such as weirs. A simple test for assessing sub- or supercritical flow is to test whether a wave will propagate upstream. The flow is subcritical if a wave will propagate upstream. If the flow jets up the sampling device, the flow is supercritical. Flows usually go critical in the exit section of a channel with upstream subcritical flow. Hydraulic jumps occur when flows move from a steep channel to a much flatter channel. One may substitute the channel depth for hydraulic depth in wide trapezoidal channels.

Energy Equation

The energy equation describes the conservation and dissipation of energy in a flow. Flow energy arises from three components. Elevation energy, energy due to depth or pressure, and kinetic energy describe the energy at a point. The units of the energy equation are *L-F*/unit weight or *L*. Thus, fluid energy is expressed as length units, or *head.* Refer to a fluid mechanics text (e.g., Streeter 1971) for an in-depth derivation of the energy equation. One may relate energy loss to the surface slope. This enables the energy relationships for a constant channel geometry shown in equations 8.4a–b:

$$E_1 = E_2 + losses_{1\text{-}2} \tag{8.4a}$$

$$E_i = v^2/2g + y_i + z_i \tag{8.4b}$$

where

E_i represents the kinetic, depth, and elevation energy components;
y_i is the depth of flow at i;
z_i is the elevation of point i above a datum;
$losses_{1\text{-}2}$ is the change in energy per unit travel length, may be expressed as the surface slope per unit length as shown in equations 8.4a–8.6. Figure 8.6 shows the nomenclature.

The energy equation is also often expressed as

$$e_1 + z_1 = e_2 + z_2 + losses_{1\text{-}2} \tag{8.4c}$$

where E_i is the kinetic and pressure depth energies. One sometimes calls the parameter e the specific energy.

Specific energy e measures energy from the bottom of the channel. Specific energy is a useful concept when flows are approximately parallel and when one may neglect energy

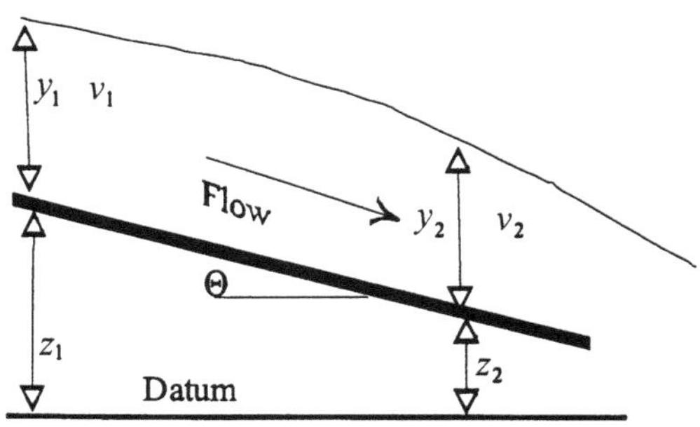

Figure 8.6. Basic nomenclature for open channel flow energy computation.

dissipation. Losses are then due to flow transitions, hydraulic jumps, or other phenomena under study. The energy loss rate is described by the surface slope, computed using

$$surface\ slope = s_f = \frac{\left(\frac{n^2 v_1^2}{\phi_{dim}^2 R_1^{4/3}}\right)_1 + \left(\frac{n^2 v_2^2}{\phi_{dim}^2 R_2^{4/3}}\right)_2}{2} \quad \textbf{(8.4d)}$$

where

s_f is the slope of the water surface (dimensionless);
n is the Manning roughness (discussed in detail in uniform-flow analysis);
v_i is velocity (L/T);
R_i is hydraulic radius (L);
ϕ_{dim} is a dimensions-related constant.

One may also write slope relationships as

$$S_o = (z_1 - z_2)/fx = tan\,(\acute{A}) \quad \textbf{(8.5)}$$

$$fe/fx = S_o - S_f \quad \textbf{(8.6)}$$

where

S_o is the bottom slope;
S_f is the water surface slope.

The losses are equal to the elevation fall per unit horizontal length with uniform flow. The bottom and water surface slopes provide a way to visualize losses. Although total energy changes when elevation changes, *the specific energy does not change in uniform flow.* Thus, knowing the conditions at the outfall of a channel, one can integrate upstream until $\Delta e/\Delta x$ approximates zero (uniform flow). Upstream integration also applies for a backwater curve, given the depth at a point behind a dam. Upstream integration is applicable in any subcritical flow situation. Downstream integration is applicable to supercritical flow problems, though it is rarely used.

The energy equation is useful for describing transitions from reservoirs to channels, from culverts to channels, or related situations. As is the custom in pipe flow analyses, one may express losses as $K_{tran}v_2^2/2g$. The transition geometry determines K_{tran}, the loss coefficient.

The energy equation, coupled with a flow equation such as the Manning equation, is the basic relationship used in analyzing transition flows, including emergency spillways and culverts. The energy equation is the basic relationship used in developing backwater curves. *In uniform flow, the fluid loses elevation potential at the same rate as the flow dissipates energy. Open channel flows dissipate energy via friction at the wetted perimeter and turbulence within the flow. We will be mainly concerned with uniform flows.*

Momentum Equation

The momentum equation comes directly from application of Newton's second law. One equates the sum of external forces to the fluid momentum change to arrive at the momentum equation. Equation 8.7a and Figure 8.7 show the momentum balance:

$$\sum force = mass\frac{\Delta v}{\Delta T} = \frac{v_1\,\gamma v_1\,A_1}{g\ unit\ time} - \frac{v_2\,\gamma v_2\,A_2}{g\ unit\ time} = M_1 - M_2 \quad \textbf{(8.7a)}$$

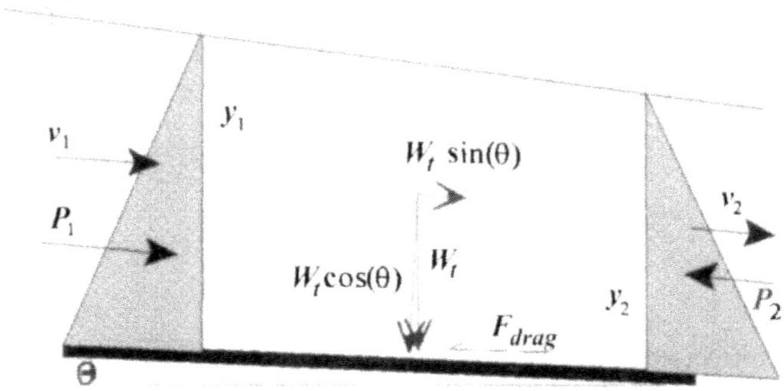

Figure 8.7. Schematic of channel showing terms of the momentum balance.

Substituting $q = A_1 v_1 = A_2 v_2$, one can also write equation 8.7a as

$$\sum force = mass\frac{\Delta v}{\Delta T} = \frac{q\gamma}{g}(v_1 - v_2) = \frac{q^2\gamma}{gA_1} - \frac{q^2\gamma}{gA_2} = M_1 - M_2 \qquad \textbf{(8.7b)}$$

where

pressure force P_1 is $\gamma y_1^2/2$ (F);
pressure force P_2 is $\gamma y_2^2/2$ (F);
W_t sinθ is the horizontal weight component (F);
F_{drag} is the drag force induced by channel bottom and wall resistance (F);
ΣForce $= P_1 - P_2 + F_{drag} + W_t \sin\theta$; $(q\gamma/g)(v_1 - v_2)$ is the momentum change (F);
q is flow rate (L^3/T);
v_i is velocity at locations $I = 1$ or 2 (L/T);
A_i is the cross-sectional area at $i = 1$ or 2 (L^2);
γ is the unit weight of the fluid (F/L^3);
g is the acceleration of gravity (L/T^2).

In words, the momentum equation states

pressure force$_1$ + horizontal component of weight − pressure force$_2$ − drag
= change in momentum

This statement of the momentum equation assumes a small slope. We will now use the momentum equation to develop a relationship describing the hydraulic jump in a rectangular channel. Starting with equation 8.7b we have

$$M_1 - M_2 = \Sigma Force$$

where

$M_1 = \gamma q^2/gA_1$;
$M_2 = \gamma q^2/gA_2$.

With a rectangular channel of width b, substituting $q = V_1A_1 = V_2A_2$, $A_1 = by_1$, and $A_2 = by_2$ and letting pressure force components cancel and neglecting drag results in

$$\frac{q_{uw}^2}{g}\left(\frac{1}{y_1} - \frac{1}{y_2}\right) = \frac{1}{2}(y_2^2 - y_1^2) \qquad \textbf{(8.8)}$$

where

q_{uw} is the flow per foot of width (= q/b); units are L^2/T;
g is gravity (L/T^2).

Substituting the definition of the Froude number $Fr = q_{uw}/y(gy)^2$ allows us to write

$$\frac{y_2}{y_1} = \frac{1}{2}(\sqrt{1 + 8Fr_1^2} - 1) \quad \textbf{(8.9)}$$

The momentum principle is useful when energy dissipation is extremely great over a short distance and bottom drag is negligible. The hydraulic jump is one such example. For nonrectangular channels, use the appropriate hydraulic element expressions and use the hydraulic depth D in the Froude number. The hydraulic jump equations for nonrectangular channels can be difficult to solve due to the number of roots required. Refer to works such as Chow 1959 and French 1985 for more details.

The hydraulic jump is important for energy dissipation. Energy dissipaters are used at the foot of spillways and at other transitions from steep to mild slopes.

Example 8.1: Assume a flow of 80 cfs in a 3 ft wide rectangular channel down a steep slope; the depth is 2 ft. The slope makes a sudden transition to a lower slope. Will a hydraulic jump occur?

The channel cross-sectional area is 3 ft × 2 ft = 6 ft^2. The hydraulic depth for a square channel is A/t = 6 ft^2/3 ft = 2 ft; $v_1 = q/A_1$ = (80 ft^3/s)/6 ft^2 = 13.33 ft/s

The Froude number $Fr_1 = v/(gy)^{0.5}$ = (13.33 ft/s)/[32.2 ft/s^2 (2 ft)]$^{0.5}$ = 1.66 and flow is supercritical.

From equation 8.9, $y_2 = \left(\frac{2\text{ ft}}{2}\right)\{[1 + 8(1.66)^2]^{0.5} - 1\}$ = 3.8 ft with conservation of momentum. The depth would be somewhat less in practice because some momentum is lost due to drag forces on channel bottom and sides.

From continuity, v_2 = (80 ft^3/s)/[(3.8 ft)(3 ft)] = 7.01 ft/s. It is assumed that the channel is steep enough to maintain this velocity. Fr_2 = (7.01 ft/s)/[32.2 ft/s^2 (3.8 ft)]$^{0.5}$ = 0.63 (subcritical). Expect a hydraulic jump.

Example 8.2: How much energy is lost in the hydraulic jump of Example 8.1?

The kinetic energy is

$$v_1^2/2g = (13.33\text{ ft/s})^2/[(2\text{ ft})(32.2\text{ ft/s}^2)] = 2.759\text{ ft}$$
$$v_2^2/2g = (7.01\text{ ft/s})^2/[2(32.2\text{ ft/s}^2)] = 0.767\text{ ft}$$

The static energy is

$$y_1 = 2\text{ ft}$$
$$y_2 = 3.8\text{ ft}$$

Combining static and kinetic energy leads to

$$e_1 = v_1^2/2g + y_1 = 4.759\text{ ft}$$
$$e_2 = v_2^2/2g + y_2 = 4.567\text{ ft}$$

The lost specific energy was $e_1 - e_2$ = 4.759 ft − 4.567 ft = 0.192 ft. This is not surprising since the hydraulic jump is an energy dissipater.

The hydraulic jump provides energy dissipation in structures, further discussed in Chapter 11.

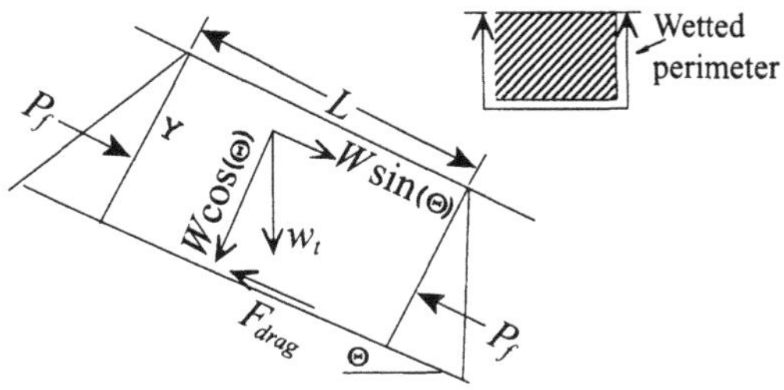

Figure 8.8. Simplified forces bearing upon uniform flow in an open channel.

Chezy Uniform Flow Equation

Engineers outside the United States employ the Chezy uniform flow equation for channel design. The Chezy equation has theoretical roots. Consider the free-body diagram in Figure 8.8. Evaluating forces along the direction of flow, one can write equations 8.10 and 8.11, assuming uniform flow:

$$F_{drag} = \frac{1}{2}\rho C_D A_d v^2; \quad A_d = PL \tag{8.10}$$

$$W_t \operatorname{Sin}(\theta) + P_1 - P_2 = F_{drag} \tag{8.11}$$

where

F_{drag} is a resisting force;
P_1, P_2, is the pressure force at points 1 and 2.

The pressure forces cancel and thus do not contribute. The component of the weight parallel with the channel bottom counters the resisting fluid drag force. Writing F_{drag} as $C_d PL\rho v^2$ and expressing W_t as ρgAL and substituting into equation 8.11 gives the first two terms of equation 8.12:

$$\rho gAL\sin(\theta) = C_D PL\rho v^2 \quad [\approx \gamma RS = t = C_D \rho v^2] \tag{8.12}$$

where

L is an arbitrary length;
C_d is a drag coefficient
$\sin(\theta) \sim \tan(\theta)$ represents the slope;
ρ is the water density;
A is the cross-sectional area;
g is gravity.

From the first two terms of equation 8.12, recall the definition of the hydraulic radius R as A/P and assume $\sin(\theta)\ \tan(\theta)$ = slope S_o. Next, solve for v and consolidate constants. We then have

$$v = C_{Chezy}\sqrt{RS_o} \tag{8.13}$$

where

R is the hydraulic radius (L);
S_o is the slope (−);

v is the velocity (L/T);
C_{Chezy} is the roughness coefficient, sometimes called the Chezy coefficient ($L^{0.5}/T$).

The two terms on the right-hand side of equation 8.12 show the forces expressed as stress on the channel side.

Manning Uniform Flow Equation

The Manning uniform flow equation is the standard uniform flow equation in the United States. According to Henderson (1966), an Irish engineer by the name of Manning discovered that the Chezy equation multiplied by $R^{1/6}$ provided consistently better agreement with experimental data. It is shown as

$$v = \frac{£_{dim}}{n} R^{2/3} S^{1/2} \quad \textbf{(8.14a)}$$

$$v = \frac{\phi_{dim}}{n} R^{0.67} S_o^{0.5} \quad \textbf{(8.14b)}$$

where ϕ_{dim}=1.486. This is (3.28 ft/m)$^{0.333}$ for English (ft/s) units and 1 for SI (m/s) units. The constant n is the channel roughness. The roughness factor n is a straightforward modification of the Chezy roughness C where $1/C$ was found experimentally to depend on $R^{1/6}$. The unit for n is s/ft$^{0.33}$. In the flow form, found by multiplying both sides of the equation by area, we have equation 8.14b.

Frequently one sees the term *conveyance*. King and Brater (1963) define conveyance (K_{conv}) as

$$K = \frac{\phi_{dim}}{n} A R^{2/3} \quad \textbf{(8.15a)}$$

$$q = K_{conv} \sqrt{S_o} \quad \textbf{(8.15b)}$$

Solving for velocity or flow with known depth and channel parameters is straightforward with the Manning equation. Solving for depth requires a rearrangement of the equation to have terms containing depth on the left and all other known parameters on the right, as shown in equation 8.16:

$$AR^{2/3} = \left(\frac{nq}{\phi_{dim}\sqrt{S}} \right) \quad \textbf{(8.16)}$$

Substitute appropriate channel elements into the left-hand side of the above relationship and put the known values into the right-hand side. Then solve for the y value satisfying the equation.

Tables 8.2 and 8.3 summarize values for n with naturally and artificially lined channels.

Channel Design and Construction

A downhill gradient must exist for an open channel to be an option for transport. The constancy of the flow, the amount of the slope, and to some extent the applications govern the nature of the channel design. The overarching goal is to satisfy the water conveyance requirements economically.

Table 8.2. Permissible velocity and Manning roughness values for naturally lined channels.

Material	Approximate roughness n values for stability design[a]	Clear-water permissible velocity v_p values (m/s)[b]	Water-transporting colloidal silt permissible velocity v_p values (m/s)[c]
Fine sand, colloidal	0.02	0.46	0.76
Sandy loam, noncolloidal	0.02	0.53	0.76
Alluvial silts, noncolloidal	0.02	0.61	0.92
Ordinary firm loam	0.02	0.61	1.07
Volcanic ash	0.02	0.76	1.07
Shales and hardpans	0.025	0.76	1.83
Stiff clay, very colloidal	0.025	1.14	1.53
Alluvial silts, colloidal	0.025	1.14	1.53
Fine gravel	0.02	1.83	1.83
Graded loam to cobbles when noncolloidal	0.03	0.76	1.53
Graded silts to cobbles when colloidal	0.03	1.14	1.68
Coarse gravel, noncolloidal[d]	0.025	1.22	1.83
Cobbles and shingles[d]	0.035	1.22	1.68

Source: Adapted from Lane 1955.

[a]Suggested n value for capacity design of intermittent channels is 1.25 times the table value, not to exceed 0.06. For soils not listed, use judgement. If soil is known to be easily erodible, use sandy loam values. If soil is erosion resistant, treat as shale/hardpan.

[b]Convert velocities to ft/s by multiplying above values by 3.28.

[c]Use velocities with silt unless runoff comes from lined surface such as a parking lot or flow is designed for specialized purposes such as aquaculture; convert velocities to ft/s by multiplying above values by 3.28.

[d]See references such as Simons and Senturk 1992 or Goldman et al. 1986 for design procedures for linings composed of larger aggregate materials.

Table 8.3. Typical Manning equation roughness n values for a variety of artificially lined channels or conduits.

Lining material	Minimum n	Maximum n[a]
Artificially lined channel:		
Asphaltic concrete	0.012	0.018
Concrete	0.017	0.03
Smooth metal	0.011	0.015
Corrugated metal	0.021	0.026
Plastic	0.012	0.014
Shotcrete	0.016	0.017
Planed wood	0.01	0.015
Unplaned wood	0.011	0.015
Pipe materials:		
Cast iron pipe	0.011	0.015
Clay or concrete drain	0.011	0.02
Clay or vitrified sewer	0.01	0.017
Corrugated plastic tube	0.015	0.02
Corrugated metal ring	0.021	0.026
Corrugated metal helical	0.013	0.015

Source: Adapted from Chow 1959.

[a]Use the maximum n in cases of intermittent flows; use an average of the maximum and minimum n values for continuous flows.

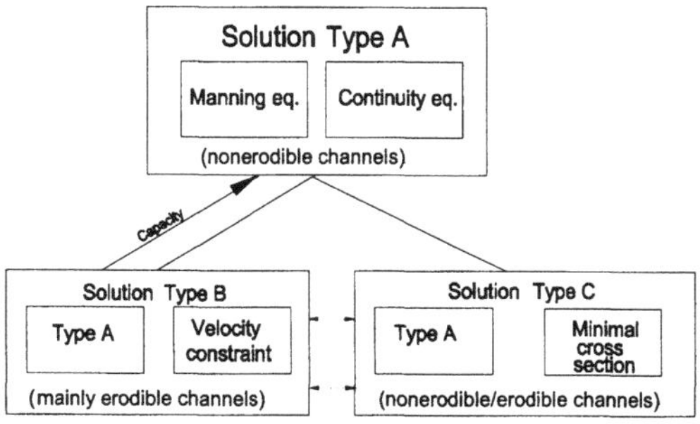

Figure 8.9. Interrelationships among the channel design problem solution types discussed in this chapter. Satisfaction of Type B may or may not lead to satisfaction of Type C and vice versa. After satisfying Type B, one may revisit Type A, particularly when flow is occasional and a capacity evaluation is appropriate.

This text evaluates three channel flow problem solution types:

- Type A, where stability is not an issue due to the presence of an artificial or natural rock lining, and one desires to compute depth with other geometric elements known
- Type B, where natural lining stability is important, and one simultaneously solves for depth with velocity fixed
- Type C, where optimal cross section is important, and stability may or may not be important depending on whether or not the lining is bare soil or other natural lining that may be susceptible to erosion

Figure 8.9 shows the interrelationship between these problem categories. The Type A solution approach is foundational to Type B and C solutions. A given channel design may or may not satisfy Type B and Type C solution criteria simultaneously. One must be able to apply each design approach in creative ways to determine the best engineering solution.

Channel design requires the channel to carry the design flow without eroding. One may stabilize by placing a nonerodible lining or by satisfying velocity constraints in erodible linings. Channels having erodible linings require evaluation under multiple criteria, especially when the flows are occasional.

Type A: Solutions Involving the Manning Equation and Continuity Equation without Additional Constraints

Solution Type A applies where channels are naturally lined with rock or artificially lined with a nonerodible material; key geometric elements except one are fixed; straightforward solution for q. One may use the Type A solution procedure to solve problems involving:

- channels where all elements except depth are fixed by external constraints (e.g., the channel construction machinery determines bottom width and side slopes; see ASAE 1997a, c)
- trapezoidal drainage ditches with entering tile lines requiring specified depth (solve for required bottom width given a required depth and side slope), roughness, and flow rate
- determination of existing channel performance when the flow rate changes or roughness increases due to age or other conditions

- determination of channel performance in capacity evaluation when other criteria (e.g., Type B) set the bottom width (trapezoid) or side slope (triangular)

External circumstances often dictate channel, slope, roughness, flow rate, side slope, and bottom width. One can solve for depth iteratively. One may then readily compute other hydraulic elements. The solution steps are as follows:

1. Write the Manning equation as $v = (\varphi_{dim}/n)\, R^{2/3}S^{1/2}$ or $q = (\varphi_{dim}/n)\, AR^{2/3}S_o^{1/2}$. Select n for the material and increment upward if the flow is occasional.
2. If all geometry is known, and if slope, roughness, v, or q is unknown, solve one equation in step 1 to get the unknown directly.
3. If the roughness, slope, and flow are known and one other geometric entity is unknown, use the flow form of the Manning equation and isolate the known quantities.
4. For a general trapezoid with fixed bottom width (common), solve for the left-hand side; put in hydraulic elements for a trapezoid; substitute iterative depth values into the right-hand side until the sides agree to within 5%. Add freeboard amount to depth to get the depth with freeboard. If the geometry is other than a trapezoid (e.g., parabola or triangle with fixed top width), solve in a similar way, using appropriate hydraulic elements. Add freeboard.
5. Check the Froude number. In case of near critical flow (e.g., $Fr = 1\ 0.1$), increase roughness by considering other materials, change slope if possible by considering other routes, change bottom width, or use some other combination of methods to avoid the critical condition. Critical flows may develop destructive waves. Supercritical flows require energy dissipation approaches such as plunge pools, baffles, or hydraulic jumps; therefore, one usually desires to decrease slope or increase bottom width. Designs involving supercritical flow, beyond the scope of this text, should also incorporate energy dissipation devices and prevent standing waves by adding turns and constrictions (see Chow 1959).

Example 8.3: Given a trapezoid channel, $n = 0.025$, slope = 0.1%, steady flow = 28.3 m³/s, bottom width = 6.1 m, $z = 2$, freeboard factor = 20% of design depth, find the normal depth. Substitute into $nq/\varphi S_o{}^{0.5} = AR^{2/3}$ or

$$0.025(28.3\ \text{m}^3/\text{s})/(1)0.001^{0.5} = 22.37 = (6.1\ \text{m}\ y + 2y^2)\{(6.1\ \text{m}\ y + 2y^2)/[6.1\ \text{m} + 2y(1 + 2^2)^{0.5}]\}^{0.67}$$

Estimate depth for starting the iteration by making the wide channel assumption where hydraulic radius R is equal to the depth and $z = 0$ in equation 8.16. Then equation 8.16 appears as

$$22.37 = (6.1\ \text{m}\ y)y^{0.67} \quad \text{and} \quad y = 2.18\ \text{m}$$

Start the iteration with this value for depth.

An iterative solution for y yields $y = 1.9$ m and $D = 2.286$ m. For occasional flows, increase the n by 25% before calculating y. The example is adaptable for other channel cross sections by using appropriate hydraulic elements.

Check the Froude number: $Fr = v/(gD_h)^{0.5}$; compute A to be 18.85 m²; $v = 1.5$ m/s; $t = 13.7$ m; $D_h = 18.85\ \text{m}^2/13.7\ \text{m} = 1.38$ m; $Fr = (1.5\ \text{m/s})/(9.81\ \text{m/s}^2 \times 1.38\ \text{m})^{0.5} = 0.41$. Therefore, the flow is subcritical.

Example 8.4: A trapezoid channel has $n = 0.025$, slope = 0.1%, steady flow = 28.3 m^3/s, $z = 2$, and freeboard factor = 20% of design depth. Tile lines enter the channel at a depth of 2 m, requiring the design depth to be specified as 2.5 m (without freeboard). What is the required bottom width?

Starting with equation 8.16 and substituting known quantities gives

$$0.025(28.3\ m^3/s)/(1)0.001^{0.5} = [b(2.5\ m) + 2(2.5\ m)^2]$$
$$\{[b(2.5\ m) + 2(2.5\ m)^2]/[b + 2(2.5\ m)(1 + 2^2)^{0.5}]\}^{0.67}$$

An iterative solution gives a bottom width of approximately 2.3 m. The Froude number is 0.40 and the depth with 20% freeboard is 3 m.

Channels having parabolic and triangular cross sections have straightforward iterative solutions. For a parabolic channel, one may substitute functions of hydraulic radius and area into equation 8.16 and solve for top width or depth assuming one or the other is known. For a triangular channel, one may usually solve for depth with known side slope z. For a circular channel, where diameter determination is the goal, one must insert functions of hydraulic radius, wetted perimeter, and cross-sectional area versus depth (see Figure 8.4) into equation 8.16 to iterate for depth assuming diameter is known. An alternative approach for the Type A evaluation when the Manning n is not constant with depth is given in the vegetated waterway analyses for each of these cross sections.

Type B: Manning Equation and Continuity Equation Subject to Velocity Constraints; Erodible Channel Stability Design

Type B solution approaches are used where lining stability is at issue. The Type B procedure is useful in situations where stability requirements such as permissible velocity are implied. Examples include

- terraces and diversions (manipulate slope usually; discussed in Chapter 10)
- grassed waterways (manipulate hydraulic radius usually; discussed in Chapter 9)
- naturally lined surface drainage channels
- channels are usually short (e.g., field scale) and spoil material handling is usually not an issue
- channels usually flow intermittently; one designs for stability and capacity; changing the Manning n to a higher value or, with vegetated waterways, assuming unmowed vegetation is usually all that is required for the capacity design.

Stability of a natural channel depends on the erosivity of the flow. Erosivity arises from drag forces acting on the wetted perimeter. Channel drag forces cause linings to become unstable. One can estimate drag forces with

$$\tau = C_d \frac{\rho}{2} v^2 \quad \textbf{(8.17)}$$

where

τ is the shear stress ($F/L^2 = M/LT^2$);
C_d is a drag coefficient ($-$);
ρ is the mass density (M/L^3);
v is the flow velocity (L/T).

Since shear stress is the mechanism causing natural linings to fail and since shear stress is proportional to the square of velocity, one can specify a maximum flow velocity. This velocity is the maximum permissible velocity for the lining.[2] The two most popular design criteria for channel stability are the tractive force method and the permissible velocity method. The text will cover the permissible velocity method, leaving the tractive force method to more advanced texts such as French 1985.

With the Manning equation, one may manipulate hydraulic radius, slope, or surface to force a design to conform to a permissible velocity requirement. The hydraulic radius may be manipulated (reduced) by forcing a channel to flow broad and shallow, for example. One may vary the slope sometimes. Restrictions on roughness (e.g., using channel linings of various sizes to vary the Manning n) are possible; discussion of varying roughness is deferred to more advanced treatments. Restrictions on hydraulic radius and slope are the common approaches for velocity control.

This chapter focuses mainly on velocity control with management of hydraulic radius. Usage of slope for velocity control is deferred to Chapters 9 and 10. Set up the permissible velocity method–hydraulic radius control by performing the following steps:

1. Solve the Manning equation for $R = [(nv_p)/(\varphi_{dim} S_o{}^{1/2})]^{1.5}$. Dimensions are ft-s or m-s.
2. Solve the continuity equation for A from $A = q/v_p$.

3a. For a trapezoid: Write A and wetted perimeter equations in terms of bottom width, z, and depth. One usually knows z. See equation 8.18:

$$A = by + zy^2 \qquad \textbf{(8.18)}$$

Then solve for d and b after substituting and rearranging the relationships into a form solvable with the quadratic formula as follows:

$$R = A/[b + 2y(Z^2+1)^{0.5}] \qquad \textbf{(8.19)}$$

Solve the relationship for b as a function of y:

$$b = (A - zy_s^2)/y_s \qquad \textbf{(8.20)}$$

Substitute back into the relationship for R:

$$R = A/\{[(A - Zy^2)/y] + 2y(Z^2 + 1)^{0.5}\} \qquad \textbf{(8.21a)}$$

$$A = R\{[(A - Zy^2)/y] + 2y(Z^2 + 1)^{0.5}\} \qquad \textbf{(8.21b)}$$

Rearranging equation 8.21b as a quadratic equation in y results in

$$y^2[Z - 2(Z^2 + 1)^{0.5}] + (A/R)y - A = 0 \qquad \textbf{(8.22)}$$

Using the quadratic formula (note the sign change in equation 8.22) results in

$$y = [-b_1 \pm (b_1^2 - 4a_1c_1)^{0.5}]/(2a_1) \qquad \textbf{(8.23)}$$

where

$a_1 = 2[(z^2 + 1)^{0.5}] - z;$
$b_1 = -A/R;$
$c_1 = A.$

Put the sensible value (usually the smallest positive value unless other factors such as need to accommodate tile drain outlets force a deep depth) y into equation 8.20 to solve for bottom width b. Appendix B contains a closed form of the solution shown in equation 8.23.

Sometimes the trapezoid may degenerate to the case where $b = 0$. A negative under the radical in the above quadratic formula suggests a degenerating trapezoid. If so, one may fix the bottom width to zero or to a bulldozer blade width (e.g., 1.2 m). *The negative radical implies that the stream will flow at a less-than-critical velocity.* Design for capacity by increasing n by 25% using Type A approaches. In other cases, one may compute a negative bottom width. Other approaches to the negative problem are discussed in Chapter 9. Warner (1998) found this approach to work well for computerized solutions.

3b. For a parabola: One can use the approximate relation between hydraulic radius and depth (see Table 8.1) to solve for depth, then use the top width relation to go immediately to top width. This is suitable for design purposes. Solve for T. Add freeboard. Note: Capacity design for parabolic cross sections is discussed in Chapter 9.

3c. For a triangle: Using the approximate relation between hydraulic radius and depth (see Table 8.1), solve for depth. This is suitable for design purposes. Check the solution by solving the detailed relation for depth, get top width, and add freeboard. The triangle solution shown in the chart is one valid solution; however, another valid solution may exist that is apparent only from the mathematical solution. Complete the capacity design as described in step 3a.

4. For the capacity design when flows are occasional, increase n and resolve as a Type A problem, using the same bottom width as determined with stability and solve for depth. With trapezoids, use the bottom width just determined. With triangles, use the z just determined. Parabolic channel capacity design is discussed in Chapter 9.

Example 8.5: Given an occasional flow of 100 cfs (2.83 m^3/s) over a 2% slope with a natural trapezoidal channel with a $z = 4$ and where the loamy soil permissible velocity is 2 ft/s (0.61 m/s) and $n = 0.03$, determine the flow depth.

$$A = (100\ \text{ft}^3/\text{s})/(2\ \text{ft/s}) = 50\ \text{ft}^2$$
$$R = [0.03(2\ \text{ft/s})]/[1.49(0.02)^{0.5}]^{2/3} = 0.15\ \text{ft}\ (0.046\ \text{m})$$

Compute the depth using the quadratic equation (equation 8.23):

$$a_1 = 4.24$$
$$b_1 = -333.33\ \text{ft}$$
$$c_1 = 50\ \text{ft}^2$$

This gives $y = 0.1538$ ft, 78.46 ft. Reject the second solution as impractical.[3] Solving for b with the first solution using equation 8.20 gives

$$50\ \text{ft}^2 = by + zy^2$$
$$0.15\ \text{ft} = R = A/P = 50\ \text{ft}^2/[b + 2y(z^2 + 1)^{0.5}]$$

One may solve for b using equation 8.20:

$$b = [50\ \text{ft}^2 - (2)(0.15\ \text{ft}^2)]/0.15\ \text{ft} = 333\ \text{ft}\ (101.5\ \text{m})$$

The bottom width is quite large (unless this is a constructed overland flow waste treatment system!). Consider lining the channel with coarse gravel, having a $v_p = 1.83$ m/s (6 ft/s) and an $n = 0.025$. Proceeding as above,

$$A = 16.66\ \text{ft}^2$$
$$R = 0.6\ \text{ft}$$

$$a_1 = 4.24$$
$$b_1 = -27.77 \text{ ft}$$
$$c_1 = 16.66 \text{ ft}^2$$

This leads to $y = 0.67$ ft and $b = 22.2$ ft before freeboard. This design is more tenable.

If the expression under the radical above ($b^2 - 4ac$) were negative, the trapezoid would have been impossible. One could then set the bottom width and move onto the capacity design phase. One would design for capacity by increasing n by, say, 25% and using the bottom width determined in the stability design. Capacity design requires a Type A solution.

The capacity design

$$AR^{0.67} = 1.25nq/\varphi S^{0.5} = 1.25(0.03)(100 \text{ cfs})/1.486(0.02)^{0.5} = 17.844$$
$$= (by + zy^2)\{(by + zy^2)/[b + 2y(z^2 + 1)^{0.5}]\}^{0.67}$$

Substituting $b = 22.2$ ft and iterating gives $y = 0.75$. Adding freeboard of 20% gives $y_f = 0.9$ ft.

Selection of the appropriate permissible velocity is critical. *In cases when the radical is not negative, computed bottom width for stability is sensitive to the chosen permissible velocity. Do not use an overconservative permissible velocity.*

Example 8.6: Suppose Example 8.5 (loamy soil lining) had specified a triangle.

$$R = y/2$$

from Table 8.1; therefore,

$$y = 2(0.15) = 0.3 \text{ ft}$$
$$A = zy^2$$

therefore,

$$z = 50/0.3^2 = 555$$

$$T = 2(0.3)(555) = 333.3 \text{ ft}$$

This is an acceptable approximation.

Calculations for a channel lined with coarse gravel are left to the student, as are freeboard calculations. Increase n by 25% and carry the stability z value over to the capacity design.

Example 8.7: Suppose Example 8.5 (loamy soil lining) had specified a parabolic shape.

Using relationships for a parabolic cross section results in a depth of $y = 1.5R$ (see Table 8.1) = 0.225 ft. $T = 50 \text{ ft}^2/[(0.67)(0.225 \text{ ft})] = 332$ ft. Calculations for a channel lined with coarse gravel are left to the student, as are the capacity design and freeboard calculations. We discuss capacity design for parabolic channels in Chapter 9.

The example problem requires a wide, shallow channel to convey the flow at the specified velocity. In practice, we would look for an alternative route requiring a lower slope or an alternative lining without the velocity constraint.

Occasional Flow Channels

These channels usually accumulate debris between runoff events. This increases the resistance to flow. To compensate, one may increase the Manning n value. Increasing the Manning n value by 25% (but not to exceed $n = 0.06$ for typical field-scale problems) is suggested. This suggestion applies to solution Types A and C as well. For trapezoidal channels, the capacity design would follow the stability design. The purpose of the stability design is to set the bottom width. Use the determined bottom width along with a Manning n of 0.06 when determining the final depth and freeboard. With triangular channels, one may find the z in the stability calculation and determine the final depth using an adjusted Manning n. A rapidly converging trial-and-error approach for parabolic channels is given in Chapter 9.

Type C: Solve the Manning Equation and Continuity Channel Subject to Minimal Wetted Perimeter with or without Side Slope z Constraint; the Economical Channel

The Type C solution procedure is useful when one wants to minimize excavation and lining costs. Examples include

- lined channels requiring minimal excavation, such as long irrigation canals or other water delivery channels
- long, unlined channels such as drainage channels provided slope is low (e.g., <0.003) and permissible velocities are not exceeded

Often the economic channel is the one carrying the flow with minimum cross section. For naturally lined channels, the slope will often be low (e.g., 0.002 or less) in order to be stable. One can show that the best choice for the trapezoidal channel bottom width to depth (b/y) ratio is given by

$$\frac{b}{y} = 2(\sqrt{1 + z^2} - z) \tag{8.24}$$

It can be shown that the optimum z is 0.5774 (giving a $b/y = 1.15$). This is half an equilateral trapezoid (Streeter 1971). The analysis assumes concrete (or other lining) costs in the bottom section are the same as costs in the side sections. More advanced treatments account for variation in materials cost with position on sides or bottom (e.g., see French 1985). One can usually achieve a stable optimum channel if the slope is less than 0.002 or 0.003.

Flow, roughness, and slope are usually known with lined channels where velocity is not limiting. The goal here is to minimize the removed area with constraint on z. *Again, the optimum b/y ratio with no constraint on z is $b/y = 1.15$.* Advanced open channel flow texts further explore the optimum z question. Some typical z values are given in Table 8.4. To solve for the design depth:

1. Get b as a function of y from equation 8.24 using z values suggested in Table 8.4. Substitute into the hydraulic elements relations R and A in Table 8.1.
2. Write the Manning equation, substituting in the above relations. Solve iteratively for depth and back-solve for the bottom width. Add freeboard.

Table 8.4. Constraining z values for general trapezoidal channels, natural linings as specified; or optimum/ y/b values in the case of other geometries.

Soil or lining type	Side slopes—Horizontal:vertical = z :1 In trapezoidal channels no deeper than 1.2 m, z should not be less than . . .	In trapezoidal channels 1.2 m and deeper, z should not be less than . . .
Peat and muck	0[a]	0.25[a]
Heavy clay	0.5[a]	1
Clay or silt loam	1	1.5
Sandy loam	1.5	2
Loose sandy soil	2	3
Channel lined with gravel, cobbles, etc.	4 to 6[b]	4 to 6[b]
Any channel to be routinely crossed	4 or greater	4 or greater
Optimum (stabilized sides) trapezoid	$z = 0.5774$	
Optimum (stabilized sides) triangle	$z = 1$	
Optimum (stabilized sides) rectangle	$b/y = 2$	
Lined (stable) circular channel	$y = D/2$	

Sources: Adapted from Schwab et al. 1993 and Streeter 1971.

[a]In the case of an optimized geometry, use the optimal Z if it equals or exceeds the tabulated amount.

[b]Side slopes suggested for gravel and cobble lining are approximate; more detailed design approaches are required for determining stability of these linings. Simons and Senturk (1992) discuss design approaches for aggregate-lined channels.

3. If naturally lined, be sure the permissible velocity does not exceed that for the wetted surface. If artificially lined, check the permissible velocity if applicable and check the Froude number.

Example 8.8: Given a flow rate of 40 cfs (1.14 m^3/s), $n = 0.025$, trapezoid cross section, slope = 0.25%, silty clay ($z = 1$ if depth less than 1.2 m or $z = 1.5$ otherwise), assume d is less than 1.2 m, so $z = 1$ initially.

$$b = 2y[(1 + 1^2)^{0.5} - 1] = 0.84y$$

(equation 8.24).

$$R = (0.84y^2 + (1)y^2)/[0.84y + 2y(1 + 1^2)^{0.5}] = 0.501y$$

(Table 8.1 and equation 8.24).

$$A = [0.84y + (1)y]y = 1.84y^2$$

(Table 8.1 and equation 8.24).

$$q = (1/0.025)(0.501y)^{0.67}\,(1.84y^2)(0.0025)^{0.5} = 1.14\ \text{m}^3/\text{s}$$

(Manning with R and A). Solving for y gives $y = 0.77$ m.

Check z: $z = 1$ is acceptable since y is less than 1.2 m. If y were greater than 1.2 m, one would select the new z and resolve for y.

Velocity check: $v = (1/0.025)[(0.501)(0.77\ \text{m})]^{0.67}(0.0025)^{0.5} = 1.1$ m/s, less than permissible velocity for bare soil. If the velocity had been above, say, 2 m/s, then one would either consider stabilizing the lining or redesign the channel based on permissible velocity. Economic factors would determine the final design.

For the lined optimum triangle, use $z = 1$ in the hydraulic elements and do not use the b/y relationship. For the optimum lined rectangle, $b = 2y$. The optimum circular cross section is defined by $y = D/2$. In cases where the optimum channel carries occasional flows, increase the Manning n by 25% before starting the calculation, as with the Type A solution.

How does one determine which solution type is applicable? Channels with slopes greater than 0.003 usually require the Type B approach unless an artificial lining is used wherein velocity is not constrained. Channels with slopes less than 0.002 can usually be designed using Type C solutions because they will not usually exceed the critical velocity. One should check the economics of channel cross sections designed with lining (Type A) and without lining (Type B and Type C) when the slope lies between 0.002 and 0.003. Aesthetic factors and client desires should be evaluated before making the final decision.

Flow Transition Analyses

A flow transition occurs when flows move from a reservoir to a channel, from a channel to a culvert, or from one channel geometry to another. We will study two types of problems. Texts on open channel flow such as French 1985 cover flow transitions in detail. Losses in transitions may be expressed as $K_{tran}v_2^2/2g$ where v_2 is the exit velocity. Table 8.5 shows values of K_{tran} for several transition types.

The transition from a reservoir to a spillway calls for a combination of the energy equation to describe energy conservation between the reservoir and the spillway and the Manning equation to define the uniform flow depth in the spillway. This analysis is also useful for describing flow through a partially submerged culvert inlet.

Table 8.5. Suggested coefficients (K_{tran}) associated with selected channel transitions.

Flow situation	K_{tran} value
Sharp transition from reservoir to outlet channel	1.0
Flow from channel into culvert	0.7–1.0
Smooth transition from reservoir to outlet channel	0.5–0.6
Smooth transition from one geometry to a narrow geometry (flow convergence)	0.1
Sharp transition from narrow geometry to a wide geometry (diverging flow)	$0.3\ (v_u^2 - v_d^2)/2g$
Flow around smooth barrier	0.05–0.2

Example 8.9: Suppose a flow is entering into a 0.6 m wide concrete ($n = 0.015$) rectangular channel at a 5% slope from an impoundment, where the depth puts the surface level at 0.2 m above the channel bottom at the entrance. Losses can be expressed as $K_{tran}v_2^2/2g$ where as K_{tran} is a coefficient of 0.75. Velocity v_1 can be taken as zero. Velocity $v_{chan} = v_2$ is the uniform velocity expected in the channel (e.g., use the Manning equation). Depth $y_1 = 0.2$ m and y_{chan} (y_2) must be calculated. We can sketch the problem as shown in Figure 8.10.

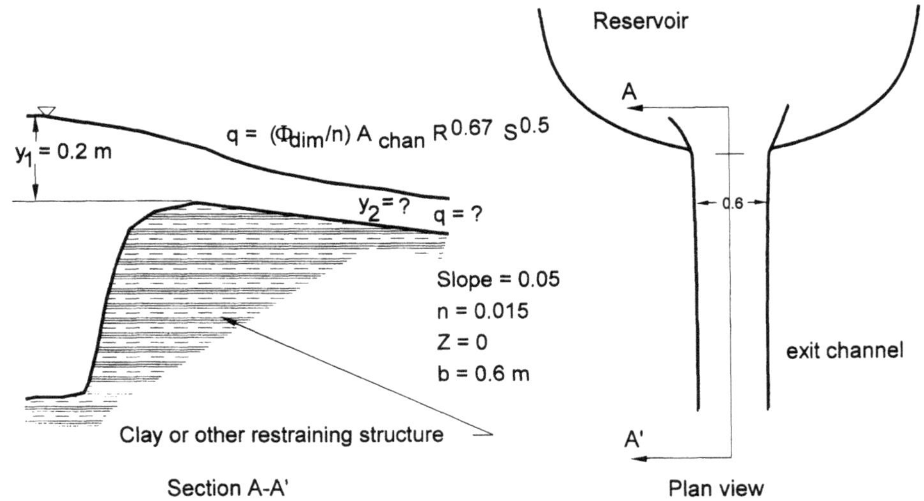

Figure 8.10. Schematic of a flow transition problem.

The energy equation, written in terms of specific energy (recall specific energy is constant with uniform flow), coupled with the Manning equation, appears as

$$y_1 = \frac{v_2^2}{2g} + y_{chan} + \frac{K_{tran}v_2^2}{2g}; \quad v_2 = v_{chan} = \frac{\phi_{dim}}{n_{chan}} R_{chan}^{2/3} \sqrt{S_o} \qquad \textbf{(8.25a,b)}$$

Substitute the expression for channel hydraulic radius $R = (0.6 \times D_2)/(0.6 + 2 \times D_2)$, slope, and roughness into the Manning equation. Then substitute the Manning equation for velocity v_{chan} in the energy equation. Refer to the nomenclature in Figures 8.6 and 8.10.

One can iteratively solve equation 8.25 for y_2. Substitute values for y_2 into the Manning equation expressed using hydraulic elements of the channel to solve for flow depth. Substitute this depth back into the Manning equation to get the velocity. Calculate the flow by applying the continuity equation. For the above condition, the depth is 0.0623 m and the flow is 0.0465 m^3/sec. One may increase or decrease the channel bottom width to meet flow requirements.

Hydrograph Routing in a Channel or Stream

Streams have noticeable effects on hydrograph shapes. Modeling the effect of streamflow on hydrograph shape is important in modeling complex watersheds. The complex watershed requires superposition of subwatershed hydrographs to find the complex watershed composite hydrograph (e.g., see Warner 1998). Consider a stream reach in equilibrium flow from a volume basis. As a hydrograph passes through the inlet of a channel reach, the stored volume

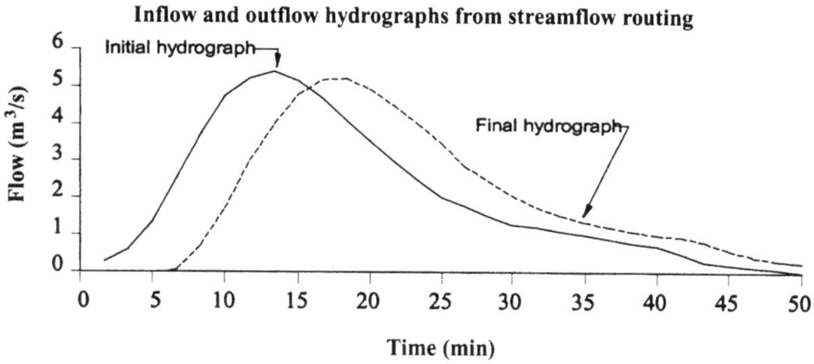

Figure 8.11. Influence of a 1000 m channel on a storm hydrograph with no inflow along the reach.

is increasing in the rising stage since inflow exceeds outflow. After hydrograph passage at the inlet, there is some point in time where outflow exceeds inflow. The flow is designated as unsteady. If flow is admitted along the stream reach, the flow is designated as unsteady, gradually varied flow. Several approaches may model these flows based on a digital solution of the continuity alone, ignoring momentum effects. Results of a storage approach known as Muskingum-Cunge (see Haan et al. 1994) are shown in Figure 8.11 for a simple stream reach. One may use combinations of the continuity equation and the momentum equations. French (1985), Chow (1959), and others give detailed discussions of unsteady flow modeling.

Channel Seepage Losses

Channel losses due to seepage may be analyzed using approaches developed by the U.S. Bureau of Reclamation (USBR 1952) where unit losses are proportional to the channel cross-sectional area using equation 8.26:

$$losses = c\sqrt{A_{cross}} \qquad \textbf{(8.26)}$$

where

losses is the flow rate lost per unit mile of length (cfs/mi);
A_{cross} is the cross-sectional area (ft^2);
c is a constant equal to 0 for sealed lining, 0.08 for compacted clay and clay loam lining; 0.13–0.2 for sandy loam and volcanic ash; 0.34−0.44 for sandy soil with rock or gravel.

Losses are minimized with proper compaction.

Construction

Some factors to consider in open channel design are the following:

A channel must carry the flow with adequate freeboard at the given channel slope.

Channel flow may be the peak flow of the runoff hydrograph. In the case of an irrigation canal, flow must be adequate to supply plant water needs plus 10–15% to account for seepage and other losses. In the case of drainage ditches, flow is determined using ASAE (1997b) standards (see Chapter 5).

One usually determines the channel slope from local topography. The decision about how to route the channel is largely economic and/or aesthetic.

Appropriately line the channel. Bare soil, stabilized soil (e.g., large rock), plastic liners, and asphalt all have a place. ASAE 1997d discusses installation of plastic liners in channels.

A channel must be stable (naturally lined channels). The limiting velocity method is one of several methods. The limiting velocity can be chosen from Table 8.2, depending on soil and flow conditions.

Table 8.4 gives guidance for side slope selection in naturally lined channels.

The channel must be functional. For example, a drainage channel must have adequate depth for drainage discharge, and a waterway outlet must be deep enough for terrace channels to drain. On vegetated outlets, see Chapters 9 and 10. On tile drainage outlets, design tile drain capacity, first setting depth within the berm.

A channel must be economical if feasible and if it is long. Often the economical channel is the one carrying the flow with minimum cross section. See the Type C solution method discussion.

The channel must be constructable with equipment on hand (graders, bulldozers, scrapers with blades made to give certain profiles, etc.).

The channel spoil must be considered. Spoil is often placed in a trapezoidal spoil bank having slope z values of 3 or 4 to facilitate driving on the spoil bank. The bank must be high and wide enough to contain the volume. One should provide a berm of 3 m between the spoil bank toe and the channel for access purposes. With on-farm construction one usually spreads the spoil over adjacent lands. With public construction or deep on-farm cuts a spoil bank is a part of the design. For a spoil bank, provide a berm for protection against sloughing and consider equipment operation and access when siting it. Berm width depends on channel z (if $z = 1$, then berm width $= 2d$; if $z = 2$, then berm width $= d$) and channel depth and there is a minimum width of 3 m. There is an increasing trend to separate topsoil from subsoil in earthmoving operations to facilitate revegetation of the soil following construction. Especially with irrigation canals, one may transport spoil along the canal length for filling depressions. The material must have appropriate well-graded texture to achieve adequate soil sealing. Chapter 12 further discusses soil sealing and compaction topics. Another alternative for spoil placement is to haul it to a fill area.

Channel curvature must be considered. Channels with constant slope usually change directions in regions with much topographic relief. Chow (1959) shows how curvature can affect the effective roughness of a channel. Channel curvature is defined by the intersection angle (deflection of one straight segment from an adjacent segment) and the radius of curvature of the curve. Degree of curvature is defined in Figure 8.12. Maximum curvatures as a function of slope and channel width are given in Table 8.6. Equipment used in construction of channels is shown in Table 8.7.

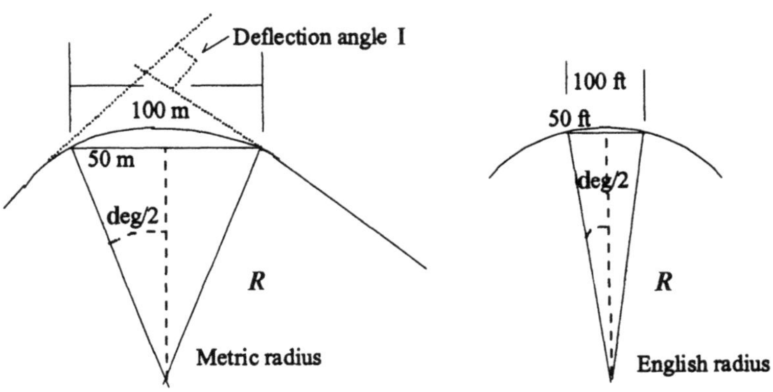

Figure 8.12. Schematic showing how degree of curvature is defined for English and metric units, where I is the deflection angle and deg/2 is half the degree of curvature.

Table 8.6. Maximum curvatures as a function of slope and channel width.

		Approx. maximum curvature (degrees)[a]	
Channel top width (m)	Slope (%)	English	Metric
4 to 6	< 0.06	19	62
4 to 6	0.06 to 0.12	14	46
5 to 11	< 0.06	11	36
5 to 11	0.06 to 0.12	10	33
11	< 0.06	10	33
11	0.06 to 0.12	7	23

Source: USDA-NRCS 1973.

Notes: [a] Degree of curvature in SI units is 3.28 times degrees of curvature in English units. The degree of curvature is not important in channels having low (e.g., less than the permissible velocity of bare soil) velocities and top widths smaller than those listed above.

Table 8.7. Construction machinery for channels.

Function	Action	
	Continuous (excavation or transport; generally used on long channels) short drainage channels or diversions)	Intermittent (alternate excavate and transport steps; generally used on
Excavation	Wheel excavator: deep, narrow; used for subsurface drainage installation	Dragline: These operate from a crane-type machine
	Plow-type ditcher: small shallow such as terraces	Clamshell
	Excavating conveyors: defined cross section; conveys spoil material to waiting transport or to channel side; large irrig. or dng canals	Backhoe: general purpose; wide variety of hoe attachments; pivoting hoes very good for finish work
	General-purpose motor blade grader: small, shallow channels	Crawler shovel: wide variety of shovel attachments; pivoting hoes also available
	Elevating grader: general; tillage for loosening soil followed by elevating conveyor belt	Motor scraper: general use; limited except for shallow channels
	Hydraulic dredge: removal of submerged sediments	Bulldozer: general purpose; wide variety of blade attachments and rear spades available
	Rotary ditcher: deep, narrow	Dozer w/pullback blade: good for finish
		Motor scraper: fairly good for excavation when fitted with spade attachment
		Crawler/wheel loaders: excavation and transport
Spoil Spreading	Blade grader: general transport	Bulldozer: fairly good for transport
	Pans: excavation and transport	Hauler: transport
	Tillage machines: finishing	Motor scraper: fairly good for transport
	Terracing machines: specialized for terrace channels	Pull-back blade: short-distance transport
	Sheepfoots roller/wheel compactor: used for compaction of clayey/sandy soils.	

Note: See Appendix E and for pictures of selected equipment.

Table 8.8. Customer requirements prompt sheet for application to open channel design problems.

Customer requirements	Customer requirements
Protect area	Water value
Aesthetics	Available $$ for excavation and spoil storage/transport
Water quality objectives	Floodplain impacts
Incoming water quality	Cost of land in the area
Drain upland area	Relation to other watershed BMPs
Minimal hydraulic impact on lower areas?	The larger hydraulic picture of the region
Clear water needed or silts allowable?	State/federal legislation/agency rules
Water conveyance efficiency requirement?	Client's concept of long and short; is it important to obtain an optimal cross section?
Land use in watershed	Availability of maps, surveys and aerial photos of water-shed and potential channel routes
Livestock or other wildlife in area?	
Safety requirements	
Sensitivity to blasting and other noise	
Water to be sold to customers along channel?	

The design question prompt sheets in Tables 8.8 and 8.9 may be helpful in focusing questions to clarify the above points in specific situations.

Design ditches and mark their location in the field with centerline and slope stakes. Slope stakes must be properly set according to the local topography. Channel construction is as much art as science. Short channels are usually constructed with intermittent equipment as shown in Table 8.7. Long channels such as drainage channels and irrigation canals are usually constructed with continuous equipment. Refer to works such as Harris 1994 for comprehensive treaments of channel construction techniques. Associations such as the Land Improvement Contractors Association (http://ourworld.compuserve.com/homepages/delmarva/staff.htm) also provide helpful information.

Table 8.9. Engineering factors prompt sheet for application to open channel design problems.

Engineering factors	Engineering factors
Grade direction	Energy extraction or dissipation methods
Life	Side slope limits
Curves allowed?	Bottom width limits
Roughness	Soil conditions
Leakage allowed?	Spoil storage/placement
Water treatment within?	Surrounding land use
Sediment load?	Baseflow vs. storm of other flow
Channel capacity	Temporary measures needed during construction
Soil stability	Flow measurements to be made?
Liner strength/endurance	Required precision?
Materials specifications	Interfaces to other channels/structures/drainage (surface or underground)
Special velocity requirements	

Shape and design requirements are influenced by the type of equipment available. Machinery should be selected to fit the requirements of the individual job; however, for smaller jobs available machines often dictate design parameters. Selection of equipment also depends on soil and water conditions, accuracy required, shape and dimensions of the channel and spoil bank, moving requirements, volume of work, and other financial factors. Explosives require much care and proper permits, experience, and training.

Channels deteriorate due to poor design, improper construction, and lack of adequate maintenance. Major causes for deterioration are sedimentation in the channel, excessive vegetation growth, channel and bank erosion, burrowing animals, high sediment load, poor location and alignment, improper depth and width, inadequate culvert and bridge capacity, failure to agree ahead of time on who should maintain the channel, and the general lack of interest in maintenance by the public, landowners, and officials. One should not routinely cross channels with vehicles or foot traffic.

Vegetate or otherwise protect spoil banks and maintain vegetation in vegetated channels. Check for scouring around bridges and culverts. Keep cattle from damaging banks. Look for sediment buildup and try to identify the source before the problem gets serious. Maintenance costs can be as low as 1–5% of the installation cost annually. One should anticipate this cost ahead of time.

Outlook

Design of a channel involves some or all of the Type A, B, and C solution approaches. A simplified decision aid for simple open channel design problems is provided in Figure 8.13. First, we considered the Manning-continuity equations wherein one usually solves for depth or, in some cases, channel width. Next, we considered solution to Manning-continuity with velocity constraints wherein one develops a cross-sectional geometry meeting hydraulic radius and area targets in erodible channels. Finally, we considered solutions to Manning-continuity with constrained wetted perimeter, and possibly the side slope z minimizing the channel cross-sectional area. For introductory channel design, Figure 8.13 will help one decide when various checks are important.

The introductory principles of channel design given here should not be considered the final word in channel design. Composite linings add to the complexity of channel design. Alternative procedures for design for stability (e.g., tractive force) are available. More advanced texts treat these problems. Chapters 9 and 10 treat two channel types, grassed waterways and terraces/diversions, in detail.

The equations of channel design may be cubic or even quartic. Blind application of equation-processing software to channel design equations can lead to imaginary solutions or, worse yet, real but impractical solutions. The engineer must determine the relevant solution. An Internet search will yield links to several commercially available channel design software packages (e.g., see Warner 1998).

Quality function deployment (QFD) can help one develop site-specific designs. Tables 8.8 and 8.9 give QFD prompt sheets. Tables 8.8 and 8.9 are not intended to serve as QFD sheets for a particular design. Rather, the tables list key items to be considered. You may modify these tables as you gain experience.

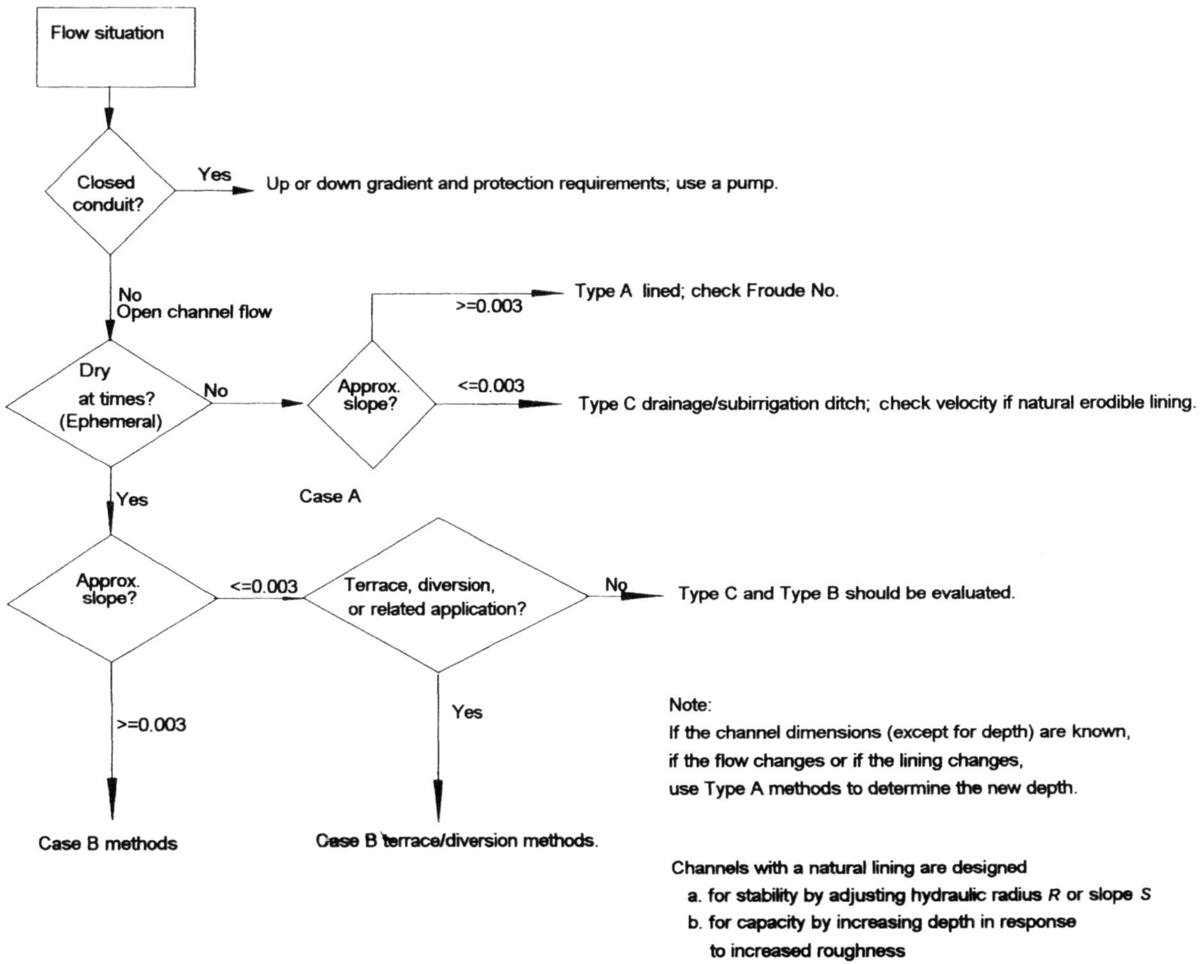

Figure 8.13. Decision aid for relating open channel flow design approaches. The conditional evaluations involving slope are approximate and should not be taken as rigorous breakpoints. Roughness, slope, cross-section type, and flow are assumed to be known, and other geometric elements are assumed unknown and to be determined.

Problems and Questions

Draw sketches of all completed designs.

1. What is the depth of flow in a rough (n = 0.02), concrete-lined trapezoidal channel, z = 3, slope = 0.002, bottom width = 1 m, while the flow is 2 m^3/s? Check the Froude number. What would be the depth and Froude number if the lining were resurfaced with a smooth concrete surface (n = 0.011)? Which surface would you recommend based on the Froude number?
2. Design a trapezoidal channel with z = 4 with a noncolloidal alluvial silt lining to move a 25 cfs flow down a 2% slope. *Hints:* Type B for stability, then increment n by 20% (this simulates effects of aging and debris accumulation) for capacity using Type A approach; add freeboard.
3. Compute the most efficient bottom width for an open trapezoidal channel with a flow rate of 2.5 m^3/s in a silt loam soil. The slope is 0.9%. (*Hint:* earth lining, low slope, Type C solution based on "efficient.") Next, compute the depth if n increases to 0.035 due to age and/or debris accumulation. Assume slope, flow, and channel width remain the same. (*Hint:* Use Type A approach.)
4. (a) What is the design capacity of a steady-flow irrigation canal with a total depth of 1.5 m including freeboard, bottom width of 1.2 m, 2 : 1 side slopes, and hydraulic gradient of 0.09%? Assume the freeboard is 20% of the depth and n = 0.04.
   (b) Repeat for the case where the roughness degrades to n = 0.045 over time.
   (c) Estimate the seepage losses if the canal is in sandy loam soil and is 2 mi long.
5. How deep is the flow in a trapezoidal channel, z = 4, lined with smooth plastic, capacity of 15 m^3/s, steady flow, gradient of 1%, freeboard factor of 20%, bottom width of 2 m? Is the resulting flow subcritical, critical, or supercritical? What if the roughness factor increases by 25% (this usually happens with age)? (*Hint:* Type A.)
6. A trapezoidal, concrete-lined irrigation ditch is to be built along a 0.5% slope. The ditch capacity is to be a steady 6 m^3/s. It is important to minimize the cross-sectional area so as to minimize the concrete required as well as the excavation costs. What is the resulting depth of the trapezoidal channel, including a 20% freeboard? Compute the Froude number. (*Hint:* Select the n value from the charts.)
7. Design a trapezoidal channel to carry the runoff from a 3600 ha watershed in the delta and coastal ricelands. The slope along the ditch route is 0.27%, the soil is heavy clay, and the average watershed slope is 0.45%.
   (a) Compute runoff using the ASAE drainage formula in Chapter 5; use the flowchart (Figure 8.13); ricelands implies long-term wetness; the slope is less than 0.003 (e.g., almost flat!); drainage ditches tend to be long so one should first optimize the cross section (Type C). Increase the n by 25% to account for the occasional nature of the flow before doing the Type C solution. Since the lining is soil (none other mentioned), one should check the final velocity to ensure it is less than the permissible velocity for the soil. If the channel velocity is greater than permissible, recommend a lining. Recalculate, being sure to use the appropriate side slope z with the lining. Assume a trapezoidal cross section. Don't forget the standard 20% freeboard.

(b) Try using the Type B solution method, considering both the deep and the shallow depths. A large depth and narrow width may be practical here since drain tube outlets will be intercepted by the channel. Could one intercept drain tubes at the 3.5 m depth using this design? If one of the Type B stability designs is suitable, increase n by 25% and solve for depth at capacity.

(c) Assume the drain tubes intersect the channel at the 2 m depth. Find the depth of flow using a bottom width of 1.5 ft (dug with a backhoe). (*Hint:* Assuming a trapezoidal channel in earth, use the minimum z for the given soil condition and then solve as Type A, unknown depth.) Freeboard is probably not necessary here because the depth needed to intersect the drain lines will ensure that channel depth is equal to or greater than 1.2 times the normal depth.

Note: Channel design gets involved! This design effort would be completed after these and perhaps additional analyses based on site data were performed. Economics coupled with aesthetics would dictate the design to handle the flow.

8. There is a need to move a 55 cfs occasional flow down a 7% slope over a silty clay loam soil. Since land is available for a wide channel, it was decided to not pay the cost to line but rather to go with a natural channel. Design a trapezoidal channel that must be crossed and must handle this flow rate. (*Hint*: This is a Type B problem.) Next, since the flow is occasional, design the channel for capacity by increasing the n value used in the stability analysis by 25%. (*Hint:* Capacity is a Type A problem, using the bottom width determined in the stability analysis.)
9. A high water mark in a natural unlined parabolic channel in a silt loam soil suggests a 1 m deep peak flow for a recent runoff event. The top width at the high water mark was 1.75 m. The channel slope is found by surveying to be 2.5%. Estimate the flow. Was scouring (erosion) likely under the peak flow condition, assuming a fine sandy bottom?
10. A concrete-lined irrigation ditch is to be built along a 0.5% slope. The ditch capacity is to be 6 m^3/s. It is important to minimize the cross-sectional area so as to minimize the concrete required as well as the excavation costs. Use (a) triangular channel, (b) rectangular channel, and (c) circular channel cross sections. Assume concrete lining in all cases. Which of the optimal cross sections has the smallest wetted perimeter? Which has the smallest cross-sectional area? Note that wetted perimeter and cross section determine which channel is the most economical to install, all other factors being equal.
11. Compute the most efficient bottom width for an open trapezoidal channel with a flow 2.5 m deep in a silt loam soil. What are the velocity and channel flow rate if the gradient is 0.09% and $n = 0.035$?
12. A reservoir serves as a source of water for an irrigation canal. The static height of the reservoir is 0.5 m relative to the concrete channel bottom entrance to the reservoir. The entrance is smooth and rounded. If the channel has a slope of 1% and is trapezoidal, $z = 1$, $b = 1.5$, what is the normal depth of flow existing in the channel? What is the flow rate? If the flow was to be 5 m^3/s, find the required static height. If the channel were optimum, the entrance rounded, and the flow 5 m^3/s, what is the required static height at the channel entrance?
13. Develop a computer pseudocode flowchart for solving for the depth of a parabolic channel with a known slope, soil conditions, roughness, and given a permissible velocity.
14. A 100 m^3/s flow needs to be moved down a 1000 m direct distance with a fall of 100 m. Flow events are occasional. An alternative path of 10,000 m is available (e.g., a longer way around). The soil is a shale. If excavation costs are \$2 per m^3 soil and

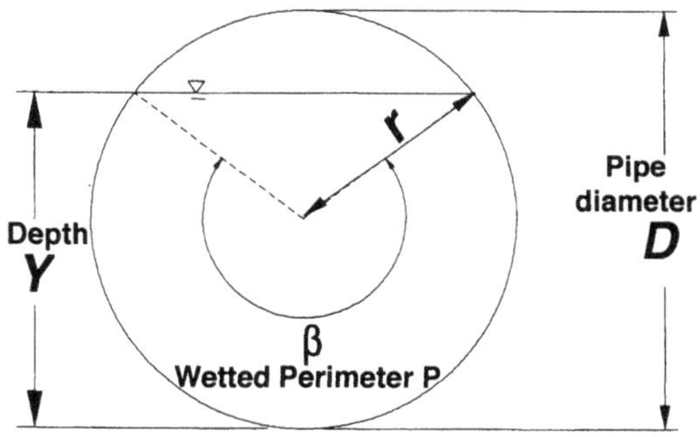

Figure 8.14. Definition sketch for a circular channel.

an additional \$1.50 per lineal meter per unit wetted perimeter of concrete lining or \$0.20 per lineal foot per unit wetted perimeter of a rock/cobble lining, which is the preferred choice based on initial investment?

15. Chow (1959) gives the hydraulic elements of a circular section as shown in Figure 8.14, where area A and wetted perimeter P are given by

$$A = 0.125(\beta - \sin\beta)D^2 \quad \textbf{(8.27a)}$$

$$P = 0.5\beta D \quad \textbf{(8.27b)}$$

where angle β is in radians.

Using trigonometry, relate depth Y to hydraulic radius r. See Figure 8.14.

16. Derive the hydraulic elements (A, P, R) for a roadside curb, shown in Figure 8.15. Express the hydraulic elements in terms of y and z.
17. Construct a QFD table using factors included in Tables 8.8 and 8.9. What additional factors would you add to the customer requirements or engineering requirements tables?
18. Show that the quadratic equation solution to the Type B problem is identical to the closed form solution.
19. Define the following channel elements: hydraulic radius, side slope ratio, cross-sectional area, freeboard, and roughness.
20. What is the difference between the Manning and Chezy equations?
21. What relationships are required for solving for flow rate and depth of a transition between a reservoir and a channel?
22. Define the components of fluid energy. How may one combine the energy equation and the Manning equation to solve a flow transition problem?
23. What are the basic relations for fluid pressure and fluid statics?

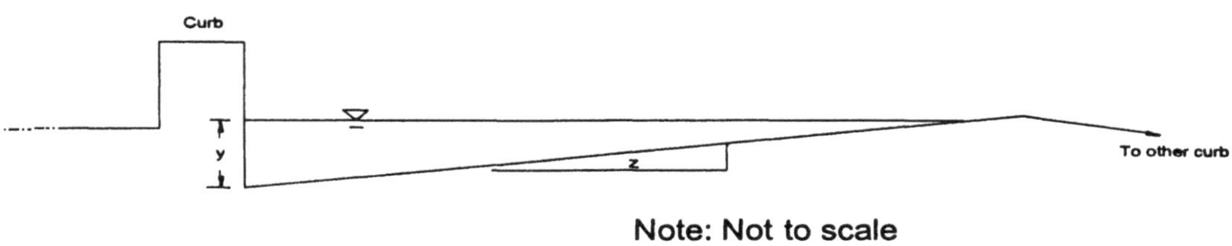

Figure 8.15. Schematic of roadside curb.

24. Derive the Manning equation; define all terms.
25. Name each of the three categories of problems and describe what types of channel design problems each category is most applicable to.
26. Specify the main use of the following equations in this chapter: (a) Manning, (b) energy, (c) continuity, (d) momentum.

Notes

1. The steady-state CE given in equation 8.1 was apparently first used by Hero of Alexandria prior to A.D. 150 (Biswas 1970).
2. The tractive force method incorporates cross-sectional geometry effects and allows for introduction of gravitational forces, in addition to water-induced shear. Thus the method is somewhat more rigorous than the permissible velocity method. French (1985) reconciles the two methods, showing near equivalence in most cases.
3. Computations with open channel design can be greatly facilitated using equation-processing software; however, beware of the pitfalls. Symbolic math packages produce elegant solutions of the equations, but many of these solutions may be extraneous and/or impractical. For example, the quadratic equation can lead to two real solutions of the trapezoid yielding a permissible velocity. One will be broad and shallow; the other, narrow and deep. Both have the same hydraulic radius and area. One would choose the shallow and broad channel for construction ease. In the case of the triangular channel, we can get four roots when using the exact definition of hydraulic radius. The designer must beware!

References

ASAE. 1997a. Concrete slip form canal linings. S289.1 SEPT92. American Society of Agricultural Engineers, St. Joseph, MI.

———. 1997b. Design and construction of surface drainage systems on agricultural land in humid areas. EP302.4 AUG93. American Society of Agricultural Engineers, St. Joseph, MI.

———. 1997c. Agricultural drainage outlets—open channels. EP407.1 DEC96. American Society of Agricultural Engineers, St. Joseph, MI.

———. 1997d. Installation of flexible membrane linings. EP340.2 DEC96. American Society of Agricultural Engineers, St. Joseph, MI.

Beltran, J. M. 1999. Land drainage. In *CIGR Handbook of agricultural engineering: Land and water management.* American Society of Agricultural Engineers, St. Joseph, MI.

Biswas, A. K. 1970. *History of hydrology*. North-Holland, Amsterdam, Holland.

Chow, V. T. 1959. *Open channel hydraulics.* McGraw-Hill, New York, NY.

FAO. 1998. The state of world fisheries and aquaculture. United Nations Food and Agricultural Organization, Rome, Italy.

French, R. H. 1985. *Open channel hydraulics.* McGraw-Hill, New York, NY.

Goldman, S. J., K. Jackson, and T. A. Bursztynsky. 1986. *Erosion and sediment control handbook.* McGraw-Hill, New York, NY.

Haan, C. T., B. J. Barfield, and J. C. Hayes. 1994. *Hydrology and sedimentology of small catchments.* Academic Press, New York, NY.

Harris, F. 1994. *Modern construction and ground engineering equipment and methods.* 2d ed. Longman, White Plains, NY.

Henderson, F. M. 1966. *Open channel flow.* Macmillan, New York, NY.

Hudson, N. 1971. *Soil conservation.* Cornell University Press, Ithaca, NY.

King, H. W., and E. F. Brater. 1963. *Handbook of hydraulics.* McGraw-Hill, New York, NY.

Lane, E. W. 1955. Design of stable channels. *Transactions of the ASCE* 120:1234–60.
Schwab, G. O., D. D. Fangmeier, W. J. Elliot, and R. K. Frevert. 1993. *Soil and water conservation engineering.* 4th ed. John Wiley & Sons, New York, NY.
Simons, D. B., and F. Senturk. 1992. *Sediment transport technology: Water and sediment dynamics.* Water Resources Publications, Littleton, CO.
Streeter, V. L. 1971. *Fluid mechanics.* 5th ed. McGraw-Hill, New York, NY.
USBR. 1952. Linings for irrigation canals. U.S. Bureau of Reclamation, Washington, DC.
USDA Natural Resource Conservation Service (USDA-NRCS). 1973. *Drainage of agricultural lands.* Water Information Center, Port Washington, NY.
Warner, R. C. 1998. SEDCAD[7] 4 for Windows95/NT: Notes for 5–7 Aug. 1998 short course. Department of Biosystems and Agricultural Engineering, University of Kentucky, Lexington.

9

Vegetated Waterways

Vegetated, or grassed, waterways are open channels with intermittent flows. Between 1996 and 1997, the U.S. land area protected by grassed waterways increased by 20,700 acres, for a total of 1,665,500 acres (NRCS 1998). Grassed waterways are extensively used around the world, although design methods are somewhat different (Hudson 1971). As with other open channels (Chapter 8), we will focus here on design methods commonly used in the United States.

Vegetated waterways are used in rural and urban runoff drainage situations. Uses include renovation of gullies, conveying water from field terrace systems, and conveying water from urban areas such as parks. Vegetated waterways are useful where being "ecologically friendly" is advantageous (see Goldman et al. 1986). Figure 9.1 shows a vegetated waterway in an agricultural setting. Grassed waterways with steeper side slopes, usually called *diversions,* are used to divert surface runoff away from areas below the diversion.

Vegetated waterway design employs the open channel Type B stability solution methodology followed by a capacity evaluation (Chapter 8). Hydraulic radius is the preferred manipulation; however, one may also increase slope to achieve the permissible velocity. Vegetated waterways are a special kind of open channel. The roughness coefficient changes drastically between the mowed and unmowed vegetation conditions. The Manning roughness n demands a special design methodology. In the mowed condition, the roughness is least and the potential for erosive velocities is greatest. The design must also consider the limited capacity of the channel when unmowed.

The roughness coefficient in the Manning equation varies in vegetated waterways, depending on vegetation type and height. One can categorize vegetated waterways according to the type and condition of the vegetation. The Manning n tends to vary consistently with velocity within these vegetal categories, known as retardance classes. Charts and analytical relationships are available to solve the Manning equation for each retardance class.

Figure 9.1. Small vegetated waterway in an agricultural field. Vegetated waterways protect naturally occurring draws and serve as discharges for surface outlet terraces.

Goals

- Identify situations where vegetated (grassed) waterways are most appropriate.
- Design vegetated waterways with trapezoidal, triangular, and parabolic cross sections for specified peak runoff.
- Develop channel maintenance strategies.

Background

A grassed waterway is a natural or constructed channel lined with erosion-resistant grasses. Grassed waterways are usually broad and shallow. A grassed waterway is typically 10–35 m wide and flows at depths less than 1 m. The usual design storm recurrence interval is 10 yr. The grassed waterway requires careful selection of vegetation type. Vegetation depends on the climate and on the channel flow frequency. Topography is also a limitation. Steep slopes require impractically wide channels, taking much land from agricultural or urban uses. Advantages of grassed waterways include potential forage production in the waterway and aesthetic enhancement in urban landscape design. Grassed waterways, though more expensive than a bare-earth channel, are much less expensive than lined channels per unit length.

The design of vegetated waterways begins with the Manning equation (equation 8.14).

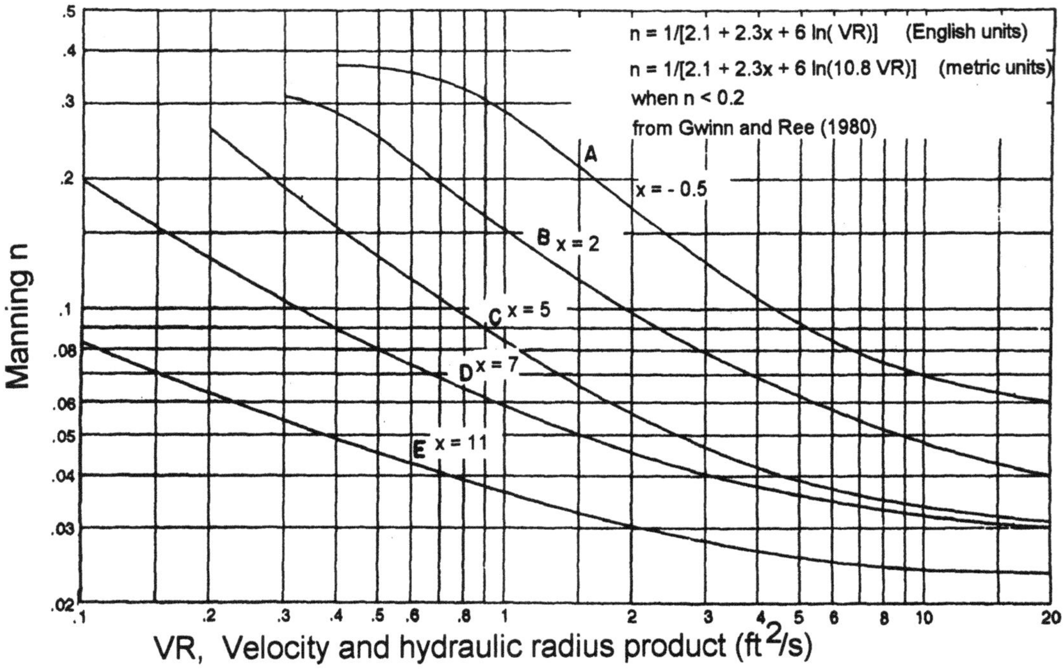

Figure 9.2. Relationship between Manning roughness *n* and product of velocity and hydraulic radius for each of five vegetation roughness classes ranging from very high (A) to very low (E). An approximate relationship between the parameters that was developed by Gwinn and Ree (1980) and is useful for computer-modeling purposes is also shown.

Vegetated waterways are designed using the permissible velocity method outlined by equations 8.18–8.23. The Manning *n* varies with vegetation height, as shown in Figure 9.2. Grasses are classified into the retardance classes shown in Figure 9.2, depending on the growth characteristics of the grass. One may use the Manning equation and Figure 9.2 in iteratively solving for velocity.

Vegetated channels, like other intermittent-flow channels, must be designed for both stability and capacity. Stability design follows a Type B stability solution with step 1 modified to enable the use of charts for variable Manning *n*. Determining capacity is a trial-and-error process. The following examples show several vegetated channel design scenarios.

Conservationists have classified grasses into retardance classes depending on height. Class A represents very dense, high vegetation, and at the other extreme, class E represents burned or very closely mowed vegetation. A given grass will usually be in one retardance class when unmowed and in another when mowed. One may effectively solve the Manning equation for each retardance class by using the graphs given in Figures 9.3–9.7. Alternatively, the *n* versus *vR* relationship (Figure 9.2) enables computer iteration for direct solutions of the Manning equation.

Channel Planning

Vegetative waterway planning is in many respects similar to planning for any channel, with the addition of vegetation selection. Specific elements include the following.

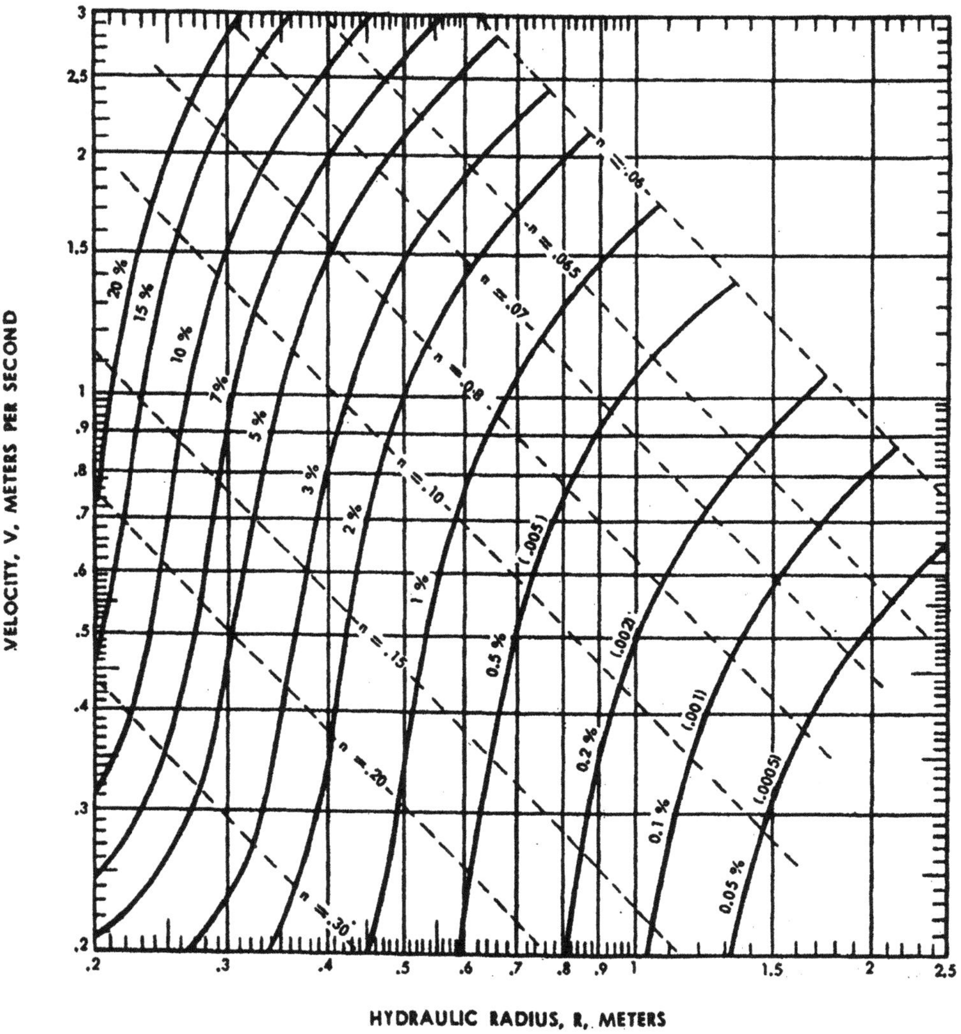

Figure 9.3. Graphical solution of the Manning equation for vegetated waterways: very high retardance, class A. (From NRCS 1966.)

Runoff

The design storm depends on the upstream and downstream situations. Traditionally the 10 yr return period with storm duration equal to the catchment time of concentration defines the design storm.

Channel Slope

Determine the channel slope from the local topography. Select the slope based on topography and desired channel routing. Slopes exceeding 5% frequently lead to large widths and shallow depths. Vegetation selection is critical when the slope exceeds 5%.

Channel Cross Section

The cross section of the channel is usually trapezoidal, triangular, or parabolic. The chosen cross section often depends on available construction equipment and on client desires. Trapezoidal

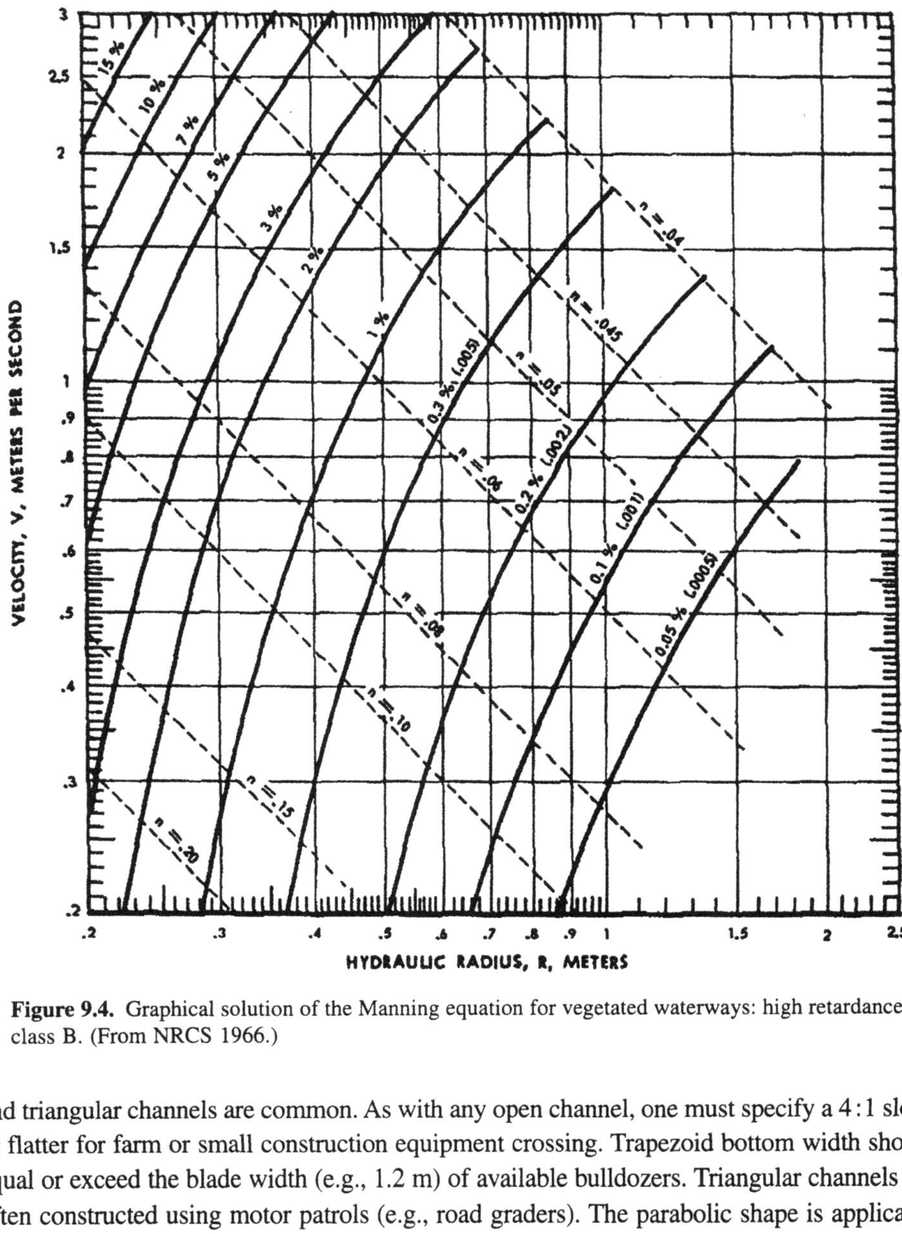

Figure 9.4. Graphical solution of the Manning equation for vegetated waterways: high retardance, class B. (From NRCS 1966.)

and triangular channels are common. As with any open channel, one must specify a 4 : 1 slope or flatter for farm or small construction equipment crossing. Trapezoid bottom width should equal or exceed the blade width (e.g., 1.2 m) of available bulldozers. Triangular channels are often constructed using motor patrols (e.g., road graders). The parabolic shape is applicable with a natural channel or a designed channel with natural aesthetics. Triangular and trapezoidal channels approach the parabolic shape when maintenance is minimum. Parabolic channels can be difficult to cross with farm and small construction equipment.

Vegetation Selection

The selection of a sustainable vegetation depends on climate and soil conditions plus tolerance to wet soils. Characteristics of commonly used vegetation are tabulated in Tables 9.1 and 9.2. Vegetation may be either unmowed or mowed during a runoff event. Retardance classes A, B, C, D, or E categorize the vegetation roughness. We usually work with B (unmowed condition) or D (mowed condition). Local experience may dictate other retardance class selections. One must design for stability (critical in the mowed condition) and capacity

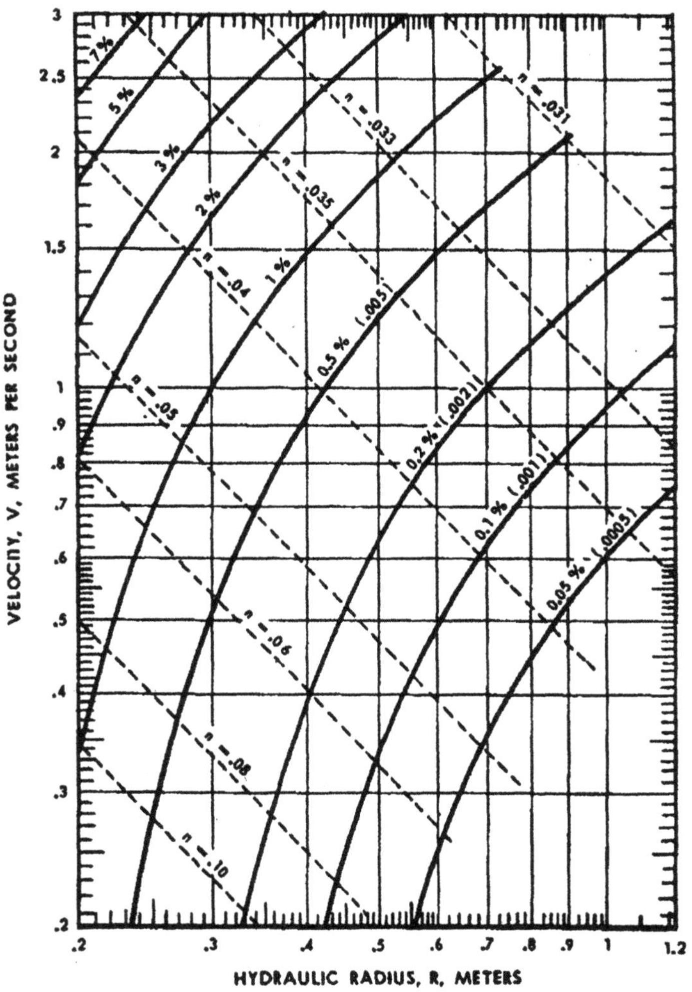

Figure 9.5. Graphical solution of the Manning equation for vegetated waterways: moderate retardance, class C. (From NRCS 1966.)

Table 9.1. Vegetation selection guide.

Geographical region of United States	Suggested vegetation
Northeastern	Kentucky bluegrass, redtop, tall fescue, white clover, bird's-foot trefoil
Southeastern	Kentucky bluegrass, tall fescue, Bermuda grass, reed canary, centipede
Western Gulf	Bromegrass, bluestem, native grass mixture, tall fescue
Southwestern	Wheatgrass, western and tall wheatgrass, smooth bromegrass
Northern Great Plains	Smooth bromegrass, wheatgrass, redtop, switch grass, native bluestem mixture

Source: Adapted from Schwab et al. 1993.

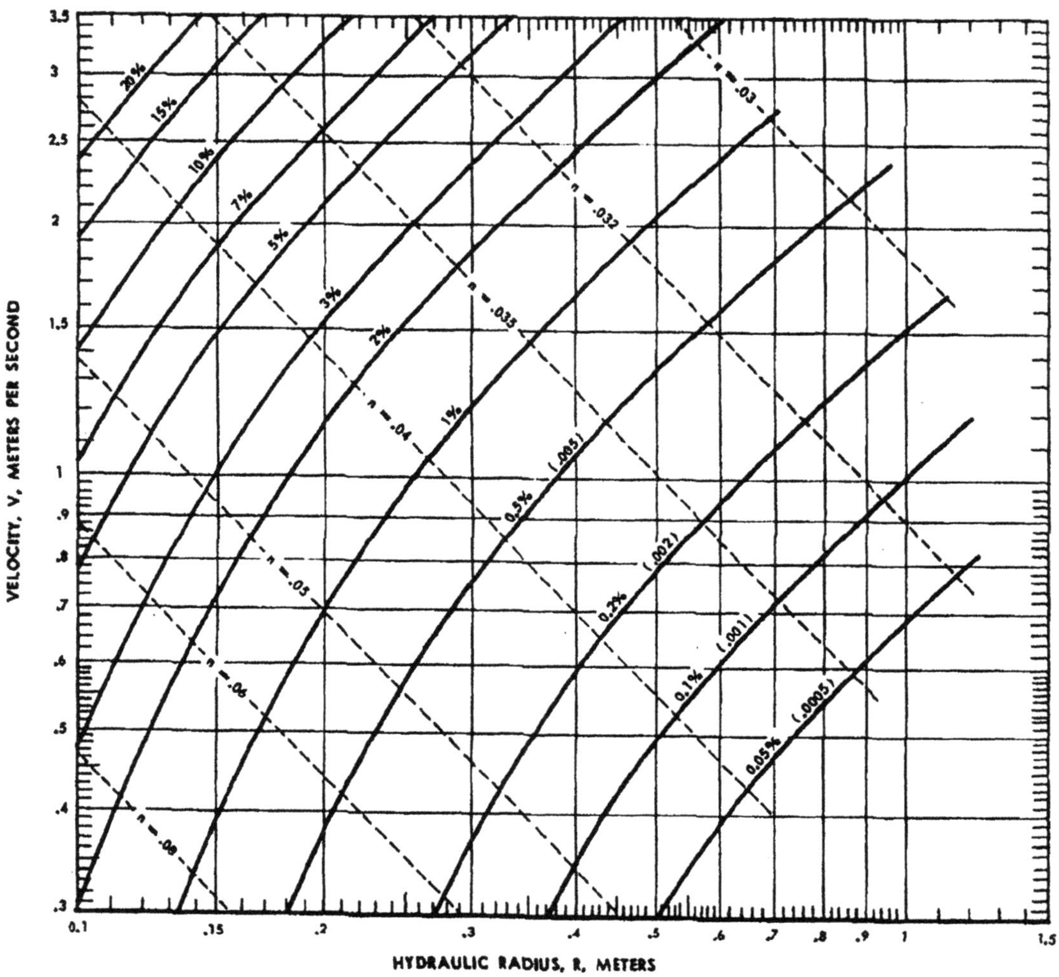

Figure 9.6. Graphical solution of the Manning equation for vegetated waterways: low retardance, class D. (From NRCS 1966.)

(critical in the unmowed condition). One cannot anticipate storm timing. Figures 9.3–9.7 provide graphical solutions to the Manning equation for the respective retardance classes.

Design Velocity for Stability

The design velocity depends on soil erosivity, vegetation type, vegetation condition, and channel slope. Velocity values are tabulated in Table 9.3. These data came from experimental studies. Ree 1949 and references cited therein give additional details concerning design velocity determination.

Channel Side Slopes

Design for crossing with equipment requires a $z = 4$ or more. If crossing is not necessary, a parabolic channel or lower z values in the trapezoid or triangle are acceptable. The minimum z for the soil (refer to Table 8.4) must be less than or equal to the selected z.

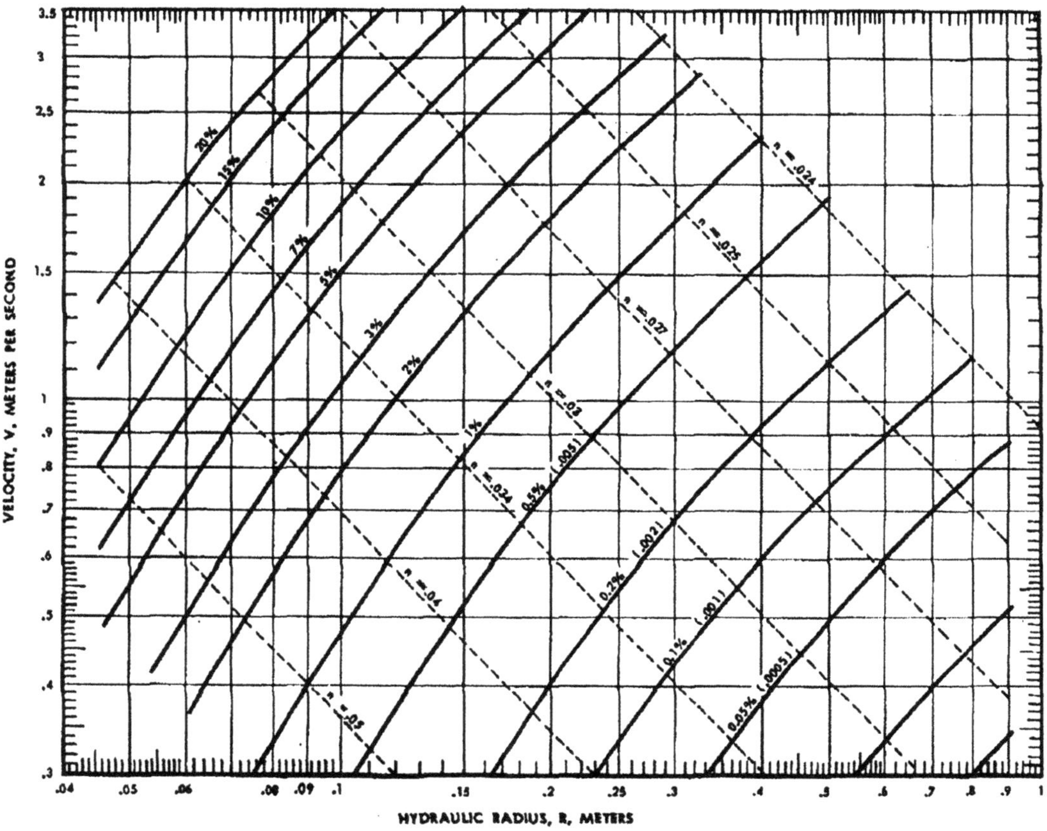

Figure 9.7. Graphical solution of the Manning equation for vegetated waterways: very low retardance, class E. (From NRCS 1966.)

Many agencies have in-house guidelines for the choice of z. Hydraulic elements for common grassed waterway cross sections are given in Table 9.4.

Design Procedures

The design of vegetated waterways requires knowledge of local conditions to select the grass. One must have knowledge of the maintenance level to anticipate possible mowing of the channel. Because the roughness of the channel depends on the grass selected (and grass maintenance), researchers have developed special design procedures. Ree (1949) proposed the initial methodology, later improved by Larson and Manbeck (1960) and by Gwinn and Ree (1980). One designs channels for *stability* and *capacity,* manipulating hydraulic radius.

Stability

Design for maximum permissible velocity for the soil-vegetation condition. Apply the following modifications of the approach outlined in equations 8.18–8.23. Use the hydraulic elements in Table 9.4 with the s subscript.

Pick maximum velocity (v_p) based on vegetation and soil conditions.

Choose a hydraulic radius from a low-retardance class (usually D or E; could be C in certain situations).

If the cross section is trapezoidal, calculate $A = q/v_p$; solve for bottom width knowing z, A, and hydraulic radius R from hydraulic elements. Proceed using the quadratic formula (equation 8.23) as developed in the Type B channel in Chapter 8, where

$$a_1 = [2(z^2 + 1)] - z$$
$$b_1 = -A/R$$
$$c_1 = A$$

Therefore, repeating equation 8.23:

$$y_s = [-b_1 \pm (b_1^2 - 4a_1c_1)^{0.5}]/(2a_1) \quad \textbf{(8.23)}$$

Put the sensible y_s value (usually the smallest positive) into the area equation above to solve for b. Denote y as y_s indicating depth at the stability criterion. Calculate bottom width from equation 8.20:

$$b = (A - zy_s^2)/y_s \quad \textbf{(8.20)}$$

Table 9.2. Vegetal cover classification according to retardance.

Retardance	Cover	Condition	Height (m)
A	Reed canary	Excellent stand, tall	0.9
A	Yellow bluestem		
B	Smooth bromegrass	Good stand, not mowed	0.3–0.4
B	Bermuda grass	Good stand, tall	0.3
B	Native grass mix	Good stand, unmowed	
B	Tall fescue	Good stand, unmowed	0.5
B	*Lespedeza sericea*	Good stand, not woody, tall	0.5
B	Grass-legume mixture	Good stand, unmowed	0.5
B	Reed canary	Good stand, mowed	0.3–0.4
B	Tall fescue and bird's-foot trefoil or ladino clover	Good stand, unmowed	0.5
B	Blue grama	Good stand, unmowed	0.3
B	Other class B mixtures	Good stand, unmowed	0.3
C	Bahia grass	Good stand, unmowed	0.2
C	Bermuda grass	Good stand, mowed	0.15
C	Redtop	Good stand, headed	0.4–0.5
C	Grass-summer legume mix	Good stand, unmowed	0.2
C	Centipede grass	Very dense cover	0.15
C	Kentucky bluegrass	Good stand, headed	0.3–0.5
D	Smooth bromegrass	Good stand, mowed	0.1–0.2
D	Bermuda grass	Good stand, mowed	0.1–0.2
D	Red fescue	Good stand, headed	0.3–0.5
D	Buffalo grass	Good stand, unmowed	0.1–0.2
D	Grass-legume fall mixture	Good stand, unmowed	0.2
D	*Lespedeza sericea*	After mowing	0.1
D	Most class B grasses after mowing	After mowing	0.1
E	Bermuda grass	Good stand, closely mowed	0.05
E	Burned stubble		
E	Very poor stand of vegetation		

Source: U.S. Soil Conservation Service (NRCS) 1979.

Table 9.3. Permissible velocities v_p for vegetated channels.

Grouped cover grasses	Slope range[a]	Permissible velocity (m/s)[b] Erosion resistant soils	Easily eroded soils[c]
Bermuda grass	0 to 5	2.4	1.8
	5 to 10	2.1	1.5
	> 10	1.8	1.2
Bahia grass, Buffalo grass, Kentucky bluegrass, Smooth brome, Blue gramma, Tall fescue	0 to 5	2.1	1.5
	5 to 10	1.8	1.2
	> 10	1.5	0.9
Grass mixtures, Reed canary	0 to 5	1.5	1.2
	5 to 10	1.2	0.9
Lespedeza sericea, Weeping lovegrass, Yellow bluestem, Redtop, Alfalfa, Red fescue or mowed fescue	0 to 5[d]	1	0.8
Common lespedeza[e], Sudan grass[e]	0 to 5[d]	1	0.8

Source: U.S. Soil Conservation Service (NRCS) 1979.

[a] Do not use on slopes steeper than 10% except for vegetated side slopes in combination with a stone, concrete, or highly resistant vegetative center section.

[b] Use velocities exceeding 1.5 m/s only where there are good covers and where proper maintenance is assured.

[c] Easily eroded soils are those with USLE K values exceeding 0.2 (English) or 0.0258 (metric). Soils with very low organic matter or severely disturbed soils should be considered easily erodible. Warner (1998) makes these recommendations for use in SEDCAD.

[d] See footnote a, limiting the slope to 5%.

[e] Use on mild slopes or as temporary protection until permanent covers are established.

If the cross section is triangular, calculate $y_s = 2R$. Calculate parameters needed for computing the z value; $A = q/v_p$. Compute the z value from $z = A/y_s^2$. One can then compute the other parameters from hydraulic elements. If z is smaller, leave z at original value (this ensures stability). If z is larger, increase z to larger value.

If the cross section is parabolic, calculate $y_s = 1.5R$; $A = q/v_p$; $t_s = 1.5A/y_s$, where t_s is the top width determined for stability and y_s is the depth in the stability design.

Capacity

Capacity designs require trial-and-error calculations. Use the hydraulic elements in Table 9.4 with the *cp* subscript. The goal is to select a D (greater than y) meeting the retardance conditions in retardance class A, B, or possibly C. Retardance class B is the usual choice unless the vegetation dictates otherwise in the unmowed state. Low maintenance would dictate the roughest condition expected for the vegetation, and frequent, regular mowing may enable use of a higher retardance class.

Table 9.4. Approximate hydraulic elements for the stability, capacity, and capacity with freeboard added cases for channel geometries commonly used for vegetated channels.

	Cross-sectional area	Wetted perimeter	Hydraulic radius	Top width
Trapezoid			$A(y_a)/P(y_s)$	$t_s = b + 2y_s z$
	$by_s + zy_s^2$	$b + 2y_s\sqrt{z^2+1}$	$A(y_{cp})/P(y_{cp})\ y_*$ (wide channel)	$T_{cp} = b + 2y_{cp}z$
			$y_*/2$	$T_f = b + 2y_f z$
	$by_{cp} + zy_{cp}^2$	$b + 2y_{cp}\sqrt{z^2+1}$	(near zero bottom width and $z > 3$ where $_*$ refers to $_{s,cp}$ or $_f$ as appropriate)	
Triangle				$t_s = 2y_s z$
	zy^2	$2y_s\sqrt{z^2+1}$	$\approx \frac{1}{2}y_s$	$T_{cp} = 2y_{cp}z$
			$\approx \frac{1}{2}y_{c_1}$	$T_f = 2y_f z$
	zy_{cp}^2	$2y_{cp}\sqrt{z^2+1}$	*when* $z \geq 3$	
Parabola	$\frac{2}{3}T_s y_s$	$T_s + \frac{8y_s^2}{3T_s}$	$\approx \frac{2}{3}y_s$	$t_s = \frac{a}{0.67y_s}$
	$\frac{2}{3}T_{cp}y_{cp}$	$T_{cp} + \frac{8y_{cp}^2}{3T_{cp}}$	$\approx \frac{2}{3}y_{cp}$	$T_{cp} = t_s\left(\frac{y_{cp}}{y_s}\right)^{1/2}$
				$T_f = T_{cp}\left(\frac{y_f}{y_{cp}}\right)^{1/2}$

Note: Nomenclature; y = depth for stability; D = depth for capacity; D_f = depth with freeboard; z = side slope ratio; t_s = top width before freeboard; T_{cp} = top width at capacity; T_f = topwidth after freeboard; b = bottom width. Subscripts; s refers to stability, cp refers to capacity and f refers to the freeboard condition. Freeboard conditions are not shown for all elements but follow from the use of the freeboard depth and topwidth.

Estimate y_{cp} as follows (see Larson and Manbeck 1960). If parabolic or trapezoidal and t or b for stability is between 2 and 6 m, add 0.1 m to the depth for stability (y_s) if the slope is greater than 5%, or add 0.17 m to the depth if the slope is 1% or less. For t or b between 6 and 27 m, reduce the above depth increments to 0.05 m (slope 5%) and 0.15 (slope > 5%). If t or b is greater than 27 m, reduce the depth increments to 0.05 m with all slopes.

Calculate parameters using the guess R_{cp}.

If the cross section is trapezoidal, use b as determined above, z as set initially, and your estimate of y_{cp}. Estimate A_{cp} and R_{cp}, where y_{cp} is the depth under the capacity criterion.

If the cross section is triangular, use the z and y_{cp} determined above to estimate R_{cp}; estimate capacity.

If the cross section is parabolic, calculate T_{cp} guess $= t_s$ (guess y_{cp}/y_s; calculate guess R_{cp}; compute guess A_{cp}.

Enter retardance class B, Figure 9.4 (or appropriate rough condition retardance class), on R axis, go to bottom slope, pick guess v from v axis.

Compute the guess q = guess $A_{cp} \times$ guess v_{cp}.

If guess q is less than q, increase guess depth and recalculate hydraulic parameter. If guess q is greater than q, decrease guess d and recalculate hydraulic parameter. If guess q is within 10% of q, add freeboard.

Freeboard

A freeboard value is added to account for freeze/thaw and to provide a margin of dry channel. One does not want incipient overtopping at the usual operating level. For earthen channels, the freeboard value is usually 0.1 m or 20%. Modify the capacity depth by adding freeboard to the capacity depth. Modify the top width by using appropriate hydraulic elements (see Table 9.4, f subscripts). If the vegetated waterway is serving as an outlet for a terrace system, design for flow as above. Achieve the depth by setting the channel to conform with terrace outlets.

Example 9.1 (trapezoidal channel design): Given a 500 cfs (14.17 m^3/s) flow over a Bermuda grass waterway at a 2% slope in erosion-resistant soils, the desire for a z = 4 for crossing ease, and v_p = 2.4 m/s, find b.

$$A = 14.17\ \text{m}^3/2.4\ \text{m} = 5.905\ \text{m}^2$$

From the class D retardance read R = 0.38 m.

Using equation 8.23,

$$a_1 = 2\,(17) - 4 = 4.246$$
$$b_1 = -15.54\ \text{m}$$
$$c_1 = 5.905\ \text{m}^2$$
$$y = 0.428\ \text{m},\ 3.23\ \text{m}$$

Use the shallow depth here.

Then from equation 8.20,

$$b = [5.905\ \text{m}^2 - 4(0.428\ \text{m})^2]/0.428\ \text{m} = 12.08\ \text{m}$$

Design for capacity: b = 12.08 m from the stability analyses above.

Add 0.1 m to the depth of 0.43 m found above for d = 0.53 m; check as shown in Table 9.5.

Then add 0.1 m freeboard to get a depth of y_f = 0.68 m, b = 12.08 m, z = 4, T_f = 17.52 m.

Table 9.5. Calculation table for capacity check, trapezoidal grass waterway, Example 9.1.

Guess y_{cp} (m)	$R_{cp} = \dfrac{by_{cp} + Zy_{cp}^2}{b + 2y_{cp}(Z^2 + 1)^{0.5}}$ (m)	$A_{cp} = zy_{cp}^2 + by_{cp}$ (m^2)	v_{cp} (class B) (m/s)	$q = A_{cp}\, v_{cp}$ (m^3/s)
0.53	0.46	7.52	1.7	12.78
0.58	0.49	8.35	1.8	15.03 OK

Example 9.2 (trapezoidal channel attempt, negative radical; triangular channels and other approaches): Given a 50 cfs (1.42 m^3/s) flow over a Bermuda grass waterway at a 2% slope in erosion-resistant soils, the desire for a $z = 4$ for crossing ease, and $v_p = 2.4$ m/s for Bermuda grass on the 2% slope, find b.

$$A = (1.42\ \text{m}^3/\text{s})/(2.04\ \text{m/s}) = 0.59\ \text{m}^2$$

From Figure 9.4, $R = 0.38$ m. Then

$$a_1 = 2(17) - 4 = 4.246$$
$$b_1 = -0.59\ \text{m}^2/0.38\ \text{m} = -1.55\ \text{m}$$
$$c_1 = 0.59\ \text{m}^2$$

The term $b^2 - 4ac$ is negative. Practically, the negative radical tells us the 4:1 trapezoid, even with a bottom width of zero, cannot physically move this flow down a 2% slope at the permissible velocity. This is the degenerate trapezoid. This provides an opportunity to explore several solution strategies before recommending an approach for dealing with the degenerate trapezoid. One may decrease z (e.g., consider compromising the ability to easily cross the channel) and use triangular geometry. One may increase the channel slope if topography allows, using a triangular cross section or using a trapezoidal cross section with set bottom width. Finally, one might design only for capacity since the channel will be stable, although sedimentation could be problematic. Examples are shown for each case.

A. *Triangular channel design* (compromise the easy-crossing requirement). $R = y_s/2$; therefore, $y_s = 0.76$ m. $A = zy_s^2$; therefore, 0.59 $m^2 = z[(0.76\ \text{m})^2]$, yielding $z_s = 1.02$, or round z to 1. Now we must design for capacity. Set up the computation table as shown in Table 9.6. Add 20% freeboard to the depth of 0.9 m for a total depth of $y_f = 1.08$ m, $T_f = 2.16$ m (see Table 9.4 for hydraulic elements), and $z = 1$ (see above). The method shown in Table 9.6 is appropriate for the triangular grassed waterway in general. The approximation of the hydraulic radius by $y/2$ becomes inaccurate when $z < 3$. If the resulting z is less than 3, try another approach.

B. *Triangular cross section, increase channel slope.* The bottom width may be zero, resulting in a triangular cross section. Suppose that we allow an increase in slope and a triangular geometry, $z = 4$, $A = 0.59\ \text{m}^2$. From the area relationship 0.59 $m^2 = zy_s^2 = 4y_s^2$; therefore, $y_s = 0.38$ m and $R = y_s/2 = 0.19$ m (see Table 9.4). From the class D curve we find a slope of 7% gives the required $v_p = 2.4$ m/s at $R = 0.19$. Now design for capacity as shown in Table 9.7, in which the format of Table 9.6 is used. Add freeboard to the depth of 0.48 m to have a total depth of $y_f = 0.576$ m, $z = 4$, $T_f = 4.61$ m.

Table 9.6. Calculation table for channel capacity for the case of degenerate trapezoid where a triangular channel was selected with the channel z decreased to achieve permissible velocity v_p in the stability design phase, Example 9.2A.

Guess y_{cp} (m)	$R_{cp} = y_{cp}/2$ (m)	$A_{cp} = z_s y_{cp}^2$ (m^2)	v_{cp} (class B) (m/s)	$q = A_{cp} v_{cp}$ (m^3/s)
0.85	0.425	0.7225	1.5	1.08
0.95	0.475	0.90	1.8	1.64
0.9	0.45	0.81	1.7	1.38 OK

Table 9.7. Capacity design calculation table for the case of a degenerate trapezoid where triangular channel was selected with the channel slope increased to achieve permissible velocity v_p in the stability design phase, Example 9.2B.

Guess y_{cp} (m)	$R_{cp} = y_{cp}/2$ (m)	$A_{cp} = z\, y_{cp}^2$ (m^2)	v_{cp} (class B) (m/s)	$q = A_{cp}\, v_{cp}$ (m^3/s)
0.45	0.225	0.81	1.2	0.97
0.5	0.25	1.0	1.6	1.6
0.48	0.24	0.92	1.5	1.38 OK

Increasing the slope may allow for a shorter, more direct routing of the channel. The available slope is limited at many sites.

C. *Trapezoidal cross section, increased slope.* If the geometry is trapezoidal and b is, say, 1.2 m (e.g., bulldozer blade width), then one could solve for y_s from the trapezoidal area relationship. Then, knowing b, y_s, and z, solve the trapezoidal area equation for y, and compute the hydraulic radius. As with the triangular case, select the slope where v_p and R_s intersect on the class D (or C or E) curve. From the area relationship $0.59\ m^2 = by + zy^2 = 1.2y + 4y^2$. This equation can be solved using the quadratic formula $y = [-b\,(b^2 - 4ac)]/2a$ where $a = z = 4$, b = bottom width = 1.2 m, $c = -A = -0.59\ m^2$. The realistic value for y_s is 0.262 m. Using the trapezoidal hydraulic radius formula gives $R = 0.175$ m. On the class D chart (Figure 9.4) $v_p = 2.4$ m/s and $R_s = 0.175$ m intersect where the slope is 7%. We next design for capacity with slope = 7% and bottom width = 1.2 m. As before, we use the class B chart. Table 9.8 is the capacity calculation table. The first guess depth is 0.1 m + 0.262 m = 0.36 m. Add 20% freeboard to the depth of 0.36 m to get $y = 0.46$ m, $z = 4$, $b = 1.2$ m, $T_f = 4.88$ m (see Table 9.4).

D. *Reduce permissible velocity, maintain trapezoidal geometry.* We could also relax the permissible velocity (which by definition results in stability), use triangular (or parabolic) geometry, and hold $z = 4$ and solve for capacity, leaving the slope at 2%. One could set z and b and find a slope, proceeding as in approach C above. *Solution for stability is not necessary because velocity is by definition less than the maximum permissible.* The bottom width may be zero, or another value may be dictated by available construction equipment. Bulldozers are often used in construction of grassed waterways and blade widths less than 1.2 m are unlikely.

Table 9.8. Trapezoidal waterway capacity calculation where slope was adjusted to achieve permissible velocity v_p in the stability design phase, bottom width $b = 1.2$ m, Example 9.2C.

Guess y_{cp} (m)	$R_{cp} = \dfrac{by_{cp} + Zy_{cp}^2}{b + 2y_{cp}(Z^2 + 1)^{0.5}}$ (m)	$A_{cp} = z\, y_{cp}^2 + by_{cp}$ (m^2)	v_{cp} (class B) (m/s)	$q = A_{cp}\, v_{cp}$ (m^3/s)
0.36	0.23	0.95	1.2	1.14
0.42	0.26	1.21	1.5	1.82
0.38	0.24	1.03	1.3	1.34 OK

Table 9.9. Trapezoidal waterway capacity calculation, $b = 1.2$ m, velocity in stability design phase less than permissible v_p, Example 9.2D.

Guess y_{cp} (m)	$R_{cp} = \dfrac{b\,y_{cp} + Z\,y_{cp}^2}{b + 2\,y_{cp}(Z^2 + 1)^{0.5}}$ (m)	$A_{cp} = z y_{cp}^2 + b y_{cp}$ (m^2)	v_{cp} (class B) (m/s)	$q = A_{cp} v_{cp}$ (m^3/s)
0.45	0.275	1.35	0.58	0.78
0.55	0.326	1.87	0.85	1.59
0.52	0.31	1.71	0.8	1.37 OK

Suppose we desire a trapezoidal cross section with bottom width of 1.2 m. Then the capacity design proceeds as shown in Table 9.9. Add 20% freeboard to the class B design depth of 0.52 m to get $y = 0.62$ m, $z = 4$, $b = 1.2$ m, and $T_f = 6.16$ m (see Table 9.4).

One should determine the velocity in the mowed condition (i.e., use the same approach for capacity except use the mowed-condition chart, Figure 9.6) in order to assess the impact of sedimentation. One may use methods identical to the capacity design in Table 9.9 for these computations. Channels designed to flow at less than the permissible velocity could require high maintenance due to high sedimentation. Consider other options if sediment-laden runoff is present. Other approaches, including the parabolic cross section, are applicable. *Recommendation: Use approach D unless sedimentation is an obvious limitation. Use approach D unless site conditions indicate that increasing the slope might be possible. If slope can be increased, consider approaches B and C. If a high sediment load is expected, consider approaches A, B, or C, based on specific site conditions.*

Example 9.3 (parabolic channel design): Given a 500 cfs (14.17 m^3/s) flow over a Bermuda grass waterway at a 2% slope in erosion-resistant soils, design a vegetated waterway having a parabolic cross section.

Velocity $v = 2.4$ m/s and stability is designed using class D. From the class D retardance read $R = 0.38$ m. $A_s = q/v = (14.17\ m^3/s)/(2.4\ m/s) = 5.9\ m^2$. For the parabolic cross section, $R = 0.67 y_s$ or $y_s = 0.567$ m; $t_s = A_s/(0.67 y_s) = 15.54$ m. Design for capacity following Table 9.10.

Table 9.10. Parabolic channel capacity calculation table, Example 9.3.

Guess y_{cp} (m)	$R = 0.67\,y_{cp}$ (m)	$T_{cp} = t_s(y_{cp}/y_s)^{0.5}$ (m)	$A_{cp} = 0.67\,T_{cp}\,y_{cp}$ (m^2)	v_{cp} (class B) (m/s)	$q = A_{cp}\,v_{cp}$ (m^3/s)
0.6	0.4	15.98	6.42	1.4	8.99
0.7	0.47	17.26	8.09	1.8	14.57
0.68	0.455	17.01	7.75	1.8	13.95 OK

Choose depth = 0.69 m. Add freeboard of 20% to get $y_f = 0.82$ m, $T_f = (15.54\ m)(0.82\ m/0.567\ m)^{0.5} = 18.69$ m.

In Example 9.3, going to retardance class B from class D will change the velocity from 2.4 to 1.7 m/s. What if one designs for only the unmowed condition and then mows the channel? This usually leads to a narrower bottom width and a deeper channel. Mowing vegetation reduces roughness and leads to excessive velocities. Thus, the channel would be unstable.

Other Design Approaches

One may determine bottom width for the channel based on available construction equipment blade width. Check for stability in the mowed condition. Check for capacity in the unmowed condition. If stability is not satisfied, then one must increase width or decrease slope. Capacity design is the next step. Warner (1998) found this approach to work well in computerized solutions.

Construction and Maintenance

Between construction and grass establishment, the vegetated waterway is vulnerable to erosion by storms much smaller than the design event. One should divert water from the channel during the construction phase if possible. Otherwise, temporary linings or nettings, shown in Figure 9.8, are appropriate for protection during this phase. More advanced texts (e.g., see Haan et al. 1994) specify non-Manning-based methods for design of channels with temporary linings. Protection during the construction phase is important in urban settings where adverse aesthetic and financial consequences are likely. Straw mulch is sometimes used; however, straw is not resistant to erosion.

Selection of an adequate vegetation is key to avoiding maintenance problems throughout the channel life. If low continuous flows develop during the useful life of a vegetated waterway, one may have to put a small lined section in the channel bottom.

As with open channel flow in general, works such as Harris 1994 give extensive discussion on earth construction techniques. Associations such as the Land Improvement Contractors Association (http://ourworld.compuserve.com/homepages/delmarva/staff.htm), ASAE (http://www.asae.org), and ASCE (http://www.asce.org) also provide helpful information.

Maintenance of vegetated waterways begins with maintaining proper fertility for the chosen grass lining. Apply mulch or surface coverings as shown in Figure 9.8 to ensure good stand development and minimal injury to the stand during the establishment phase. Prolonged droughts or prolonged wet seasons may result in stand deterioration. Machine or animal traffic, overgrazing, and rodent-induced damage necessitate maintenance. Reshaping and reseeding may be necessary. Annual inspections and inspection after major storms will ensure control of problems related to sediment deposition, scouring, gullying, and traffic damage. Channels flowing at less than the permissible velocity in the stability phase may be more vulnerable to sedimentation.

Outlook

Our discussion of vegetated waterways has further developed the concept of stability design followed by capacity design. Terraces and diversion channels (discussed in Chapter 10) expand on this design approach. The trapezoidal channel is the "standard" design cross section for vegetated waterways. Using z values near 4 is typical since these can be crossed. In some

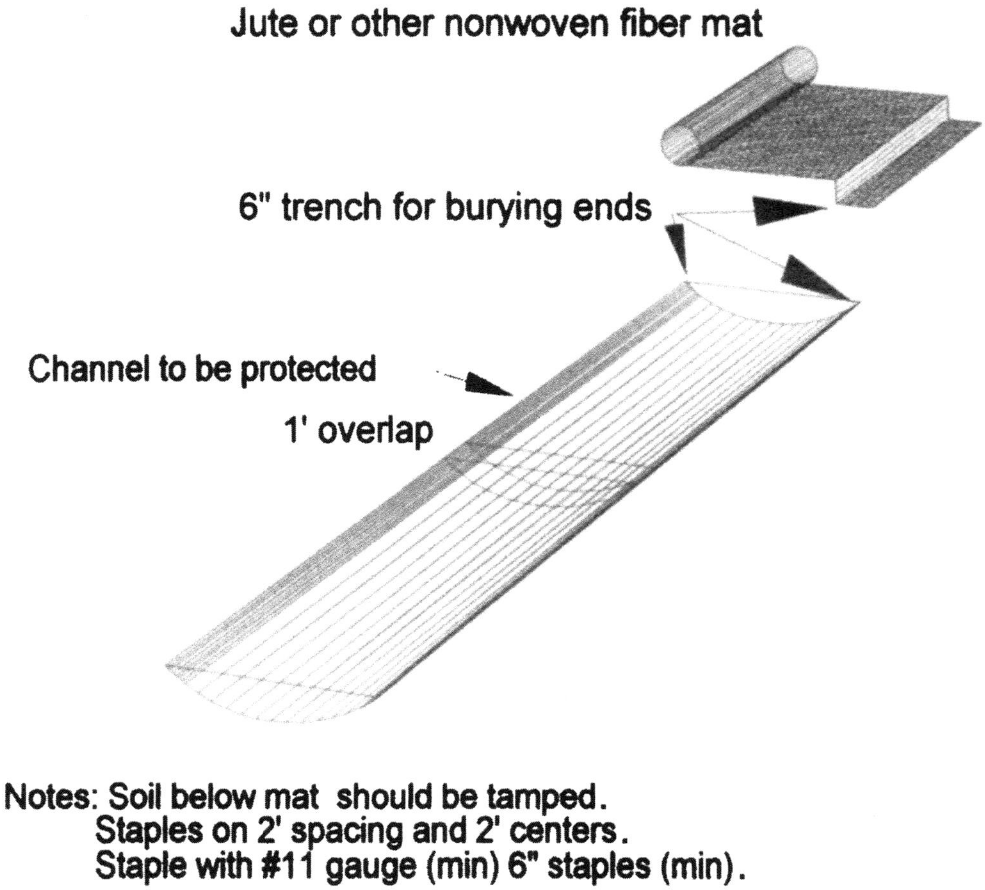

Figure 9.8. Simple procedure for installing jute mat to stabilize a newly constructed grassed waterway. (Adapted from Goldman et al. 1986.)

low-slope, high-flow scenarios, the bottom width of a desired trapezoid with 4:1 side slopes can approach zero. The general recommendation is to set the bottom width at the width of the excavator to be used, then design for capacity. Specific site conditions may warrant changes in channel routing to increase slope. Channels originally constructed as triangular or trapezoidal tend in time to become ever more parabolic under low-maintenance programs.

The grassed waterway is an aesthetically attractive option for urban settings. The grassed waterway may also be incorporated into some biological treatment schemes for nutrient pollution removal. Appendix C further discusses maintenance and construction.

The requirements of designing vegetated waterways assuming hydraulic radius manipulation are summarized in Figure 9.9. Do the low-roughness condition first to design the bottom width of the trapezoidal section, the side slope z of the triangle, or the top width of the parabolic section. This satisfies stability. Then design for capacity in the high-roughness condition, varying depth or top width to give the area satisfying the continuity equation. Finally, provide for freeboard. Conclude the design with a sketch showing key channel elements. Lastly, periodic inspections and maintenance are important for maintaining functional grassed waterways. An Internet search will reveal links to several commercially available software packages (e.g., see Warner 1998) for designing channels, including vegetated waterways.

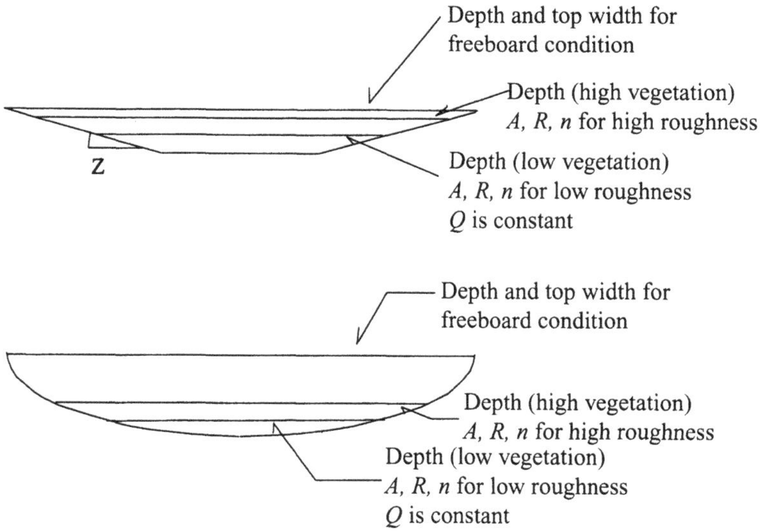

Figure 9.9. Design requirements summary for vegetated trapezoidal and parabolic channels. Note the general similarity with other channel types.

Problems and Questions

Draw sketches of all completed designs.

1. Design a parabolic-shaped grassed waterway to carry 1.5 m^3/s in an easily eroded soil, with Bermuda grass in good condition and 4% slope.
2. Design a trapezoidal waterway with 4:1 side slopes to carry 0.6 m^3/s from a terraced field where soil is resistant to erosion and the channel slope is 4%. The vegetation choice is bromegrass.
3. Design a triangular waterway to carry 1.8 m^3/s from a field resistant to erosion and 6% channel slope with fescue lining. Can this channel be easily crossed by farm or small construction equipment?
4. Design a trapezoidal channel to carry 25 m^3/s flow over a 1% slope in an easily erodible soil. Fescue is to be used. The channel must be crossed with farm equipment.
5. Design a parabolic channel and a triangular channel to carry 25 m^3/s flow over a 1% slope in an easily erodible soil. Fescue is to be used. Compare the resulting channels by superimposing scaled drawings of both channels.
6. A trapezoidal channel, in an easily erodible soil, has been in the unmowed state (fescue) for many years and has accumulated sediment from recent development activity. The channel slope is 4%, there is a 4:1 side slope condition, and the bottom width is 2 m. The documented channel flow depth is 1 m. Determine the flow velocity. Calculate the flow. Would the channels contribute sediment downstream (e.g., be stable) if mowed?
7. Design a channel to carry 50 m^3/s in a case where the Bermuda grass vegetal lining may be burned as part of a controlled forest burn from time to time. The channel slope is 3% and the soil is easily erodible. The channel must be crossed by farm equipment. Compare this channel to an identical nearby channel with mowed, unburned Bermuda grass.
8. A 100 m^3/s flow is to be conveyed down a 1000 m direct distance with a fall of 100 m. An alternative path of 10,000 m is available (i.e., a longer way around). If excavation

costs are \$2 per m^3 soil and an additional \$5 per lineal m of concrete lining, or an additional cost of \$1 per lineal m of fescue vegetation lining, which is the preferred choice based on initial investment? If annual maintenance is 5% of the initial investment and the life of the project is 20 yr, what is the choice?

9. Prove that the simplifications for hydraulic radius with the triangular and parabolic cross sections given in Table 9.4 are within 5% of the exact values given in Table 8.4, provided the limiting conditions of Table 9.4 are satisfied.
10. What customer requirements concerning water conveyance may point toward a vegetated waterway recommendation?
11. In what situations would one favor a parabolic channel cross section rather than a triangular or trapezoidal cross section?
12. In cases where the slope is fixed and one initially specifies a trapezoidal channel, what options does one have when exactly meeting the stability condition is impossible?

References

Goldman, S. J., K. Jackson, and T. A. Bursztynsky. 1986. *Erosion and sediment control handbook.* McGraw-Hill, New York, NY.

Gwinn, W. R., and W. O. Ree. 1980. Maintenance effects on the hydraulic properties of a vegetation lined channel. *ASAE Transactions* 23:636–42.

Haan, C. T., B. J. Barfield, and J. C. Hayes. 1994. *Hydrology and sedimentology of small catchments.* Academic Press, New York, NY.

Harris, F. 1994. *Modern construction and ground engineering equipment and methods.* 2d ed. Longman Publishing Group, White Plains, NY.

Hudson, N. 1971. *Soil conservation.* Cornell University Press, Ithaca, NY.

Larson, C. L., and D. M. Manbeck. 1960. Improved procedures for vegetated waterways. *Agricultural Engineering* 41:694–96.

NRCS. 1998. USDA national resources inventory, 1997 update (http://www.nrcs.usda. gov). Natural Resource Conservation Service, Washington, DC.

Ree, W. O. 1949. Hydraulic characteristics of vegetation for vegetated waterways. *Agricultural Engineering* 30:184–87, 189.

Schwab, G. O., D. D. Fangmeier, W. J. Elliot, and R. K. Frevert. 1993. *Soil and water conservation engineering.* 4th ed. John Wiley & Sons, New York, NY.

U.S. Soil Conservation Service (NRCS). 1979. *Engineering field manual for conservation practices.* USDA-NRCS, Washington, DC.

Warner, R. C. 1998. SedCad[7] 4 for Windows95/NT: Notes for 5–7 Aug. 1998 short course. Department of Biosystems and Agricultural Engineering, University of Kentucky, Lexington.

10

Terraces and Diversions

The terrace is a topographic modification that may serve as a combination channel and impoundment. The diversion is similar to a terrace except it typically drains a larger area and is used mainly for flow control and does not usually impound water. Terraces are used mainly for erosion control. It should be noted that conservation tillage is becoming the preferred approach for erosion control with most important row crops. However, conservation tillage options are not yet available for many specialty crops. Terracing and related practices increased by 72,300 acres between 1996 and 1997 (NRCS 1998). Terraces are used worldwide for erosion control (Hudson 1971; De Wrachien and Chesci 1999). Figure 10.1 shows an ancient bench terrace system.

Terracing reduces slope length and peak flow and thus decreases soil erosion. Erosion terrace channels and diversions with surface outlets are designed using Type B slope manipulation solutions. Besides stability, the design of terraces, diversions, and related intermittent-flow channels includes a capacity design phase. Many of the ideas involved in agricultural terrace design are applicable to urban erosion control situations. We will integrate principles of channel design with erosion control, water conservation, and harvesting strategies.

Goals

- Focus hydrological analyses, erosion principles, and waterway design principles covered thus far on the design of terraces, diversions, and their outlets. Type B slope manipulation stability solutions are of particular interest. Recall that the Type B problem is the solution of the Manning equation with velocity constraints; a Type A capacity solution then completes the channel design.
- Classify terraces according to profile, channel cross section, field orientation, and outlet type.
- Calculate terrace spacing, channel slope, and channel depth.
- Identify similarities and differences between terraces, diversion terraces, and diversions.
- Design diversion terrace channels and diversion ditches.
- Calculate storage volumes and drainpipe sizes for terraces with tile outlets.

Background

Why study erosion terraces if other approaches to erosion control are more widely used? Three reasons: (1) Conservation tillage is not possible for all crops. Vegetable crops are

Figure 10.1. Ancient bench terraces at Delphi, Greece.

examples. (2) With long slopes, surface cover alone may not be sufficient for erosion control. (3) Many aspects of terrace design are readily transferable to urban situations, erosion control at construction sites, and bioremediation and other farm- or field-scale environmental problems. Diversion terraces are commonly used in construction erosion control. Diversions are used for water management. The forest products industry is looking to terracing as an erosion control strategy during stand replenishment following harvesting. One can modify the principles used in constructing a water conservation terrace to give a method for water harvesting. This may be useful for some bioremediation projects away from a water source. Terracing concepts continue to be germane to erosion control strategies.

Nonparallel terracing and, later, parallel terracing (see Figure 10.2) proliferated in the age of conventional tillage. Conservation tillage schemes represent another approach to erosion control in agriculture. The arrival of conservation tillage and large equipment explains the abandonment of many terrace systems in the southeastern United States.

The major differences between a terrace and a diversion channel are the side slope, possible length, spacing, flow capacity, time of concentration computation, and possibly outlet type. Terrace channels are triangular with z values often greater than 4. Diversions are often triangular or trapezoidal and usually have z values of 4 or smaller. We may design an agricultural terrace to fit machines of prescribed widths, whereas we design diversions with fixed z values. Lengths of diversions may exceed those of conventional terraces. Flows in diversions are higher than in conventional terraces. Time of concentration is set at 5 min for terraces and computed for diversions. Diversions usually have surface outlets, while terraces may have surface or tile outlets.

The emphasis here will be on single terraces and diversions. For the usual 10 yr storm, surface waterway outlet, one needs to determine (1) spacing of the terrace/diversion, (2) slope causing excessive velocities, and (3) depth of the channel for capacity evaluation. For the terrace with a tile outlet, the storage berm height and terrace drainage system must be determined. Diversions find wide application in soil erosion management plans for urban construction sites. One may refer to works such as Beasley et al. 1984 for more comprehensive discussions of field terrace systems.

Figure 10.2. Comparison between parallel and nonparallel broadbase terrace channels. Parallel terraces follow the contour only approximately and require cut-and-fill analyses. Parallel terraces are much more farmable than nonparallel terraces. Primary design goals are spacing, channel capacity, and channel slope.

Terrace Definitions and Types

A terrace is a ridge and channel, constructed across the slope at a suitable location to intercept surface runoff water. One may construct it with an acceptable grade to the outlet or with a level channel and ridge. Figures 10.3–10.5 show common terrace types and nomenclature. Several types of terraces are common, especially in the lower Midwest region of the United States. Powell and Steichen 1981a and ASAE 1997a discuss terrace types and designs. Terraces are classified based on alignment, cross section, or outlet type. Nonparallel terraces follow the contour of the land. Parallel terraces have equal spacing throughout their length. They require more cut and fill than nonparallel terraces. However, parallel terraces make farming practices easier.

Bench terraces are sometimes used on steep (>20%) slopes (e.g., see Figures 10.1 and 10.5). They are also an option for improving irrigation efficiency on gently sloping soils. The conservation bench terrace incorporates a wide, flat channel to enhance infiltration. The grassed backslope terrace has a steep backslope (2 : 1 or greater) kept in perennial grass.

Broadbase and bench terraces may have a variety of channel cross sections. The terrace channel cross section as used in this text implies a high z value, which may be arbitrarily

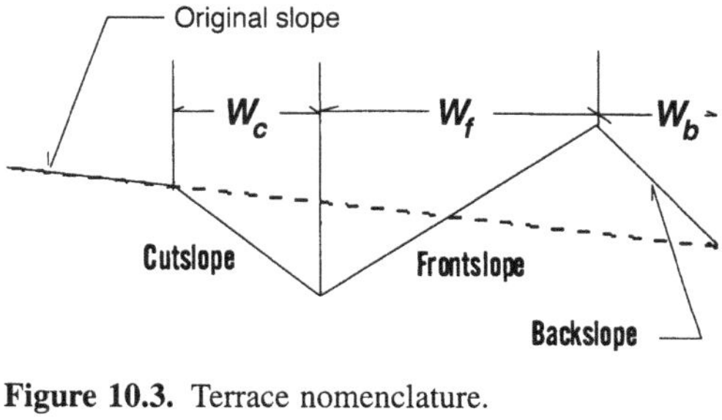

Figure 10.3. Terrace nomenclature.

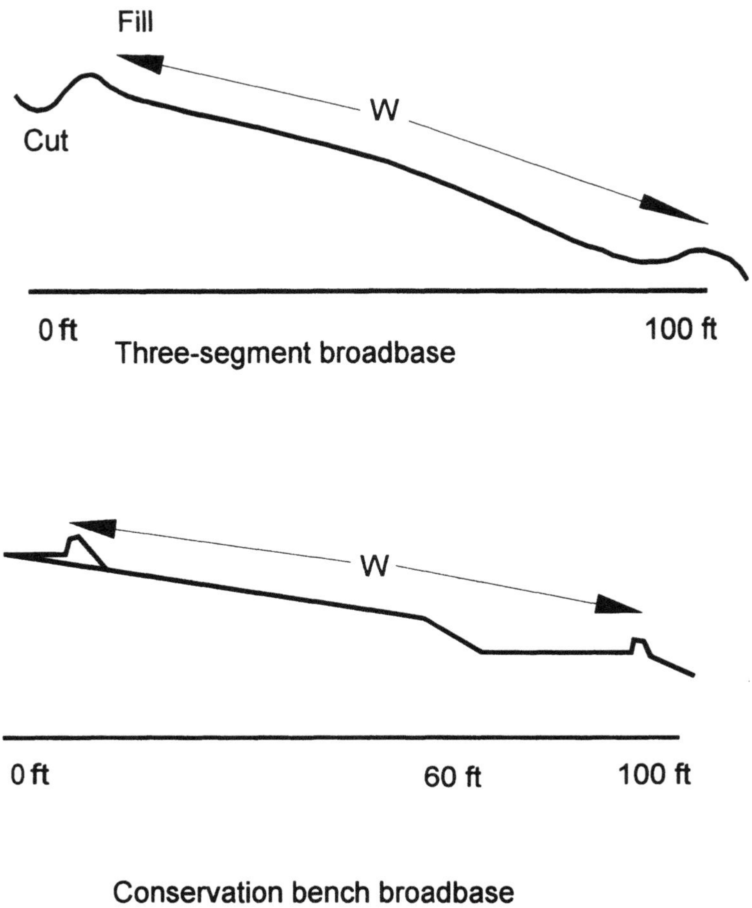

Figure 10.4. Selected examples of common terrace systems. The three-segment broadbase terrace is generally used for erosion control and may have a tile or vegetated outlet. The conservation bench broadbase usually has a blocked outlet (not shown) for water conservation purposes.

set or may be designed to conform to machine width. The diversion terrace has z values lower than the broadbase terrace and may have a nonsoil lining. Cutslopes, frontslopes, and backslopes of the diversion terrace are not intended for farming. The diversion channel is similar to the diversion terrace; however, it serves flow control purposes, whereas terraces and diversion terraces are primarily for erosion control.

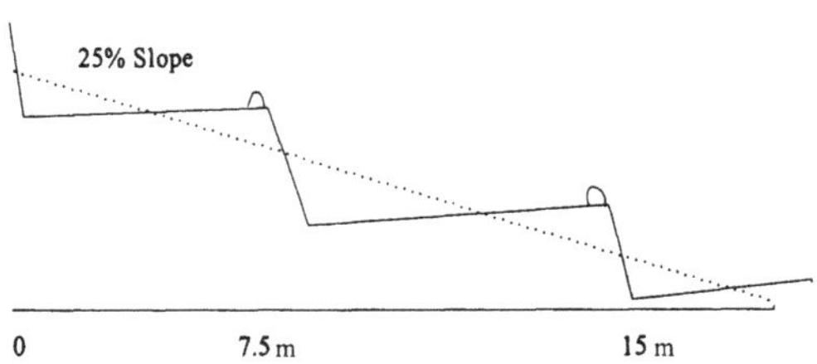

Figure 10.5. Old-style bench terrace used on steep slopes. This terrace is very expensive and requires deep soils.

Terraces may have vegetated outlets, natural draw outlets, subsurface-drained outlets, or blocked outlets depending on the purpose of the terrace and/or the site topography (Powell and Steichen 1981b). Terrace outlets are further discussed following terrace system design. They may also be parallel or nonparallel, as shown in Figure 10.2.

Function of Terraces

Properly placed, constructed, and maintained terraces reduce runoff and soil losses. They prevent formation of rills and gullies. They contribute toward the reclamation of badly gullied fields by intercepting runoff before it becomes concentrated and attains an eroding velocity. Terraces allow sediment to settle from runoff water, improving the quality of surface water leaving the field. Some terrace systems modify the land slope, improving farmability. By reducing runoff rate and soil losses and by enhancing infiltration, they help maintain the productivity of the soil.

Terraces encourage contour farming as a conservation practice. Well-planned terraces add value to practically all soils. The exceptions are stony, sandy, or shallow soils not permitting economical construction and maintenance.

Land slope affects the practicality of terraces. As land slope increases, soil loss from erosion increases. However, the cost of terrace construction and maintenance, and associated farming difficulties, increase with slope. Narrow-base terraces with grassed backslopes are used in more steeply sloped fields. Beasley et al. (1984) elaborate on the cost-benefit ratios of terracing for conventional agriculture.

Erosion control is usually the prime objective in terracing in humid regions. In arid regions, water conservation can replace erosion control as the prime objective. Terrace designers should account for factors such as allowable soil loss, most intensive use expected for the land, and the expected level of management. Farmability is a second important objective in conventional agriculture. Terraces must be as farmable as possible to have them accepted and maintained. Terraces with good farmability will

- be parallel wherever possible to reduce point rows, save time in farming operations, and reduce crop damage caused by machinery turning between terraces;
- be spaced to fit the equipment employed at the location;
- be spaced as wide as possible without having erosion exceed the allowable soil loss;
- be planned and constructed with a farmable cross section;
- provide good access roads to all parts of a terraced field (waterways should never serve as roads).

One may sacrifice some design requirements for non-row-crop situations such as forests and construction sites.

Terraces can enable moisture conservation. In areas of low rainfall and permeable soils, level terraces with closed ends are sometimes used for moisture conservation and erosion control. The moisture conservation benefit occurs because terraces increase opportunity time for infiltration.

Erosion Terrace Spacing for Erosion Prevention

Broadbase and bench terraces are widely used for erosion control. Erosion reduction primarily and equipment dimensions secondarily factor into the terrace spacing decision.

We first discuss the minimum spacing calculation based on erosion reduction. We then adjust the spacing to fit equipment width for high field efficiency. ASAE 1997a provides two approaches for minimum terrace spacing computation. Figures 10.2 and 10.4 define the spacing (W).

The *vertical interval map* (VI map), based primarily on erosive rainfall, is the oldest technique for terrace spacing specification. The vertical interval is calculated using

$$VI = Xs + Y \tag{10.1}$$

where

VI is vertical interval between consecutive terraces, m;
X is a constant for geographical locations (see ASAE 1997a terrace standard and Figure 10.6);
s is land slope in m per 100 m;
Y is a constant based on soil erodibility, cropping systems, and crop management (0.3–1.2, with lower value for easily erodible soil and higher value for conservation tillage with mulch).

Compute the horizontal interval by dividing the land slope into the vertical interval. One may approximate the horizontal interval with the slope length due to the typically low slopes.

The RUSLE terrace spacing method enables an accounting of rainfall, soil, and anticipated cropping in the terracing spacing decision. The RUSLE method is as follows. Repeating equation 6.1, the USLE, one may write

$$A_{Ero} = R_{rf}K_{sl}L_fS_fC_{prac}P_{prac} \tag{6.1}$$

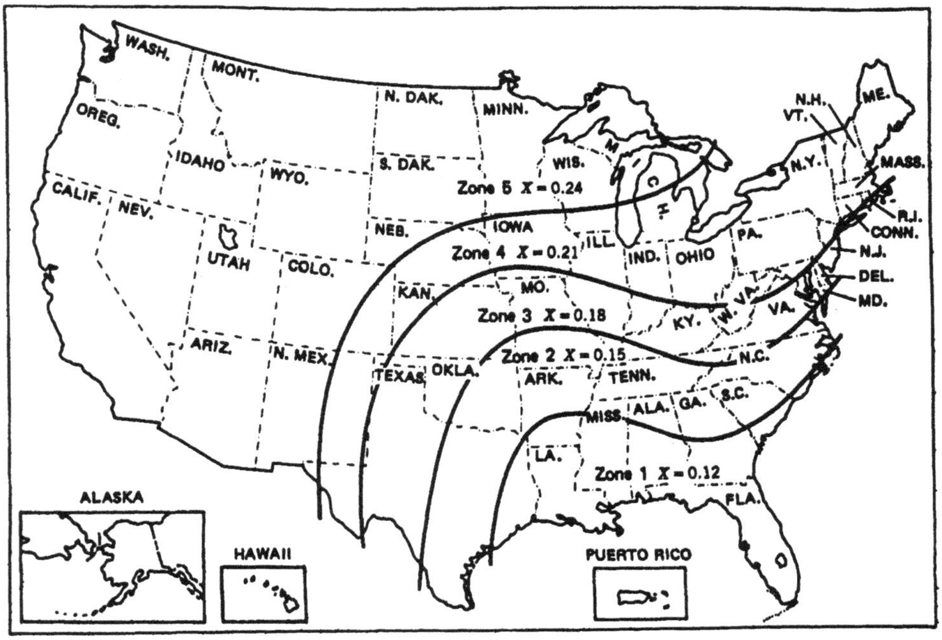

Figure 10.6. Terrace spacing parameters for the VI map terrace spacing standard. (ASAE 1997a; used with permission.)

where

A_{ero} is the soil loss tolerance (Mg/ha); commonly 5–11 Mg/(ha-yr);
R_{rf} is a rainfall factor (average annual, Figure 6.9);
K_{sl} is a soil factor (see equation 6.3);
L_f is a length factor;
S_f is a slope factor (see equations 6.7a–c);
$C_{prac} = C_{avg\ annual}$; practice factor for the rotation being planned (see Tables 6.2, 6.4, and 6.5 for computation details);
$P_{prac} = P_c \times P_s \times P_t$ where contouring $P_c = 0.6$ from Chapter 6 (see Table 6.3); strip-crop P_s from Chapter 6; terrace $P_t = 1.0$. (*Note:* We do not terrace the zone between terraces! Use the P_t value for terracing when characterizing erosion from the field. Refer to equation 6.8 and associated discussion.)

Solve the USLE for the length factor using the acceptable loss rate. Form a ratio to determine the maximum value of the length factor when staying within the soil loss tolerance as shown in equation 10.2:

$$\frac{Tol}{A_{ero}} = \frac{required\ L_f}{actual\ L_f} \quad \textbf{(10.2)}$$

where L_f is the USLE L factor. Recalling equation 6.2 for L_f (noting notation change of L to terrace spacing W_i), and auxiliary equations 6.5–6.7, yields

$$R_{rfstorm} = \frac{\Sigma(EI_{30})}{100} = 0.29(i_{30\ \max})\sum_{i=1}^{N} P_i(1 - e^{-0.05i_i}) \quad \textbf{(6.2)}$$

$$m_{er} = \frac{\sin(\theta)}{\sin(\theta) + 0.269[\sin(\theta)]^{0.8} + 0.05} \quad \textbf{(6.5)}$$

$$\theta = \tan^{-1}\left(\frac{s}{100}\right) \quad \textbf{(6.6)}$$

$S_f = 3.0[\sin(\theta)]^{0.8} + 0.56$ if the slope length 1 is shorter than 4 m **(6.7a)**

$S_f = 10.8\sin(\theta) + 0.03$ if the slope length 1 is longer than 4 m and $s < 9\%$. **(6.7b)**

$S_f = 16.8\sin(\theta) - 0.5$ if the slope length 1 is longer than 4 m and $s \geq 9\%$. **(6.7c)**

where

θ is the slope angle;
s is the land slope as a decimal;
L_f is the slope length measured along the land slope.

First, solve for the L_f factor before terracing. Second, solve for the required length factor from the spacing-tolerance ratio. Then solve for the spacing ($1 = W_i$) along the slope length after terracing. Vertical interval then is equal to the slope times the slope length.

For optimum farmability with modern equipment, one should space terraces at 36, 18, and 9 m (120, 60, 30 ft) (ASAE 1997a). Beasley et al. (1984) present an extensive listing of terrace spacings accommodating modern farm equipment. Select the tabulated spacing less than the calculated spacing. Sometimes one may reevaluate the rotation so the minimum spacing above will be adequate. For nonagricultural applications, construction or other regulatory considerations may dictate the required interval.

Example 10.1 (ASAE vertical interval method): Given soil on 5% slope in Athens, Georgia, K_{sl} = 0.02 Mg-ha-hr/(ha-MJ-mm), C_{prac} = 0.2, *Tol* = 7 Mg/ha-yr, R_{rf} = 5000 MJ-mm/ha-hr-average annual storm, and a 100 by 100 m, or 1 ha, field. Compute the maximum allowable VI map spacing.

$$VI = XS + Y$$

$$X = 0.15 \text{ for Athens from the } X \text{ map}$$

$$Y = 0.3 \text{ m (easily erodible soil)}$$

$$VI = 0.15(5\%) + 0.3 = 1.05 \text{ m}$$

$$\text{slope length } W_i = 1.05 \text{ m}/0.05 = 21 \text{ m}$$

Example 10.2 (USLE procedure for Example 10.1):

$$A_{ero} = R_{rf} \text{ (MJ-mm/ha-hr-average annual storm) } K_{sl} \text{ [Mg-ha-hr/(ha-MJ-mm)]}$$
$$L_f S_f C_{prac} P_{prac} \text{ Mg/ha-yr}$$

We need $L_f = (100/22)m_{er}$. θ is found to be 2.86 degrees (equation 6.6), leading to m_{er} = 0.4 (equation 6.5) and L_f = 1.83 (equation 6.2). S_f = 10.8 sin θ + 0.03 = 0.56 (equation 6.7b). $P_{prac} = P_t P_c P_s$ where P_t = 1 (no terraces between the terraces!), P_s = 1, and P_c = 0.6 for P = 0.6 (Table 6.3, choose 0.6 because we are barely over the length limit). Then A_{ero} = 12.29 Mg/ha-yr, which is excessive. (See Example 6.1 for more USLE calculations.) We need to reduce L_f proportionally as follows: 7 Mg/ha-yr/12.29 Mg/ha-yr = $L/1.83$, leading to target L of 1.04. Terrace spacing W_i is then $1.04^{1/0.4}$ (22 m) = 24.27 m, and VI = 0.05(24 m) = 1.25 m. Based on general machinery width, select a spacing of 18 m.

One could solve directly for the L_f factor, then solve for spacing. The USLE result is the preferred method of spacing calculation, though both results are close in this instance. In both cases, the 18 m spacing is the choice.

The VI map and USLE methods of spacing calculation may lead to greatly different spacings. One should then reconsider values used for terms in both spacing relationships. Carefully evaluate the crop rotation efficacy (i.e., C_{prac} factor of the USLE spacing), particularly if the spacing of the USLE is much shorter than the VI map spacing. The USLE spacing is more appropriate.

Terrace Channel Length, Cross Section, Maximum Grade, and Capacity

Channels associated with bench and broadbase erosion control terraces are intermittent-flow channels, designed for stability and capacity. One then controls the channel slope for stability (maximum permissible velocity) and channel depth for capacity. *The design return period is usually 10 yr. The time of concentration for terraces is set at 5 min.* For terraces,

Figure 10.7. Broadbase erosion terrace with contour row cropping following the row near the left bottom with large side slope z. (Courtesy of NRCS.)

set the overland flow time at 5 min instead of computing a value. Figure 10.7 shows a typical terrace channel. The channel in Figure 10.7 is triangular with a very large z (e.g., $z = 10+$), and one farms all parts of the channel and embankment.

Length

Agricultural terrace lengths depend on field shapes and outlet locations. One should use the short lengths to avoid excessive flows at the discharge. Maximum practical terrace lengths are about 300–500 m. Diversion terraces can be deeper, wider, and longer.

Channel Cross Section

One must design the broadbase (and sometimes bench) terrace system with the present or anticipated future farm equipment of the landowner or grower in mind. Size the cutslope, frontslope, and backslope to allow one pass of the most critical (i.e., usually widest) tillage or harvesting equipment, usually a planter or harvester. This affects the z values for the various slopes. Equating equipment width to a slope requires trial-and-error solution. Figure 10.8 shows the relationships between the z values, land slope, and machine widths. For design purposes, one assumes a triangular channel. In practice, ASAE (1997a) mandates a minimum channel width of 0.9 m and a minimum ridge top width of 0.9 m.

An alternative approach involves designing at several, say, three, z values, assuming $z =$ 5, 10, 20 on all slopes. The computed depths including freeboard enable interpolation of a z giving the required slope face width for a predetermined machine width.

Maximum-Minimum Allowable Slope

Terrace channel grades should usually be less than 1%. The grade is selected based on the ability to remove excess precipitation with a velocity constraint of less than 0.6 m/s (0.45 m/s in highly erosive soils). Terrace grades can be either uniform or variable.

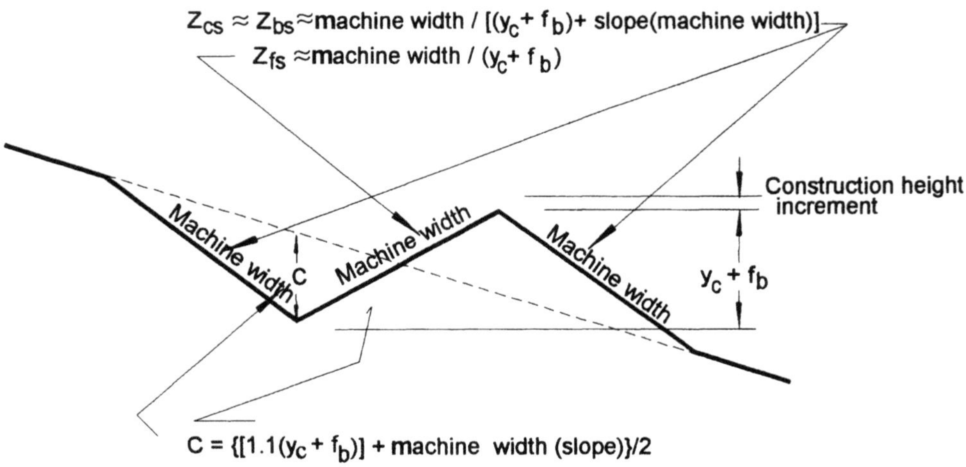

Figure 10.8. Farmability of terrace systems is ensured when one pass of the largest equipment will be accommodated by the cutslope, frontslope, and backslope. An alternative is to set z_{fs} to a large value (e.g., 20 or greater so machinery width is not a serious constraint). The large-z assumption is possible because of the relative insensitivity of the result to the z value used in the computations. ASAE (1997a) recommends a minimum channel width of 0.9 m and a minimum ridge top width of 0.9 m. These are ignored in channel design. A 10% additional construction height increment is estimated to account for settling.

One computes peak flow from the rational equation, stated originally as equation 3.7:

$$f_{GA} = -K_c \frac{\Delta H}{\Delta L} \tag{3.7}$$

where

A is the area (= terrace spacing $W \times$ length L);
C_{rat} is the rational coefficient;
I is the design rainfall intensity computed from the 10 yr–t_c storm, where for terraces $t_c = 5$ min.

Beasley et al. (1984) suggest a C_{rat} value of 0.7 for field terraces. If the terrace is a diversion terrace, use an appropriate C_{rat} for the land above the terrace. One must satisfy the Manning equation (equation 8.14) with the appropriate permissible velocity (i.e., Type B solution):

$$q = \frac{\phi_{dim}}{n} A R^{2/3} S^{1/2} \tag{8.14}$$

where

S is slope (dimensionless);
R is hydraulic radius (L);
v_p is permissible velocity (L/T);
n is the Manning roughness coefficient (0.035 recommended for slope determination; 0.06 for capacity);
ϕ_{dim} is the units coefficient (1 for metric and 1.49 for English units).

The hydraulic radius for the triangular channel is given by

$$R = \frac{area}{wetted\ perim} = \frac{Zy^2}{2y\sqrt{Z^2+1}} = \frac{Zy}{2\sqrt{Z^2+1}} \approx \frac{y}{2} \tag{10.3}$$

Note that we can also use the simplified relation $R = y/2$ since $z > 3$ for the triangular terrace/diversion channel. By the continuity equation, one can write

$$\underset{1}{q_p} = \underset{2}{\frac{C_{rat}iW_iL}{43{,}560}} = \underset{3}{v_pA_c} = \underset{4}{v_pZy^2} = \underset{5}{\frac{0.0028}{10{,}000}C_{rat}iLW_i} = \underset{6}{\frac{\phi_{dim}}{n_s}R^{2/3}\sqrt{S}Zy^2} \tag{10.4}$$

where

v_p is the permissible velocity (2 ft/s [0.61 m/s] for most soils; 1.5 ft/s [0.46 m/s] for extremely erodible soils);
n_s is the Manning n for stability ($n_s = 0.035$);
A_c is the cross-sectional area of the terrace channel;
W_i is the spacing between terraces (ft, m);
S is the slope;
L is the terrace length (ft, m).

The product W_iL is the area intercepting the runoff.

Solving the continuity equation (first and fourth terms in equation 10.4) for depth gives

$$y_s = \sqrt{\frac{q_p}{Zv_p}} \tag{10.5}$$

where we introduce y_s in place of y to denote depth of flow at the stability limit.

One may then arrange equation 10.4 for a solution for maximum slope:

$$\underset{1}{S_{max}} = \underset{2}{\frac{v_p^2n_s^2}{\phi_{dim}^2R^{4/3}}} = \underset{3}{\frac{v_p^2n_s^2}{\phi_{dim}^2\left(\dfrac{Zy_s^2}{2y_s\sqrt{1+Z^2}}\right)^{4/3}}} \approx \underset{4}{\frac{v_p^2n_s^2}{\phi_{dim}^2\left(\dfrac{y_s}{2}\right)^{4/3}}} \tag{10.6}$$

The above equation follows from the continuity equation and the area of the triangular cross section, solved for depth. Solution for S_{max} is straightforward after assuming a trial z value ranging from 10 to 20 or larger. If the machine width is not an issue from the cultural practice viewpoint, assume a z of 20 and solve for S_{max}.

ASAE (1997a) suggests stability maximum slopes and drainage minimum slopes. The stability maximum slope provides a cap for the maximum slope predicted above. The drainage minimum slope provides a minimum for sedimentation and drainage purposes. The channel slope can operate within the window defined by the maximum and minimum slopes in Table 10.1.

Terrace Capacity and Depth

Relationships given above are useful for terrace depth calculation. Use a Manning $n_{cp} = 0.06$ for the depth calculation. Use the maximum grade calculated above. This time, slope

Table 10.1. Suggested maximum and minimum terrace grades.

Terrace length or length from upper end of long terraces (m)	Stability maximum slope defined by the minimum equation 10.6 value or the value given below (%)
30	2
31–60	1.2
61–150	0.5
151–365	0.35
366	0.3
	Drainage minimum slope (%)
Soils with slow internal drainage	0.2
Soils with good internal drainage	0

Sources: Adapted from Beasley 1963 and ASAE 1997a.

is an input value. Slope should exceed 0% on extremely well drained soils and 0.2% on typical soils. Velocity must not exceed the permissible velocity. One may then calculate depth. Then add a freeboard of 0.5 ft (0.15 m). Solve the relations above for depth Y_{cp} to arrive at the following relationship:

$$\underset{1}{\frac{\phi_{dim}}{n_{cp}} Z y_{cp}^2 \left(\frac{y_{cp}}{2}\right)^{0.67} \sqrt{S_{usd}}} = \underset{2}{\frac{C_{rat} i W_i L}{43{,}560}} = \underset{3}{\frac{0.0028 C_{rat} i W_i L}{10{,}000}} \tag{10.7}$$

$$= \underset{4}{q_p} \approx \underset{5}{\frac{\phi_{dim}}{n_{cp}} \frac{W_m}{(y_{cp} + 0.15)} y_{cp}^2 \left(\frac{y_{cp}}{2}\right)^{0.67} \sqrt{S_{usd}}}$$

The slope S_{usd} is the selected slope. It should be between S_{max} of Table 10.1 (or the S_{max} computation) and S_{min} of Table 10.1. (Selecting halfway between is one approach.) One can solve the above relationship (the right-hand side) for depth. The new approximate z is $W_m/(y_{cp} + 0.15\text{ m})$ if one is contemplating fitting machine width to the terrace slopes. Recheck the maximum slope at this new z. On finding a combination of z and depth, add 0.15 m freeboard. Finally, account for 10% settling by multiplying the freeboard height by 1.1. The 1.1 factor adds 10% construction height to allow for settling. Estimate the cut depth using the formulas in Figure 10.8.

Example 10.3 (critical machine width equals channel slope face width): Given a 100×9 m (328×30 ft) terrace system, 5% slope, machine width of 6.1 m (20 ft) and $v_p = 0.6$ m/s (2 ft/s), size the terrace channels.

Assume $z_{fs} = 15$; find that the 5 min rain was 1.1 in (27.9 mm) and the intensity was 336 mm/hr; area = (100 m)(9 m)/(10,000 m^2/ha) = 0.09 ha.

$$q_p = 0.0028CiA = (0.0028)(0.7)(336\text{ mm/hr})(0.09\text{ ha}) = 0.06\text{ m}^3\text{/s (2.12 cfs)}$$

Compute y_s using equation 10.5 and S_{max} using equation 10.6 (first and fourth terms).

$$y_s = \{(0.06\text{ m/s}^3)/[(15)(0.6\text{ m/s})]\}^{0.5} = 0.082\text{ m}$$

$$S_{max} = [(0.6\text{ m/s})^2(0.035^2)]/(0.082\text{ m}/2)^{1.33} = 0.0309$$

Select a design channel slope of 0.005 (Table 10.1 maximum controls). Solving equation 10.7, fourth and fifth terms, yields

$$0.06\ m^3/s = (1/0.06)6.1\ m/(y_{cp} + 0.15)(y_{cp}^2)(y_{cp}/2)^{0.67}(0.005)$$

The depth $y_{cp} = 0.122$ m + 0.15 m freeboard = 0.272 m. The frontslope = 6.1 m/0.272 = 22 compared to a z_{fs} of 15. Check y_s. Go back and check the maximum slope at z_{fs} = 22; $y_s = [(0.06\ m^3/s)/(22)(0.6\ m/s)]^{0.5} = 0.067$ m; the maximum slope was $S_{max} = [(0.06\ m^3/s)^2\ (0.035^2)]/(0.067\ m/2)^{1.33} = 0.02$, still well over 0.005, so channel gradient maximum of 0.005 is acceptable.

Figure 10.8 provides relations for backslope, cutslope, and construction depth. Backslope = cutslope = z_{bs} = 6.1 m/[0.272 m + 0.02(6.1 m)] = 15. The construction height y_{const} is 1.1 × 0.272 m = 0.3 m.

Cut depth = [0.3 m + 6.1 m (0.02)]/2 = 0.21 m.

Example 10.4 (large side slope *z* such that machine width not required to fit on side slope face width): Given a 100 × 9 m (328 × 30 ft) terrace system, 2% field slope, v_p = 0.6 m/s (2 ft/s), size the terrace channels. Assume the terrace need not accommodate a critical machine width. Note that the 2% field slope corresponds to a *z* value of 50. The chosen *z* must be smaller. Set z_{fs} = 20; use the flow rate computed in Example 10.3, $y_s = [(0.06\ m^3/s)/(20)(0.6\ m/s)]^{0.5} = 0.07$ m (equation 10.5). $S_{max} = (0.6\ m/s)^2(0.035^2)/(0.07\ m/2)^{1.33} = 0.0381$ (equation 10.6). Select a design slope of 0.005 (Table 10.1 maximum controls). Solving equation 10.7, fourth and first terms, yields

$$0.06\ m^3/s = (1/0.06)(20)(y_{cp}^2)(y_{cp}/2)^{0.67}(0.005)^{0.5} \text{ or } y_{cp} = 0.13\ m$$

Adding 0.15 m of freeboard gives a depth of 0.28 m, then adding a 10% construction height brings the depth at construction to 0.31 m. To estimate the cut depth, use the actual or slope face width and relationships in Figure 10.8. If the terrace was designed with a set z value (e.g., *z* = 20), figure the face width of the cutslope at the computed depth at capacity.

Back slope = cutslope = z_{bs} = *x* m/[0.31 m + 0.02(*x* m)] = 20. Solving for *x* yields face width *x* = 10.3 m. Cut depth = [0.31 m + 10.3 m(0.02)]/2 = 0.26 m.

Study the similarity of the results in Examples 10.3 and 10.4. This corroborates the lack of sensitivity of the design process to the *z* value when *z* gets large.

Broadbase and Bench Erosion Terrace Design Summary

1. Determine the terrace spacing by RUSLE or VI map methods.
2. Determine the largest machine width having the most critical height-depth requirements (e.g., usually a tillage machine, planter, mower, or harvester).
3. Estimate the maximum gradient with a trial *z* ranging from 5 to 20. One may incorporate this *z* relation into the iterative solution as was done in Example 10.3. Select a channel slope between the table minimum and the computed maximum. Alternatively, one may set the *z* to a large value such as 20 following Example 10.4.

4. Estimate the capacity depth (y_{cp}) near the terrace exit for the longest terrace (or terrace with the largest flow), then estimate the backslope and frontslope according to machinery width requirements. Repeat the range of z values (single large z for approximate work). Compute $y_{const} = 1.1y_{cp}$:

   - The depth of cut for a balanced terrace $= [y_{const} + W_m(S_{field})]/2$; repeat for each z_{fs} if applicable if the terrace is designed to accommodate the critical machine width. If z_{fs} is fixed (terrace does not accommodate a critical machine width), add freeboard of 0.15 m and compute y_{const}.
   - The frontslope $z_{fs}\ W_m/(y_{const} + fb)$; find the z_{fs}, y_{cp} combination compatible with the required W_m.

$$\text{cutslope } z_{cs} = W_m/[y_{const} + fb + (S_{field}W_m)].$$
$$\text{backslope } z_{bs} = W_m/[y_{const} + fb + (S_{field}W_m)],$$

   where fb = freeboard; S_{field} = field slope; W_m = machine width or equivalent machine width.

Diversion Terraces and Diversions in Agricultural and Urban Applications

Diversion Terraces

The diversion terrace is an erosion control approach using channels with low z values and possibly linings other than soil. Otherwise, the diversion terrace is similar to a broadbase or bench terrace system in terms of spacing. *The diversion terrace is a terrace whose catchment area is largely bare soil and whose channel and catchment area are involved in uses other than agriculture.* The z is fixed at 4 or lower for a diversion terrace. Diversion terraces may have soil or other linings and the stability-capacity design follows guidelines for the lining. Fixing z at less than 4 can result in cheaper installations. Diversion terraces are not usually broadbase or bench terraces. Diversion terraces have steep cutslopes and frontslopes and may have steep backslopes. For a diversion terrace, assume the critical machinery width does not need to fit on the slope face. The methodology discussed above to solve for depth with soil linings is suggested. For other linings, design for stability at the design n, then increase n by 25% for the capacity design. If the terrace channel is vegetated, use methods for vegetated channels (stability and capacity). To complete the solution, add freeboard, add 10% for construction height, and compute the cut depth using the width of the excavator blade. Flows are computed using the method for terraces discussed previously.

Diversions

A diversion has a steep frontslope and steep cutslope channel constructed across the slope to divert excess water to a safe discharge area. Diversions do not have an overt erosion control purpose. *The major difference between a terrace and a diversion channel is the side slope, possible length, spacing and flow capacity, time of concentration computation, and possibly outlet type (diversions are not usually tile drained). A diversion has a generalized catchment area that may be forested, pastured, or put to other use, whereas the diversion terrace and terrace are designed assuming a tilled soil catchment. Spacing is not an issue with diversions.* Terrace channels are triangular and have z values greater than 4, while diversions are either triangular or trapezoidal and have z values of 4 or smaller.

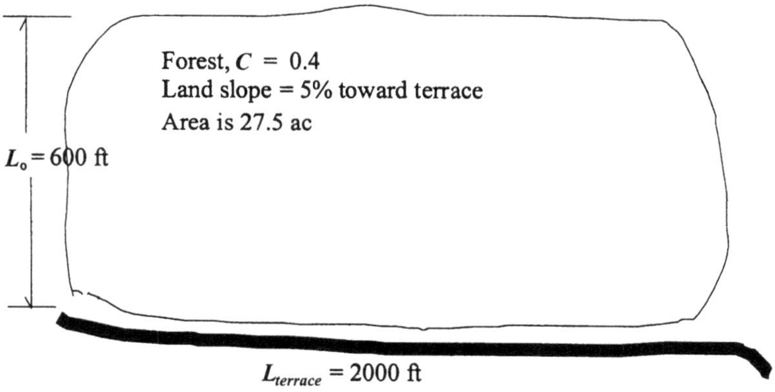

Figure 10.9. Diversion terrace situation.

One may design an agricultural terrace to accommodate machines of prescribed widths, but one designs diversions with fixed z values. Flows in diversions are higher than in conventional terraces. Times of concentrations are set at 5 min for terraces and computed for diversions. Diversions may have any lining type. Length or slope is not arbitrarily constrained. Figure 10.9 shows a diversion channel and its associated drainage schematically.

One designs the triangular diversion channel as a terrace channel, discussed above. One may design the bottom width of the trapezoid using Type B stability methods. One may hold the slope constant and design for a stable bottom width as was done in Chapters 8 and 9. One may sometimes fix the bottom width at 1.2 m or so (e.g., bulldozer blade width) and adjust the slope to keep from exceeding the permissible velocity. One follows stability design with capacity design, setting the Manning n to $n_{cp} = 0.06$ as with terrace design or, with a vegetated waterway, satisfying the tall grass.

The diversion terrace is an integral part of many erosion control plans. Figure 10.10 shows a typical diversion channel. Compute Q with the rational equation as for a terrace channel or with the TR55 method, except one computes the time of concentration t_c using

Figure 10.10. Diversion channel with a vegetated lining. Design methodology for vegetated lining is appropriate here.

the combination of overland flow to ditch plus travel time in the ditch. Typically, a 10 yr storm frequency with duration equal to time of concentration defines the design storm. One may compute the overland flow velocity using the overland flow method time-of-concentration calculation in Chapter 5. The C factor in the rational equation is appropriate to the land use above the terrace. Haan et al. (1994) discuss diversion design in more detail. Examples 10.5–10.7 show the process.

Example 10.5 (soil lining, slope determined): Suppose one has an unlined diversion with a permissible velocity of 0.46 m/s (1.5 ft/s), z of 4, triangular, in the situation shown in Figure 10.9. Design the channel for a 10 yr storm.

First, compute the flow rate based on the 10 yr storm with the rational method.

$$t_c = [600 \text{ ft}/(2.5\ (0.05)) + (2000 \text{ ft}/1.5 \text{ ft/s})] = 2406.6 \text{ s} = 0.67 \text{ hr}$$

A 10 yr–0.67 hr storm in Athens is 2.2 in.

$$I = 2.2/0.67 = 3.28 \text{ in/hr}$$

$$q = 0.4\ (3.28 \text{ in/hr})(27.5 \text{ ac}) = 36 \text{ ac-in/hr} = 36 \text{ cfs} = 1.02 \text{ m}^3/\text{s}$$

Since $z = 4$ and $v_p = 0.46$ m/s, we can compute the depth (equation 10.5) and then the maximum slope (equation 10.6):

$$y_s = (q/v_p z)^{0.5} = \{(1.02 \text{ m}^3/\text{s})/[(0.46 \text{ m/s})(4)]\}^{0.5} = 0.75 \text{ m}$$

Compute S_{max} using the Manning equation with $n_s = 0.035$:

$$S_{max} = [(n_s v_p)/(y_s/2)^{0.67}]^2 = [(0.035)(0.46 \text{ m/s})]/[(0.75 \text{ m}/2)^{0.67}]^2 = 0.00096.$$

Do not impose the limits of Table 10.1 on diversion slopes; use the computed S_{max}.

Next we design depth y_{cp} for capacity using $n_{cp} = 0.06$. From equation 10.7, first and fourth terms,

$$q = 1.02 \text{ m}^3/\text{s} = 4y_{cp}^2(1/0.06)(y_{cp}/2)^{0.67}(0.00096)^{0.5}$$

$$y_{cp}^{2.67} = 0.7857$$

$$y_{cp} = 0.91 \text{ m}$$

Adding a 0.15 m freeboard results in a depth of 1.06 m. The construction height is 1.1×1.06 m = 1.17 m. Assuming a balanced diversion, one may calculate cut depth and z values as shown in Figure 10.8.

Example 10.6 (vegetation lining, slope determined): Assume Example 10.5 had an easily erodible trapezoidal cross section, vegetated with fescue, minimum bottom width of 1.2 m, $z = 3$. Design the diversion channel for a 10 yr storm. Assume the grass is rapidly established.

First, determine the slope to give permissible velocity of the fescue, mowed (class D, Figure 9.6) condition. $v_p = 1.5$ m/s assuming slope is less than 5%.

$$T_c = \text{overland flow time} + \text{ditch flow time} = t_c = 600 \text{ ft}/[2.5(0.05)]$$
$$+ (2000 \text{ ft} \times 0.3048 \text{ m/ft})/1.5 \text{ m/s} = 1480 \text{ s} = 0.41 \text{ hr}$$

The calculated rainfall intensity (Weiss method) was 114 mm/hr, leading to a peak flow of 1.4 m^3/s from the rational equation. Find hydraulic radius R by first calculating $A = q/v_p = (1.4\ m^3/s)/1.5\ m/s = 0.93\ m^2$. Given the minimum bottom width $b = 1.2$ m, one may write equation 5.20 for y unknown to get $A = by + zy^2$ or $0.95 = 1.2y + 3y^2$. The nonnegative answer is $y = 0.39$ m using the quadratic formula.

$$R = A/[b + 2y(z^2 + 1)^{0.5}] = 0.93/[1.2 + 2(0.39)(3^2 + 1)] = 0.26$$

One may refer to Appendix B, equation B.3, for a closed form solution for depth when the bottom width is fixed. Use the vegetation retardance class D chart to find the slope of 0.02 at the intersection of the hydraulic radius and permissible velocity values. *Note:* In some cases one may determine stability assuming bare soil, since vegetation may be slow to establish. Use the bare soil minimum velocity and solve the Manning equation for critical slope, using R as computed above.

Holding slope constant in class B (Figure 9.4) conditions, following Example 9.1 capacity phase, we have the results shown in Table 10.2.

Table 10.2. Calculations for capacity in Example 10.6.

Guess y_{cp} (m)	$R_{cp} = \dfrac{1.2y_{cp} + 3y_{cp}^2}{b + 2y_{cp}(z^2 + 1)^{0.5}}$ (m)	$A_{cp} = 3y_{cp}^2 + 1.2y_{cp}$ (m^2)	v_{cp} (class B) (m/s)	$q = A_{cp} v_{cp}$ (m^3/s)
0.53	0.32	1.47	0.95	1.39 OK

Find the depth in the unmowed channel to be 0.53 m + 0.15 m freeboard = 0.68 m. If the slope were fixed, compute as a vegetated waterway, discussed in Chapter 9.

Example 10.7 (soil lining, fixed slope): Assume Example 10.5 had an easily erodible trapezoidal channel, bare silt loam soil lining, and a design $z = 3$. Suppose the channel slope is fixed at 4%. Design the diversion channel for a 10 yr storm.

From Example 10.5, the flow was 1.02 m^3/s. Using Type B channel solution methodology,

$$n_s = 0.035$$

$$A = q/v_p = 1.02\ m^3/s/0.46\ m/s = 2.22\ m^2$$

$$R = (nv_p/\varphi_{dim}S^{1/2})^{1.5} = [0.035(0.46\ m/s)/(1(0.04)^{1/2})]^{1.5} = 0.023\ m$$

$$a_1 = 2(z^2 + 1) - z = 3.32$$

$$b_1 = -A/R = -2.22\ m^2/0.023\ m = -96.52\ m$$

$$c_1 = 2.22\ m^2$$

$$y_s = [-b_1 \pm (b_1^2 - 4a_1c_1)^{0.5}]/2a_1 = \{96.52 - [96.52^2 - 4(2.22)(3.22)]^{0.5}\}/2(3.32) = 0.023\ m$$

$$\text{bottom width } b = (A - zy_s^2)/y_s = [2.22\ m^2 - 3(0.023\ m^2)]/0.023\ m = 96.45\ m$$

Designing for capacity use $n_{cp} = 0.06$. Then the Manning equation with trapezoidal elements and bottom width substituted appears as

$$1.02\ m^3/s = (1/0.06)(96.45y_{cp} + 3y_{cp}{}^2)\{(96.45y_{cp} + 3y_{cp}{}^2)/[96.45 + 2y_{cp}(3^2 + 1)]\}^{0.67}(0.04)^{0.5}$$

Iterating for y gives $y_{cp} = 0.032$ m. Add the freeboard of 20% to get $y_{fb} = 0.04$ m.

A better lining choice would be a cobble and shingle ($n = 0.035$) with $v_p = 1.68$ m/s. Use the flow found in Example 10.6 (where v_p was 1.5 m/s), thus $q = 1.4$ m^3/s. $A = 1.4/1.68 = 0.83$ m and $R = 0.16$ m. Using the quadratic formula leads to $y_s = 0.18$ m and $b = 4.07$ m. Designing for capacity at this bottom width and $n_{cp} = 0.06$ leads to $y_{cp} = 0.19$ m. Adding 20% freeboard gives $y_{fb} = 0.23$ m.

Diversion Terraces for Construction Site Erosion Control

Terraces for construction site erosion control are sometimes referred to as diversion terraces. One may apply the approaches outlined above for the diversion to temporary terraces for slope length reduction on excavated slopes. Figure 10.11 shows construction diversions integrated with other erosion control practices for slope length reduction and erosion control.

The design of the construction diversion proceeds as follows. Compute the design runoff capacity based on the 10 yr–5 min storm or based on travel time plus overland flow over the excavated slope. Use the value giving the highest runoff flow rate. Set the channel z value to a specified value (e.g., from 2 to 20), depending on construction equipment available. Solve for maximum grade as with terraces. Solve for capacity with $n = 0.06$ for bare soil. If the channel is lined, use methods in open channel flow, Type A. Recall that the Type A methodology is the Manning solution without additional constraints.

Figure 10.11. Erosion control strategies for construction sites include concrete-lined diversion channels, open-weave wood fiber impregnated with grass seed (shown growing), silt fencing, and rock linings.

Example 10.8: Size the terrace channels given an excavated slope requiring a terrace system 100 m long and 9 m spacing for adequate erosion control (328 × 30 ft), 5% slope, and machine equipped to excavate a 3:1 slope (e.g., $z = 3$). Lining is natural, and $v_p = 0.6$ m/s (2 ft/s). *Note:* Spacing could be determined using methods of Examples 10.1 and 10.2.

This example is identical to the field terrace system except we constrain z to 3. See Example 10.4. Find the 5 min rain of 1.1 in (27.9 mm) and the intensity of 1.02 m³/s:

$$\text{area} = 100 \text{ m } (9 \text{ m})/(10{,}000 \text{ m}^2/\text{ha}) = 0.09 \text{ ha}$$

$$q_p = 0.0028$$

$$C_{rat} iA = 0.0028(0.7)(336 \text{ mm/hr})(0.09 \text{ ha}) = 0.06 \text{ m}^3/\text{s } (2.12 \text{ cfs})$$

$$y_t = (0.06 \text{ m}^3/\text{s})/[(0.6 \text{ m/s})3]^{0.5} = 0.18 \text{ m}$$

$$S_{max} = (0.6 \text{ m/s})^2(0.035)^2/(0.18 \text{ m}/2)^{1.33} = 0.011$$

Select a maximum slope of 0.01. Design for capacity

$$0.06 \text{ m}^3/\text{s} = (1/0.06)(3)y_{cp}{}^2(y_{cp}/2)^{0.67}(0.01)^{0.5}$$

$$y_{cp} = 0.23 \text{ m} + 0.15 \text{ m freeboard}$$

for a total depth of 0.38 m. Construction height is 1.1 × 0.38 = 0.42 m.

Example 10.9 (diversion terrace, steep slope, concrete-lined channel): A highway embankment, slope of 30%, through a sensitive wetland area is to be protected using concrete-lined trapezoidal diversion channels ($z = 3$, bottom width = 0.3 m). The embankment is 200 m long and is located in Athens, Georgia. Figure 10.11 shows a typical scenario. The soil loss tolerance is 5 Mg/(ha-yr). The Athens USLE R_{rf} factor is 5000 MJ-mm/(ha-hr-average annual storm), and the K_{sl} factor for the soil is 0.042 Mg-ha-hr/(ha-MJ-mm), computed as in Example 10.2. A seed-impregnated open-weave mat having a $C_{prac} = 0.01$ is to protect the embankment surface. Determine the channel spacing and channel design.

Spacing may be determined using the USLE method. The P_{prac} factor is 1 since this is a construction situation. The slope angle is 16.7 degrees and the slope exponent is 0.66. Estimate a slope length value and compute the average annual soil loss. Adjust the guess until a value of 5 Mg/(ha-yr) obtaining the tolerance within 5% (one could also use the approach in Example 10.2). The solution converges at 8.8 m. For the 200 m length, the catchment area is 0.18 ha. Using the 10 yr–5 min storm for Athens leads to a 146 mm/hr intensity (Weiss method for precipitation). From the rational equation, we have a design runoff of 0.05 m³/s. For the channel design, use a Manning n = 0.015(1.25) = 0.02 for the Type A channel design. The 1.25 factor increases the concrete surface Manning n for the capacity design. A slope value of 0.005 is selected. The design flow depth is 0.12 m. Adding 20% freeboard gives a channel depth of 0.14 m. *Note:* Choosing a channel slope of 0.01 resulted in a Froude number close to 1, whereas the slope of 0.005 resulted in a Froude number of 0.7. The extremely low flow going critical would not be serious; however, it is good practice to avoid critical flow situations in general. *In addition, the computed spacing and, consequently, the channel flow are sensitive to the C_{prac} factor, soil K_{sl} factor, and the R_{rf} factor. One must carefully select these parameters when designing for steep slopes.*

Terrace Outlet Selection

The terrace system may use vegetated waterway outlets, tile outlets, or natural wooded draws or, in the case of the water conservation terrace, have no outlets. The natural wooded draw consists of brushy growth with few or no large trees. A dense canopy and grasses are present. The major advantages of the natural draw are low expense and excellent wildlife habitat (Powell and Steichen 1981b). Using natural draws may impede travel and increase field transit times. The vegetated waterway (Chapter 9) is the most popular outlet alternative.

The tile (sometimes called pipe) outlet is advantageous when field topography does not lend itself to parallel terraces. The outlet structure is vulnerable to damage by machines and animals. Sediment accumulation is also a problem. Plowing away from the outlet gives sediment control. One may reduce machine and animal damage by securing the vertical outlet with fence posts.

In Missouri, the NRCS has adopted the practice of placing tile outlets at runoff accumulation points more than 60 m from the field boundary. Accumulation points within 30–60 m of the field boundary are drained with vegetated waterways. The NRCS has found this practice to provide economical parallel terraces.[1]

Calculations for Tile Outlet Terrace or Sedimentation Pond

Tile outlet terraces and erosion terraces are identical in purpose but have different design criteria. With tile outlet terraces, erosion of the channel bottom is not a major issue. The tile outlet terrace is a relatively slow draining basin designed to contain a volume instead of a peak flow. Design focuses on berm height for volume control and on drain hydraulics to achieve drainage before flood injury occurs. One also designs the drainage pipe to avoid scouring problems along the drain when using a perforated drain.

The hydraulics of the tile outlet terrace and the sedimentation pond outlet are identical. The residence time in a sedimentation pond must be equal to or greater than a set time to achieve adequate sediment entrapment. The residence time of the tile outlet terrace must be equal to or less than a set time to avoid flood injury to crops. The tile outlet terrace design process seeks to reduce both the storage berm height and the size of the outlet drainage system. The tile outlet terrace is useful for situations where parallel terraces would not be feasible due to irregular slopes. Figure 10.12 shows a tile outlet installation. Chapter 12 further discusses sedimentation ponds.

Tile outlet terraces satisfy two objectives. First, the terrace should not overflow during the design storm. Second, the terrace must remove the water within a specified time. Many crops should not stand in water more than 24 hr. Drainage time reduces to 6 hr for specialty crops vulnerable to root damage by flooding. One may invoke other criteria for other specialized applications. Figure 10.13 shows nomenclature for a tile outlet terrace.

Design using 10 yr–24 hr storms for agricultural applications. For urban or bioremediation purposes, one may invoke other criteria specific to the nature of the problem. Water volumes are of interest. A modified rational equation for runoff volume calculation is applied since the watershed is small, sloped, and assumed relatively impervious. The runoff volume estimation assumes a triangular rainfall distribution over 24 (or 6) hr and uses the

Figure 10.12. Typical tile outlet terrace. The vertical riser has little protection and is vulnerable to machine damage.

excess rainfall coefficient discussed in Chapter 5. The time required for the water to flow to the terrace is negligible. Thus, one has

$$\text{runoff volume} = 10^{-4} C_{excess} I_{24} L W_i \text{ (metric units)} \tag{10.8}$$

where

C_{excess} is the runoff fraction (C_{excess} = 0.6 typically);
I is the depth (m) of the 10 yr–24 hr storm;
L is the run length of the terrace (m);
W_i is the width between terraces (m).

One must remove this volume of water within 24 hr for most crops (6 hr for some specialty crops) to avoid flood injury. Quick water removal is desirable to reduce berm height.

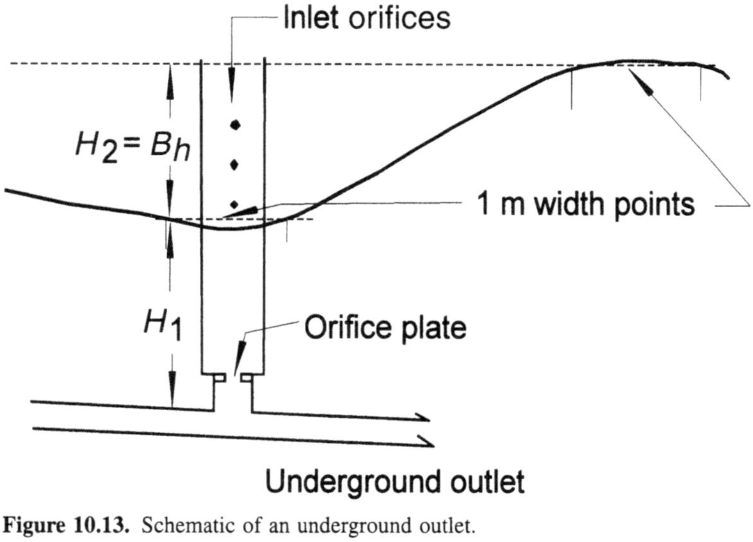

Figure 10.13. Schematic of an underground outlet.

Set a berm height and compute storage (ha-m). Beasley et al. (1984) give a simple relationship for estimating volume:

$$V_{storage} = 0.0617 B_h A_{surface}\ (ha\text{–}m) \quad \textbf{(10.9)}$$

where

B_h is the berm height (m);
$A_{surface}$ is the surface area of the water at the berm height (ha).

Convert to ac-ft by multiplying ha-m by 8.11.

Next, form a ratio of storage volume/runoff volume. Enter the horizontal axis of the underground outlet routing curve (Figure 10.14) and estimate the peak flow ratio.

Figure 10.14 also applies when routing flows through principal spillways of ponds and related structures. Chapters 11 and 12 further cover these topics.

Assuming a triangular hydrograph with equal rise and recession times, estimate peak storm runoff as

$$q_{ip} = 2V_{runoff}/D_{storm} \quad \textbf{(10.10a)}$$

where

V_{runoff} is expressed in m^3;
D_{storm} is storm duration (24 hr or 6 hr).

The peak outflow q_{op} is found using Figure 10.14 and

$$q_{op} = q_{ip} f(V_{store}/V_{runoff}). \quad \textbf{(10.10b)}$$

Adjust storage (berm height) to get a satisfactory release time. An orifice plate controls the outlet flow rate of a pipe drain system (Beasley et al. 1984). Calculate the orifice area using

$$q_{op} = 0.6 A_{orf}\sqrt{2g(H_1 + 0.7B_h)} \quad \textbf{(10.11a)}$$

where

A_{orf} is the orifice area (m^2) to pass q_{op};
H_1 is the distance from the orifice plate to the bottom 1 m width point;
B_h is the distance between the bottom and top 1 m berm width measures.

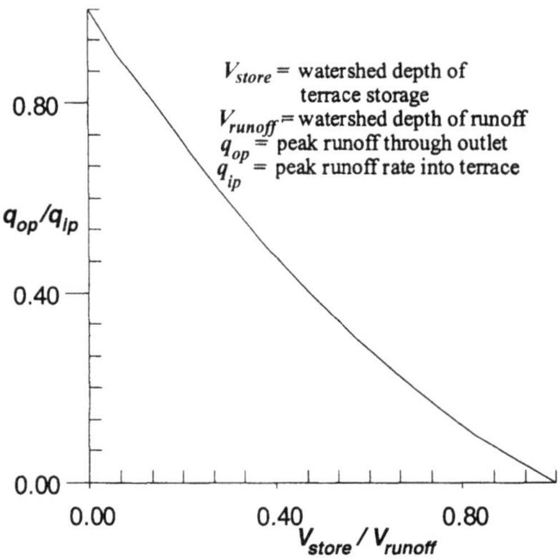

Figure 10.14. Underground outlet routing curve. (Redrawn from Beasley et al. 1984.)

See Figure 10.13. H_2 is also used as the berm height.

The orifice is necessary to ensure that outflow will not exceed specified flows downstream. Applying the 0.7 factor gives the effective average portion of B_h representing an effective average depth for flow computation.[2]

The number of holes in the upper filter can be determined using equation 10.12:

$$N_h = \frac{0.3q_{op}}{a_h\sqrt{B_h}} \tag{10.12}$$

where

N_h is the number of holes required;
a_h is the area of one filter hole (suggested 1 in^2 = 0.00545 ft^2);
B_h is the maximum elevation of water; effective berm height (ft);
q_{op} is the flow rate (ft^3/s).

Round up to the next number to specify minimum holes required. Equation 10.12 requires the indicated units.

One may design the underground drainage lines assuming perforated pipe with velocity limits or sealed pipe without velocity limits. For sealed pipe, write the Manning equation as shown in equation 10.13:

$$q_{op} = \frac{\phi_{dim}}{n}\left(\frac{\pi D^2_{sealed}}{4}\right)\left(\frac{D_{sealed}}{4}\right)^{2/3}\sqrt{S_p} \tag{10.13}$$

Solving equation 10.13 for diameter D_o yields

$$D_{sealed} = \left(\frac{3.22q_{op}n}{\phi_{dim}\sqrt{S_p}}\right)^{3/8} \tag{10.14}$$

where

S_p is the pipe gradient;
D_{sealed} is the minimum pipe diameter for full pipe flow based on the pipe gradient;
n is the Manning n for the pipe material (usually 0.011 for smooth pipe, 0.016 for corrugated pipe);
the constant 3.22 is $4(4)^{0.67}/\pi$.

There are limiting velocities in tile outlets when using drainpipe with unsealed joints or bell-end tongue-and-groove pipe. These limiting velocities depend on the surrounding soil type. If velocity must be limited, use the following procedure.

First, estimate the full flow expected in the pipe using the Manning equation, assuming a commercially available pipe diameter D_{ou}:

$$v_{full} = \frac{\phi_{dim}}{n}\left(\frac{1.21D_{ou}}{4}\right)^{2/3}\sqrt{S_p} \tag{10.15a}$$

$$q_{full} = \frac{™D^2_{ou}}{4}v_{full} \tag{10.15b}$$

where

D_{ou} is the trial perforated pipe diameter;
v_{full} is the maximum expected flow velocity in the perforated pipe;

Table 10.3. Maximum velocity values ($v_{pktable}$) in underground tiles from terraces.

Soil texture	Drain tile with unsealed joints (use for perforated pipe)		Drain tile with wrapped joints or bell-end tongue-and-groove pipe	
	ft/s	m/s	ft/s	m/s
Sand and sandy loam	3.5	1.07	6	1.83
Silt and silt loam	5	1.54	6	1.83
Silty clay loam	6	1.83	7	2.13
Clay and clay loam	7	2.13	9	2.74

Source: Adapted from Griessel and Beasley 1971.

q_{full} is the flow rate at full pipe flow at the gradient S_p;
other nomenclature defined previously.

The constants appearing in the hydraulic radius term of equation 10.14 and in the area term of equations 10.15a–b are selected for estimating maximum flows. They are based on Figure 8.4. Using maximums instead of full flow values provides for conservative checks on the pipe flow velocity. Use the same values of n as above. D_{ou} should be commercially available and may be somewhat larger than D_o computed for the nonlimiting velocity analyses.

Next, calculate q_{op}/q_{full}, enter Figure 8.4 to estimate y_{ou}/D_{ou}. One may then estimate R_{op}/R_{ou} from Figure 8.4 and estimate v_{pk} using

$$v_{pk} = \frac{\phi_{dim}}{n}\left[\frac{f(R_{ou}/R_o)D_{ou}}{4}\right]^{0.67}\sqrt{S_p} \qquad \textbf{(10.16)}$$

Select the maximum velocity $v_{pktable}$ for the appropriate soil condition from Table 10.3; check that v_{pk} is less than the $v_{pktable}$. If v_{pk} exceeds $v_{pktable}$, then perforated pipe should not be used.

Design each interval following each inlet as one goes downslope, noting that Q changes with the number of drain line inlets.

Consult ASAE 1997b for rodent control, outlet protection, and pressure relief suggestions with underground conduits.

Example 10.10 (tile outlet design): Design an underground outlet for a field terrace system, sandy loam soil, 268 m long, spaced 50 m apart. Corn is grown. Assume the outlet is in the middle of the field.

The area drained by the terrace is 268×50 m/(10,000 m^2/ha) = 1.34 ha.

For a berm of 1 m at the outlet and estimated surface area of 1.34 ha, the storage from equation 10.9 is $S = 0.0617(1)(1.34) = 0.083$ ha-m. Use correct units in this empirical equation.

The 10 yr–24 hr rain for Athens, Georgia, is 142 mm = 0.142 m (Weiss method). The runoff volume from equation 10.8 is

$$V_{runoff} = C_{excess} iLW_i = 0.6(0.142\text{ m})(268\text{ m})(50\text{ m})/(10{,}000\text{ m}^2/\text{ha})$$
$$= 0.1142\text{ ha-m} = 1142\text{ m}^3$$
$$q_{ip} = 2(1142\text{ m}^3)/[(24\text{ hr/day})(3600\text{ s/hr})] = 0.0264\text{ m}^3/\text{s}$$

(equation 10.10a).

Referring to the Beasley et al. (1984) routing curve in Figure 10.14, find the following:

$$q_{op}/q_{ip} = f(V_{stor}/V_{runoff}) = f(0.083/0.1142) = f(0.727) = 0.17$$

from Figure 10.14. Then

$$q_{op} = 0.17(0.0264) = 0.0045 \text{ m}^3\text{/s}$$

One could reduce the berm height and recalculate the q_{op}. Since this berm is small already, we will leave it at the current size. Assume the underground outlet is 1 m below the surface and the berm height is 1 m (not counting freeboard and construction height).

From equation 10.11a,

$$0.0045 \text{ m}^3\text{/sec} = 0.6\, A_{orf}\{2(9.81)[1 + 0.7(1)]\}^{0.5}$$
$$A_{orf} = 0.0013 \text{ m}^2$$

Note that 0.6 is the orifice coefficient. The diameter D_{orf} is then

$$D_{orf} = (4A_{orf}/\pi)^{0.5} = 4(0.0013 \text{ m}^2/3.14159) = 0.0407 \text{ m} = 41 \text{ mm}$$

The number of 1 in (25 mm) diameter holes in the inlet device from equation 10.12 is

$$N_h = [0.3(0.0045 \text{ m}^3\text{/s})(35.29 \text{ ft}^3\text{/s/m}^3\text{/s})]/\{(0.00545 \text{ ft}^2)[(1 \text{ m})(3.28 \text{ ft/m})]\}$$
$$= 3.28 \text{ or } 4$$

Suppose the underground outlet slopes downward at 5% for 100 m. Consider the case of no pipe perforations. Thus, limiting velocity is not an issue.

Solve equation 10.14 for

$$D_{sealed} = [3.22\ (0.0045 \text{ m}^3\text{/s})(0.011)/(0.05)^{0.5}]^{3/8} = 0.066 \text{ m} = 66 \text{ mm}$$

This is below the smallest standard drain tubing of 4 in diameter. Thus, the 4 in (101 mm) is recommended.

Suppose 101 mm diameter perforated pipe were available with $n = 0.016$. The maximum flow v_{full} in the 101 mm pipe would be

$$v_{full} = (1/0.016)(1.21(0.101 \text{ m}/4))^{0.67}(0.05)^{0.5} = 1.35 \text{ m/s}$$

leading to

$$q_{full} = (3.14)((0.101 \text{ m})^2/4)(1.35 \text{ m/s}) = 0.011 \text{ m}^3\text{/s}$$

(equation 10.13). Thus

$$q_{op}/q_{full} = 0.0045 \text{ m}^3\text{/s}/0.011 \text{ m}^3\text{/s} = 0.41$$

and from Figure 8.4, $y_{ou}/D_{ou} = 0.45$ and $R_{op}/R_{ou} = 0.95$. Then using the Manning equation,

$$v_{pk} = (1/0.016)[0.95(0.101 \text{ m}/4)]^{0.67}(0.05)^{0.5} = 1.14 \text{ m/s}$$

The calculated velocity is slightly higher than the $v_{pktable} = 1.07$ m/s value tabulated in Table 10.3 for the sandy loam soil. Thus, the 101 mm perforated pipe should not be used. One may, using Figure 8.4, find that a 152 mm (6 in) pipe will give acceptable velocities.

Parallel Terrace System Planning and Design

Factors influencing terrace location include land slope; soil conditions such as degree and extent of erosion; proposed land use; presence of boulders, trees, gullies, farm roads, fences, and other impediments to cultivation; row layout; type of terrace and outlet type (ASAE

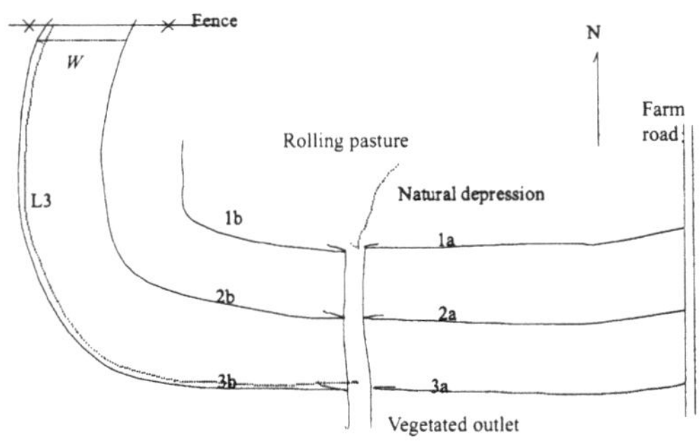

Figure 10.15. Hypothetical terracing situation involving three terraces with vegetated waterway outlet.

1997a; Schwab et al. 1993; Beasley et al.,1984). Minimum maintenance, ease of farming, and adequate control of erosion are the criteria for terrace location. Placing the terrace ridge just above eroded spots, gullies, and abrupt changes in slope simplifies alignment. Satisfactory locations for roads and fences are on the ridge, on the contour, or on spoil locations beside the outlet. Consider the situation shown in Figure 10.15.

One usually installs the top terrace first, starting from the outlet end. Starting with other terraces may lead to sequential failure due to the excessive drainage area resulting during the construction phase. General rules for laying out the top terrace are (1) the drainage area above the top terrace ordinarily should not exceed 1 ha; (2) if the top of the hill comes to a point, one may increase the spacing to 1.5 times the regular spacing; (3) on long ridges, where the terrace approximately parallels the ridge, one should specify the regular spacing; and (4) if short, abrupt changes in slope occur, the terrace should be placed just above the break.

Obstructions or topographic features below the top terrace, or the need to make terraces parallel, may require placing a terrace at another point downslope. This terrace is designated the *key* terrace because other terraces are set at normal spacing from this terrace.

Level terraces (water conservation/harvesting terraces) are used in arid regions. On flat slopes in the Plains, one sometimes constructs level terraces with alternate ends open, allowing water to flow from one terrace to the other. Beasley et al. (1984) give additional information on water conservation terrace layout and planning. Water harvesting is presented as an example of how one may apply agricultural terracing knowledge to bioremediation problems.

The desire to maintain parallel terraces in fields with nonuniform topography requires soil transport along the terrace. Unbalanced terraces imply cutting parts of a given terrace to the entire flow depth or deeper while other sections of the terrace depth may be constructed wholly of fill material. One may vary the channel grade between the maximum and minimum allowable grades. Many terrace designs are unbalanced in that the cut depth is not constant along the channel and one transports soil along the length of the channel. Unbalanced terrace design requires cut-and-fill computations to accomplish a balance in soil volume cut and soil volume filled. Unbalanced terrace design is an exercise in balancing cuts and fills along an individual terrace or in a field terrace system. Channel slopes may vary according to flow, whereas with the balanced design procedure, only one channel slope is calculated. Economical designs require minimal transport of spoil.

The construction and layout follows. For each terrace, one should plot a profile of the original ground surface versus distance from the outlet, starting at the elevation of the outlet. Next, plot the channel bottom versus distance. Compute a range of maximum allowable slopes as a function of length along the terrace where the flow rate varies as a function of length because the terrace-drained area changes as the length changes. Vary the slopes (between limits set by stability maximum slope and drainage minimum slope) to reduce altered profiles. Next, plot the ridge height (i.e., equals the flow depth plus freeboard).

Where the ground surface is less than the depth of cut, one must alter the profile to increase depth. Resulting spoil is transported to locations where the ground surface is more than the depth of cut below the ridge height. Ideally, one would try to balance cuts and fills, but transport distance should be kept in mind. Appendix C and Beasley et al. 1984 present additional discussion and give pragmatic advice concerning construction and installation.

Terrace maintenance is important. Powell and Kok (1994) suggest that plowing from the top down to the channel bottom may maintain terrace shape over time. The plow throws the soil upslope, leaving a dead furrow at the channel bottom. One may plow backslopes from the top of the backslope down to the bottom of the backslope for the same effect. Local offices of the NRCS may also provide additional information.

NRCS engineers began computerizing terrace design in the late 1960s. Gaddis and Steiner (1988) presented a terrace design program solving for channel depth for both balanced and unbalanced conditions. The program accounts for cuts and fills with unbalanced terraces. Channel options include the broadbase, steep backslope, and flat channel terrace styles. One may specify tile outlets or surface outlets if erosion control is the objective. Likewise, one may specify water impoundments when water conservation is the primary objective.

Outlook

This chapter brings the channel design to a convenient stopping point. Table 10.4 summarizes the channel design issues presented in this and the previous two chapters. Many options have been presented that may be applied for storm water management and erosion control in a variety of situations.

Terraces are a viable strategy for erosion control when conservation tillage may not be an option. Closely related to agricultural terraces are diversion terraces, differing from terraces only in that they may have smaller z values. Diversions are applicable in both urban and agricultural settings. They are primarily designed for flow control and not erosion benefits. Diversions are designed according to the time of concentration of the supplying watershed. Water conservation terraces originally described for agricultural use may have application for harvesting water for remote bioremediation applications. This introduction to terrace design has focused mainly on single balanced terraces and diversions. Beasley et al. (1984) discuss unbalanced terrace systems in more detail. New applications such as forest products may require appropriate modifications to the erosion terrace design procedure to suit client needs.

Erosion control terrace channel design (vegetated outlets) involves stability and capacity design similar to vegetated waterways. Indeed, stability and capacity design methodology holds for any channel that is dry at times and has a natural lining. With terrace and diversion channels, the roughness n is constant in both cases. In stability, one manipulates either the hydraulic radius or the slope. Grassed waterways usually manipulate the hydraulic radius to achieve a permissible velocity. Terrace channel design manipulates the

Table 10.4. Selected channel design solution types and attributes.

Attribute:	Type A	Type B	Type C	Vegetated waterway	Machine-width terrace	Diversion
Computation:	Capacity	Stability	Optimal	Type B, Type A	Type B, Type A	Type B, Type A
Place discussed:	Chapter 8	Chapter 8	Chapter 8	Chapter 9	Chapter 10	Chapter 10
Primary use:	Performance change, capacity evaluation	Stability in erodible channels	Optimal cross sections	Conveyance	Collection	Collection
Rural or urban?	Both	Both	Both	Both	Rural mainly	Both
Design relations:	Basic Manning continuity equation; iterate for one unknown; usually in trapezoidal channel with known depth	Basic Manning continuity equation; solve for *R* or *S* and *A* and plug into hydraulic elements to solve; solve for *y* at capacity	Basic Manning continuity equation, *b*/*y* or optimal *Z* or other parameter	Manning variable *n*; retardance charts, continuity equation; stability and capacity	Basic Manning; approx. hydraulic elements, continuity equation; solve for slope and then *y* at capacity usually	Basic Manning; approx hydraulic elements, continuity equation; solve for slope and then capacity usually
Spacing?	Not applicable	Not applicable	Not applicable	Not applicable	VI or RUSLE	Not usually
Cross section:	Any	Any	Any	Trapezoidal, triangular, parabolic	Triangular	Triangular, trapezoidal
Flow type:	Constant or occasional	Frequently occasional	Usually constant	Occasional	Occasional	Usually occasional
Side slope Z[a]	Any	Depends on material	Usually optimal	Typically 4 or more	Typically 4 or more[b]	4 or less
Lining?	Usually nonerodible (e.g., concrete)	Erodible (e.g., cobbles or bare soil)	Nonerodible unless slope less than about 0.002	Vegetation	Bare soil	Bare soil, vegetation, erodible, nonerodible
Velocity control	None	Restrict hydraulic radius usually	None	Restrict hydraulic radius usually	Restrict slope	Restrict hydraulic radius or slope
Froude number needed?	Yes unless channel is erodible	Not needed when stability is an issue	Yes if non-erodible, else, no	No	No	No, unless non-erodible

Attribute:	Type A	Type B	Type C	Vegetated waterway	Machine width terrace	Diversion
Stability?	Not applicable	Yes	Not applicable; if erodible one should check permissible velocity	Yes: Class D or E retardance usually	Yes	Yes, unless nonerodible. If vegetation lined, do as for veg. waterway
Capacity?	Only when occasional, if so, increase *n* by 25%	Yes when occasional, usually the case; if yes, increase *n* by 25% or other spec.	No unless flow is periodic; if yes, increase *n* by 25% or other spec.	Yes: Class A or B retardance	Yes; n = 0.06	Yes; $n = 0.06$ for bare soil lining or 25% increase for other linings; if vegetation lined, do as for vegetated waterway
Outlet type	Channel	Channel	Channel	Channel	Channel, tile, none[d]	Channel usually
Associated flow and time of concentration:	TR55 or rational and Kirpich[c]	TR55 or rational and Kirpich	TR55 or rational and Kirpich	TR55 or rational and Kirpich	Use rational with time of concentration = 5 min	Overland time plus channel travel time and rational

[a]Z value implies trapezoidal or triangular cross section, not applicable if cross section is parabolic or other non-generalized trapezoidal shape.
[b]For a diversion terrace in agricultural applications one may restrict the *Z* to less than 4.
[c]Kirpich or any other approach for estimating time of concentration on a watershed as a whole. TR55 recommended unless watershed is 5 ac or less in size.
[d]Channel methodology is not used with the water conservation terrace.

channel slope toward the same end. One may simplify the design process because of depth insensitivity to the side slope z value.

Design summaries for terraces and diversions channels follow.

Erosion Control Terraces

Slopes may or may not accommodate critical machine width; see Examples 10.3 and 10.4.

1. Determine spacing using ASAE or RUSLE approaches (see Examples 10.1 and 10.2).
2. Determine the runoff using the rational method ($C = 0.7$), area defined by terrace length and spacing.
3. Select an approximate z.
4. Compute depth at stability based on q, the continuity equation, and maximum velocity for bare soil condition ($v = 0.61$ m/s for most soils, $v = 0.46$ m/s for easily erodible soils) using equation 10.5.
5. Compute maximum slope from minimum (value from equation 10.6, maximum in Table 10.1). Use $n = 0.035$.
6. Compute the actual frontslope z value based on computed depth; go back to step 4 if the difference is more than 10%.
7. Determine capacity using equation 10.7 (first and fourth terms) and $n_{cp} = 0.06$.
8. Add freeboard and construction settlement allowance.
9. Compute frontslope and backslope z along with depth to cut using relationships in Figure 10.8.

Diversion Terraces

Cross section is triangular, low z; see Example 10.4.

1. Determine spacing using USLE approaches (see Examples 10.2 and 10.8)
2. Determine the runoff using the rational method ($C = 0.7$), area defined by terrace length and spacing.
3. Set z to 4 or lower for the (usually) triangular channel.
4. Compute depth at stability based on q, the continuity equation, and maximum velocity for bare soil condition ($v = 0.61$ m/s for most soils, $v = 0.46$ m/s for easily erodible soils) using equation 10.5. (Skip this step if concrete lining used.)
5. Compute maximum slope from minimum (value from equation 10.6, maximum in Table 10.1). Use $n = 0.035$ for soil or the appropriate value for the lining used. Set the slope if concrete lining is used.
6. Determine capacity using equation 10.7 (first and fourth terms) or appropriate Type A channel, and $n_{cp} = 0.06$ (or appropriate value for lining used).
7. Add freeboard and construction height if applicable.
8. Compute frontslope and backslope z along with depth to cut using relationships in Figure 10.8.

Diversions

May be triangular, parabolic, or trapezoidal; z is low; bare soil, rock, etc., or vegetated channels; no spacing; see Examples 10.5, 10.6, 10.7). If bare soil:

1. Compute time of concentration from overland flow, and compute channel permissible velocity.

2. Compute the rational q using weighted rational coefficient for catchment area.
3. Design for stability: Type B channel design unless lined channel used.
4. Design for capacity by increasing $n = 0.06$ (or appropriate value): Type A. Add freeboard.
5. If a vegetal lining is used and slope is fixed, see vegetated waterway examples.

Tile Outlet Terraces

See Example 10.10.

1. Propose a berm height and get the runoff volume based on excess rain $C = 0.6$ (equation 10.8) from a 10–24 hr storm. Compute storage volume.
2. Find the design peak inflow from equation 10.10a.
3. Find the design outflow using Figure 10.14.
4. Find required orifice area (equation 10.11a) and number of filter holes (equation 10.12).
5. Compute full-flow diameter with solid pipe (equation 10.14)
6. Compute the velocity and flow in commercially available perforated pipe (equations 10.15a–b).
7. Using the commercially available diameter of perforated pipe where full flow exceeds the design flow; use equation 10.16 and Figure 8.4 to compute actual velocity. If velocity recommended in Table 10.3 is exceeded, go to next larger pipe diameter and recheck the velocity.

Some software for terrace design is published (Gaddis and Steiner 1988) that incorporates channel design and associated cut-and-fill balances. Appendix C further discusses terrace construction and layout. Organizations such as the Land Improvement Contractors Association also provide helpful information related to terrace construction.

PROBLEMS AND QUESTIONS

Draw sketches of each completed design.

1. Given a terrace system with a 27 m spacing and a 400 m length and easily erodible soil, compute the required channel slope and flow depth at capacity. Assume the terrace channel has a $z = 20$, construction height is 10% of depth, and a freeboard of 20%. The 10 yr storm intensity is 110 mm/hr.
2. A system of graded terraces in Tifton, Georgia, is parallel and farmed with four-row equipment with 30 in rows. The soil is a Tifton sandy loam, 1% organic matter, 60% sand (30% very fine), 15% silt, and 25% clay, with a moderate permeability and granular structure. The land slope is 5%, erodibility is high, and the soil loss tolerance is 7 Mg/ha-yr. The field is roughly rectangular, 500 ft down (e.g., from the high point in the field to the channel) and 800 ft across the slope (e.g., parallel to the terraces). The terrace outlet is on the side of the field. The C_{prac} factor is 0.14 for the conventional tillage rotation commonly used. An alternative rotation using conservation tillage giving a $C_{prac} = 0.10$ is also acceptable to the client if needed. (Recall the C_{prac} computation discussion in Chapter 6.) The terrace outlet type is a grassed waterway. Sketch and label the field and channel, showing the designed channel cross section.

(a) Compute the ASAE VI map terrace spacing, adjusting for machinery width.
(b) Compute the USLE spacing, adjusting for machinery width. Is it practical from an erosion standpoint to farm this field with the conventional tillage rotation? Would you use the ASAE VI or the USLE spacing? Why? What C_{prac} factor would you recommend as a target?
(c1) Assume a trial depth and tentatively determine the channel z based on machinery width.
(c2) Use the machine width not hydraulically significant design approach, assuming $z = 20$. Compute the depth for stability.
(d1) Using the spacing from (b) and the z from (c1), estimate the maximum allowable grade of the channel.
(d2) Using the spacing from (b) and the z and depth from (c2), estimate the maximum allowable grade of the channel.
(e) What is the design depth? *Note:* If one is designing for machinery width, check z; if z is not adequate, redo (d1). If assuming machine width not significant, go to step f.
(f) What is the depth of cut, including freeboard? Use the stated machine width as needed. Note that the equation for depth of cut presumes the cutslope face to be equal to the machine width. Depth of cut is approximate when using the constant-z approach.

3. Rework Problem 2 assuming the soil is in Athens, Georgia, Cecil sandy loam, 15% very fine sand, 25% silt, 50% clay, 1% organic matter, massive structure, land slope of 3%.
4. In designing a grassed waterway, why not routinely vary S instead of R for a stable channel? Would varying S for diversions be more likely to work with diversions as opposed to existing streams that were being improved?
5. Compare and contrast the methods for terrace channels and those for grassed waterway design.
6. A diversion in Athens, Georgia, is needed to protect a playground on the lower side of conventionally tilled field. The land slope is 5%, and erodibility is high. The field is roughly rectangular, 500 ft downslope and 800 ft in the direction of the channel. The bare-soil-lined diversion is on the lower side of the field. Design the diversion for a 50 yr storm. Use a triangular cross section, $z = 3$.
   (a) What is the time of concentration?
   (b) What is the flow rate?
   (c1) What is the maximum channel slope if the diversion is rapidly established in fescue (assume the mowed condition)?
   (c2) What is the maximum channel slope if the diversion is in bare soil? (Here we use the stability design approach of Example 10.5 to determine the maximum slope because the channel is being constructed in bare soil, which will prevail until the vegetation gets established.)
   (d) What is the design depth after freeboard (use the capacity design method of Example 10.6 to design the channel for capacity)? Indicate which condition (c1 or c2) was used to determine the slope.
7. A diversion in Maysville, Kentucky, is needed to protect bottomland acreage from hillside runoff. The hillside slope is 8% and is covered in established hardwood forest. Soil erodibility is high. The forest is roughly rectangular, 800 ft downslope, and 1500 ft in the direction of the channel. The fescue-lined diversion is on the lower side of the field. Design the diversion for a 50 yr storm. Use a trapezoidal cross section, $z = 3$.

Provide a sketch of the final design with the following assumptions:
(a) Assume the slope is fixed at 5%, design the bottom width and depth.
(b) Assume the channel slope is not fixed, design the channel slope and depth using a fixed bottom width of 1.2 m.
(c) Assume the channel slope is not fixed and the cross section is triangular, design the depth and side slope value.

8. Design the berm height and an underground outlet for a field terrace system 400 m long, spaced 100 m apart. Corn is being grown and the drainage system will slope at 10%. The location is Lexington, Kentucky.
9. Design the berm height and an underground outlet for a field terrace system 100 m long and spaced 100 m apart. Broccoli is being grown in Clemson, South Carolina, and the drainage system will slope at 2%. Assume the outlet is near the middle of the field.
10. Design a water conservation terrace to be constructed in western Kansas with a spacing of 200 m on a field with a 0.75% slope.
11. The Department of Transportation in your state is building a road through a sensitive wetland area where the EPA requires erosion control. The soil is 25% clay and 5% very fine sand, moderate structure, low permeability, no organic matter. The slope is 1 : 2 (i.e., $z = 2$). The embankment run is 300 m long. Assume a mat with $C_{prac} = 0.01$ can be used. A slip-form device for constructing a concrete-lined channel with a 0.2 m bottom width is commonly used by contractors for these channels. Design the channel spacing and the channel cross section.
12. What are the pros and cons of the vegetated waterway outlet? The tile outlet?
13. How do terraces reduce soil erosion?
14. What are the essential differences between a terrace and a diversion?
15. What are the differences between a soil conservation terrace and a water conservation terrace?
16. In designing terrace channels for stability, why not vary hydraulic radius R instead of channel slope S?
17. Define three types of terrace systems (do not confine yourself to channel cross sections).
18. Define *frontslope, cutslope,* and *backslope*.
19. From the hydraulic elements for a circle, derive equations 10.12, 10.13, 10.15, and 10.16.
20. Compare and contrast drainage times and effective flow rates by evaluating equations 10.11a and 10.11b. Assume constant surface area with depth in the terrace. Use a time of 24 hr. Be sure units are consistent when using equation 10.11b. Is the factor of 0.7 reasonable in equation 10.11a?
21. Compare and contrast *diversion, terrace,* and *diversion terrace*.

NOTES

1. Personal communication from Dr. Donald Pfost, extension engineer, University of Missouri.
2. One could also use the emptying time equation given by equation 10.11b:

$$t = \frac{1}{C_a A_o \sqrt{2gy}} \int_{y_1}^{y_2} A(y) y^{-1/2}\, dy \qquad \textbf{(10.11b)}$$

where C_d is the discharge coefficient; A_o is the orifice area; $A(y)$ is the surface area as a function of depth y (in the volume calculation $A(y)$ was assumed constant); y_2 is the initial head; and y_1 is the final drained head. Texts such as Streeter 1971 give additional details.

References

ASAE. 1997a. Design, layout, construction, and maintenance of terrace systems. ASAE Standard S268.4 DEC96. American Society of Agricultural Engineers, St. Joseph, MI.

———. 1997b. Design and construction of subsurface drains in humid areas. ASAE Standard EP260.4 DEC96. American Society of Agricultural Engineers, St. Joseph, MI.

Beasley, R. P. 1963. *A new method of terracing.* Missouri Agr. Expt. Station Bulletin 699. University of Missouri, Columbia.

Beasley, R. P., J. M. Gregory, and T. R. McCarty. 1984. *Erosion and sediment pollution control.* 2d ed. Iowa State University Press, Ames.

De Wrachien, D., and G. Chesci. 1999. Soil conservation: Erosion control. In *CIGR Handbook of Agricultural Engineering: Land and water engineering*, ed. H. N. van Lier, L. S. Pereira, and F. R. Steiner. American Society of Agricultural Engineers, St. Joseph, MI.

Gaddis, R. J., and D. Steiner. 1988. *Nebraska terrace program C version, rev. 1.0.* Report to the U.S. Soil Conservation Service (NRCS), Lincoln, NE.

Griessel, O., and R. P. Beasley. 1971. *Design criteria for underground terrace outlets.* University of Missouri Extension Division Sci. Tech. Guide 1525. University of Missouri, Columbia.

Haan, C. T., J. C. Hayes, and B. J. Barfield. 1994. *Hydrology and sedimentology of small catchments.* Academic Press, New York, NY.

Hudson, N. 1971. *Soil conservation.* Cornell University Press, Ithaca, NY.

NRCS. 1998. *USDA national resources inventory, 1997 update* (http://www.nrcs.usda.gov). Natural Resource Conservation Service, Washington, DC.

Powell, G. M., and H. Kok. 1994. Maintaining terraces. KSU Extension Bulletin no. C-709 Rev. Kansas State University, Manhattan, KS.

Powell, G. M., and J. M. Steichen. 1981a. Terrace system options. KSU Extension Bulletin no. AF-74. Kansas State University, Manhattan, KS.

———. 1981b. Terrace outlet alternatives. KSU Extension Bulletin no. AF-70. Kansas State Uni versity, Manhattan, KS.

Schwab, G. O., D. D. Fangmeier, W. J. Elliot, and R. K. Frevert. 1993. *Soil and water conservation engineering.* 4th ed. John Wiley & Sons, New York, NY.

Streeter, V. 1971. *Fluid mechanics.* McGraw-Hill, New York, NY.

11

Hydraulics of Water Management Structures

Structures help stabilize actively eroding areas, control and monitor flows, and convey storm water runoff. Structures were needed to make the measurements leading to the runoff coefficients in Chapter 5. Various structures are integral components in best management practices (BMPs) for sediment control in the United States and developed countries. Ancient civilizations used hydraulic structures before 600 B.C.E. for flow measurements and flow controls (Biswas 1970). Flow-controlling mechanisms include orifice, weir, open channel, and closed conduits. Pipe slope, length, and degree of inlet/outlet submergence determine the flow category.

Structures covered in this chapter find application in waste treatment plants and in a wide range of food-processing plants. Worldwide, increases in aquaculture are creating a continuing demand for new structures associated with fish ponds (Gopalakrishnan and Coche 1994). Figures 11.1–11.5 illustrate structures for gully stabilization, conveyance under roadways, elevation drop structure, runoff control, and irrigation management.

Channels often require stabilization to maintain gradients (and therefore flow velocities) below erosive levels. Structures enable slope reductions and energy dissipation at slope changes. Structures adapt channels to elevation, where gullies are often problems. Structures are used to route water over or around dams. Structures enable water flow measurement and control for open channels in irrigation applications. Conservation structures consist of concrete, wood, metal, or other materials.

Goals

- Visualize structures with respect to various inlets, conveyances, and outlets.
- Identify the structure hydraulics based on inlet and conveyance hydraulics and appreciate the need to dissipate energy at the outlet.
- Apply appropriate weir, flume, and culvert hydraulics as required for arriving at hydraulic structure designs.

Figure 11.1. Small dam for flow control in a canal.

Structure Types

Figures 11.1–11.5 show a variety of structure types. Figure 11.6 (see Schwab et al. 1993) shows schematics of additional types. Standard designs are available from Brakensiek et al. 1979 and USBR 1987. This text is not concerned with structural design. Hydraulic design is the primary interest.

The basic features of a water management structure are the *inlet*, *conduit*, and the *outlet*. One names and classifies structures according to these features. Inlet hydraulics and sometimes conduit or outlet hydraulics limit flow through weirs, flumes, chutes, and drop inlets. Hydraulics of the inlet and conduit can limit flow in pipe-type spillways. Erosion

Figure 11.2. Culvert outlet showing riprap for flow energy dissipation.

Figure 11.3. Weir for controlling flow in a water delivery channel for irrigation. (Courtesy of ASAE.)

minimization is usually the major design consideration for outlets. Erosion or scour is a serious problem when the free outfall condition exists. Common designs consist of various sill and block arrangements. Additional components include wing walls, side walls, head wall extensions, cut off walls, and toe walls. Figure 11.7 shows some structural examples.

Figure 11.4. Flow control structure in an urban area for controlling and measuring runoff flow rates. Increasing impermeable surface area results in higher peak flows, which in turn lead to downstream flooding unless measures such as flow rate–limiting structures are placed upstream. Sharp crested weir hydraulics apply to this structure.

Figure 11.5. Parshall flumes used for measuring flow rates in irrigation channels.

Good foundations are essential to the success of permanent structures. One should remove surface soils. Pipes should have seepage collars to prevent seepage-induced erosion from causing eventual failure. Much of the design information for structures comes from testing of models in the laboratory.

Rules of thumb are available for selecting structures, depending on the head and the maximum anticipated flow rate. Beasley et al. (1984) give an excellent elaboration of flow selection, shown in Figure 11.8.

Hydraulic Concepts

Specific Energy

The specific energy of a flow, depth y plus velocity energy ($v^2/2g$), was defined in Chapter 8 as flow energy relative to the channel bottom. Figures 11.9a–b show specific energy plots. A flow may possess the same specific energy at two different depths and corresponding velocities. Critical depth is the depth at which water flowing in a channel has the lowest energy head. Flow at a Froude number of unity enables unique stage discharge relationships.

Hydraulic Jump

A hydraulic jump, shown in Figure 11.9a, occurs when the flow of water changes from a depth less than critical depth to a depth greater than critical. Hydraulic jumps convert kinetic (velocity) energy to potential (elevation) energy. Energy is lost during the transition due to friction and turbulence in the jump zone. The hydraulic jump provides energy dissipation. Blocks and sills in the outlet portion of the structure control the location of the hydraulic jump. Example 8.2 discussed energy loss in hydraulic jumps. Because the hydraulic jump position depends on depth and flow rate above and below the jump, use stilling basins to hold the jump at a fixed location.

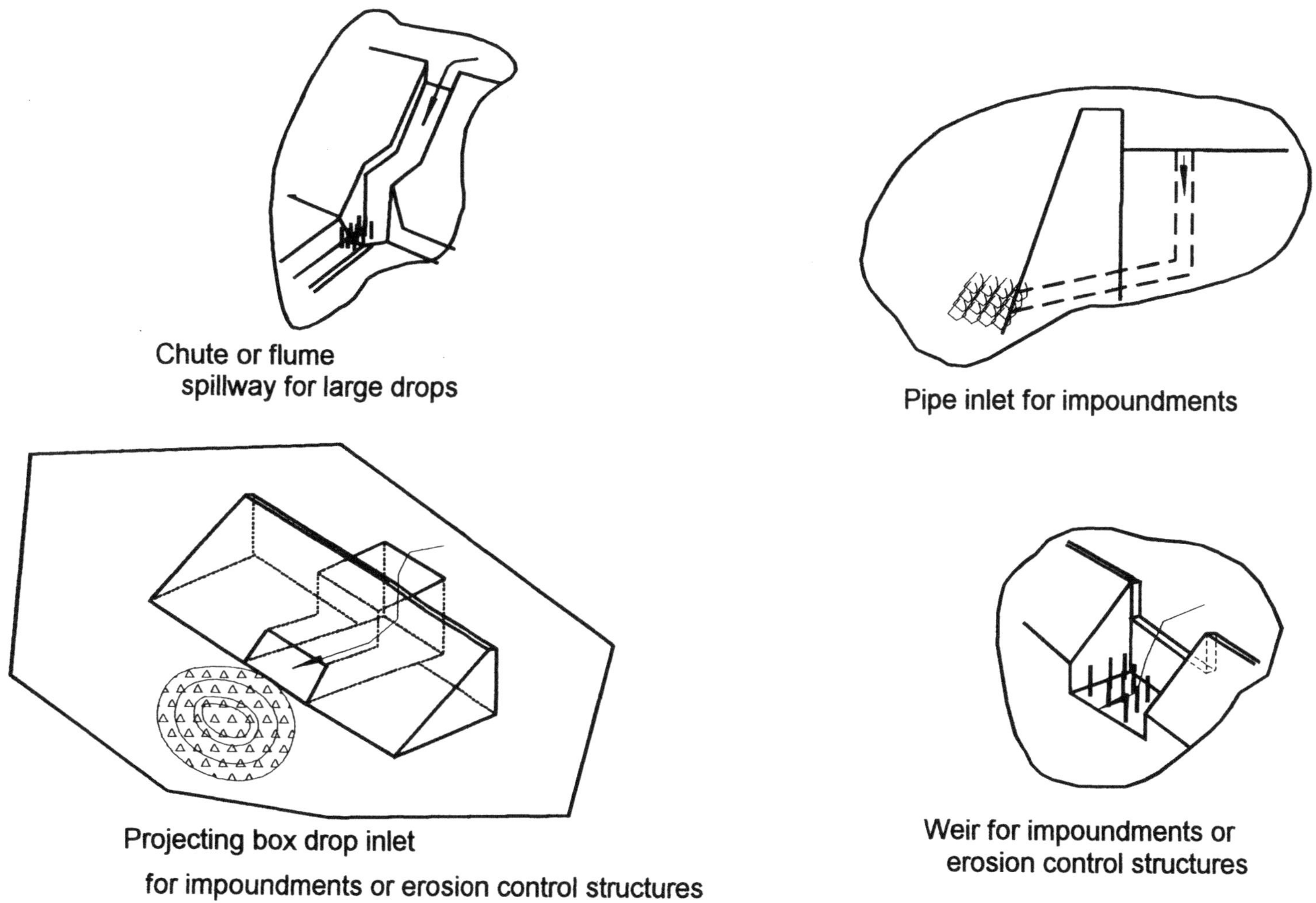

Figure 11.6. Schematic showing four common spillway types and some uses.

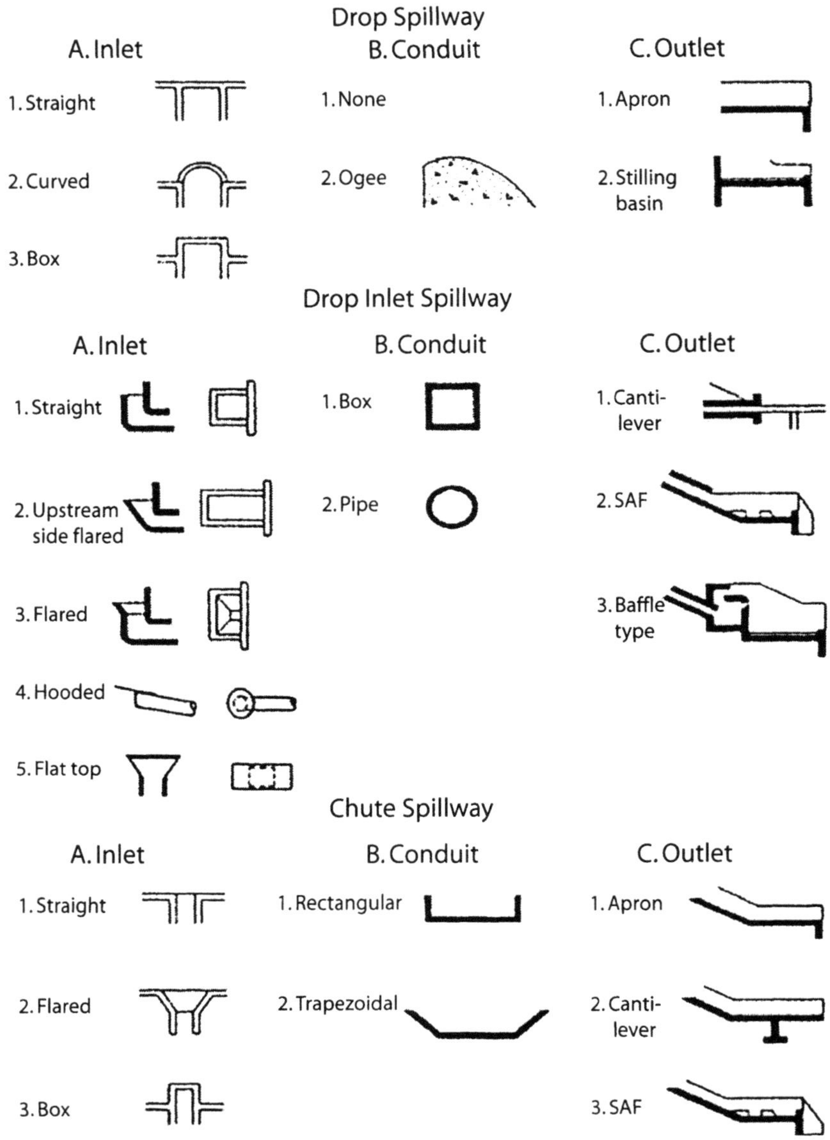

Figure 11.7. Common variations of drop, drop inlet, and chute spillways. One should know how to compute weir and orifice flow through each inlet and how to compute channel or pipe flow through the various conduits. (From Schwab et al. 1993; used with permission of John Wiley & Sons.)

Subcritical-Supercritical Flow Transitions

Important hydraulic concepts related to the function of hydraulic structures are the critical depth and specific energy. Figure 11.9b shows a specific energy curve for a subcritical-supercritical flow transition. Structures for discharge measurement work because the structure causes a transition from subcritical (depth > critical, $Fr < 1$) to supercritical flow (depth < critical, $Fr > 1$). Thus, in the flow structure the Froude number equals 1. Coupling with the continuity equation and setting the Froude number (Fr) to 1 enable a unique relationship between stage and discharge.

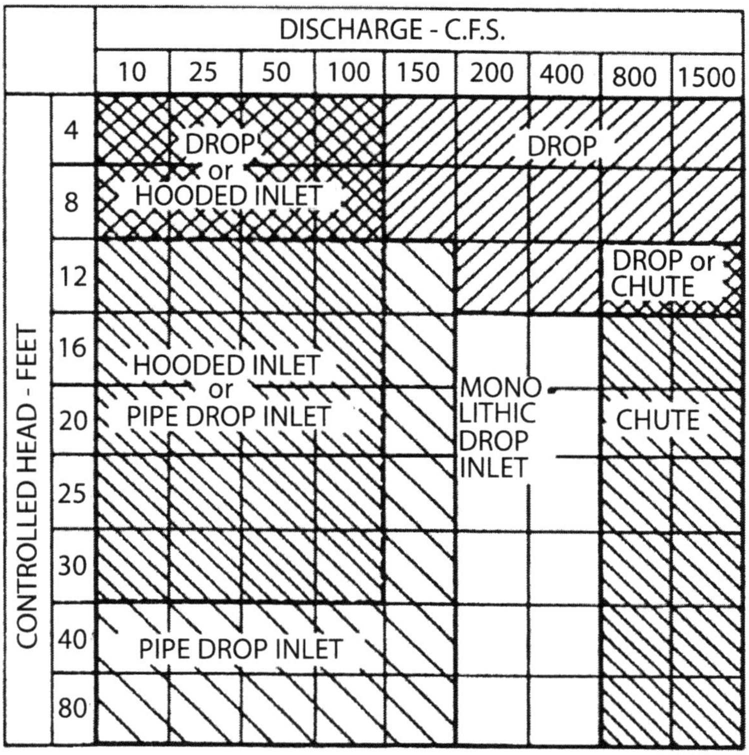

Figure 11.8. Suggested structures for a range of head and discharge rates. (From Beasley et al. 1984; used with permission of Iowa State Press.)

Stage-Discharge Relationships of Weir Inlets and Flumes

The drop spillway, box-inlet drop spillway, and formless flume (Blaisdell 1948; Blaisdell and Donnelly 1951; and Wooley et al. 1941) behave according to weir hydraulics. A simplified derivation of the weir formula follows. At critical depth the Froude number, defined as $Fr = v/(gD_h)^2$, is equal to 1; therefore $v = (gD_h)^{1/2}$. For a rectangular (top width T_w = bottom width L) weir flow, $D_h = A/T_w = LH/L$ = depth = H. One may insert the

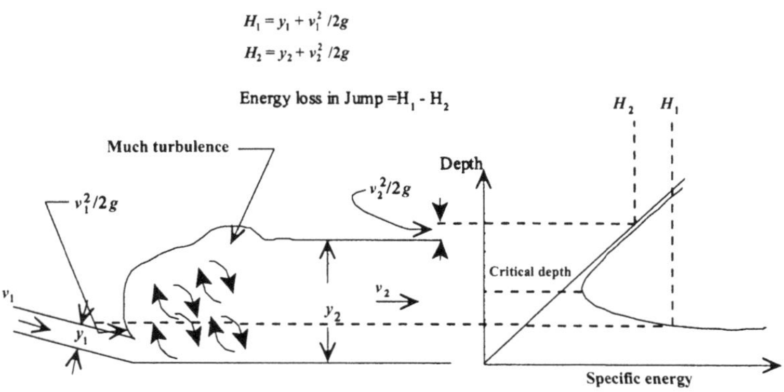

Figure 11.9a. Hydraulic jump schematic showing energy relationships (not to scale).

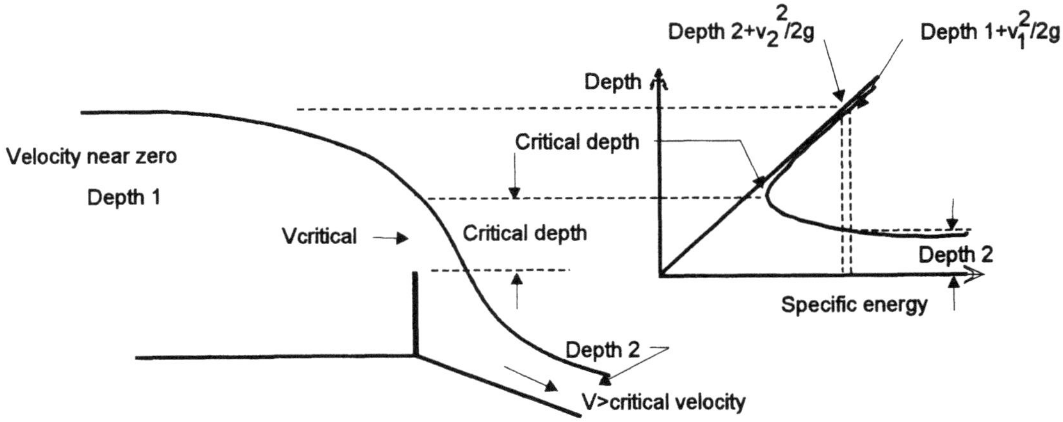

Figure 11.9b. Energy relationships for a subcritical-supercritical flow transition used in quantifying discharge (not to scale).

appropriate definition for D_h for other cross sections. Coupling the Froude number relationship with the continuity equation ($q = Av = LHv$ for a rectangular weir) creates the basis for a heuristic stage-discharge relationship for many structures. One can write $q = \text{Const} \times g^{1/2}AH^{1/2} = \text{Const} \times g^{1/2}LHH^{1/2}$, where Const is a discharge coefficient approximately equal to 0.6. This leads directly to the weir equation:

$$q = C_w LH^{3/2} \quad \text{English, ft, ft}^3\text{/s} \qquad \textbf{(11.1)}$$
$$q = 0.55C_w LH^{3/2} \quad \text{Metric, m, m}^3\text{/s}$$

where C_w represents the weir constant.

The constant C_w corrects for converging flow effects and energy loss in the weir cross section and includes the square root of the gravitational constant g. Other, more formal derivations of equation 11.1 proceed directly from the energy equation and do not invoke the Froude number. Streeter 1971 and other basic fluid mechanics texts give the derivation for the rectangular, triangular, and trapezoidal weirs. Using the energy equation–based derivation enables us to measure the flow depth upstream where the velocity is nearly zero. This also frees us from having to pinpoint precisely where the Froude number is exactly 1.

Many references use English units. Be careful to check consistency of units before proceeding. For stage measurement devices, follow established procedures for measurement location, allowable upstream flow conditions, and allowable exit conditions (e.g., outlet submerged or free discharge). Haan et al. (1994) summarize hydraulic relations for several types of weirs in Figure 11.10. This figure also presents relationships for sluice gates and orifices.

Drop spillways can control the fall of water and prevent further erosion. Flow passes through a weir, drops to an apron or stilling basin, and passes into the downstream channel.

For the free flow one calculates capacity for a drop spillway with the weir formula. The length L is the length of the three sides of a box inlet, the arc length of an arch inlet, and the crest length of a straight inlet. We measure head from the weir or flume bottom near the point of critical flow.

For many situations, a value of $C_w = 3.2$ will produce satisfactory results for design purposes. One can describe other structures using the following C_w values: box inlets,

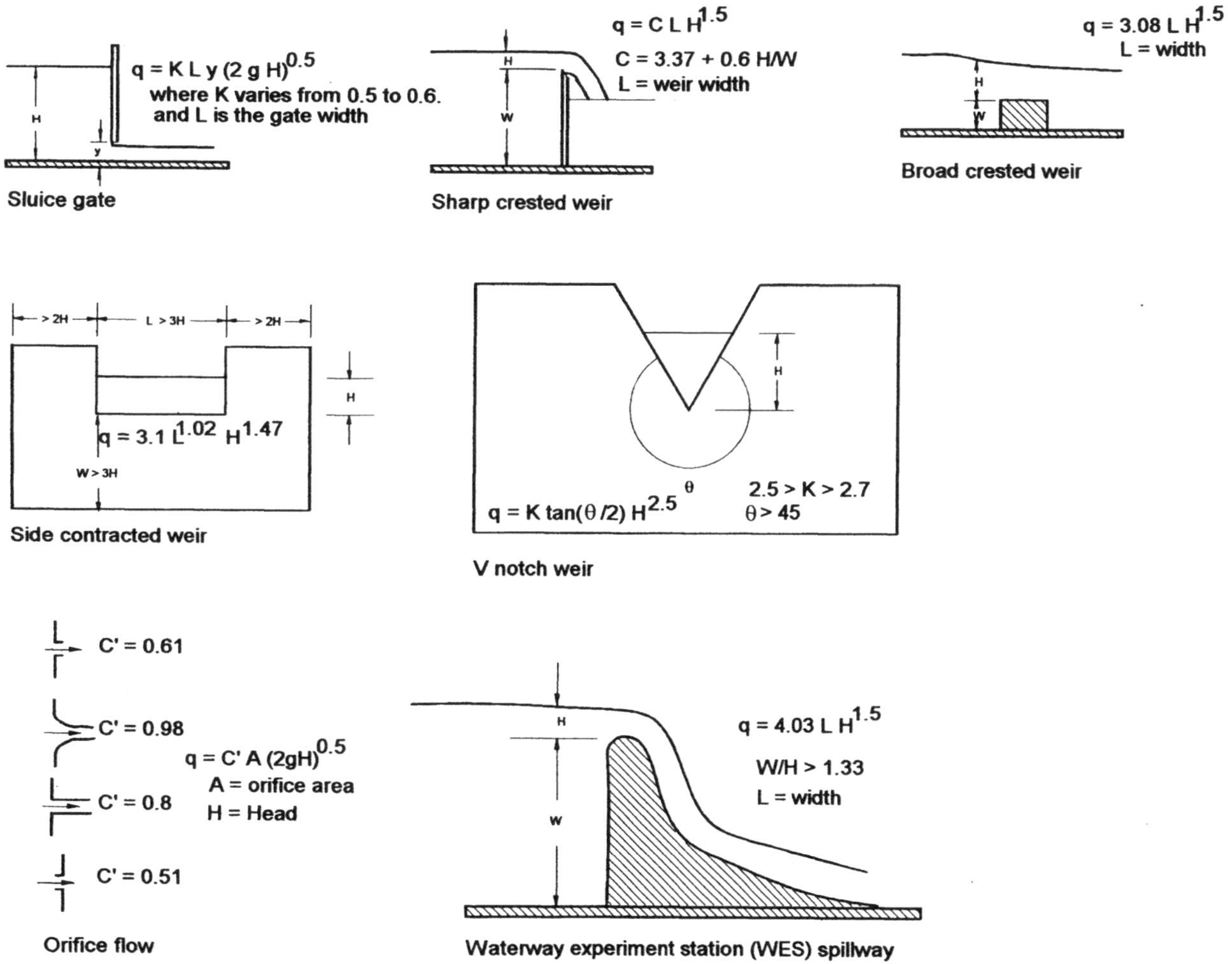

Figure 11.10. Flow relations for common flow control and inlet devices in English (FPS) Units.

$C_w = 3.2$ (when the ratio of head to box width is 0.2 or greater; otherwise, refer to Blaisdell and Donnelly 1951 or similar references); drop spillways, $C_w = 3.2$; formless flume, $C_w = 3.9$. Figure 11.10 also gives accepted values of the weir coefficient. For weirs designed for flow measurement, use these relationships and C_w values.

Brater and King (1976) give additional information concerning weir constants. The design relation allows a direct calculation of flow rate from depth. One can determine expected depth at a given flow rate. Size the structure to keep minimum flow between 10% and 100% of full depth. The approach velocity must be near zero above the structure for the simple relation above to be valid. Advanced courses cover weir selection, placement, and sizing.

> **Example 11.1:** What is the required length of a sharp-crested weir ($C_w = 3.2$) for moving 0.5 m^3/s at an available head of 0.85 m? Solve for the case of a broad-crested weir ($C_w = 3.9$).
>
> Begin with equation 11.1 : $0.5\ m^3/s = 0.55(3.2)(L)(0.85\ m)^{1.5}$ leading to $L = 0.36$ m. For the broad-crested weir we have $0.5\ m^3/s = 0.55(3.9)(L)(0.85\ m)^{1.5} = 0.30 = L$. If this were a projecting box inlet spillway, then the length of each of the three flowing sides of an equal-sided box would be $L/3 = 0.10$ m.

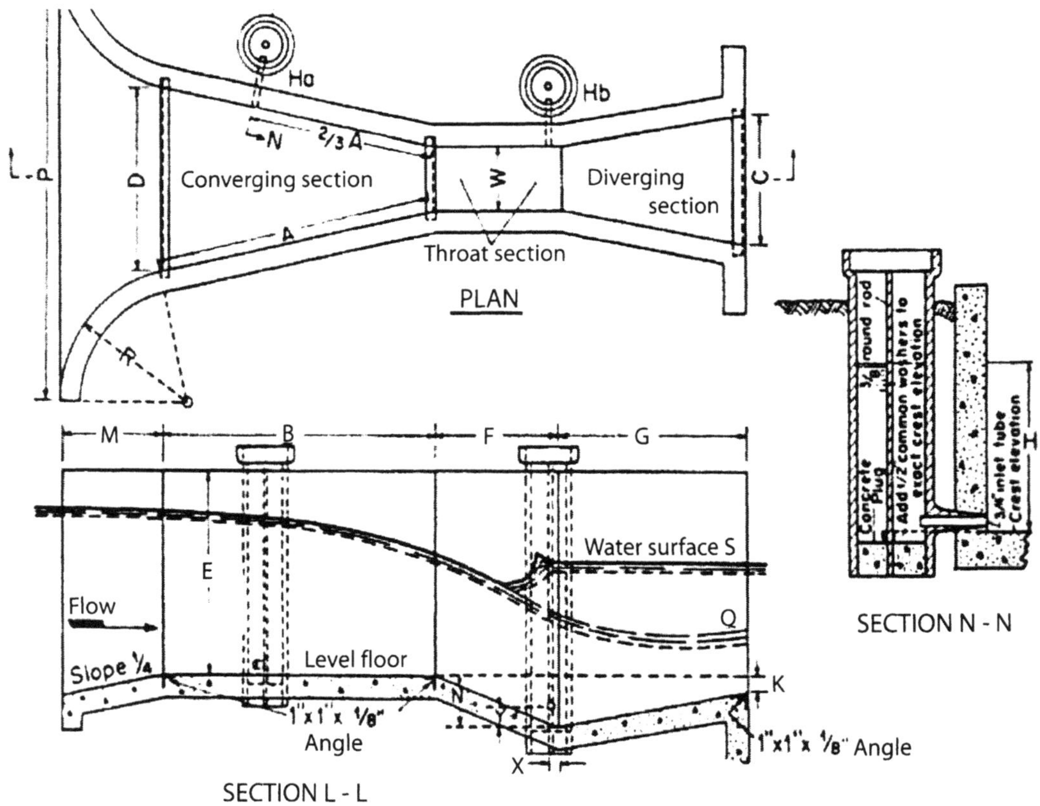

Figure 11.11. Schematic of Parshall flume, showing the critical flow section where the critical section may be submerged. (Courtesy of NRCS.)

The Parshall flume, shown in operation in Figure 11.5, enables measurement of relatively clean flows with low head loss. Figure 11.11 shows a schematic of the Parshall flume. The Parshall flume is used in surface irrigation applications. It is also employed for monitoring flows in many processing applications involving open channel flows. Equation 11.2 gives the stage discharge:

$$q = AH_a^b \tag{11.2}$$

where

A ranges from 0.99 (3 in throat), 2.06 (6 in throat), 3.07 (9 in throat), $4W$ (1–8 ft throat), to $3.6875W + 2.5$ (10–50 ft throat); W is the throat width (ft);
b ranges from 1.55 (3 in throat), 1.58 (6 in throat), 1.53 (9 in throat), $1.522W^{0.026}$ (1–8 ft throat), to 1.6 for throat widths from 10 to 50 ft;
H_a is the inlet head (ft).

Note the similarity with the basic weir equation. The unusual shape of the critical flow section explains the (empirically determined) deviation from the weir coefficient of 3.2 and the exponent deviation from 1.5.

The critical flow section is often submerged in a Parshall flume. Submergence affects the flow when the water level is from 60% to 100% (or greater) of the inlet head. Refer to works such as Chow 1959 and Brakensiek et al. 1979 for factors and diagrams giving submergence correction factors. They also give additional construction details. Parshall flumes perform best with low sediment concentrations.

Example 11.2: Given a Parshall flume, 50% submerged, throat width of 1 ft, and inlet head of 0.5 ft, compute the flow.

Since submergence is less than 50%, no correction is needed. From equation 11.2 and coefficients for the flume with a 1 ft throat,

$$q = 4(1\text{ ft})(0.5\text{ ft})^{1.522(1\text{ ft})^\wedge(0.026)} = 1.98\text{ cfs}$$

Refer to Chow 1959 or Brakensiek et al. 1979 if submergence is greater than 50%.

The H_x flume, where subscript $x = s, m, l$ (i.e., small, medium, large), is a critical flow device shaped to prevent the trapping of sediment from sediment-laden flows. Figure 11.12 shows a field view of an H_m flume. Figure 11.13 shows the relative proportions of the H_s flume. (See Brakensiek et al. 1979 for similar drawings of H_m and H_l flumes.) The H_x flume is based on the trapezoid that projects to a surface normal to the flow direction (see the front view in Figure 11.13).

Using the heuristic approach of equation 11.1 and substituting the hydraulic elements of the projected trapezoidal section, one can expand the weir equation as shown:

$$q = 10^{-6}\sqrt{g}\left[c_0 + c_1\frac{y}{2B_0} + c_2 Log_{10}\left(\frac{y + 6.092\,\text{mm}}{0.5\,Des}\right)\right] \left(\frac{[2B_0 + (B_1 + b_1)(y + h)^2]^{1.5}}{\sqrt{2B_0 + [2(B_1 + b_1)(y + h)]}}\right) \tag{11.3}$$

Figure 11.12. Flume and sampling equipment for measuring runoff hydrographs, sedimentation concentration vs. time, and pesticide concentrations vs. time. A mechanical stage recorder is partially visible under the raised cover on the right. The sediment-sampling apparatus is not visible in this photograph. The installation is located at the USDA-ARS station at Watkinsville, Georgia.

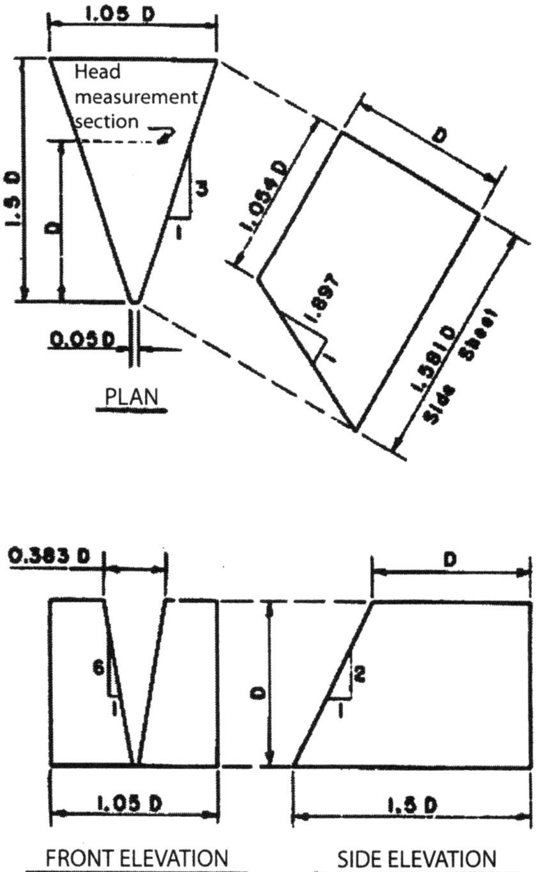

Figure 11.13. Three views of an H_s flume. The H_m and H_l flumes are similar in style but have modified proportions to better accommodate larger flows. (From Brakensiek et al. 1979.)

where

B_o is one-half the bottom width of the flume at its outlet (in Figure 11.13, plan view, B_o is $0.025D$, mm);
Des is the design depth of the *H* flume (*D* in Figure 11.13, mm);
y is the depth of flow in the flume (mm);
g is gravity (9810 mm/s^2);
q is flow rate (l/s);
B_1 is the projected side slope value for the flume as seen from the front elevation (e.g., in Figure 11.13, front view, $B_1 = 1/6 = 0.167$);
b_1 is a projected slope offset found using nonlinear regression;
h is a depth offset found to be nonzero using nonlinear regression (the offset is attributed to "virtual" projection);
c_0, c_1, c_2 are components of the weir C_w coefficient.

The lowercase *b* coefficients are estimated using nonlinear regression approaches. The author wrote the *c* coefficients following an expansion of a formula for weir losses from King and Brater (1963).

Setting the Froude number equal to 1 and applying the continuity equation with the projected trapezoidal geometry leads to equation 11.2. Offsets for fitting purposes were then added as shown. The first large bracket in the above equation is the C_w value. Table 11.1 gives selected geometric elements for the flumes. Table 11.2 gives values for the coefficients of equation 11.3. Using equation 11.2 presumes a determined H_x flume geometry. One may first solve equation 11.3 using coefficients for all flow values for the approximate flow rate calculation. Refinements to the calculated flow are then possible from using appropriate coefficients corresponding to the approximate flow. Equation 11.3 is useful for modeling stage discharge of H_x flumes with geometric characteristics bounded by the standard models given by Brakensiek et al. (1979).

Equation 11.3 fails when the depth y is less than $0.1Des$ because the subcritical to supercritical transition does not happen consistently with shallow depths. It is also implied

Table 11.1. Selected geometric elements for *H* flumes.

Type	$2B_o$ (mm)	D (mm)	B_1 (−)	H_{max} (mm)	q_{max} (l/s)
H_S	6.096	121.92	0.1667	118.87	2.274
H_S	9.14	182.88	0.1667	179.83	6.258
H_S	12.19	243.84	0.1667	240.79	12.941
H_S	15.24	304.8	0.1667	301.75	22.738
H_M	15.24	152.4	0.5	149.35	9.373
H_M	22.86	228.6	0.5	228.6	27.102
H_M	30.48	304.8	0.5	301.75	54.368
H_M	45.72	457.2	0.5	454.15	150.928
H_M	60.96	609.6	0.5	606.55	310.919
H_M	76.2	762.0	0.5	758.95	543.683
H_M	91.44	914.4	0.5	911.35	860.832
H_M	137.16	1371.6	0.5	1368.55	2378.6
H_L	243.84	1219.2	1.0	1219.2	3313.1

Source: Brakensiek et al. 1979.

Table 11.2. Summary of estimated coefficients for the general critical flow model at selected flow conditions.

	All flumes in Brackensiek et al. 1979, ($0.1Des \leq H < Des$) and flow rate q is . . .				
Term	q = all	$q \leq$ 115 l/s	$q \leq$ 30 l/s	$q \leq$ 15 l/s	$q \leq$ 8 l/s
h (mm)	−0.57	−0.39	−0.21	0	0
b_1	0.059	0.083	0.075	0.078	0.0793
c_0	0.673	0.684	0.689	0.685	0.684
c_1	0.003	−0.0023	−0.0022	−0.0025	−0.0025
c_2	0.027	0.0875	0.0924	0.0895	0.0901
Residual range (l/s)	−22 to +14	−2.6 to 0.6	−0.26 to 0.15	−0.05 to 0.07	−0.03 to 0.05

Source: Tollner, unpublished data, 1998.

that the flow depth in the flume is less than the flume height. Brakensiek et al. 1979 contains detailed rating tables and construction drawings for H_x flumes.

Example 11.3: A sediment-laden flow not expected to exceed 0.1 m^3/s (100 *L/s*) is expected from a watershed. Select an H_x flume and compute the maximum expected depth.

Select an H_m flume with D = 457.2 mm from Table 11.1. Use coefficients with q = 115 l/s from Table 11.2 in equation 11.3 to determine the maximum flow depth of 388 mm, which lies between 0.1*Des* and *Des*.

Discharge Relations of Orifices and Sluice Gate Inlet Devices

Consider the energy, shown as equation 11.4, assuming negligible losses:

$$\frac{p_1}{\gamma} + \frac{v_1^2}{2g} = \frac{p_2}{\gamma} + \frac{v_2^2}{2g} \qquad \textbf{(11.4)}$$

where the datum is the channel bottom. One can assume velocity v_1 and p_2 to be near zero. This yields

$$\frac{p_1}{\gamma} = \frac{v_2^2}{2g} = y \approx v = \sqrt{2gy} \qquad \textbf{(11.5)}$$

Substituting $q = Av$ into equation 11.5 leads directly to the orifice equation. The inclusion of the C_{orf} coefficient accounts for energy losses and momentum effects due to convergence. Sluice gate analysis proceeds along similar lines. These equations are valid in any consistent unit system.

The computation for sluice gates is a straightforward application of the orifice equation. An orifice constant of 0.6 is adequate in most standard design situations. Refer to Figure 11.10 for accepted orifice coefficients. King and Brater (1963) also give orifice constants for many orifice configurations.

Flow Hydraulics of Closed Conduits

Either the inlet section or the conduit may control the flow through closed conduits such as culverts and other pipe structures. Inlet flow control is either weir limited (unsubmerged) or orifice limited (submerged). Conduit flow control is either open channel (unsubmerged inlet and outlet) or pipe limited (inlet and outlet submerged or unsubmerged). Figure 11.14 shows culvert nomenclature associated with culvert analyses, and Figure 11.15 shows possible flow categories depending on inlet and outlet levels and various pipe parameters.

Weir flow, orifice flow, open channel flow, or pipe flow may control the flow rate, depending on the stage of water at inlet and outlet. This text discusses only the case where the exit water level does not control (except with full pipe flow, which does not require the simplification). A more detailed analysis of exit water conditions is deferred to advanced courses.

Submerged Inlets—Free Discharge

If the inlet is submerged, orifice or pipe flow dictates the flow rate. The orifice flow (category B; see Figure 11.15) condition is

$$q = A_{orf} C_{orf} \sqrt{2gH} \quad \text{with} \quad H = H_{inlet} - 0.5 D_{orf} \; (\text{pipe laying near horizontal}) \qquad \textbf{(11.6)}$$
$$H = H_{inlet} \; (\text{pipe standing vertically})$$

where

q is flow capacity (L^3/T);
A_{orf} ($= A_p$) is conduit cross-sectional area (L^2);
H is head causing full pipe flow (L; see Figure 11.15);
D_{orf} ($= D_p$) is the diameter of the orifice or pipe;
H_{inlet} is the (submerged) inlet water surface elevation relative to the inlet invert;
g is gravity (L/T^2);
C_{orf} is the orifice coefficient, normally 0.6.

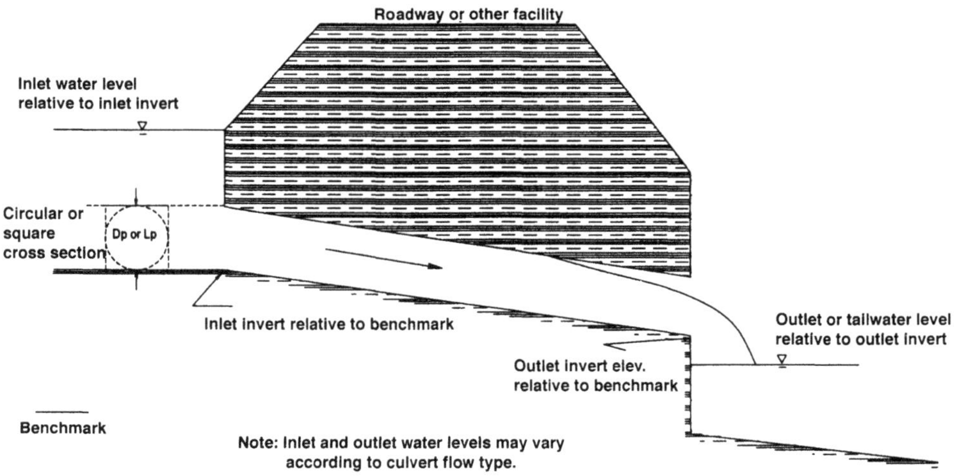

Figure 11.14. Culvert nomenclature.

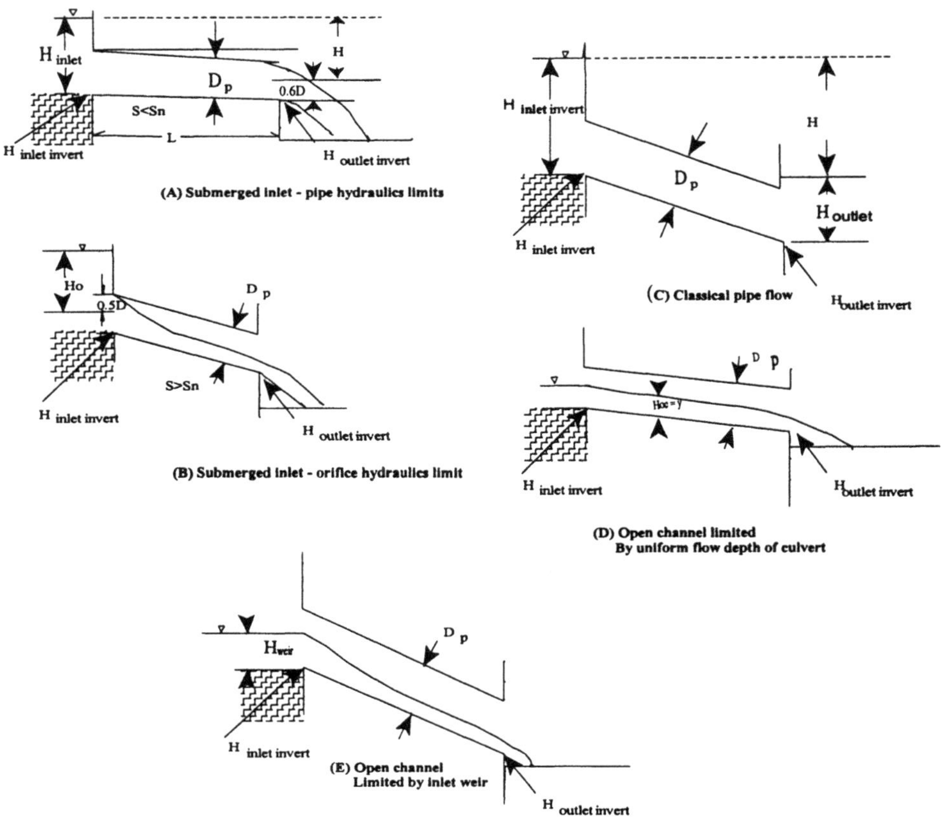

Figure 11.15. Some scenarios of flow through a closed conduit. Consider the definition of head in categories A through C. Categories A and B are required when designing a free discharge culvert to pass a peak discharge subject to head limitations. These problems require a test (calculation of flow for each type and selection of minimum flow) to determine which of A or B controls. Category C is generally unique to communicating bodies of water, using the category A equation with a modified definition for head. Categories D and E are solved using the weir equation at the entrance and the Manning equation plus energy equation as in the transition problem. The controlling category is the one providing minimum discharge. Categories D and E are helpful in stage-discharge relationships for spillways and related structures. Culverts whose outlets are submerged to a varying extent require advanced treatment beyond the scope of this text.

Definition: Invert elevation refers to the elevation of the bottom of the pipe relative to some benchmark elevation datum.

Example 11.4: What is the orifice area required to move 0.5 m^3/s at a head of 2.85 m, as measured from the center of the orifice to the free water level behind the orifice?

One may begin with equation 11.6, substituting known quantities. $0.5\ m^3/s = 0.6A\,[2(9.8\ m/s^2)(2.85\ m)]^{0.5}$ leading to $A = 0.1114\ m^2$. The diameter is $[4(0.1114)/\pi]^{0.5} = 0.376$ m.

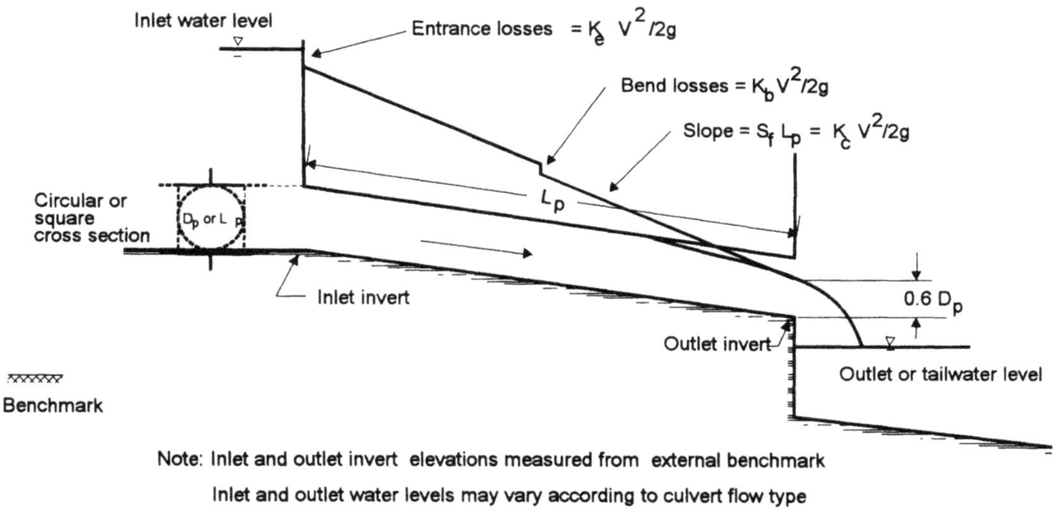

Figure 11.16. Culvert situation of Figure 11.14 showing the energy grade line.

One may derive the orifice equation as shown in equation 11.5, setting the datum in the center of the orifice. The pipe flow (category A) condition is described by writing an energy balance:

$$H = y_1 - y_p = \frac{v_{pip}^2}{2g} + K_e\frac{v_{pip}^2}{2g} + K_b\frac{v_{pip}^2}{2g} + S_f L \qquad \textbf{(11.7)}$$

where

v_{pip} is the velocity in the pipe (m/s);
y_p is the pressure head at the pipe exit (m).

Figure 11.16 shows the components of the energy balance along with the energy grade line.

One may derive the friction slope S_f from the Manning equation (equations 8.14a–c), taking φ_{dim} as 1, giving

$$S_f = v_{pip}^2 \frac{n^2 4^{4/3} 1000^{4/3}}{D_p^{4/3}} \qquad \textbf{(11.8a)}$$

where D_p is the pipe diameter (mm). This approach works well with rough pipe. This is because large ε/D ratios and high Reynolds numbers lead to constant friction factors. This can be verified by looking at a Darcy-Weisbach friction factor diagram (see any fluid mechanics text). Multiplying both sides by $2g$ gives

$$2gS_f = v_{pip}^2\, 2g\frac{n^2 4^{4/3} 1000^{4/3}}{D_p^{4/3}} = K_c v_{pip}^2 \qquad \textbf{(11.8b)}$$

where K_c is introduced for simplification. Substituting the right-hand side of equation 11.8b into equation 11.7, solving for v_{pip}, and multiplying by pipe cross-sectional area

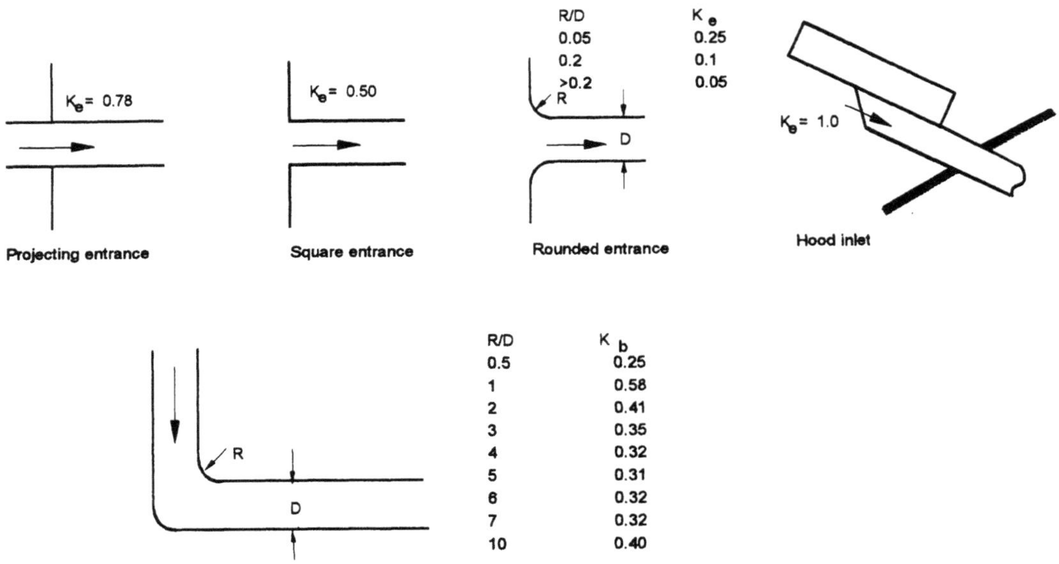

Figure 11.17. Entrance and bend loss coefficients.

(A_p) leads directly to the discharge relationships summarized in

$$q = \frac{A_p\sqrt{2gH}}{\sqrt{1 + K_e + K_b + K_c L}} \tag{11.9a}$$

$$H = (H_{inlet} + H_{inlet\ invert}) - H_{outlet\ invert} - 0.6D_p \tag{11.9b}$$

where

L is conduit length (m);
K_b is the bend coefficient (Figure 11.17);
K_e is the entrance coefficient (Figure 11.17);
K_c is the conduit loss coefficient (equations 11.9c or 11.9d below);
H_{inlet} is the (submerged) inlet water surface elevation relative to the inlet invert;
$H_{inlet\ invert}$ is the pipe bottom elevation at the inlet relative to a surveying benchmark;
$H_{outlet\ invert}$ is the pipe bottom elevation at the outlet relative to a surveying benchmark.

Note: Other minor losses can usually be neglected in culvert pipe flow analyses.

Charts (see Figure 11.17) from Schwab et al. 1993 or formulas (equations 11.9c–d) provide coefficient estimates. The culvert pipe flow equation is a modification of the orifice equation.

The coefficient K_c, the conduit loss factor for circular pipes, is

$$K_c = \frac{Const_{dim1}\, n^2}{D_p^{4/3}} \tag{11.9c}$$

where

K_c is the conveyance coefficient (m^{-1} or ft^{-1});
D_p is the pipe diameter in mm or inches;
$Const_{dim1}$ is 1,244,522 (D_p in mm; K_c in m^{-1}) or 5087 (D_p in inches; K_c in ft^{-1});
n is the Manning roughness factor for the pipe or conduit material.

Equation 11.9c may vary depending on round-off errors. For square conduits the loss factor is

$$K_c = \frac{Const_{dim2}\, n^2}{R^{4/3}} \tag{11.9d}$$

where

K_c is the conveyance coefficient (m^{-1} or ft^{-1}) as above;
R is the hydraulic radius in m or ft;
$Const_{dim2}$ is 19.6 (R in m; K_c in m^{-1}) or 29.16 (R in ft; K_c in ft^{-1}).

One may derive the dimension factors in equations 11.9a–d starting with equations 11.8a–b using the appropriate φ_{dim} in the Manning equation.

Pipe flow may be differentiated from orifice flow (test of category A or category B) with a submerged inlet and unsubmerged outlet using

$$s_n = K_c \frac{v_{pip}^2}{2g} \tag{11.10}$$

where s_n is the neutral slope.

The rationale for equation 11.10 is apparent in Figure 11.16. The neutral slope condition occurs when the slope of the energy grade line is the slope of the pipe. For actual pipe slopes greater than the computed neutral slope, orifice flow prevails. For pipe slopes less than the neutral slope, pipe flow hydraulics prevails. The more straightforward approach for testing whether category A or B controls flow is to compute the flow for both conditions and take the minimum.

Category B and category A designs are the most used design methodologies for culverts. The usual design goals are to (1) size the culvert pipe such that a specified flow will not exceed a given head; (2) determine the head required by a given flow and pipe size; and (3) compute the flow for a given pipe and head. The problem of determining the required pipe for a given discharge requires iteration or trying several pipe sizes. The problem of determining the required head for a given discharge and a known pipe solution involves a one-pass solution. The problem of determining required flow for a given pipe and a known head involves a direct solution. Compute orifice flow for the final design to check the flow-controlling mechanism.

Write equations 11.9a–d and substitute known values. Write the equation for head H and substitute known values. If flow q is the only unknown, simply "plug and chug." If H_{inlet} is unknown, solve equations 11.9a–d for H, then compute H_{inlet} from the appropriate head equation. If the pipe diameter is unknown, one must iterate.

Note: If H_{inlet} is unknown and one knows diameter and flow, assume initial inlet submergence. Find the head H in equations 11.9a–d. Put the found H in the head equation to solve for H_{inlet}. If H_{inlet} is negative, then the inlet is not submerged. This suggests weir control or open channel within the pipe control. *One would miss this possible outcome if one immediately solved for neutral slope using continuity.*

This is an approximate design procedure. Results are good most of the time. Advanced methods are available for comprehensive culvert flow analyses. Consult a text on open channel flow for more advanced methods. Excellent computer software for culvert design is available. Similar techniques are applicable to the hood inlet and other discharge spillways for structures (Beasley et al. 1960).

Example 11.5 (solving for flow with known head and known pipe size): Pipe size is 762 mm (30 in); a corrugated culvert is 12 m (39.36 ft) long. Inlet invert elevation is 127.9 m; outlet invert elevation is 127.7 m. Headwater elevation is 129.5 m (i.e., H_{inlet} = 1.6 m = 129.5 m − 127.9 m), and water elevation at the pipe exit is 126.8 m. What is the discharge?

The discharge is a free discharge since the water level at the pipe exit is below the outlet invert. The actual culvert slope is (127.9 m − 127.7 m)/12 m = 0.0167. Assume pipe flow holds. Use equations 11.9a–d. The head H = 129.5 m − 127.7 m − 0.6(2.5 ft/3.28 ft/m) = 1.342 m; K_e = 0.5, K_c = 0.112 m^{-1} (n = 0.025 and equation 11.9c), K_b = 0, A_p = 0.456 m^2. Then, from the pipe flow equation, one has

$$q = \{0.456\,\text{m}^2\,[(2)(9.8\ \text{m/s}^2)(1.342\ \text{m})]^{0.5}\}/[1 + 0.5 + (0.112\ \text{m}^{-1})12\ \text{m}]^{0.5}$$
$$= 1.386\ \text{m}^3/\text{s}$$

Now we must check the neutral flow condition:

$$v_{pip} = q/Ap = (1.386\ \text{m}^3/\text{s})/0.456\ \text{m}^2 = 3.03\ \text{m/s}$$
$$S_n = [0.112\ \text{m}^{-1}\,(3.03\ \text{m/s})^2]/[2(9.8\ \text{m/s}^2)] = 0.053$$

Since the actual slope is less than the neutral slope (0.0167 < 0.053), pipe flow is valid.

If the neutral slope had been exceeded, one would use the orifice flow equation to compute the flow rate. One could also use the Mavis plot (Figures 11.18–11.19) to estimate the flow in the orifice mode as follows. The head for the orifice is

$$H_{inlet} - 0.5D_p = 1.6\ \text{m} - [0.5(0.762\ \text{m})] = 1.22\ \text{m}$$
$$H/D = 1.22\ \text{m}/0.762\ \text{m} = 1.6$$

From Figure 11.19, $q/D_p^{5/2} = 2.3$ and $q = 2.3(0.762\ \text{m}^{2.5}) = 1.15\ \text{m}^3/\text{s}$. This flow is somewhat less than pipe flow; thus the Mavis-predicted orifice flow could be argued to be the minimum. Note that plots such as Figures 11.18 and 11.19 are approximations. The student is encouraged to use the orifice equation to find orifice flow.

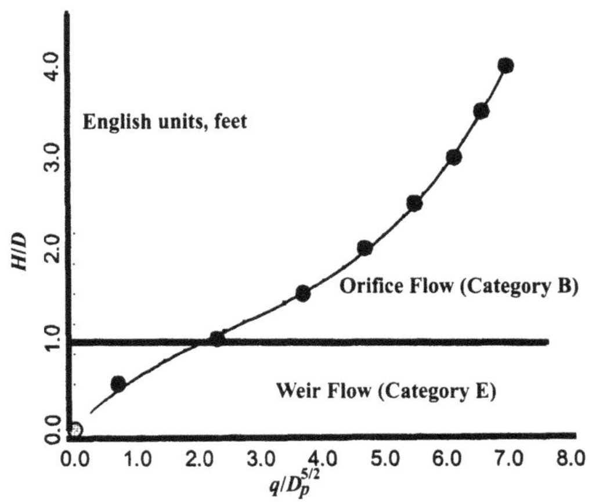

Figure 11.18. Mavis plots for unsubmerged and submerged flow through a pipe entrance, assuming no pipe effects such as an open channel or closed pipe (English units of cfs for flow and feet for head and diameter).

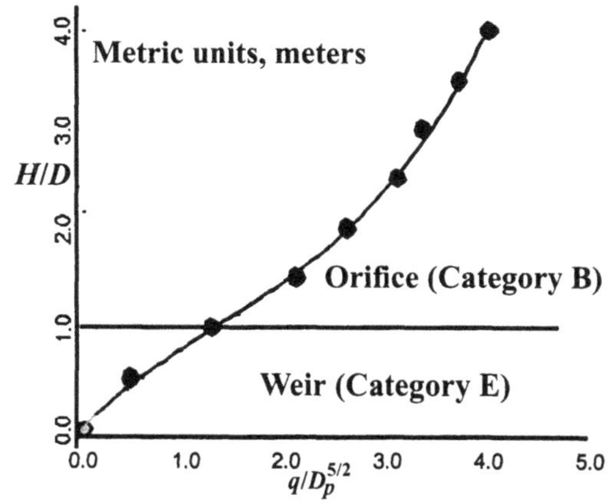

Figure 11.19. Same as for previous figure except units are meters for head and diameter and m^3/s for flow.

Example 11.6 (solving for head for given discharge and known pipe size): A culvert is 12 m (39.36 ft) long, no bends. The flow is 5 m^3/s discharge. Inlet invert elevation is 127.9 m; outlet invert elevation is 127.7 m. Water level at exit is 126.8 m (i.e., free discharge). What inlet water elevation is required to convey the flow through a 1 m diameter corrugated pipe? Assume free outfall. Starting with equation 11.9b,

$$H = (H_{inlet} + 127.9\text{ m}) - 127.7\text{ m} - 0.6(1\text{ m}) = H_{inlet} - 0.4\text{ m}$$

where H_{inlet} is the distance from the inlet invert to the water surface at the inlet.

$$a = \pi(1\text{ m})^2/4 = 0.785\text{ m}^2$$
$$K_c = 0.0796\text{ m}^{-1}\ (n = 0.025)$$
$$K_e = 0.5$$
$$K_b = 0$$

Using equation 11.9 gives

$$5\text{ m}^3/\text{s} = \{0.785\text{ m}^2\ [2(9.81\text{ m/s}^2)(H_{inlet} - 0.4\text{ m})]^{0.5}\}/[1 + 0.5 + 0.0796\text{ m}^{-1}\ (12\text{ m})]^{0.5}$$

Note that

$$[1 + 0.5 + 0.0796\text{ m}^{-1}\ (12\text{ m})] = 1.57$$
$$[5\text{ m}^3/\text{s}(1.57)/0.785\text{ m}^2]^2 = 2(9.81\text{ m/s}^2)(H_{inlet} - 0.4\text{ m})$$
$$H_{inlet} = 5.47\text{ m}$$

Therefore,

$$V_{pip} = (5\text{ m}^3/\text{s})/0.785\text{ m}^2 = 6.37\text{ m/s}$$
$$S_n = 0.0796[(6.37\text{ m/s})^2/2(9.81\text{ m/s}^2)] = 0.16$$

Pipe flow is valid.

Example 11.7 (solving for pipe size for known discharge and known head): Find the straight corrugated pipe size to convey 5 m^3/s with an inlet depth of 3 m. Length of culvert is 12 m. Inlet and outlet elevations, n, and inlet conditions are as in Example 11.6.

Starting with equations 11.9a–d, assuming $K_e = 0.5$ and $K_b = 0$, we can write

$$5\text{m}^3/\text{s} = ((3.14D_p^2/4)\ \{2(9.81\ \text{m/s}^2)[(3\ \text{m} + 127.9\ \text{m}) - 127.7\ \text{m} - 0.6D_p]\}^{0.5})/$$
$$\{1 + 0.5 + (12)[1{,}244{,}522(0.025^2)/(1000\ \text{mm/m}\ D)^{1.33}]\}^{0.5}$$

Using an iterative solver results in $D_p = 1.17$ m pipe diameter. To check neutral slope

$$K_c = 0.0645$$
$$A_p = 1.077\ \text{m}^2$$
$$V_{pip} = 5\ \text{m}^3/\text{s}/1.077\ \text{m}^2 = 4.64\ \text{m/s}$$
$$S_n = 0.0645(4.64\ \text{m/s})^2/[2(9.81\ \text{m/s}^2)] = 0.071$$

Pipe slope $= (127.9 - 127.7)/12 = 0.017$
Pipe flow is the correct flow mode.

Example 11.8 (orifice flow prevails): Find the flow with a straight 1 m diameter corrugated pipe 4 m long on a 65% slope. The inlet invert is at 100 m, the inlet depth is 10 m, and free-fall conditions exist. Use the inlet entrance conditions given in Example 11.6.

$$A_p = 0.785\ \text{m}^2$$
$$K_e = 0.5$$
$$K_c = 1{,}244{,}522(0.025^2)/1000^{1.33} = 0.079$$
$$H = (100\ \text{m} + 10\ \text{m}) - [100\ \text{m} - 0.65(4\ \text{m})] - 0.6(1\ \text{m}) = 12.0\ \text{m}$$

Equation 11.9a appears as

$$q = 0.785\ \text{m}^2\ \{2[9.81\ \text{m/s}^2(12\ \text{m})]\}^{0.5}/[1 + 0.5 + 0.079(4)]^{0.5} = 8.94\ \text{m}^3/\text{s}$$
$$V_{pip} = 8.94/0.785 = 11.4\ \text{m/s}$$
$$S_n = 0.079(11.4^2)/[2(9.81)] = 0.52$$

Pipe flow is not valid since neutral slope is less than the actual slope of 0.2. Thus, one must recalculate using orifice flow. Using equation 11.6,

$$H_{orifice} = 10 - 0.5(1) = 9.5\ \text{m}$$
$$C_{orf} = 0.6$$
$$q = 0.785\ \text{m}^2(0.6)\ \{2[9.81\ \text{m/s}^2(9.5\ \text{m})]\}^{0.5} = 6.43\ \text{m}^3/\text{s}$$

Submerged Inlet and Submerged Outlet (Category C)

Treat this situation as a normal pipe flow problem. The category C equation is identical to the category A as above, except head H is now the difference between the upper and lower water surfaces. The design relationship is

$$q = \frac{A_p\sqrt{2gH}}{\sqrt{1 + K_e + K_b + K_cL}} \qquad \textbf{(11.11a)}$$

$$with\ H = H_{inlet} + H_{inlet\ invert} - (H_{outlet\ invert} + H_{outlet\ (subm)})$$

where $H_{outlet\ subm}$ is the water level elevation at the pipe outlet; other nomenclature defined as for equations 11.9a–d.

If the distance from the inlet and outlet culvert inverts to a head-limiting reference above the culvert (e.g., a roadway) is known, one may compute the head H as shown in equation 11.11b:

$$H = (H_{road} - H_{upper\ water\ surf}) - (H_{road} - H_{road\ water\ surf}) \qquad \textbf{(11.11b)}$$

where

H_{road} is the road surface (or other head-limiting condition) elevation relative to a surveying benchmark;

$H_{upper\ water\ surf}$ is the distance from H_{road} down to the upper water surface level;

$H_{lower\ water\ surf}$ is the distance from H_{road} down to the lower water surface level.

Example 11.9 (solving for flow with submerged inlet and outlet): Suppose a horizontal 24 in culvert crosses a 5 ft wide trail 30 ft below the water surface on the upper side and 2 ft below the water surface on the lower side. The culvert pipe length is 30 ft. The submerged corrugated steel pipe connects the two sides. Will the culvert transfer 50 cfs without flooding? The driving head H is the difference in water elevations, relative to the trail surface: 30 − 2 = 28 ft. Using the pipe flow equation and loss factors and converting to metric units, write equation 11.11a (using equation 11.11b for head computations) as

$$50\ \text{cfs}/(35.28\ \text{cfs/m}^3) = \{0.292\ [(2)\ (9.8\ \text{m/s}^2)H]\}^{0.5}/\{(1 + 0.5 + 0.15[30\ \text{ft}/(3.28\ \text{ft/m})]\}^{0.5}$$

Solve for $H = 3.45$ m $= 11.32$ ft. Thus, a water surface differential of 11.32 ft will transfer 50 cfs. Under these stated conditions the culvert would actually transfer

$$q = 0.292\ \text{m}^2\ \{(2(9.8\ \text{m/s}^2)[28\ \text{ft}/(3.28\ \text{ft/m})]\}^{0.5}/\{1 + 0.5 + 0.15\ \text{m}^{-1}\ [30\ \text{ft}/(3.28\ \text{ft/m})]\} = 2.22\ \text{m}^3/\text{s} = 79\ \text{cfs}$$

If the problem is to determine the pipe diameter that will pass a specified flow at a specified head, propose pipe diameters and solve for flow at set head. Continue until the selected diameters give a range of flows bracketing the needed flow. Select the next largest available diameter.

Unsubmerged Inlets and Unsubmerged Outlets (Categories D and E)

When the inlet and outlet are not submerged, then either open channel flow (category D) or weir flow (category E) controls.

First, consider category E weir flow. Mavis (1943) developed dimensionless relationships describing the weir and orifice flow of a culvert with a short pipe and free outfall. Our major concern here is with the weir portion of the curve. For a square-edged entrance inlet equations 11.12a–b describe the flow in situations where weir and orifice flow (category B) prevail (this regression relationship fit the Mavis 1943 data with an R^2 of 0.98):

$$\frac{H_{inlet}}{D_p} = 0.68\frac{q}{D_p^{5/2}} - 0.1392\left(\frac{q}{D_p^{5/2}}\right)^2 + 0.01756\left(\frac{q}{D_p^{5/2}}\right)^3 (\textit{English units}) \quad \textbf{(11.12a)}$$

$$\frac{H_{inlet}}{D_p} = 1.06\frac{q}{D_p^{5/2}} - 0.3005\left(\frac{q}{D_p^{5/2}}\right)^2 + 0.0722\left(\frac{q}{D_p^{5/2}}\right) \quad \textbf{(11.12b)}$$

Figures 11.18 and 11.19 show equations 11.12a–b graphically. Flow is in units of cfs or m^3/s. Diameter and head are expressed in feet or meters. As an approximation, one may also use the weir equation with C of 3.9.

Open channel flow (category D) is now considered. This problem is identical to the flow transition problem of Chapter 8. With open channel flow, the Manning equation coupled with the energy equation is used to describe the transition hydraulics. Refer to the previous discussion of hydraulic elements (Table 8.1), the Manning equation, and flow transitions. First, calculate the flow neglecting the transition effect, calculating the flow and velocity by the Manning equation. Then, using the velocity, compute $v^2/2g$ to account for change in velocity head. Subtract this head from the original depth to approximate the transition. One should use an energy loss $K_{tran} = 1–2$ where energy loss is $K_{tran}v^2/2g$ to account for energy losses at the inlet. The test of whether open channel flow or weir hydraulics controls when both inlet and outlet are unsubmerged is to compute the flow in both conditions and select the mechanism producing the least flow.

Example 11.10 (inlet and outlet unsubmerged): Suppose a 30 in corrugated pipe at a 1% slope was partially submerged at a depth of 15 in. Compute the flow by weir flow and open channel flow.

Using Figure 11.18 for weir flow, for $H_{inlet}/D_p = 0.5$, read $q/D^{5/2} = 0.8$ (ft). Since $D_p = 2.5$ ft, $q = 0.8(2.5^{5/2}) = 7.9$ cfs.

For open channel flow (refer to Figure 8.4 for channel elements for a circle), write the Manning equation. The energy equation and the Manning equation must be satisfied. Assume zero velocity above the inlet. We will first solve the Manning equation, then calculate the acceleration and loss term, adjust head downward, recalculate the Manning equation, and adjust until the losses equal the initial adjustment. Consider that $(1 + K_{tran})v_{pip}^2/2g$ is the acceleration and transition head loss with $K_{tran} = 1$. Select a guess depth of 1.25 ft. $R/R_O = 1$ and $A/A_O = 0.5$ from Figure 8.4. Perform a first pass at the Manning equation where

$$R = 1(2.5 \text{ ft}/4) = 0.625 \text{ ft}$$
$$A = 0.5(3.14)(2.5 \text{ ft})^2/4 = 2.45 \text{ ft}^2$$

Then

$$v = (1.49/0.025)(0.625 \text{ ft})^{0.67} (0.01)^{0.5} = 4.35 \text{ ft/s}$$
$$q = 4.35 \text{ ft/s}(2.45 \text{ ft}) = 10.67 \text{ cfs}$$

The acceleration and losses = 2(4.35 ft/s)^2/[2(32.2 ft/s^2)] = 0.58 ft. Then the depth of flow is about 0.67 ft instead of 1.25 ft. Recalculate the hydraulic radius to be 0.6(2.5 ft/4) = 0.375 ft and area to be 0.15(3.14)(2.5 ft)^2/4 = 0.74 ft^2. (Verify these estimates using Figure 8.4.) Then

$$v = (1.49/0.025)(0.375)^{0.67}\ (0.01) = 3.08\ \text{ft/s}$$

$$q = 3.08\ \text{ft/s}(0.74\ \text{ft}^2) = 2.27\ \text{cfs}$$

Losses and acceleration are now

$$2[(3.08\ \text{ft/s})^2]/[2(32.2\ \text{ft/s}^2)] = 0.3$$

Let depth = 0.95 ft. Then

$$R = 0.7(2.5\ \text{ft}/4) = 0.44\ \text{ft}$$
$$A = (0.35)(3.14)(2.5\ \text{ft/s})^2/4 = 1.72\ \text{ft}^2$$
$$v = (1.49/0.025)(0.44\ \text{ft}^{0.67})\ (0.01) = 3.43\ \text{ft/s}$$
$$q = 3.43\ \text{ft/s}(1.72\ \text{ft}^2) = 5.91\ \text{cfs}$$

Losses and acceleration are then 0.36 ft, acceptably close to the assumed 0.3 ft. Continue the iteration until solutions converge to within 15%. Since open channel flow results in the minimum flow, open channel flow is limiting. Had the pipe been shorter and/or steeper, weir flow would likely be the controlling mechanism.

The category D and category E situations are specifically useful in developing stage-discharge curves for culverts and spillways. Routing curves for culvert structures are developed using stage-discharge curves.

Culvert Design When Outlet Is Partially Submerged (not shown in Figure 11.15)

What happens when the exit water level can affect the flow through the pipe? Assuming a submerged inlet and physical slope less than the neutral slope (pipe flow hydraulics), one may modify the relationship for computing head in the pipe flow equation. As the culvert moves from free discharge to submerged outlet, a simple approach is to proportionally increase the $0.6D_p$ term in equation 11.9b as the pipe outlet moves from no submergence to full submergence. For example, if the outlet is halfway submerged, one may use the free-discharge approach, changing the 0.6 coefficient to 0.8. When the outlet is submerged, the coefficient is 1.0. A more rigorous analysis of partial outlet submergence is beyond the scope of this text.

The pipe diameter proved to be the most influential in a sensitivity analysis. The diameter is usually precisely known. The Manning roughness was the next most influential. It varies by as much as 50%.

In summary, culvert problems usually involve category A or category B, culvert for passing a peak flow; category C, communicating water bodies; or category D or E, stage-discharge relationships for culverts or spillways.

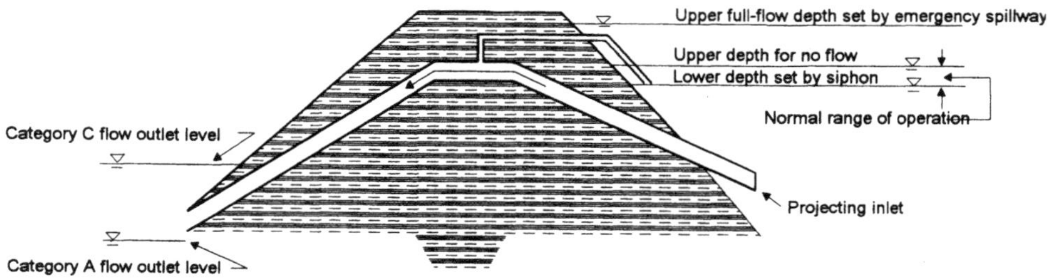

Figure 11.20. Schematic of a siphon spillway with projecting inlet. Tail water elevations defining category A and category C flow types are indicated. Inlet flow control levels are noted. Drawing is not to scale.

Stage-Discharge Curves for Culverts and Spillways

How can one describe the flow-depth relation of a culvert or spillway structure? The compound structure is any flow path with an inlet and a closed conduit such as a pipe. Figure 11.8 provides suggestions on spillway type as a function of flow rate and head. The following technique enables generation of stage-discharge curves for the selected structure.

One simple spillway structure is the siphon, shown in Figure 11.20. Treat the siphon as a variation of full-flow culvert flow, category A or C. The most distinguishing feature of the siphon is the high middle pipe section. Pressures less than atmospheric may exist within the middle pipe section. The inlet is usually a projecting-style inlet. A siphon spillway in a pond may appear as shown in Figure 11.20. The siphon pipe must withstand negative pressure values.

Other spillways involve weir flow, followed by orifice flow, and finally pipe flow. Considering the problem before and after inlet submergence is useful. Before submergence, weir flow or open channel flow mechanisms control the flow. After inlet submergence, orifice flow or pipe flow could prevail. The condition giving the minimum flow is in control. In situations involving additional spillways carrying runoff, add the contributions of each spillway at each value of depth. This often happens with a pond. Culvert and pond spillway analyses and modeling bring together all the material in hydraulic structures discussed in this chapter.

Example 11.11 (stage-discharge computation for spillways: vertical riser): Compute the stage-discharge curve for both the principal spillways shown in Figure 11.21 (case A, vertical riser; and case B, horizontal pipe) where the pipe is corrugated and is 0.8 m in diameter.

Equation-processing software or a spreadsheet is useful for solving these problems. The TK© Solver rule sheet details the mechanisms involved in the compound spillway. The principal pipe spillway flows at the minimum possible of the pipe, orifice, and weir conditions. The emergency spillway adds in once the stage reaches the flood storage height. The solution given below was developed using an equation-processing package. Variables, solution rules, and results are shown in Tables 11.3–11.5.

Table 11.3. Variable sheet and rule sheet for TK® solver model for a principal spillway, case A, vertical riser.

St	Input	Name	Output	Unit	Comment
	3	FStor		m	Depth from top of riser pipe to bottom
	10	Lemerg		m	Crest width of emerg spillway
	0.8	Dp		m	Pipe diameter
	1	Kb		-	Bend coefficient
	1	Ke		-	Entrance coefficient
		Kc	.107095917	-	Conveyance loss coefficient
	5	Lpv		m	Vertical length of pipe riser
	30	Lph		m	Horizontal run of pipe
	0.025	n			Manning roughness of pipe
L	5	Hinlet		m	Depth from invert to water surface at I
	100	Hininv		m	Inlet invert elevation
	90	Houtinv		m	Outlet invert elevation
	0.6	C		-	Pipe weir coefficient
L		qtot	0	m^3/s	Total pipe and emerg spillway flow
L		qw	83.433805	m^3/s	flow through pipe as weir
L		qp	2.09017787	m^3/s	flow through pipe
L		qorf	0	m^3/s	flow through pipe as orifice
L		qemerg	0	m^3/s	flow through emerg spillway
		Ap	0.321698816	m^2	Cross sectional area of pipe
		H	14.52	m	Head driving pipe flow
		q	0	m^3/s	Flow through pipe spillway (min of orif
		Hemerg		m	Elev of emerg spillway above pipe
	9.81	g		m/s^2	Gravity

-

Rule sheet

S	Rule
Comment; STAGE DISCHARGE CURVE -PRINCIPLE SPILLWAY WITH VERTICAL PIPE RISER	
Comment; Save the function sheet as a import transfer for hydrout1*.tk	
* Undetermined condition	if Hinlet >Lpv then qp = $Ap*\sqrt{(2*g*H)}/\sqrt{(1 + Kb + Ke + (Kc*(Lph)))}$ else qp = 0; flow from pipe flow-free discharge
Comment; One may in some cases wish to include Lpv in the pipe flow equation length determination.	
* Undetermined condition	if Hinlet ≥ Lpv then $Kc=(1244522*n^2)/((1000*Dp)^{1.33})$ else Kc = 0; riser same diameter as horizontal member
* Undetermined condition	if Hinlet ≥ Lpv then H = (Hinlet+Hininv) − (Houtinv + (0.6*Dp)) else H = 0; free discharge assumption
* Unsatisfied	$Ap = (3.14159*Dp^4)/4$
* Undetermined condition	if Hinlet ≥ Lpv then $qw = C*(3.1416)*Dp*(Hinlet\text{-}Lpv)^{1.5}$ else qw = 0; weir flow
* Undetermined condition	if Hinlet ≥ Lpv then $qorf = 0.6*Ap*\sqrt{(2*g*(Hinlet\text{-}Lpv))}$ else qorf = 0; orifice flow
* Unsatisfied	q = min(qp, qw, qorf)
* Undetermined condition	if Hinlet ≥ (Lpv+FStor) then $qemerg = 3.2*Lemerg*(Hinlet\text{-}(Lpv+FStor))^{1.5}$ else qemerg = 0; emerg spillway as weir
* Unsatisfied	qtot=q+qemerg; total flow

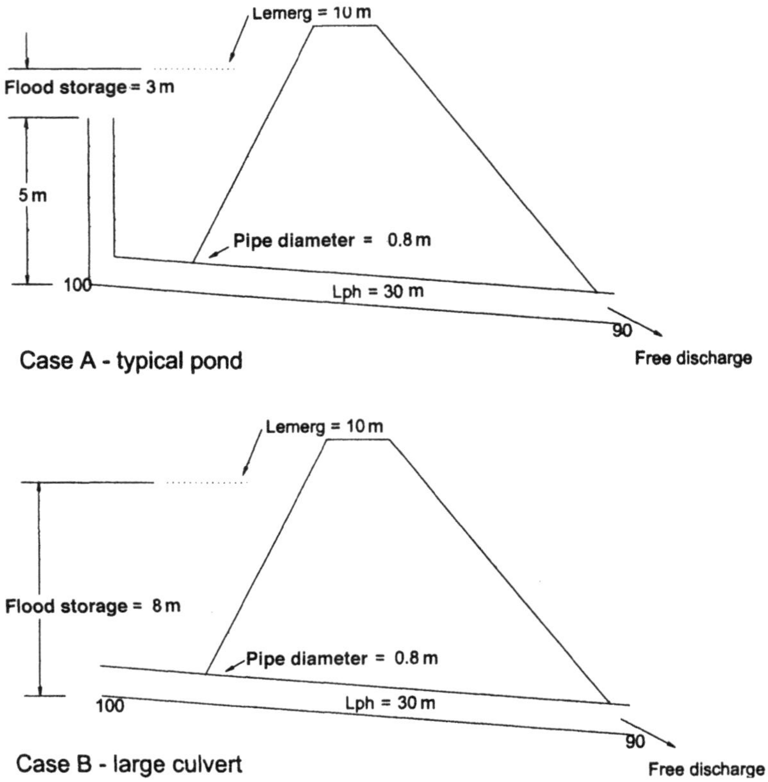

Figure 11.21. Spillway cases for Example 11.11.

Table 11.4. Results table showing the different components of the discharge (qtot) for the model in Table 11.3.

Note: all flows expressed in m^3/s

	Hinlet	qtot	qw	qp	qorf	qemerg
2	5	0	0	0	0	0
3	5.25	0.188496	0.188496	2.197064547	0.4274843487	0
4	5.5	0.5331471993	0.5331471993	2.215580504	0.6045541637	0
5	5.75	0.7404246114	0.9794539471	2.233942997	0.7404246114	0
6	6	0.8549686974	1.507968	2.25215578	0.8549686974	0
7	6.25	0.9558840631	2.107449347	2.270222456	0.9558840631	0
8	6.5	1.047118527	2.770314111	2.288146487	1.047118527	0
9	6.75	1.131017276	3.490994774	2.305931198	1.131017276	0
10	7	1.209108327	4.265177594	2.323579788	1.209108327	0
11	7.25	1.282453046	5.089392	2.341095337	1.282453046	0
12	7.5	1.351824206	5.960766898	2.358480807	1.351824206	0
13	7.75	1.417805188	6.876875571	2.375739056	1.417805188	0
14	8	1.480849223	7.835631577	2.392872835	1.480849223	0
15	8.25	5.541316739	8.835215912	2.4098848	1.541316739	4
16	8.5	12.91320847	9.874024311	2.426777512	1.599499971	11.3137085
17	8.75	22.44024945	10.95062803	2.443553445	1.655639763	20.78460969
18	9	33.70993739	12.063744	2.460214987	1.709937395	32
19	9.25	46.48392267	13.21221161	2.476764447	1.762563123	44.72135955
20	9.5	60.60141632	14.39497438	2.493204057	1.813662491	58.78775383
21	9.75	75.94439779	15.61106529	2.509535976	1.863361076	74.08103671
22	10	92.42143612	16.85959478	2.525762292	1.911768126	90.50966799

Table 11.5. Rules for stage-discharge determination with pipe mechanical spillway, no vertical riser.

```
:STAGE DISCHARGE CURVE FOR PIPE MECHANICAL SPILLWAY – NO RISER
; Save the function sheet as an import transfer for HYROUT1*.tk
if Hinlet >Dp then qp=Ap*√(2*g*H)/√(1 + Kb + Ke + (Kc*(Lph))) else qp=0; flow from pipe
   flow free discharge
if Hinlet ≥ Dp then Kc=(1244522*n^2)/((1000*Dp)^1.33) else Kc=0;
if Hinlet ≥ Dp then H=(Hinlet+Hininv)-(Houtinv+(0.6*Dp)) else H=0; free discharge assumption
Ap=(3.14159*Dp^2)/4
if (AND(Hinlet ≥ Lpv, Hinlet<Dp)) then qw=C*Wpw*(Hinlet-Lpv)^1.5 else qw=0; weir flow; use
   weir equation and wetted perimeter as length
if Hinlet < Dp then Wpw=3.1415*Dp*WP%WPo(Hinlet/Dp) else Wpw=C*3.1415*Dp
if Hinlet ≥ (0.5*Dp) then qorf=0.6*Ap*√(2*g*(Hinlet- (0.5*Dp))) else qorf=0 ; orifice flow
if Hinlet < Dp then Hinlet=(ychan) +(1+Ktran)*((Vchan^2)/(2*g)) ; open channel flow before inlet
   submergence
Vchan=(1/n)*(Rp^0.67)*(√(Sp))
Sp=(Hininv-Houtinv)/Lph
if Hinlet < Dp then Rp=(Dp/4)*R%Ro(ychan/Dp) else Rp=Dp/4
if  Hinlet < Dp then Apo=Ap*Ap%Apo(ychan/Dp) else Apo=Ap
if Hinlet <Dp then qoch=Apo*Vchan else qoch=0
if Hinlet > Dp then q=min(qp,qorf) else q=min(qw, qoch); determine the flow in the pipe
if Hinlet ≥ (FStor) then qemerg=3.2*Lemerg*(Hinlet-(Lpv+FStor))^1.5 else qemerg=0; emerg spill-
way as weir
qtot=q+qemerg; total flow
```

Models for the vertical riser solution are shown in Figure 11.20. The method for the horizontal pipe is identical to that of the vertical pipe except for the possible flow mechanisms. Categories D and E culvert flows are possible with the horizontal pipe, no vertical riser. The rule sheet for the horizontal pipe case is shown in Table 11.5.

If the pond level is below the vertical riser elevation, there is no flow. The pipe always functions as an orifice due to the steep slope, as one may see from the table of vertical riser results (Table 11.4). The pipe edge acts first as a weir and then as an orifice. Finally, pipe hydraulics becomes applicable if the physical slope is less than the neutral slope. The mechanism resulting in the minimum flow is the controlling mechanism. Thus, one may calculate flow by each mechanism and take the minimum. One measures the head for the weir and orifice from the top of the riser. One measures the head for pipe flow following Figure 11.15, case A assuming free discharge. Culvert flows of categories D and E are possible with the horizontal pipe.

The horizontal pipe will experience open channel and/or weir flow before inlet submergence. Open channel flow is modeled by applying the energy equation coupled with the Manning equation (see Example 11.14). Model the flow area using circular channel elements (Figure 8.4). List functions based on Figure 8.4 may be used to model the hydraulic elements of circular sections before submergence with the open channel mechanism. Because of the need to iterate for the open channel mechanism, one must set the depth of flow in the channel (ychan) as a guess.

Weir flow is modeled measuring head from the inlet bottom. Use circular channel elements to estimate weir length (wetted perimeter). The Mavis (1943) charts (Figures 11.18 and 11.19) could also have been used.

Pipe flow and orifice flow are possible once submergence occurs. One measures orifice head from the center of the pipe. Measure the pipe flow head as shown in Figure 11.15 (case A assuming free discharge).

We enter the stage variable (called Hinlet) as an input list. One should include several points before submergence occurs. Solve for the flow (Qtot) at each stage. Then develop a function with a range of Qtot and a domain of Hinlet.

Example 11.12 (stage-discharge computation for spillways: horizontal pipe): Compute the stage-discharge curve for an 0.457 m (18 in) diameter corrugated metal pipe riser connected to a 15.24 m (50 ft) pipe of the same material and diameter. The pipe contains a 90-degree bend and has no physical slope. The riser top is 3.05 m (10 ft) above the outlet bottom. Use a spreadsheet for the computations. Weir flow will first occur. Using a weir coefficient $C_w = 3$, we can write from equation 11.1

$$q = 0.55(3)(L)(H^{3/2}) = 0.55(3)(\pi d)(H^{3/2}) = 0.55(3)(1.435 \text{ m})(H^{3/2}) = 2.37(H^{3/2})$$

where H is the distance of the water from the top of the spillway.

Orifice flow (see equation 11.6) is then possible and is described using

$$q = C_{orf}A\,(2gH)^{0.5} = 0.6(0.164)\,[2(9.81)(H)]^{0.5} = 0.436H^{1/2}$$

where

A = area $= 3.1415(0.457^2)/4 = 0.164 \text{ m}^2$;
H = distance of water from top of spillway.

Full pipe flow is also possible at deeper depths. The culvert equation (equation 11.9a) may be used to describe full pipe flow. Assuming an $n = 0.025$, $K_e = 1$, $K_b = 0.5$, and $K_c = 1{,}244{,}522(0.025)^2/(457 \text{ mm})^{1.33} = 0.225 \text{ m}^{-1}$, we may then write

Stage-discharge evaluation

Stage (m)	Weir flow (m^3/s)	Orifice flow (m^3/s)	Pipe flow (m^3/s)	Flow (m^3/s)	Type (m^3/s)
0	0	0	0.496418	0	
0.1	0.074945981	0.137875306	0.505284	0.074946	Weir
0.2	0.211979244	0.194985128	0.513996	0.194985	Orifice
0.25	0.29625	0.218	0.518297	0.218	Orifice
0.5	0.837921536	0.308298557	0.539289	0.308299	Orifice
0.75	1.539360155	0.377587076	0.559495	0.377587	Orifice
1	2.37	0.436	0.578995	0.436	Orifice
1.25	3.312175692	0.487462819	0.59786	0.487463	Orifice
1.5	4.353968018	0.533988764	0.616147	0.533989	Orifice
2	6.703372286	0.616597113	0.651183	0.616597	Orifice
2.25	7.99875	0.654	0.668012	0.654	Orifice
2.5	9.368247568	0.68937653	0.684428	0.684428	Pipe
2.75	10.80805104	0.723024204	0.700458	0.700458	Pipe
3	12.31488124	0.755174152	0.716131	0.716131	Pipe
3.25	13.88587935	0.786010178	0.731467	0.731467	Pipe
3.5	15.51852401	0.81568131	0.746489	0.746489	Pipe
3.75	17.21056974	0.844310369	0.761214	0.761214	Pipe
4	18.96	0.872	0.775659	0.775659	Pipe
4.5	22.62388146	0.92489567	0.803772	0.803772	Pipe
5	26.49740553	0.974925638	0.830934	0.830934	Pipe

Figure 11.22. Spreadsheet for the approximate stage-discharge curve for the Example 11.12 spillway.

the culvert equation as

$$q = 0.164\ \text{m}^2\ [2(9.81\ \text{m/s}^2)(H_{cul})]^{0.5}/[1 + 1 + 0.5 + 0.225\ \text{m}^{-1}(15.24\ \text{m})]^{0.5} = 0.298\ (H_{cul})^{0.5}$$

$$H_{cul} = H + 3.05 - 0.6(0.457\ \text{m/s}^2) = H + 2.775$$

(Derive this from equation 11.9b!) The culvert equation for this example then appears as

$$q = 0.298(H + 2.775)^{1/2}$$

This example has no emergency spillway.

One can then evaluate these equations over a range of stages using a spreadsheet. The spreadsheet for this example appears in Figure 11.22. The calculations of the various flow mechanisms are straightforward from the above relationships. The flow column is the minimum of the flows for each mechanism at the respective stage.

The culvert/spillway pipe must stand up to the loads in question, including negative pressures found in siphons. Culverts should be at least 0.6 m below the surface. Appendix C discusses strength of pipes relative to loading and soil conditions.

Closed Conduit Systems for Urban Storm Water Collection

The category E (closed conduit, open channel flow) situation discussed with culvert flow is useful for describing flow in storm water collection systems, commonly known as storm sewers. Storm sewer networks are designed by first identifying the outlet invert elevation. Move upgrade, identifying grade change points and junctions. At the first junction, proceed upgrade along one of the branches, continuing to identify grade changes and additional junctions. Continue the grade change and junction identification until the complete network has been tentatively identified. Following the initial layout, one computes the design storm (usually 10 yr–T_C) using the rational method.

Curb gutters or other channels provide initial collection. Curb inlets then divert the flow to sewers. Variations of the weir formula or other empirical formulas are used to design inlets (see Meadows and Walski 1999). In small drainage networks, one may ignore travel time. Peak flows may be summed at flow path junctions.

Note the flows at each inlet and junction. Design the component pipes to accommodate the flows as closed-conduit open channels based on the known flow, slope, and pipe material using the category E methodology. Excellent computer software is available for designing storm sewer collection systems, including curb inlets (e.g., see Meadows and Walski 1999).

One must be cognizant of state and local codes when designing storm sewers. Pipe size should increase as one moves downgrade along a branch of the network. A decreased size could be specified if the same flow moves from a shallow gradient to a steep gradient. Local codes often specify that pipe size should not decrease because decreasing pipe diameters downgrade may be susceptible to clogging.

As urbanization continues to encroach on agricultural and forested lands, the natural resources engineer will become more involved in the development of storm water drainage systems using closed conduits. Discharge from storm sewers is becoming regulated. The increased regulation has the indirect effect of bringing increased scrutiny on all designs for storm water collection systems. One should keep abreast of new developments in the regulatory environment.

Water Measurement Structures for Irrigation and Aquaculture

Flow measurement is customary in arid areas, where growers often purchase water. The techniques are applicable to humid areas. In humid areas, one may wish to quantify flows for general information or regulatory purposes. Weirs (rectangular, triangular, trapezoidal, and several other cross section types), orifices, and various flumes (*H* flumes, Parshall flumes) are widely used for flow measurements. These devices accomplish a sub- to supercritical flow transition. The unique stage-discharge at critical flow is the core attribute of these flow measurement devices. Selection of constants and measuring position is critical for accurate measurements. Bos et al. (1984), King and Brater (1963), Pair et al. (1983), and Walker and Skogerboe (1987) provide additional information regarding device selection, placement, and field use.

A variety of permanent structures are needed to control irrigation water. Adjustable gates control flows from irrigation canals into farm ditches. These are similar to sluice gates. Distribution boxes with appropriate baffling are used to pass water from pump outlets into multiple channels. Weirs and flumes are widely used to measure flow rates. Figure 11.23 shows a flow distribution box used when delivering irrigation water via open channels. Distribution boxes incorporate weirs and sluice gates structures.

Yoo and Boyd (1994) and Coche and Muir (1992) discuss construction of weirs for measuring and controlling water flows in ponds for aquaculture. They also discuss inlet and outlet devices for fish ponds.

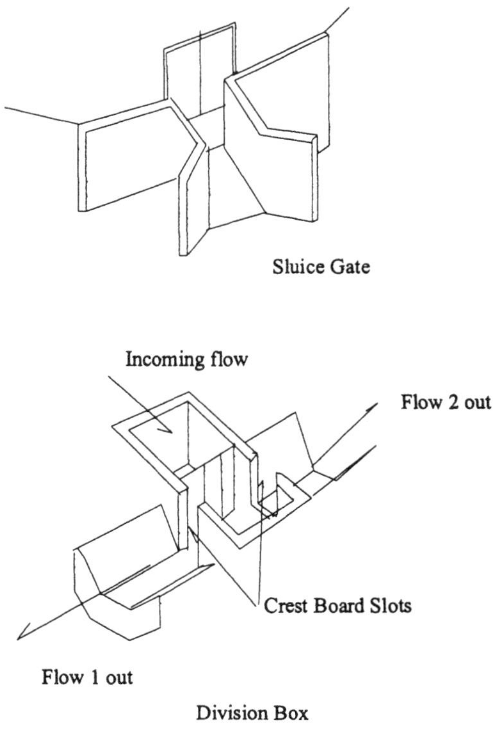

Figure 11.23. Schematic of irrigation canal distribution box.

Outlook

The process for structure selection and design is as follows:

1. Compute design peak flow.
2. If structure is a principal spillway or gully stabilization structure, see Figure 11.8 to select structure based on peak flow.
3. If structure is a weir, inlet, orifice, or sluice gate, see Figure 11.10 for stage-discharge relationship. If a weir is involved, see Figure 11.7 for the length of a rectangular weir, box inlet, or other structure. See Examples 11.1 and 11.2. For a Parshall flume or H_x flume see equations 11.2 and 11.3.
4. If the structure is a culvert, identify the flow class from Figure 11.15. Draw a picture! Most design problems involve selection of road height (i.e., H_{inlet}) with a free discharge or submerged discharge. To solve for H_{inlet} when q is known and there is a free discharge, solve as weir (see Figures 11.7 and 11.8), orifice (equation 11.6), and pipe (equations 11.9a–d). The condition giving the minimum H_{inlet} controls.

   See Figure 11.16 for K_b and K_e; equation 11.9c–d for K_c.

   See Examples 11.5–11.9 for culvert design with indicated knowns and unknowns.

   If one is solving for Hinlet and the outlet is submerged, one need only analyze the pipe flow case (see equations 11.11a–b).

   If one knows distances from a head-limiting reference above the culvert (e.g., a roadway), use equation 11.11b.

   If one is solving for diameter to pass a given flow and given Hinlet (free discharge), use equation 11.9a, iterating for D. If the discharge is submerged, use equations 11.11a–b and iterate for D.
5. Other cases in Figure 11.15 are used for stage-discharge calculations, required for hydrologic routing of storm water through culvert structures.
6. The principal spillway may act as a weir, orifice, or pipe. Pipe flow is adequate if the discharge is submerged. See Examples 11.11 and 11.12.

The inlet and conduit define the flow hydraulics of a structure.

Energy dissipation is accomplished in outlets. The inlet section is the Froude number transition region (e.g., weirs). Inlet transition hydraulics describes the transition from channel to culvert with unsubmerged inlet and unsubmerged outlet. The conduit is the pipe in a culvert or a channel on a large broad-crested weir. Open channel and/or pipe flow describe the hydraulics. Exit water level and energy dissipation devices define the outlet. Exit water level affects the nature of the flow in the conduit and the inlet (e.g., the full pipe culvert flow with inlet and outlet submerged). Energy dissipation design is the subject of more advanced courses. *Successful analysis of the structure lies in knowing which flow mechanism is controlling and how to measure the operational head for the flow control mechanism.*

Control structures must have sufficient capacity to pass the design discharge and safely dissipate the kinetic energy of the flow within the structure without destroying the structure. Total forces on structures are often quite large. Static loads can also be quite large. Information herein is good for flow estimates. Accurate flow measurements entail too many structural details for discussion here; good information is available in the references cited above.

Design of structures is important to avoid overflow failures and/or to obtain accurate flow measurements. Plan installations to avoid freeze-thaw, seepage, and rodent-induced

Table 11.6. Prompt sheet for assessing customer requirements. Users may add additional issues.

Purpose?
Peak flow reduction?
Sediment content and other nutrients?
Handle flows over the flow range adequately
Flow measurements and/or head regulation?
Energy dissipation within structure?
Stability of flows over time
Head loss in line with available head and flow
Discharge regulation or erosion stabilization/prevention?
Pump storage?
Consequences of short- or long-term flooding near structure
Discharge from particular location along the depth profile?
Land use/value near structure and upstream
Is leakage a problem?
Maximum depth?
Storage/surface area needed if impoundment
Life?
Available funding
Environmental/regulatory Issues
. . .

failures. The primary reasons for failure of conservation structures are

- insufficient capacity
- insufficient energy dissipation, including unanticipated supercritical flows
- improper installation, leading to seepage failures that undermine the structure

Table 11.7. Prompt sheet for technical requirements. Users may add additional issues.

Withstand flow now and through time given water quality
Energy dissipation at discharge of structure?
Capacity/volume
Accurately measure flows over required flow range?
Foundation requirements?
Freeze/thaw issues?
Materials specifications
Energy dissipation approaches
Soil stability
Sediment issues?
Hydraulic conductivity of surrounding soils
Surrounding soil physical and chemical characteristics
Sand/silt/clay/rock content (e.g., texture) of surrounding soils
Stability of soil to piping failure
Plastic/bentonite feasible if sealing needed?
Anticipated pressure heads (positive or negative)
Costs of pipe, excavation, liners, concrete work, etc.
. . .

Table 11.6 gives a prompt sheet for customer requirements related to structures. A prompt sheet for technical requirements is given in Table 11.7.

The U.S. Federal Highway Administration (1985) has developed state-of-the-art culvert design procedures. An excellent method for determining peak flows based on observations at culverts is one by Bodhaine (1976). Engineering design software for sediment basin design such as SedCad7 (Warner 1998), the TR20 software of the NRCS (1983), and the SWMM software of the USEPA (1994) aid in developing stage-discharge relations and do hydrologic routing of storms through the structures. An Internet search will reveal several excellent commercial software packages for hydraulic analyses of culverts and other structures.

Problems and Questions

Understand how to measure the effective weir length for all the structures in Figure 11.7! Draw complete sketches of problem situations.

1. Given the flow of 1.5 m^3/s and a rectangular weir with a bottom width of 1 m, what is the depth of flow in the weir throat? Assume a weir coefficient of 3.1. Draw a sketch of the weir.
2. Determine the design dimensions of a box inlet (1.5 $\times$ 1.5 m, see Figure 11.24); we need the depth here) to carry 7 m^3/s. Assume broad-crested weir control prevails.
3. What is the maximum capacity of a straight, broad-crested drop spillway having a crest length of 3 m and a depth of flow of 0.9 m?
4. Determine the crest length for a straight inlet drop spillway to carry 7.5 m^3/s. The depth of flow is not to exceed 0.9 m. What are the dimensions of a square box under the same conditions?
5. A spillway has the Waterways Experiment Station (WES) spillway shape (see Figure 11.10). What is the length required to accommodate a 3 m^3/s flow rate if one constrains the head to not exceed 0.4 m?
6. Determine the maximum culvert flow expected in the following situation: a concrete pipe culvert 2 ft in diameter and 50 ft long, one 90 degree bend, entrance flush with retainer wall, with inlet invert at 101 ft and outlet invert at 98 ft, inlet height maximum at 5 ft above the inlet invert, and exit water level at 90 ft.
7. (a) A 0.25 m diameter corrugated pipe trickle tube spillway (designed as a culvert) projects into a pond with the water level 2 m above the inlet invert. The discharge is a free discharge. If the spillway lies on a 5% slope and is 5 m long, what is the discharge flow rate?
   (b) If the spillway lies on a 0.05% slope and is 100 m long, what is the flow rate?
   (c) If the water level is 20 m above the inlet invert and the pipe is 5 m long at a 20% slope, what is the discharge?
8. A flow of 3 m^3/s must be passed under a roadway, and the vertical distance from the road surface to the inlet invert is 4 m. The pipe is 15 m long and has no slope. Assume a free discharge. Assume a square-edged inlet. What diameter of smooth concrete pipe is required to pass the flow without flooding?
9. A road at 20 m elevation has a straight culvert 100 m in length. The culvert is to be placed at an inlet invert elevation of 15 m. The culvert must carry an estimated peak runoff of 0.4 m^3/s. The outlet invert is 14.0 m. Maximum exit water level elevation is 10.09 m. The entrance is square with a concrete head wall flush with the end of the pipe. Corrugated metal pipe with $n = 0.025$ is used. Compute the pipe diameter required to avoid flooding.

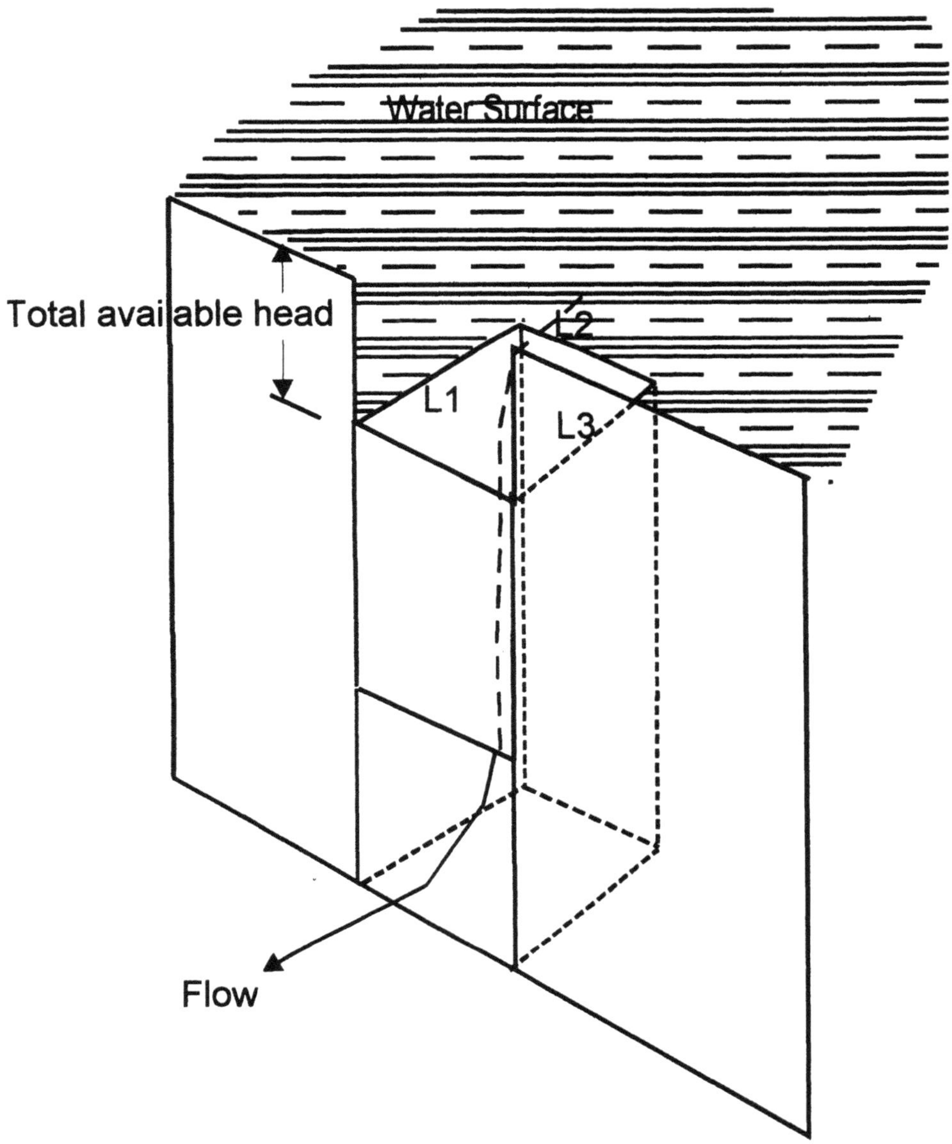

Figure 11.24. Projection box inlet.

10. A 0.3 m diameter smooth concrete pipe connects two reservoirs, one with a 3.5 m level above the inlet invert, the other 2.5 m above the outlet invert. The pipe lies on a 0.01% slope and is 50 m long. Assume a square-edged inlet. What is the flow through the pipe?
11. A culvert is 1 m in diameter and 25 m long with one bend, is made of corrugated pipe, and is on a 0.5% slope. The inlet pool is 10 times the culvert diameter for a distance of 10 m upstream. Assume a square-edged inlet. Estimate the flow rate through the culvert when the inlet depth is 0.1, 0.3, 0.5, and 0.7 times the pipe diameter, assuming the discharge is free.
12. Compute the stage-discharge curve for an 0.6 m diameter corrugated metal pipe riser connected to a 25.24 m pipe of the same material and diameter. The pipe contains a

90 degree bend and has no slope. Assume a square-edged inlet. The riser top is 5.05 m above the outlet bottom.

13. Develop the stage-discharge curve for the structure in Example 11.12, case B.
14. A flow of 10 m^3/s must pass down a steep slope with a total head drop of 11 m. Suggest a structure type and size the structure.
15. Compute the size of a 20 m straight corrugated pipe culvert needed to conduct 10 m^3/s if the road surface is 8 m above the upper inlet invert, the entrance is projecting, and the exit is a free discharge.
16. The purpose of this problem is to explore the effect of culvert situations on controlling flow mechanisms. Be sure to evaluate orifice/pipe and weir/channel possibilities when appropriate. The pipe diameter is fixed because it happens to be available. In each situation below, (1) compute H_{inlet}, (2) identify the controlling flow category, and (3) show the energy grade line on a situation drawing.
    (a) A 457 mm diameter straight culvert 100 m in length is to be placed under a farm road to carry an estimated peak runoff of 0.28 m^3/s. The invert elevation of the pipe inlet is 15.24 m, and the outlet invert is 14.0 m. Maximum exit water level elevation is 10.09 m. The entrance is square with a concrete head wall flush with the end of the pipe. Corrugated metal pipe with n 5 0.025 is used.
    (b) Rework case a when the exit water level elevation is 18 m, all other variables the same.
    (c) Rework case a with a 10 m long culvert, all other variables the same.
    (d) Rework case b with the 10 m long culvert.
17. From what point does one measure the upper invert elevation in a culvert pipe? Lower invert elevation?
18. What categories of flow are possible with culvert pipe and submerged or nonsubmerged inlet having a free discharge?
19. Derive the culvert equation for category B flow conditions for the square conduit case. *Hint:* Follow the steps outlined in the derivation of the circular conduit case.
20. What flow control mechanism operates for each structure shown in Figures 11.6 and 11.7? How does one calculate the head for each mechanism on the various structures?

References

Beasley, R. P., J. M. Gregory, and T. R. McCarty. 1984. *Erosion and sediment pollution control.* 2d ed. Iowa State University Press, Ames.

Beasley, R. P., L. D. Meyer, and E. T. Smerdon. 1960. Canopy inlet for closed conduits. *Agricultural Engineering* 41:226–28.

Biswas, A. K. 1970. *History of hydrology.* North-Holland Publishing Co., Amsterdam, Holland.

Blaisdell, F. W. 1948. Development and hydraulic design: Saint Anthony falls stilling basin. *Transactions of the ASCE* 113:483–520.

Blaisdell, F. W., and C. A. Donnelly. 1951. *Hydraulic design of the box inlet drop spillway.* USDA-SCS-TP-106. USDA Soil Conservation Service, Washington, DC.

Bodhaine, G. L. 1976. Measurement of peak discharge at culverts by indirect methods. In *Techniques of water resources investigations of the United States Geological Survey,* book 3, chap. A3, U.S. Geological Survey, Washington, DC.

Bos, M. G., J. A. Replogle, and A. J. Clemmens. 1984. *Flow measuring flumes for open channel systems.* John Wiley & Sons, New York, NY.

Brakensiek, D. L., H. B. Osborne, and W. J. Rawls. 1979. *Field manual for research in agricultural hydrology.* Agricultural Handbook 224. USDA-SEA, Washington, DC.

Brater, E. F., and H. W. King. 1976. *Handbook of hydraulics.* 6th ed. McGraw-Hill, New York, NY.

Chow, V. T. 1959. *Open channel hydraulics.* McGraw-Hill, New York, NY.

Coche, A. G., and J. F. Muir. 1992. *Pond construction for freshwater fish culture: Pond-farm structures and layouts.* FAO Training Series 20/2. Food and Agriculture Organization of the United Nations, Rome.

Gopalakrishnan, V., and A. G. Coche. 1994. Handbook on small-scale freshwater fish. FAO Training Series no. 24. Food and Agriculture Organization of the United Nations, Rome.

Haan, C. T., J. C. Hayes, and B. J. Barfield. 1994. *Hydrology and sedimentology of small catchments.* Academic Press, New York, NY.

King, H. W., and E. F. Brater. 1963. *Handbook of hydraulics.* 5th ed. McGraw-Hill, New York, NY.

Mavis, F. T. 1943. *The hydraulics of culverts.* Pennsylvania Engineering Experiment Station Bulletin 56. Pennsylvania State University, College Station.

Meadows, M. E., and T. M. Walski. 1999. *Computer applications in hydraulic engineering.* 3d ed. Haestad Press, Waterbury, CT.

NRCS. 1983. *Computer program for project formulation: Hydrology.* SCS Technical Release 20. Soil Conservation Service (Natural Resource Conservation Service), Washington, DC.

Pair, C. H., W. H. Hinz, K. R. Frost, R. E. Sneed, and T. J. Schlitz. 1983. *Irrigation.* 5th ed. Irrigation Association, Silver Springs, MD.

Schwab, G. O., D. D. Fangmeier, W. J. Elliot, and R. K. Frevert. 1993. *Soil and water conservation engineering.* 4th ed. John Wiley & Sons, New York, NY.

Streeter, V. L. 1971. *Fluid mechanics.* 5th ed. McGraw-Hill, New York, NY.

U.S. Bureau of Reclamation (USBR). 1987. *Design of small dams.* 3d ed. Government Printing Office, Washington, DC.

U.S. Environmental Protection Agency. 1994. *SWMM: Storm water management model.* EPA 823-C-94-001. USEPA, Washington, DC.

U.S. Federal Highway Administration. 1985. *Hydraulic design of highway culverts.* Hydraulic Design Series no. 5, Report no. FHWA-IP-85-15. Federal Highway Administration, Washington, DC.

Walker, W. R., and G. V. Skogerboe. 1987. *Surface irrigation: Theory and practice.* Prentice-Hall, New York, NY.

Warner, R. C. 1998. *SedCad[7] 4 for Windows95/NT: Notes for 5–7 Aug. 1998 short course.* Department of Biosystems and Agricultural Engineering, University of Kentucky, Lexington.

Wooley, J. C., M. W. Clark, and R. P. Beasley. 1941. *The Missouri soil saving dam.* Missouri Agricultural Experiment Station Bulletin 434. University of Missouri, Colombia.

Yoo, K. H., and C. E. Boyd. 1994. *Hydrology and water supply for pond aquaculture.* Chapman & Hall, New York, NY.

12

Impoundments and Embankments

Since antiquity, embankments and ponds have provided water supplies and flood control around the world. Soils engineering as a discipline did not develop until the early twentieth century (Jumikis 1984). Natural resources engineers rely on knowledge of soil properties for many field-scale projects. This chapter provides a brief introduction to field-scale impoundment and embankment engineering.

Virtually all soil-water management projects entail calculations of soil and water volume. Understanding of soil properties is crucial for locating roads and structures related to natural resources engineering. Design of ponds and embankments is crucial for surface water supply development, waste treatment and storage, and some bioremediation schemes. Slope modifications on fields for agricultural and other purposes entail earthmoving operations. Impoundments such as ponds, sediment basins, and lagoons are widely used for crop irrigation or for containment of runoff or wastes. Increasing aquacultural operations in many developing countries are creating a demand for new pond construction (Gopalakrishnan and Coche 1994). The cost of impoundments arises mainly from earthmoving expenses.

Site factors such as soils, seepage, topography, evaporation, etc.; structure objectives such as flood control, storage, fish, recreation, waste treatment, etc.; incoming water quantity, quality, and distribution; and safety aspects such as embankment, downstream flood hazards, and access by people and livestock are all involved in structure hydraulic design. Contemporary designs of safe and effective earthen impoundments and embankments bring together the disciplines of biology, soil physics, and soil mechanics. Quantifying the impact on a structure of an inflowing storm hydrograph brings together many hydrologic and hydraulic principles discussed earlier. Some of these features are apparent in Figure 12.1, which shows a small levee with inlet and outlet structures.

Goals

- Identify soil properties included in soil survey publications relevant to soil volume computations.
- Determine spillway size and flood storage capacity from outflow peak hydrographs based on stage-discharge and stage-storage relationships.
- Design agricultural ponds, waste storage structures, sediment basins, and aquacultural structures.
- Identify conditions known to cause embankment.

Figure 12.1. Levee pond under construction showing an inlet flow measurement (H flume) structure and a drop inlet structure for flow removal. The levee pond typically has no emergency spillway.

Soil Fundamentals for Embankment Construction

Soil Survey

The USDA-NRCS soil survey is an excellent starting point for determining suitable sites for proposed impoundments or embankments. The soil survey publication contains detailed maps (aerial photographs) of regions, showing codes for each soil class (called mapping units). A soil survey typically gives the USDA and other classification schemes along with pH, permeability, available water capacity (used in irrigation design), textural analyses, shrink-swell potential, Atterburg limits (plasticity index, plastic limits, and liquid limits), optimum compaction density, and optimum compaction moisture. Soil surveys give data for nearly every soil-related parameter required for general runoff and erosion analyses.

Soil surveys discuss ponds, embankments, terraces, agricultural drainage, irrigation, terraces, waterways, septic tank filter fields, lagoon impoundments, and utility as topsoil and road fill. Surveys discuss limitations for agricultural and other uses such as light industry, campsite and intensive play areas, picnic grounds, and trafficways.

Thus, the soil survey can be a significant resource in planning ponds, embankments, and other structures or land uses. One must first understand the engineering terms. State and local NRCS offices, Farm Services Administration (FSA) offices, and land grant university libraries provide soil surveys for most counties of the United States.

Soil Properties

A soil contains solids, liquids, and gaseous phases. The solids can be mineral or organic. Minerals are by far the most common component of the solid phase. Organic soils may occur in bogs or swampy areas and require special testing and handling

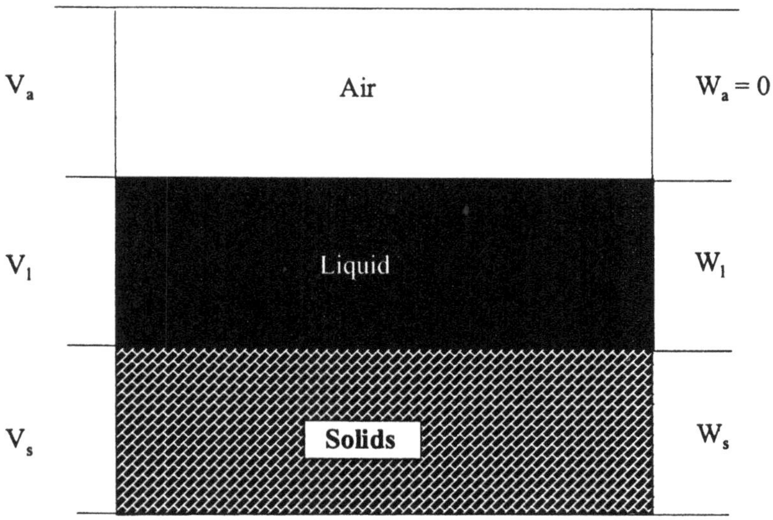

Figure 12.2. Phase diagram for soil.

for urban or agricultural construction or tillage. Consider the diagram in Figure 12.2. Let

V_a = volume of gas;
V_s = volume of solids;
V_l = volume of liquid (usually water);
V_T = total volume;
W_a = weight of gases = 0;
W_s = weight of solids;
W_l = weight of liquid;
W_T = total weight.

The weights and volumes of the phases define common parameters of interest.

Definitions of some commonly used parameters to describe soil conditions follow.

Dry bulk density is the weight of dry solids (W_s) over the total volume. This value for agricultural soils can range from about 1 Mg/m^3 (tilled soil) to greater than 1.7 Mg/m^3 (compacted sandy loam).

Particle density (G) is the weight of dry solids (W_s) over the volume of soil solids; it ranges from 2.5 to 2.7 Mg/m^3 for mineral soils and from 1.0 to 1.5 Mg/m^3 for organic muck soils. A value of 2.6 Mg/m^3 represents most soils encountered in agricultural and construction applications. Use measured particle density data when soils are from a swamp or bog.

Proctor moisture-density (B) is the optimum moisture content for compaction and the value of the bulk density when compacted at optimum moisture.

Porosity (n_p) is $(V_l + V_a)/V_T$.

Void ratio (e) is $(V_l + V_a)/V_s$.

Gravimetric water content (w_g) is the weight of water divided by the weight of dry solids.

Liquid limit (LL) is the moisture content at which a grooved sample will flow together in a specialized cup raised 1 cm and dropped repeatedly for 25 times.

Plastic limit (*PL*) is the moisture content at which 1/8-inch-long threads of soil will maintain integrity while being rolled between one's hands.

Shrinkage limit is the water content that, when decreased, will not cause further shrinkage (i.e., expansion of cracks) of the soil. Some clays have a shrinkage limit approaching zero, depending on the clay chemical constituents. Cohesive soils change from dark to light color near the shrinkage limit.

Clay activity is the $LL - PL/(\%\text{clay})$.

Plasticity index (*PI*) is the $LL - PL$.

Equations 12.1–12.3 provide some useful relationships:

$$n_p = 1 - \frac{BD}{PD} \qquad \textbf{(12.1)}$$

$$\theta_V = w_g BD / SG_w \qquad \textbf{(12.2)}$$

$$e_v = \frac{n_p}{1 - n_p} \qquad \textbf{(12.3)}$$

where

n_p = porosity;
θ_V = volumetric water content;
BD = dry bulk density (M/L^3);
PD = particle density (M/L^3), usually ranges from 2.5 to 2.7 Mg/m^3;
e_V = void ratio, volume of voids (V_a and V_l) over volume of solids.

One usually ignores the last term (SG_w) of equation 12.2, the specific gravity of water (1 Mg/m^3). Units for equation 12.2 are derived as follows:

$$m^3_{\text{water}}/m^3_{\text{dry soil}} = (Mg_{\text{water}}/Mg_{\text{dry soil}})(Mg_{\text{dry soil}}/m^3_{\text{dry soil}})/(Mg_{\text{water}}/m^3_{\text{water}})$$

Soil mechanics texts such as Jumikis 1984 provide some additional relations between phase weights and volumes.

Soil Classification

Soil classification begins with size analyses and is presented in a percent passing curve. All particles in a soil are judged passing "*x* size" if they are smaller than *x*. The percent passing curve is identical to the cumulative probability curve discussed earlier (e.g., Figure 2.7). Figure 12.3 shows percent passing curves for two soil materials.

Particle size is the sole criterion for the USDA soil classification scheme. The combinations of sand, silt, and clay sizes determine soil texture. Figure 12.4 shows the USDA textural triangle. Jumikis (1984) lists the particle size limits and shows some typical particle size curves. The clay fraction and associated ions in the liquid phase affect soil behavior. Two types of clay are the 1-1 clays (e.g., kaolinite) and 2-1 clays (bentonite). The 1-1 clays are usually nonexpansive, while the 2-1 clays are expansive.

The primary classification for soils for engineering applications is the Unified Soil Classification (USC) system. This system lumps silt and clay fractions into "fines." The USC system also incorporates liquid limit (*LL*) and plastic limit (*PL*) along with percent passing curves. The plasticity index (*PI*), *LL*, and *PL* are the *Atterburg limits*. Along with sand and fines, Atterburg limits determine the engineering properties of soils.

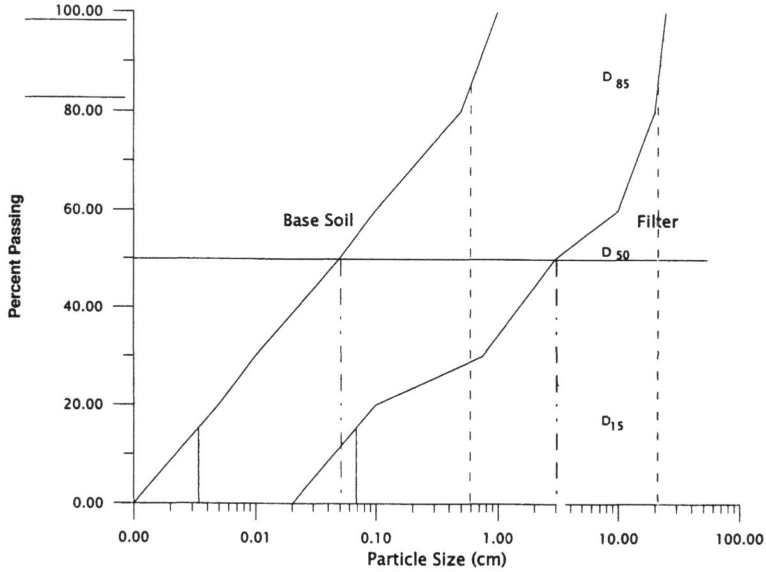

Figure 12.3. Percent passing curve for two soil materials. The D_{15}, D_{50}, and D_{85} values are significant for soil filter design.

The percent passing curve allows us to assess the gradation. Gradation is quantified using the uniformity index $U = D_{60\%}/D_{10\%}$ where $D_{60\%}$ is the diameter passing 60% of the soil and $D_{10\%}$ is the diameter passing 10% of the soil. Well-graded soils ($U < 5$) usually have a linear percent passing curve while poorly graded soils ($U > 15$) may have a curvilinear percent passing curve. Soil mechanics texts (e.g., Jumikis 1984) provide test details.

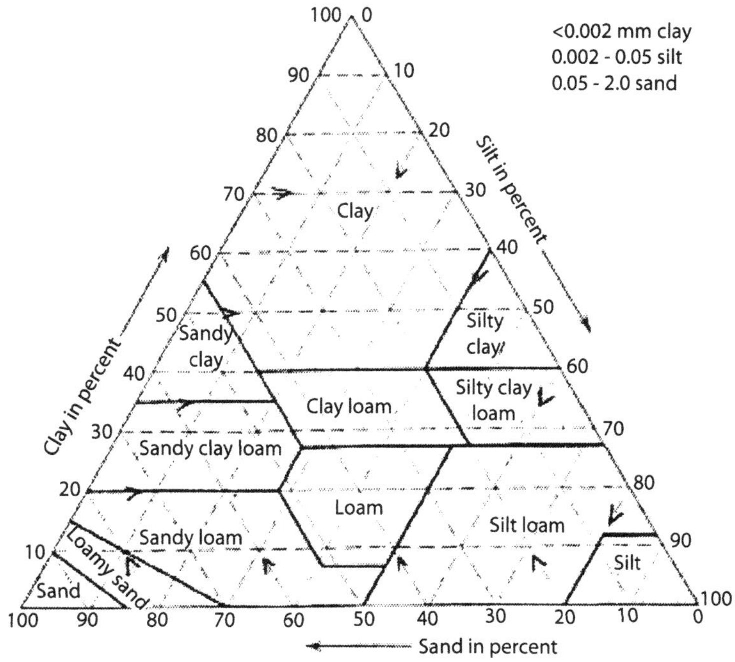

Figure 12.4. The USDA soil textural triangle and soil classification scheme.

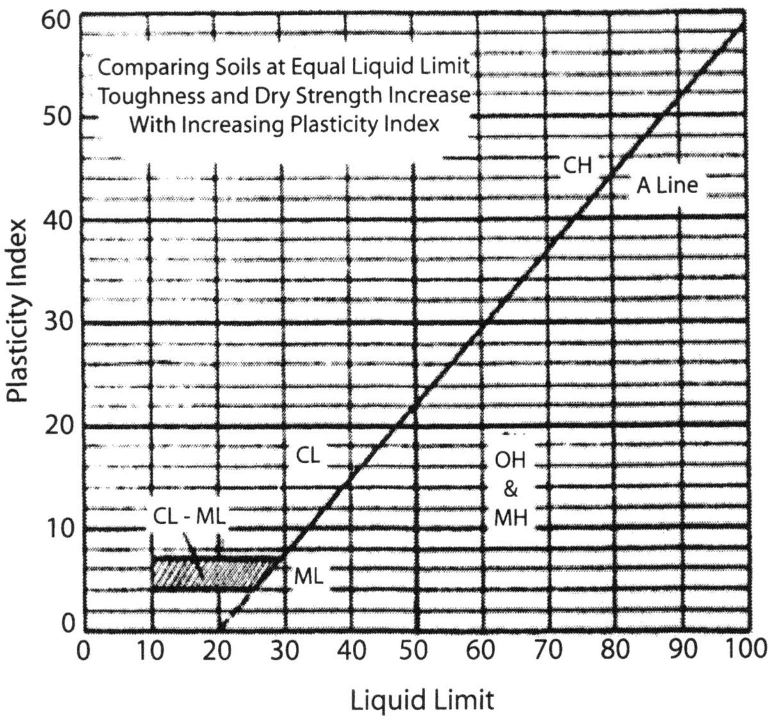

Figure 12.5. Unified Soil Classification system.

Values for Atterburg limits account for clay type and the liquid ionic composition. Soil surveys report USC data. One may infer a variety of engineering properties from the USC shown in Figure 12.5.

Knowledge of the USC for a sample enables us to determine the potential of various sites for impoundment seals, load-bearing foundations, roadways, septic tank leach fields, etc. Figure 12.6 shows the USC system correlated to various use categories. NRCS 1998 gives detailed information concerning soil classification with the USC system.

The amount and composition of the clay fraction have large influences on the soil classification because they impact Atterburg limits. The USDA classification system reflects the effect of clay percentage but not the clay activity. Hillel (1998) refers to seven clay mineral groupings. The most important are the kaolin group, common in the southeastern United States and China; and the montmorillonite group, found in many locations in the western United States, western Europe, and in "black belt" soils. Montmorillonitic soils have a large shrink-swell potential, and kaolinitic soils have a low shrink-swell potential.

Figure 12.7 summarizes other available soil classification systems. They include the ASTM (American Society for Testing Materials), MIT (Massachusetts Institute of Technology), and the International Soil Classification system.

Soil classification is the starting point for project feasibility assessments. Soil survey publications provide information on a county-by-county basis. Close cooperation with soil scientists, NRCS personnel, local architectural and engineering soil testing laboratories, and structural engineers is essential.

Major Divisions (1)	(2)	Symbol Letter (3)		Name (4)	Value as Subgrade when Not Subject to Frost Action (5)	Value as Subbase when Not Subject to Frost Action (6)	Value as Base when Not Subject to Frost Action (7)	Potential Frost Action (8)	Drainage Characteristics (9)
Coarse-grained soils	Gravel and gravelly soils	GW		Well-graded gravels or gravel-sand mixtures, little or no fines	Excellent	Excellent	Good	None to very slight	Excellent
		GP		Poorly graded gravels or gravel-sand mixtures, little or no fines	Good to excellent	Good	Fair to good	None to very slight	Excellent
		GM	D	Silty gravels, gravel-sand-silt mixtures	Good to excellent	Good	Fair to good	Slight to medium	Fair to poor
			U		Good	Fair	Poor to not suitable	Slight to medium	Poor to practically impervious
		GC		Clayey gravels, gravel-sand-clay mixtures	Good	Fair	Poor to not suitable	Slight to medium	Poor to practically impervious
	Sand and sandy soils	SW		Well-graded sands or gravelly sands, little or no fines	Good	Fair to good	Poor	None to very slight	Excellent
		SP		Poorly graded sands or gravelly sands, little or no fines	Fair to good	Fair	Poor to not suitable	None to very slight	Excellent
		SM	D	Silty sands, sand-silt mixtures	Fair to good	Fair to good	Poor	Slight to high	Fair to poor
			U		Fair	Poor to fair	Not suitable	Slight to high	Poor to practically impervious
		SC		Clayey sands, sand-clay mixtures	Poor to fair	Poor	Not suitable	Slight to high	Poor to practically impervious
Fine grained soils	Silts and clays *LL* is less than 50	ML		Inorganic silts and very fine sands, rock flour, silty or clayey fine sands or clayey silts with slight plasticity	Poor to fair	Not suitable	Not suitable	Medium to very high	Fair to poor
		CL		Inorganic clays of low to medium plasticity, gravelly clays, sandy clays, silty clays and lean clays	Poor to fair	Not suitable	Not suitable	Medium to high	Practically impervious
		OL		Organic silts and organic silt clays of low plasticity	Poor	Not suitable	Not suitable	Medium to high	Poor
	Silts and clays *LL* is greater than 50	MH		Inorganic silts, micaceous or diatomaceous fine sandy or silty soils, elastic silts	Poor	Not suitable	Not suitable	Medium to very high	Fair to poor
		CH		Inorganic clays of high plasticity, fat clays	Poor to fair	Not suitable	Not suitable	Medium	Practically impervious
		OH		Organic clays of medium to high plasticity, organic silts	Poor to very poor	Not suitable	Not suitable	Medium	Practically impervious
Highly organic soils		Pt		Peat and other highly organic soils	Not suitable	Not suitable	Not suitable	Slight	Fair to poor

Figure 12.6. Unified Soil Classification system and associated uses. (From Jumikis 1984; used with permission of Krieger Publishing Co.)

Flood Routing through Reservoirs and Related Structures

Storm routing techniques are the most general tool for pond inlet/outlet structure design. Impoundments must hold design volumes, and outlet structures must satisfy certain design flow rates. Thus far, we have assumed that flows must satisfy the peak flow of a runoff event. The structure must accommodate the storm volume as a whole or part. It must discharge the portion not retained in some period. Flood routing allows us to assess the tradeoff between storage and discharge.

Storage in impoundments can mitigate downstream peaks. The usual procedure is to propose a culvert or structure, then route the hydrograph through the structure using the compound stage-discharge curve for the structure. If the storage is not well used, the structure is oversized. If the storage is exceeded, the structure is undersized. Continue iterations until a satisfactory solution is reached. The storage-depth relation is determined using the average end area formula discussed in Appendix C.

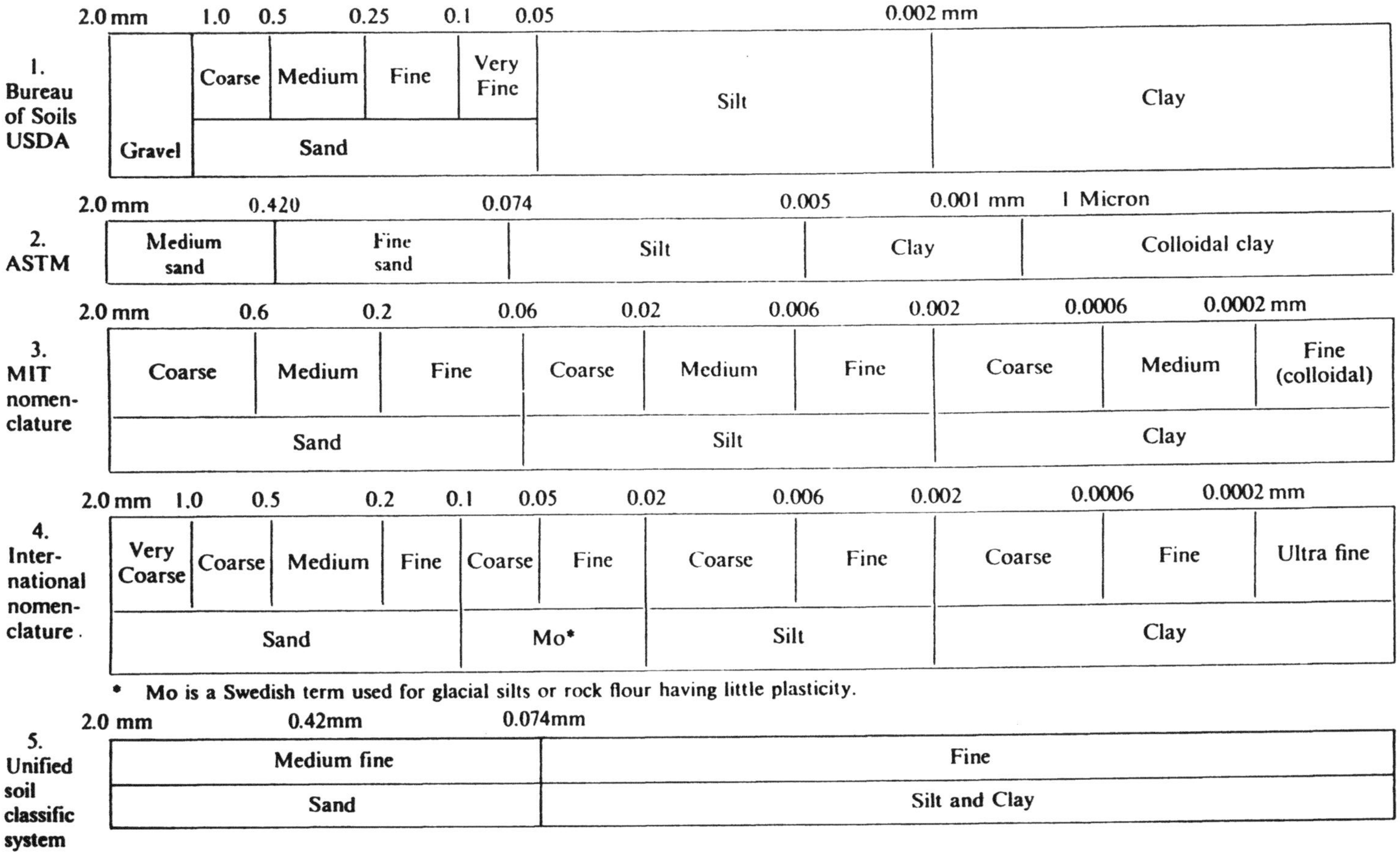

Figure 12.7. Comparison of available soil classification systems. Other schemes have replaced the gravel classification of the USDA. The American Association of State Highway and Transportation Officials (AASHTO; not shown) is similar to the systems shown except that a gravel class is added (greater than 2 mm and up to 75 mm) and silt and clay are lumped together. (From Jumikis 1984; used with permission of Krieger Publishing Co.)

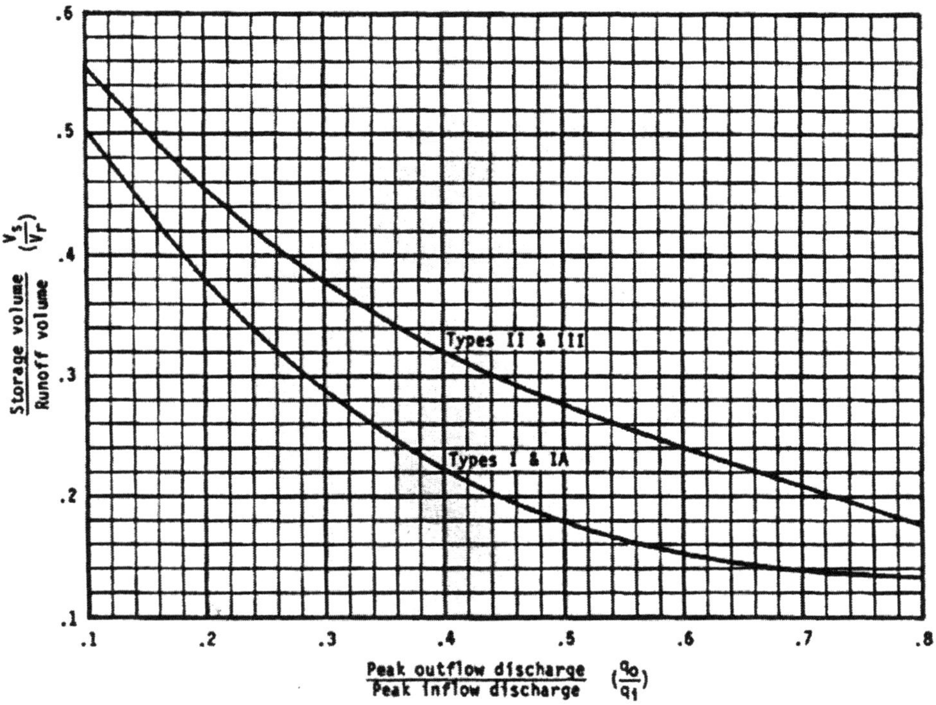

Figure 12.8. NRCS approximate detention basin routing for indicated rainfall types. (Courtesy of NRCS.)

Two approaches are presented for reservoir routing: the NRCS TR55 method and storage routing.

NRCS TR55 Routing

The NRCS developed the TR55 method as an aid for sizing principal spillways and detention ponds in urban applications. One should not use the method when errors exceed 25%. The NRCS biased the method to prevent undersizing of outflow components. The incoming hydrograph should be similar to the basic hydrograph shown in Figure 5.3. The spillway should be a basic pipe or weir. The reservoir storage and runoff volume must be expressed in equivalent units (e.g., watershed inches). A watershed inch is the volume in cubic feet (meters) of reservoir storage divided by the watershed area in consistent units. Similarly, one must express peak inflow and outflow in the same units. Figure 12.8 shows a design plot for the NRCS TR55 routing method. NRCS 1986 provides additional details.

Example 12.1: A watershed in the Type II area of 123 ac has a pond with an available storage of 4.5 ac ft corresponding to 4.5 ac ft/123 ac = 0.036 ft = 0.44 in of storage in watershed inches. The runoff volume from a 10 yr storm is 2.93 in. The peak runoff rate is 127 cfs.

The storage volume ratio is 0.44 in/2.93 in = 0.15 in. We extrapolate the outflow to inflow ratio of 0.85. Thus, the design outflow for the spillway is (0.85)(127 cfs) =

107.9 cfs. The NRCS TR55 method is approximate. Use this method only for designing "typical spillways." A rigorous design requires a storm routing analysis, which in turn requires a storage-depth relation and a stage-discharge relation.

Storage Routing

Storage routing is a generalized routing approach applicable even to stream routing (see Haan et al. 1994). Storage routing is also called hydrologic routing. Storage routing is not dependent on hydrograph shape. Storage routing is useful for any type of principal and emergency spillway configuration if one knows the stage-discharge relations. Storage routing imposes no limits on storm return period. Storage routing is defined as

$$infl\ dt = out\ dt + \Delta St \tag{12.4}$$

where

infl is the inflow (L^3/T);
out is the outflow (L^3/T);
St is the storage (L^3); sometimes expressed in mixed units such as hr-m^3/s or cfs-hr;
ΔSt indicates a change in storage (L^3/T);
dt is the differential time interval (T).

The task is to solve the differential equation in equation 12.4. One begins by discretizing equation 12.4 as shown in equation 12.5:

$$\Delta t\frac{infl_1 + infl_2}{2} = \Delta t\frac{out_1 + out_2}{2} + St_2 - St_1 \tag{12.5}$$

where the subscript 1 indicates the current time and 2 indicates the time at the next step. The solution begins at time zero. The inflow value at the next time (2) is also known. To use this equation one must isolate the unknown values at time step 2 on the left-hand side. Thus, one can write

$$\frac{St_2}{\Delta t} + \frac{out_2}{2} = \frac{infl_1 + infl_2}{2} + \left(\frac{St_1}{\Delta t} + \frac{out_1}{2}\right) - out_1 \tag{12.6}$$

One may compute the left-hand side knowing the value of all terms on the right-hand side. Given the inflow hydrograph, one may step through the solution of equation 12.6. Some preparatory calculations are necessary to carry out the storage routing procedure. Take inflow values from the inflow hydrograph.

A routing term may be defined using

$$routing\ term = \frac{St}{\Delta t} + \frac{out}{2} \tag{12.7}$$

One can precalculate the routing term as a function of storage and stage. Table 12.1 shows the analysis steps for computing the routing term. Example 12.2 shows the procedure.

The outflow-stage relationship is developed by calculating outflow versus stage through the principal and emergency spillways (if applicable). Appendix C discusses storage-stage relationship development. One may also use the Beasley et al. (1984) storage-stage formula for ponds (see the discussion on terrace underground outlets in Chapter 10 and see equation 10.9, where B_h is depth above the pond bottom).

Table 12.1. Storage routing curve calculation template.

Stage (m) (Col. A)	Outflow or spillway discharge (m^3/s) (Col. B)	Storage *St* (m^3) (Col. C)	$St/\Delta t$ (m^3/s) (Col. D)	*out*/2 (m^3/s) (Col. E)	$St/\Delta t$ + *out*/2 (m^3/s) (Col. F)
0	0	0	0	0	0
. . . (Select stage at several intervals)	(from compound stage-discharge curve)	(from average end area or prismoidal formula calculations; see Appendix C)	(computed by dividing storage in col. C by Δt)	(computed by taking half of the outflow in col B)	(sum of col. D and col. E)
. . .					
Max					

Note: Develop a plot of col. F vs. col. B for subsequent use. Choose Δt so that three to five time increments occur prior to the inflow hydrograph peak time.

Example 12.2: Prepare the routing preliminary computations from the storm inflow hydrograph shown in Figure 12.9. The structure of interest is the reservoir with the outlet structure shown in Example 11.11, case A. We define the reservoir stage-storage relationship for this example by Storage = $H_{inlet} \times 200 + 500 \times H_{inlet}^2 + 100 \times H_{inlet}^3$ (We discuss development of stage-storage relationships in later sections.)

Pick a time step of 0.2 hr since this will allow several computations before reaching the time to peak. We also know the inflow hydrograph at this interval.[1]

Develop a plot of column F versus column B of Table 12.1 (routing parameter vs. outflow). Plotting the stage-discharge and the stage-volume relationships is helpful. One may also plot the outflow (y axis) versus the storage volume on log-log paper (not shown). Spreadsheets or equation-processing software packages are useful for making these computations.

One can then arrange flood routing computations as shown in Table 12.2. Results for Example 12.2 are shown in Tables 12.3 and 12.4. They are plotted in Figure 12.10.

Figure 12.10 indicates that reservoir storage contained nearly all the storm volume without causing a significant rise. The emergency spillway never flowed in this storm event. Reducing the storage-depth relation or increasing the pipe diameter would have increased the outflow peak.

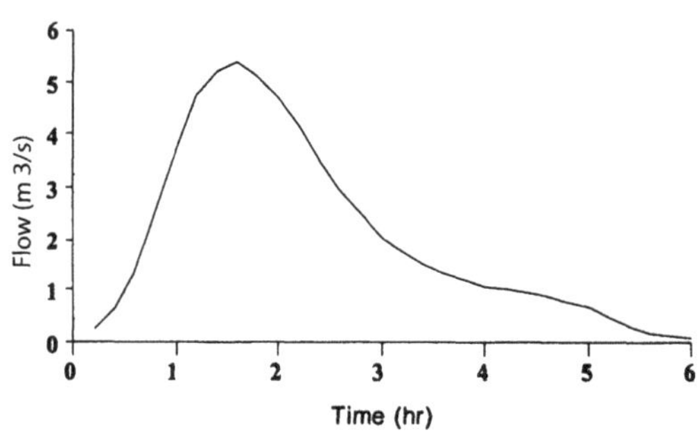

Figure 12.9. Inflow hydrograph for Example 12.2.

Table 12.2. Hydrologic flood routing computations template.

Label (Col A)	Time $\Delta t = XX$ (hr) (Col B)	*Inflow* $infl_i$ (m^3/s) (Col C)	$(infl_1 + infl_2)/2$ Each row is i and the prior row is $i - 1$ (m^3/s) (Col D)	$(St/\Delta t) + (out/2)$ Each row is i and the prior row is $i - 1$ (m^3/s) (Col E)	*Outflow out* Each row is i and the prior row is $i - 1$ (m^3/s) (Col F)
$i = 0$	0	0	0	0	0
$i = i + 1 = 1$	Δt (select Δt so the incoming hydrograph peak is at least 5 increments into the storm)	(from inflow hydrograph)	{Col C(i) + Col (C($i-1$))}/2	(compute from routing equation; Col D(i) + Col E($i-1$) − Col F($i-1$))	(Enter Col E (i) value into the routing parameter vs outflow plot made from Table 12.1. Record the outflow discharge here. Now index to next i and repeat.)
$i = i + 1 = 2$	$2\Delta t$				
$i = i + 1 = 3$	$3\Delta t$				
...	...	...	...	...	...
$i = N$	Max. time (end of inflow hydrograph)	...	...	...	...

Table 12.3. Results of storage function and routing parameter for Example 12.2.

Element	Hinlet	qtot	STORS	OUTFLPAR
1	5	0	26000	36.11111111
2	5.25	.188496	29301.5625	40.79086258
3	5.5	.5331471993	32862.5	45.90893471
4	5.75	.7404246114	36692.1875	51.33158383
5	6	.8549686974	40800	57.09415102
6	6.25	.9558840631	45195.3125	63.24920939
7	6.5	1.047118527	49887.5	612.81175371
8	6.75	1.131017276	54885.9375	76.79597739
9	7	1.209108327	60200	84.21566527
10	7.25	1.282453046	658312.0625	92.08436888
11	7.5	1.351824206	71812.5	100.4154954
12	7.75	1.417805188	781212.6875	1012.2223575
13	8	1.480849223	84800	118.5182024
14	8.25	5.541316739	91832.8125	130.3162313
15	8.5	12.91320847	99237.5	144.2864653
16	8.75	22.44024945	107023.4375	1512.8637879
17	9	33.70993739	115200	176.8549687
18	12.25	46.48392267	123776.5625	195.1538537
19	12.5	60.60141632	132762.5	214.6930693
20	12.75	75.94439779	142167.1875	235.426626
21	10	92.42143612	152000	257.3218292

Table 12.4. Outflow computations for Example 12.2.

Element	TIME	INFLOW	AVG INFLOW	ROUTING PA	OUTFLOW
1	.2	.28	.14	.14	.0001938462
2	.4	.62	.45	.5898061538	.0008166547
3	.6	1.36	.99	1.578989499	.0021862932
4	.8	2.55	1.955	3.531803206	.0048901891
5	1	3.74	3.145	6.671913017	.0092380334
6	1.2	4.76	4.25	10.91267498	.0151098577
7	1.4	5.21	4.985	15.88256513	.021991244
8	1.6	5.38	5.295	21.15557388	.0292923331
9	1.8	5.15	5.265	26.39128155	.0365417745
10	2	4.67	4.91	31.26473977	.0432896397
11	2.2	4.11	4.39	35.61145013	.0493081617
12	2.4	3.54	3.825	312.38714197	.1469532619
13	2.6	2.97	3.255	42.49518871	.3032653954
14	2.8	2.46	2.715	44.90692332	.4656717084
15	3	2.01	2.235	46.67625161	.5624774136
16	3.2	1.78	1.895	48.00877419	.6134122661
17	3.4	1.5	1.64	412.03536193	.6526529479
18	3.6	1.3	1.4	412.78270898	.6812198272
19	3.8	1.2	1.25	50.35148915	.7029610978
20	4	1.1	1.15	50.79852805	.7200488837
21	4.2	1	1.05	51.12847917	.7326610612
22	4.6	.9	.95	51.34581811	.7407075499
23	4.8	.8	.85	51.45511056	.7428799851
24	5	.7	.75	51.46223057	.7430215116
25	5.2	.5	.6	51.31920906	.7399515935
26	5.4	.3	.4	50.97925747	.7269571539
27	5.6	.2	.25	50.50230032	.7087257614
28	5.8	.15	.175	412.96857455	.6883244239
29	6	.1	.125	412.40525013	.6667916959

It is intriguing to note the intersection of the inflow hydrograph with the maximum point of the outflow hydrograph. Storage is highest at maximum depth and therefore maximum outflow. Before the maximum outflow occurs, storage is increasing over time. After maximum outflow, storage is decreasing. Thus at maximum outflow the derivative of storage with respect to time is zero. From equation 12.4, inflow must equal outflow when ΔSt equals zero.

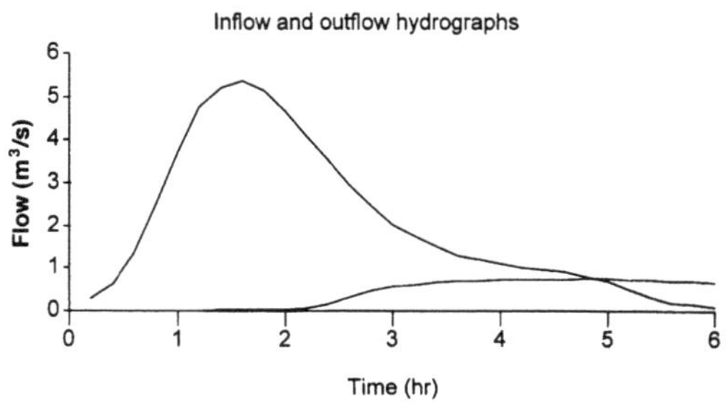

Figure 12.10. Inflow and outflow hydrographs for the pond of Example 11.11, case A, with the output hydrograph of Example 12.2.

A simulation package such as STELLA is helpful for doing reservoir routing if one knows the depth-storage function and the depth outflow function. The NRCS has excellent software for routing floods through structures (NRCS 1983). The SedCad7 software described by Warner (1998) also does reservoir routing. One can identify excellent software for routing with an Internet search. The simple storage routing approach assumes the water in the reservoir and the incoming water to be massless and have no momentum. Momentum can be important with long, narrow reservoirs. More advanced texts cover approaches accounting for dynamic effects.

General Pond Design

Pond design builds upon basic soil mechanics and hydraulic topics discussed above. Ponds may be levee, excavated, hillside, or watershed styles as shown in Figure 12.11. Ponds store water or provide water for site-specific uses. Pond design begins with obtaining or preparing a topographic map of the proposed site.

Pond Siting

Mapping the site is followed by a soil survey of the region to evaluate the need for sealing with natural or artificial sealants (e.g., bentonite, plastic liner). Soil limitation ratings

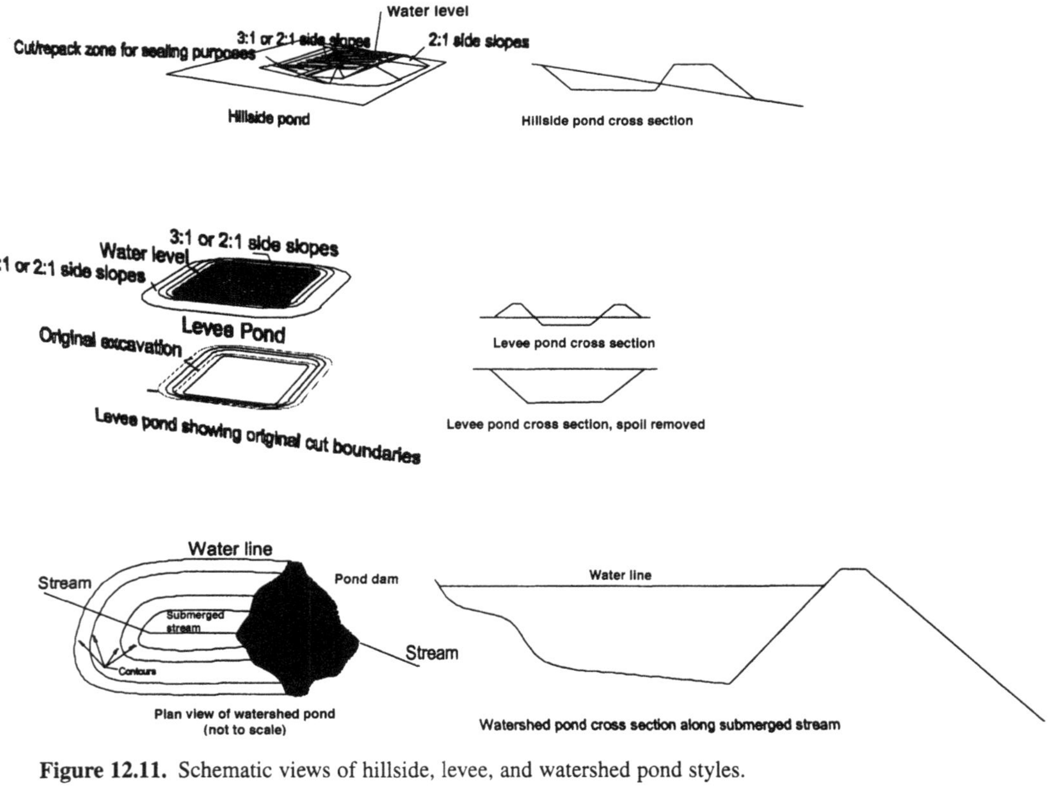

Figure 12.11. Schematic views of hillside, levee, and watershed pond styles.

Table 12.5. Selected soil limitations for ponds with significant excavation.

	Limitation rating			
Property	Slight	Moderate	Severe	Restrictive feature
Depth to sulfidic or sulfuric layer (ft)	> 3	1.5–3	< 1.5	Potential acidity or toxicity (e.g., aquaculture)
Thickness of organic soil (ft)	< 1.5	1.5–2.5	> 2.5	Seepage
Lime requirement (tons/acre)	< 2	2–10	> 10	Mineral acidity (e.g., aquaculture)
pH of 1.5 to 2 ft layer of pond bottom	> 5.5	4.5–5.5	< 4.5	Too acid (e.g., aquaculture)
Clay content (%)	> 35	18–35	< 18	Excessive seepage
Terrain slope (%)	< 2	2–5	> 5	Slope
Depth to water table (ft)	> 2.5	1–2.5	< 2.5	Hard to drain (e.g., aquaculture)
Frequency of flooding[a]	None	Rare/ occasional	Normal/ periodic	Flooding
Small stones (%)	< 50	50–75	> 75	Small stones
Large stones (%)	< 25	25–50	> 50	Large stones
Organic matter (%)				
Low clay content soil (<60% clay)	< 4	4–12	> 12	Excessive humus
High clay content soil (>60% clay)	< 8	8–18	> 18	Reducing environment
Depth to rock (ft)	> 5	3–5	< 3	Shallow, seepage

Source: Adapted from Hajek and Boyd 1990.

[a] None = no possibility; rare = up to 5 times per 100 years; occasional = 5–50 times per 100 years; frequent = more than 50 times per 100 years.

for aquaculture ponds are suggested in Tables 12.5 and 12.6 (Hajek and Boyd 1990). These limits apply to most ponds. After site selection, one then proposes an embankment height enabling adequate water storage while not submerging critical areas. Various embankment heights must be evaluated until the desired working storage volume plus flood storage is obtained. The design also includes principal and emergency spillways and inlet structures. Embankment height in this discussion is for practical purposes less than 3 m high. One should engage a soils engineer for larger embankments.

Because of the historic importance of these impoundments to agriculture, we will give an extended discussion of watershed pond design. Principles are easily extended to levees, excavated ponds, and hillside ponds. Coche and Muir (1992) and Yoo and Boyd (1994) discuss pond design for aquiculture. Figure 12.11 shows a generic watershed pond. The design process begins with a topographic map. Select the basin with the potential to yield the maximum water storage for minimum embankment volume.[2] Choice locations for ponds are naturally occurring constrictions along drainage paths enabling maximal water storage, all other factors being equal. Be sure to abide by all regulations dealing with wetland preservation.

Using a relevant cross section, sketch the embankment on the plan view. One normally specifies embankment slopes with $z = 3$ upstream and $z = 2$ downstream. The choice of $z = 3$ upstream is usually sufficient to avoid sloughing of small side slopes (e.g., 3 m or

Table 12.6. Selected soil limitations for levees and embankments.

Property	Limitation rating Slight	Moderate	Severe	Restrictive feature
Average clay content at depth greater than 3.3 ft	> 35 Clayey	18–35 Loamy	< 18	Too sandy
Depth to sulfidic or sulfuric layer (ft)	> 3	1.5–3	< 1.5	Potential acidity or toxicity (e.g., aquaculture)
Lime requirement (tons/acre)	< 2	2–10	> 10	Mineral acidity (e.g., aquaculture)
pH of 1.5 to 2 ft layer of pond bottom	> 5.5	4.5–5.5	< 4.5	Too acid (e.g., aquaculture)
Engineering classes	—	—	All G and S classes	Too stony
Slope (%)	< 8	8–15	> 15	Slope
Thickness of organic soil (in)	< 6	6–18	< 18	Subsides, hard to compact
Depth to water table (ft)	> 3	1.5–3	< 1.5	Wetness
Fraction >3 in diameter	< 25	25–50	> 50	Large stones
Depth to bedrock	> 3	1.5–3	< 1.5	Depth to rock
Shrink-swell	Low	Medium to high	Very high	Shrink-swell cracking
Erodibility *K* (English)	< 0.1	0.1–0.3	> 0.3	Erosion

Source: Adapted from Hajek and Boyd 1990.

less in height on fast drawdown), and the choice of $z = 2$ downstream results in a stable backslope for small embankments. Deviations from these z values are allowable only if alternative z values satisfy a bank stability analysis (beyond the scope of this text). Locate the embankment cross section at each 30 m station along the top of the embankment. One may estimate embankment top width from

$$W_{TW} = 0.4H_{embmx} + 1 \tag{12.8}$$

where

W_{TW} is the top width (m);
H_{embmx} is the maximum embankment height (m).

On the site map, tentatively locate the embankment. Develop a relation between volume of pond and depth using the volume formulas in Appendix C. Find the required storage volume based on design needs, and determine the elevation of the principal spillway satisfying the design requirement.

Other matters related to siting and construction are important for success or failure. Keep livestock out of the pond and off the embankment. Keep trees off the embankment. Keep burrowing animals (e.g., muskrats) in check. Use antiseep collars, properly installing them with appropriate compaction. Scarify the pond basin to a 0.25 m depth and compact at optimal moisture. Try to identify zones of permeable material and avoid or seal. Beware of ledge rock; ledge rock is troublesome with respect to seal development. When placing sealing materials, use thin layers and compact at Proctor moisture conditions.

Pond and Embankment Seepage Prediction and Control

Seepage reduction through the embankment and pond bottom is important. In practical terms, an adequate seal is one with seepage less than or equal to 0.125 in/day (3.67×10^{-6} cm/s), the federally mandated value. Seepage reduction may require measures ranging from compaction to application of bentonite clay or chemical additives. One must do a soil analysis to investigate the suitability of soil for sealing. The analyses include Atterburg limits, Proctor density, and hydraulic conductivity on the soil compacted to Proctor density. *Soils that are well graded, do not have over 20% gravel, have 20–50% sand, do not have over 30% silt, and have 15–25% clay are good candidates for sealing with compaction to Proctor density* (Schwab et al. 1993). Also, refer to Tables 12.5–6. Organic matter should be negligible. The compacted soil should have a K_{och} value of 10^{-7} cm/s or smaller if possible. Soils outside the above limits may have excess seepage losses due to consolidation (coarse sands and gravels, fine uniform sands below critical density) or fail in shear (plastic clays). One must correct the actual measured pond level decline to account for evaporation when evaluating seepage. A class A evaporation pan with an appropriate pan coefficient and a computed (e.g., Blaney-Criddle; consult a climatologist) evaporation using data from a nearby meteorological station are two possible approaches. Climatologists prefer the computed evaporation due to difficulties in estimating site-specific pan coefficients.

The seepage through a clay liner is shown by

$$v = K_c \frac{y_{lag} + d}{d} \tag{12.9}$$

where

K_c is hydraulic conductivity (L/T);
y_{lag} is the depth of water in the structure (L);
d is the liner thickness (L);
v is the specific seepage rate (L/T) per unit bottom area.

One may use equation 12.9 to determine liner thickness or adequacy with known head or hydraulic conductivity. Details showing how equation 12.9 is a straightforward application of Darcy's law are left to the student.

Failure to find soils with adequate sealing potential may require the importation of soils or the use of sealants such as Wyoming bentonite. Dispersants such as soda ash, tetra sodium pyrophosphate (TSPP), and sodium tetra phosphate (STP) are used as sealing aids with flocculated clays often occurring in limestone or high loess soils. Soil cements and asphalts may also reduce seepage to acceptable levels. Biological based sealing mechanisms passively seal many waste impoundments. Most regulatory agencies desire more demonstrably active control measures. Scenarios B and C in Figure 12.12 are applicable when suitable soils are not available. A plastic liner (ASAE 1997c) is also a possibility. Soils in regions of high gypsum are troublesome. Gypsum may dissolve and cause piping. Soils with permeable strata require clay or plastic liners. Soils containing ledge rock are sealed with clay or plastic liners. Leakage may occur via ledge rock fractures. One must provide adequate cover above and below linings to avoid puncture failure. Clay and plastic liner treatments may cost $100,000 or more per hectare (Reed et al. 1995). If soil with sealing potential is available within a mile of the site, the

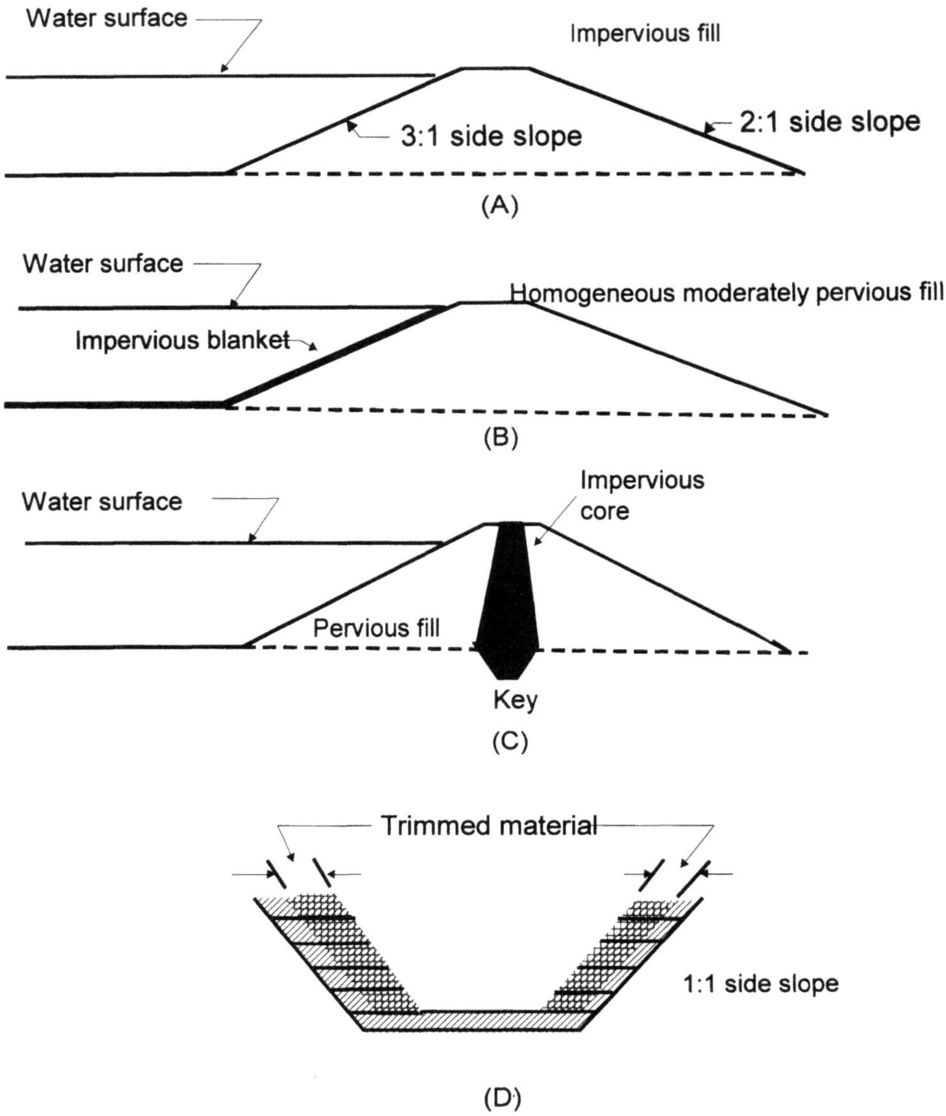

Figure 12.12. Approaches for sealing earthen structures: (A) largely impervious fill; (B) pervious fill and blanket seal; (C) impervious core with pervious fill; and (D) clay liner installed in lagoon with steep sidewalls where seal was stair-stepped in lifts one machine width up the wall initially, then trimmed, with cuttings becoming part of the bottom seal.

usual practice is to compact the pond basin and transport the suitable soil for building a core seal.

Soil seals can be 90% adequate and leak at the 90% rate of an unsealed soil. This leads to the question: How does one ensure that the structure is 100% sealed without excessive soil testing? Construction engineers ensure an adequate soil seal by almost continuous observations at the job site. Contractors who have well-maintained and reliable equipment and who operate in a methodical manner are the most successful. Soil lifts should not exceed 30 cm in thickness before compaction. Successful contractors make moisture checks before each water application and ensure uniform water applications. One then compacts in a timely manner. Smooth rollers are used in sands and sheepsfoot rollers are used in more clayey soils. Well-managed compaction programs coupled with adequate soil sampling ensure adequate clay liner development. Similar and appropriate workmanlike approaches also ensure the successful application of plastic liners.

Soil structure modification and sealing are well-developed disciplines. Renfro (1968), Cedergren (1977), and Emerson et al. (1978) each present various physical, chemical, and installed-liner approaches to soil sealing. NRCS 1997 and 1998 give further detailed information and example problems related to sealing impoundments with clay liners and plastic liners.

Imported clay liners must satisfy size requirements with respect to underlying material to maintain liner integrity. Similarly, rock toe filters must satisfy size requirements in relation to the embankment fill to maintain integrity of the embankment. The rock toe filter (example shown in Figure 12.13b) protects the embankment from effects of residual seepage. Seepage concentrates near the bottom of the embankment on the downstream side of an impoundment. The graded sand and gravel filter must meet certain size requirements, given in equations 12.10a–c, to reduce soil movement under the influence of hydraulic gradients:

$$\frac{D_{15}(filter)}{D_{15}(base)} \geq 5 \tag{12.10a}$$

$$\frac{D_{15}(filter)}{D_{85}(base)} \leq 5 \tag{12.10b}$$

$$\frac{D_{50}(filter)}{D_{50}(base)} \leq 25 \tag{12.10c}$$

where D_{xx} refers to the particle diameter at xx percent passing.

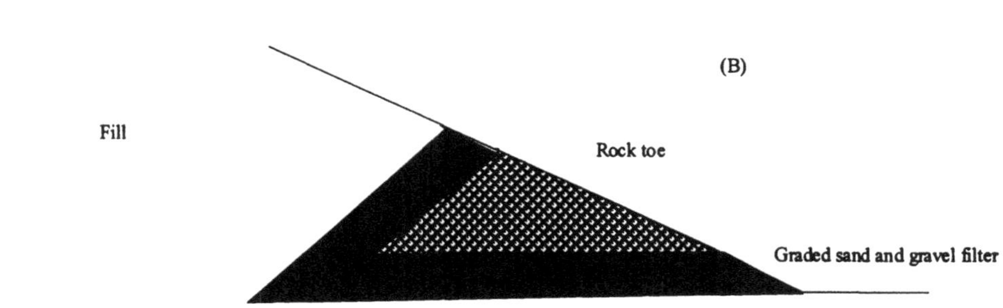

Figure 12.13. (a) Nomenclature for the approximate seepage relation and (b) detail of a rock toe filter arrangement recommended for the downstream side of a pond embankment.

The filter material and base material must satisfy equation 12.10a provided that the filter does not contain more than 5% of particles finer than 0.074 mm and provided that the particles have a maximum size of 76 mm.

The percent passing curve for the filter should be roughly parallel to that for the base soil. The ratio should not exceed 25, as shown in equation 12.10c.

For pipe drains, the D_{85} of the filter should exceed 2 times the openings in the drain pipe. The particle size should not exceed 75 mm. The D_{15} and D_{85} are the particle diameters at which 15% and 85% are smaller on a mass basis. USBR 1973 and NRCS 1994 further discuss filter criteria. NRCS 1994 presents detailed filter design examples. Refer to the percent passing plots for the base soil and filter media to verify filter media suitability.

Example 12.3: Does the filter shown on the percent passing example in Figure 12.3 pass the equation 12.10 criteria? The D_{15} ratio is 0.07 mm/0.025 mm = 0.28, which is acceptable. The D_{15}/D_{85} ratio is 0.07 mm/0.5 mm = 0.14, which is acceptable. The D_{50}/D_{50} ratio is 2 mm/0.05 mm = 60, slightly high. The D_{85} of the filter is about 11 mm, larger than the 2 mm openings usually found in pipe drain line, and thus is acceptable. The example roughly meets the maximum particle size of 76 mm. The filter has excessive fine material. The filter should consist of somewhat larger material in the smaller size zone. The fraction of larger particles is acceptable. Because criteria are (nearly) satisfied, this filter is accepted.

Embankment Profile, Protection, and Volume

One may estimate slope widths using the slope stakes relationship:

$$d_{ss} = \Delta El\, Z + \frac{w}{2} \qquad \textbf{(12.11)}$$

where

d_{ss} is the distance from centerline to edge of fill or cut (L);
ΔEl is the difference in elevation from the stake to the bottom of the ditch or the top of the fill (L);
Z is the side slope ratio;
w is the top width of the embankment (W_{TW}).

See Appendix C for additional discussion on setting slope stakes. The nomenclature for equation 12.11 is shown in Figure 12.14.

Side slope z values for ponds are usually 3 for the submerged embankment side and 2 for the back of the embankment. The z value of the back side of a high-clay embankment should be 3. One should perform soil stability analyses on large structures to ensure embankment stability. Soil stability analysis is beyond the scope of this text.

Construct an embankment profile along the centerline as well. The profile should extend into the bank as far as the emergency spillway. The emergency spillway bottom should be a minimum of 0.3 m above the top of the principal outlet. The emergency spillway

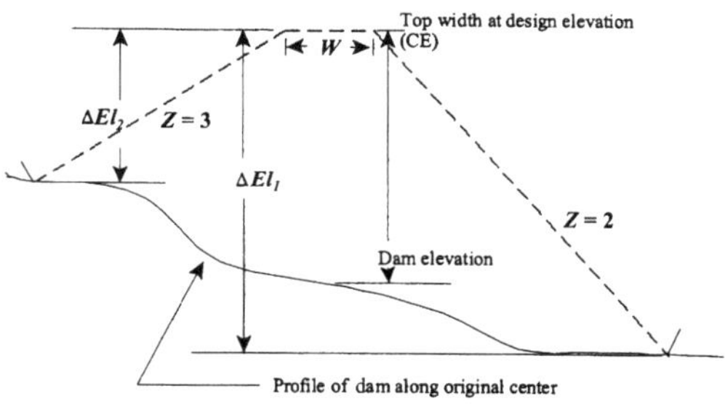

Figure 12.14. Applications of slope stakes equation in the case of a fill over an arbitrary bottom, showing the finished embankment with a dotted line.

may be higher above the principal outlet if one needs more flood storage for peak flow reduction downstream. Embankment height (settled) is flood storage plus maximum allowed depth in emergency spillway plus wave height and frost depth. Constructed height is 1.1 × (design height − elevation at embankment base on profile). Figures 12.15 and 12.16 show details relating to principal spillway installation.

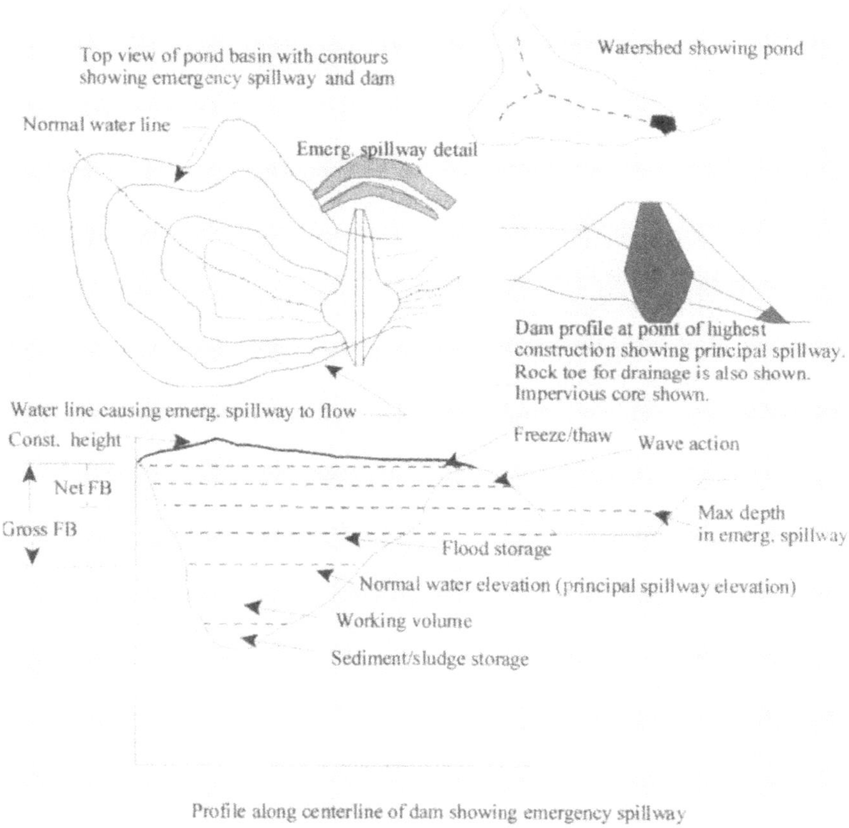

Figure 12.15. Plan view and embankment cross section for a watershed pond impoundment with key water elevations shown. (Redrawn from Schwab et al. 1993.) Other impoundments for specialized purposes may not contain one or more of the following: flood storage, wave action allowances, an emergency spillway. Embankment height is determined by the liquid volume storage/depth requirements shown.

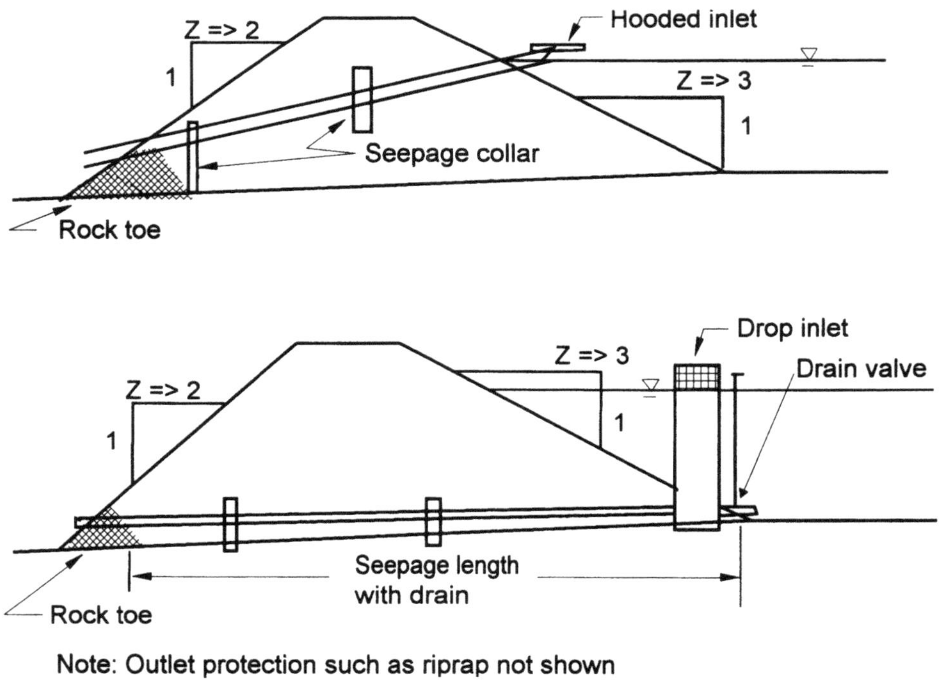

Figure 12.16. Typical principal spillway types showing seepage collars. Other spillway types, such as the siphon, flumes, and weirs, may also be used as principal spillways.

One should specify seepage collars such that seepage length is 10–15% larger than the length of the spillway conduit. Seepage collars are critical in soils where piping failures are known to occur. Piping failure occurs in soils with low cohesion and clay particle sizes. Soils engineers coined the term *piping* to describe water conduits formed in saturated soils. Piping failures become catastrophic if not stopped quickly. Periodic inspections are essential. Piping failures may develop near installed conduits since compaction is more difficult near the conduit. Highly weathered kaolinitic clays such as found in the southeast piedmont are strong candidates for piping failure. Emerson et al. (1978) discuss piping failure in detail.

One may estimate the number of seepage collars using equation 12.12:

$$N_{collar} = \frac{C_{collar} L_p}{L_c} \tag{12.12}$$

where

C_{collar} is a coefficient (0.1–0.15);
L_p is the pipe length that is saturated;
L_c is the length of the flow path around the seepage collar or twice the radius (less the pipe radius) of the collar;
N_{collar} is the number of seepage collars required.

Seepage collars should not be more than 25 ft (7.62 m) apart. Figure 12.17 shows seepage collars on the conduit of the principal spillway for a small pond. Proper compaction management around the conduit is critical to control piping failure.

Figure 12.17. Disks serving as seepage collars on the principal spillway conduit for a small pond.

Compute the embankment volume using the average end area formula discussed in Appendix C (see equations C.1a–e). Compute the volume of solids of compacted soil. This volume of suitable soil solids must exist on the site or be transported from another location.

Example 12.4: A soil volume of 144.45 m^3 having a void ratio of 1.1 was found in a valley proposed for pond development. The soil would be suitable for use in a pond embankment if it were compacted to a void ratio of 0.6. How much volume would this soil occupy in the pond embankment?

Soil solids volume in the valley must equal the soil solids volume in the compacted structure. First, compute the volume of solids:

$$V_{T1} = 144.45 \text{ m}^3 = V_s + V_v = V_s + V_s e_1 = V_s + 1.1V_s = 2.1V_s$$

This yields a $V_s = 68.79 \text{ m}^3$. The final volume is given by

$$V_{T2} = 68.79 \text{ m}^3 + e_2(68.79) = 68.79 \text{ m}^3 + 0.6(68.79 \text{ m}^3) = 110.06 \text{ m}^3$$

Suppose we hauled the soil from the borrow site to the embankment site in a 2 m^3 hauler. Further, the material had a dry bulk density of 1.1 Mg/m^3 while in the hauler. How many trips are required to move the soil?

$$\text{porosity } n_p = 1 - BD/PD = 1 - (1.1 \text{ Mg/m}^3)/(2.65 \text{ Mg/m}^3) = 0.58$$
$$\text{void ratio} = n_p/(1 - n_p) = 0.58/(1 - 0.58) = 1.39$$

The haul volume is then $68.79 \text{ m}^3 + 1.39(68.79 \text{ m}^3) = V = 164.4 \text{ m}^3$. The number of trips required is $164.4 \text{ m}^3/(2 \text{ m}^3/\text{trip}) = 83$ trips

Spillway Design

Principal Spillway Design with Flood Storage as the Primary Objective

Spillways are important in watershed ponds and hillside ponds fed by surface runoff. Principal spillway design is concerned with setting the elevation of the spillway to allow the desired working volume. One then sizes the structure using tables or by making flow estimates based on various routing procedures. Principal spillway design requires knowledge of incoming peak flow or hydrograph, available storage, treatment objective (if any), allowable or desired peak flow below the structure, and/or constraints on pipe and weir sizes. The Beasley et al. (1984) routing method provides preliminary estimates. One may also use the storage method or NRCS TR55 approximate method. Begin the spillway design process by selecting the spillway type. The structure selection guide (Figure 11.8) aids the selection process. The storm frequency-duration for runoff storage volumes is usually 2 yr–6 hr (average infiltration rate) or 2 yr–24 hr (slow infiltration rate) for rural reservoirs or ponds. The storage volume is much larger for waste treatment facilities. Agency regulations specify storage volume of waste impoundments. Calculate peak flow rates from the rational equation $q_p = C_{rat}iA$ ($q_p = 0.0028C_{rat}iA$, metric) or the NRCS TR55 method. The NRCS method applies if the watershed area exceeds 5 ac. The flow q_p is here the incoming flow and designated as q_{ip}. Use duration equal to watershed T_c for peak calculations. Compute runoff volume from the NRCS curve number method (V_{rp}). Calculate the detention storage (V_{sp}) between principal and point of incipient flow in the emergency spillway.

Principal spillway elevation is set to achieve the desired working volume. Several approaches are available for sizing the principal spillway. One should start the iterative analysis with a 250 mm diameter. Compute head loss using methods similar to the culvert pipe flow condition, modifying the pipe size as needed until achieving convergence. Beasley et al. (1984) present charts and relations for morning glory, canopy, hood inlet, and drop inlet structures. See Beasley et al. 1984 for discussion of pipe thickness, quality, and related issues.

The NRCS TR55 (see Figure 12.8) and storage routing methods could also be used to estimate the required maximum flow through the principal spillway. The NRCS SITES computer software (NRCS 1996) enables a comprehensive analysis, using storm routing. This software is downloadable from http://www.usda.nrcs.gov. Then, size the pipe to convey the flow when the inlet flow level is at the point of incipient flow in the emergency spillway.

Emergency Spillway Design with Flood Storage as a Primary Objective

The emergency spillway is often a short grassed waterway with a broad-crested weir control section. Emergency spillway elevation depends on needed flood storage.

A suggested storage based on experience in the midwestern United States (25 yr storm, 200 mm spillway diameter) may be estimated using equation 12.13. This equation was developed from NRCS data given in Schwab et al. 1993 and represents ($R^2 = 0.96$)

suggested flood storage for ponds with a surface area of less than 1.25 ha and with watersheds yielding less than 4.2 m^3/s peak runoff (Schwab et al. 1993). Equation 12.13 assumes a typical Corn Belt pond with a mechanical spillway pipe of 200 mm diameter.

$$V_{sto} = 0.0306 + 0.478q_p - 0.0477q_p^2 - 1.192S_a - 0.1546q_pS_a + 0.7013S_a^2 \quad \textbf{(12.13)}$$

where

V_{sto} is storage depth (m);
S_a is water surface area (ha);
q_p is design peak runoff into the pond (m^3/s).

Use 0.3 m minimum for flood storage depth.

In other words, V_{sto} gives guidance on how high to place the emergency spillway bottom from the principal spillway crest. Sometimes regulations legislate storage. Perform a volume calculation for storage volume knowing the depth of the storage zone. Discharge values exceeding 4 m^3/s or surface areas exceeding 1.25 ha entail the use of other methods. One should investigate the site for reasonable storage levels and then route the incoming hydrograph for spillway design. One may compute empirical values of flood storage from Figure 12.18, developed based on contractor experience.

One may also estimate required surface area generating a required storage volume from an NRCS chart such as shown in Figure 12.19. After determining storage, select principal and emergency spillway designs accommodating incoming storms. Hydrological routing would be the appropriate test for finding workable combinations. Both approaches are appropriate when pond storage is the primary design goal and downstream flow rates are secondary.

Emergency spillway size depends on pond storage, watershed size, watershed runoff potential, and design storm. Secondary design factors may include frequency of allowed use. Difficulties of excavation or limited space may sometimes entail the use of more expensive structures for emergency spillways. Emergency spillways are designed using steady-flow hydraulics at the peak flow rate. Determine design return period for the structure as a whole: typically 25 yr for small farm ponds; up to 100 yr where embankment breach may endanger life and property. Duration for runoff volume determinations is 6 hr for average infiltration or 24 hr for slow soils.

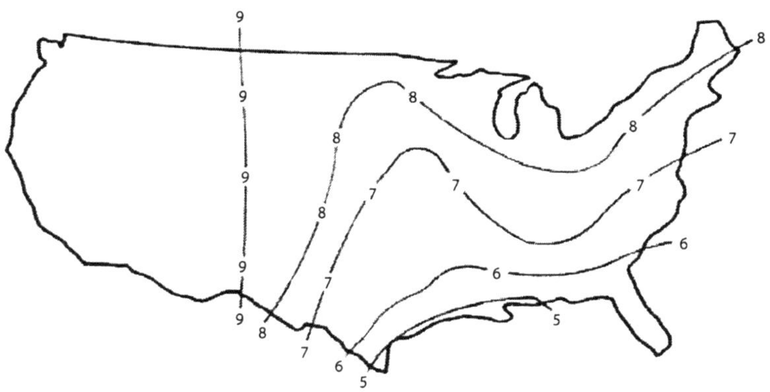

Figure 12.18. Empirical experience factors for computing storage for small reservoirs in the absence of runoff data. Divide the surface area (acres) of the pond by the factor above to estimate flood storage (feet). Storage here is not a factor of incoming flow. (Courtesy of LICA.)

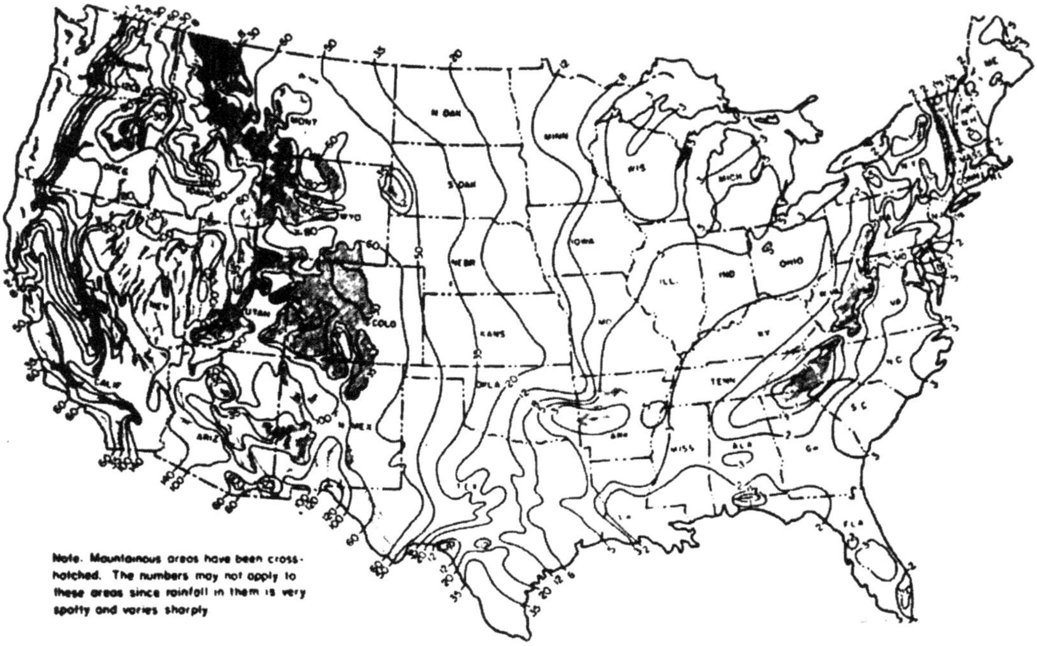

Figure 12.19. Map showing approximate size of watershed necessary for long-term storage of 1 ac-ft of runoff. The contour values are roughly proportional to the ratio of evapotranspiration to rainfall for the region. One ac-ft is 0.1233 ha-m; 0.405 ha is 1 ac. (From Beasley et al. 1984; used with permission of Iowa State Press.)

Beasley et al. (1984) give many tables showing precalculated flows for a variety of structures. By referring to the embankment profile, one must determine whether the crest width is acceptable. If not, one must increase depth, requiring redesign of the embankment and spillway. The exit section of the spillway must be sloped so as not to exceed maximum permissible velocity. Alternatively, one may use the TR55 approximate method. For higher return period storms use storage routing (e.g., NRCS 1996) software.

Spillway Design When Flood Storage Is Not the Primary Objective

With levee, hillside, and excavated ponds the primary objective may be to accomplish sedimentation, waste treatment, fish culture, etc. With sedimentation or waste treatment, one must ensure adequate hydraulic residence times for settling or treating the runoff under the design peak inflow condition. Harvest and water quality issues are important with design for aquaculture.

Principal spillway design with hillside ponds depends on the inflow. Hillside ponds supplied with diversions may have only an emergency spillway. The spillway must pass the steady flows associated with stream or spring diversions if applicable.

Embankment Safety: Selection of Design Storm

The breaching of an embankment can have serious consequences for structures and people downstream. Therefore, one should check local, state, and federal legislation before proceeding with embankments that impound large amounts of water and/or are quite high.[3] The NRCS gives guidelines for design storm selection stemming from safety implications. Check with the NRCS or other agencies regarding embankment safety legislation.

Soil stability analyses for large structures are required where failure would jeopardize life and property downstream. Embankment safety legislation–mandated procedures override the suggestions presented in the present text. NRCS 1991 discusses NRCS criteria for embankment safety. Consult NRCS offices for information regarding state embankment safety legislation.

APPLICATIONS

Agricultural Pond

Water volume calculations are important for irrigation planning and other uses. One must match water volume to the capability of the producing watershed to generate runoff in the case of water harvesting or watershed ponds. Table 12.7 summarizes various agricultural water requirements.

For irrigation or other related purposes where one may store water in a pond, Figure 12.19 gives a crude estimate of drainage in acres needed for an acre-foot of water yield. One may use Figure 12.18 for flood storage estimation. Compute pond volumes using

Table 12.7. Water requirements (working volumes) for agricultural uses.

	Average use	
Type of use	l/day	ha-m/yr
Household, all purpose, per person	190–380	0.01
Fire protection		0.03
Dry cow or steer, per 450 kg mass	75–115	0.003
Milk cow, 450 kg, including milkhouse and barn sanitation	130–170	0.006
Horse or mule, 450 kg	45	0.0017
Turkeys per 100 head	60	0.002
Chickens per 100 head	35	0.001
Swine per 45 kg	10–20	0.0006
Sheep per 45 kg	8	0.0003
Orchard spraying per year of tree age per tree per application	4	
Humid region irrigation per ha		0.3–0.45 m
Arid region irrigation per ha		0.3–1.5 m

Source: Adapted from Midwest Plan Service 1987.

Note: Make an allowance for sediment accumulation equal to 10 to 50% depending on erosion in the watershed. Assume seepage and evaporation losses range from 40 to 60%.

Water usage rates for bioremediation projects are highly dependent on the details of specific projects.

Appendix C, equations C.1a–e or C.2 as appropriate. Schwab et al. (1993) and Beasley et al. (1984) give detailed agricultural pond design examples.

Example 12.5: For a pond in Athens, Georgia, compute the pond area required to store 0.74 ha-m (6 ac-ft) of water.

From Figure 12.19, we need approximately 1.75 ac to store each acre-foot of water near Athens. Thus the watershed acreage is 1.75(6) = 10.5 ac = 4.25 ha.

For flood control structure design, the minimum duration storm is 6 hr, according to the NRCS. Sometimes 24 or 48 hr may be the minimum, where life and property are jeopardized by structural failure. More accurate runoff estimates require more detailed watershed and climate models.

Compute the water volume using the average area formula or the cone frustum approximation shown in Appendix C, equations C.1a–e and C.2. Table 12.7 tabulates water volume requirements for selected agricultural operations. The watershed must provide the water and one must appropriately size the pond to hold the requirement plus an allowance for sediment storage.

Example 12.6: Suppose a pond was to provide 20 ha-m water for assorted uses and 1 ha-m for sediment storage. Percolation losses are 60% of the total pond volume based on local experience. The following area-stage data are available: 0 ha, 80 m; 3 ha, 90 m; 6 ha, 100 m; 12 ha, 110 m. (This pond is very deep.) Set the principal spillway elevation.

The total stored water required is stored volume/(1 − losses) or (20 ha-m + 1 ha-m)/(1 − 0.6) = 52.5 ha-m. Using the average area volume formula, the stage-volume relation is 85 m, 15 ha-m; 95 m, 45 ha-m; 105 m, 90 ha-m. Interpolate to place the principal spillway at 96.67 m.

In practice, one would analyze the selected design using NRCS TR55 routing methods or possibly storage routing. One should use routing methods when the design contains nonstandard structures. Routing methods should be part of the analysis if structure failure could jeopardize life or property.

Sedimentation Basins, Traps, and Barriers

Sedimentation basins and traps are routinely used in construction sediment capture. Levee, excavated, hillside, and watershed pond configurations are used as sites permit. These ponds usually have trickle tube principal spillways and should have emergency spillways. *Sedimentation basins* or *silt basins* refer to engineered structures designed to treat runoff from drainage areas. If the drainage area exceeds 2 ha, the basin has the following characteristics (Goldman et al. 1986):

- compacted embankments
- one or more inflow points carrying polluted runoff

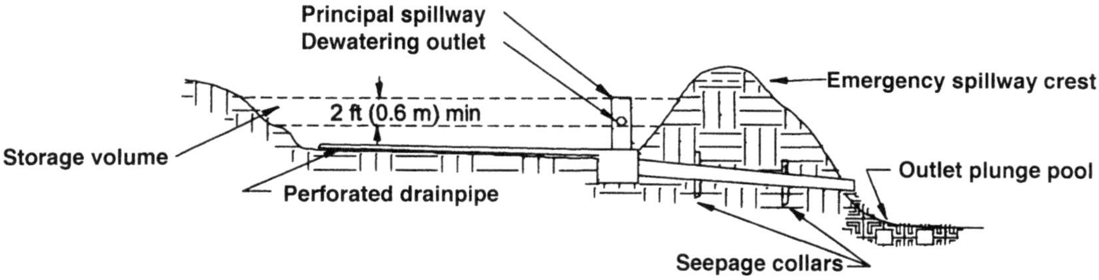

Figure 12.20. Schematic cross section of a sedimentation basin.

- baffles to spread the flow throughout the basin
- a securely anchored pipe riser as the principal spillway
- an emergency spillway
- outlet protection to reduce erosion at the outlet if the outlet is into a natural drainage way (outlet protection is not a major concern if the outlet is into a storm drain)

Figure 12.20 shows a schematic of a sedimentation basin. Sedimentation basins are designed with some empirical factors introduced to account for non-ideal aspects. Figure 12.21 shows the operation of an ideal sedimentation basin.

One configures the settling basin so the particles with set size or larger will fall through the settling zone before the horizontal velocity carries the particles to the exit riser. The storage zone must have adequate capacity to store the settled soil fraction. Soil particles (specific gravity = 2.5 Mg/m^3) settle according to the relationship given in Figure 12.22.

In a properly designed basin, the time to fall to the storage zone T_V will equal the time to flow horizontally from the inlet to the exit. The vertical fall time (T_V) may be expressed as

$$T_V = \frac{y}{v_s} \quad \textbf{(12.14)}$$

where

y is the depth of the settling zone (L);
v_s is the settling velocity (L/T).

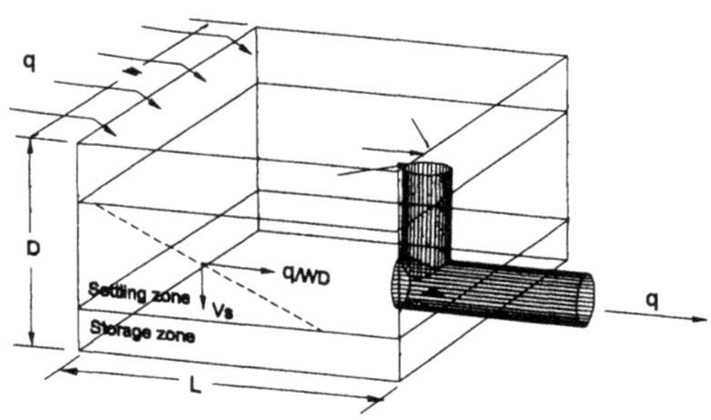

Figure 12.21. Ideal settling basin.

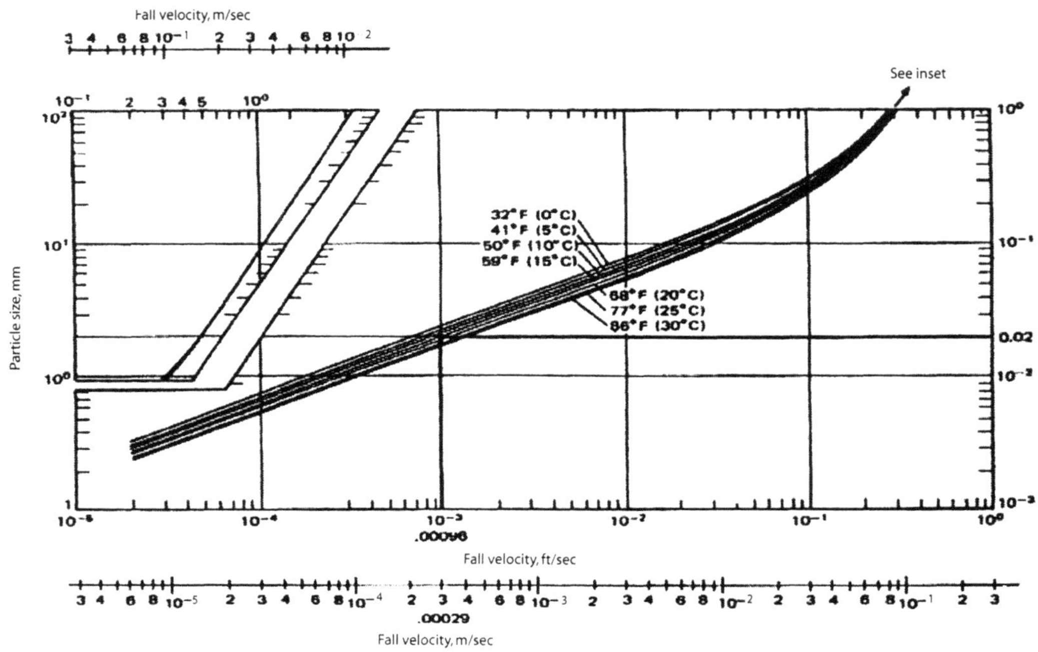

Figure 12.22. Fall velocity for spherical sand particles as a function of sediment size and temperature.

One may express horizontal flow time as

$$T_H = \frac{L}{q_s/W_b y} \tag{12.15}$$

where

q_s is the average inflow rate (L^3/T);
L is the basin length (L);
W_b is the basin width (L).

Since T_v must equal T_H, one may set the right-hand sides of the above two relationships equal to each other. Noting WL is the surface area A_s, the ideal basin surface area is described by

$$A_s = W_b L = \frac{q_s}{v_s} \tag{12.16}$$

Turbulence in the settling zone hinders particle settling. Turbulence effects on settling are modeled using a trapping efficiency coefficient. Figure 12.23 shows basin sediment-trapping efficiency as a function of a surface area adjustment factor. The specific gravity of the particles is 2.5 Mg/m^3. Stokes's law describes the fall velocity for particles less than 0.1 mm. Turbulent drag begins to dominate with larger particles.

The plug flow assumption implied in Figure 12.21 requires some adjustment due to basin shape. Figure 12.23 shows the effect of shape on trapping efficiency. The US Environmental Protection Agency recommends an adjustment factor of 1.2 (Mills and Clar 1976). The surface area formula then appears as

$$A_s = \frac{1.2 q_s}{v_s} \tag{12.17}$$

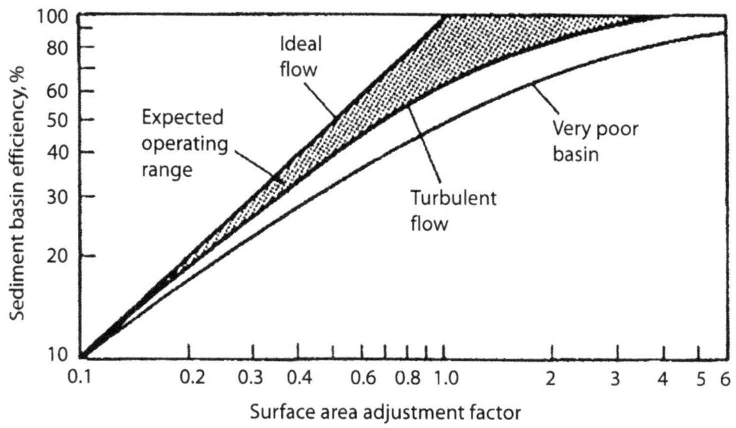

Figure 12.23. Impact of turbulence on basin sediment-trapping efficiency.

For a typical range of particle sizes having settling velocities given in Table 12.8, one can compute the surface areas per unit inlet flow from the above formula as shown in Table 12.8. One designs sedimentation basins to trap particles greater than or equal to 0.02 mm. All soil particles less than 0.02 mm pass through.

One may distribute the flow from the inlet over the sedimentation basin width using baffling. This prevents large dead zones, as shown in Figure 12.24. The depth of the settling zone must be sufficient to prevent scour along the top of a full storage zone. Goldman et al. (1986) suggest a length to settling zone depth ratio of 200 : 1 and a settling zone depth of 2 ft (0.6 m).

The design runoff rate q_s for computing the surface area of a sediment basin is the average runoff of a 10 yr–6 hr storm. This results in a lower surface area than expected using the peak runoff for design. Goldman et al. (1986) contend that the lower surface area does not significantly reduce the basin efficiency.

Table 12.8. Surface area requirements of sedimentation basins.

		Settling velocity		Surface area requirements	
Particle size (mm)	Soil texture	(ft/s)	(m/s)	ft^2 per ft^3/s	m^2 per m^3/s
0.5	Coarse sand	0.19	0.058	6.3	20.7
0.2	Medium sand	0.067	0.020	17.9	58.7
0.1	Fine sand	0.023	0.007	52.2	171.0
0.05	Coarse silt	0.0062	0.0019	193.6	635.0
0.02	Medium silt	0.00096	0.00029	1,250.0	4,101
0.01	Fine silt	0.00024	0.000073	5,000	16,404
0.005	Clay	0.00006	0.000018	20,000	65,617

Source: Adapted from Goldman et al. 1986.

Note: Goldman et al. (1986) recommend 0.02 mm as a design particle size in many applications. The temperature is 20°C.

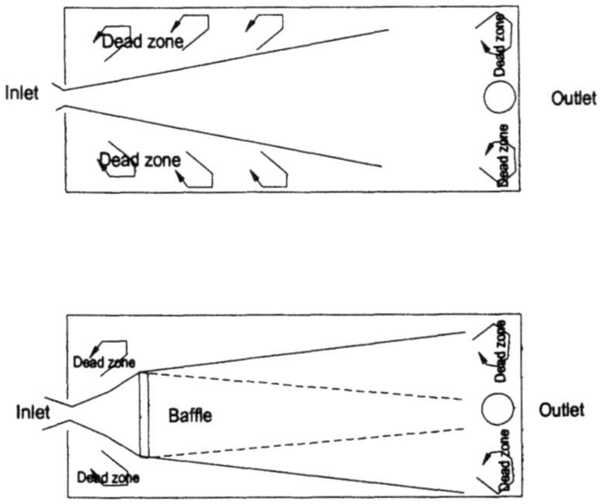

Figure 12.24. Effect of baffling to diffuse fast-moving flows over the basin width for increased effectiveness. The unbaffled basin (top) has dead space and flow velocity much greater than the design velocity in the settling zone. Baffling (bottom) slows the incoming velocity and eliminates much of the dead space. Outlet flows are typically slow, not requiring outlet baffling.

One may dewater sedimentation basins by placing an orifice on the outlet riser at the bottom of the settling zone. Dewatering may also be accomplished by placing a tile drain covered with sand and gravel in the excavation bottom. The tile drain connects to the bottom of the riser pipe. Practitioners have tried a variety of other approaches with differing degrees of success. The orifice area required to drain the settling zone is given by

$$A_{orf} = \frac{A_s\sqrt{2y_s}}{3600t_{sd}C_{orf}\sqrt{g}} \quad \textbf{(12.18)}$$

where

A_{orf} is the orifice area;
A_s is the basin surface area (m^2);
y_s is the settling zone depth(m);
t_{sd} is the time (hr) needed to drain the basin settling zone;
C_{orf} is the orifice coefficient (usually 0.6);
g is gravity (m/s^2).

The drainage time t_{sd} is usually 24 hr.

The riser diameter may range from 45 cm (30 cm high riser) to 228 cm (152 cm high riser). The gauge thickness ranges from 16 gauge (30 cm high riser) to 10 gauge (213 cm high riser). Goldman et al. (1986) give additional details concerning riser design. The principal spillway, with a head up to the emergency spillway, handles peak flow from storms exceeding the 10 yr – 6 hr runoff. One designs emergency spillways as for ponds, except flood storage may be the same as the dewatering zone volume or may include some increment above the top of the principal spillway. Emergency spillways are customarily designed to flow at the 2 yr interval. Packages such as SedCad7 (Warner 1998) model the sedimentation process in a sedimentation basin.

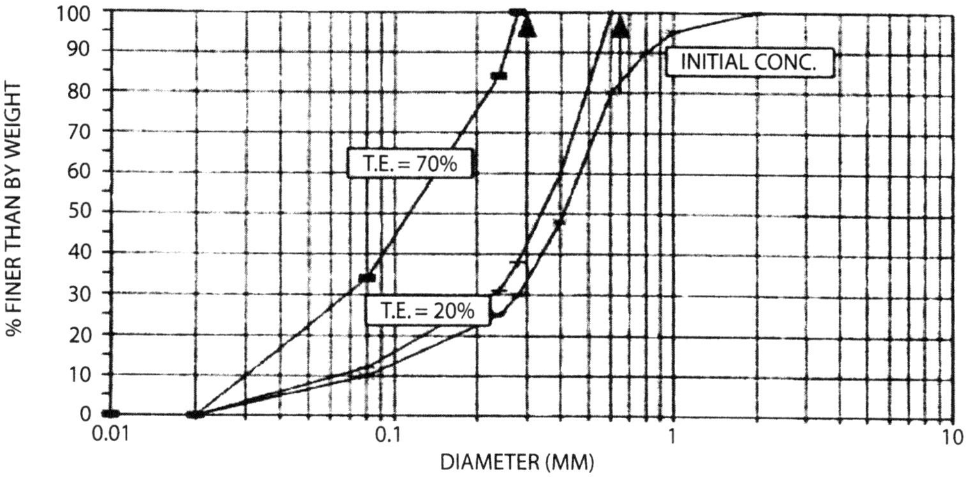

Figure 12.25. Sediment distributions resulting from passage of incoming sediment distribution through structures having the indicated trapping efficiencies. The final size distributions are plotted as follows: (1) Determine the trapping efficiency TE. (2) Take the point on the initial percent finer (InitialPF) distribution representing (1−TE) to 100%. For example, if TE is 70%, plot the diameter of particle at 30% finer at 100%. (3) Other points on the final curve (FinalPF) may be represented by FinalPF = InitialPF/(1−TE) where FinalPF ≤ 100. *Percent finer is identical to percent passing.* (From Haan et al. 1994; used with permission of Academic Press.)

Leakages from the basin bottom and seepage through the basin embankment are not issues. Rock toes are not ordinarily used in sedimentation basins. Principal and emergency spillways are of the same critical importance in sedimentation basins as with ponds. Return periods for principal and emergency spillway design storms are not as large as with ponds. Freeboards are similar to those discussed for ponds. One may compute dam and water volumes using Appendix C, equations C.1a–e or C.2 as appropriate.

Passage of a hydrograph and sedigraph through a sedimentation basin or trap structure is treated identically to the process outlined in Chapter 6. The delivery ratio D in that discussion is identical to 1 − trapping efficiency (TE). The process consists of (1) evaluating the cumulative size distribution at the structure inlet for the event; (2) evaluating the total mass trapped during a storm event; (3) and deducting the sediment mass trapped from the larger particles of the cumulative size distribution as shown in Figure 12.25, which shows a hypothetical initial size distribution and resulting size distributions for structures with trapping efficiencies of 20 and 70%. One may compute outflow sedigraphs from structures using methods outlined in Chapter 6.

Example 12.7: A construction site has a drainage area of 5 ha in Atlanta, Georgia. The site is on a hillside with a 10% slope. Diversion terraces are spaced every 20 m along the original slope length of 225 m. All runoff will move to a single sedimentation basin. The soil K = 0.011. A sieve analysis suggests 80% of the soil is larger than 0.02 mm.

Using methods in Chapter 6, the average annual erosion R_{rf} = 5000, L_f = 0.95, S_f = 1.17, C_{prac} = 1, and P_{prac} = 0.6. The average annual soil loss is 38 Mg/ha-yr. The

10 yr–6 hr storm is 102 mm or 17 mm/hr. Using a rational C_{rat} of 0.6, we find q_s = 0.14 m^3/s. From Figure 12.22 the settling velocity of the 0.02 mm particle is 0.00029 m/s. The minimum required surface area is A_s = 1.2(0.14 m^3/s)/0.00029 m/s = 579 m^2 (equation 12.17). The storage depth must store the average annual erosion × percent greater than 0.02 mm × the area. Storage weight is 38 Mg/ha-yr(0.8)(5 ha) = 152 Mg/yr. The density of uncompacted dry soil is approximately 1.06 Mg/m^3. Thus, the storage volume requirement is 152 Mg/yr/1.06 Mg/m^3 = 143 m^3/yr or 143 m^3 with annual cleanouts. The storage depth is 143 m^3/579 m^2 = 0.25 m. Set the depth of the settling zone to 0.6 m. The length must be less than 200 m × 0.6 = 120 m. Set length = 100 m. The width must be 579 m^2/100 m = 5.8 m.

The unobstructed orifice size at the bottom of the settling zone (y_{sd} = 0.6 m) to dewater in 24 hr is $A_{orf} = \{579\ m^2\ [2(0.6\ m)]\}/[(3600\ s/hr)(24\ hr)(0.6)\ (12.81\ m/s^2)]$ = 0.00394 m^2 (equation 12.18). Two 50 mm diameter holes will provide the required orifice area.

Sedimentation basins frequently fail due to lack of timely cleaning. Unusually heavy rains can also contribute to basin failure. Sedimentation basins treating drainage areas less than 2 ha are called sediment traps. Sediment traps may not have all the above features. Sediment traps are short lived compared to basins. Traps may have pipe risers as above, straw bales, or filter fabric (e.g., silt fence) to define the sedimentation zone. Storage is usually not an issue with sediment traps. Straw bale dams or silt fences across swales and storm sewer inlets are examples of sediment traps if appropriately engineered. Sediment trap design follows principles similar to sedimentation basin design with some simplifications. Goldman et al. (1986) and ASAE (1997a; although this reference uses the term *basin,* the term *trap* is more appropriate) give additional details relating to sediment trap design.

Sediment barriers are nonengineered temporary structures. Silt fences or straw bales placed on the contour are examples of sediment barriers. Sediment barriers may also be placed at slope transitions, along streets, and around storm water inlets and sidewalks. Figure 6.25 shows a straw bale barrier placed to retain sediment.

One may place sediment barriers across small swales. Goldman et al. (1986) recommend not using a barrier if the flow in a swale exceeds the capacity of a grass waterway defined by the swale. Instead, use an engineered basin or trap.

Sediment ponds are often hillside ponds, levee ponds, or excavated ponds. Approximate volume calculations for excavation and spoil banks are shown in Figure 12.26. These ponds are assumed square. Excavation is computed to give the desired water volume. Outlet structures are not shown. Excavation/bank ratios are assumed equal. In practice, this ratio will likely exceed 1 slightly. In many cases, soil near the surface may not be as desirable as deeper soil for sealing purposes. Contractors should partition soil during construction, enabling placement of desirable soil near the water boundary for sealing purposes.

Construction and Safety Issues

Works such as Stephens 1991, NRCS 1991 and 1996, and Harris 1994 discuss pond and structure installation. Appendix C further discusses installation of channels and structures.

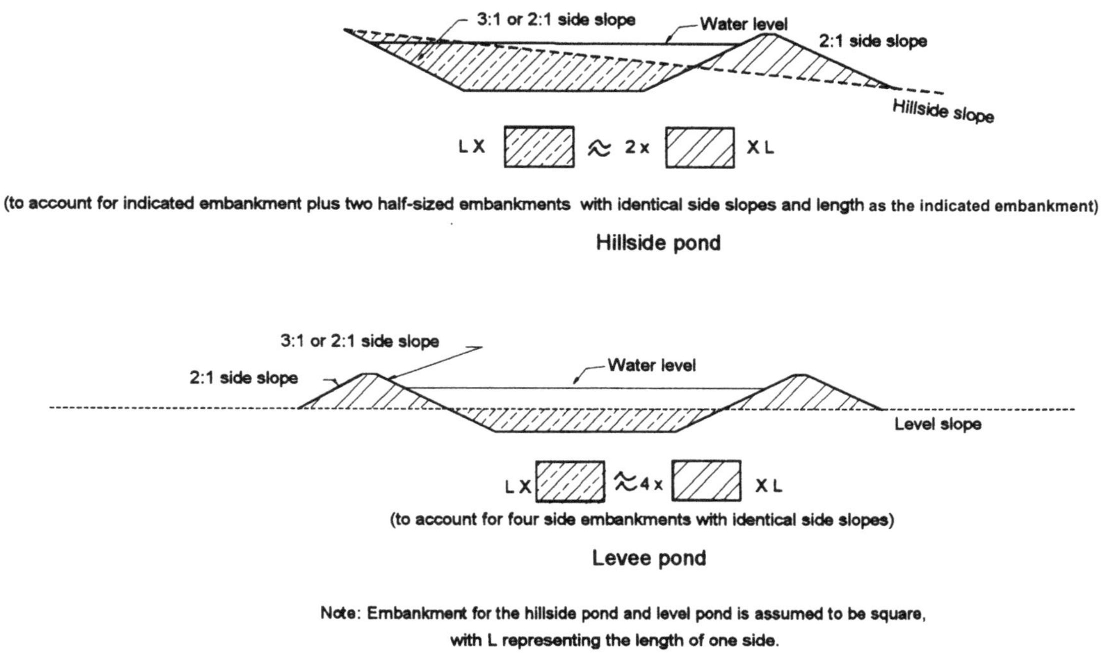

Figure 12.26. Volume calculations for levee ponds and hillside ponds.

Many states have regulations regarding embankment design for large reservoirs, modeled after details given in NRCS 1996. Embankment safety evaluation requires a soil stability analysis for strength, a topic beyond the scope of this text. Use adequate linings and proper rock toe filtering. These issues are most critical for large impoundments.

Check with NRCS or other local authorities for site requirements related to personnel safety. These may vary depending on the purpose of the impoundment. Excellent NRCS publications are available at http://www.usda.nrcs.gov. Proper fencing and signage is essential. For waste lagoons and related bodies requiring periodic sampling, appropriate steps and platforms with handrails should be provided. One should provide for equipment for mowing and/or spraying banks or harvesting fish.

Structures must rest on an appropriate soil foundation. The soil must be adequate from surface hydraulic and seepage perspectives. The soil survey is an excellent starting point for detailed discussions on the suitability of different soils for different uses. Land forming operations are well developed for agricultural applications and for other civil applications as well. Pond design and sedimentation design software packages incorporate many ideas covered in this chapter.

Outlook

Designs of ponds, lagoons (or other levees), or watershed impoundments should be completed only after an appropriate engineering analysis of customer and technical requirements. Table 12.9 provides guidance for the analysis. Each item represents a broad category

Table 12.9. Customer and technical requirement prompt sheet for impoundments.

Customer requirements	Technical requirements
Recreation?	Flood storage limitations based on topography
Aesthetics?	Required design storm return periods
Fish and wildlife?	Watershed characterization
Peak flow reduction needed?	Water quality limits (sediment, nutrients)
Working volume requirements set by known storage volume or hydraulic residence time?	Principal spillway
Flood storage high priority?	Emergency spillway types based on topography
Sediment removal objective?	Seepage and other soil limits?
Waste treatment objective?	Soil strength parameters for embankments
Waste storage objective?	Is primary objective flood storage and/or peak flow reduction?
Water supply objective?	Is primary objective to satisfy mass balance and/or hydraulic residence time?
Agricultural objective (e.g., irrigation)?	Population density: urban or rural?
Aquacultural objective?	Buffers or other setbacks for odors/air quality issues?
Personnel safety	Embankment safety issues
Impact on neighbors?	Other regulatory issues
	Fencing, signage, and other safety features
	See Tables 12.3 and 12.4 for potential siting limitations.

encompassing many subtopics appropriate in each site-specific application. Revisiting the customer and technical requirement prompt sheets in Chapter 8 (Tables 8.8 and 8.9) would be useful also. The watershed pond was used as the primary focus for discussion, but slopes, inlet/outlet structures, volume calculations, seepage control, and soil topics are directly relevant to other pond types.

Generalized pond design moves through the following steps:

1. Clearly state the purpose for the pond, lagoon, or other impoundment. Revisit engineering design methodology when making this determination.
2. Site the pond based on topographic data or site-specific factors to accomplish desired working volume. Define the pond type (see Figure 12.11). One should site watershed impoundments to impound as much water as possible with the least embankment volume. Use $z = 3$ for the submerged face and $z = 2$ for the back face of the embankment. Compute top width using equation 12.8. Plot the embankment to scale on a topographic map (see Figure 12.15). Ensure that the watershed size is adequate (Figure 12.19). Estimate embankment and water volumes from Appendix C, equations C.1a–e. With a lagoon or waste treatment structure, site as dictated by the need for storage or treatment.
3. Examine the soil for seepage potential. Plan to line the pond if appropriate soil is not available.
4. Determine the working volume and flood storage volume needed (Table 12.7 is useful for rural ponds). Determine availability of flood storage. Check to see if the site will accommodate the required storage (see Table C.1 in Appendix C for the storage calculation procedure) and allow a suitable emergency spillway. For routine ponds, relationships between storage, surface area, and design incoming peak flows are available (equation 12.13, Figure 12.18). For special cases such as sedimentation waste treatment, the

hydraulic residence time is critical and will dominate the hydraulic design. For storage, the mass balance must be satisfied, and storage must be available under scenarios specified by regulation. Size embankments and excavations for the appropriate pond style.

5. Design the principal spillway. Design outlets and seepage collars as needed. One may compute flows through spillways of small ponds using the Beasley et al. 1984 routing procedure (Figure 10.14) or Culp (1948) equations. These relations assume ordinary weir or pipe spillways. The NRCS TR55 method (see Figure 12.8) may be used for ponds in urban settings. If flood storage is limiting, spillway structures are not ordinary pipes or weirs; if one is meeting a peak flow reduction goal, one should use storage routing (see equation 12.6). NRCS software (TR20) is available for this purpose. Spillway design for ponds should apply to sedimentation basins as well, accounting for the specialized nature of these spillways (see Example 12.7). With a lagoon or storage impoundment, there may be no principal spillway.
6. Design the emergency spillway for ponds based on a 25 (or greater) yr storm and available flood storage. One may use the NRCS TR55 routing procedure (Figure 12.8) for estimating the emergency spillway size of small reservoirs. Designs where loss of life or significant property is likely on failure should be checked using storage routing. Treat emergency spillways for sedimentation basins similarly. A 2–10 yr return storm is commonly used. Emergency spillways on waste storage structures are not commonly specified, especially when there is limited catchment area draining to the pond. Consult NRCS or other agencies regarding emergency spillways on waste impoundments.
7. The embankment height for ponds must allow for all freeboards shown in Figure 12.15. Freeboards for sedimentation basins are limited to sediment storage, sedimentation working volume, flood storage, freeze/thaw, and the construction height increase to allow for settling. Freeboards for waste impoundments are commonly limited to sediment/sludge storage, working volume, freeze/thaw, and the construction height increase.
8. Compute embankment volume, including the core for sealing if necessary. Show all allowances for storage, wind, and settlement after construction. Estimate width and height. Tables C.1 and C.2 in Appendix C provide a guide for volume calculations.
9. Consider embankment safety issues.
10. Address all other safety issues such as fencing and signage.

Objectives for ponds and lagoons may span the gamut from providing recreational benefits, irrigation, water for a bioremediation operation, peak flow reduction for flood control, sedimentation, or waste treatment. The approach to the problem depends on objectives and constraints. One must use a routing method (e.g., NRCS 1996) to ensure comparability of the mechanical spillway, flood storage capacity, and emergency spillway to the incoming storm runoff hydrographs. In cases where there is a treatment objective such as sedimentation or waste treatment, residence time must be appropriate. Pond/lagoon storage should handle specified return period storms.

Principles covered here extend to any field- or farm-scale project involving the siting of structures with environmental or conservation implications. One cannot overemphasize the necessity of good topographic maps or survey data. As global positioning technology improves, survey data acquisition will become easier. Elevation-based control of excavating machinery is now a reality.

These farm- and field-scale operations are useful in small bioremediation applications as well. We can now address comprehensive erosion control plans and other regional watershed management questions. The ASAE 1997d standard presents many channel- and structural-based approaches for use in erosion control. ASAE developed this standard for mapping purposes. Ponds and embankment calculations are applicable to many other structures such as waste lagoons, drainage and irrigation sumps, etc. Although objectives are different, the principles are still valid. ASAE publishes excellent guides for waste structures. What if an embankment is situated in an area with little or no downstream development and designed accordingly, then downstream development occurs? This is a societal dilemma.

Software packages such as SedCad and PONDPACK7 integrate runoff models and various pond outlet structures with digitized topographical descriptions enabling easy evaluation of various pond and outlet scenarios at given locations.

Problems and Questions

1. Given a soil with a dry bulk density of 1.5 Mg/m^3, what is the porosity? What is the void ratio?
2. Given the above soil, what is the new void ratio following tillage given a new dry bulk density of 1.0? What is the new porosity?
3. Calculate the volume in the embankment shown in Figure 12.27 using the average end area formula. Calculate the top widths of the cross sections using $W = 1 + 0.4H_{embmx}$; allow 10% settlement.

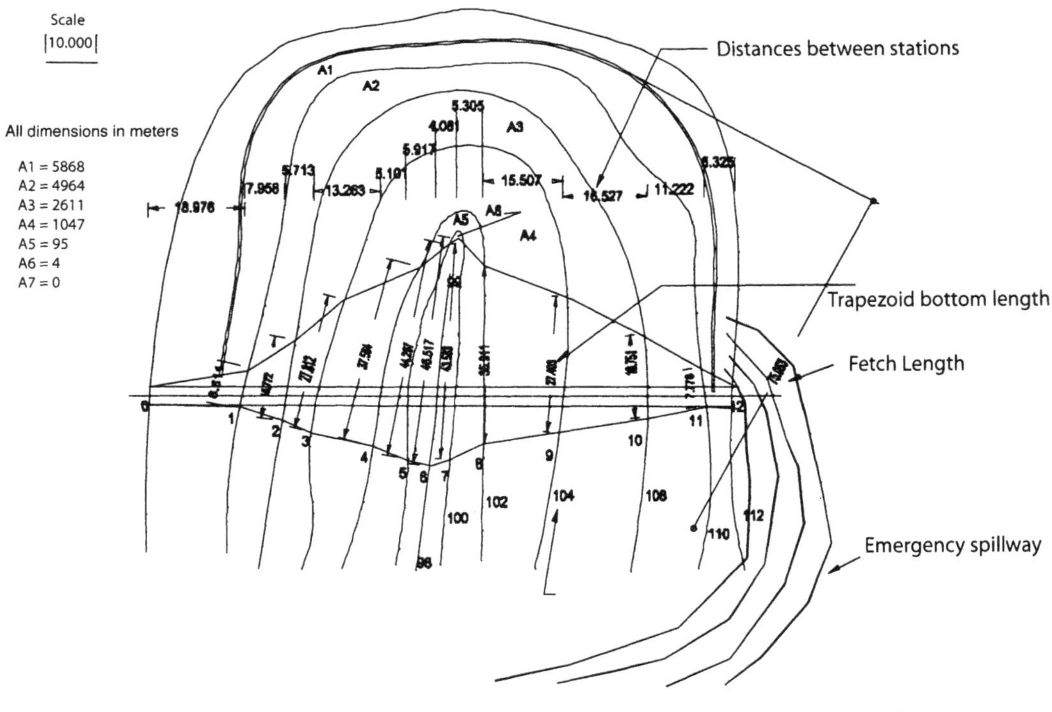

Figure 12.27. Pond and embankment for selected homework problems. A1 through A7 denote areas enclosed by contours above the embankment (water surface area as a function of stage). Distances between the cross-sectional trapezoids are dimensioned as are the trapezoid bottom lengths.

4. Calculate the water volume in the pond shown in Figure 12.27 using the average end area formula. What is the total water volume assuming all embankment material came from the impoundment? Further, assume the void ratio of the compacted embankment is 0.5. Assume the total solids volume of the borrow area is available to become water volume.
5. A pond embankment has 15,000 m^3 and is compacted at an average void ratio of 0.55. The soil has good sealing properties and has an undisturbed void ratio of 0.12. Compute the volume of soil one must take from this area to construct the embankment.
6. Develop a 10 yr–1 hr storm hydrograph for a 50 ac watershed in Knoxville, Tennessee. Route the flow through the culvert of Example 11.11 assuming the storage of case B is available.
7. A watershed of 50 ac has a pond with an available flood storage of 17 ac-ft. The runoff volume from a 10 yr storm is 50.15 ac-ft. The 10 yr peak runoff rate is 127 cfs. The 50 yr peak runoff is 500 cfs, and the storage with an operating emergency spillway is 10 ac-ft. The 50 yr runoff volume is 140 ac-ft. Design the principal spillway using the NRCS TR55 method.
8. Route the storm of Table 12.10 through the reservoir of Example 11.11 using the structure of case B. Assume the stage-storage relationship of Example 11.11 holds. A 9 m wide emergency spillway is situated 3 m above the pipe. The pipe is 20 m long and has a 90 degree bend, no rounding. Compute the modified stage-discharge relationship and determine the peak outflow from the sediment pond with this pipe.
9. Calculate the wave height, gross freeboard, and net freeboard for the embankment in Figure 12.27. At what elevation should the emergency spillway be situated, assuming a flood storage of 0.3 m? At what elevation should the mechanical spillway be placed?
10. Assuming a 1 ha garden requiring 1 ha-m of water over a season, would the full pond of Figure 12.27 provide adequate water? Assume 10% sediment storage and assume 60% seepage and evaporation losses.

Table 12.10. Stage storage and inflow data for Problem 12.8.

Stage M	Storage M^3	Discharge M^3/s	Time Hr	Inflow M^3/Sec	Time (cont.)	Inflow (cont.)
0	0	0	0.2	0.28	3.6	1.3
0.075	360	0.17	0.4	0.62	3.8	1.2
0.15	612	0.28	0.6	1.36	4	1.1
0.2	1,224	0.57	0.8	2.55	4.2	1
0.31	1,728	0.85	1	3.74	4.4	0.9
0.35	2,232	1.13	1.2	4.76	4.6	0.8
0.4	2,736	1.42	1.4	5.21	4.8	0.7
0.45	3,348	1.7	1.6	5.38	5	0.5
0.6	4,068	1.81	1.8	5.15	5.2	0.3
0.91	8,172	1.93	2	4.67	5.4	0.2
1.2	13,968	2.04	2.2	4.11	5.6	0.15
1.5	20,000	2.2	2.4	3.54	5.8	0.1
1.8	30,000	2.6	2.6	2.97		
2	45,000	2.7	2.8	2.46		
2.1	55,000	2.9	3	2.01		
2.2	70,000	3.2	3.2	1.78		
2.3	85,000	3.5	3.4	1.5		

Table 12.11. Data for Problem 12.12.

Contour elevation (m)	Length of water surface (m)	Accumulated storage (ha-m)
24.4	0	0.0
25.6	30	0.025
26.8	91	0.123
28	183	0.395
29.2	244	0.888

11. If the 25 yr storm peak flow was 2 m^3/s and the runoff volume was 0.4 ha-m, how wide would the emergency spillway of the pond in Figure 12.27 need to be? What would the size of the mechanical spillway need to be?
12. A farm pond is to provide 0.42 ha-m of water for 100 milk cows; seepage and evaporation losses are 45% of the stored volume, settlement allowance for fill is 5%, and the flood spillway depth is 0.3 m. Determine the elevation of normal water surface, flood spillway, and top of embankment before and after settlement where the base of the embankment is 24 m. Assume a peak flow of 1.5 m^3/s, frost depth of 0.15 m, and water surface area of 0.45 ha. Table 12.12 gives additional data.
13. A holding pond is to store 0.5 ha-m for use in a bioremediation project. A continuous flow rate of 0.2 m^3/s is expected. The pond receives no additional surface runoff and has a plastic liner. The pond is to have a principal spillway and is in a flat location. The spoil is to be used in a nearby location and thus will not be near the pond. Design the pond and size the principal spillway.
14. Given a location by your instructor, find the USDA and Unified Soil Classifications systems from the soil survey.
15. With the completion of this chapter, nearly all elements used in erosion control planning have been presented. Consult the ASAE standard entitled *Mapping Symbols and Nomenclature for Erosion and Sediment Control Plans for Land Disturbing Activities.* Identify the hydraulic principles associated with the runoff control measures, outlet protection, and sediment traps/barriers.
16. Give one strategy for reducing the peak flow from a watershed.
17. Compute the uniformity coefficients of the soils with percent passing curves shown in Figure 12.3. Are these soils well graded?
18. What is *dry bulk density, void ratio, gravimetric moisture content* (dry basis), and *volumetric moisture content?* Define these using a phase diagram.
19. Given an elevation of a principal spillway, what factors does one consider in setting the elevation of an emergency spillway and a constructed embankment height?
20. Explain how to calculate the volume of a pond embankment.
21. Explain how to calculate the water volume contained by a pond.
22. A zoning board proposes a site for conversion from agricultural to residential zoning. They propose septic tanks for sewerage handling and disposal. For a generalized location, what steps would one take to determine the adequacy (or lack thereof) of septic tanks for waste treatment?

Notes

1. TK®Solver was used for this routing. Some TK® advanced features (List functions, ELEMENT function, Save function sheet, Merge function sheet) were used in the solution. A spreadsheet may also be used; however, one must fit a function to relate the routing parameter to discharge (Table 12.1, col. F vs. col. B).
2. Refer to Appendix C for a discussion of techniques for computing earth and water volumes.
3. For example, the Georgia Safe Dams Act requires all embankments impounding 50 or more acre-feet or having embankments 25 feet or higher to be inventoried. Requirements vary among other states.
4. Available from Civil Software Design, Ames, Iowa and Haested Methods, 37 Brookwood Road, Waterbury, CT. Software package names are provided as a convenience and their mention is not to be construed as an endorsement.

References

ASAE. 1997a. Water and sediment control basins. S442 DEC92. American Society of Agricultural Engineers, St. Joseph, MI.

———. 1997b. Design of anaerobic lagoons for animal waste management. EP403.2 AUG93. American Society of Agricultural Engineers, St. Joseph, MI.

———. 1997c. Installation of flexible membrane linings. EP340.2 DEC96. American Society of Agricultural Engineers, St. Joseph, MI.

———. 1997d. Mapping symbols and nomenclature for erosion and sediment control plans for land disturbing activities. S422 MAR95. American Society of Agricultural Engineers, St. Joseph, MI.

Beasley, R. P., J. M. Gregory, and T. R. McCarty. 1984. *Erosion and sediment pollution control.* 2d ed. Iowa State University Press, Ames.

Cedergren, H. R. 1977. *Seepage, drainage, and flow nets.* John Wiley & Sons, New York, NY.

Coche, A. G., and J. F. Muir. 1992. *Pond construction for freshwater fish culture: Pond-farm structures and layouts.* FAO Training Series 20/2. Food and Agriculture Organization of the United Nations, Rome.

———. 1995. *Pond construction for freshwater fish culture: Building earthen ponds.* FAO Training Series 20/1. Food and Agriculture Organization of the United Nations, Rome.

Culp, M. M. 1948. The effect of spillway storage on the design of upstream reservoirs. *Agricultural Engineering* 29:344–46.

Emerson, W. W., R. D. Bond, and A. R. Dexter. 1978. *Modification of soil structure.* John Wiley & Sons, New York, NY.

Goldman, S. J., K. Jackson, and T. A. Bursztynsky. 1986. *Erosion and sediment control handbook.* McGraw-Hill, New York, NY.

Gopalakrishnan, V., and A. G. Coche. 1994. *Handbook on small-scale freshwater fish.* FAO Training Series 24. Food and Agriculture Organization of the United Nations, Rome.

Haan, C. T., B. J. Barfield, and J. C. Hayes. 1994. *Design hydrology and sedimentology for small catchments.* Academic Press, New York, NY.

Hajek, B. F., and C. E. Boyd. 1990. *Rating soil and water information for aquaculture.* Fisheries Research and Development Project, Central Research Institute for Fisheries, USAID/Indonesia, Jakarta, Indonesia.

Harris, F. 1994. *Modern construction and ground engineering equipment and methods.* Longman Publishing Group, White Plains, NY.

Hillel, D. 1998. *Environmental soil physics.* Academic Press, New York, NY.

Jumikis, A. R. 1984. *Soil mechanics.* Robert E. Krieger Publishing Co., Malabar, FL.

Midwest Plan Service. 1987. *Private water systems.* MPS, Iowa State University.

Mills, T. R., and M. L. Clar. 1976. *Erosion and sediment control, surface mining in the eastern US.* U.S. Environmental Protection Agency, Washington, DC.

NRCS. 1983. *Computer program for project formulation: hydrology.* SCS Technical Release 20. Soil Conservation Service (Natural Resource Conservation Service), Washington, DC.

———. 1986. *Urban hydrology for small watersheds.* 2d ed. SCS Technical Release 55. Soil Conservation Service (Natural Resource Conservation Service), Washington, DC.

———. 1991. *Earth dams and reservoirs.* Technical Release 60. USDA-NRCS, Washington, DC.

———. 1994. *National engineering handbook.* Chap. 26, Gradation design of sand and gravel filters. USDA-NRCS, Washington, DC.

———. 1996. *SITES: Water resource site analyses computer program.* Version 96.1. USDA-NRCS, Washington, DC.

———. 1997. *Agricultural waste management field handbook.* 651.1080. Appendix 10D: Geotechnical design and construction guidelines. USDA-NRCS, Washington, DC.

———. 1998. *National engineering handbook.* Appendix 7B: Identifying soils for engineering purposes. USDA-NRCS, Washington, DC.

Reed, S. C., R. W. Crites, and E. J. Middlebrooks. 1995. *Natural systems for waste management and treatment.* 2d ed. McGraw-Hill, New York, NY.

Renfro, G. 1968. Sealing leaking ponds and reservoirs. SCS-TP-150. USDA-SCS, Washington, DC.

Schwab, G. O., D. D. Fangmeier, W. J. Elliot, and R. K. Frevert. 1993. *Soil and water conservation engineering.* 4th ed. John Wiley & Sons, New York, NY.

Stephens, T. 1991. *Handbook on small earth dams and weirs: A guide for siting, design, and construction.* Paul & Company Publishers Consortium, Concord, NH.

U.S. Bureau of Reclamation (USBR). 1987. *Design of small dams.* 3d ed. Government Printing Office, Washington, DC.

Warner, R. C. 1998. *SedCad 4 for Windows95/NT: Notes for 5–7 Aug 1998 short course.* Department of Biosystems and Agricultural Engineering, University of Kentucky, Lexington.

Yoo, K. H., and C. E. Boyd. 1994. *Hydrology and water supply for pond aquaculture.* Chapman & Hall, New York, NY.

13

Shallow Groundwater Management with Drains and Wells

Surface and subsurface drainage strategies have been the major means of removing excess water from soils. Dideriksen et al. (1978) reported excess water to be a major problem on 25% of U.S. cropland. Surface and subsurface drainage is a major concern in many developing countries (Beltran 1999). Worldwide, excess salinity, often resulting from lack of adequate leaching and drainage, affects some 60 million irrigated hectares out of the estimated total 260 million irrigated hectares (Jensen 1993).

Many of the world's most productive soils need drainage to (1) remove excess water during the growing season; (2) increase timeliness of tillage, planting, and harvest operations; and (3) in arid regions remove excess salts from the plant root zone. Figures 1.2, 1.3, 1.6, and 1.7 show the effects of lack of drainage. Worldwide, crop growth effects, salinity management, water quality issues, and drainage technology continue to be important (see Evans 1998). Wetlands conservation interests are currently slowing—but not stopping—the growth of the agricultural drainage industry in the United States; however, nonagricultural applications such as waste treatment are currently outpacing agricultural drainage activity (NRCS 1998). Drainage of turf in parks and athletic facilities continues to gain in importance.

Three important aspects of drainage and groundwater management are (1) surface drainage, (2) pipe drainage (called tile or tube drainage in the older literature), and (3) well development and drainage (see Figures 13.1–13.3). This chapter will examine applications of the Dupuit-Forchheimer assumptions of groundwater flow.

Goals

- Predict surface drainage flows and design surface drainage ditches.
- Determine pipe drain spacing for subsurface drainage systems.
- Identify drainage effects on water quality for salinity and nitrogen management.
- Predict shallow-well flow rates.
- Apply the Dupuit-Forchheimer (DF) assumptions in the design of simple drainage-infiltration systems.

Surface Drainage

Surface drainage systems seek to divert excess rainfall before infiltration or flooding. Surface drainage systems, exemplified in Figure 13.4, are used for slowly permeable soils. Surface systems may include bedding channels, parallel field drains, parallel lateral ditches,

Figure 13.1. Drainage channel in North Carolina coastal plain providing nutrient management (denitrification through winter months), surface and subsurface drainage, and subirrigation capability during growing season.

and cross-slope ditch systems. In fields with random depressions, random field drains are a possibility.

One usually uses bedding channels on soils of low conductivity that have a slope less than 1.5%. Pipe drainage is not usually feasible in these instances. The bedding channels may be created with field tillage equipment. Spacing between channels may range from 7 to 28 m (Schwab et al. 1993). The channels often have side slopes of 8 or greater, enabling easy crossing with farm equipment. The side slopes are farmed.

Parallel field drains are excavated ditches with steep side slopes, as shown in Figure 13.1. One does not cross these ditches in ordinary field operations. One may space parallel

Figure 13.2. Drainage collection sump with a diesel-powered and belt-driven low-head/high-capacity pump.

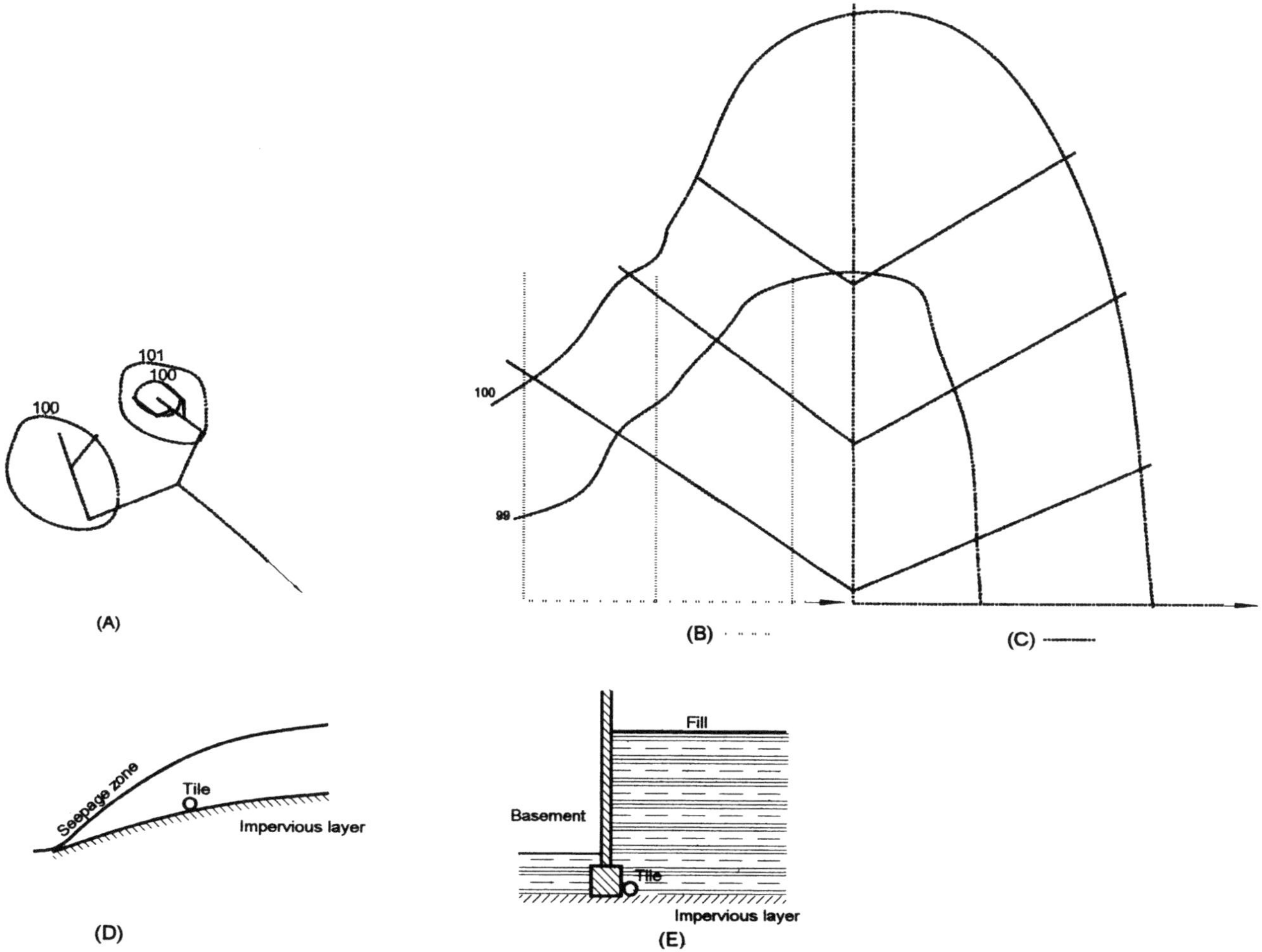

Figure 13.3. Scenarios for placing drainage pipe: a) random pattern for draining random spatial depressions; b) gridiron pattern for draining a uniformly sloped space; c) herringbone pattern for draining a large shallow basin; d) individual drains for intercepting seepage; and, e) individual pipe for providing drainage around a structural foundation. (Modified from Schwab et al. 1993.)

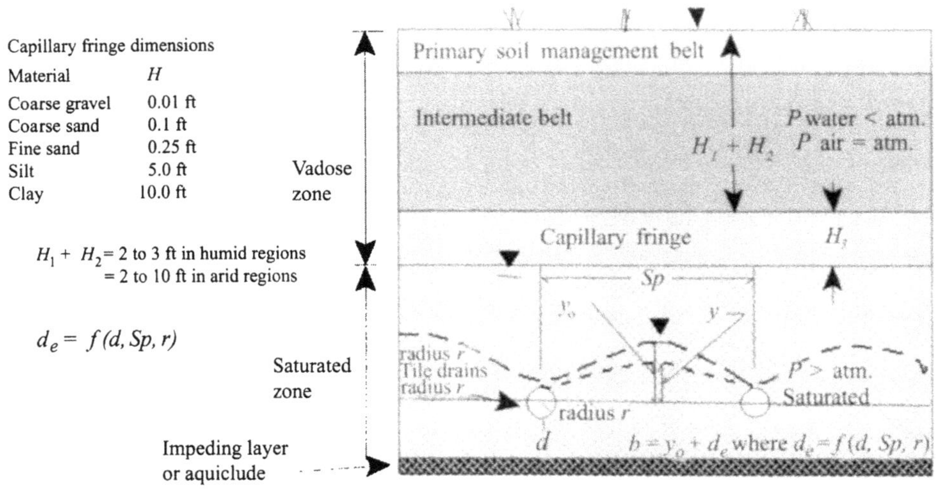

Figure 13.4. Schematic of vadose zone showing pipe drainage lines overlying impeding layer and showing nomenclature for steady-state and non-steady-state flow equations. Note that the intermediate zone is usually lumped with the primary soil management zone in drainage applications. Depth d_e is the equivalent depth, not the actual depth. Equivalent depth d_e depends on actual depth *d*, spacing *Sp*, and to a minor extent the pipe radius *r*.

drains as shown in Table 13.1. Parallel field drains both intercept surface runoff and provide for subsurface outflow into the channels. Spacing criteria in Table 13.1 reflect both aspects. Cross-slope ditches are essentially terraces designed following terrace design procedures. Conservation engineers developed the random ditch field drain concept for draining individual field depressions. The ASAE (1997a) drainage ditch formula, discussed in Chapter 5, is a published standard for calculating runoff needed for drainage ditch design. This is a specialized approach specifically for surface drainage outlets in areas with land slope less than 0.5% and watershed areas less than 4000 ha (10,000 ac).

ASAE (1997a) presents the 20-40 rule for designing drainage ditches following a ditch junction. When the watershed of the smaller of two ditches is less than 20% of the total area drained by both ditches, design the outlet ditch using the formula (see equations 5.13a–c)

Table 13.1. Parallel surface drain channel specifications for water table control.

	Sandy soil	Other mineral soils	Organic soils, peat and muck
Maximum spacing (m)[a]	200	100	60
Minimum side slope *z*	1	1.5	0 to 1
Minimum bottom width (m)	1.2	0.3	0.3
Minimum depth (m)	1.2	0.8	0.9

Source: Adapted from Schwab et al. 1993.

[a]Spacing criteria reflect crop yield concerns; potential pollution impacts due to nutrient loss are not considered in these recommendations.

with M equal to the sum of the areas for each ditch. If the area is 40% or more, apply the above formula to each contributing area separately, then sum the totals for the design flow rate. For areas between 20 and 40%, compute the flow both ways and interpolate to compute the design flow rate.

ASAE (1997a) has other standardized procedures for drainage ditch design. Surface drainage channels, shown in Figure 13.1, are maintained like any open channel.

Example 13.1: Compute the runoff from a drainage basin of 1000 km^2 located in the coastal plain ($C_{ASAE} = 30$).

From the ASAE drainage ditch formula, $q_{pASAE} = 0.013(30)(1000\ \text{km}^2)^{0.833} = 123\ \text{m}^3/\text{s}$.

Surface and subsurface drainage enterprises often require pump stations, such as shown in Figure 13.2, for lifting water small distances. Pumps employed for this purpose are typically the mixed or axial flow centrifugal pumps having low-head/high-flow characteristics. Appendix E further discusses pumps and pumping topics.

In the western United States the drainage coefficient is not the basis for design. Instead, one may use the 5 yr rainfall runoff. One should check local requirements. Many states have provisions for water management districts or drainage districts.

The Manning equation Type B (stable cross section) or Type C (optimal cross section) provides the basis for channel design. The depth of these channels must exceed the pipe drain outlet depths.

Introduction to Subsurface Flow

The primary reason for subsurface drainage is to lower the piezometric water level or water table. Figure 13.1 shows an open ditch for drainage, subirrigation, and nutrient control. Maintaining high water tables in winter months encourages denitrificaton in nitrate-rich soils.

The basic relationship at the core of subsurface drainage and well design is Darcy's law, stated as

$$f = -K_c \frac{\Delta H}{\Delta L} \quad or \quad q = -AK_c \frac{\Delta H}{\Delta L} \tag{13.1}$$

where

f is the moisture flux (L/T);
q is the volumetric flow rate (L^3/T);
K_c is hydraulic conductivity (L/T), often equal to saturation conductivity K_{sat};
ΔH is change in potential or head compared with a datum (L);
ΔL is the length over which the head change occurs (L).

The right-hand expression reflects application of the continuity equation to get flow q in L^3/T.

Often one must calculate the travel time for a pollutant to move from point a to point b, knowing the Darcian flux. Is it distance divided by speed? *Not exactly!* One must divide the Darcian velocity by soil effective porosity, then divide the corrected velocity into the distance to get the time, giving

$$T_{Tvl} = \frac{Ln_p}{q} \tag{13.2}$$

where

T_{Tvl} is travel time;
L is the distance;
n_p is the soil porosity;
q is the Darcian flux.

Review the discussion in Chapter 3 for additional details.

The Dupuit-Forchheimer (DF) assumptions provide the basis for many approximate subsurface flow analyses. The DF assumptions are

- horizontal flow, uniform with depth
- no vertical flow component

The DF assumptions are foundational for pipe drainage, well flow, and other specialized analyses involving flow through porous media.

Pipe Drainage

Pipe drainage has extensive application to agricultural settings. Drainage using perforated pipe is also an important tool in many waste application systems. Irrigation applications requiring a leaching fraction are usually drained using pipe drainage. The leaching fraction in agriculture for salinity control may range from 0 to 0.1. It approaches 1 when leaching salts or other contaminants (see Chapter 14). Figure 13.3 shows several approaches for using pipe drains to remove water from fields.

Subsurface drainage results in soil aeration and maximum plant root development. Subsurface drainage favors aerobic microbes and plant root growth. Drainage increases the growing season by enabling earlier planting via better trafficability. It increases tilth benefits because wet soils cannot be tilled. Drainage in conjunction with irrigation has historically been used to remove salts and other pollutants from agricultural soils (Rhoades 1974). Van Schilfgaarde (1974b) and NRCS (1973) provide additional drainage discussion. Pipe drainage goes hand in hand with agricultural irrigation of arid soils and with soil remediation involving leaching. Figure 13.4 provides a definition sketch of pipe drainage showing nomenclature.

Pipe Spacing

ASAE (1997b, c) has produced pipe drainage standards for humid and arid areas. Design parameters for pipe systems include spacing, depth, grade, design drainage rate, and pipe size. One may determine drain spacing from soil hydraulic properties and depth to restricting layers.

Derivation of the ellipse equation for steady-state conditions begins with equation 13.3:

$$q_x = -yf_x = -K_{sat}y\frac{dy}{dx} = \left(\frac{Sp/2 - x}{Sp/2}\right)\frac{i_d}{2} \tag{13.3}$$

Term: 1 2 3 4

where

q_x is the rate of flow per unit drain length (measured along z) toward the drain (L^2/T) over any distance x (L);
Sp is the spacing of the drains (L);
x is measured between the drain lines (L);
y is the depth from the surface (L);
z is the length along a drainpipe;
K_{sat} is the hydraulic conductivity (L/T);
i_d is the rate of inflow per unit drain length (L^2/T).

The third and fourth terms imply that the Darcian rate of flow out one pipe is equal to half the total inflow into the space between two pipes. Equating terms 3 and 4 in equation 13.3 and separating variables, substituting $i = i_d/Sp$, integrating, and solving for Sp gives

$$Sp = \left[\frac{4K_{sat}(b^2 - d_e^2)}{i}\right]^{1/2} \tag{13.4}$$

where

b is the equivalent depth to the barrier d_e plus height of the water table above the drain;
i is the rate of inflow (L/T);
Sp is the computed spacing (L);
d_e is the equivalent depth to the impeding layer, corrected for convergence at the pipe.

The corrected depth d_e is found knowing the actual depth to the impeding layer d using equations 13.5a–c. If $0 < d/Sp < 0.31$, use

$$d_e = \frac{d}{1 + \frac{d}{Sp}\left[2.55 \ln\left(\frac{d}{r}\right) - c_{eq}\right]} \tag{13.5a}$$

where

$$c_{eq} = 3.55 - 1.6\frac{d}{Sp} + 2\left(\frac{d}{Sp}\right)^2 \tag{13.5b}$$

If $d/Sp > 0.31$, then use

$$d_e = \frac{Sp}{2.55\left[\ln\left(\frac{Sp}{r}\right) - 1.15\right]} \tag{13.5c}$$

To solve the ellipse equation, first assume a tile spacing and then compute the equivalent depth to the impermeable layer. Adjust the spacing until the ellipse equation converges to the correct value. Example 13.2 shows the procedure.

Example 13.2: Suppose we drain an irrigated area to a depth of 1.8 m; the minimum depth to the water table is 1.5 m. The hydraulic conductivity is 0.5 m/day. Depth to an impeding layer is 6.7 m. Excess irrigation is 1.2 mm/day. Assume a 0.127 m pipe (diameter). Find the drain spacing.

Assume a spacing of 55 m. Then $d = 6.7$ m $- 1.8$ m, or $y = 4.9$ m and $d/Sp =$ 4.9 m/55 m = 0.09, less than 0.31. Then we must use equations 13.5a–b:

$$C_{eq} = 3.55 - 1.6(4.9/55) + 2(4.9/55)^2 = 3.42$$

Note that 1.2 mm/day is 0.0012 m/day.

$$d_e = 4.9/\{1 + [(4.9/\text{sp})\ 2.55 \ln (4.9/0.127) - (4.9/55)\ 3.42]\} = 3.28 \text{ m}$$
$$Sp = (\{(4)(0.5 \text{ m/day})[(3.58 \text{ m})^2 - (3.28 \text{ m})^2]\}/(0.0012 \text{ m/day}))^{1/2} = 58.6 \text{ m}$$

If Sp was more than 10% different from the initial assumed Sp value, resolve using the most recent Sp value as the next guess Sp. Convergence usually occurs in two or three iterations.

Parameters in the above equation are variable. In actuality, the spacing decision for a field is a tradeoff between the theoretical value and the economic value.

Non-steady-state drainage is usually more relevant than steady-state drainage in humid areas. The presence of sensitive crops may require dropping the water table by some distance to ensure that submergence lasts for no more than, say, 12 hr. The Van Schilfgaarde (1974a) relationship (equation 13.6a) is one way of solving for spacing in these circumstances:

$$Sp = \left[\frac{9K_{sat}td_e}{n_{pd}\,(\ln[y_0(2d_e + y)] - \ln[y(2d_e + y)])}\right]^{1/2} \qquad \textbf{(13.6a)}$$

where

t is the time (T) for the water table to drop from y to y_0, measured at the centerline between two pipes (see Figure 13.4);
n_{pd} is the drainable porosity of the water-conducting soil expressed as a fraction, often defined as voids draining due to gravity; Table 3.1, equilibrium moisture content, provides some guidance for drainable porosity;
y_0 is the initial water table height above the drain line;
y is the final water table height above the drain line;
Sp is the computed drain spacing.

Van Schilfgaarde (1974a) derived this equation following an analogy with the one-dimensional transient heat conduction equation. Other nomenclature is the same as for the steady-state case. The dependence of spacing on equivalent depth makes this case an iterative process as well. Example 13.3 shows the use of equation 13.6a.

Example 13.3: Calculate the drain spacing for a falling water table assuming K = 0.3 m/day, n_{pd} = 0.01, drain depth to centerline = 0.9, y_0 = 0.9 m, y = 0.6 m, and t = 1 day for water to drop from y = 0.9 m to 0.6 m. The drain is 0.61 m above a claypan barrier.

Assume a spacing Sp of 20 m. From equations 13.5, d/Sp = 0.03. If pipe diameter = 0.127 m, then the pipe r = 0.064 and d/r = 9.37. Then c_{eq} = 3.504 and d_e = 0.56 m. From equation 13.6a,

$$Sp = [((9)(0.3 \text{ m/day})(1 \text{ day})(0.56 \text{ m}))/(0.01)[\ln\{0.9 \text{ m } (2{*}0.56 \text{ m} + 0.6 \text{ m}) - \ln\{0.6 \text{ m}(2{*}0.56 \text{ m} + 0.9 \text{ m})\}]]^2 = 18 \text{ m}$$

which is reasonably close (e.g., 10%) to the assumed 20 m.

Table 13.2. Suggested spacing and drain depths for humid areas for optimum crop yield.

Soil	Hydraulic grouping	Conductivity (m/day)	Spacing (m)	Depth (m)
Clay and Clay loam	Very slow to slow	0.001 to 0.005	9 to 21	0.9 to 1.1
Fine sandy loam	Moderate	0.02 to 0.06	30 to 37	1.2 to 1.4
Sandy loam	Moderately rapid	0.06 to 0.13	30 to 60	1.2 to 1.5
Peat and muck	Rapid	0.13 to 0.25	30 to 90	1.2 to 1.5

Source: Adapted from ASAE 1997.

Note: These recommendations do not apply when the objectives include the reduction of nitrates in effluent.

Small errors in the initial and final water table height cause large errors in predicted time. Hydraulic conductivity is influential. The solution is not sensitive to pipe radius, drainable porosity, or depth to the impermeable layer. The low dependence on pipe radius effectively enables decoupling of the pipe spacing determination from the pipe size determination. Pipe diameter determination depends on the pipe flow.

The Van Schilfgaarde (1974a) equation is not applicable if the drain lies on the barrier (e.g., $d = d_e = 0$). Van Schilfgaarde developed equation 13.6b for the case $d_e = 0$:

$$Sp = \left[\frac{9K_{sat}t}{(2n_{pd}y_0y)/(y_0 - y)}\right]^{1/2} \tag{13.6b}$$

Nomenclature is the same as for equation 13.6a. He assumed an initial elliptical water table with equations 13.6a–b. See Van Schilfgaarde 1974b for relationships derived using other initial water table configurations.

As with steady-state drainage, actual spacing is usually an economic tradeoff. One can also design deep drainage ditches using both the steady and the non-steady-state approaches. For humid areas, Schwab et al. (1993) suggest spacings shown in Table 13.2. Spacings in Table 13.2 do not reflect nutrient control objectives.

Pipe Depth

Pipe depth should not be deeper than the restricting layer and should not be less than 0.6 m when there is vehicle traffic over the surface. Skaggs (1998) suggested that profitability from a crop yield point of view was largely insensitive to depth and spacing. Table 13.2 shows target depths for optimum crop yields. Along with the other depth constraints, the field slope will often dictate depth. Drainage lines in irrigated soils are usually deeper in arid applications to collect saline leachate. Narrow spacing and shallow placement minimized nitrate leaving fields (Skaggs 1998). It was hypothesized that better water table control afforded by narrowly spaced and shallow drains encouraged denitrification.

For nonagricultural applications where drainage pipe must support heavy construction equipment, one must match the strength properties of the drainpipe to the loading conditions. Appendix C provides a pipe loading discussion.

Pipe Slope

Table 13.3 presents minimum slopes to avoid sand/silt buildup in drains. Grades steeper than 2–3% can lead to undermined supports, causing possible misalignments in the tile line and resulting in sand entry and blockages.

Table 13.3. Minimum slopes (%) for tile drains.

Nominal diameter (mm)	Fine sand or silt not present[a]		Fine sand or silt may enter the drain[b]	
	Tile	Tubing	Tile	Tubing
	Grade (%)			
75	0.08	0.10	0.60	0.81
100	0.05	0.07	0.41	0.55
125	0.04	0.05	0.30	0.41
150	0.03	0.04	0.24	0.32

Source: Adapted from ASAE 1997b.

[a]Minimum clearing velocity of 0.15 m/s.
[b]Minimum clearing velocity of 0.42 m/s.

Design Drainage Rates

Drainage fractions based on years of experience are tabulated in Table 13.4 for humid conditions. For spacing with steady-state flow, one may substitute these drainage rates into the steady-state spacing equation. Do not substitute that portion of drainage entering the pipe via a surface opening. Table 13.4 gives drainage coefficients (determining flow rate) for hydraulic capacity design of the tile. The drainage coefficients in Table 13.4 are also applicable for tile line hydraulic capacity with non-steady-state flows. ASAE (1997c) describes drain design processes for unsteady flow systems.

Drainage is crucial for salinity control in arid regions. For arid regions, drainage and seepage discharge from irrigated lands may be estimated using

$$D_{c\,arid} = \frac{I(P + Seep)}{100t_{inte}} \tag{13.7}$$

Table 13.4. Drainage coefficients (D_c) for humid areas.

Crops and degree of surface drainage	Drainage coefficient (D_c)[a]	
	Mineral soil (clay and silt) (mm/day)	Organic soil (mm/day)
Field crops		
Normal[b]	10 to 13	13 to 19
With blind inlets	13 to 19	19 to 25
With surface inlets	13 to 25	25 to 38
Truck crops		
Normal[b]	13 to 19	19 to 38
With blind inlets	19 to 25	38 to 51
With surface inlets	25 to 28	51 to 102

Sources: Adapted from ASAE 1997b.

[a]Where available follow local recommendations.
[b]Adequate surface drainage with outlets to other drains or ditches provided.

where

$D_{c\ arid}$ is a drainage coefficient for arid regions in mm/day;
P is deep percolation expressed as a percentage of irrigation application;
Seep is field canal seepage loss expressed as a percentage;
I is irrigation depth in mm;
t_{inte} is the time between irrigations in days.

Deep percolation would include leaching plus (100 − application efficiency)%. Canal seepage S would be (100 − conveyance efficiency)%. The drainage coefficient usually ranges from 3 to 6 mm/day. In arid regions, the maximum flow is governed mainly by leachate collection. Flows vary from 0.2 to 70 l/s per 100 m of drain length, with an average of 9.3 l/s per 100 m. Consult local expertise in these situations. Salinity and similar conditions might need to be taken into account in waste irrigation approaches, soil remediation schemes, and perhaps some bioremediation schemes. Example 13.4 applies equation 13.7 to drainage coefficient prediction.

Example 13.4: Given an irrigation depth of 25 mm, 5 mm leaching and deep percolation losses, 10% seepage losses in the irrigation lateral, and a 5-day irrigation interval, estimate the arid drainage coefficient.

$$P = (5/25)100 = 20\%$$

From equation 13.7,

$$D_{c\ arid} = 25 \text{ mm } (20\% + 10\%)/(100\% \times 5 \text{ days}) = 1.5 \text{ mm/day}$$

Drain Size (Diameter)

The Manning equation, full flow, is customarily used for tile line sizing. Using the drainage coefficient to express flow rate as depth drained times area over time and expressing hydraulic radius as $r/2$, the Manning equation appears as

$$d = 51.7(D_cAn)^{0.375}\, s_d^{-0.1875} \quad \textbf{(13.8)}$$

where

A is area in hectares;
n is the roughness coefficient;
s_d is the drain slope;
D_c is the drainage coefficient in mm/day;
d is the tile diameter in mm.

For clay tiles, $n = 0.013$. For corrugated tubing, use $n = 0.015$ for diameters ranging from 76 to 203 mm; 0.017 for diameters ranging from 203 to 305 mm; and 0.02 for diameters greater than 305 mm. Figures 13.5 and 13.6 show nomograph solutions for this equation. Examples 13.5 and 13.6 show the computations.

Example 13.5: Determine the diameter of corrugated plastic tubing and flow rate where the slope is 0.4%, the drainage area is 12 ha, and the drainage coefficient is 13 mm/day.

From equation 13.8,

$$d = 51.7\ (13 \text{ mm/day} \times 12 \text{ ha} \times 0.017)^{0.375}\ (0.004)^{-0.1875} = 210 \text{ mm}$$

The next largest commercially available size is 254 mm.

The pipe flow rate is

$$q = 12 \text{ ha} \times 13 \text{ mm/day} \times 10{,}000 \text{ m}^2\text{/ha} \times 1000 \text{ l/m}^3 \times 0.001 \text{ m/mm} \times 0.0000116 \text{ day/s} = 18.1 \text{ l/s}$$

Example 13.6: Determine the corrugated pipe size to carry the design flow from 910 m length of pipe spaced 30 m apart if drainable porosity is 0.04 and the water table must drop 0.24 m/day. The drainage pipe gradient is 0.0015.

$$q = 910 \text{ m} \times 30 \text{ m} \times 0.24 \text{ m/day} \times 0.04 \times (1000 \text{ l/m}^3)/(86{,}400 \text{ s/day}) = 3 \text{ l/s}$$

$$\text{area} = 910 \text{ m} \times 30 \text{ m}/(10{,}000 \text{ m}^2\text{/ha}) = 2.73 \text{ ha}$$

$$D_c = 0.24 \times 0.04 \times 1000 \text{ mm/m} = 9.6 \text{ mm/day}$$

$$d = 51.7(9.6 \text{ mm/day} \times 2.73 \text{ ha} \times 0.015)^{0.375}(0.0015)^{-0.1875} = 123 \text{ mm}$$

Select 127 mm as the next largest size available.

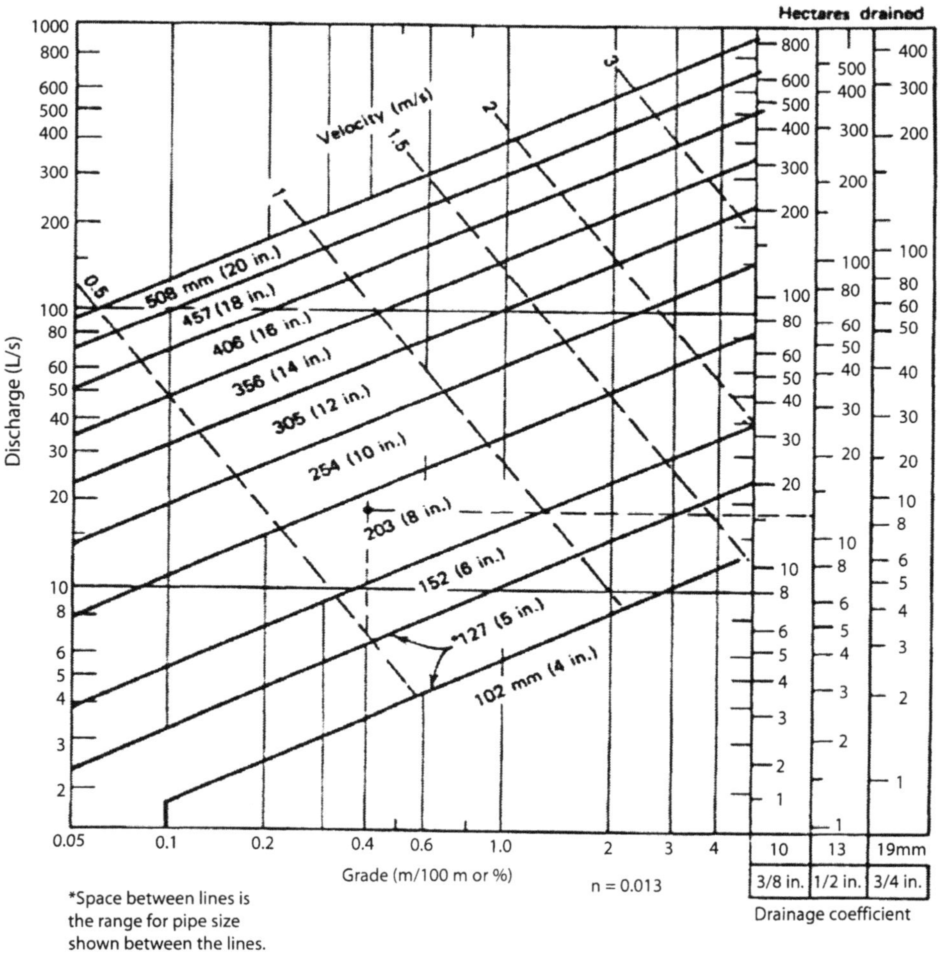

Figure 13.5. Nomograph for solving the tube drain capacity equation, smooth pipe with $n = 0.013$. (From ASAE 1997b; used with permission.)

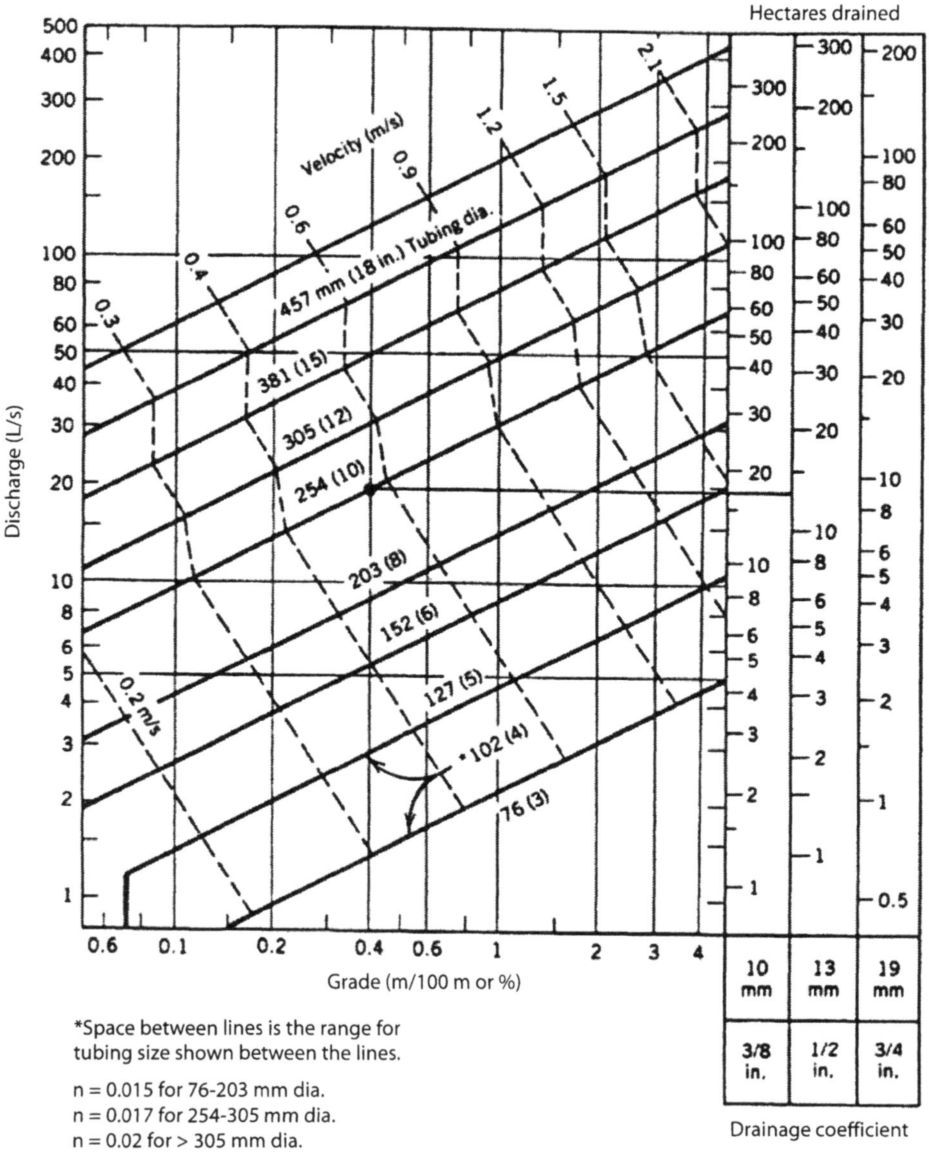

Figure 13.6. Nomograph for solving the tube drainage equation for corrugated plastic tubing with *n* varying as indicated. (From ASAE 1997b; used with permission.)

Compute the flow rate for a main as for a lateral, using the field area above the discharge point. Allow for expansion and provide for surface inlets as needed.

Drainage System Design and Installation

A design starts with an accurate topographic field survey, along with determinations of soil texture, soil layering, and impermeable layer presence. Select the relevant drainage coefficient. Determine the optimum water table depth for optimum production and/or optimal nutrient management in consultation with an agronomist. Determine the depth and spacing, using the tabulated recommendations above for humid

Figure 13.7. Plastic drain tube installation with laser grade control.

areas, unless precise knowledge of soil properties is available, in which case use the steady-state or non-steady-state calculations suggested above. Spacing design may be made using a nominal tube radius of 101–151 mm. Tube radius has little effect on the spacing calculation. For arid to semiarid areas, set spacing based on calculations outlined in ASAE (1997c) standards. Use surface inlets to drain any potholes, etc. Determine pipe size as discussed above, being sure to use the appropriate field area. Do not forget relief pipes, breathers, and sedimentation basins as required. Relief pipes and breathers are placed at locations where pipeline slope changes. Sedimentation basins are placed at junctions as needed to enable sediment cleanout. Use flap valves at outlets for rodent control. Figure 13.7 shows the installation of a flexible drain tube.

The drain line should be installed on an appropriate bed for support and to keep excess sediment from entering the pipeline. The bed should fit the form of the pipeline as shown in Figure 13.8. This reduces the possibility of crushing the pipe. Concerning filter design, the D_{85} of the bed (filter) material should be greater than or equal to the maximum openings

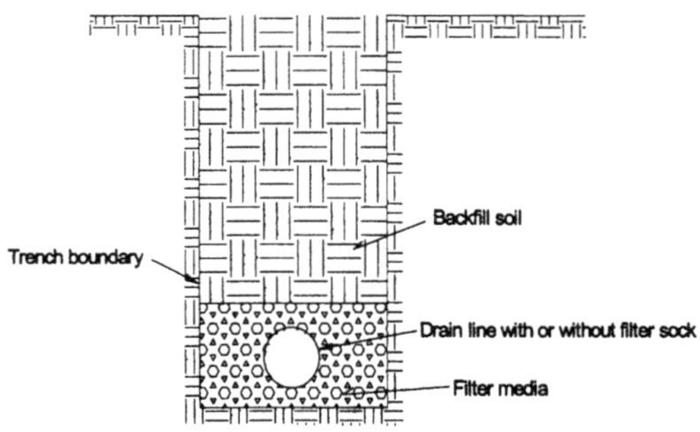

Figure 13.8. Cross section of a properly installed pipe drain.

in the drain line. Ratios defined by equations 13.9a–c should hold (NRCS 1973) for drain line filters:

$$\frac{d_{50}(filter)}{d_{50}(base)} = 12 \ through \ 58 \tag{13.9a}$$

$$\frac{d_{15}(filter)}{d_{15}(base)} \geq 12 \ through \ 40 \tag{13.9b}$$

$$\frac{d_{15}(filter)}{d_{85}(base)} \leq 12 \ through \ 40 \tag{13.9c}$$

No more than 10% of the filter material should pass the 25 mm sieve, and 85% of the filter material should be larger than the holes in the drain line.

The filter sock is an alternative approach to excluding sediment from the pipelines. The filter sock is a fabric material inhibiting movement of soil material into the pipe while maintaining adequate conductivity. The use of the filter sock is reducing the importance of the filter aspects of the bed. One may purchase drain line with the filter sock preinstalled.

Example 13.7: Evaluate the drain pipe filter with sizes given below for suitability for a soil with the specified sizes:

D_{15} of the filter is 0.03 mm
D_{15} of the soil base is 0.007 mm
D_{50} of the filter is 1.5 mm
D_{50} of the soil base is 0.07 mm
D_{85} of the filter is about 10 mm
D_{85} of the soil base is about 0.5 mm

Is this filter material compatible with the soil base?

The D_{50} ratio is 1.5 mm/0.07 mm = 21.5, which is within the acceptable range. The D_{15} ratio is 0.03 mm/0.007 mm = 4.28, which is somewhat low. The D_{15}/D_{85} ratio is 0.03 mm/0.5 mm = 0.06, which is acceptable. The filter D_{85} is larger than the 2 mm openings usually found in drain line, which is acceptable. Approximately 10% of the proposed filter material passes the 25 mm sieve, which again is acceptable. The filter should consist of somewhat larger material in the smaller size zone to be adequate. The fraction of larger particles is acceptable. One should plot these points for the filter and the base on a percent passing curve.

This introduction to pipe drainage included steady-state drainage and some simple non-steady-state cases. In practice, one analyzes drainage problems over an entire season or seasons. ASAE (1997c) discusses a strategy based on similar equations for performing a seasonal analysis in arid regions. In arid and humid areas, one can study seasonal drainage problems using software such as DRAINMOD (Skaggs 1980; Skaggs et al. 1988).

Most drainage literature views the drainage spacing-depth question from a crop yield standpoint. Using DRAINMOD, Skaggs et al. (1998) have shown that crop profitability is insensitive to the depths and spacing usually found in humid regions. In other words, crop yield does not vary greatly with spacing for commonly used spacings. Thus, design of drainage systems for yield is forgiving in the face of choices made by owners and contractors. Deep drains with wide spacings accomplish the greatest level of water removal and are preferred based on installation cost, ranging from \$1.30 to \$3.60 per meter of length. Skaggs et al. (1998) have shown that the addition of an environmental objective—minimization of nitrate losses in drain effluent—is not inherently forgiving. Total nitrate in drain effluent is much greater from the deep, wide installations. Thus, a shallower, narrow installation can significantly reduce nitrogen loss (by encouraging denitrification in winter) without significantly impairing crop yields.

Accessories

Proper specification of drainage system accessories may ensure the effectiveness and design life usefulness of a drainage system. Surface inlets are intake structures for surface water removal from potholes, road ditches, or other depressions. One should remove water with surface inlets whenever practical. The major functions of the structure are to admit water without admitting debris. Various grates or blind inlets filter out debris. The blind inlet consists of 150–200 mm aggregate near the drain and 250–350 mm aggregate near the surface. Aggregate is uniformly graded in size between the drain and soil surface. One must maintain inlets to ensure adequate performance.

One should specify sedimentation basins at junctures where slope decreases with the flow. These basins are approximately $1 \times 1 \times 1.5$ m (L, W, D) concrete or masonry structures covered with a solid cover or grate. These may serve as junction boxes as well.

Relief pipes and breathers are small vertical risers extending from the drain line to the soil surface. Relief pipes provide pressure relief, and breathers allow vacuum breaks. One usually requires breathers where the slope changes from low to high values, and pressure relief where the slope changes from high to low values. Negative pressures resulting in pipe failure can develop just above steep sections. In addition, large positive pressures resulting in blowout can develop at the bottom of steep sections. Relief pipes and breathers solve these problems.

Pipe outlets should have gates for rodent exclusion. One should spread soil as needed to ensure at least 0.6 m pipe coverage. Use drainage envelopes consisting of uniformly graded rock and gravel or corrugated tubing with a filter sock for excluding fine soils as needed. To prevent root intrusion, brush and trees should not grow within 30 m of a pipeline. To prevent outlet blockage, outlets should project several centimeters from collection channel banks. One should take precautions to minimize erosion under drain outlets. Installations in muck soils are especially prone to bacterial deposits, which cause clogging. Advanced texts (e.g., Luthin 1973; NRCS 1973) give additional discussion on system accessories and maintenance of drainage ways.

Subirrigation

Figure 13.9 shows a scheme for tile drainage/subirrigation with pipe drains. Figure 13.1 shows an open channel drain that is also used for subirrigation. Both pipe and channel systems for subirrigation require appropriate control structures.For example, crest boards placed in vertical slots maintain desired levels in subirrigation systems (e.g., water table control systems). Sumps are common drainage structures. A sump is a structure for pump inlet and storage at a drainage pump station. Figures 13.10–13.14 show some generic drainage control structures for pipe systems having the dual drainage/subirrigation function.

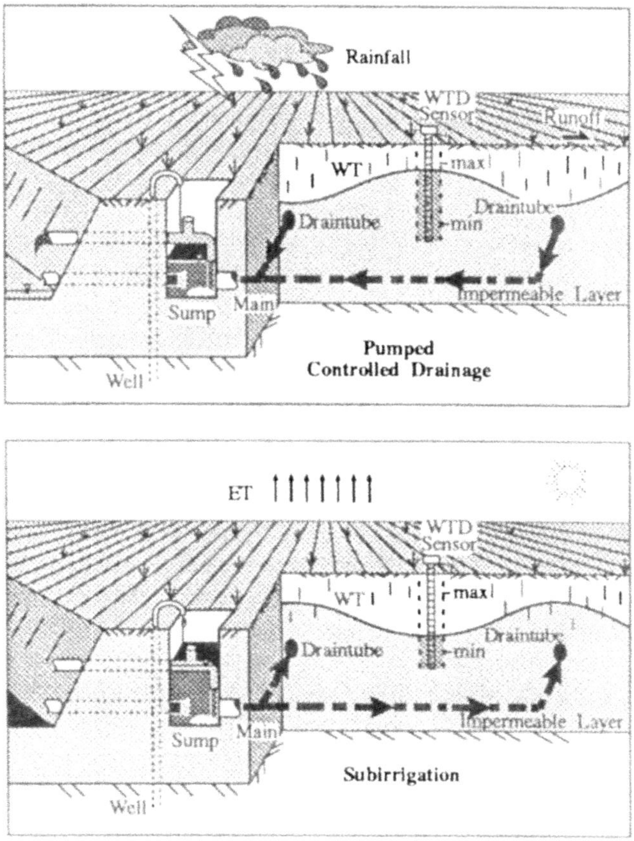

Figure 13.9. Combination drainage/subirrigation system. (From Fouss et al. 1990; used with permission of ASAE.)

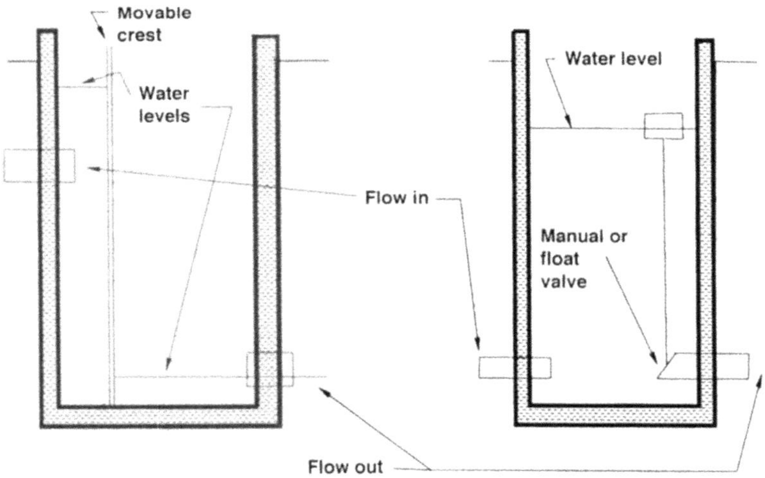

Figure 13.10. Drainage/subirrigation control structure schematic, with and without removable crest boards.

Figure 13.11. Water table control structure. (Courtesy of Dr. Sharyl Walker, University of Illinois.)

Materials

Drainage structures are fabricated from polyethylenes (PE), polyvinylchlorides (PVC), various epoxies and other plastics, metal, and concrete. Most of the PVC and related plastic materials are prefabricated. Manufacturers can provide much information on corrosion resistance and field durability. Manufacturers should provide test results (American Society of Testing and Materials, http://www.astm.org) on request. One should include test results and expectations into installation contracts . Corrosion, durability in the presence of sunlight and extreme temperatures (especially with PVC and PE materials), pressure requirements, brittleness, expansion-contraction with temperature, and crushing

Figure 13.12. Interior of T valve for flow control in subirrigation applications. (Courtesy of Dr. Sharyl Walker, University of Illinois.)

Figure 13.13. Exterior of T valve used for flow control in subirrigation applications. (Courtesy of Dr. Sharyl Walker, University of Illinois.)

resistance are important issues. One must specify minimum allowable hydraulic conductivity for pond liners (e.g., clay blankets, high-density polyethylene [HDPE], other geosynthetic materials). Relevant ASTM material specifications should be included in design reports.

Metals employed in drainage structures range from various steels to galvanized metal products. Vendors and manufacturers can provide ASTM test results when needed, and one should clearly write expectations of these results into project contracts. The American Institute of Steel Construction (http://www.asci.org) provides excellent publications relating to reinforcing steel design and materials specification.

Figure 13.14. Water table control structure viewed from above, used for nutrient management and subirrigation in the North Carolina coastal plains.

For concrete structures built on site, one should specify air content and consistency, mix design, inspection and testing, consolidating and curing, preparation and removal of forms, and fabrication and placing of reinforcement. The American Concrete Institute (http://www.aci-int.org) and Portland Cement Institute (http://www.portcement.org) provide excellent publications and guidance.

Standards organizations such as ASAE (http://www.asae.org) and ASCE (http://www.asce.org) are excellent resources on hydraulic structure installation and maintenance. Local offices of the NRCS will also provide assistance.

Shallow Wells

The shallow well is an important drainage tool in bioremediation projects and involves a straightforward application of DF hydraulics. Shallow wells also serve as water sources for small-scale irrigation projects. Well hydraulics is practically a discipline in its own right (e.g., see Bouwer 1978), and this chapter only introduces the topic and concerns only gravity and confined wells in homogeneous soils bounded by an aquiclude (an impervious layer). The discussion is limited to steady-state withdrawal scenarios.

Groundwater table, piezometric surface, intake and recharge areas, water-bearing formation or aquifer, cone of depression in both gravity and confined wells, and cone of injection are common terms. Figures 13.15–13.16 illustrate some terminology.

The U.S. Geological Survey site at http://www.usgs.gov provides extensive information about groundwater aquifer status in the United States. The USGS data base covers only large aquifers.

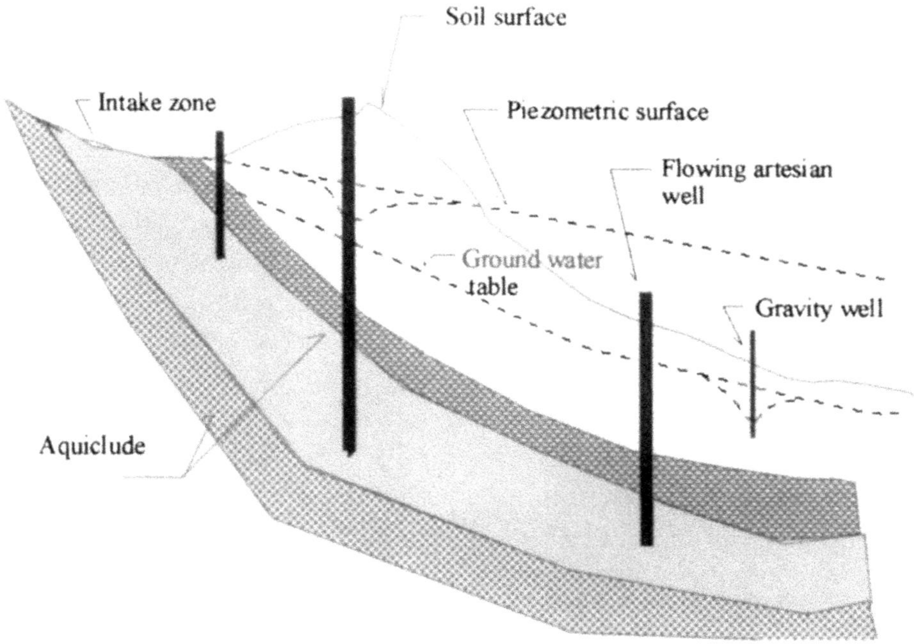

Figure 13.15. Schematic of an aquifer showing recharge zone and various flow conditions.

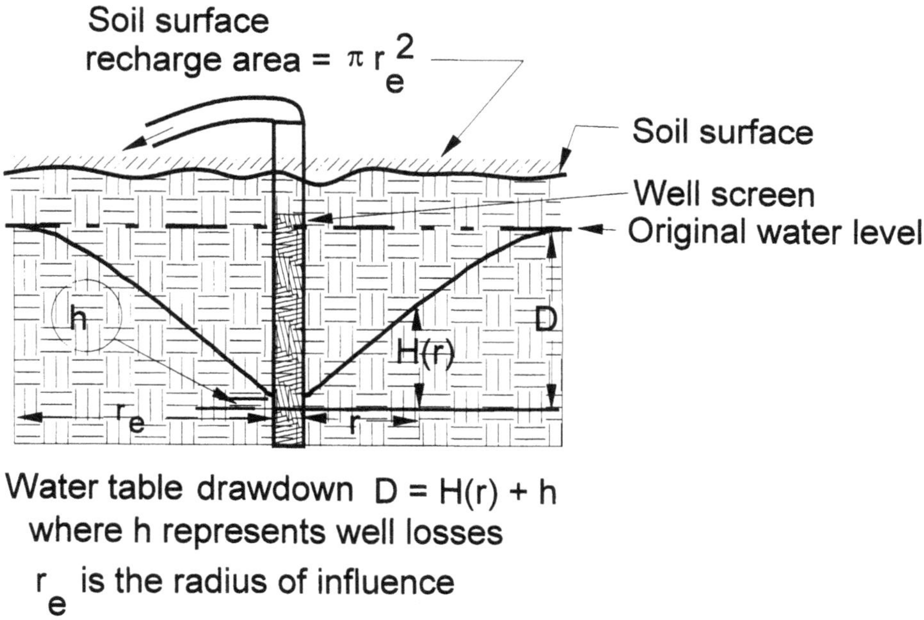

Figure 13.16. Schematic of gravity well with recharge.

One may define an aquifer as a soil formation conveying water. In Figure 13.15 the zone between the aquicludes is a confined aquifer. An aquiclude is impervious to water movement. Shallow groundwater tables are often found between the soil surface and an aquiclude. Rainfall percolation often recharges shallow groundwater tables. Confined water tables receive recharge on higher elevations having permeable soils. The piezometric surface is the height to which water will rise in a system. The flowing artesian well is a discharging well because the actual well elevation is below the piezometric surface. The gravity well by definition penetrates the shallow groundwater table. The cone of depression is a lowering of the shallow groundwater table near a gravity well. Active wells in confined aquifers have a cone of depression in the piezometric surface. A cone of injection occurs when one pumps water into a well. The analyses of this chapter will be limited to fully penetrating wells operating under steady-state conditions. Well losses are not considered. One should identify the terms defined above in Figures 13.16 and 13.17, which show elements of flow analyses in unconfined and confined aquifer systems.

Flow from a gravity well with no recharge may be described as follows. One may start with a simple mass balance on a cylinder at radius r from the well as denoted in case A in Figure 13.17. Flow q is negative since flow is leaving the system. The q is found from applying Darcy's law. The datum for defining the heads is the bottom of the well. Area times velocity is the well flow from the continuity equation. This analysis assumes a fully penetrating well. The unconfined flow relationship derivation begins with the flow (q) equation in Figure 13.16. Separating the variables leads to

$$\frac{q\,dr}{2\mathrm{p}K_{sat}r} = y\,dy \qquad \textbf{(13.10)}$$

Integration from the well radius (r_w) to the radius of influence (r_e) and from depth at

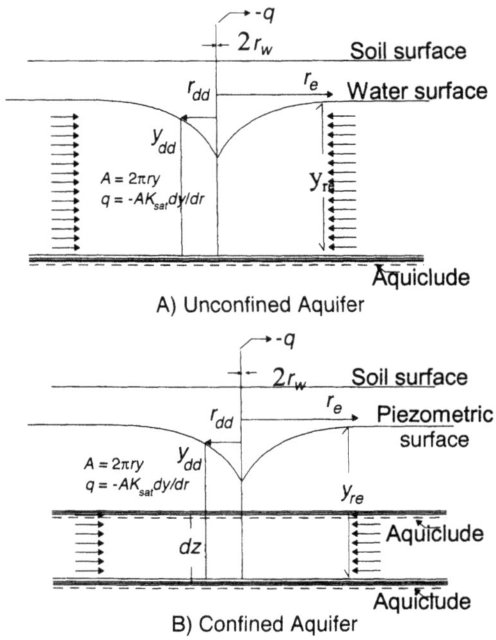

Figure 13.17. Definition sketches for DF analyses of (a) unconfined and (b) confined wells.

well (h) to static water table height (H) then results in

$$q_{ucon} = \frac{\pi K_{sat}(y_{re}^2 - y^2)}{\ln(r_e/r_w)} \qquad \textbf{(13.11a)}$$

where

K_{sat} is the hydraulic conductivity (L/T), understood to be at saturation conditions;
r_w is the radius of the well (L);
r_e is the equivalent radius of influence (L), which is the distance away from the well such that the water surface is essentially flat;
y_{re} is the water table height above the lower barrier before drawdown (L); that is, water table height at or beyond radius r_e;
y is the water table height at the well after drawdown (L);
q is q_{ucon}, the steady-state unconfined well flow rate (L^3/T).

One may also find the drawdown ($y_{re} - h_{dd}$) at some radius r_{dd} by solving equation 13.11a for drawdown as shown in equation 13.11b:

$$Q = \frac{\pi K_{sat}(y_{re}^2 - y_{dd}^2)}{\ln(r_e/r_{dd})} \qquad \textbf{(13.11b)}$$

where y_{dd} is the water table height at r_{dd}. The solution is left to the student.

The gravity equation does not account for well losses. The relation is based on the DF assumptions. These assumptions are not true near the well.

Depth to the impermeable layer is influential in the problem. The conductivity is also influential. Although the radius of influence is imprecise, its effect is small, allowing less precise knowledge to suffice.

For the confined aquifer with complete well penetration, the flow is

$$q_{conf} = \frac{2\pi d\, K_{sat}\,(y_{re} - y)}{\ln(r_e/r_w)} \tag{13.12a}$$

where

q_{conf} is the flow rate from a fully penetrated confined aquifer;
d is the thickness of the confined zone layer;
y_{re} is the height of the piezometric surface above the top of the confining layer before pumping, or the height of the piezometric surface at or beyond r_e after equilibrium;
y is the effective pressure head of water in the well relative to the top of the confining layer.

One may derive the confined aquifer equation following the procedure shown for the unconfined aquifer. As for the unconfined case, the well is assumed to fully penetrate the aquifer. This derivation is left to the student (see case b in Figure 13.17).

Solving equation 13.12a for $(y_{re} - y_{dd})$ gives drawdown at any point between the well radius and the radius of influence:

$$q = \frac{2\pi d\, K_{sat}\,(y_{re} - y_{dd})}{\ln(r_e/r_{dd})} \tag{13.12b}$$

The radius of influence r_e depends on soil formation and texture. Table 13.5 provides estimates of radius of influence and approximate soil properties for well design approximations. Theoretically, the existence of the radius of influence with no recharge is impossible. The use of the radius of influence notion implies some limited recharge. Examples 13.8 and 13.9 show uses of the shallow well equations.

Table 13.5. Soil properties for use in simple water well design problems.

Soil formation and texture	Total porosity (%)	Drainable porosity n_{dp} (%)	Typical K_{sat} (m/day)	Radius of influence r_e (m)
Silty clay or finer	40 to 45	0.01 to 0.03	10^{-6} to 10^{-4}	< 30
Fine sand with some clay or silt	40 to 45	5 to 15	0.0001 to 1	30 to 90
Fine to medium sand formations fairly clean and free from clay and silt	10 to 20	15 to 25	0.011 to 5	90 to 180
Coarse sand and fine gravel formations free from clay and silt	20 to 30	20 to 25	50 to 5000	180 to 300
Coarse sand and gravel, no clay or silt	30 to 40	25	100 to 1000	300 to 600

Sources: Adapted from Bennison 1947 and Bouwer 1978.

Example 13.8: One drills a 0.304 m diameter well to a 100 m deep confined fine sand/silt formation with a K_{sat} = 0.01 m/day. Use 30 m as the radius of influence. Confining layers are 50 m and 100 m below the surface. If the total piezometric drawdown is 50 m, what is the pumping rate? What is the drawdown 10 m away from the well center?

Solving for the flow using equation 13.12a yields

$$d_z = 50 \text{ m} = \text{the zone thickness between the confining layers}$$
$$y_{re} - y = 50 \text{ m}$$

Thus

$$q = 2(3.14)(50 \text{ m})(0.01 \text{ m/day})(50 \text{ m})/(\ln 30 \text{ m}/0.152 \text{ m}) = 29.71 \text{ m}^3\text{/day}$$
$$= 0.00034 \text{ m}^3\text{/s} = 0.01 \text{ cfs} = 5.44 \text{ gpm (gallons per minute)}$$

The drawdown at 10 m is found from equation 13.12b:

$$29.71 \text{ m}^3\text{/day} = 2(3.14)(0.01 \text{ m/day})(50 \text{ m})(30 \text{ m} - y_{dd})/(\ln(30 \text{ m}/10 \text{ m}))$$
$$30 - y_{dd} = 10.4 \text{ m} \qquad y_{dd} = 19.6 \text{ m}$$

Example 13.9: Assume the above case was unconfined and the actual drawdown at the well was 50 m. The level was within 25 m of the surface before the well began operating. The bottom confining layer is present. The unconfined analysis is applicable.

One must first establish the drawdowns with respect to the well bottom.

The y at the well is 100 m − 25 m − 50 m = 25 m. The level y_{re} at r_e is 100 − 25 = 75 m.

$$q = 2(3.14)(0.01 \text{ m/day})[(75 \text{ m})^2 - (25 \text{ m})^2]/\ln(30 \text{ m}/0.152 \text{ m}) = 59.41 \text{ m}^3\text{/day}$$

(equations 13.11a–b). The drawdown 10 m from the well is

$$59.41 \text{ m}^3\text{/day} = 2(3.14)(0.01 \text{ m/day})[(75 \text{ m})^2 - h_{dd}^2]/\ln(30 \text{ m}/10 \text{ m})$$

or

$$1039.31 \text{ m}^2 = (75 \text{ m})^2 - y_{dd}^2$$

This gives y_{dd} = 67.71 m. The drawdown is 75 m − 67.71 m = 7.28 m.

The DF assumptions enable simple analyses of seepage from open channels, subsurface runoff over an impermeable layer, uniform infiltration, and drainage to a stream. Bouwer (1977) describes other DF analyses potentially useful for bioremediation. Figure 13.18 shows a deep well drilling rig.

One could irrigate over some surface above the well and continuously circulate the water in some bioremediation schemes. We may also pump and inject as shown in Figure 13.19. The cone of injection may at first approximation be treated as a mirror image of the cone of depression. One may derive equations for confined and unconfined injection wells, noting q is positive in the injection case. With the injection well, the DF assumption is theoretically adequate but practically less realistic because it is not accurate near

Figure 13.18. Rotary drilling rig.

the start of injection and the time to attain a realistic radius of influence is unknown. Nyer et al. (1996) give numerous applications of injection and withdrawal wells for bioremediation applications. Tube drainage applications may also be feasible in some instances.

Increasing recharge may reduce the radius of influence to some extent. The horizontal flow assumption then becomes more problematic and the equation is less accurate. True horizontal flow is also questionable when the K value is spatially variable.

Some bioremediation schemes involve gas injection or pumping. One may compute intrinsic permeability (dimensions of length squared; see equation 3.2a) for the porous media

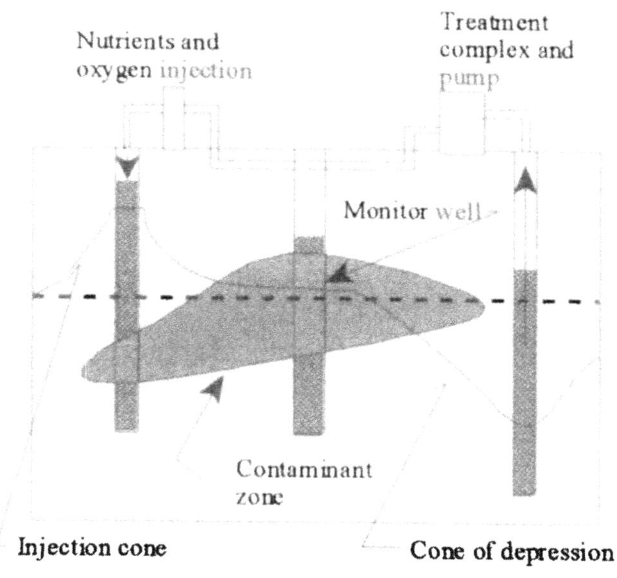

Figure 13.19. Bioremediation scenario involving pumping, treating, and injection. (Redrawn from Nyer et al. 1996.)

and conductivity computed for the gas. Use a radius of influence as with water in these computations. One may then use confined well hydraulics to model the problem. This assumes laminar flow and complete well penetration. These analyses neglect density effects. They assume horizontal movement of the gas over a zone at least as large as the radius of influence.

Situations with no confining layer require a more advanced analysis because the DF assumptions fail. One should regard this solution as a limiting case, as situations rarely satisfy these assumptions.

Example 13.10: Suppose one desires to inject and withdraw 10 m^3/s air from an unsaturated porous medium using the approach shown in Figure 13.19. A 0.304 m diameter well is installed to the top of a confining layer 10 m below the surface. Neglect losses in the well column. Assume the surface is impermeable. Estimate the gauge pressure required at the wellhead.

From Table 13.5, estimate the K_{sat} to be about 0.05 m/d (5.787×10^{-7} m/s) and the radius of influence for water to be about 100 m. From equation 3.2a the intrinsic conductivity is

$$5.787 \times 10^{-7} \text{ m/s } (102 \text{ N/m}^3)/0.001 \text{ N-s/m}^2 = 0.59 \text{ m}^2$$

The gas conductivity is then

$$K_{air} = 0.59 \text{ m}^2(0.1223 \text{ N/m}^3)/1.8 \times 10^{-5} \text{ N-s/m}^2 = 4009 \text{ m/s}$$

See any fluids text for air density and viscosity values. One must use the force units of dynamic viscosity. Then

$$10 \text{ m}^3\text{/s} = 2(3.14)(10 \text{ m})(4009 \text{ m/s})(y_{re} - y_d)/\ln(100 \text{ m}/0.304 \text{ m}) = 2 \times 10^{-4} \text{ m} = y_{re} - y_d$$

The gauge pressure is 0 at the surface. The gauge pressure at the well column is then

$$2 \times 10^{-4} \text{ m air } 1.3 \text{ kg/m}^3/9.81 = 0.1325 \text{ N/m}^3 \times 2 \times 10^{-4} = 2.65 \times 10^{-5} \text{ N/m}^2$$

This is a very low pressure and is not surprising due to the large well and low air flow rate. This should be regarded as a first order estimate due to the assumptions made in this simplistic analysis.

Well Installation and Maintenance

One may drill wells by a variety of methods. Jetting is good in shallow sand formations. Boring is used in shallow clay formations. The cable tool, which is the historical method, and the more modern rotary method perform well for deep well development in a variety of formations. Well drilling and maintenance are special disciplines. Drilling contractors are the best source of expertise for water supply wells, sampling wells, and test wells in a locality. Protection of the wellhead from pesticides and other potential groundwater contaminants is essential. Campbell and Lehr 1973 is an excellent source for various well drilling and development technologies.

Uniform Infiltration and Drainage to a Nearby Stream

Land application waste treatment/disposal schemes may be modeled as a steady application of water over a site with an aquiclude at some distance below the surface. Flow to a

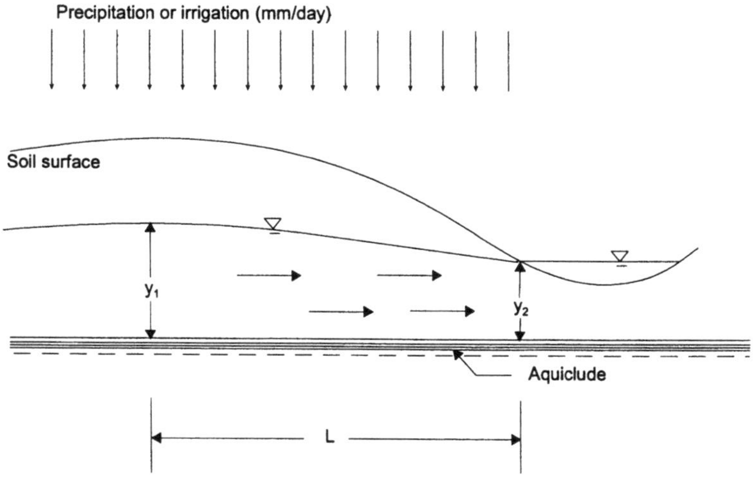

Figure 13.20. Infiltration of rainfall or irrigation and drainage to a stream.

nearby surface stream may be estimated using DF as follows, referring to Figure 13.20. One can write Darcy's law as

$$f_x = -K_{sat}\frac{dy}{dx} \tag{13.13}$$

where *dy/dx* is the water table surface slope. The datum for these analyses is the confining layer. One can write the flow per unit distance *x* as

$$q_x = -K_{sat}y\frac{dy}{dx} = Px \tag{13.14}$$

The term *Px* on the right is an expression of the rainfall (*P*) per unit distance *x* and per unit distance *y* (into the paper). Separate variables, integrate *x* from 0 to *L*, integrate *h* from h_2 to h_1, and solve for h_1 as shown in equation 13.15:

$$y_1 = \sqrt{y_2^2 + \frac{PL^2}{K_{sat}}} \tag{13.15}$$

Example 13.11: Given a soil with $K_{sat} = 0.01$ m/day, $h_2 = 10$ m, a precipitation rate of 0.005 m/day, and a distance from the stream to the top of the mound of 200 m, find the steady-state height of the groundwater mound. Assume the soil surface at the top of the mound is 200 m above the confining layer.

From equation 13.15,

$$y_1 = ((10\text{ m})^2 + \{(0.005\text{ m/day})[(200\text{ m})^2]/(0.01\text{ m/day})\}) = 142\text{ m}$$

Thus, the groundwater mound height is 200 − 142 = 58 m below the soil surface. Making a sketch of these conditions reveals significant departure from horizontal flow. Define the heads with respect to the bottom confining aquiclude. One can develop many other analyses involving the DF assumptions in a similar way. Texts such as Bouwer 1977 and Crites and Tchobanoglous 1998 give additional examples based on straightforward application of DF groundwater flow.

Outlook

Worldwide, removal of excess water from land area for crop production, desalinization for crop production, and the growth in nontraditional applications requiring water removal bode well for the future of a drainage industry. Drainage needs vary greatly from country to country. In the United States, the continuing need for surface drainage as development progresses and the growth of nontraditional drainage applications such as bioremediation provide continuing demand for drainage engineering services. Evaluation of existing drainage systems for additional benefits such as nutrient management will continue to stimulate research in the coming years. The Dupuit-Forchheimer assumptions will likely lie at the heart of new approaches in agricultural and nontraditional drainage system design.

One should attempt drainage only after ascertaining that the operation will not endanger wetlands. If possible, remove excess water with surface drainage techniques. Flat, permeable soils underlain with slowly permeable soils usually require subsurface pipe drainage. These systems can be managed for subirrigation. Modeling regional aquifers requires advanced techniques. Drainage and groundwater are each the topic of several advanced texts.

The addition of environmental objectives to the crop production objectives necessitates ongoing research. For example, Skaggs (1998) showed, using 40 yr simulations with DRAINMOD (Skaggs 1980; Skaggs et al. 1988) in many soils and climates, that shallow, narrow spacings reduce nitrate loss in drain effluent. He reported crop yield to be insensitive to spacing and depth with common drainage practices. Thus, prospects for nitrate control in drain effluent seem good if one applies proper management. The addition of the nutrient control objective to crop production/water control objectives will likely cause future revisions to subsurface drainage practices. Drainage effluents are often saline, especially in arid regions having limited water available for leaching excess salts (see Chapter 14).

Specification of drainage and subirrigation systems must include all materials specifications to be complete. This includes the specification of piping, envelope materials, accessory fittings such as vacuum breakers and pumps, flow measurement structures, concrete, etc. Appendix C further discusses materials specifications for structures. Texts such as Luthin 1973, Van Schilfgaarde 1974b, and Schwab et al. 1993 provide detailed information as do professional societies such as ASAE (http://www.asae.org), ASCE (http://www.asce.org), and ICID (http://cirrus.hrwallingford.co.uk/g7icid.htm). Contractors in many regions may be determined using Internet sites such as http://www.drainpro.com and via the Land Improvement Contractors Association (http://ourworld.compuserve.com/homepages/delmarva/staff.htm). An Internet search will reveal several commercial drainage design software packages.

Designs should include ASTM specifications for all construction materials. The National Ground Water Association (http://www.h2o-ngwa.org) publishes information on technical materials standards. Specifications for bioremediation wells follow closely those for water. Consult works such as Nyer et al. 1996 for case study reports.

Problems and Questions

Draw sketches of all completed designs where applicable.

1. In Michigan a 1000 ha location with heavy clay and with a 0.1% slope is to be surface drained by one drainage canal. There are no subsurface drains. Compute the runoff rate and design an optimal unlined canal to carry the discharge.

2. A clay loam with an impermeable layer is 5 m below the surface, and 101 mm diameter drains are installed at approximately 1 m below the surface. The soil conductivity is 1.1 m/day. The desired water table height is 1.3 m above the impermeable layer. A sustained rainfall of 25 mm/day is falling. Compute the steady-state drain spacing to achieve the desired water table.
3. A bioremediation project requires water application to leach a contaminant. Drains (101 mm diameter) were put 0.05 m over an impermeable layer 4 m deep. The water table can remain 3.5 m above the impermeable zone. The K is 3 m/day. The needed leaching rate is 20 mm/day. Compute the required drain spacing to maintain the desired water table level.
4. In a dry area, an irrigation system applies 30 mm irrigation plus a 5% leaching requirement on a 7 day cycle with seepage losses of 15% from the supply canal. The soil is a loam and is impermeable at 3 m. Find the drainage coefficient. If one treated the water application as a "quasi-steady-state" situation (e.g., compute the average daily equivalent application rate), how widely spaced would the 127 mm diameter drainage pipes need to be?
5. Given a soil with a drainable porosity of 0.1 in a clay loam having an impermeable layer 4 m below the surface and an original water table height of 3 m above the impermeable layer, compute the time required to drop the water table 0.3 m.
6. Water is used to transport a bioremediation nutrient broth (similar to water) over a site prepared for bioremediation. Drains are to be installed to complete the preparation. Drains will be placed on an impermeable zone 5 m below the surface. Conductivity of the soil is 8 m/day. Drainable porosity is 0.14. Soil above the impermeable zone will be saturated with the broth; however, the site must become essentially aerobic from the surface to a 4.5 m depth within 8 hr for the microbe to do its job (i.e., the water table must fall from the surface to 4.5 m). The process would cycle many times to complete the bioremediation. What is the drain spacing to drop the water table in the 8 hr interval?
7. Your client has a 20 ha field in a humid region with vegetable production. Size the corrugated plastic main carrying all the drainage effluent from this region. The slope is 1%. What is the design flow?
8. A 300 × 300 m fine sandy area is being bioremediated by leaching water at a rate of 25 mm/day and recovering the leachate via shallow wells. The soil K is 24 cm/day. The wells fully penetrate to a 50 m deep boundary. What is the spacing of the wells to pump the effluent back to maintain a steady-state condition, assuming the wells are spaced at half the radius of influence?
9. A well is to supply 100 gpm. The aquifer has a K value of 1 cm/day. The aquifer is 50 m thick, is fully penetrated by a well, and is overlain by an impermeable layer plus 100 m of surface soil. The piezometric head causes water to come within 10 m of the surface. At steady state the height can drop to the top of the aquifer but no farther. Can the well meet the 100 gpm flow rate requirement? If the flow is 50 gpm, what is the drawdown 5 m from the well? Present sketch(s) with your solution.
10. Develop prompt sheets for engineering design methodology customer requirements and technical specifications encompassing surface and subsurface drainage problems.
11. What are the Dupuit-Forchheimer assumptions? How are they applied in reference to Figures 13.4, 13.16, and 13.20?

12. What are the advantages and disadvantages of surface drainage approaches? Pipe drainage approaches?
13. What is meant by *radius of influence*? How can one compute a realistic estimate for flow from a shallow well using this imprecise term?
14. Figure 13.4 identifies generic components of the vadose zone in a pipe drainage situation. Figure 14.3 (Chapter 14) identifies key components in the water management zone important for irrigation planning. Compare and contrast Figure 13.4 with Figure 14.3. Relate the generic components of these figures to features identified in Figure 13.20. What common attribute of the intermediate zone is ignored in Figure 13.20 and, by extension, in Example 13.11?

References

ASAE. 1997a. *Design and construction of surface drainage systems on agricultural lands in humid areas.* EP302.4 AUG93.American Society of Agricultural Engineers, St. Joseph, MI.

———. 1997b. *Design and construction of subsurface drains in humid areas.* EP260.4 DEC96. American Society of Agricultural Engineers, St. Joseph, MI.

———. 1997c. *Design, construction, and maintenance of subsurface drains in arid and semiarid areas.* EP463.1 AUG95. American Society of Agricultural Engineers, St. Joseph, MI.

Beltran, J. M. 1999. Land drainage. In *CIGR handbook of agricultural engineering: Land and water engineering.* American Society of Agricultural Engineers, St. Joseph, MI.

Bennison, E. W. 1947. *Ground water, its development, uses, and conservation.* E. E. Johnson, St. Paul, MN.

Bouwer, H. 1978. *Groundwater hydrology.* McGraw-Hill, New York, NY.

Campbell, M. D., and J. H. Lehr. 1973. *Water well technology: Field principles of exploration, drilling, and development of groundwater and other selected minerals.* National Water Well Association, McGraw-Hill, New York, NY.

Crites, R., and G. Tchobanoglous. 1998. *Small decentralized wastewater management systems.* WCB McGraw-Hill, New York, NY.

Dideriksen, R. I., A. R. Hildebaugh, and K. O. Schmude. 1978. *Erosion inventory-sheet and rill erosion.* ASAE Technical Paper no. 78-2514. American Society of Agricultural Engineers, St. Joseph, MI.

Evans, R. O., ed. 1998. *Drainage for the 21st century: Food production and the environment.* ASAE Publication 02-98. American Society of Agricultural Engineers, St. Joseph, MI.

Fouss, J. L. 1990. Drain tube materials and installation. In *Drainage for agriculture,* ed. J. Van Schilfgaarde. Agronomy Monograph no. 17. American Society of Agronomy, Madison, WI.

Jensen, M. E. 1993. The impacts of irrigation and drainage on the environment. Fifth N. D. Gulhati Memorial Lecture. International Commission on Irrigation and Drainage, The Hague.

Luthin, J. N. 1973. *Drainage engineering.* John Wiley & Sons, New York, NY. Reprint, R. E. Kreiger, Huntington, NY.

NRCS. 1973. *Drainage of agricultural land.* National Engineering Handbook Section 16. Reprint, Water Information Center. Syosset, NY.

———. 1998. *USDA national resources inventory, 1997 update,* http://www.nrcs.usda.gov. Natural Resource Conservation Service, Washington, DC.

Nyer, E. K., S. Fam, D. F. Kidd, F. J. Johns II, P. L. Palmer, G. Boettcher, T. L. Crossman, and S. S. Suthersan. 1996. *In situ treatment technology.* Lewis Publishers, Boca Raton, FL.

Rhoades, J. D. 1974. Drainage for salinity control. In *Drainage for agriculture,* ed. J. Van Schilfgaarde. Agronomy Monograph no. 17. American Society of Agronomy, Madison, WI.

Schwab, G. O., D. D. Fangmeier, W. J. Elliot, and R. K. Frevert. 1993. *Soil and water conservation engineering.* 4th ed. John Wiley & Sons, New York, NY.

Skaggs, R. W. 1980. *DRAINMOD reference report.* USDA-SCS (NRCS), South National Tech. Center, Fort Worth, TX.

———. 1998. Notes from presentation given in February 1998, Graduate Research Seminar, Athens, GA.

Skaggs, R. W., J. E. Parsons, and K. D. Konyha. 1988. DRAINMOD version 4.0: An overview. ASAE Technical Paper no. 88-2563. American Society of Agricultural Engineers, St. Joseph, MI.

Van Schilfgaarde, J. 1974a. Nonsteady flow to drains. In *Drainage for agriculture,* ed. J. Van Schilfgaarde. Agronomy Monograph no. 17. American Society of Agronomy, Madison, WI.

———. ed. 1974b. *Drainage for agriculture.* Agronomy Monograph no. 17. American Society of Agronomy, Madison, WI.

14

Introduction to Irrigation Design

Irrigation is critical for crop production in many parts of the world. Irrigation makes up shortfalls in precipitation during critical growth periods in humid regions. Seasonal applications in arid regions may exceed 1 m. Irrigation is increasing in importance for land disposal of wastewater and is an integral component of some bioremediation schemes. Water can be an efficient nutrient carrier and pollutant remover from sites undergoing remediation. Nonagriculturally motivated irrigation applications along with conventional agricultural applications continue to increase in market share.

Agricultural applications have driven the development of irrigation methods and techniques. Irrigation enabled civilizations in the arid Mediterranean basin and in the Orient to establish long-term residency. Irrigated agriculture reduces the effects of climate fluctuations on our lifestyle. Seasonal water applications in humid areas may exceed 150 mm over a growing season. Crops in arid areas often receive up to 2 m of irrigation. In the farm and ranch irrigation survey of the Census of Agriculture, nearly 25% of all U.S. farmland was irrigated. Worldwide, irrigated area is increasing by 30% per year (van der Leeden et al. 1990). Asian countries now have 50% of the world's irrigated land (FAO 1999). Agricultural irrigation is increasing in importance worldwide (van Lier et al. 1999).

To specify irrigation for a particular site—nontraditional application or for agricultural crops—one needs to answer most or all of the following questions:

- What is the volume of the available soil storage reservoir?
- What is the available water in the soil storage reservoir?
- What is the peak rate of water use?
- What is the intake rate of the soil?
- What system options are available?
- What pumping rate is necessary?
- What is the quality of water required?

Answers will depend on the application and/or type of irrigation system used. Agricultural applications usually try to meet peak evapotranspiration (ET) needs, while nonagricultural applications apply as much water as possible without exceeding hydraulic, nutrient, or biochemical oxygen demand (BOD) limits as specified by regulatory agencies.

Irrigation system design is a diverse discipline. An introductory text in natural resources engineering cannot address all methods of applying water (and nutrients, pesticides, etc.)

to crops or treatment areas. Nontraditional irrigation applications are viewed as extensions of agricultural irrigation, as these applications usually are concerned with a subset of the agricultural questions. The text first addresses core questions such as those listed above in an agricultural context and then introduces some important water delivery systems and some nontraditional applications.

Goals

- Overview irrigation approaches and systems.
- Determine the parameters (e.g., answers to above questions) affecting irrigation design.
- Calculate seasonal water requirements for given crops, irrigation interval and peak flow, and irrigation pumping capacity.
- Design basic sprinkler laterals for agriculture; suggest modifications of the agricultural approach for some nontraditional applications; specify a sprinkler-based irrigation machine (center pivot) approach; and design a simple microirrigation system.

Irrigation Systems Overview

Irrigation systems convey and apply water for meeting plant transpiration and evaporation needs. They distribute nutrients (fertigation) and pesticides (chemigation). Sprinkler systems also provide frost protection and crop cooling. Irrigation is moving beyond agricultural and ornamental applications.

Figures 14.1a–f show an array of approaches to agricultural irrigation. Surface systems are widely used in areas where one may impound water at high elevations and distribute it by gravity. Perhaps the most common irrigation approach used in developed countries is the movable or solid set sprinkler lateral. The need to reduce the labor required to move sprinkler laterals motivated the development of the side roll and hose-tow systems. Linear move and pivot systems are popular due to their potential for energy and labor savings. Microirrigation approaches are increasing in popularity due to their potential for labor, energy, and water savings.

Table 14.1 summarizes and classifies irrigation systems according to application pros and cons. Jensen (1981), Finkel (1983), James (1988), Cuenca (1989), Kruse et al. (1990), Hoffman et al. (1990), Stewart and Nielsen (1990), and Rochester (1995) also summarize various irrigation systems for crops and specialized applications such as turf.

One may classify irrigation approaches as open surface, closed conduit surface, or subirrigation. As of 1990, open surface irrigation and closed surface irrigation were the most prevalent, accounting for 58% of U.S. irrigated land area. Most surface irrigation occurs west of the Mississippi River. Surface irrigation theory strongly depends on infiltration phenomena, closely coupled with the hydraulics of surface transport in a conveyance channel. Surface irrigation is becoming less important in the agriculture of developed countries. Other approaches, such as microsprinklers and drip, are superior with respect to water application efficiency (discussed later in this chapter). Microsprinklers are miniature sprayers with a wetted radius on the order of 15 m. Drip systems use drip emitters which depend on soil capillarity to distribute moisture over plant root zones. Texts such as Walker and Skogerboe 1987 provide detailed treatment of classical and modern surface irrigation approaches that are beyond the scope of this chapter.

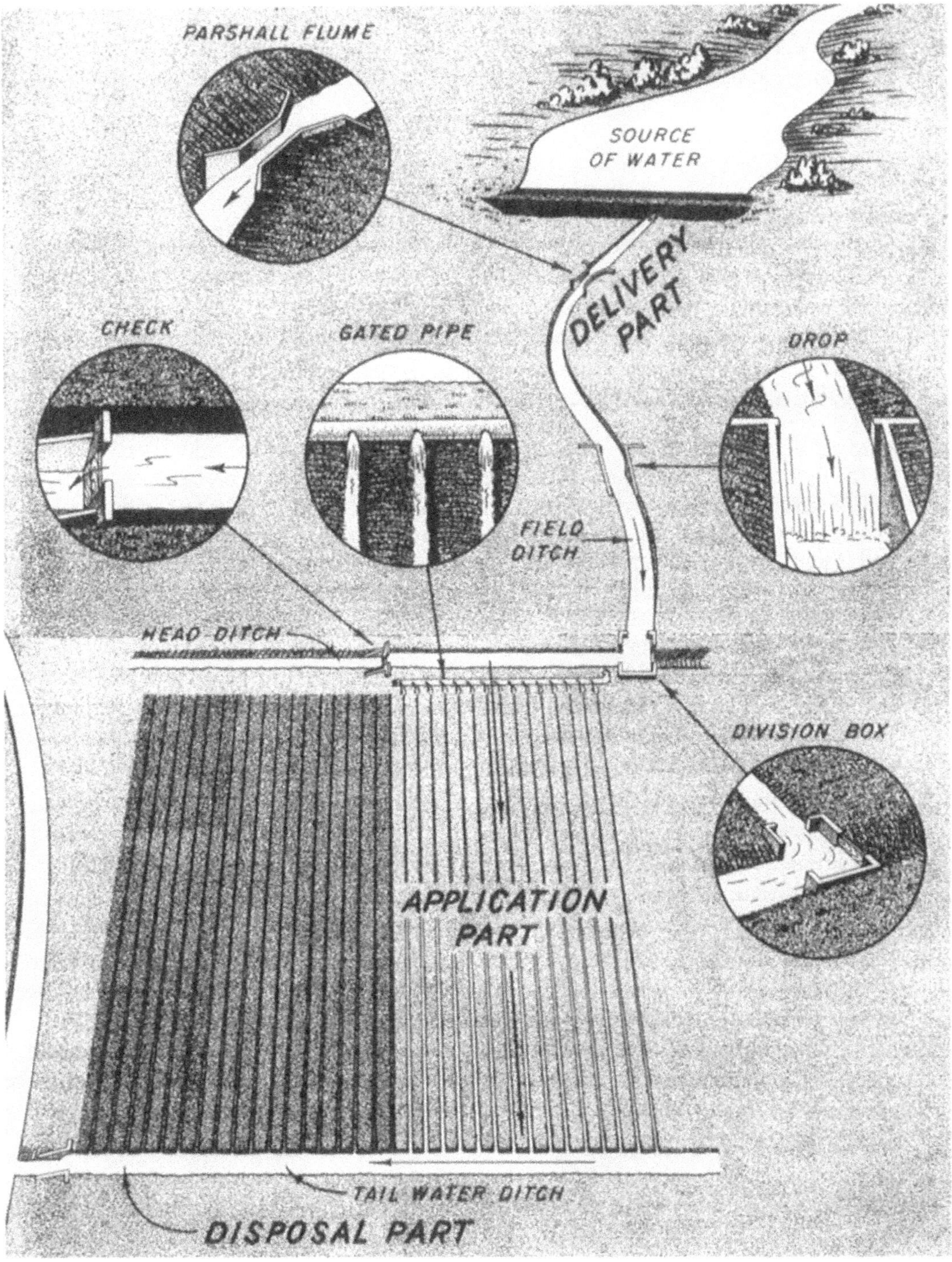

Figure 14.1a. Improved surface irrigation system showing a variety of structures for water measurement and management. (Courtesy of NRCS.)

Figure 14.1b. Movable lateral shown with a gun sprinkler. Laterals may contain a variety of sprinkler styles.

One may classify closed conduit irrigation systems as movable or solid set. Movable systems include hand-move sprinkler laterals, center pivots, traveling guns, and linear move systems. Solid set systems stay in place over a growing season and include sprinklers and a variety of microirrigation systems. Systems are also classified as intermittent or continuous application. Open conduit surface systems and sprinkler systems are usually intermittent application systems, with irrigation replenishing the soil storage reservoir periodically. Continuous (microirrigation; possibly subirrigation as well) systems provide daily ET demands with small reliance on the soil storage reservoir.

Figure 14.1c. Side roll lateral showing close-up of the powered movement mechanism. (Courtesy of NRCS.)

Figure 14.1d. Hose-tow irrigation system with big gun nozzle. Guns are also used on center pivots ends, cable towed travelers, and on hand-moved laterals.

Water table irrigation (subirrigation) is a special case involving management of a subsurface drainage system. Water movement with subirrigation is upward. One may practice subirrigation in localities having plentiful fresh water and a confining soil layer (e.g., parts of the Atlantic coastal plain). These systems provide irrigation, drainage, and some nutrient management (e.g., denitrification) needs. See Skaggs et al. 1988 and references contained therein for additional discussion of subirrigation.

Many sprinkler and microirrigation options exist and are worthy candidates for specific sites. The intermittent movable sprinkler system and the solid set sprinkler system meet a variety of objectives ranging from crop watering, nutrient application, frost prevention, and cooling. Thus, sprinkler and microirrigation systems are most emphasized.

Soil-Water-Plant Relations

Soil Storage

Agricultural irrigation approaches (except drip/trickle) strive to fill the soil storage reservoir as fast as the soil will accept water, move to other field locations, and then return when peak ET has drawn down the soil storage reservoir to the management-allowed deficit (MAD). The *water release curve* is crucial to understanding the soil water storage reservoir. A water release curve is the relation between volumetric water content (x axis)

Figure 14.1e. Linear move lateral irrigation system. (Courtesy of NRCS.)

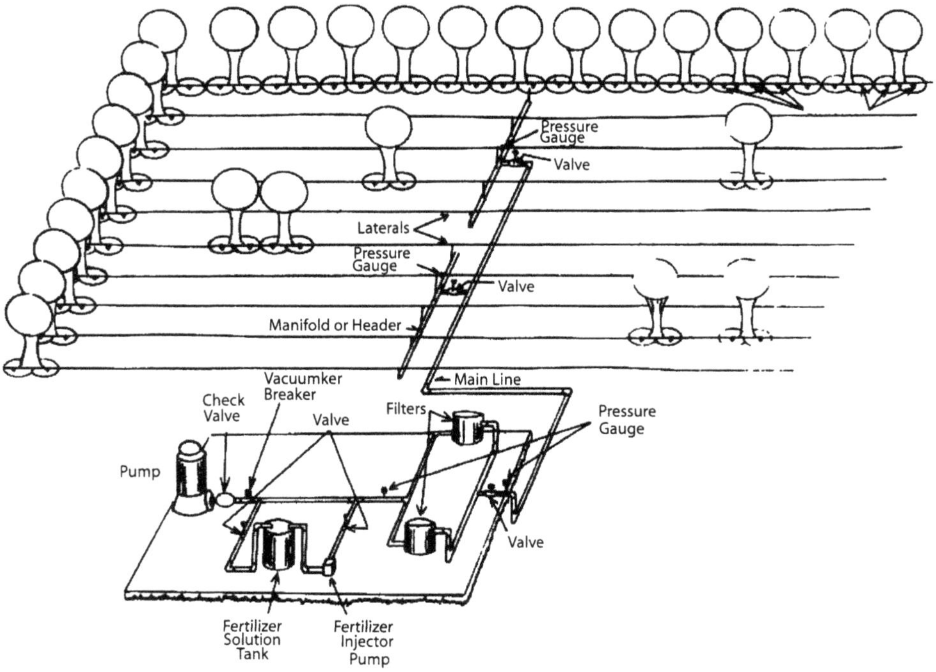

Figure 14.1f. Typical drip system layout. (From Howell et al. 1981; used with permission of ASAE.)

Table 14.1. Summary of selected irrigation system attributes as affected by site and situation factors.

	Improved surface systems		Sprinkler systems			Micro irrigation	Subirrigation (see Chapter 15 for more on this topic)
Site and situation factors	Redesigned surface system	Level basins	Intermittent mechanical move	Continuous mechanical move	Solid set and permanent	Emitters and porous tubes	
Infiltration rate	Moderate to low	Moderate	All	Medium to high	All	All	All
Topography	Moderate slopes	Small slopes	Level to rolling	Level to rolling	Level to rolling	All	Small slopes
Crops	All	All	Generally shorter crops	All but trees and vineyards	All	High value generally	All except very shallow rooted
Water supply	All but high salts	All	Salty water may harm plants	Salty water may harm plants	Salty water may harm plants	All: can potentially use high-salt waters	Salty water may harm
Application efficiency	60–70%	Average 80%	Average 70–80%	Average 80%	Average 70–80%	Average 80–90%	?
Capital required	Low to moderate	Moderate	Moderate	Moderate	High	High	Moderate
Energy requirement	Low	Low	Moderate to high	Moderate to high (LEPA can reduce)	Moderate	Low to moderate	Low
Management skill	Moderate	Moderate	Moderate	Moderate to High	Moderate	High	Low
Machinery operations	Medium to long fields	Short fields	Medium field length, small interference	Some interference, circular fields unless linear move	Some interference	May have considerable interference	Minimum interference
Duration of use	Short to long	Long	Short to medium	Short to medium	Long term	Long term	Long term
Weather	All	All	Poor in windy conditions	Better in windy conditions than sprinklers	Windy conditions reduce performance, good for cooling	All	All
Chemical application	Fair	Good	Good	Good	Good	Very good if clogging can be reduced	No
Long-term salinity management	Fair	Good	Good	Good	Good	Poor	No
Frost prevention/ crop cool objectives	No	No	Fair	Fair	Good	Fair if micro-sprayers used, else no	No

Source: Adapted from Schwab et al. 1993, and NRCS 1983.

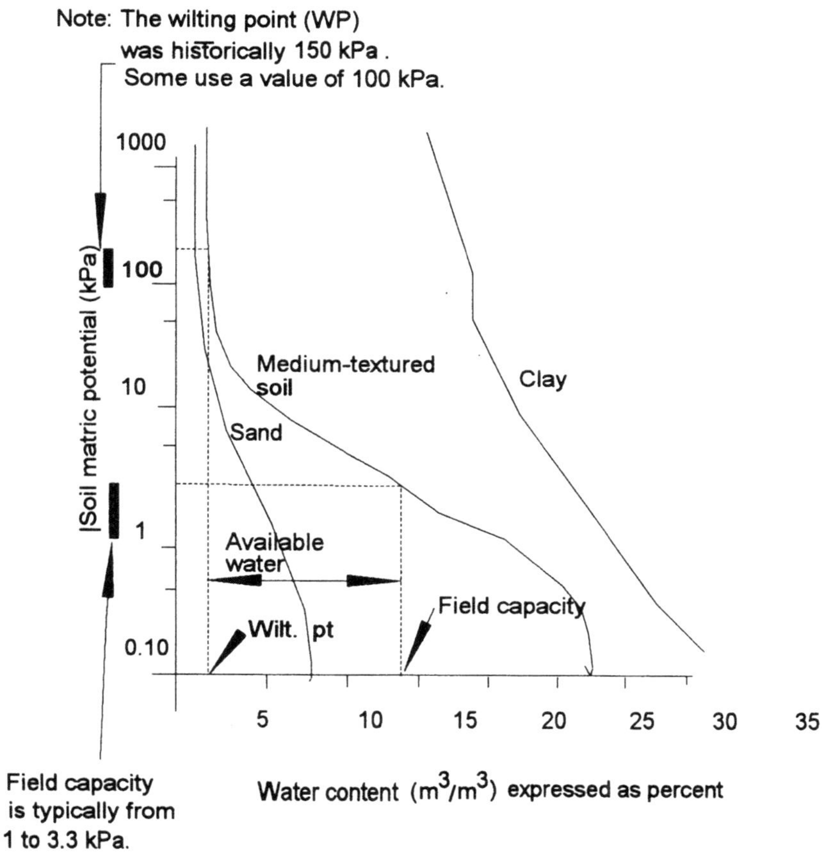

Figure 14.2. Generic water release curves. The range of available water for the medium-textured soil is on the left.

and potential (y axis). Figure 14.2 shows a water release curve. Volumetric water content is a product of soil dry bulk density (grams dry solids/cm^3 volume), gravimetric moisture content (grams water/grams dry solids), and specific gravity of water (grams water/volume). The matric potential is the pressure (expressed as cm water) that one must exert to cause moisture equilibration at specified water content.

Field capacity (*FC*) on the wet side and wilting point (*WP*) on the dry side bound the soil water storage reservoir. Field capacity is the value that a well-drained soil will hold, resisting gravity. The wilting point is the value that completely shuts down transpiration and causes permanent plant wilting. The available water (*AW*) is the amount between these extremes. Figure 14.2 illustrates *FC* and *WP* for a medium-textured soil.

How does a soil develop such low (negative) potentials (called matric or capillary potentials) as water content goes down? Surface tension provides the key. One may compute soil matric potential using

$$y_{rise} = -\frac{2\sigma_{st}}{\rho g r_p} = -\frac{2\sigma_{st}}{\gamma r_p} \tag{14.1}$$

where

h is the capillary or matric potential (L);
σ_{st} is the surface tension of water (F/L, 72.75 dynes per cm at 20°C);

ρ is the density of water (M/L^3);
g is gravity (L/T^2);
γ (= ρg) is the unit weight ($M/L^2T^2 = F/L^3 = 981$ dynes/cm^3)
r_p is the radius (L) of the largest pore holding water at pressure potential y_{rise}.

Equation 14.1 was discussed in Chapter 3 (see Figure 3.3). Clay soils, having a preponderance of smaller pores, hold water at very low potentials. Sandy soils, on the other hand, drain at higher potential levels. Zhang (1997) and Carsel and Parrish (1988) summarize approaches and needed parameters necessary to estimate water retention curves for various soil types. Appendix F discusses mechanisms plants use to counter the soil water potential.

Plant Available Water

The available soil water is partitioned into two components: the readily available water (*RAW*) and the remaining water (*RW*). Usage of all the available water (*AW*) entails severe plant wilting. The purpose of crop irrigation is not only to maintain crop viability but to maintain optimal yields. Experience has shown that 40–65% of the *AW* is consumed before yield declines occur. This fraction of *AW* is designated the *MAD*.

The *RAW* is computed using

$$RAW = MAD \sum_{i=0}^{layers} (AW)_i y_i \tag{14.2}$$

where

i is the layer number of effective rooting depths;
AW_i is the available water in layer i (L);
y_i is the depth of layer i (L);
MAD typically ranges from 0.4 to 0.6 and may be lower for high-value crops.

Calculate AW_i using

$$AW_i = FC_i - WP_i \tag{14.3}$$

where

FC_i is the field capacity (volumetric moisture at 3.3 kPa, 0.33 bar, 330 cm water);
WP_i is the volumetric moisture at the wilting point as read from the water release curve. The wilting point potential is historically 147 kPa, 15,000 cm water, and 15 bars. Some advocate reducing this value to as low as 99 kPa, 10 bars, 10,000 cm water.

In deep soils one may estimate the effective rooting depth by dividing the maximum rooting depth by 2. Plants extract 70% of water from the upper 50% of the root zone in a nonimpeding soil under well-watered conditions (Hansen et al. 1980). The lower 50% becomes more important as water stress increases. The upper 50% of the rooting zone is the effective soil water management zone. In soils with hardpans impeding root growth, the soil water management zone may extend to the impeding layer if the layer is within 0.6–1 m of the surface. Using 1 ft or 0.3 m thickness in these calculations is customary. One should consult local experience or a state irrigation guide for detailed information relating to rooting depth and *RAW*. Approximate soil properties, including *AW*, are tabulated in Table 14.2. Table 14.3 tabulates effective rooting depths for selected plant species.

Table 14.2. Summary of selected soil properties.

Soil texture	Sat. cond. (mm/hr)	Total pore Space (% vol.)	Field cap. (% vol.)	Perm wilt pt (%vol.)	Bulk density (Mg/m^3)	Avail. water (% vol)	Avail. water (mm/m)
Sandy	50 (25–250)	38 (32–42)	15 (10–20)	7 (3–10)	1.65 (1.55–1.80)	8 (6–10)	80 (70–100)
Sandy loam	25 (12–75)	43 (40–47)	21 (15–27)	9 (6–12)	1.5 (1.4–1.6)	12 (9–15)	120 (90–150)
Loam	12 (8–20)	47 (43–49)	31 (25–36)	14 (11–17)	1.4 (1.35–1.5)	17 (14–20)	170 (140–190)
Clay loam	8 (3–15)	49 (47–51)	36 (31–42)	18 (15–20)	1.35 (1.3–1.4)	18 (16–22)	190 (170–220)
Silty clay	3 (0.25–5)	51 (49–53)	40 (35–46)	20 (17–22)	1.3 (1.25–1.35)	20 (18–23)	210 (180–230)
Clay	5 (1–10)	53 (51–55)	44 (39–49)	21 (19–24)	1.25 (1.2–1.3)	23 (20–25)	230 (200–250)

Source: Adapted from Hansen et al. 1980; used with permission of John Wiley & Sons publishers.

Table 14.3. Summary of effective rooting depths and peak water use for selected crops.

Crop	Effective rooting depth in the absence of impeding layers (m)	Peak rate of water use (mm/day) Cool	Moderate	Hot
Alfalfa	0.9–1.2	5	6	9
Beans	0.3–0.9	3	4	6
Corn	0.6–1.2	5	6	8
Pasture	0.5–0.8	5	6	8
Cotton	0.8–1.2	5	6	9
Annual Fall vegetation	0.1–0.3	1	2	3
Potatoes	0.3–0.6	4	5	6
Peanuts	0.6–0.9	4	6	8
Sorghum	0.5–0.8	3	5	6
Annual Summer vegetation	0.1–0.3	3	4	6
Strawberries	0.3–0.5	3	4	6
Tobacco	0.3–0.6	3	4	6

Source: Adapted from Schwab et al. 1993.

Notes: If the impeding layer is shallower than indicated above, use the impeding layer depth. See the irrigation guide for your location for specific recommendations for crops at your location. Irrigation guides are available from the local NRCS or Cooperative Extension Service.

Example 14.1: Suppose cotton, with a maximum rooting depth of 2 m (thus an effective rooting depth of 1 m), is growing in a soil that is sandy for the top 0.33 m, sandy loam for the next 0.33 m, and clay for the next 0.33 m. The *MAD* is 0.5. From equation 14.2, using values from Table 14.2, one may write

$$RAW = 0.5[(80 \text{ mm/m} \times 0.33 \text{ m}) + (120 \text{ mm/m} \times 0.33 \text{ m}) + (230 \text{ mm/m} \times 0.33 \text{ m})] = 64.5 \text{ mm}$$

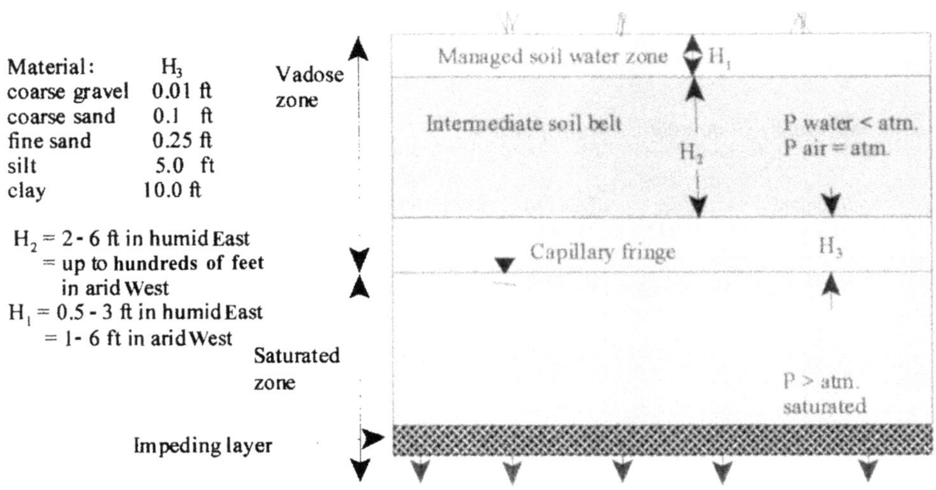

Figure 14.3. Vadose zone features.

The Vadose Zone

Figure 14.3 shows a generic vadose zone for a typical soil. The saturated zone contains pressures greater than atmospheric as depth increases. The capillary fringe depends on the soil texture. Pore pressures decrease as one moves above the water table. Gravity drains large pores above the capillary fringe.

Between the capillary fringe and the primary water management zone is the intermediate zone. The intermediate zone is essentially the transition between the realm of agriculture or surface activity and the geologic zone. In the humid eastern United States, the intermediate zone often includes one or more impeding layers. The intermediate and primary zones are often the most biologically active zones. The intermediate and geologic zones are often the targets of bioremediation activities. Many soils in the Southeast may have an impeding layer at the bottom of the managed soil water zone. In wetland situations or in other high water table soils there is sometimes no intermediate zone. Drainage lowers the water table from the primary soil management zone (see Figure 13.4).

The managed soil water zone represents the managed portion of the vadose zone in most cases and corresponds to the upper 50% of the plant root system (e.g., that portion which contains effective plant available water). Depth to water table, applicable soil water release curve, and the presence of impeding layers in the intermediate zone determine the vadose zone for a specific location.

Figure 14.4 shows how soil capillary potential and mechanical impedance affect plant root growth in the managed and intermediate soil zones. Increasing mechanical impedance, as indicated by increasing bulk density values, reduces root growth. Impeding layers have higher bulk density values. One often observes the mechanical impedance effect in soils with high sand percentages. Some advocate the use of moisture to reduce mechanical impedance effects. Moisture for mechanical impedance reduction deserves consideration, based on Figure 14.4. Maintaining the water table where the capillary fringe overlaps the root zone is ideal. Subirrigation schemes (used in some Atlantic coastal plains soils) achieve this goal. Water table management is further discussed in Chapter 13.

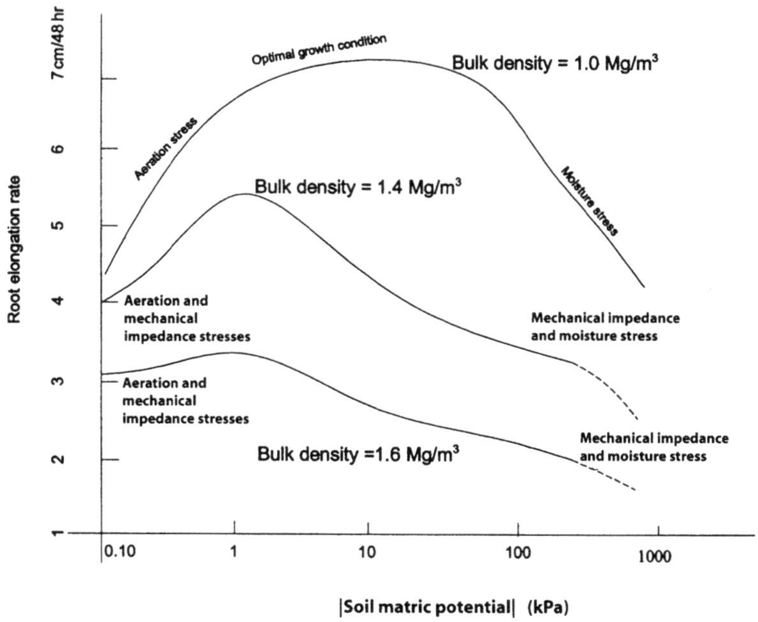

Figure 14.4. Mechanical impedance as a function of soil matric potential (absolute value), aeration, and compaction for a generic plant root. (Redrawn from Baver et al. 1972.)

Soil Intake Rate

Recall the infiltration function for a soil, exemplified in Figure 14.5. The irrigation intake rate for a soil is essentially the steady-state infiltration rate.

One can simulate passage of an irrigation nozzle over a location (e.g., treat the sprinkler pattern as a hyetograph) and compute effective runoff and infiltration using the Green-Ampt approach. Works such as Cuenca 1989 provide detailed information.

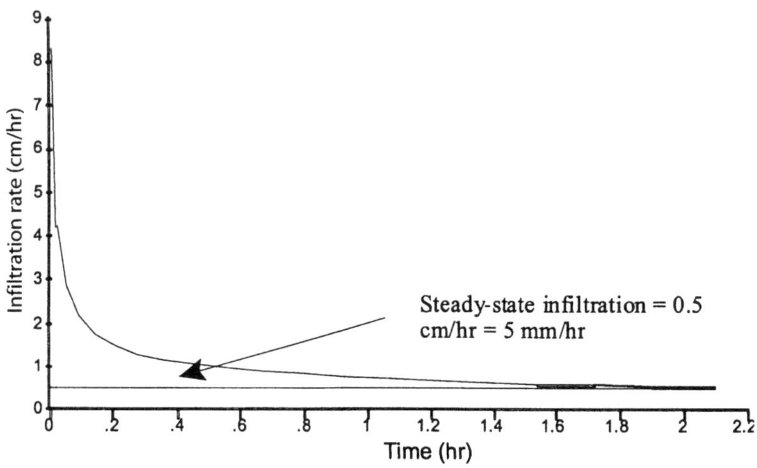

Figure 14.5. Intake rate vs. time for a soil as predicted using the Green-Ampt (see equation 3.10) infiltration equation.

Table 14.4. Approximate water intake rates as a function of topography and cultural practices.

Soil texture and profile conditions	Maximum water application rate for slope and cultural conditions (mm/hr)			
	0% Slope		10% Slope	
	With cover	Bare	With cover	Bare
Light sandy loams uniform in texture to 2 m	32	20	19	11
Light sandy loams over more compact subsoils	25	15	13	8
Silt loams uniform in texture to 2 m	16	10	10	6
Silt loams over more compact subsoil	10	6	5	3
Heavy-textured clays or clay loams	5	3	3	2

Source: NRCS 1983.

The steady-state infiltration rate should be field measured for irrigation design purposes. Table 14.4 summarizes for planning purposes approximate infiltration rate values for selected soils.

Example 14.2: One designs an irrigation system to apply water at the maximum intake rate of a uniform light sandy loam soil. The soil is sloping and has no cover. The available water holding capacity is 64 mm. How long should the system take to replenish this amount of water?

Since soil is sloping and has no cover, pick the intake rate of 11 mm/hr from Table 14.4.

$$t_{set} = 64 \text{ mm}/11 \text{ mm/hr} = 5.8 \text{ hr}$$

Water Quality Issues and Leaching Requirement for Crop Production

Inorganic anions and cations, organic compounds (e.g., pesticide residues, organic wastes), and biological and/or microbiological contamination presence/absence determine water quality. Salinity, determined mainly by sodium, calcium, and magnesium, is often the determining quality factor for irrigation water suitability. Salinity is an issue when applying liquid wastes or reclaimed water to soils even in humid areas. The white crust in Figure 14.6 is due to precipitated salts, rendering the soil useless for agricultural purposes. Appropriate management with water of suitable quality coupled with judicious crop selection may alleviate the difficulties of salinity.

In wastewater treatments involving nutrient and BOD application, one must not exceed monthly or annual constituent additions. For example, in land application design for animal or other waste, one must evaluate the water storage capacity and infiltration capacity of the soil. These provide the upper weekly limit on allowed water application; total nitrogen applied annually coupled with the N uptake by the crop may set a limit on total annual application, and BOD applied in a week's or month's time may limit total application. These and other limits vary by state.

Figure 14.6. Salinity, indicated by the light areas, is a product of natural seepage, canal seepage, and excess irrigation.

The limiting factor in arid region irrigation is salinity. Salinity may also be a factor in humid regions when wastewater effluents are applied. Specific ions may also limit crop growth. For example, selenium and boron are now known to limit crop yields in California. Hoffman et al. (1981) present other elements causing yield limitations. The limiting concentration of the various pollutants varies by crop.

One should assess chemical contamination, sediment concentrations, pesticides, and other organics before irrigating with a given water supply. Sediments and organics leading to algae production cause problems for microirrigation systems. Selected chemical constituents such as selenium and boron may pose threats to wildlife. High concentrations of sodium, calcium, and magnesium salts can render water unsuitable for irrigation. The ratio of sodium to calcium and magnesium is also significant. Salinity is significant in the West. It is becoming more important in the East with waste applications.

For salinity measurements, one does not always have to do an expensive chemical analysis. The electrical conductivity of water is linearly proportional to salt content. Thus, one can express water quality as electrical conductivity, expressed in decisiemens per meter, or dS/m. One can compute the leaching requirement to avoid salt buildup by

$$LR = \frac{EC_w}{5EC_e - EC_w} \tag{14.4}$$

where

EC_w is the electrical conductivity of the irrigation water source (dS/m);
EC_e is the saturated soil-water extract EC, which should not exceed tolerance values in Table 14.5 (dS/m);
LR is the leaching requirement (L).

Table 14.5 gives the salt tolerance for selected crops in electrical conductivity units. In instances where the BOD or other biological parameter or specific nutrients control application volume, consult regulatory agencies for detailed design and testing procedures. The total water applied is therefore $(1 + LR) \times$ soil water deficit. To maintain optimal plant and microbial activity, soil water deficits usually should not exceed the *MAD* (e.g., 50%) of the available water.

Table 14.5. Critical salinity conductivity values for selected crops.

	Yield potential[a]				
	100%		90%		
Crop	EC_e	EC_w	EC_e	EC_w	Max. EC_e
Field crops:					
Corn	1.7	1.1	2.5	1.7	10
Cotton	7.7	5.1	9.6	6.4	27
Sorghum	4	2.7	5.1	3.4	18
Soybean	5	3.3	5.5	3.7	10
Wheat[b]	6	4	7.4	4.9	20
Vegetable crops:					
Beans	1	0.7	1.5	1	7
Lettuce	1.3	0.9	2.1	1.4	9
Potato, sweet potato	1.6	1.1	2.5	1.7	10
Tomato	2.5	1.7	3.5	2.3	13
Forage crops:					
Alfalfa	2	1.3	3.4	2.2	16
Bermuda grass	6.9	4.6	8.5	5.7	23
Sudan grass	2.8	1.9	5.1	3.4	26
Fruit crops:					
Date palm	4	2.7	6.8	4.5	32
Grape	1.5	1	2.5	1.7	12
Orange, grapefruit, lemon	1.7	1.1	2.3	1.6	8

Source: Adapted from Ayers and Westcot 1985.

[a]All values in dS/m at 25°C.

[b]During germination and seedling stage, EC^e should not exceed 4 or 5 dS/m. Data may not apply to new semidwarf varieties of wheat.

The white crust in Figure 14.6 reflects a long-term absence of excess water for leaching purposes. The application of a leaching fraction presumes a drainage system, discussed in Chapter 13, to remove the leachate. Effluent from drainage in irrigated areas is a significant water quality problem in the arid West. Salinity levels, nutrients, and pesticide residues are often high enough to be of environmental concern in arid regions and with waste application systems in humid regions.

Irrigation Efficiency

There are three basic efficiency notions: conveyance efficiency, water application efficiency, and water use efficiency.

Conveyance Efficiency

Conveyance efficiency is percentage of water delivered to the point of use in a delivery system. Conveyance efficiency is used for assessing seepage losses in irrigation canals. It should approach 100% for good pipe systems. Conveyance efficiency on a statewide basis ranges from 75 to 90% in western states (Solley et al. 1998). One may estimate losses per unit length using the USBR channel seepage relationship (equation 8.26).

Example 14.3: We pump 15 m^3/s water into a distribution canal and recover 11 m^3/s at the canal end, 5 km downstream. What is the conveyance efficiency of this canal?

$$100 \times [(11\ m^3/s)/(15\ m^3/s)] = 73.3\%$$

Water Application Efficiency

One may define water application efficiency (WAE) as water stored in the root zone compared with applied water. Water not beneficially used goes into surface runoff, evaporates, or moves below the root zone. WAE does not treat leaching as a beneficial use. This value is most useful when applied to field-sized and smaller areas. WAE may range from less than 50% for subirrigation, 60% for surface systems, 70% for sprinkler systems, and up to 90% for some microirrigation systems.

Example 14.4: A center pivot pumps 1000 gpm (2.23 cfs; 0.065 m^3/s) onto a field. The runoff was channeled and measured, leading to a runoff rate of 0.012 m^3/s.

The application efficiency is then

$$100 \times (0.065\ m^3/s - 0.012\ m^3/s)/0.065\ m^3/s = 81.5\%$$

(assuming no deep percolation and no evaporation).

Even when applying water at rates less than the theoretical intake rate, expect some runoff. Achieving truly uniform distributions is difficult. The above example does not account for evaporation losses, which are often considerable.

Water Use Efficiency

Water use efficiency (WUE) is water beneficially used divided by water delivered. Beneficially used water includes transpired water and water for leaching. Water not beneficially used may run off or percolate *beyond the leaching requirement* below the root zone. This value is most useful when applied to field-sized and smaller areas. WUE is plant mass produced per unit of water applied: for sorghum, 250 kg water per kg sorghum; for corn, 350 kg water/kg corn; for wheat, 500 kg water/kg wheat; for potatoes, 640 kg water/kg potatoes; and for alfalfa, 900 kg water/kg alfalfa hay.

These efficiency notions are important for evaluating cropping systems for specific sites. There is usually a tradeoff made between high efficiency and initial investment.

Other Irrigation Effectiveness Indicators

One can also characterize effectiveness of irrigation with the uniformity coefficient (UC). One may assess uniformity using a series of catch cans distributed over the field in a grid pattern. Uniformity is computed by first calculating the average water depth caught over a specified test interval. Then one determines the deviation of each catch can from the calculated field average. Take the absolute value of each deviation. Compute the average deviation value. One can then compute *UC* as

$$UC = 1 - \frac{y_{devavg}}{y_{avgapl}} \quad \textbf{(14.5)}$$

where

y_{devavg} is the average of the absolute values of deviation, in depth of water infiltrated or caught in surface cans (*L*);

y_{avgapl} is the average applied depth (*L*).

UC values above 0.8 are usual.

Related to *UC* is the distribution uniformity (*DU*). To calculate *DU,* find the average depth as above, then sort the catch can values in descending order and separate the lower 25% of the data into a separate data set. Compute the average amount caught in the lower 25% ($d_{25\%}$). One may then calculate *DU:*

$$DU = \frac{d_{25\%}}{y_{avgapl}} \quad \textbf{(14.6)}$$

where

$d_{25\%}$ is the depth value not exceeded by 25% of the cans (*L*);

y_{avg} is the average depth applied as above.

DU values above 0.7 are generally acceptable. *UC* and *DU* values are important in surface system designs and are mainly used to evaluate existing microirrigation and sprinkler systems. Low *UC* and/or *DU* values suggest the likelihood of poor application efficiency. Burt and Styles (1994) give an excellent overview of irrigation efficiency and uniformity evaluation methods for modern irrigation approaches. Agricultural applications form the background for most discussions of efficiency; however, the concepts are relevant for nontraditional applications as well.

If one is applying nutrients or wastes through the irrigation system, efficiency and timing aspects are even more critical. For example, nutrients or wastes applied inefficiently can result in serious crop injury and significant runoff or pollution.

Example 14.5: Suppose one finds the following depths of water in catchment pans spaced uniformly under a sprinkler system under test. The pans were collected immediately after the test, eliminating evaporation effects. Create a table (see Table 14.6) with rows and columns corresponding to the field layout.

Table 14.6. Arrangement of calculations for distribution uniformity determination for Example 14.5.

Depths (mm) and deviation from the mean value of 41 mm				
50 (+9)	44 (+3)	42 (+1)	40 (−1)	47 (+6)
44 (+3)	42 (+1)	40 (−1)	39 (−3)	44 (+3)
43 (+2)	37 (−4)	32 (−9)	30 (−11)	35 (−6)
48 (+7)	42 (+1)	40 (−1)	36 (−5)	45 (+4)

The mean depth is 41 mm. The average of the five lowest readings (lower quarter) is (37 mm + 30 mm + 36 mm + 32 mm + 35 mm)/5 = 34. The $DU = 0.83$. The uniformity coefficient is 1 − 4.05 mm/41 mm = 0.90, where 4.04 mm is the average of the absolute value of the deviations.

Efficiency values defined above assume the only beneficial use of water is for meeting transpiration needs. Other efficiencies are required when engaging in crop cooling, pesticide application, bioremediation, etc. Heerman et al. (1992) provide a methodology for relating UC to irrigation economics. They generalize the method to include center pivot machines.

Irrigation Scheduling

One may schedule using a water balance. Assuming no upward movement of water into the root zone, one may write the water balance as

$$y_{ri} = y_{ri-1} + (ET - P_e) \tag{14.7}$$

where

y_{ri} *and* y_{ri-1} are the total depth of water removed from the soil root zone at the end of day i and day $i-1$ respectively;

ET is the daily ET;

P_e is the effective rainfall for day i.

A conservative estimate of P_e is one-third the rain gauge reading. One may change this factor based on local field experience.

The decision to irrigate is based on projecting forward in time considering recent water status indicators. When projecting $y_i - y_{i-1}$ to exceed the remaining soil storage, one should commence irrigation. The projection time (e.g., 1 or 2 days) should enable planning and preparation for irrigation at the time the deficit is anticipated. Example 14.6 shows the procedure.

Example 14.6: A loam soil growing cotton has an effective root zone depth of 1 m and a *MAD* of 50%. Daily *ET*, irrigation, and effective precipitation are given in Table 14.7. The *AW* is 170 mm/m or 170 mm. Applying the *MAD* results in 170 mm/2 = 85 mm. One would irrigate on day 7 based on a projection from day 6. One must make this calculation/projection daily as shown in Table 14.7.

Table 14.7. Information for Example 14.6 irrigation scheduling problem.

Date (July)	*ET* (pan × *ts* pan coef. × crop coef.; could also use calculated *ET* × crop coef.) (mm)	P_e = 0.33 raingauge (mm)	Irrigation (mm)	Deficit D_i (assume 52 mm brought forward from June 30) (mm)
1	8	0		60
2	6	7		59
3	7	0		66
4	8	0		74
5	9	0		83
6	5	7		81[a]
7	5	0	85	1
8	3	0		4

[a]Indicates one would plan to irrigate the next day based upon a projected deficit. Judgment, experience and knowledge of the water balance are required to make the best irrigation decision.

Scheduling for continuous (e.g., microirrigation) systems that operate on a daily or every other day basis through the season depends solely on the *ET* demands anticipated. When actual *ET* is less than peak, one varies system hours of operation for each block as needed. One should augment the above scheduling approach in humid regions with soil moisture or soil water tension measurements.

COMPUTING *ET* REQUIREMENTS: PEAK *ET*

Peak *ET* is important in any system containing growing plants. State irrigation guides provide peak design *ET* rates for various crops and areas. Values shown in Table 14.3 came from irrigation guides. One can also estimate peak use from equation 14.8 (Hansen et al. 1980):

$$ET_{pk} = 0.034ET_{mon}^{1.09}P_i^{-0.09} \qquad \textbf{(14.8)}$$

where

ET_{mon} is the average monthly *ET* for the region in mm (available from climatological data);
P_i is the net irrigation application in mm;
ET_{pk} is the peak *ET* rate in mm/day.

Peak *ET* from equation 14.8 is not adjusted for the crop coefficient (see Chapter 4). One should consult local guides for specific design values. The irrigation period is the *AW* (mm), adjusted for the *MAD*, divided by peak water use rate (in mm per day).

Soil moisture measurement is a basis for irrigation scheduling. The tensiometer or electrical instruments mimicking the tensiometer measure the soil water potential. These devices are useful for irrigation scheduling. When pressure potentials at predetermined locations reach predetermined low values (high levels of suction), one commences irrigation. Once one sufficiently increases potential (e.g., low suction readings), irrigation ends. Stewart and Nielsen (1990) discuss beginning/ending trigger points and tensiometer placement for many crops.

EFFECTIVE RAINFALL FOR AGRICULTURAL IRRIGATION

Rainfall provides for some plant water needs for humid regions as noted in Example 14.6. However, not all is "effective"[1] because some runs off, some rainfall percolates below the root zone, and some evaporates. One must estimate effective rainfall to adequately estimate seasonal irrigation requirements. Equations 14.9a–b (Hansen et al. 1980; NRCS 1970) can be used to estimate monthly effective rainfall. Equation 14.9a gives the effective calculated rainfall:

$$P_{eff\,calc} = f(Def)(1.25P_m^{0.824} - 2.93)(10^{0.000955ET_m}) \qquad \textbf{(14.9a)}$$
$$\text{where } f(Def) = 0.53 + 0.0116Def - 8.94 \times 10^{-5}Def^2 + 2.32 \times 10^{-7}Def^3$$

where

Def is soil water deficit or irrigation depth in mm;
f(*Def*) is an adjustment factor for deficits or net irrigation depths and equals 1 if *Def* = 75 mm;
ET_m is the average monthly *ET* in mm;
P_m is the mean monthly rainfall in mm;
$P_{eff\,calc}$ is the estimated effective rainfall for a 75 mm soil water deficit; limited to the lower of ET_{mon} or $P_{eff\,calc}$.

The effective precipitation for irrigation planning purposes must not exceed *ET*, as shown in equation 14.9b:

$$P_{eff} = Min(ET_m, P_{eff\,calc}) \quad \textbf{(14.9b)}$$

For a given rainfall event, the effective precipitation is the portion not running off. Assuming runoff was the dominant component, one could use the *CN* method to make such an estimate. One could also perform an effective rainfall analysis using an infiltration function such as Green-Ampt.

Effective precipitation is more appropriately analyzed from a frequency point of view. In other words, what is the minimum effective precipitation 1 year in 10, 50, 100, etc.? This is a topic for a more advanced text.

Intake rate is a significant factor in selecting sprinkler-based systems. State irrigation guides provide estimates of intake rates for various soils, slope, and cover conditions. Depending on the system type, relations are available for checking intake rate. Conforming to intake rate for the given soils reduces runoff. Conforming to soil intake rates is more problematic for point or line systems than for sprinkler laterals. Relations are available for intake rate calculations with the above-mentioned systems. Soil intake rates have less effect on microirrigation system design.

With furrow and related surface irrigation systems, intake rate affects the rate of water advance and resulting application efficiency. The NRCS has categorized soils into intake families. Field infiltration measurement determines the intake family. Irrigation guides tabulate the intake family for soils. Subirrigation systems use the hydraulic conductivity of the soil as the primary intake constant.

Computing *ET* Requirements: Seasonal Water Use

The text covered the Penman, Jensen-Haise, Blaney-Criddle, and pan evaporation methods for predicting evaporation in Chapter 4. Texts such as Cuenca 1989 present additional techniques. The customary approach for estimating crop *ET* is to apply a crop coefficient to the wet surface *ET* value.

Burman et al. (1981) discuss the calculation of crop water use coefficients. They separate the *Kc* value into a basal coefficient *Kcb*, *Ka*, and *Ks*, where $Kc = Kcb(Ka) + Ks$. The basal coefficient is the well-watered value and assumes no evaporation from the soil surface. The *Ka* value is found from $Ka = \ln(\%AW + 1)/\ln(101)$. Parameter *Ks* depends on soil characteristics and on days after an irrigation or rainfall event. For a soil in Kimberly, Idaho, they reported $Ks = 0.8\ (0.9 - Kcb)$ for 1 day after rain; $0.5\ (0.9 - Kcb)$ for 2 days after rain; and $0.3\ (0.9 - Kcb)$ for 3 days after rain. If *Kcb* exceeds 0.9, then one can neglect *Ks*. Burman et al. (1981) tabulate *Kcb* values for selected crops. This approach is useful for estimating seasonal water requirements. Published values for each major crop are available from NRCS regional irrigation guides.

Seasonal irrigation requirements supply the *ET* needs less the effective rainfall. Because of salt buildup associated with evaporation, an extra portion is included to leach salts downward. The efficiency of the irrigation method is also included. One can compute seasonal irrigation requirements (IR) using the water balance shown in equation 14.10:

$$IR = \sum_{i=0}^{No\ months} \left[\frac{(ET_m - P_e)(1 + LR)}{E_a} \right]_i - \Delta W \quad \textbf{(14.10)}$$

where

ET_m is the monthly evapotranspiration for each month of the growing season;
P_e is the monthly effective precipitation;
LR is the leaching requirement;
E_a is the system application efficiency, depending on the chosen system;
ΔW is the allowable change in soil zone storage over the time increment.

Equation 14.10 stems from the basic water balance arranged as shown in equation 14.11:

$$P + P_i - Q = E + Perc + \Delta St_{soil} \qquad \textbf{(14.11)}$$

where

P is gross precipitation;
P_i is total irrigation;
Q is total runoff;
E is total transpiration and soil evaporation;
$Perc$ is deep percolation, including leaching requirements;
ΔSt_{soil} is change in soil water status over some time increment.

The effective precipitation accounts for runoff. The deep percolation term accounts for the leaching fraction. The presence of the application efficiency stems from some excess water percolating deeply and some running off the site.

Example 14.7: Suppose cotton is grown and is sprinkler irrigated with water having an electrical conductivity of 1.2 dS/m; the average depth of irrigation is 25 mm and the application efficiency is 70%. Assume a zero change in soil storage. Assume negligible deep percolation. The monthly crop *ET* (see Chapter 4) and rainfalls are shown in Table 14.8. Now, one must compute the leaching requirement. From equation 14.4 and Table 14.5 for cotton with 100% yield potential, salinity must not exceed 7.7 dS/m.

$$LR = 1.2 \text{ dS/m}/[5(7.7 \text{ dS/m}) - 1.2 \text{ dS/m}] = 0.032$$

(assume leaching for 100% yield)

$$IR = [(395 \text{ mm} - 218 \text{ mm})(1 + 0.032)]/0.7 = 261 \text{ mm}$$

(equation 14.10). This is the average depth of water one must apply not to have yield reductions. Water volume required is this depth times field area.

Table 14.8. Irrigation water balance for irrigation scheduling in Example 14.7.

Month	ET (mm)	Rainfall (mm)	Eff Rain (Eqs. 14.9a–b) (mm)
June	100	100	65
July	110	90	60
August	110	90	60
September	75	50	33
Total	395	330	218

As in the effective precipitation discussion, a most relevant question is the maximum irrigation requirement for, say, 9 years in 10 instead of 5 years in 10. This requires a frequency analysis on effective precipitation and *ET*. Pair et al. (1983) provide further discussion.

Irrigation Pumping Rate for Meeting *ET* Requirements

One may compute maximum irrigation pumping rate for planning purposes using

$$q = \frac{166.7 \times area\,(hectares) \times net\ application\,(mm)}{irrigation\ period\,(days) \times oper.\ hours\ per\ day \times app.\ eff} \tag{14.12}$$

where q is pumping rate in l/min. The gross application efficiency is (*ET* over the interval + total leaching requirement)/system. Operating hours per day depends on system type (see Table 14.1).

Irrigation supplies all *ET* needs during the peak demand time in this formulation. Derivation of equation 14.12 is left to the student. Appendix E discusses power requirements. Pumping rates for special-purpose irrigation applications depend on the nature of the application.

Water Rights

Just because water may happen to be available, can one legally use it? It depends on the legal water rights, which vary from state to state. The allocation of water and rights to water varies by country, by region within a country, and by water source. Water rights are closely associated with government policy. State and federal law (or various agency regulations) may mandate conservation by certain dates. They may mandate "use it or lose it" by certain dates. Historically, two major approaches to water rights have evolved.

Riparian

The owner of land adjacent to a stream or water body can use that water. In other words, ownership of water is tied to the land. Variants of this approach are prevalent in the eastern United States.

Appropriation

Water rights are separated from the land. One can sell and purchase water access apart from the land. The appropriation doctrine is more prevalent in the western United States, where water is historically scarce. Due to the increased importance of water, one should approach each case as unique. States are moving toward more regulation of water withdrawal quantities from natural surface bodies and groundwater. Flood (1990) presents an overview of water rights in the 50 U.S. states and territories. The maps given in Figures 14.7–14.10 show that states treat groundwater and surface water differently. The possibility of selling water rights separately from the land, historically true for the western United States, is increasing in the eastern United States. Public concern for the adequacy of water supplies increased greatly with the droughts of the mid-1980s. In Florida, water management districts

Figure 14.7. Surface water rights systems in the United States. (From Flood 1990; used with permission of the American Water Works Association.)

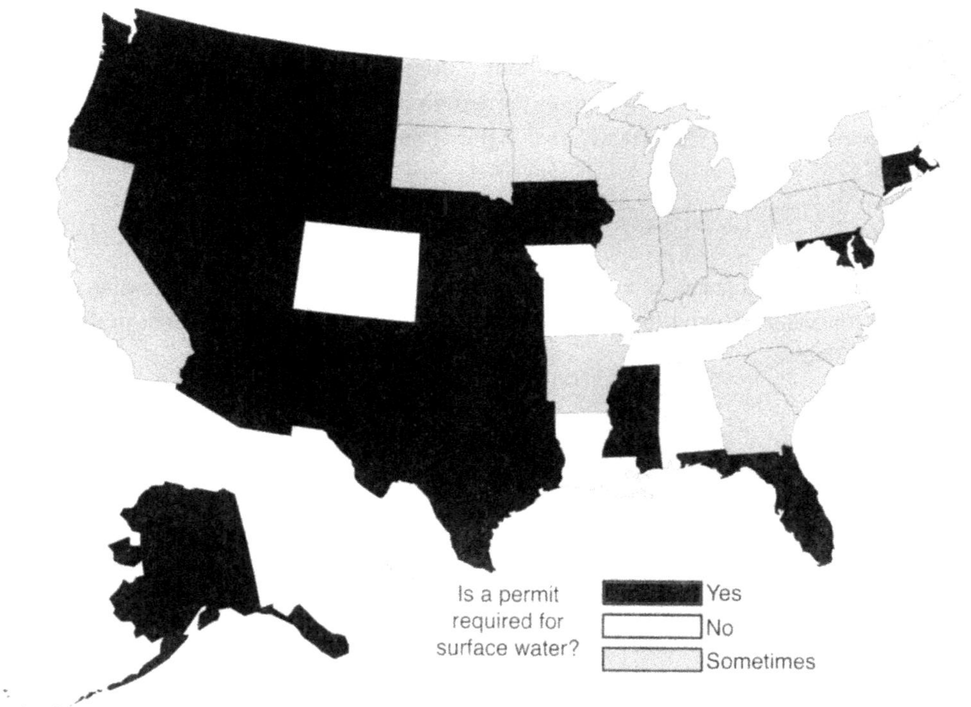

Figure 14.8. Surface water permit requirement status in the United States. (From Flood 1990; used with permission of the American Water Works Association.)

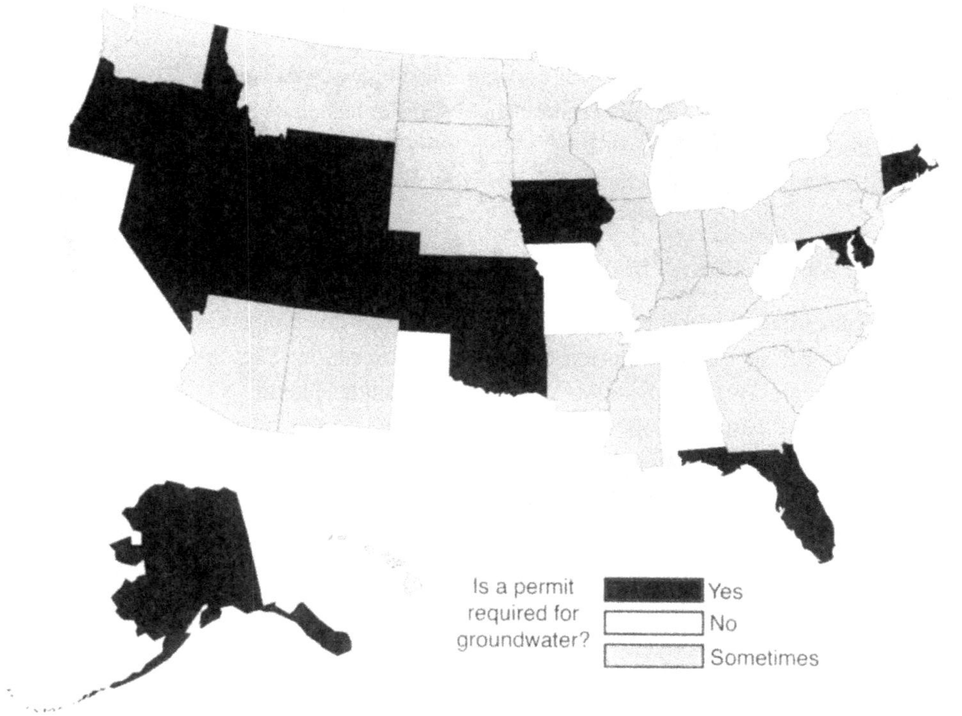

Figure 14.9. Groundwater permit requirement status in the United States. (From Flood 1990; used with permission of the American Water Works Association.)

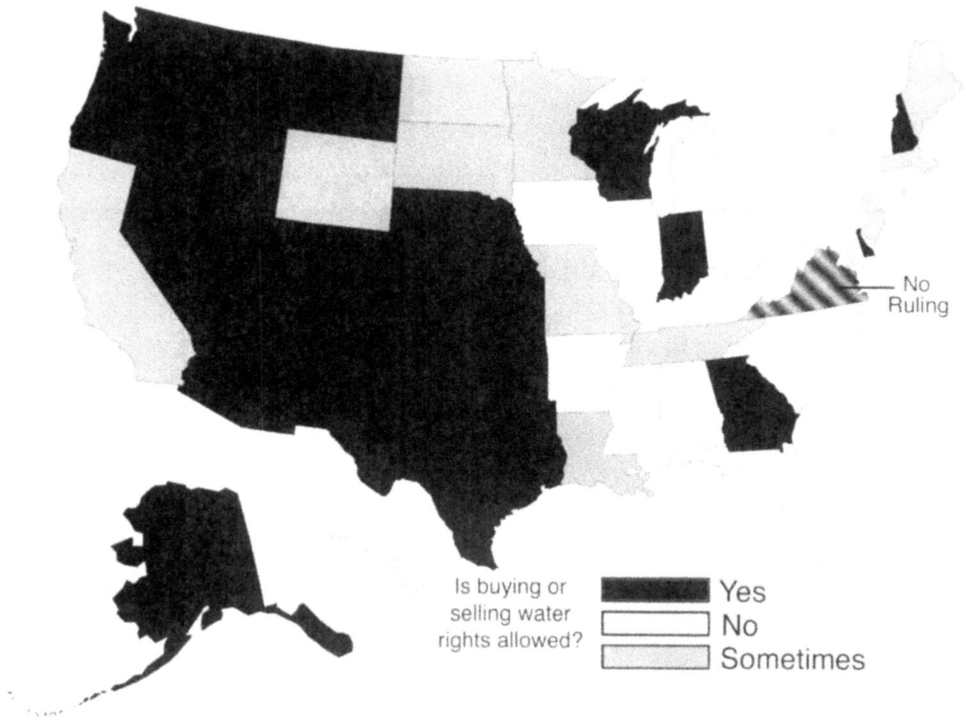

Figure 14.10. Possibility of buying or selling of water rights in the United States. (From Flood 1990; used with permission of the American Water Works Association.)

and state regulations control surface and groundwater withdrawals for irrigation through permitting. Georgia now requires monitoring of groundwater withdrawals via large irrigation wells. Regulations and laws about water rights are changing in many states. Thus, one should consult appropriate authorities early in the planning of large projects.

SPRINKLER LATERAL DESIGN FOR AGRICULTURAL APPLICATIONS

The sprinkler lateral is the basis of solid set and hand-moved pipe systems. Sprinkler laterals are useful on soils having reasonably good infiltration. They may provide uniform coverage. Figure 14.11 shows several operating sprinkler laterals.

The sprinkler lateral is labor intensive to operate. Center pivots, traveling guns, and other sprinkler-based systems can overcome the labor problems associated with moving sprinkler laterals. Discussion of more automated approaches is left to advanced texts.

A sprinkler lateral(s) system affords an excellent introduction into the design of irrigation systems. Lateral design invokes all facets of irrigation design. Pair et al. (1983), Walker and Skogerboe (1987), Jensen (1981), Cuenca (1989), and Burt and Styles (1994) treat all aspects of sprinkler, microirrigation, and surface irrigation design. Any design must live within constraints of soil intake rate and allowable soil water depletion. *Restated, the system application rate should not exceed the soil intake rate and must not take longer than a design interval or cycle time.* One must size hardware to give adequate uniformity and not have excessive friction losses or excessive pipe sizes. The text exemplifies the design process with a basic sprinkler lateral. Appendix E includes additional information on centrifugal pump selection and friction loss calculation.

The design proceeds along the following steps.

Field Data

Map the field, showing the water source, proposed laterals, and main lines. Note prevailing winds and elevations. Try to lay mains uphill and downhill and have laterals operate

Figure 14.11. Solid set sprinkler system cooling potatoes.

on the contour. For multiple laterals, try to use one or at most two diameters. Have the water supply in the center if possible. Maintain a balanced lateral operation. Also, note water rights and water quality of available water. Make appropriate chemical analyses to document water quality or waste constituents if applicable. A complete field mapping includes a soils map. If possible, determine the intake rate using an infiltrometer.

One desires to arrange lateral movements to achieve uniform flow delivery and relative easy lateral movements across the field. For example, Figure 14.12 shows possible schemes where one may pump from a stream along one boundary. It also shows other schemes with the water source in the center of the field or on the field edge. If possible, try to maintain a set number of sprinklers and a constant lateral pipe diameter. System layout is inherently an iterative process. Consider booster pumps. Adapt to soil differences when possible.

Irrigation Interval

Determine the minimum irrigation interval for the field based on peak ET and on allowed soil depletion; determine the maximum set time. The lateral cycle time must be equal to or less than cycle time computed from equation 14.13:

$$T_{cycle} = \frac{\text{allowable depletion}}{\text{peak } ET} \tag{14.13}$$

The allowable depletion is precisely the readily available water (RAW) discussed previously. The lateral(s) must cycle over the field within $T_{cycle.}$ Start with the required time for the lateral to operate at one location (set time) and add some moving and maintenance time for the total set. Total hours of operation per day should not exceed values in Table 14.9.

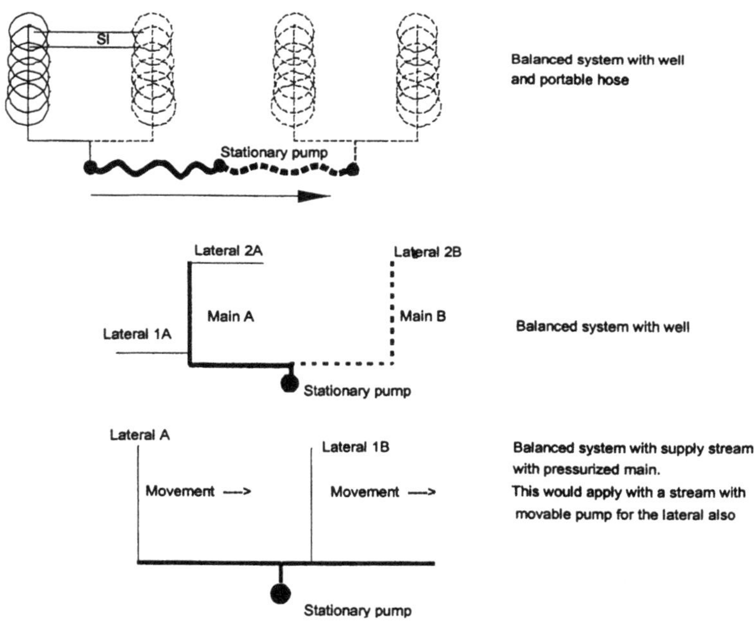

Figure 14.12. Sprinkler lateral layouts. (Modified from Schwab et al. 1993.)

Table 14.9. Comparison of selected sprinkler irrigation equipment.

Type of system	Relative cost	Relative labor cost	Practical hours of daily operation
Hand-moved lateral with standard sprinklers	0.4	4–5	12–16
Hose tow	0.5	1.5–3	16
Boom-type sprinklers, trailer mounted	0.6	4	12–16
Center pivot	1	1	24
Solid set	3–7	1	24

Source: Modified from Schwab et al. 1993.

Note: Based on 65 ha field, 63 l/s pump, 80% application efficiency.

Lateral Set Time

Estimate the required lateral set time by

$$T_{set\ time} = \frac{\text{allowable depletion}}{r_{intake}} + MandM \tag{14.14}$$

where

$T_{set\ time}$ is the set time;
MandM is the time required to move the system plus some time for maintenance;
r_{intake} is the soil intake rate from Table 14.4.

The total cycle time must be less than or equal to the total of the set times and should allow maintenance. The constraint on total operating hours per day must also be satisfied.

Nozzle and Lateral Spacing

Spacings such as 7–9 m are typical starting values. These numbers can be larger or smaller depending on the situation. See texts such as Cuenca 1989 for additional details. Try to achieve a balanced design such as is shown in Figure 14.12. Varying soils with varying intake rates and storage capacities, steep slopes, etc., require creativity in the layout. To adapt to these realities one may evaluate varying lateral spacing; however, such variations can complicate system moving and maintenance. The lateral design is still tentative. Final selection of the lateral spacing depends on sprinkler selection.

Sprinkler Nozzle Selection

The choice depends on the diameter of coverage required, available pressure, capacity of sprinkler, and nature of the crop canopy. Very low and low pressure nozzles (34–210 kPa) are mostly used on high-intake soils. The wetted radius is 6–10 m, and large droplets are the rule. High to very high pressure nozzles (350–830 kPa) give a high application rate and small droplet diameters. Small droplet diameters are subject to wind-induced variation. They perform well on all field crops and most irrigatable soils but require the most pumping energy. Medium pressure nozzles (210–415 kPa; 30–60 psi) give small water

Table 14.10. Typical nozzle performance data for medium-pressure sprinklers.

Nozzle pressure		Nozzle diameters in millimeters (in) 3.97 × 3.18 (5/32 × 1/8)		4.76 × 3.97 (3/16 × 5/32)		6.35 × 3.97 (1/4 × 5/32)	
(kPa)	(psi)	Dia[a]	l/s	Dia[a]	l/s	Dia[a]	l/s
207	30	25	0.37	26	0.52	28	0.76
276	40	27	0.43	28	0.61	31	0.9
345	50	28	0.47	30	0.68	34	1
414	60	30	0.52	31	0.74	36	1.1

Source: Adapted from Schwab et al. 1993.

Note: 1 l/s = 15.85 gpm; 1 cfs=28.32 l/s; 448.8 gpm=1 cfs; 1 m^3/s=35.28 cfs; 1 m^3/s=1000 l/s; and 1 m = 3.28 ft. Consult manufacturers for additional specifications.

[a]Diameter of coverage in meters.

droplets and a reasonably good moisture distribution pattern. Table 14.10 shows typical data for general-purpose sprinkler nozzles. Special nozzles for use in orchards and on turf are also available. Typical sprinkler heads give distributions as shown in Figure 14.13.

Pressure range, orifice diameter, and maximum diameter of coverage are usually sufficient for sizing a nozzle, as shown in Table 14.10. The optimal pressure typically results in a triangular water distribution pattern. One determines nozzle spacing along the sprinkler laterals and lateral spacing along the main, accounting for required overlap. The overlap is required to achieve a uniform application of water over the field. An overlap of 50% or so will give a uniform coverage. Wind distorts the pattern and one usually increases overlap to compensate for wind distortion. Based on experience, overlap should correspond to recommendations in Table 14.11. When wind is variable (e.g., no dominant direction or velocity) use the 0 or less than 6 km/hr conditions or substitute your judgment. One can compute sprinkler capacity from

$$q_{spr} = S_{lateral} S_{main} r_{intake} \quad \textbf{(14.15)}$$

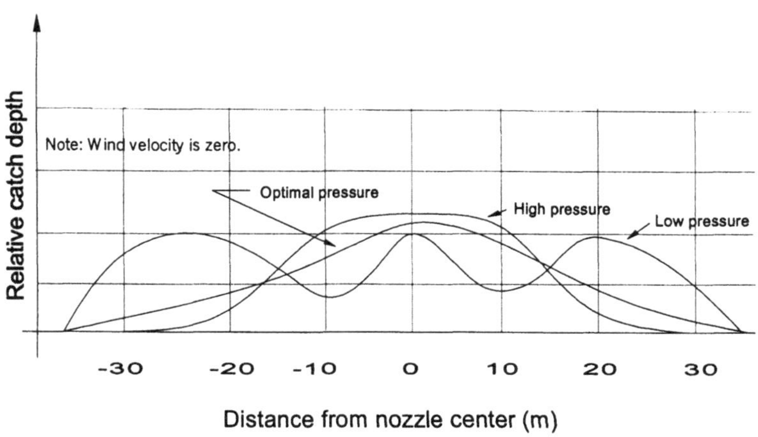

Figure 14.13. Water distribution patterns from a typical sprinkler nozzle operated at low, optimal, and high pressure.

Table 14.11. Approximate effects of wind velocity on sprinkler and lateral spacing.

	Lateral spacing in percent of the coverage diameter[a]	
Wind velocity (km/hr)	Spacing along lateral (%)	Spacing along the main (%)
0	50	65
Less than 6	45	60
7 to 12	40	50
13 or more	30	22

Source: Adapted from Schwab et al. 1993 and Pair et al. 1983.

[a]For laterals normal to the wind direction only.

where

q_{spr} is the individual sprinkler discharge;
$S_{lateral}$ is spacing along the lateral (apply % in Table 14.11 to the catalog diameter);
S_{main} is lateral spacing along the main;
r_{intake} is the soil intake rate or application rate.

The basic orifice equation (see equation 11.6) describes sprinkler discharge. One can write the orifice equation with SI irrigation units as

$$q_{spr} = 0.00111 C d_n^2 P^{1/2} \quad \textbf{(14.16)}$$

where

C (usually 0.95) is the coefficient of discharge;
d_n is the nozzle diameter in mm;
P is pressure in kPa.

Some sprinklers are actually combinations of orifices. One could solve for nozzle diameter. In practice, manufacturers tabulate flow rate as a function of pressure and nozzle size. In practice, one usually selects the closest available model and uses its value of q_{spr} for subsequent calculations.

Number of Sprinklers per Lateral

Compute the number of sprinklers needed on the lateral. Compute the total flow in one lateral by adding individual sprinkler contributions. Compute the total flow by adding flows of all laterals operated simultaneously.

Size the Lateral

Size laterals using the friction loss computed using the Scobey (1930)[2] equation. The Scobey equation is

$$H_f = F_{mo} \frac{K_{sc}\, L\, q^{m_{sc}} (4.1 \times 10^6)}{D^{3+m_{sc}}} \quad \textbf{(14.17)}$$

Table 14.12. Christiansen's factors for multiple outlets for smooth and somewhat rough pipe.

	Correction Factor F_{mo}			
	First sprinkler one sprinkler interval from main		First sprinkler one-half sprinkler interval from main	
No of sprinklers	$m_{sc} = 1.9$	$m_{sc} = 1.75$	$m_{sc} = 1.9$	$m_{sc} = 1.75$
1	1	1	1	1
2	0.63	0.65	0.51	0.53
4	0.48	0.5	0.41	0.43
6	0.43	0.45	0.38	0.4
8	0.41	0.43	0.37	0.39
12	0.39	0.42	0.36	0.38
16	0.38	0.4	0.36	0.37
20	0.37	0.39	0.35	0.37
30 or more	0.36	0.36	0.35	0.37

Source: Adapted from Christiansen 1948.

Notes: Data are based on Scobey flow exponents of $m_{sc} = 1.9$ and 1.75 for Q, where m_{sc} is the Scobey exponent. Use values in the $m_{sc} = 1.9$ columns for rough (e.g., sprinkler lateral) pipe. Use values in the $m_{sc} = 1.75$ columns for smooth plastic (e.g., drip tubing) pipe.

where

K_{sc} the Scobey constant for the pipe material;
L is the length of pipe in meters;
q is the discharge in l/s;
D is the inside diameter of the pipe in mm;
F_{mo} is the multiple outlet adjustment factor (Table 14.12);
H_f is the head loss in meters of water;
m_{sc} is a dimensionless exponent (1.9 for irrigation pipe and 1.75 for smooth plastic or PVC pipe; the 1.75 value is from the Blasius equation from fluid mechanics).

The K_{sc} factor ranges from 0.32 (transite plastic) to 0.40 (steel pipe or portable aluminum pipe with couplers) to 0.42 (portable galvanized steel pipe with couplers). Scobey showed that increasing the Blasius exponent values to 1.9 gave good agreement with measurements (Scobey 1930) when using these pipe types.

The concept of the multiple outlet (F_{mo} in equation 14.17) is not limited to irrigation lateral analyses. The multiple outlet factors, shown in Table 14.12, were determined by Christiansen (1948). He applied the Scobey (1930) equation to small pipe segments between laterals using the actual flow through the segment. Head loss over the lateral was then summed. He then computed the head loss assuming the total flow was conveyed along the entire lateral. The factors in Table 14.12 were found by taking a ratio of the head losses. Begin by taking 20% of the nozzle operating pressure for H_f. Solve the Scobey equation for D, then select the next largest commercially available pipe diameter. Select the nozzle operating pressure from a manufacturer's catalog.

Pressure at Lateral Inlets

One can compute the pressure required at the lateral inlet from equation 14.18:

$$H_n = H_a + 0.75H_f \pm 0.6H_e + H_{rp} \tag{14.18}$$

where

H_n is the pressure head at the lateral entrance;
H_a is the average nozzle pressure head (L) computed or selected from the manufacturer's catalog;
H_f is the friction loss;
H_e is the elevation difference between lateral inlet and the farthest sprinkler riser; H_e is positive when the lateral lies upslope and negative if it lies downslope;
H_{rp} is the riser height.

These pressure head components are shown schematically in Figure 14.14.

The criterion that $H_f + H_e$ not exceed 20% of H_a is a useful approach for specifying pressure extremes. If H_f exceeds this limit, distribution along the lateral is poor due to pressure variation. A lateral lying upslope is at a major disadvantage; positioning the lateral to lie downslope can be advantageous in that gravity may partially offset friction loss. One usually plans to have minimal elevation change along the lateral. Using an arbitrarily large diameter pipe reduces H_f and results in excellent uniformity; however, the design may be expensive. Pipe expense per unit length increases geometrically with pipe diameters.

Sizing the Main

Use the Scobey relationship with one or two inlets (or the number of laterals operating simultaneously) for friction calculations. The goal is to get the smallest pipe diameter carrying the flow with a reasonable friction loss. Try to keep friction losses less

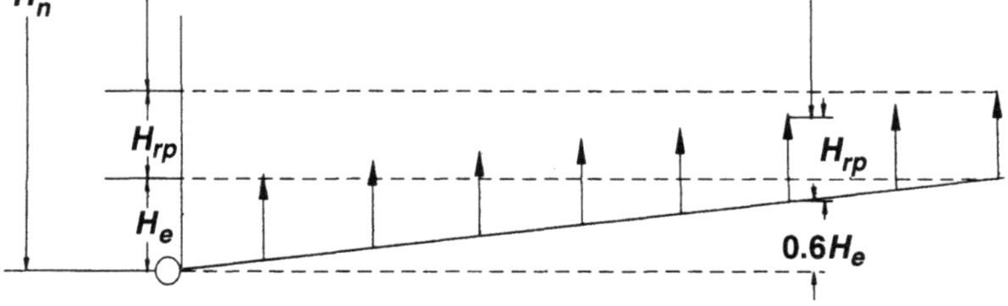

Figure 14.14. Schematic of pressure head components needed for sprinkler lateral design.

than 40 kPa per 100 m of pipe length. In meters of water, this is 4 m head loss per 100 m of length, the economical friction loss for typical irrigation pumping systems. Another criterion for irrigation mains design is that the velocities be limited to 5 ft/s (1.4 m/s). Experience has shown this limit to reduce problems due to water hammer.

Sizing the Pump

Compute pump capacity and power requirements. Appendix E covers pump selection in more detail.

Water Rights and Water Quality

Be aware of water rights and water quality issues. Make appropriate chemical analyses to document water quality or waste constituents if applicable. Effective available water when using saline water is usually less than that found using above methods. Total wastewater applied may be less than total *ET* demand. If so, the BOD and/or nutrients must not be overapplied; one would specify other water for the area if needed.

Design Summary

Draw a map showing soil type and wind and elevation information; show main plus one or two laterals.

Determine the soil intake rate (Table 14.4).

Determine root zone (Table 14.3), soil storage capacity (Table 14.2), *MAD*, and, after dividing by irrigation efficiency, the total applied after each pass.

Determine peak *ET* (Table 14.3); get set time by dividing *ET* into soil storage capacity; compute irrigation interval by dividing the peak *ET* into the total applied.

Select a sprinkler, noting operating pressure and wetted diameter (Table 14.10). Adjust the wetted diameter for lateral and main connection spacings based on wind (Table 14.11); this can be unavoidably frustrating the first few times through. The approximate nozzle data given in Table 14.10 are good for many standard applications.

Compute lateral length based on field dimensions and number of sprinklers required to get the specified overlap based on wind condition. Compute the number of connections to the main based on spacing requirements and field dimensions.

Determine area irrigated by one lateral set; compute numbers of hectares to be covered in one day to enable satisfaction of irrigation interval. Compute total sets per day. Based on set time and move time (estimate 1 hr per lateral per set), estimate the number of laterals one must operate and not exceed maximum operating hours per day.

Calculate lateral flow and pressure; select lateral diameter so pressure distribution across lateral is less than 20% of nozzle operating pressure.

Calculate the main diameter giving losses less than 4 m per 100 m length.

Size pump and suction line if applicable (see Appendix E).

Compute pump power requirement using methods in Appendix E.

Be sure water quality and water rights are acceptable. Alternative systems may be necessary (see Cuenca 1989).

Example 14.8: Design a wheel sprinkler irrigation system to irrigate a 20 ha (square) peanut field on a sandy loam bare soil over a compact subsoil in south Georgia (hot climate). The field slope is flat. Determine the limiting rate of application, the irrigation period, the net depth of water per application, the depth of water pumped per application, and the required system capacity in ha/day. The available water holding capacity of the soil is 150 mm/m (see Table 14.2) and the effective depth of the root zone is taken to be 0.6 m (see Table 14.3). Suppose 23 hr per day is available for irrigation. Take the management-allowed deficit to be 0.45. The maximum intake rate of the soil is 15 mm/hr (see Table 14.4). Winds during irrigation are 7–12 km/hr. Use a water efficiency of 0.7, motor efficiency of 0.75, and a drive efficiency of 0.95.

The total available water capacity is 0.9 × 150 mm = 135 mm. See Table 14.2. The net depth of application is 0.45 × 135 mm = 61 mm. Assuming a water application efficiency of 70%, the total application is 61 mm/0.7 = 87 mm. From Table 14.3 peanuts use 8 mm per day in a hot climate. Thus, the irrigation period is 61 mm/8 mm/day ~7 days (equation 14.13). Thus, the system must pump and discharge 87 mm on 2.8 ha/day to irrigate the field over the cycle time. The field map and proposed system layout appear in Figure 14.15. Lateral length assumes 12 m from main to first sprinkler and 8 m to the boundary from the last sprinkler. The layout assumes a hypothesized sprinkler spacing of 12 m and lateral spacing of 17.18 m.

Number of lateral sets per day: (2.8 ha × 10,000 m^2/ha)/(20 × 12 m × 18 m) = 6.48 or 8 sets per day total. Use two laterals with 4 sets per day.

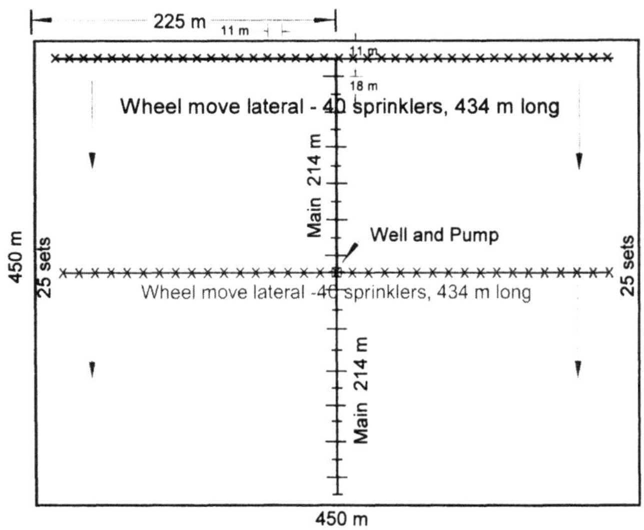

Notes: Four laterals are shown, 2 in black and 2 in grey.
Two laterals are operated at a time and 2 are ready to begin operation with minimal time loss at switchover.
One lateral set operates on the even numbered main connections while the other set operates on the odd numbered main connections.

Figure 14.15. Field layout for Example 14.8. Arrows show the movement of the laterals over time.

Discharge required per sprinkler (equation 14.15):

q_{spr} = (11 m × 18 m × 15 mm/hr × 1000 l/m3)/(1000 mm/m × 3600 s/hr) = 0.825 l/s

Round to 0.9 l/s. The intake rate came from Table 14.4.

Check compatibility of sprinkler discharge with lateral spacing and mains: Note from a manufacturer catalog (sample excerpts shown in Table 14.10) for a discharge of 0.89 l/s, a 6.35 mm × 3.97 mm sprinkler, with a diameter of 31 m, and operating pressure of 276 kPa (40 psi), will suffice. Assuming the lateral is normal to the prevailing wind of 7–12 km/hr, the sprinkler spacing is 31 × 0.40 = 12.4 or 12 m along the lateral. The spacing along the main is 0.5 × 29 = 16 m. The spacing of 18 m used is close enough. If the calculated spacings are off by more than 25%, then recompute spacings and nozzle flow and redo the above steps. The flow rate is 10% greater, which will compensate for the larger spacing.

Discharge per lateral: 20 sprinklers × 0.9 l/s = 18 l/s.

Daily operating time: gross application = 91 mm/application rate = 87 mm/15 mm/hr = 5.8 hr. Four sets per day × 5.8 hr/set = 23.2 hrs., near the limit. One wheel lateral operates at a time as shown in Figure 14.15. An alternate set wheel lateral, which is connected to every other lateral connection along the main, enables near-continuous operation, thus the system is considered acceptable.

Size the lateral and main: Assume a riser height of 0.6 m or 2 ft. The maximum pressure variation is 20% of the operating pressure or 0.2 × 310) = 62 kPa.

The allowable variation due to friction is 62/9.8 − 0.6 = 6 m.

Use portable aluminum pipe with a K_{sc} = 0.40. For 20 sprinklers, the first sprinkler should be one spacing away from the main; the F factor is 0.37. Using the Scobey equation (equation 14.8), the friction losses are calculated as shown in Table 14.13 using a length of half the lateral length. The 127 mm diameter lateral and the 178 mm diameter main give a pressure variation within the computed allowable variation of 6 m. The 178 mm diameter for the main allowed pressure loss to be less than the 4 m per 100 m required.

The pressure head at the junction of lateral and main (equation 14.18) is

$$H_n = 31.6 + 0.75(4.8) + 0.6(0) + 0.6 = 35.8 \text{ m}$$

(The average operating pressure is 310 kPa = 31.6 m; 9.78 kPa = 1 m.)

Table 14.13. Summary of friction losses with indicated lateral and main diameters.

	Friction loss (m)			
Tubing Outside Diameter (mm)	217 m lateral × F_{mo}	Lateral loss (m per 100 m)	214 m main with two laterals operating	Main loss (m per 100 m)
76	19.4	8.9	196	92
101	4.8[a,b]	2.2	48.7	23
127	1.6	0.7	15.8	7.4
152	0.7	0.3	6.7[a]	3.1
178	0.3	0.14	3.02	1.41

[a]Select the 101 mm lateral diameter that gives 4.8 m loss, which is within the computed allowable loss of 6 m. Select the 152 mm main that gives a pressure loss of 3.1 m/100 m, within the 4 m/100 m criterion.

[b]Add or subtract elevation changes to friction loss (if there was a positive or negative elevation change, respectively) when making the 20% operating pressure test.

One computes the system head using (see Appendix E):

$$H_s = SL + DL + DD + H_{ls} + M_{ls} + H_{ld} + M_{ld} + H_o + VH \qquad \textbf{(14.19)}$$

where

H_s is total system head;
SL is suction side lift;
DL is the discharge side lift;
DD is the well drawdown;
H_{ls}, M_{ls} are the pipe friction and minor suction losses;
H_{ld}, M_{ld} are the pipe friction and minor discharge losses;
H_o is the operating pressure at the end of the discharge line;
VH is the velocity head.

All heads are expressed in units of length.

For this problem, take $SL = 5$, $DL = 1$, $DD = 0$, $H_{ls} + H_{ld}$ (main losses) = 6.7, $M_{ls} + M_{ld}$ (minor losses) = 0, and $H_o = 35.8$ m, resulting in $H_s = 48.5$ m. NPSH is the net positive suction head, provided by the pump manufacturer. Minor losses are sometimes estimated by taking 5% of the friction head plus average sprinkler operating head.

The power is then 0.0098×36 l/s $\times$ 48.5 m/(0.7 $\times$ 0.75 $\times$ 0.95) = 35 kW.

Pipe network software, in which one may vary the lateral diameter over the length, is handy for large designs. One can specify the pump once the operating pressure and flow are determined.

Center Pivot System Specification

The center pivot is perhaps the most common agricultural irrigation machine in use. Approaches for specifying a center pivot are similar to those for other irrigation machines such as linear move systems. The center pivot irrigation system is a continuously moving sprinkler lateral pivoting on one end. Center pivots are extensively used for agricultural irrigation. System designers arrange the pivot nozzle package to uniformly apply water

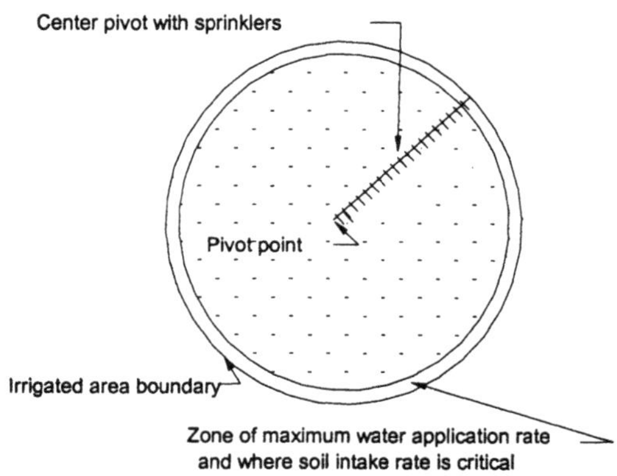

Figure 14.16. Schematic of center pivot irrigation system showing area irrigated and zone of maximum application rate.

Figure 14.17. Large center pivot system. Sprinklers may be along the pivot pipe top or on drops from the main pivot pipe. Center pivot systems rotate about a fixed point (pivot) and have a radius of coverage ranging from less than 1 ha to 50 or more ha.

over the irrigated area on each pass. As with sprinkler laterals, one must apply the required amount of water to fill the soil storage reservoir. The system must rotate with an adequate speed to replenish the soil storage zone. Figures 14.16–14.17 show center pivot systems. Pivots apply water at high rates near the end away from the pivot point. Center pivot systems are usually restricted to soils having high intake rates. These soils include sands, sandy loams, and well-structured loams. Finer textured soils usually have intake rates too low for center pivots. Unlike a sprinkler lateral, center pivots and other irrigation machines must be designed not to exceed allowable soil intake rates.

Irrigation machine design is performed at the manufacturer level. Manufacturers design the propulsion system and devise nozzle combinations for uniform water application over a range of pivot length and soil condition combinations. They determine required operating pressures. Field engineers work with clients to customize irrigation machines for specific locations or to determine site feasibility for an irrigation machine. Refer to texts such as Jensen 1981 or Cuenca 1989 for hydraulic design discussions. Field engineering to specify a system for a client is emphasized. Field engineers are concerned with required system flow rates, the soil storage reservoir, and the peak *ET*. One must specify the system size, rotation interval, and system flow. Soil intake rate should equal or exceed the maximum application rate.

System Size

The circle diameter containing the required field area must be determined. One may compute pivot length from

$$R_{pivot} = \sqrt{\left(\frac{A_{field}}{\pi}\right)} \tag{14.20}$$

where

R_{pivot} is the pivot length (m);
A_{field} is the required field area (m^2).

Equation 14.20 assumes no big gun on the end of the pivot. The pivot does not swivel at any point along the pivot length.

System Rotation Interval

The center pivot must rotate often enough to replenish the soil storage reservoir before *ET* demands empty the soil storage. The system rotation interval is identical to the cycle time defined in equation 14.3.

System Flow Rate

Flow rate for a center pivot system is computed to replenish *ET* in the soil storage reservoir for the cycle time days. One may compute flow rate using

$$q_{cp} = 1.1574 \times 10^{-5} \frac{d_{application} A_{field}}{T_{cycle}} \tag{14.21}$$

where

q_{cp} is the system flow rate (l/s);
$d_{application}$ is the application amount (mm);
A_{field} is the coverage area (m^2);
T_{cycle} is the cycle time (days).

Maximum Application Rate

The water application rate is highest at the end of the pivot (see Figure 14.16). Assuming an elliptical water application pattern, NRCS 1983 gives the maximum water application rate as

$$I_{max} = \frac{28800 q_{cp}}{R_{cp} d_{wttd}} \tag{14.22}$$

where

q_{cp} is the system flow rate, computed using equation 14.21 (l/s);
R_{cp} is the length of the pivot (m);
d_{wttd} is the wetted diameter of the sprinklers used in the nozzle package (m).

Values computed using equation 14.22 should equal or exceed the Table 14.4 value for the corresponding soil condition. One may compute the sprinkler wetting radius to handle the maximum condition. Feasibility would then depend on availability of systems (nozzles, booms, etc.) to achieve the wetted diameter.

Values for sprinkler wetted diameters are assumed constant along the length of the pivot. NRCS 1983 gives additional background about maximum soil intake rates. See texts such as Cuenca 1989 and Jensen 1981 for additional design information.

Irrigation machine design becomes challenging with soils containing high textural and topographic variability. One may design the entire system for the worst (minimum) soil storage case or (lowest) intake rate case; or one may take a weighted average of intake rate and soil storage based on area. The soil properties of the dominant soil mapping unit are often used. Field obstacles also provide challenges requiring site-specific modifications. Variability may reduce application efficiency by 5–10% or more.

Example 14.9: Design a center pivot irrigation system to irrigate a 16 ha circle. The maximum application rate is 15 mm/hr for the flat silt loam soil under pasture. Assume a *MAD* of 0.45 storage capacity and an application efficiency of 0.7.

The peak *ET* is 8 mm/day. The soil depth is 0.6 m and the soil storage is 150 mm/m. Assume one revolution in the cycle time interval. Compute the system size, cycle time, and sprinkler wetted diameter needed to not exceed the maximum soil intake rate.

The size of the pivot is R_{cp} = [(16 ha) (10,000 m^2/ha/π)] = 226 m (equation 14.21). The allowable depletion is 0.45(0.6)(150 mm) = 41 mm. The cycle time is then 41 mm/8 mm/day = 5 days.

The pivot flow rate is given by

$$q_{cp} = 41 \text{ mm} \times 160{,}000 \text{ m}^2 \times 1000 \text{ l/m}^3/1 \text{ rev.} \times 5 \text{ days/rev.} \times 86{,}400 \text{ s/day} \times 1000 \text{ mm/m} = 15.2 \text{ l/s}$$

Equation 14.21 gives the same answer.

The maximum water application rate is

$$I_{max} = 15 \text{ mm/h} = 28{,}800(15.2 \text{ l/s})/\{226 \text{ m } [d_{wttd}(3.14)]\}$$

from equation 14.22. Therefore, d_{wttd} = 41 m, very large.

One would work with the system manufacturer to determine the nozzle package and operating pressure needed at the pivot point. Design the pumping plant as for a sprinkler system.

Design for Nonagricultural Applications

Bioremediation approaches may involve water application as a nutrient carrier for maintaining microbial populations. It may involve irrigation for leaching constituents from the soil. These are examples of novel irrigation applications. Many systems are available. One may use sprinkler laterals for frost protection and bloom delay in high-value crops. Nonagricultural applications often use sprinkler systems. Sprinkler systems were determined to be good for bioremediation (Sims et al. 1986; Riser-Roberts 1992). Sprinkler systems serve other purposes involving water or temperature management. These include waste application, moisture management on logs, and urban irrigation. Microirrigation systems also have advantages, ranging from crop or turf watering to bioremediation applications.

The emphasis on the environment has raised interest in how other fluids behave in soils. The interest in bioremediation is prompting development of release curves for many other fluids. Most organic contaminants are poorly soluble in water and most have lower values for surface tension than water does. These nonaqueous-phase liquids may form pockets of undiluted compound.

Water is usually present in these applications as well, resulting in multiphase fluids with phases more dense (DNAPL, dense nonaqueous phase liquids, e.g., some organic polychlorides) and less dense (LNAPL, light nonaqueous phase liquids, e.g., gasoline) than water.

Figure 14.3 showed a generic vadose zone for a typical soil. In the humid eastern United States, the vadose zone may go all the way to the water table and may include one or more impeding layers. In bioremediation projects, the zone below the soil water management zone and into the water table may be of more interest than the soil water

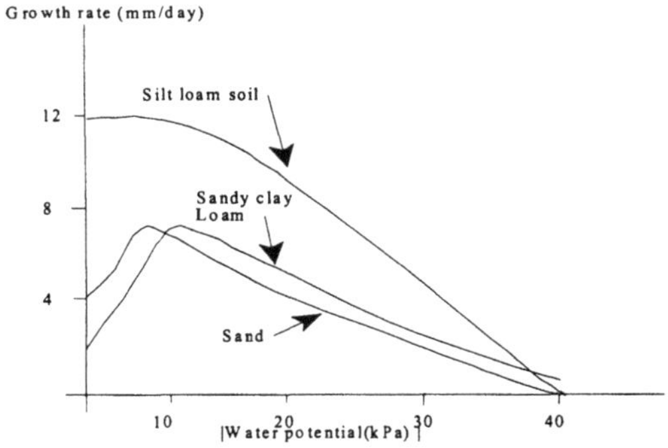

Figure 14.18. Growth rate as a function of water potential (absolute value) for *Phytopathora cinnamoni* in three soil types. (Redrawn from Baver et al. 1972.)

management zone. By managing the water table, one can influence the concentration of DNAPLs or LNAPLs and also the oxidation status. Figure 14.18 shows the influence of water potential on fungal activity.

Bacteria and fungi grow at varying rates depending on soil water potential levels. This is true for plant roots as well. The response depends on the organism's aeration status (aerobe or anaerobe). At higher (e.g., near zero on the negative side) matric potentials, the soils approach saturation, resulting in low oxygen and limited growth of aerobes. At lower matric potentials (e.g., high negative values), low water film thickness limits the growth rate. Figure 14.18 shows fungal growth data as a function of matric potential. Aerobes function well between −100 and −2000 cm water. This corresponds well to the available soil water zone (see Baker 1994). Some fungi grow well in extremely dry soils. If a bioremediation process requires an anaerobic process, we would usually want a saturated soil. If water table adjustment is a viable option (e.g., in subirrigation schemes), one may achieve denitrification by saturating soils for a period.

It helps to have a firm understanding of agricultural irrigation before designing nonagricultural systems. Some of the ways in which nonagricultural irrigation differs from agricultural irrigation are as follows:

1. Soil storage and depletion are not the only determining factors in system specification; one or more of the following should be considered: moisture content maintenance, BOD and other nutrient loading, along with hydraulic loading.
2. In the case of waste disposal, one starts with a known quantity of waste and calculates the irrigated area instead of computing the quantity of water required for a known crop area.
3. Entire systems usually operate simultaneously instead of a programmed rotation of lateral movement.
4. Salinity, usually a nonissue in humid climates, may become limiting.
5. Systems may lie over rough terrain requiring pressure regulation. Allowed variation along the lateral is much less than the 20% common with agricultural approaches.
6. New concepts of irrigation efficiency may be required to evaluate system efficacy.
7. Laterals with more than one pipe diameter are common.
8. Seasonal effective rainfall may not figure into the design.
9. Failure to adequately maintain waste irrigation systems may result in severe fines.

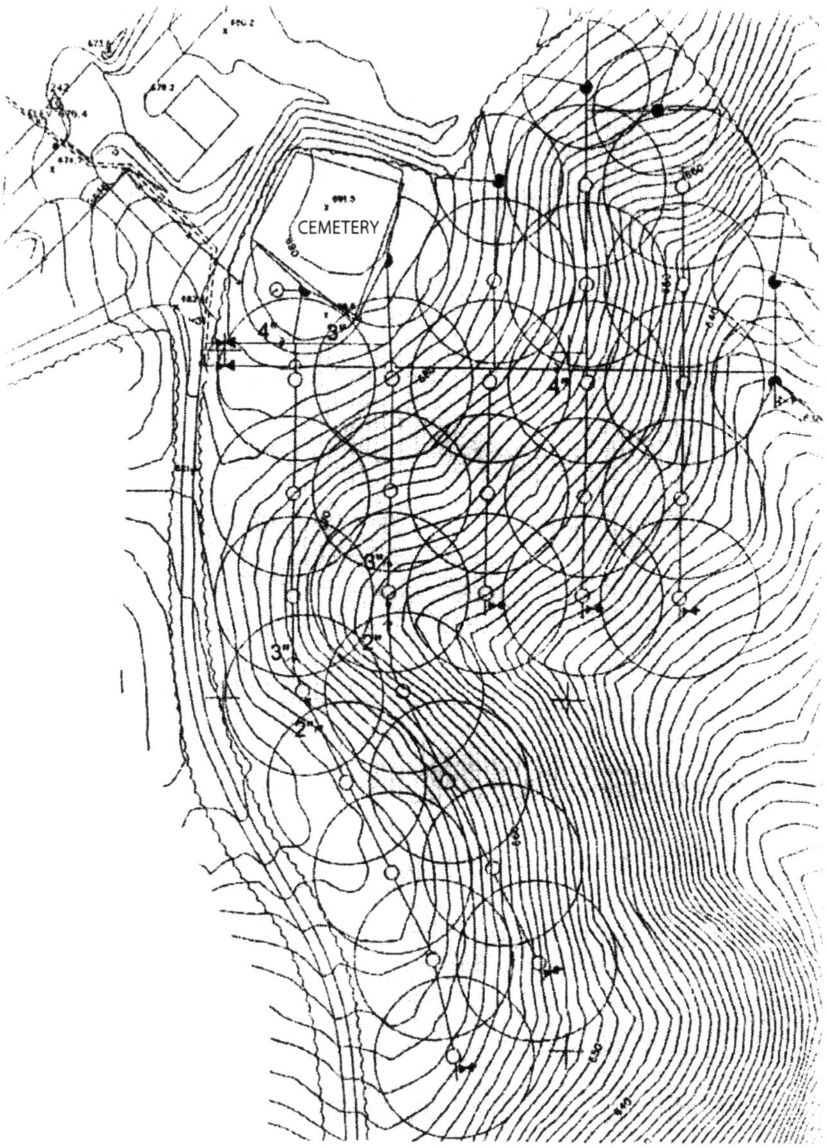

Figure 14.19. Solid-set sprinkler layout for a small land application system in Clarke County, Georgia.

10. Abrasion resistance may be as significant as corrosion resistance.
11. Regulatory agencies often specify design approaches.

Figure 14.19 shows a solid set system for land application of wastes. Observe the sprinkler overlap of about 40%. The low overlap was chosen because the trees protect the system from wind. This system, with little modification, could also serve a frost protection/bloom delay role, though additional overlap may be required due to greater wind exposure.

Several accessories (not shown) were used. The pressure/vacuum relief valve, located at the high point in the system, enables drainage in both directions. The automatic drain valves at the end of the laterals empty the laterals in cold weather. Manual shutoffs are located at lateral ends. These are closed during summer months. The system is equipped with valves enabling supply line drainage from the pressure/vacuum relief at the pump. An important accessory not shown is the pressure regulator, located on each sprinkler riser.

Not shown on Figure 14.19 is the pumping station. Approximately 250 m of 150 mm (6 in) PVC pipe bring water to the pressure relief valve. There are four 90-degree elbows with $r/R = 1$; two 45-degree elbows with $r/R = 5$; one flap check valve; one globe valve; and one gate valve on the pump discharge side. The suction side has one foot valve and one 90-degree elbow ($r/R = 1.5$) and 16 ft of 203 mm (8 in) line. Several problems at the end of the chapter lead one through simplified aspects of irrigation design for waste application.

Pipe network software is convenient for identifying economical designs. One can specify the pump once the operating pressure and flow are known.

Example 14.10: Suppose a 500 ft (155 m) long HDPE (e.g., smooth plastic) lateral with 9 sprinklers lies on a variable slope with the end riser base being 20 ft (6.1 m) higher than the main connection. The equivalent riser height is 10 ft (3.05 m), which accounts for a riser ball valve and pressure regulator losses. Pressure-regulated (50 psi, 115.5 ft water, 35.2 m) sprinklers delivering 10 gal/min (0.63 l/s) are used. A pressure of 65 psi (150 ft water, 45.6 m) is available at the lateral main connection. Size the lateral.

Substituting known values into equation 14.18 gives

$$45.6 \text{ m} = 35.2 \text{ m} + 0.75\,(H_f) + 6.1 \text{ m} + 3.05 \text{ m or } H_f = 1.67 \text{ m}$$

This is the maximum possible friction loss. One must determine a diameter giving no more than 1.67 m head loss.

Next, calculate the friction loss for several lateral diameters (51, 76, 102 mm). The total lateral flow is 9 × 0.63 l/s = 5.67 l/s. The Christiansen factor for nine sprinklers, $m = 1.75$, first sprinkler at full interval from the main, is taken as 0.37. One may use a Scobey coefficient of 0.4 even though the pipe is smooth (connections protrude into the lateral and create extra roughness). Using the Scobey equation, one has

$$H_f = [0.37(0.4)(155 \text{ m})(5.67 \text{ l/s})^{1.75}]/(Dp \text{ mm})^{4.75} = 15.2 \text{ m } (D_p = 51 \text{ mm});$$
$$2.28 \text{ m } (D_p = 76 \text{ mm}); \text{ and } 0.56 \text{ m } (D_p = 102 \text{ mm})$$

The 102 mm (4 in) diameter lateral loses are less than the allowable loss and thus meets the condition.

Microirrigation System Design with Pressure-Compensated Emitters

Microirrigation systems are widely used in agricultural applications and may have a place in bioremediation applications. Burt and Styles (1994) provide detailed drip system design methodology. For this brief introduction, the focus is on a simple bioremediation application. Figure 14.20 shows a series of tensiometers placed to monitor the wetting front from a microirrigation emitter.

The pressure-compensated emitter enables application of a well-controlled rate even when there is topographic variation. The hydraulics of microirrigation system design is

Figure 14.20. Microirrigation lateral with emitter instrumented to measure soil water status vs. distance from emitter.

in many ways similar to that of sprinkler systems. Consider the situation shown in Figure 14.21.

Suppose a pressure-compensated emitter delivers 4.2 l/hr over a 10–30 psi (69–207 kPa) pressure range. This implies the pressure at the farthest point in Figure 14.21 must not be less then 69 kPa and the pressure at the closest emitter to the well must not exceed 207 kPa.

The Blasius equation is used for design of microirrigation laterals and mains and is similar to the Scobey equation except that the exponents are lower:

$$H_f = K_{sc} L_q^{1.75} D^{-4.75} \tag{14.23}$$

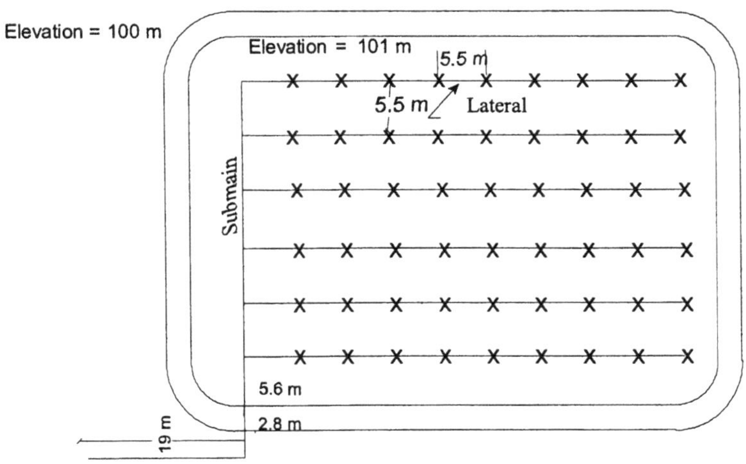

Figure 14.21. Microirrigation system in a hypothetical bioremediation application. Each junction has an emitter also.

where

K is a constant equal to 7.89×10^5 for indicated SI units; this factor contains a friction loss factor for plastic tubing.
H_f is the friction loss in m;
L is the pipe length + equivalent length (due to emitters, fittings, etc.) in m;
q is the total flow in l/s;
D is the inside pipe diameter in mm.

The Christiansen multiple outlet factors (Table 14.12) may be used to estimate pressure in tubes with multiple flows. In microirrigation systems with plastic tubing the column for $m_{sc} = 1.75$ is used. Multiplying the computed head loss by the Christiansen factors gives actual head loss.

Schwab et al. (1993) suggest the equivalent length for in-line emitters is 1–3 m per emitter; for on-line emitters the equivalent length is 0.1–0.6 m per emitter; and for solvent welded connectors the equivalent length is 0.3–1 m per emitter. They also suggest allocating 55% of allowable head loss to laterals and 45% of losses to the submain.

Example 14.11: A bioremediation application calls for using solvent welded pressure-compensated emitters each applying 4.2 l/hr over 69–207 kPa on a 5.1 m grid as shown in Figure 14.21. There is an emitter at each junction. The emitters are placed on a level surface 1 m higher than the pump. Size the laterals and the submain and determine the pressure required at the nearest submain-lateral junction.

Each lateral has nine emitters for a total flow of 37.8 l/hr and actual length of 50 m. Include an equivalent length of 0.5 m per emitter, or 4.5 m, for a total length of 54.5 m. The total length of the submain is $5.1 \times 5 + 5(0.5) = 28$ m. The Christiansen factor for nine emitters, one spacing from the main, $m_{sc} = 0.75$, is 0.43 (see Table 14.12). The factor for the submain is 0.47 (both values interpolated). The maximum allowable head loss over the system is $207 - 69 = 138$ kPa (13.8 m). Any elevation variations occurring over time must also fall within this range.

From equation 14.23, one can write the friction loss for the lateral as

$$H_{flat} = (0.43)\,(7.89 \times 10^5)\,(54.5\text{ m})\,[(37.8\text{ l/hr}/\,(3600\text{ s/hr})]^{1.75} D_{lat}^{-4.75}$$
$$= 6.37 \times 10^3 D_{lat}^{-4.75}$$

Table 14.14. Summary of friction losses with the Example 14.11 microirrigation system.

Diameter (mm)	Loss in lateral (m)	Loss in submain (m)	Total loss (m)
5	3.05	34.4	Total loss is 10.01 m (100 kPa).
7	0.62	6.96	Lateral and submain losses are
10	0.11	1.28	roughly in line with the Schwab
15	0.02	0.19	et al. (1993) loss proportions.

Likewise, one can write the friction loss for the submain as

$$H_{fsub} = (0.47)\,(7.89 \times 10^5)\,(28\ \text{m})\,[(5 \times (37.8\ \text{l/hr} + 4.2\ \text{l/hr})/(3600)]^{1.75}$$
$$D_{subm}^{-4.75} = 7.19 \times 10^4 D^{-4.75}$$

Table 14.14 shows losses calculated using several diameters and highlights the selected choice.

Assume a subsidence variation of 0.5 m (site was only 1 m above stable soil).

The pressure required at the operating junction is 69 kPa + 100 kPa + 5 kPa = 174 kPa. Flow at the junction is 4.2 × 10 × 6 = 252 l/hr = 0.07 l/s.

Approaches discussed previously may be used to design the pump station.

Microirrigation system design for agricultural applications is much more involved than this example shows at first reading. There are special approaches for determining *ET*. Non-pressure-compensated emitters are commonly specified due to lower initial cost. Uniformity of application then becomes an issue. Specification of the required zone of wetting and determination of emitter size are required. Cuenca (1989) gives additional details.

Water quality is crucial for agricultural and nonagricultural microirrigation systems. One should include back washable screens or sand filtration units to remove potentially clogging particles. Pressure gauges are placed before and after the filter to monitor the head loss. Excessive heads across the filter signal the need to backwash or change the filter. Maintenance should include periodic flushing of the entire system with chlorine or other disinfectants. This inhibits the growth of slime-producing microorganisms. Avoid water from ponds if possible.

Safety and Maintainability

Irrigation systems must be designed for safe operation. Well-designed systems also contain operational strategies for addressing maintenance issues. Figure 14.22 illustrates components commonly found in irrigation systems ranging from sprinkler laterals to trickle systems. Burt et al. (1995) give detailed discussions related to application of nutrients through irrigation systems. Hansen et al. (1980), Pair et al. (1983), and Cuenca (1989) also present practical discussions of safety and operational issues.

The injection of nutrients requires precautions ensuring that nutrients do not accidentally contaminate freshwater supplies. Approved backflow prevention valves are required in the supply line just ahead of the pump. Backflow prevention is required for any system attaching directly to a public water supply. Figure 14.23 shows additional safety strategies for safe nutrient injection (ASAE 1997).

Filtration is necessary with microirrigation to avoid emitter clogging. Filtration devices are not shown in Figures 14.22 and 14.23. Vacuum breakers, pressure regulators, and pressure gauges enable precise pressure regulation when needed. Secondary filters are sometimes placed in microirrigation laterals, especially when injecting nutrients. One usually equips laterals with flush valves. This enables flushing with bleach or other chemicals to control microbial-induced clogging, to clean injected chemicals from the system, and to facilitate winterization.

In specifying materials and accessories for irrigation systems, one must give detailed pipe specifications. Organizations such as ASAE (http://www.asae.org) and the Irrigation Association (http://www.irrigation.org) provide excellent guidance for materials specification

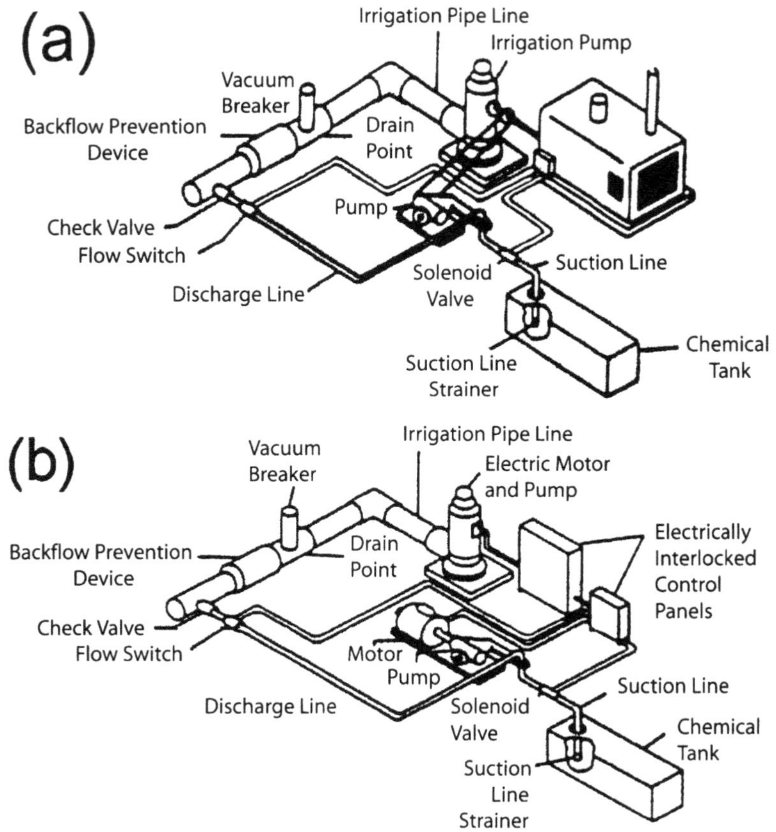

Figure 14.22 Details of chemical injection systems for irrigation with (a) diesel-powered system and (b) electrically powered system. (From ASAE 1997; used with permission.)

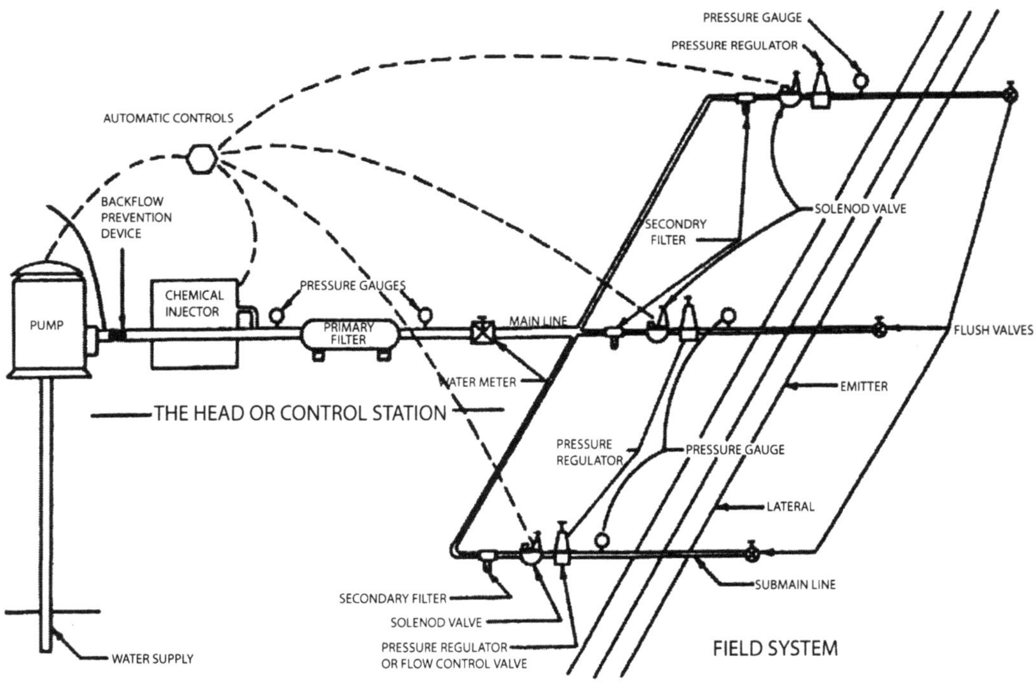

Figure 14.23. Schematic of trickle irrigation system showing important components for safety and system maintenance. (From Pair et al. 1983; used with permission of the Irrigation Association.)

and installation. Local offices of the NRCS may provide helpful information. As with any engineered project, ASTM or other standards agency test results should guide materials specifications. Corrosion, durability in the presence of sunlight and temperature (with plastic materials), pressure requirements, brittleness, expansion-contraction with temperature, and crushing resistance are examples of important issues.

Outlook

Irrigation design is a well-developed discipline, briefly introduced in this text. All irrigation systems have some issues in common. For any irrigation system to work for plant production or other application, the quality, quantity, and delivery rate of water must be satisfactory. Only then can one meet the seasonal and peak *ET* plus leaching requirement. Waste application and/or nutrient applications must be within appropriate standards to maintain downstream water quality. The system for delivering the water should be chosen to satisfy soil available water and soil intake rate constraints. Uniform application is important. Available labor and expertise must also be considered. Good-quality water is essential for crop irrigation. Nonagricultural applications carry forward the idea of not exceeding water balance parameters while achieving uniform application. The parameter most often overlooked in specifying these systems is the soil intake rate. Application rate often exceeds infiltration rate due to lack of adequate design. Ignoring limiting soil intake rates reduces water application efficiency by increasing surface runoff. Choice of appropriate materials and system safety must be high priority.

Proceedings of conferences sponsored by the Irrigation Association and/or the ASAE (e.g., Heerman and Solomon 1990) are excellent resources for tracking developments in irrigation, including nonstandard (i.e., nonfarm) irrigation case studies. An Internet search will lead to several commercially available irrigation design software packages.

Problems and Questions

Draw sketches of all completed designs where applicable.

1. Determine the evapotranspiration and irrigation requirement for small grain where the monthly *ET* coefficient $k = 0.65$ in the Blaney-Criddle formula. The average monthly temperatures for the 3-month growing season are 21.6, 23.4, and 24.6°C; the percentages of daytime hours are 8.7, 9.3, and 9.5; the average rainfall is 33, 49, and 54 mm. Assume a water application efficiency of 60%. The soil rooting zone is 1 m, and the available water holding capacity is 100 mm per meter.
2. Given a soil profile that is sandy for the first foot of depth and silty clay for the next 3 feet and where corn is to be grown, what is the available water to the corn plant? If sprinkler irrigation is used, what is the shortest time the sprinkler can operate over the soil to supply the required water? What is the cycle time of the system?
3. Determine the leaching requirement and depth of water application to an alfalfa field if a 10% yield reduction is acceptable. The salinity of the irrigation water is 1.2 dS/m, 15 days have elapsed since the last irrigation, and the average alfalfa *ET* was 9 mm/day. Assume no runoff from the irrigation.
4. Design a sprinkler irrigation system for a square 16 ha field to irrigate the entire area within a 14-day period. Not more than 16 hr per day are available for moving the pipe and sprinkling. Depth of application is to be 60 mm at each setting, not to exceed a

Table 14.15. Data for Problem 14.4.

Drawdown + suction lift (m)	Discharge (l/s)
1	8
5	10
10	12
15	15
20	19

rate of 9 mm/hr. A well 25 m deep in the center of the field will provide the water. The drawdown discharge is given in Table 14.15.

Assume a 276 kPa nozzle pressure, highest point in field is 2 m above the well site, 1 m riser height, 60% pump efficiency, and 70% engine-drive efficiency.

5. A sprinkler main with a maximum equivalent length of 400 ft (122 m) is an irrigation design scenario under evaluation. The required flow is 250 gpm (= 15.77 l/s), and the required pressure at lateral connection is 60 psi. What pipe diameter for the main is needed to keep the loss within tolerance? What pressure must be delivered to this main?
6. A lateral has 25 sprinklers, each applying 10 gpm at 50 psi. The lateral is 800 ft long. The lateral lies downslope on a 0.5% slope. Riser height is 1 m. There are no pressure regulators or other fittings on the riser. Size the lateral.
7. Four irrigation laterals, each having 10 sprinklers applying 0.567 l/s at 60 psi, operate simultaneously. Assume the four laterals effectively join at one point that is 10 m higher than the water level being pumped. Approximately 250 m of 150 mm (6 in) PVC pipe bring water to the lateral connection point. There are four 90-degree elbows with $r/R = 1$; two 45-degree elbows with $r/R = 5$; one flap check valve; one globe valve; and one gate valve on the pump discharge side. The suction side (203 mm; 8 in) has one foot valve and one 90 degree elbow ($r/R = 1.5$). What PVC pipe size is required to deliver the flow at a pressure of 65 psi to the lateral connection, within customary loss values? What pressure is required at the pump discharge? (This is similar but not identical to the situation shown in Figure 14.19.)
8. Resize the pump in Example 14.8 with both wheel laterals operating simultaneously. If the cycle time does not change, what are the implication on the required operating hours per day?
9. A HDPE smooth plastic lateral 480 ft long with eight sprinklers requiring 0.567 l/s, 50 psi, lies downhill with the base of the lateral being 12 m below the lateral connection. Assume a plastic pipe with pressure-regulated sprinklers requiring 40 psi. The equivalent riser height is 3 m. Size the lateral.
10. Redesign the microirrigation system shown in Example 14.10 for the case where two emitters, each delivering 4.2 l/hr, are located at each point. Assume each point can be treated as if there was one emitter delivering 8.4 l/hr. The goal is to raise the volumetric moisture by 0.2 units with the moisture addition. Estimate the time required for the system to operate assuming no deep percolation and an effective loss of 10% due to evaporation.
11. Given a deep sandy soil in Tifton, Georgia, specify a center pivot for 50 ac of corn. Assume the peak *ET* is 9 mm/day. The rooting depth is 0.9 m. Assume a nozzle

package requiring 50 psi is available. Assume a gate valve, two 90 degree elbows, a butterfly flap valve, two couplings, and 30 ft of pipe for connecting the pump discharge to the pivot. The well is located at the pivot point, and the level of the well is 100 ft below the surface. The deep well turbine pump is electrically driven with a direct drive. Specify pivot length, cycle time, and nozzle wetted diameter. Compute the pump power.

12. An approximately circular 10 ac site is to be bioremediated using a liquid carrier (e.g., water) to be applied at a rate of 20 mm/day. Assume a nozzle package requiring 50 psi is available. Assume a gate valve, two 90-degree elbows, a butterfly flap valve, two couplings, and 100 ft of pipe are to be used for connecting the pump discharge to the pivot. A water source at 30 psi is available at the pivot. Compute the pump power requirement.
13. Using information presented in Tables 14.1–14.5, prepare a prompt sheet summarizing key customer requirements and technical specifications for irrigation systems.
14. What is *available water*?
15. What is *management-allowed deficit*?
16. What is the *water application efficiency*?
17. How do you quantify the capacity of the "soil storage reservoir" in a layered soil?
18. Knowing the soil storage reservoir capacity and design peak *ET* rate, show how to compute the maximum irrigation interval.
19. Knowing the maximum infiltration rate of the soil, and given an irrigation system operating at the maximum interval, how do you calculate the required time for water application?
20. Describe how to size a centrifugal pump, given a system flow rate and system head.
21. What are the advantages of using sprinklers for irrigation? Disadvantages?

Notes

1. Observe that *effective* as used in this text is context dependent. For example, in our discussion of runoff *effective* was the runoff portion. Here in our irrigation discussion, *effective* is the portion of beneficially used rainfall. In a bioremediation context, the meaning could change yet again. *Effective* in bioremediation may be that portion of rainfall contributing to nutrient transport or pollutant removal. Much remains to be learned concerning how to specify effective rainfall in the bioremediation context.
2. The Scobey equation is related to the Blasius equation for closed conduit fluid flow. Recall the Moody diagram from fluid mechanics. The Blasius equation describes the flow in a smooth pipe having relative roughness approaching zero (Streeter 1971). One can show with the Blasius equation that the diameter exponent is 4.75 and the Q exponent is 1.75. Thus, the Scobey equation is simply describing lines of low relative roughness on the Moody diagram. Increasing the exponents models increased values of relative roughness on the Moody diagram acceptably well. The Scobey equation should not be used for very rough pipe.

References

ASAE. 1997. Safety devices for chemigation. EP490.1 DEC96. American Society of Agricultural Engineers, St. Joseph, MI.

Ayers, R. S., and D. W. Westcot. 1985. *Water quality for agriculture*. Irrigation and Drainage Paper no. 29, rev. 1. Food and Agriculture Organization of the United Nations, Rome.

Baker, K. H. 1994. Bioremediation of surface and subsurface soils. In *Bioremediation,* ed. K. H. Baker and D. S. Herson. McGraw-Hill, New York, NY.

Baver, L. D., W. H. Gardner, and W. R. Gardner. 1972. *Soil physics.* 4th ed. John Wiley & Sons, New York, NY.

Berg, I. O., and M. D. Groskopp. 1964. *Irrigation equipment in Wisconsin.* University of Wisconsin Spec. Cir. 90. Madison, WI.

Burman, R. D., P. R. Nixon, J. L. Wright, and W. O. Pruitt. 1981. Water requirements. In *Design and operation of farm irrigation systems,* ed. M. E. Jensen. ASAE Monograph no. 3. American Society of Agricultural Engineers, St. Joseph, MI.

Burt, C. M., and S. W. Styles. 1994. *Drip and microirrigation for trees, vines, and row crops.* Irrigation Training and Research Center, California Polytechnic State University, San Luis Obispo.

Burt, C. M., K. O'Connor, and T. Ruehr. 1995. *Fertigation.* Irrigation Training and Research Center, California Polytechnic State University, San Luis Obispo.

Carsel, R. F., and R. S. Parrish. 1988. Developing joint probability distributions of soil water retention characteristics. *Water Resources Research* 34(5):755–69.

Christiansen, J. E. 1948. *Irrigation by sprinkling.* California Agricultural Experiment Station Bulletin 670. University of California, Davis.

Cuenca, R. H. 1989. *Irrigation system design: An engineering approach.* Prentice-Hall, Englewood Cliffs, NJ.

Fangmeier, D. D, and E. N. Biggs. 1986. *Alternative irrigation systems.* Report 8885. Cooperative Extension Service, University of Arizona, Tucson.

FAO. 1999. *Irrigation in Asia.* FAO Water Report no. 18. Food and Agriculture Organization of the United Nations, Rome.

Finkel, H. J. 1983. *CRC handbook of irrigation technology.* 2 vols. CRC Press, Boca Raton, FL.

Flood, P. K. 1990. Water rights of the fifty states and territories. In *Water rights of the fifty states and territories,* ed. K. R. Wright. American Water Works Association, Denver, CO.

Hansen, V. E., O. W. Israelsen, and G. E. Stringham. 1980. *Irrigation principles and practices.* 4th ed. John Wiley & Sons, New York, NY.

Heermann, D. F., H. R. Duke, A. M. Serafim, and L. J. Dawson. 1992. Distribution functions to represent center pivot water distribution. *Transactions of the ASAE* 35(5):1465–71.

Heermann, D. F., and K. H. Solomon. 1990. *Visions of the future: Proceedings of the Third National Irrigation Symposium.* ASAE Special Publication no. 04-90. American Society of Agricultural Engineers, St. Joseph, MI.

Hoffman, G. J., R. S. Ayers, E. J. Doering, and B. L. McNeal. 1981. Salinity in irrigated agriculture. In *Design and operation of farm irrigation systems*, ed. M. E. Jensen. ASAE Monograph no. 3.American Society of Agricultural Engineers, St. Joseph, MI.

Hoffman, G. J., T. A. Howell, and K. H. Solomon, eds. 1990. *Management of farm irrigation systems.* American Society of Agricultural Engineers, St. Joseph, MI.

Howell, T. A., D. S. Stevenson, F. K. Aljibury, H. M. Gitlin, I. Wu, A. W. Warrick, and P. A. C. Raats. 1981. Design and operation of trickle (drip) systems. In *Design and operation of farm irrigation systems,* ed. M. E. Jensen. ASAE Monograph no. 3. American Society of Agricultural Engineers, St. Joseph, MI.

James, L. G. 1988. *Principles of farm irrigation system design.* John Wiley & Sons, New York, NY.

Jensen, M. E., ed. 1981. *Design and operation of farm irrigation systems.* ASAE Monograph no. 3. American Society of Agricultural Engineers, St. Joseph, MI.

Kruse, E. G., D. A. Bucks, and R. D. von Bernuth. 1990. Comparison of irrigation systems. In *Irrigation of agricultural crops,* ed. B. A. Stewart and D. R. Nielsen. ASA Monograph no. 30. American Society of Agronomy, Madison, WI.

NRCS. 1970. *Irrigation water requirements.* Technical Release 21. Washington, DC.

———. 1983. Sprinkler irrigation. In *National engineering handbook,* sec. 15, Irrigation, chap. 11. Washington, DC.

Pair, C. H., W. H. Hinz, C. Reid, and K. R. Frost. 1983. *Irrigation.* 5th ed. Irrigation Association, Silver Spring, MD.

Riser-Roberts, E. 1992. *Bioremediation of petroleum contaminated sites.* C. K. Smoley/CRC Press, Boca Raton, FL.

Rochester, E. W. 1995. *Landscape irrigation design.* ASAE Special Publication 8. American Society of Agricultural Engineers, St. Joseph, MI.

Schwab, G. O., D. D. Fangmeier, W. J. Elliot, and R. K. Frevert. 1993. *Soil and water conservation engineering.* 4th ed. John Wiley & Sons, New York, NY.

Scobey, F. C. 1930. *The flow of water in riveted steel and analogous pipes.* USDA Technical Bulletin 150. Washington, DC.

Sims, R., D. Sorensen, J. Sims, J. McLean, R. Mahmood, R. Dupont, J. Jurinak, and K. Wagner. 1986. *Contaminated surface soils in-place treatment techniques.* Noyes Publications, Park Ridge, NJ.

Skaggs, R. W., J. E. Parsons, and K. D. Konyha. 1988. DRAINMOD version 4.0: An overview. ASAE Technical Paper no. 88-2563. American Society of Agricultural Engineers, St. Joseph, MI.

Solley, W. B., R. R. Pierce, and H. A. Perlman. 1998. Estimated use of water in the United States in 1995 (accessible from http://www.usgs.gov). *U.S. Geological Survey Circular 1200.* U.S. Geological Survey, Washington, DC.

Stewart, B. A., and D. R. Nielsen. 1990. *Irrigation of agricultural crops.* Agronomy Monograph no. 30. American Society of Agronomy, Madison, WI.

Streeter, V. L. 1971. *Fluid mechanics.* McGraw-Hill, New York, NY.

van der Leeden, F., F. L. Troise, and D. K. Todd. 1990. *The water encyclopedia.* 2d ed. Lewis Publishers, Boca Raton, FL.

van Lier, H. N., L. S. Pereira, and F. R. Steiner. 1999. *CIGR handbook of agricultural engineering.* American Society of Agricultural Engineers, St. Joseph, MI.

Walker, W. R., and G. V. Skogerboe. 1987. *Surface irrigation: Theory and practice.* Prentice Hall, Englewood Cliffs, NJ.

Zhang, R. 1997. Determination of soil sorptivity and hydraulic conductivity from the disk infiltrometer. *Soil Science Society of America Journal* 61:1024–30.

15

Streamflow Processes and Lake Dynamics

Stream and lake systems convey most of the world's runoff to the ocean, where evaporation processes act to continue the hydrological cycle. Stream and lake systems are under increasing environmental pressure from human actions and other causes. The USEPA and state counterpart agencies mandate site-specific environmental assessment programs before issuing revised NPDES permits. The assessment determines the additional load a stream can handle without becoming impaired. Although agencies do not have direct authority through the Clean Water Act (CWA) for regulating nonpoint pollution, they have much persuasive capacity via the requirement for NPDES permit holders to prove the stream can handle the additional loadings. Thus, growth requires a consensus among nearby urban and rural constituencies, often a daunting political challenge.

Assessment programs require intensive data collection and modeling. Definitions and parameters discussed in Chapter 7 are applicable to stream water quality characterization from a nutrient-loading perspective. This chapter integrates a physical, biological, and chemical perspective, along with stream and lake dynamics, with the sediment and nutrient quality perspectives discussed in Chapter 7.

Goals

- Identify physical processes affecting nutrient dynamics in streams: stream morphology, velocity profiles and their measurement and influence on physical and biological processes, sediment transport and sediment sampling, gully and stream bank erosion, and channel maintenance and restoration.
- Identify major biological processes in streams, wetlands, lakes, and ponds.
- Identify chemical processes, including oxygen and sediment chemistry, and nutrient cycling, which affect nutrient dynamics in lakes and streams.
- Appropriately describe and model physical, biological, and chemical processes at specific locations in watersheds with the aid of multidisciplinary expertise and sophisticated computer modeling techniques.

Background

Environmental assessment has become a major activity in the United States and other developed countries. Since the 1970s, we have made much progress in our ability to associate symptoms of downstream and lake impairment with uplands erosion and water quality management practices. Many environmental agencies monitor pollutants, endangered species, and biodiversity. Watershed assessment is environmental assessment on a watershed scale. The United States has 617,806 miles of rivers and streams, an estimated 56% of which do not meet water quality requirements for fishing and recreation (NRCS 1998). Conditions are especially bad in Eastern Europe and the former Soviet Union. Stream impairment often follows high uplands erosion and poor water quality management. The UN Food and Agriculture Organization (FAO) has published a land use suitability planning guide (FAO 1993), which includes environmental assessment. Governmental agencies expect projects anticipating significant aid from developed countries to conform to this and similar planning processes.

Besides governmental agency interest in watershed assessment, courts may make an assignment of liability when damage has occurred to the health, income, property value, or general well-being in a community. Watershed assessment may thus have a part in litigation of cases involving environmental damage. Science will provide one platform for conflict resolution. For example, environmentalists often classify streams and rivers as impaired or unimpaired based on the clarity of the flow. In Georgia, Kundell and Rasmussen (1995) concluded that clarity alone was not an adequate standard. The emphasis in their study was the preservation of endangered macroinvertebrate benthic species. These species provide food for fish and thus are important for the health of streams. When development occurs upstream, nonturbid flows can carry significant bed loads, affecting biotic life. Perry and Vanderklein (1996) present the mix of time and space scales shown in Figure 7.7 when harvesting forested watersheds for timber. This mix of scales also applies to many urbanizing locations and greatly complicates efforts to understand stream ecosystems and predict watershed response.

Stream assessment is thus a multidisciplinary undertaking. The overall purpose of this chapter is to show how geologists, limnologists, hydrologists, ecologists, soil chemists, and engineers may join in solving these societal problems. Streams manifest the successes or failures of point source controls along with successes or failures of management practices imposed on uplands. Some processes in streams and associated wetlands may relieve some negative effects of uplands neglect. The USEPA mandates total maximum daily load (TMDL) for specific water bodies. The TMDL is the maximum amount of point and nonpoint pollutant a water segment can accept without impairment (Bookchin and Farnsworth 1996). Stream design processes, discussed in Chapter 8, provide a foundation for the TMDL assessment. Agencies have yet to assign TMDLs to many streams, although they know many do not meet intended use guidelines. TMDLs are only now being determined for most streams judged to be impaired.

The historical approach to treating water quality problems in streams and other water bodies was to treat the effect, as shown in Figure 15.1, where copper sulfate is being added to a pond for algal control. The emphasis is shifting toward addressing problems at the source when possible. This corroborates the current regulatory move toward nutrient management plans and comprehensive soil conservation plans as discussed in Chapter 7. Chapters 6 and 7 suggest strategies for reducing soil loss and for preserving water quality. These chapters emphasized nonpoint pollution control in upland areas. This chapter presents overviews of processes that operate downstream and that may mitigate or further aggravate water quality concerns. How management practices imposed on uplands affect runoff, erosion, streamflow, pathogens, metals, pesticides, and nutrients downstream is of interest here.

Figure 15.1. Application of copper sulfate to kill algae, evidence of eutrophication. The copper sulfate treats the effect, not the cause. The current emphasis in solving eutrophication problems is to control N and P upstream.

How Waters Are Biologically Impaired

Nitrogen and phosphorus are essential for the growth of aquatic plants in freshwaters. P frequently stimulates aquatic growth in freshwaters. N frequently stimulates growth in estuaries and oceans (see Correll 1998). Light availability, scour, and chemical transformations complicate the straightforward observation of N and P limitation effects on plant growth.

Streams are designated *impaired* when nitrate exceeds 10 mg/l N and ammonia exceeds 1 mg/l N. Algal blooms or evidence of eutrophication indicate impairment. There are aesthetic impacts when algae levels reach 150 mg/m^2. Excess P favors cyanobacteria, associated with taste and odor problems and off-flavors in fish. Eutrophic streams cause excessive plant growth, often clogging water intakes. Eutrophic streams deplete oxygen, resulting in species shift toward the undesirable. Total N and total P should not exceed 1.5 mg/l and 0.075 mg/l, respectively, to avoid algal blooms. In the following sections we will see how knowledge of physical, biological, and chemical processes aids in the assessment of stream health.

Predominant Physical Processes

Stream Morphology

Morphology is concerned with shape and size. Stream morphology is a discipline that was developed by geographers and earth scientists. Geographers relate stream morphology to topography and soil types in a region. Sediment production potential is loosely correlated to stream morphology. Likewise, morphological factors are loosely correlated to stream ecological indicators.

Rainfall, ice, and snow act on land surfaces to create several types of drainage patterns seen in nature. The particular resulting drainage pattern depends on several factors.

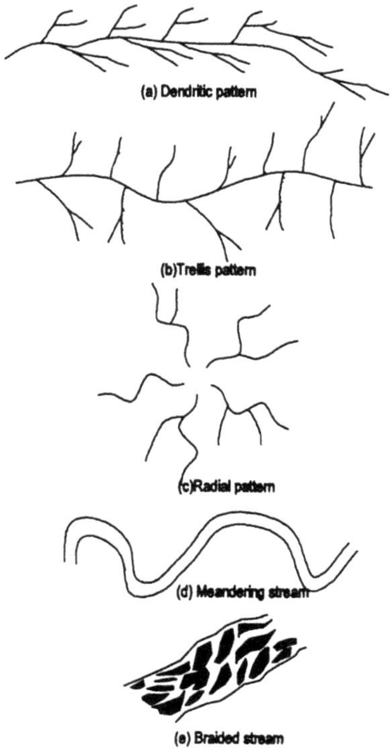

Figure 15.2. Patterns a through c are the major general patterns for smaller streams. Patterns d and e are specialized descriptions usually reserved for large streams and river systems. (Adapted from Viessman and Lewis 1996.)

Watershed slope, underlying soil and rock properties, flow history, freeze-thaw activity, and sediment transport (Viessman and Lewis 1996) affect the drainage pattern. Figures 15.2a, b, and c shows three drainage basin network types. Remote sensing techniques enable easy determination of stream morphology.

Many small tributaries characterize the most common stream pattern, the dendritic. These join into higher-order streams, eventually forming the major rivers of a region. Stream "order" describes the dendritic nature of the stream network in a watershed. A first-order stream has no tributaries. A stream of "X" order has two or more tributaries of "X − 1" order. One may describe the watershed as first, second, or higher order, depending on the stream order at the watershed outlet. Watershed modelers use these terms when linking subunits of watersheds together to model larger watersheds. Figure 15.3 shows a generic dendritic stream with classification numbers.

Long main streams intercepted by numerous shorter right-angle tributaries characterize the trellis pattern. They are common in areas where tectonic forces are active or have been active, such as the Appalachian and Rocky Mountains along the foothills. Multibasin patterns, combination of a through c in Figure 15.2, occur in flat, swampy areas with many surface depressions. Wetland processes in these systems can have a profoundly beneficial impact on water quality. Multibasin patterns occur in glaciated, windblown, and permafrost areas and are common in plains and mountain valleys. Radial drainage patterns occur in foothill areas or mountain areas.

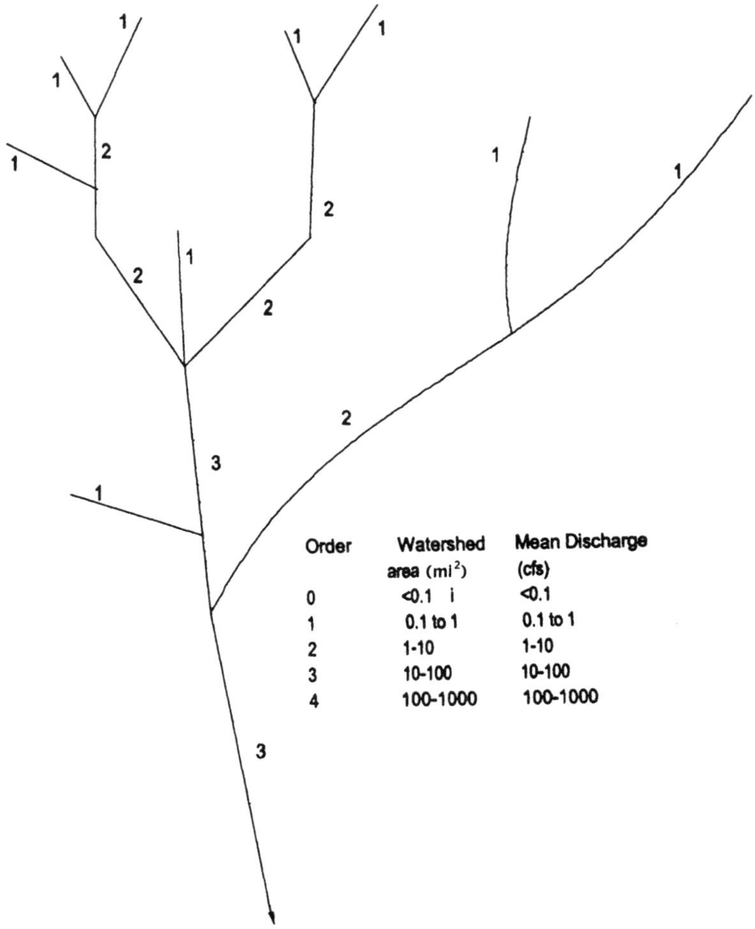

Figure 15.3. Stream order designation for generic dendritic stream network, with a subjective correspondence between watershed area and mean discharge.

Individual streams favor one or more of the two major patterns shown on Figures 15.2d and e. Streams are rarely straight except on steep slopes in homogeneous materials. Braided streams have many interconnected channels flowing around and over islands and bars, most of which are inundated during high flows. Braided streams often transport large amounts of sediment but often less than the sediment supplied. They are common in the southeastern piedmont. A related form, the deltaic, occurs with larger stream outlets. Some stream morphologists have called braided streams incipient forms of meandering streams. Many braided streams revert to meandering or other forms when sediment supplies change. Meandering in an otherwise straight channel occurs because of transverse currents arising from Coriolis forces and bed-bank shear forces.

One may also classify streams according to temporal aspects. *A perennial* stream is one that flows continuously. An *intermittent* stream is usually dry during a part of each year (dry no more than 70 days/yr). An *ephemeral* stream flows only in response to precipitation or snowmelt (dry more than 300 days/yr).

Morphologists often apply *Lane's law.* Lane's law states that the product of (1) bed slope and (2) water discharge is proportional to the product of (3) sediment size and

(4) transport rate. For example, constructing a reservoir reduces the sediment load immediately downstream. The reduced sediment load will likely cause the bottom slope to decrease (potential for gully formation) or will cause the size of particles transported to increase. By Lane's law, either slope or discharge must decrease or sediment particle size must increase. Degradation (downward cutting) of the bed of the Missouri River has occurred below some upstream reservoirs, isolating marinas and water intake structures in some locations (Viessman and Lewis 1996).

The ideas of Lane's law have been expressed in more quantitative terms by the *regime theory.* Consider the following background. An equation such as the Manning equation describes uniform flow in an open channel with essentially rigid boundaries. Three equations describe uniform flow in a channel with loose and movable boundaries. The inclusion of sediment transport adds an additional equation. One may say the natural channel is in *regime* if flow and sediment loading patterns are stable over a number of years. The *regime* concept suggests much freedom of individual behavior within a framework of laws and has no short-period connotation. The boundaries of a *regime* channel will not change unless long-term changes in the flow or sediment loading occur. Placement of reservoirs, subdivision developments, or other land use transitions affect the *regime* of streams in the drainage basin.

Graf (1971) summarized the regime approach as follows:

$$B = C_B q^{\alpha} \tag{15.1a}$$

$$D = C_D q^{\beta} \tag{15.1b}$$

$$\bar{u} = C_{\bar{u}} q^{\gamma} \tag{15.1c}$$

$$G_{ss} = p q^{\delta} \tag{15.1d}$$

$$S = 0.06 q^{-0.44} \tag{15.1e}$$

$$\alpha + \beta + \gamma = 1 \tag{15.1f}$$

$$G_B C_D C_{\bar{u}} = 1 \tag{15.1g}$$

where

q is long-term average flow rate (m^3/s);
B is average width (m);
D is average depth (m);
$\bar{u}$ is average velocity (m/s);
S is average slope;
α, β, γ, δ, C_B, C_D, $C_{\bar{u}}$, and p are empirical exponents;
G_{ss} is the suspended sediment flux (tonnes/day).

Graf (1971) presented a compilation of values for the exponents and found $\alpha = 0.26$, $\beta = 0.40$, $\gamma = 0.34$, and $\delta = 2$–3. Equations 15.1 are most useful for qualitative understanding of natural channel dynamics. The humid eastern and wet mountain areas usually had lower α values than did semiarid areas. Values of $C_{B,D,\bar{u}}$, and p vary widely, depending on the nature of flow (ephemeral, perennial, etc.).

Equations 15.1a–g suggest trends when comparing streams across drainage basins. Large basins have more discharge due to larger drainage areas. Depth, width, and velocity increase with flow, as does the total sediment load. The equation for slope describes the decreasing slope with position within a given drainage basin. Depth and width increase as

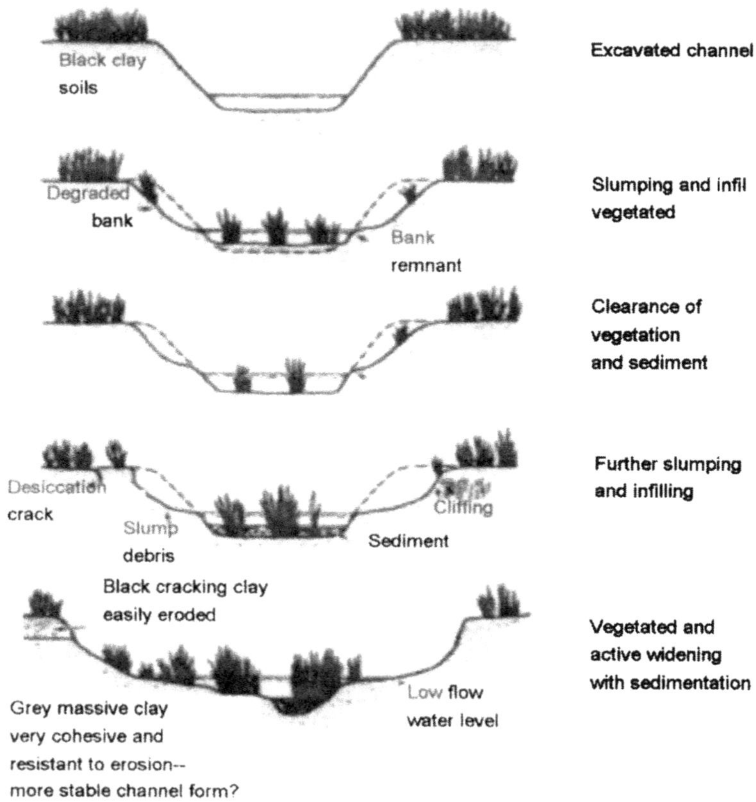

Figure 15.4. Observed sequence of channel change in an urbanized stream. Observe the change from trapezoidal to parabolic shape. (From Whitlow and Gregory 1989; used with permission of John Wiley & Sons.)

one moves downstream. Velocity frequently *decreases* in response to the lower slope. Sediment transport rate also usually decreases. Total sediment transported increases because of flow volume increases. Equations 15.1a–g are empirical generalizations and may not all hold true for specific basins. Streams may meander in lower reaches. Meandering is most likely in high-order streams of low slope. One may visualize meander length and width as the period and amplitude of a sine curve. Similar relations exist for meander length and other morphological relations.

Erodible constructed channels will show evidence of conforming to equations 15.1 and Lane's law over time. Thus, the need for periodic channel maintenance for constructed channels is not surprising. These relationships allow us to anticipate effects of introducing land use changes in regions drained by the stream. Figure 15.4 presents a sequence of channel changes without maintenance. Streams are dynamic and ever changing.

Velocity Profile

Mean velocity has been the major concern thus far. Velocity at various depths will be above or below the mean. Transport processes depend on the velocity profile and turbulent fluctuations. Knowledge of velocity profiles in open channels provides much insight into sediment and other mass transport processes. Works such as Chow 1959, Streeter 1971, and Simons and Senturk 1992 provide extensive treatment of velocity profiles in natural and artificial channels. Equation 15.2 gives the fundamental relationship for velocity versus depth in a natural channel:

$$u = \frac{u^*}{k} \ln\left(\frac{y}{y_o}\right) + C \tag{15.2}$$

where

u is the velocity (ft/s, m/s) at a point;
k is the von Karman coefficient, 0.35 for sediment-laden water, 0.4 for clear water;
C is a constant: $C = 5.5$ ft/s (smooth; 1.68 m/s) and $C = 8.5$ ft/s (rough; 2.59 m/s);
y_o is the distance from channel bottom where turbulence gives way to laminar flow;
u^* is the friction velocity, computed by

$$u^* = \sqrt{\tau_o/\rho} = \sqrt{\gamma RS/\rho} = \sqrt{gRS} \tag{15.3}$$

where

ρ is the density of water ($\approx$1000 kg/m^3);
g is gravity (9.81 m/s^2);
γ is the specific weight of the water ($\approx$1000 kg/m^3 $\times$ 9.81 m/s^2 9810 N/m^3);
R is the hydraulic radius (depth, wide channel, m);
τ_o is the channel drag;
S is the channel slope.

Values for the von Karman coefficient and the C values in equation 15.2 were determined from observations in flumes with a wide range of roughness conditions.

Fluid boundary layers play a significant role in flow hydraulics. The thickness of the laminar boundary depends on whether the boundary is smooth or rough. A smooth boundary is one where roughness elements of diameter k_s do not protrude into the turbulent flow regime. All drag results from skin friction. Skin friction is defined as shear-based resistance collinear to the flow. A rough boundary has roughness elements protruding through the laminar layer. The roughness elements exert form drag. Form drag is resistance to blunt objects oriented normal to the flow. The roughness elements in Figure 15.5 exert form drag because they protrude into the flow.

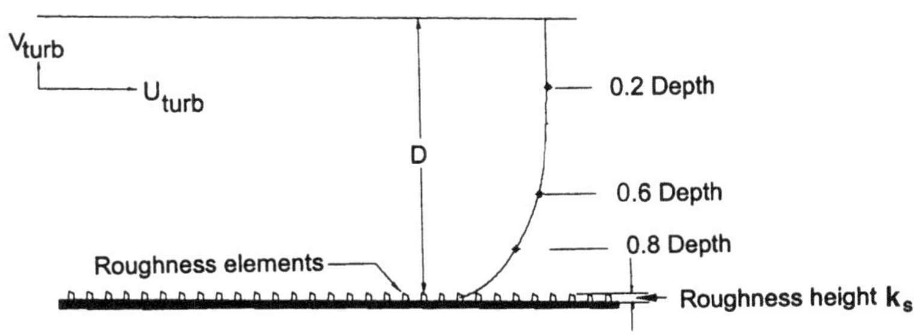

Figure 15.5. Schematic of rough boundary channel showing the log velocity profile, turbulent velocity components, and typical roughness elements. The mean velocity in the vertical direction is zero. The mean velocity in the x direction (along the channel bottom) is the velocity being profiled. The protruding roughness elements exert form drag.

Boundaries are classified using equations 15.4a–c. Equation 15.4a gives the requirement for the hydraulically smooth condition:

$$0 \leq \frac{k_s u^*}{\nu} \leq 5 \tag{15.4a}$$

The requirement for a transition boundary is given by

$$5 \leq \frac{k_s u^*}{\nu} \leq 70 \tag{15.4b}$$

Equation 15.4c gives the requirement for a hydraulically rough boundary:

$$\frac{k_s u^*}{\nu} \geq 70 \tag{15.4c}$$

Thus to determine the appropriate regime one must have a bottom material size analysis, the friction velocity u^*, and the kinematic viscosity ν. One may use the D_{50} sediment size for estimating roughness height k_s. The equations for describing the boundary limits are empirically derived from experimental observations. The rough boundary is of most interest. Figure 15.5 schematically shows rough boundaries along with a velocity profile calculated using equation 15.2.

The constant y_o in equation 15.2 may be approximated as $k_s/30.08$ with a rough channel and ν/u^* with a smooth channel, where k_s is the roughness height. Many different formulations for y_o are available in the literature (e.g., see Simons and Senturk 1992). Integrating equation 15.2 for smooth and rough (wide) channels results in the mean velocity for the smooth and rough condition (Simons and Senturk 1992):

$$\frac{\bar{u}}{u^*} = 3.0 + \frac{1}{k}\ln\left(\frac{Ru^*}{\nu}\right) \quad (smooth) \tag{15.5a}$$

$$\frac{\bar{u}}{u^*} = 6.25 + \frac{1}{k}\ln\left(\frac{30.08R}{k_s}\right) \quad (rough) \tag{15.5b}$$

In Equations 15.5 the contribution of flow from the channel bottom to level y_o (i.e., the laminar portion) to the fluid flow was neglected.

Equations 15.5a–b assume the hydraulic radius R is approximately the channel depth. Because most channels of interest are hydraulically rough, focus on equation 15.5b. Ignoring the constant 6.25 in equation 15.5b, following Chow (1959), and substituting the Manning equation will give

$$n = \frac{(R/k_s)^{1/6}}{\sqrt{g}(1/k)\ln(R/k_s)} k_s^{1/6} \tag{15.6}$$

Equation 15.6 suggests a relation between the Manning n and particle size. Simons and Senturk (1992) suggest approximations of n versus particle size as

$$n = \frac{D_{90}^{1/6}}{26} \tag{15.7}$$

where D_{90} is the particle size with 90% smaller (m). Simons and Senturk (1992) also give other, similar approximations of the Manning roughness versus particle size. They also give values of k_s for commonly used materials for designed channels. Equation 15.6

suggests a changing Manning *n* with hydraulic radius. Equation 15.7 exposes a basic problem with the Chezy equation and to some extent with the Manning equation derived in Chapter 8. Hydrologists have observed the lack of constancy of the Manning *n* in large streams.

For design purposes, the Manning equation continues to be popular. The parameter k_s is difficult to measure. The variation of *n* with *R* is real. Rivers and large streams exhibit variable Manning *n* values. There is evidence that the function of *n* versus stage depends on whether the stage is rising or falling. One may regard the *n* versus *R* relationship as *approximately* constant for engineering small streams and channels.

Example 15.1: A stream is 1 m deep and has a slope of 0.001 and a D_{50} of 0.05 m. Compare the velocity from Equation 15.5 to that predicted by the Manning equation.

One must determine hydraulic smoothness or roughness. Using equations 15.4,

$$0.05 \text{ m } [(9.81 \text{ m/s}^2) \ (1 \text{ m})(0.001)]/0.000001 \text{ m}^2/\text{s} = 4952$$

Thus, the channel is hydraulically rough.

From equation 15.5b, the average velocity is

$$u^* = (9.81 \times 1 \times 0.001)^{0.5} = 0.99 \text{ m/s}$$

$$u_{0.20} = 6.25(0.099) + (0.99/0.4)ln\left[\frac{30.08(.2)}{0.05}\right] = 1.80 \text{ m/s}$$

Comparing with the Manning formula,

$$V = (1/D_{90}{}^{1/6}/26)\ (1^{2/3})(0.001)^{1/2} = 1.77 \text{ m/s}$$

where D_{90} was taken as 0.01 m. If $D = 0.005$ is used, then $v = 2$ m/s. These differences are typical.

One may determine the mean channel velocity by taking an average of measured point velocity values at 0.8*D* and 0.2*D* or velocity at 0.6*D*, where *D* is the channel depth. Integrating the velocity profile equation over a transect and dividing by depth gives values close to the 0.6*D* value or the average of the 0.8*D* value and the 0.2*D* value. Note that depths are referenced to the *water surface*.

Streamflow Measurements

Flow and velocity in streams may be determined using several approaches. The simplest approach is to place a float on the surface and clock the time required to traverse a given distance. This enables a crude velocity measurement. To estimate discharge one requires cross-sectional area and velocity measurements.

High water marks (or existing water level) and roughness estimates enable flow determination by application of the Manning equation. This requires knowledge of roughness, channel cross-sectional area, and wetted perimeter. One may estimate roughness using

Figure 15.6. Triangular weir for measuring runoff hydrographs for a small watershed. The stage gauge is just out of the picture. This device is a broad-crested weir.

equation 15.7 or from photographic comparisons such as those provided by Chow (1959) and French (1985). Determination of wetted perimeter and hydraulic radius requires channel surveys. Most open channel flow texts (e.g., French 1985) discuss techniques for making flow estimates from measurements at multiple cross sections.

If a structure such as shown in Figure 15.6 is available, one can determine the flow directly from a stage-discharge curve for the structure. Elimination of the need to estimate roughness enhances accuracy of the flow measurement.

Culverts and bridges are useful for flow determination. If a culvert in the stream is present with a free discharge, one may use culvert hydraulics to estimate flow. Velocity may be determined knowing area and flow.

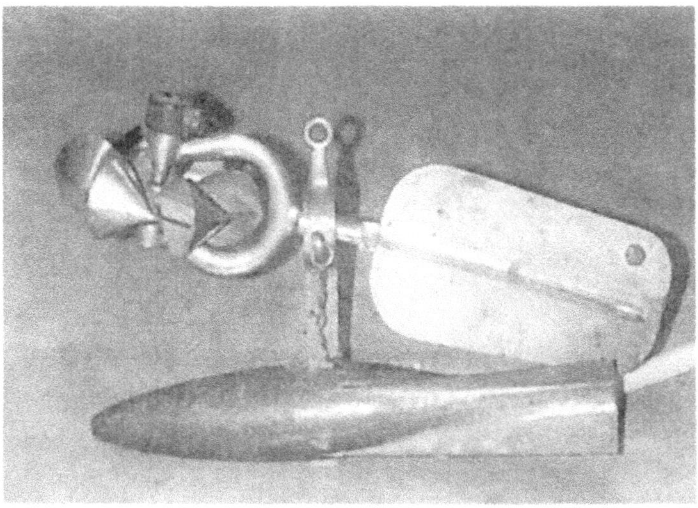

Figure 15.7. Price current meter with stabilizing weight.

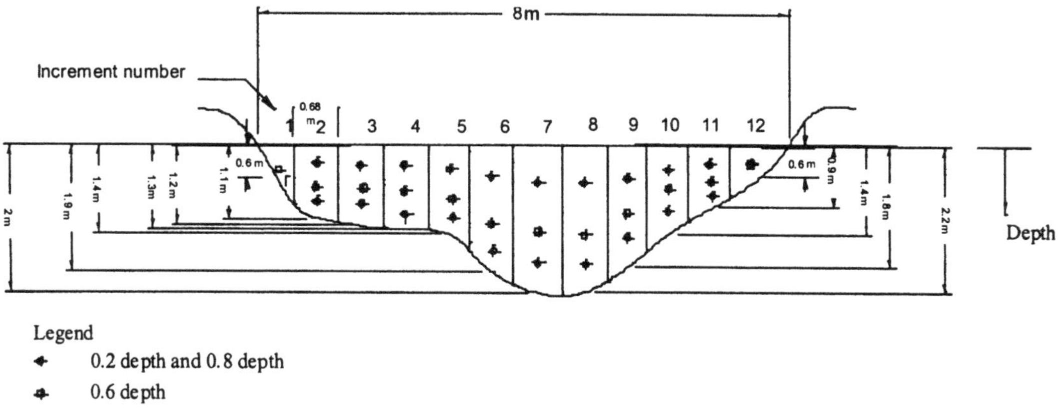

Figure 15.8. Hypothetical channel showing measurement stations for integrating velocity by area for total flow.

Large streams are difficult to survey using the methods given above. They require direct velocity measurements at various points across the channel width. One may make point velocity measurements with the Price current meter, shown in Figure 15.7. Suspending the meter from a structure such as a bridge or a boat is the norm. One may survey a large stream by collecting measurements at levels of 0.2*D* and 0.8*D* and averaging. Alternatively, one may take one measurement at the 0.6*D* level, relative to the surface, as discussed above. Figure 15.8 shows these points for a hypothetical stream. See texts such as French 1985 for additional corrections and precautions. Boat-based measurements are challenging because the stream current usually affects boat position. Acoustic Doppler techniques, discussed below with sediment sampling techniques, enable one to develop velocity profiles in streams by locating a single sensor on the stream bottom. Acoustic Doppler sensors have only recently become commercially available.

A crude approximation of stream velocity may be made by dividing the time required for a float to traverse a known distance into the travel interval and multiplying by a correction based on stream depth. Yoo and Boyd (1994) give correction factors ranging from 0.66 for a 1 ft depth; 0.7 for a 3 ft depth; 0.76 for a 6 ft depth; to 0.8 for a depth of 20 ft or more. Linear interpolation may be used. The correction factor is usually set equal to 1 regardless of depth. Cross-sectional area may be estimated by assuming a geometry in conjunction with relevant known depth, top width, side slope, and bottom width values (e.g., parabolic, trapezoidal) or by using more detailed channel survey data. The flow rate is determined by multiplying the stream velocity by the stream cross-sectional area.

One may also use pitot tubes or floats to estimate flow within a culvert pipe. Some have used pitot tubes for estimating velocity at points along two or three vertical transects. Pitot tubes are not routinely used for stream current measurements due to inadequate precision.

The U.S. Geological Survey (USGS) has published historical measured data from many streamflow measurement stations in the United States. Sources such as van der Leeden et al. 1990 give comparable data for major rivers of the world. One may find the USGS data at links within the Internet site at http://www.usgs.gov. Available data include surface water and sediment concentration plus groundwater statistics.

Example 15.2: Estimate the flow in a stream measured as shown in the shaded portion of Table 15.1. Complete Table 15.1 to compute the streamflow.

Table 15.1. Flow rate calculations from data gathered at bridge crossing at location of Figure 15.8.

Distance b from bank to meter (m)	Depth y (m)	Meter depth below surface (m) (either 0.6y 0.2y and 0.8y)	Point velocity (m/s)	Average velocity V_i (m/s)[a]	q_i (m^3/s)[b]
0.34	0.6	0.36	0.14	0.14	0.07
1.02	1.1	0.22	0.5	0.35	0.26
		0.88	0.2		
1.7	1.2	0.24	1.3	1.03	0.84
		0.96	0.75		
2.38	1.3	0.26	1.3	1.15	1.02
		1.04	1.0		
3.06	1.4	0.28	1.5	1.35	1.29
		1.12	1.2		
3.74	1.9	0.38	1.6	1.5	1.94
		1.52	1.4		
4.42	2.0	0.4	1.8	1.65	2.24
		1.6	1.5		
5.1	2.2	0.44	1.9	1.75	2.62
		1.76	1.6		
5.78	1.8	0.36	1.6	1.45	1.77
		1.44	1.3		
6.46	1.4	0.28	1.6	1.4	1.33
		1.12	1.2		
7.14	0.9	0.18	1.4	1.2	0.73
		0.72	1.0		
7.82	0.6	0.36	0.3	0.3	0.12

$q_i = 14.23$ m^3/s

[a]Average the 0.8D and 0.2D values or use the 0.6D value

[b]Figure the incremental q_i using

$$q_i = 0.5\,(b_{i+1} - b_{i-1})\, y_i\, V_i$$

where b_i are the entries in the first shaded column and y_i is the entry from the second shaded column.

Sediment Transport Processes

Chapter 6 discussed sediment yields from upland watersheds. Streams transport part of this sediment to structures and beyond. In the upper reaches of a stream having steeper slopes, streams carry sediment as wash load. Wash load is different from bed load. Wash load originates upstream. Equilibrium usually does not exist between the transported sediment and the bed materials of the channel bottom. The techniques of sediment routing discussed in Chapter 6 are useful for predicting the wash load.

Further along the reach, the generic stream becomes a depositing (or aggrading) stream. At this point, the slope has decreased. The transported sediment and the channel bed material equilibrate. One may then use techniques for bed material load transport prediction.

Figure 15.9 shows a velocity profile above a rough bottom. As the velocity increases, the rate of sediment bouncing (more properly, saltation) along the bed increases. The capacity of a flow to suspend sediment therefore increases with velocity.

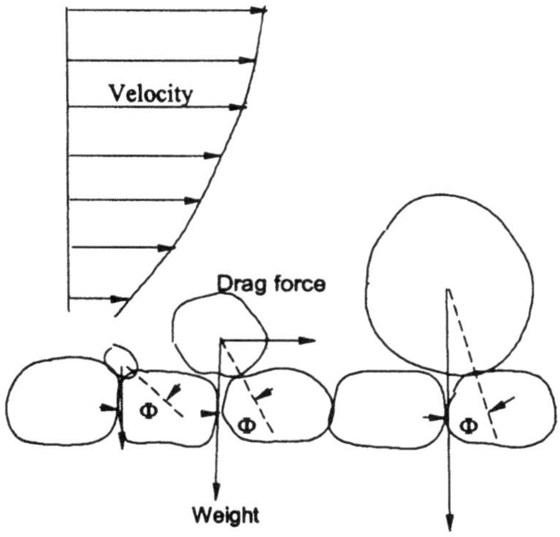

Figure 15.9. Close-up of channel bottom with log velocity profile shown. Particles A, B and C exert form drag and skin friction. Movement occurs when fluid-induced drag exceeds friction related to weight and particle support. The angle is the effective friction angle (e.g., drag force over weight at incipient particle movement).

There is a velocity below which no transport occurs for a given particle size. The Shields curve shown in Figure 15.10 defines the boundary between transport and no transport. The Shields diagram is a plot of the Shields parameter versus the friction velocity Reynolds number:

$$\frac{\tau_o}{(\rho_s - \rho)gD} \quad versus \quad \frac{u^*D}{\nu} \tag{15.8}$$

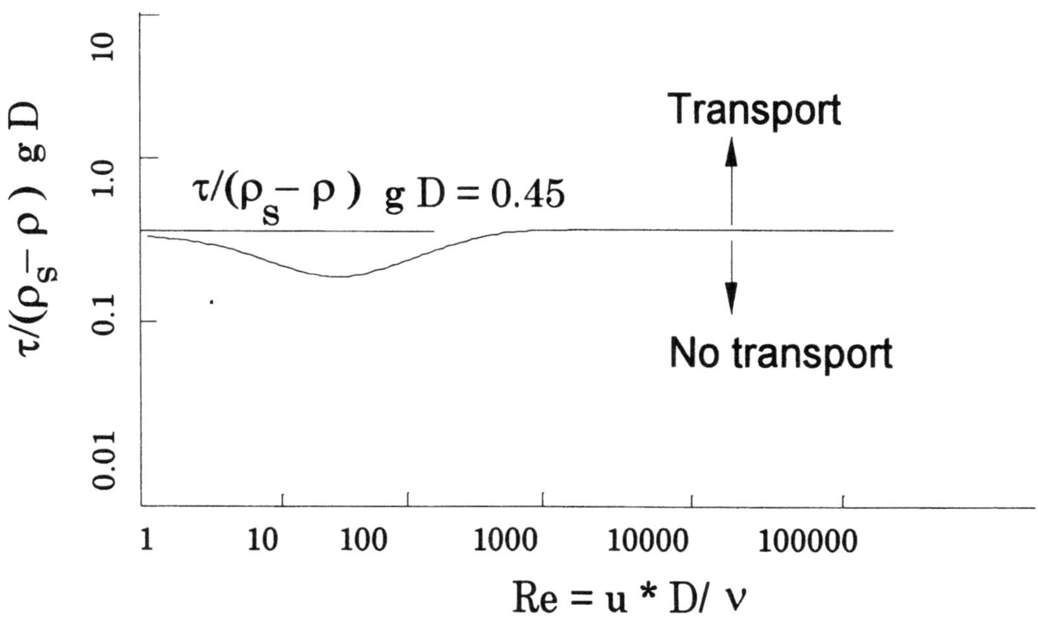

Figure 15.10. Shield's curve for predicting incipient motion of sediment particles.

where

$\tau_o = \rho gRS$, the tractive force on the channel bottom (kg/m-s^2 = N/m^2; use the mass form of units when demonstrating nondimensionality);
ρ_s is the sediment density (kg/m^3);
ρ is the fluid density (approximated using the density of water, 1000 kg/m^3);
g is gravity (m/s^2);
D is the particle diameter (m);
u^* is the friction velocity (m/s) = (gRS);
ν is the kinematic viscosity (10^{-6} m^2/s for water).

Values for the left- and right-hand sides of relation 15.8 falling above the line in Figure 15.10 indicate transport conditions exist. Values plotting below indicate stationary conditions. Increases in tractive force cause increased sediment transport.

Graf (1971), Simons and Senturk (1992), and Yang (1995) give excellent reviews of sediment transport formulas. Those texts provide derivation of total load, bed load, and suspended load sediment transport formulas. These formulas suggest that as u* (and τ_o) increase, so does the capacity to transport sediment. Many formulas treat the bed load and suspended loads separately because of the desire to quantify the effect of suspended load on other processes such as light penetration or biological processes.

Of the many sediment transport relations, this text presents the Meyer-Peter-Mueller relationship in a nondimensional form:

$$\frac{q_{bw}}{\rho u^{*3}} = 8\left(1 - \frac{0.047}{\tau_*}\right)^{3/2} \quad \textbf{(15.9a)}$$

$$\tau_* = \frac{\rho u^{*2}}{(\gamma_s - \gamma)D_{50}} \quad \textbf{(15.9b)}$$

$$q_{bw} = 8\left[1 - \frac{0.047(\gamma_s - \gamma)D_{50}}{\rho u^{*2}}\right]^{1.5} \rho u^{*3} \quad \textbf{(15.9c)}$$

where

q_{bw} is the submerged weight of the transported sediment (kg/m-s);[1]
D_{50} is the median diameter of the transported material (m);
u^* is the friction velocity = (gRS) (m/s);
γ_s is the unit weight of sediment (usually 2500–2700 kg/m^3)
γ is the unit weight of water (usually 1000 kg/m^3)
ρ is the weight density of water (usually 102 N/m^3 = kg s^2/m^4).

Equation 15.9c shows the operational form of the Meyer-Peter-Mueller relationship. Example 15.3 shows how one may apply the Meyer-Peter-Mueller relationship.

Example 15.3: The allowable sediment storage in a reservoir is 50 $\times 10^6$ m^3. For a river 30 m wide, slope of 0.05, mean particle size of 0.5 mm (0.0005 m), depth of 1.5 m, compute the time required to fill the reservoir with sediment assuming a trapping efficiency of 95%. The unit weight of water is 1000 kg/m^3 and the unit weight of sediment is 2550 kg/m^3. Compute the Shields curve parameters.

First, review the units of equation 15.9c. Note that $\gamma_s - \gamma$ is 1550 kg/m^3 and ρ is 102 kg s^2/m^4.

Compute the friction velocity:

$$u^* = (9.81 \text{ m/s}^2 \times 1.5 \text{ m} \times 0.005) = 0.271 \text{ m/s}$$

$$q_{bw} = 8(1 - (0.047(1550 \text{ kg/m}^3)(0.0005 \text{ m})/\{102 \text{ kg s}^2/\text{m}^4 [(0.271 \text{ m/s})^2]\})^{1.5}\ 102 \text{ kg s}^2/\text{m}^4(0.271 \text{ m/s})^3 = 15.4 \text{ kg/m-s}$$

The 30 m river then transports

$$15.4 \text{ kg/m-s } (30 \text{ m}) = 461 \text{ kg/s}$$

The allowable sediment storage will be filled in

$$50 \times 10^6 \text{ m}^3/\{[(461 \text{ kg/s})/1550 \text{ kg/m}^3]\ (365 \text{ days/yr})(86{,}400 \text{ s/day})(0.95)\} = 5.6 \text{ yr}$$

The reservoir in Example 15.3 will fill with sediment prematurely. One should anticipate much more sediment storage here. The Shields curve parameter

$$\tau/[\rho_s - \rho)\ gD]\ (1000 \text{ kg/m}^3)(9.81 \text{ m/s}^2)(1.5 \text{ m})(0.05))/$$
$$[(1500 \text{ kg/m}^3)\ (9.81 \text{ m/s}^2)\ (0.0005 \text{ m})] = 100$$

much greater than 0.45, consistant with the above findings.

Sediment transport formulas usually assume a quasi-steady-state supply of sediment. This is not true in many watersheds even when there are no major land use changes occurring. The variable nature of runoff causes pulses in sediment delivery. When significant land use shifts occur and sediment load greatly increases, the load predicted by equation 15.9c is questionable. This is due to nonequilibrium between suspended load and bed material loads.

Interestingly, after heavy erosion followed by reforestation and improved management, lower reaches of streams may slowly degrade. The watersheds feeding into Lake Lanier, Georgia, heavily mined for gold in the nineteenth century, continue to yield mercury-laden sediments originating from lower stream reaches (the gold-smelting process required mercury). Rivers in the Georgia piedmont are degrading in response to improved water quality as land is returned to pasture or forest following the decline in cotton growing. Sediment-free water moving into a river with fine sediments can cause channel bottoms to degrade (see Graf 1971; Yang 1995). Soil erosion processes in stream reaches may extend for time scales from 100 years to geologic epochs. The reality of sedimentation time scales requires a tentative approach to addressing sediment-related problems.

Sediment Sampling

The collection of sediment data is difficult since sediment load depends on flow, in turn depending on watershed response to rainfall events. One may collect suspended sediment and bed load using devices such as those shown in Figures 15.11a–d. As a rule, the major limitation to sediment transport model development is the collection of adequate data. The Interagency Committee on Water Resources (1963) gives extensive discussions of physical/gravimetric techniques and problems of sediment sampling in natural streams. The rising stage sampler shown in Figure 15.11a is the most modern sampling approach. One may use the sampler to collect sediment during the streamflow hydrograph rising

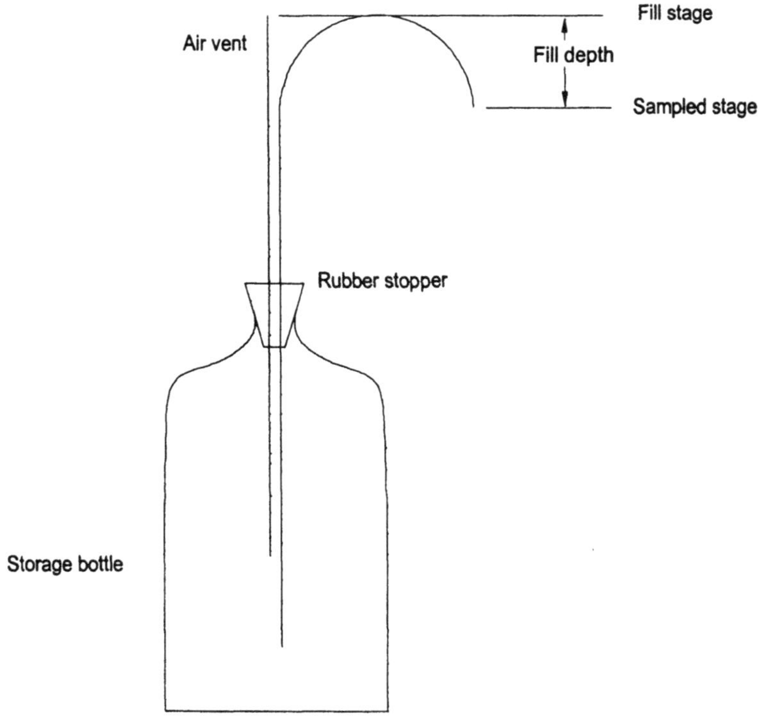

Figure 15.11a. Rising stage sediment sampler.

limb in intermittent and ephemeral streams. The devices shown in Figures 15.11b–c are standard approaches for collecting gravimetric sediment samples from the stream (Figure 15.11b) and from the streambed (Figure 15.11c).

Automated sampling devices are becoming more popular. They work best when flow passes through a structure and sediment concentrations are thoroughly mixed. This enables collection of total sediment load samples. Gravimetric analysis is the primary approach for sediment concentration determination.

Acoustic Doppler techniques are becoming popular in field environments for profile velocity measurements and suspended sediment measurement (see Figure 15.11d). Acoustic techniques also hold promise for rapid sediment load assessments. Together, acoustic Doppler velocity measurements and sediment measurements may provide easier approaches to sediment transport analyses.

Gully and Stream Bank Erosion

In gully erosion water concentrates in narrow channels and, over short periods, removes the soil from this narrow area to considerable depths, ranging from 1–2 ft to as much as 75–100 feet. Gully erosion sometimes occurs at groundwater seeps, where soils are moist and unstable for long periods. Gully and stream bank erosion are just as significant as splash and rill erosion. Sediment in streams is more likely to come from the stream bank than from uplands when the uplands are protected from excessive erosion. In a south China study, diCenzo and Luk (1997) showed that 85% of sediment in a stream originated from stream bank erosion. Gully formation processes may be described using sediment transport

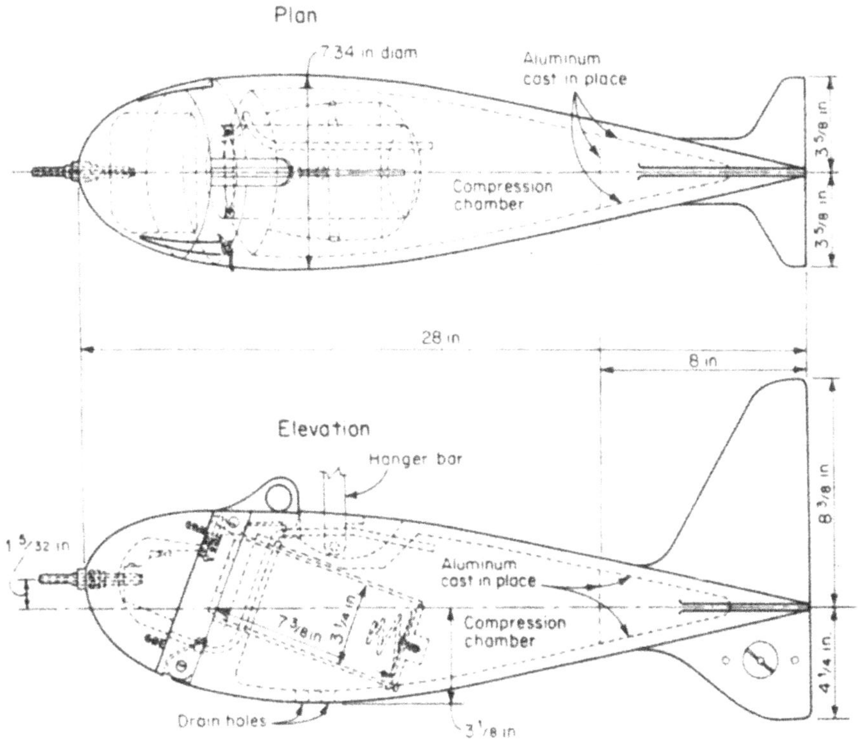

Figure 15.11b. Point integrating suspended load sediment sampler US P-61. One may integrate over depth by steadily moving the sampling device over the profile while collecting the sample. (From Interagency Committee on Water Resources 1963.)

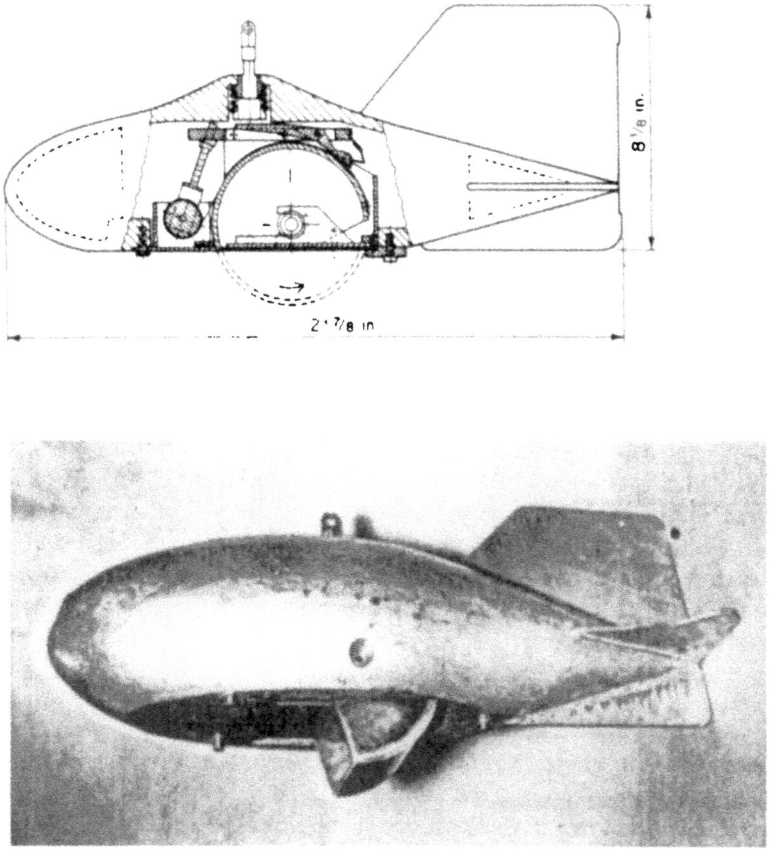

Figure 15.11c. Bed load sampler US BM-60. (Interagency Committee on Water Resources 1963.)

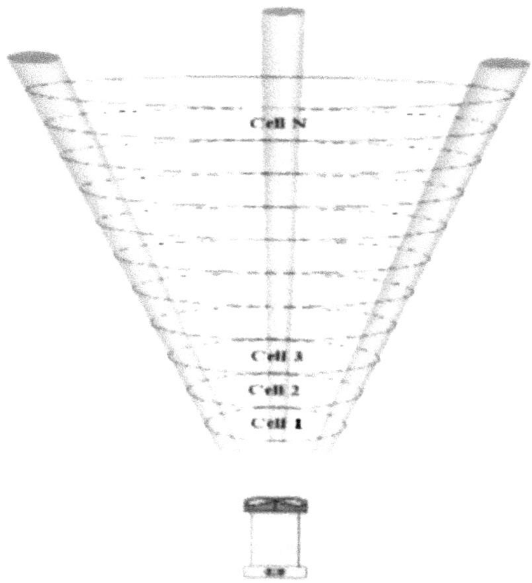

Figure 15.11d. Schematic drawing showing zones or cells wherein velocity profile information is obtained with acoustic Doppler measurements. Turbidity measurements are also obtained with the instrument.

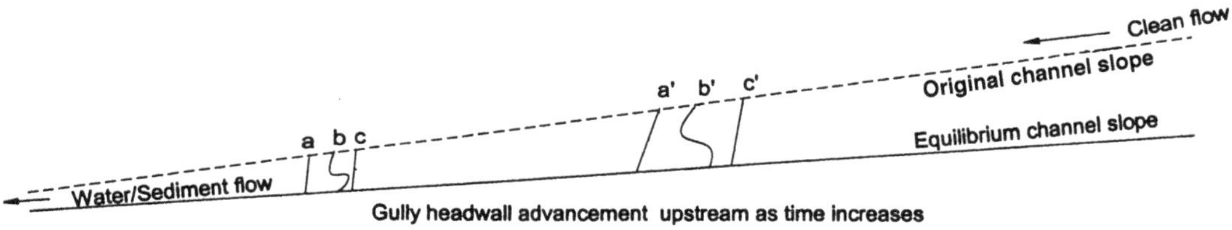

Figure 15.12. A gully headwall advancing over time.

equilibration analyses. Consider a steady flow carrying no sediment. The flow has a given capacity to transport sediment, from equations 15.9a–c. The flow will "seek" a (lower) slope and sediment load resulting in equilibrium. A discontinuity between the equilibrium slope and field slope will result in a gully headwall as shown in Figure 15.12. The headwall (a in Figure 15.12) moves upstream by a process of scour. The scour undermines the headwall (b), causing eventual sloughing of the remainder of the headwall (c). The process continues, resulting in the advance of the gully headwall. The sloughing process is difficult to model.

Figure 15.12 is a hypothetical one-dimensional portrayal of gully development. Using remote sensing techniques, Thomas and Welch (1988) documented gully headwall advancement in the Georgia piedmont. Stream erosion is a three-dimensional extension. Streams may erode sensitive bank material at low flow, undercutting the stream bank, resulting in bank sloughing. Stream and gully erosion are difficult to model. The ephemeral gully erosion model (EGEM) under development by the USDA-ARS erosion laboratory is one rigorous attempt to model stream and gully erosion. Existing gully erosion models predict the erosion due to scouring but do not predict erosion due to sloughing. Gully erosion processes related to scouring are included in the latest versions of the Water Erosion Prediction Project (WEPP) model.

Channel Maintenance and Restoration

The NRCS (1998) suggests restoration proceeding along one of two tracks: (1) a single-species restoration focusing on habitat requirements of certain life stages of a species (e.g., spawning of trout or salmon, preservation of bottom-dwelling organisms); or (2) an ecosystem restoration approach focusing design resources on the chemical, hydrologic, and geomorphic aspects. The NRCS (1998) provides details on trout and salmon stream restoration. Ecosystem restoration approaches require a watershed hydrologic response analysis; selection of preliminary right-of-way for the channel; bed material analyses of the channel bottom; hydrologic, hydraulic, and sediment transport rate analyses of the channel to determine the discharge range; and an analysis of the preferred morphological type based on a regime analysis. Morphological analyses include meander characterization. The analysis is iterative, each step depending on preceding choices.

Soil bioengineering (Schiechtl 1980) approaches are useful for controlling water velocities on slopes and in channels. Soil bioengineering is the integration of mechanical and biological materials for soil stabilization and erosion control. Live and dead plants provide stabilization. Hedge layers and brush layers (periodic rows of debris on the contour) have prevented erosion on construction slopes. Seeded mats, jute mats (degrade in 1–2 yr), or coconut fiber mats (degrade in 4–10 yr) are useful for stabilizing

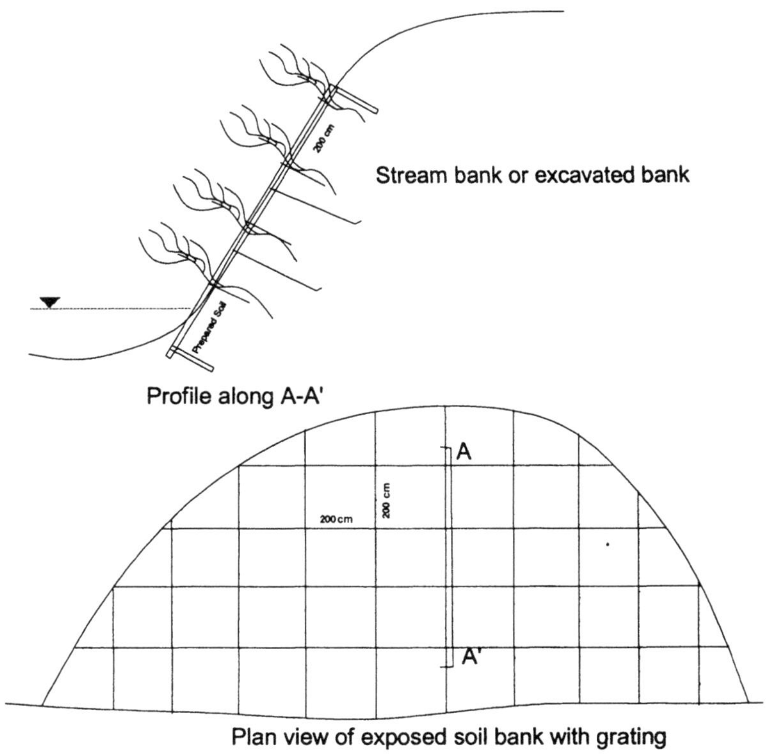

Figure 15.13a. Slope erosion control scheme for bank stabilization of stream with low flow. This approach is also applicable to other excavated banks.

slopes or channels. They are usually not placed around the entire channel cross section. They may be used as temporary linings until vegetation develops. Geotextiles, made of nondegradables such as polypropylene, have successfully controlled erosion on stream banks. Brush mattresses and seeded mats are examples of successful approaches for waterway stabilization. Many find decorative rock an aesthetically pleasing option for lining channels. Rock gabions (wire mesh bags filled with stones) and/or log cribbing with plant cuttings between the gabions (or cribbing) have proven functional and attractive.

Figures 15.13a–c show some bioengineering techniques. Soil bioengineering is labor intensive and thus costly, but the aesthetic appeal of bioengineering techniques makes the approach popular in urban areas.

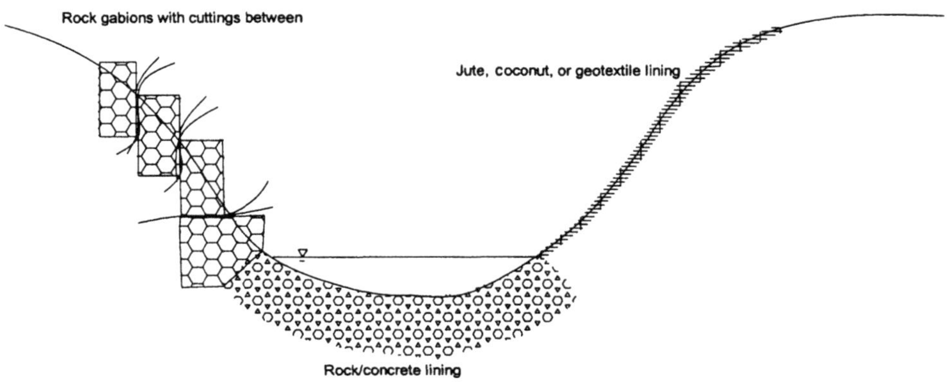

Figure 15.13b. Bioengineering techniques suggested for channel stabilization.

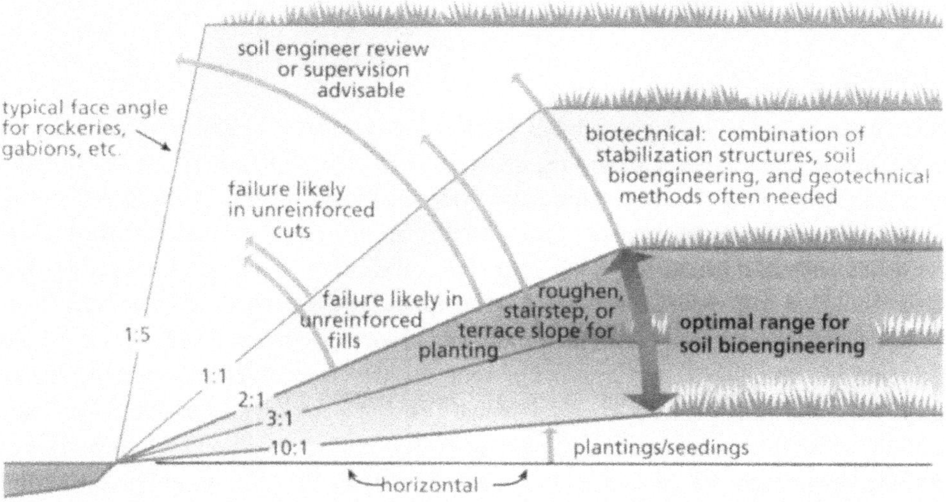

Figure 15.13c. Suggested stream bank restoration approaches as a function of bank slope. (From NRCS 1998.)

Natural riparian buffers are, not surprisingly, larger than BMP-related buffers. Natural buffers are large enough to enable substantial species diversity and a species succession from the water body to the upper buffer edge. They represent true ecotones (ecological transitions) and possess time scales of turnover ranging into thousands or millions of years (Mitsch and Gosselink 1993). NRCS 1998 provides an excellent discussion of practical approaches for stream channel and stream bank modification for rural and urban use.

Riparian buffers are proven BMPs for stream maintenance. Riparian buffers are zones of undisturbed vegetation between streams and fields. Conservationists sometimes call riparian buffers *vegetative filter strips.* They represent a BMP for agronomic and animal agriculture and silviculture. Chapter 6 discusses the capacity of a riparian buffer for sediment trapping. Chapter 7 discusses the capacity for riparian buffers to remove offending nutrients from runoff. Riparian buffers have advantages for wildlife conservation and have aesthetic advantages. One should review the BMPs for stream protection presented in Tables 7.1–7.4.

Stream corridor restoration is important because only 56% of the nation's 617,806 miles of rivers and streams supports multiple uses (NRCS 1998). Comprehensive treatment of stream restoration approaches is available at links within http://www.nrcs.usda.gov.

Physical Process Summary

Lower-order streams in dendritic, trellis, and radial systems tend to be eroding, with deposition occurring in higher-order reaches. The multibasin system usually occurs in upland swamps and wetlands. Infusions of sediment into these lower-order streams cause changes having decades-long time scales. Likewise, when sediment sources are eliminated, decades-long morphology changes commence. Streams may then evidence erosion or degrading behavior. Meandering, braded, anabranching, and reticulate streams are usually parts of high-order streams in large basins. These systems are usually depositing sediment. Velocity profiles are important to understanding some biological processes. Velocity measurement is important for stream assessment. The Manning *n* relates to roughness concepts important for predicting velocity profiles. Velocity affects sediment transport, in turn affecting biological and chemical processes.

Biological Processes

Streams

Natural streams and lakes teem with life. Habitats are extremely variable. Streams and lakes in the humid United States are among the most diverse of all temperate climates. Ecologists find the Tennessee-Cumberland River watershed to be the most biologically diverse ecosystem in a temperate climate (Anon. 1999). Various indicator organisms for ecological health such as daphnia, mayflies, stone flies, caddis flies, some minnow species, other filter-feeding crustaceans (Crites and Tchobanoglous 1998), and selected plant species are now used to assess stream ecological health and changes to stream health caused by land use changes. Developers and regulatory agencies will team engineers with ecologists, geologists, and soil scientists as development plans are proposed; thus some familiarity with these disciplines is essential. Briefly, the stream ecologist will pull a sample of some 200 organisms from the candidate stream to be impacted by development and determine the proportions of Ephemeroptera (mayflies), Plecoptera (stone flies), and Trichoptera (caddis flies). A reference stream in the area must then be sampled. If the number of organisms in the first stream is 85% of the corresponding reference stream value, the stream is designated to be unimpaired. If the value is less than 50%, the stream is seriously impaired. As one can imagine, selection of the reference stream can be controversial.[2]

Even in small streams, there can be much biological variation across the channel. One may class the biological characteristics of open waters into two broad categories: pelagic and benthic. The pelagic habitat is the free water away from the influence of shore or bottom substrates. The benthic habitat is associated with the substrate of the lake, extending from near the shoreline into deepest water. Ecologists further classify benthic as littoral or profundal. One may define the littoral habitat as the bottom zone extending from the shoreline to a distance where wave action is insignificant. The benthic profundal region is the bottom zone of the stream or lake having inadequate light penetration for bottom-attached organisms (Maitland 1990). One may measure light penetration using photometers and Secchi disk. The Secchi disk has distinct black and white features. The Secchi disk measurement is the depth where the disk is not discernable. Light penetration in freshwater streams is usually limited to the top 0.6–1.3 m (2–4 ft).

Figure 15.14 shows these habitat types. One may differentiate the benthic class based on eroding (sometimes referred to as degrading; usually in upper reaches) or depositing

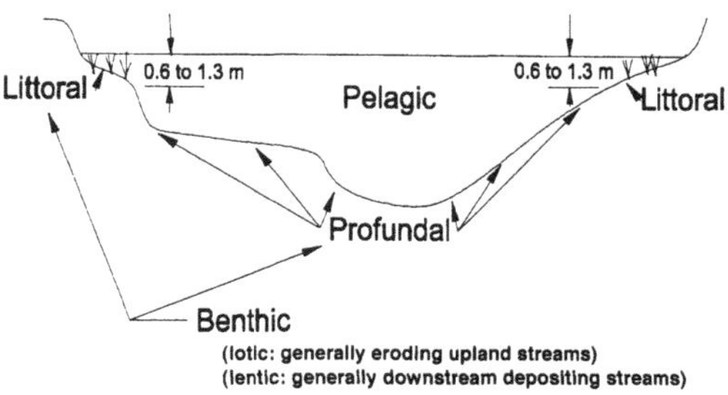

Figure 15.14. Habitat type designations common in stream ecology. Littoral and profundal zones generally occur in depositing streams (lentic or lotic ecology). Light penetration (usually from 0.6 to 1.3 m) distinguishes the littoral zone from the profundal zone. The littoral and profundal designations are not generally applicable to eroding streams.

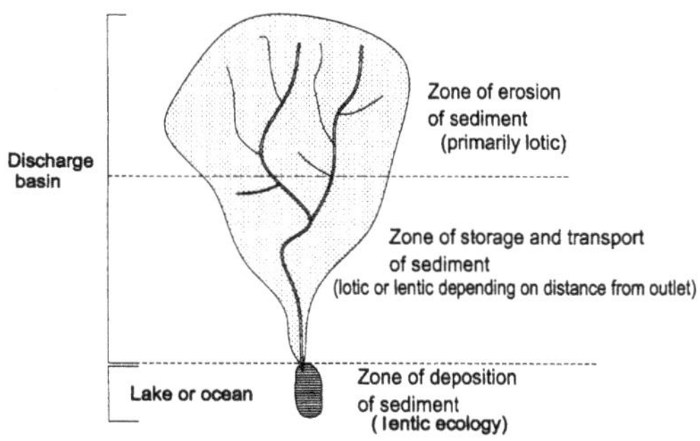

Figure 15.15. Hypothetical dendritic watershed showing predominant benthic characteristics as a function of relative distance from lake or ocean.

(lower reaches) conditions. Figure 15.15 shows the generic zones of sediment transport on a watershed scale. Limnologists refer to *lotic* ecology as that found in running water and *lentic* ecology as that found in standing water. Thus, upper reaches of streams are lotic and lower reaches having reduced velocity are *lentic*.

The pelagic region supports vertebrates and invertebrates. The benthic region is host to by far the most diverse flora and fauna associated with the stream. As streams transition from swift cascading waters to slow-flowing reaches, changes in associated biology occur. Some algae, stone flies (Plecoptera), blackflies (Simuliidae), and caddis flies (Trichoptera) are associated with the upper reaches of eroding streams with rock and gravel bottoms. Gravel bottoms provide important habitat for fish eggs. Aggrading streams and lakes usually host bottom-dwelling crustaceans, mollusks, algae, fungi, bacteria, protozoans, worms, rotifers, and nematodes. Insects, amphibians, birds, and land mammals affect the system. Lower stream reaches are usually much more ecologically diverse due to increasing nutrient levels. Endangered aquatic species are often benthic-dwelling organisms (e.g., the snail darter). Limnology texts such as Maitland 1990 and Cole 1994 provide much more detail concerning specific organisms.

Lotic and lentic ecologies have a bearing on the types of organisms contributing to chemical transformations. Lotic zones usually develop attached filamentous and other organisms facilitating useful bioremediation based on bottom and aquatic plant surface area. Lentic ecologies develop suspended organisms that occupy water volume. Wetland processes exploit the high surface area due to aquatic vegetation and the water volume.

Limnologists have studied species diversity and distribution versus distance from river discharge. Figure 15.16 provides a generic relationship. Ecologists have categorized the benthic plant communities of streams in England as follows (Maitland 1990):

Eroding or degrading (lotic habitat):
- Torrential, mainly on rocks
- Nonsilted, mainly on stones

Depositing or aggrading (lentic habitat):
- Partly silted, mainly on gravel
- Silted, mainly on silt
- Littoral (zone near the edge of quiet stream or lake), mainly on mud

These zones each host numerous species of benthic dwelling organisms.

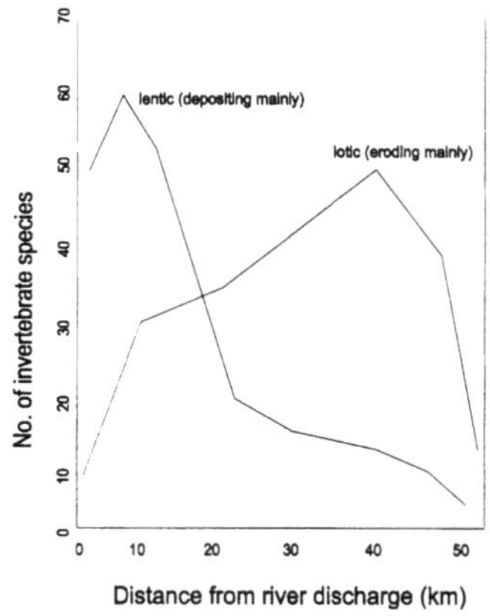

Figure 15.16. Number of species as a function of distance upstream for a river in Scotland. The aquatic survey is an important body of information for TMDL assessment. (Redrawn following Maitland 1990.)

The dendritic, trellis, and radial streams (see Figure 15.2) are often eroding but may also be depositing in their lower reaches. The braided stream is usually depositing. A wetland is perhaps the ultimate depositing stream. As streams begin to meander, the habitat diversity increases as the meanders tend toward oxbow lakes. The wetland contains the most diverse flora and fauna of any stream type. Even with changes in flow, slope, velocity, sediment transport and river shape, an incredible capacity exists for species to adapt to the environment. Land use changes resulting in extreme nutrient and sediment loadings can, however, overwhelm stream systems. The purpose of the TMDL is to define loads not causing significant human-induced ecology shifts.

Sustained sediment wash load usually occurs with large-scale land use transformations. Wash load reduces benthic species diversity and numbers in eroding and depositing stream reaches. Evidence exists that watersheds can recover to a limited extent once the sediment wash load returns to a "quasi"-normal condition.

Intermittent and ephemeral steams possess substantial benthic life, although it may not be immediately apparent. Many algae, macroinvertebrates (e.g., nematodes, rotifers), bacteria, and fungi have periods of dormancy inherent in their life cycles. Other life forms live in small pockets of persisting moisture. Some bacteria and many fungi develop spores. Some organisms form cysts or go into stasis (a suspension of life processes) until streamflow resumes and creates more favorable habitats.

Wetlands

The wetland is a transition between terrestrial and aquatic biosystems. The biological activity of the wetland is even more intense than the vigorous microbiology of littoral zones of moving streams. Good light penetration, rich nutrients, and slow-moving currents aid biological processes. Wetlands also exhibit a vigorous macrobiology (see Figure 15.17). The diversity and number of organisms during the open emergent and submergent phases explain the high water quality restoration potential of a mature wetland. Riparian zones around streams accomplish many processes occurring in wetlands but on a smaller scale.

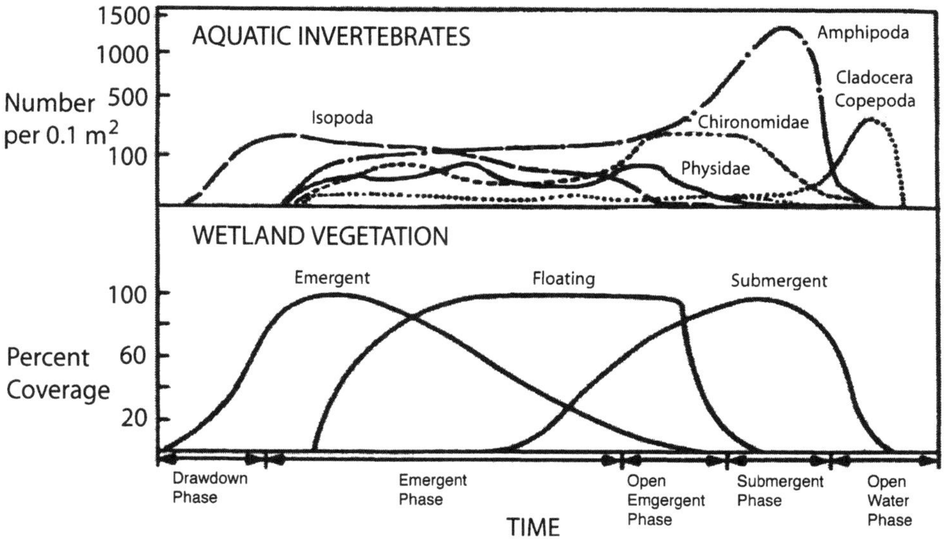

Figure 15.17. Aquatic invertebrates and wetland vegetation in prairie freshwater marshes that draw down during drought and then reemerge when the rains return, at which time floating vegetation appears that eventually submerges, bringing about open water. The cycle repeats every 15–20 years or whenever droughts cause drawdown. (From Mitsch and Gosselink 1993; used with permission of Van Nostrand Reinhold Publishers.)

The wetland functions as a bioremediation system due to its extreme biotic diversity. Ironically, the diversity greatly complicates accurate prediction of pollutant removal. The wide variability in spatial and temporal scales greatly complicates verification of relationships describing chemical and microbial process kinetics. Thus, one should treat these relationships as rough approximations.

Natural wetlands occur in every state and reflect regional differences in climate, vegetation, soil, and hydrologic condition. Ecologists divide wetlands into five types: marine (open ocean), estuarine (coastal and tidal marshes), riverine (freshwater river and stream channels), lacustrine (standing lakes), and palustrine (inland marshes, swamps, and bogs). Natural wetlands effectively remove some pollutants such as BOD, nitrogen, sulfur, and possibly phosphorus. Wetlands foster sediment deposition and affect temperature, which may lead to numerous physical, biological, and chemical transformations. Constructed wetlands mimic the natural wetland functions (van der Leeden et al. 1990). Mitsch and Gosselink (1993) provide extensive discussions of the biology of these systems.

Oxygen Transport

Oxygen transport is crucial for maintaining the aerobic status of streams. Fish and many other higher organisms inhabiting the pelagic region of streams and lakes require oxygen. Organisms of the benthic region usually prefer oxygen. Without oxygen, facultative and anaerobic microbial species may continue activity in anaerobic conditions if detritus or other carbon sources are present. Anaerobic conditions adversely affect macroinvertebrate and fish numbers. Turbulence in the flow aids the reoxygenation process. Photosynthetic algae also add oxygen; however, in times of cloudiness or darkness the algae respire or consume oxygen. Organic carbon added to the body as point waste, falling leaves, and other detritus creates oxygen sinks. Knowledge of the oxygen balance enhances our ability to predict the impact of additional organic carbon loadings on stream oxygen status.

The Streeter-Phelps equations (Tchobanoglous and Schroeder 1985) were developed for describing the oxygen mass balance in a stream reach. These relationships model oxygen consumption by BOD and benthic carbon along with reoxygenation, which depends on algal populations and streamflow characteristics. A detailed derivation of the Streeter-Phelps equations is beyond the scope of this text. Table 15.2 provides a summary of the

Table 15.2. Variable and rule sheets showing units and nomenclature for a TK© model that solves the Streeter-Phelps equations.

Variable sheet...

Status	Input	Name	Output	Unit	Comment
	3	De		M	Stream depth
	.24	u		M/S	Stream velocity
	21	T		C	Temperature
	7.5	Oinit		G/M^3	Initial oxygen concentration
	4	Mino2		G/M^3	Minimum allowable Oxygen concentration
L	3.51214261	θh		Day	Hydraulic residence time (set as list when computing plots)
F	12.9846094	Li		G/M^3	Initial BOD at discharge point after mixing with stream flow (guess when solving for θ h)
	0	rap		G/M^2 Day	Reaeration due to photosynthesis (input as positive)
	0	r r		G/M^2 Day	Oxygen uptake due to respiration (input as negative)
	.2	k20		Day^{-1}	Standard oxygen uptake rate
	−.8	rb20		G/M^2 Day	Temperature corrected benthic oxygen uptake rate
	−.26666667	rb		G/M^2 Day	Benthic oxygen uptake rate (input as negative for uptake)
	.2112	k		Day^{-1}	Oxygen uptake rate constant
	.374469785	k2		Day^{-1}	Reaeration rate constant
	.000182512	DL		M^2/D	Molecular oxygen diffusion coefficient
	.7	Di		G/M^3	Initial oxygen concentration deficit
	4.2	Dc		G/M^3	Oxygen deficit (critical)
L		DO2	4.2	G/M^3	Oxygen deficit (set as list output when computing plots)
L		O2	4	G/M^3	Oxygen concentration (set as list output when computing plots)

Rule sheet...

Status	Rule
Cancelled	Dc = (((k*Li *Exp (−k*θh)) − rb − rap − rr)/k2); Cancel when inputing θh
Cancelled	θh = (1/(k2 − k))*Ln((k2/k) − (((k2*Di*(k2 − k)) + ((rb + rap + rr) * (k2 − k)))/((k^2) * (Li)))); Cancel when inputing θh
* Unsatisfied	k2 = 294 * (Sqrt(DL * u))/$De^{1.5}$
* Unsatisfied	DL = 1.76E − 4 * $1.037^{(T-20)}$
* Unsatisfied	rb = ((rb20/De))
* Unsatisfied	k = k20*$1.056^{(T-20)}$
* Unsatisfied	Di = 8.2 − Oinit
* Unsatisfied	Dc = 8.2 − Mino2
* Unsatisfied	DO2 = (((k*Li)/(k2−k)) * (exp(−k*θh) − exp(−k2*θh))) + (Di*exp(−k2*θh)) −(((rap + rr + rb)/k2) * (1 − exp (−k2*θh)))
* Unsatisfied	O2 = 8.2−DO2

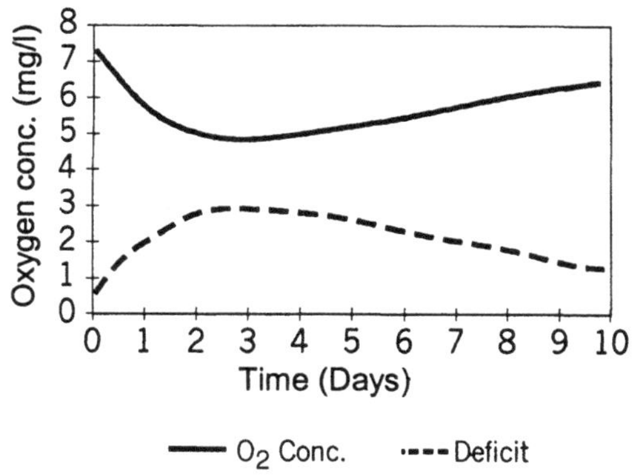

Figure 15.18. Oxygen sag curve predicted using the Streeter-Phelps equations plotted versus time (distance along stream reach). The deficit curve is a solution of the model for oxygen utilization in Table 15.2. The oxygen concentration curve is the deficit curve subtracted from 8.2 g O_2/m^3, which gives the stream oxygen status.

Streeter-Phelps model as adapted to an equation-processing software package. These equations model the oxygen consumed as water moves through a defined stream reach. Subtracting oxygen used from the equilibrium oxygen concentration in a stream (typically 8.2 mg/l) gives the oxygen concentration available for biological activity. Figure 15.18 shows a typical oxygen concentration versus distance along the reach. The enhanced stream water quality model (QUAL2E), a water quality program available from the USEPA (see links within http://www.epa.gov), also uses a mass balance. Proper use of QUAL2E and related models requires investments in time to learn the proper use of the software and in effort to determine appropriate input values.

Lakes and Ponds

Large ponds and lakes are usually biologically active. Lakes have littoral and profundal regions as discussed above and possess a lentic ecology. Ecologists divide the pelagic region of a lake into three regions: the epilimnion, into which light penetrates and in which photosynthetic algae grow; the hypolimnion, dark and cold; and the thermocline, the boundary between the epilimnion and the hypolimnion (Maitland 1990). Figure 15.19

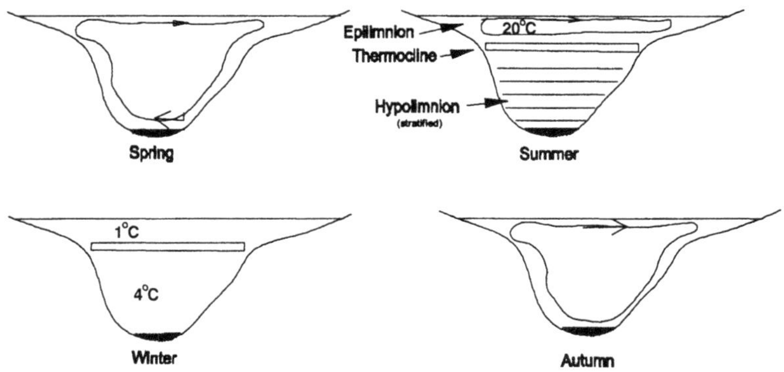

Figure 15.19. Lake circulation patterns and temperature profiles as a function of season in temperate regions.

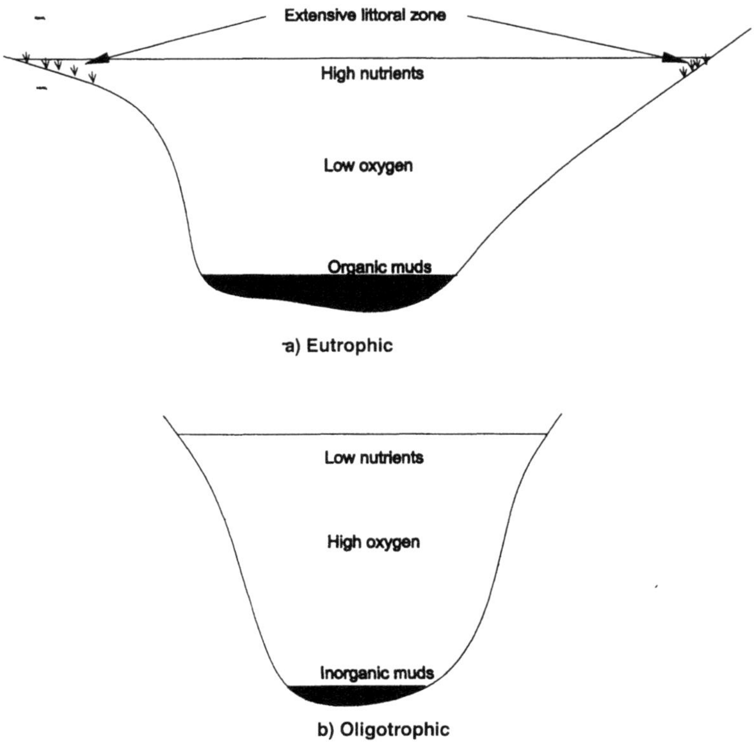

Figure 15.20. A schematic representation of the major differences between eutrophic and oligotrophic lakes.

delineates these zones. Temperate lakes and ponds as a rule cool to 4°C in winter. They cool all the way to the bottom because water at 4°C is maximum density. In the spring and autumn, the thermocline disappears and the lake circulates in response to wind-driven surface currents. Figure 15.19 shows an idealized turnover cycle. Depending on particular conditions the turnover may or may not occur. The turnover phenomenon requires an adaptable ecology. Turnover can cause taste and odor problems in drinking water originating in temperate lakes. Tchobanoglous and Schroeder (1985) discuss approaches for quantifying the lake state in the presence of incoming waters.

Warm tropical lakes are thermally stratified in the summer but may circulate the remainder of the year. Equatorial lakes occur only in warm tropical areas. Permanent stratification is normal although this may break down intermittently due to wind. Some lakes in arctic regions never rise above 4°C in the summer. They have inverse thermal stratification (e.g., water at the surface just under ice is colder than deeper water) in the winter.

Depending on the nutrient levels, the lake may be eutrophic or oligotrophic. The eutrophic lake as a rule has more biological activity. Lakes are normally oligotrophic, implying low nutrient levels. Excessive nitrogen or phosphorus will likely cause eutrophication. High algae levels suggest this condition. Figure 15.20 compares eutrophic and oligotrophic lakes. Table 15.3 gives additional characteristics. Maitland (1990) also presents data for a third category encountered less frequently, the dystrophic lake, occurring in stony and muck conditions. Limnologists further subdivide these categories (e.g., see Cole 1994).

A healthy stream or lake need not always be free of turbidity; some streams are naturally turbid at certain seasons. However, a water body may be clear and blue—to all appearances an aquatic paradise—and yet contain certain chemicals that render the stream

Table 15.3. Characteristics of eutrophic and oligotrophic lakes as a function of selected physical attributes.

Character	Eutrophic	Oligotrophic
Basin shape	Broad and shallow	Narrow and deep
Lake substrate	Fine organic silt	Stones and inorganic silt
Lake shoreline	Weedy	Stony
Water transparency	Low	High
Water color	Yellow or green	Green or blue
Dissolved solids	High, much N and Ca	Low, poor in N
Suspended solids	High	Low
Oxygen	High at surface, low under ice or thermocline	High
Phytoplankton	Few species, high numbers	Many species, low numbers
Macrophytes	Many species, abundant in shallow water	Few species, some in deep water
Zooplankton	Few species, high numbers	Many species, low numbers
Zoobenthos	Few species, high numbers	Many species, low numbers
Fish	Few species	Many species

Source: Adapted from Maitland 1990.

a biological desert (Bardach 1962). Many industrial wastes contain irritants or poisons that aquatic life forms ingest. Other pollutants such as organic carbon may result in oxygen depletion. In the gills of fishes and other stream dwellers, sometimes only one thin cell layer separates the blood and other body fluids from the surrounding water. Some chemicals diffuse through the gills to act on other organs. Other compounds clog the gills, resulting in death by suffocation. Reliance on observation alone can cause major misjudgments. Detailed studies of life forms must supplement physical and chemical measurements along a stream to correctly assess stream health.

One must view streams and lakes as living ecosystems to appreciate the potential contribution of stream ecological processes to environmental enhancement. Species diversity is a potent buffer against short-term variations in upstream nutrient loadings. However, long-term land use changes often result in changes in sediment and nutrient loadings. This causes long-term changes in numbers of species in the benthic and pelagic regions. These changes are most noteworthy when endangered species are threatened. In extreme cases, excessive nutrient and sediment loadings can turn a living stream or lake into an open sewer and lagoon. Aesthetically undesirable life forms then dominate.

Physical and Biological Process Summary

Stream morphology (channel order, arrangement, channel shape, and bottom slope) has much influence on stream velocity, which in turn influences erosion (lotic) and deposition (lentic) activity. Erosion and deposition in turn greatly influence benthic and pelagic region ecology. The littoral zone ecology differs in lotic and lentic streams. The profundal zone may not exist in lotic streams. This zone may develop extensively in lentic streams, usually deeper than 1 m. Turbidity influences the ecology in the pelagic zone. TMDL assessment requires definition and documentation of key ecological indicators in each ecological zone. Aeration correlates positively to stream velocity, as one may see via the reaeration term in the Streeter-Phelps equations. Temporal scales of ecological change may persist for years, making cause and effect hard to assess. Scale effects of ecological change in uplands discussed in Chapter 7 compound the change effects in downstream bodies. Recall the interactions between streamflow morphology and likely eroding/depositing behavior and associated ties with benthic, pelagic, littoral, and profundal ecologies.

Predominant Nutrient and Chemical Processes

Oxygen and Sediment Chemistry

The occurrence of low oxygen levels has implications beyond immediate impact on stream biology and microbiology. Low oxygen creates the likelihood of low oxidation-reduction (redox) potentials, especially in benthic regions. Many elements exist in multiple states of oxidation. Reduced forms (e.g., Fe^{2+}, Mn^{2+}) tend to be more soluble than oxidized forms (e.g., Fe^{4+}, Mn^{3+}) in certain pH zones. *Redox combined with water pH has implications concerning the solubility of many chemical compounds.* Redox is often expressed as Eh (Nernst potential in volts) or pE (logarithms of oxidized form over reduced form). *Low potentials or negative logarithm values of pE denote reduced (e.g., often more soluble) forms, and vice versa.* Stumm and Morgan (1981) provide much more discussion of redox and pH interactions.

Acid mine drainage, natural sulfide deposits, and dissolved CO_2 or other acid sources may contribute to low pH values in streams. Natural or human-induced BOD loadings or inorganic chemical reactions may deplete oxygen levels, favoring low redox states. Aquatic chemists have developed relationships between redox potential and pH for various compounds. Low pH values may intensify solubility problems. Two examples are shown for illustrative purposes in Figure 15.21. Figure 15.22 shows the effect of redox potential on common elements at a pH of 7. *A stream polluted with heavy-metal constituents that encounters heavy organic loadings downstream will likely have serious soluble-metal pollution due to the relatively high solubility of the reduced forms of many metal cations.* Soluble cations intensify transport of anion pollutants (e.g., nitrate, nitrite, several phosphorus forms).

Figures 15.21 and 15.22 are strictly applicable to idealized situations containing only those compounds on the respective graphs. Inclusion of other elements affects solubility via possible common ion effects. Development of solubility relationships requires input from aquatic chemists and is beyond the scope of this text. These relationships are presented to show the effect of aeration status on element solubility. Again, lowering the redox potential and lowering pH increases the likelihood of elemental solubility. Ammonia nitrogen (as opposed to nitrates), sulfide (as opposed to sulfates), and ferrous (as opposed to ferric) are favored at low redox potentials and are more soluble than their oxidized counterparts.

Other Processes

Oxygen processes are only one of several dynamic forces affecting chemical status. Precipitation, sorption, ion exchange, and partitioning are mechanisms employed to describe important chemical processes. Precipitation is the transformation of a material from dissolved ion phases to the solid phase. Precipitation keeps many toxic metals from being in solution. Sorption denotes adsorption onto surfaces or absorption into soil bodies, usually

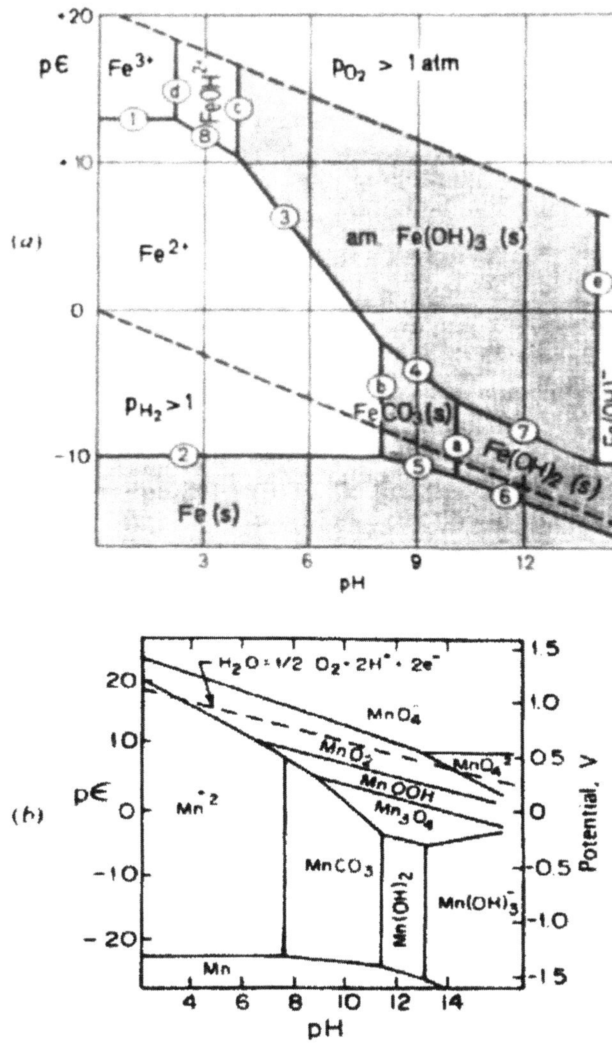

Figure 15.21. Redox (pE) potential vs. pH for (a) iron-carbonate-hydroxide system and (b) manganese-carbonate-hydroxide system. As redox drops and acid conditions develop, more soluble forms of manganese and iron predominate. (From Stumm and Morgan 1981; used with permission of John Wiley & Sons.)

in irreversible reactions. Sorption processes cause pesticides to selectively adsorb onto clay and soil organic matter. Likewise, clays tend to selectively adsorb phosphorus. The ability of a soil to adsorb phosphorus, pesticides, or other potential pollutants is finite. Ion exchange is responsible for the reversible adsorption of nutrients in soils. Ion exchange processes are reversible, whereas many sorption reactions are not completely reversible. Partition coefficients describe the selective absorption of some metals or organic materials to other organic materials. Many organic polymers selectively absorb heavy metals from liquids.

Watershed development and resulting increased sediment load may affect chemical interactions. Removal of riparian vegetation can increase temperature by exposing streams to more direct sunlight. Reactions may be kinetically faster in runoff from asphalt than

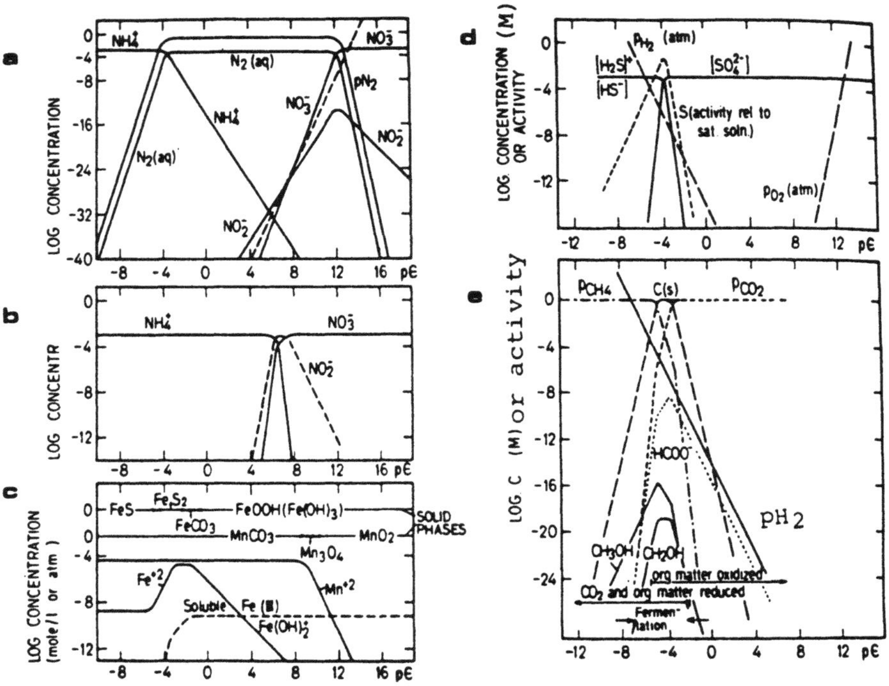

Figure 15.22. Concentration diagrams as a function of redox potential with pH = 7 for (a) nitrogen, (b) nirtrogen without N_2, (c) iron and manganese, (d) sulfur, and (e) selected carbon species. Note that reduced forms of metals, sulfur, and other compounds are frequently more soluble than oxidized forms. (From Stumm and Morgan 1981; used with permission of John Wiley & Sons.)

from vegetation because asphalt runoff is usually warmer. Increased sunlight, thus increased ultraviolet (UV) light, can cause breakdown of polycyclic hydrocarbons. Some of the UV-mediated reactions bring about toxic compounds. UV light selectively breaks down some polycyclic aromatic hydrocarbons. The influence of turbidity, temperature, and UV light penetration on all species needs research.

The littoral and profundal regions of depositing streams may be a source/sink for chemicals going into and out of solution. This is because clays and detritus may change oxidation status as loading changes. McCutcheon et al. (1993) elaborate on the chemical aspects of water quality.

The large number of chemical constituents possible in a wetland or stream, coupled with the interactions of oxygen and biomass, make theoretical wetland design difficult. Data from natural wetlands suggested zero-order and first-order kinetics for removing BOD, nitrogen, and phosphorus in constructed wetlands. Overloading a constructed wetland with BOD may depress the redox potential. This may increase soluble metals. Phosphorus may also become more mobile in heavily loaded wetlands. Miller et al. (1997) provide an extensive bibliography of sediment and chemical removal processes in natural and constructed wetlands.

Nutrient Sampling

The collection of adequate chemical constituent data for stream assessment is expensive. One must properly collect and hold samples in specified conditions to ensure valid tests. Nitrogen samples must be appropriately pretreated and held to ensure valid

Figure 15.23. Automated sampling system for stream organic and nutrient assessment.

analyses. Samples must be protected from UV light to avoid breakdown in cases of polycyclic hydrocarbons. Clesceri et al. (1998) present accepted pretreatment and analysis procedures. Automated samplers such as shown in Figure 15.23 are widely used for field sample collection.

Nutrient Cycling

Nitrogen, carbon, and sulfur cycle in streams just as they cycle in uplands. Each of these elements has numerous gaseous forms, as one may recall from Figures 7.4a–c. Sulfur is in many ways similar to nitrogen. Phosphorus cycling is interesting because it may change forms as stream pH and redox potentials vary. Redox and pH will change the amounts and rates of nitrogen and phosphorus in the various pools.

One may view the nutrient cycling process in the stream context not as a circle but as a downstream spiral. Thus, there is a stream velocity–travel distance–travel time relationship

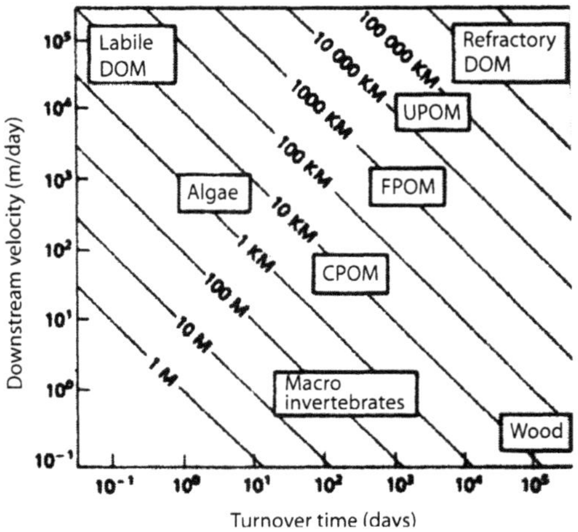

Figure 15.24. Suggested turnover lengths for carbon of various forms in a stream with a flow of 3–30 m^3/s. CPOM is coarse particulate organic matter; FPOM is fine particulate organic matter; UPOM is ultrafine particulate organic matter; DOM is dissolved organic matter. (Form Newbold 1996; used with permission of Blackwell Science Publishers.)

in the turnover time associated with nutrient cycles. Newbold (1996) suggests organic turnover as shown in Figure 15.24. Significant turnover may occur with dissolved organics and low velocities. Turnover of carbon gives interchange of solid, liquid, and gaseous carbon phases and enables carbon to leave the system. Thus, turnover implies remediation opportunities.

The USEPA has developed software in an attempt to model streamflow processes and project risk assessments. The enhanced stream water quality model (QUAL2E) is applicable to well-mixed dendritic streams and models the major reactions of nutrient cycles, algal production, benthic and carbonaceous oxygen demand, and atmospheric reaeration. It can predict up to 15 water quality constituent concentrations. The USEPA intends QUAL2E as a water quality planning tool for supporting the development of TMDLs. QUAL2E can also be used with field sampling for identifying the magnitude and quality characteristics of nonpoint sources. By operating the model dynamically, the user can study diurnal dissolved oxygen variations and algal growth. However, the effects of dynamic forcing functions, such as headwater flows or point loads, cannot be modeled with QUAL2E. The USEPA storm water management model (SWMM) also supports TMDL determination (USEPA 1994). Other software, such as WASP (Ambrose et al. 1993), is used when estuarial processes are important. Software mentioned in Chapter 7, such as AGNPS, may also be used to estimate nonpoint contributions of watersheds.

The USEPA supports the use of BASINS (Lahlou et al. 2000) for a comprehensive approach to TMDL determination. BASINS (Better Assessment Science Integrating Point and Nonpoint Sources) is a comprehensive modeling system that can simulate the response of natural and man-made drainage systems to precipitation events. BASINS models upland and stream processes using the ARCVIEW geographic information system (GIS) package as a data manager. One should periodically visit links within http://www.epa.gov and http://www.wcc.nrcs.usda.gov to see the latest in available tools for modeling streamflow and upland water quality processes.

An Internet search using key words such as *hydrologic modeling, basins, risk assessment, ecotoxicology, sediment transport models,* etc., will provide sites with additional commercially and governmentally available software packages for watershed characterization. Comprehensive modeling of ecosystems is under development at several research centers. An adequate science base for modeling macrobiological components of total watershed assessment is currently lacking.

Outlook

This chapter provides a broad overview of extremely complex physical, biological, and chemical processes. A reasonably good understanding of factors contributing to stream TMDLs exists; however, much remains to be learned about precise interrelationships. The intent here is to be introductory and general and to equip the natural resources engineer to interact with other specialists. This chapter has introduced the discipline of stream morphology, with associated relations between velocity profiles, cross section, and sediment transport, and the disciplines of limnology and hydrology as related to natural stream and lake processes. The text has shown how the friction velocity (u^*) relates to sediment transport capacity and to the reoxygenation potential of streams. Nutrient cycling from limnology perspectives was introduced, and ecological variation along a typical stream was illustrated. Stream oxygen balances play a key role via redox potential with respect to nutrient solubility. Velocity profiles and oxygen transport are two major processes with which engineers can readily identify. These processes are critical to understanding stream and lake processes. The chapter introduced nomenclature used by limnologists for describing stream ecology. One should appreciate the potential of long temporal and spatial scales when studying past, present, and proposed future stream alterations. NRCS 1998 gives an excellent overview of the scope of stream problems. It discusses multidisciplinary and multiagency approaches for achieving a political consensus on solution approaches. It provides additional background on most topics introduced in this chapter.

The overriding goal of water quality efforts is the preservation of fitness for use by relevant species. The object of point and nonpoint pollution control efforts is to prevent values for parameters in Figure 7.2 from increasing beyond those for "typical" surface and ground waters. A scientific approach to determining the TMDL values for water bodies requires watershed scale assessment and modeling. Modeling will aid in testing the impact of additional pollutant loads on streams with TMDLs because modeling is the most efficient way to link knowledge of the physical, biological, and chemical processes occurring in streams and lakes. Streamflow models such as QUAL2E and uplands water quality models such as AGNPS (discussed in Chapter 7) will find extensive application in evaluating various land management plans. Storm water management in accordance with BMPs coupled with comprehensive nutrient management planning in agricultural operations will be essential for maintaining capacity for municipal growth. Tools for storm water and agricultural BMP evaluation will be useful for watershed TMDL development.

The ability to comprehensively model natural resources engineering is tentative given the components remaining to be integrated into the overall knowledge framework. Regulation has been the primary approach used to motivate environmental compliance. Regulation often proceeds without the benefit of fully understanding the system, often leading to citizen resistance. Full understanding of spatial and temporal phenomena is exceedingly difficult. AGNPS, SWMM, BASINS, and QUAL2E are major state-of-the-art programs for predicting stream dynamics. NRCS 1998, this chapter, and exposure to models and programs such as those listed above could easily encompass one semester of study.

The emphasis on stream and lake assessment, coupled with extensive emphasis on modeling to link physical, biological, and chemical processes, will lead to even better associations of stream and lake impairment symptoms with uplands management practices. This will impact agricultural and urban activity alike. As consequences of uplands management are better understood, those bearing the brunt of the consequences will make their case and controversies will erupt. Community problem-solving approaches undergirded by high ethics and integrity are needed. Developments in ecological science and software modeling will aid the process in time.

Problems and Questions

NRCS 1998 contains an excellent listing of review questions in the introduction section.

1. A stream in a region with extensive agricultural production with extreme erosion in the nineteenth and early twentieth centuries has been under mixed pine-hardwood forest for the last 50 years. Explain what may be happening to the stream regime.
2. A stream is 2 m deep and has a slope of 0.01 and a D_{50} of 0.1 m. Estimate the velocity at 0.2*D*, 0.6*D*, and 0.8*D*. Compare these to the average velocity.
3. Compute the flow in the stream with the velocity measurements shown in Table 15.4.
4. A stream is equipped with markers 100 m apart. A float requires 1 min to traverse the distance between the markers. The average cross-sectional area is parabolic with a depth of 3 m and a top width of 5 m. Estimate the existing streamflow rate.
5. A reservoir design life is 100 yr. For a 10 m wide river, 0.0005 slope, 0.5 mm mean particle size, and 2.5 m depth, compute the required reservoir sediment storage capacity.
6. A fish farm needs to have clear water for fish production. Suppose one constructs the channels on a slope of 0.0001 and a mean particle size of 0.05 mm lines the bottom.

Table 15.4. Data for Problem 15.3.

Distance *b* from bank to meter (ft)	Depth *D* (ft)	Depth below surface (m) (either 0.6*D* or 0.2*D* and 0.8*D*)	Point velocity (ft/s)	Average velocity $\overline{u}$ (ft/s)	q_i (ft^3/s)
2	1	0.6	0.54		
4	3.5	2.8	0.98		
		0.7	1.6		
6.5	5.2	4.2	1.3		
		1	1.6		
8	6.3	5	1.3		
		1.3	1.8		
9.5	4.4	3.5	1.5		
		0.9	1.7		
12	2.2	1.3	1.1		
13	0.8	0.5	0.64		
14	0	0	0		

The side slope z is 3, the depth 1 m, and the bottom width is 3 m. Compute the flow. Use the Shields curve to determine if the flow will entrain sediment.

7. Define *lentic, lotic, littoral, profundal, pelagic, benthic, eutrophic, oligotrophic, wash load, bed material load, bed load.*
8. What are the defining characteristics of an eroding (degrading) stream versus a depositing (aggrading) stream?
9. What contributions does the limnologist, the geologist, and the hydrologist make to streamflow assessment? Define TMDL and discuss some approaches used for assessing TMDL for streams.
10. How do time scales and nutrient turnover relations explain why precise quantitative responses to sediment or nutrient pulse loadings are so difficult.
11. What are some techniques one may use for flow measurements in natural streams? Assume structures that may facilitate the measurement methods are available nearby.
12. Why would the rules of thumb presented in the text for estimating runoff based on stream size not hold in times of significant watershed development? What is meant by the concept of *regime* in the context of stream development and water quality?
13. How can comprehensive nutrient management planning on uplands aid in stream improvements downstream? Review Tables 7.1–7.4 for BMPs applicable to stream protection. Summarize relevant practices from these tables and rate their effectiveness (e.g., 1–4 rating as used in these tables) for nutrient control.
14. Four principles of BMP selection were identified in chapter 7. Review these principles and identify how they may work in a natural stream or wetland.
15. Why is the total pollutant load in a watershed not necessarily that measured in a stream at the watershed outlet?
16. Given two streams with identical BOD loadings with one being of the dendritic type and one being braided, which one will likely carry the most waste without becoming impaired?
17. During a watershed assessment, one notices very little macroinvertebrate presence and reduced fish populations along a stream reach, compared to a comparable reference stream. Physical indicators appear to be otherwise normal (e.g., no foams, colors, odors, apparent sludges). There is some evidence of soil deposition. The watershed is undergoing urbanization. From this limited evidence, what do you suspect?
18. During a watershed assessment, one notices very little biological activity of any kind, compared to a comparable reference stream. Physical indicators appear to be otherwise normal (e.g., no foams, colors, odors, apparent sludges). The watershed is in an urban area. Before making a judgment, what type of data would you collect next?
19. It has often been said that "the solution to pollution is dilution." Why does society find this "solution" increasingly inadequate?
20. Indicate the stream order of the streams in part a of Figure 15.2.
21. Why do some lakes "turn over"?
22. Define some microbiological process, chemical processes, and physical processes that have a bearing on water quality as a flow moves down an eroding stream, depositing stream, wetland, and lake.
23. Perform an Internet search to identify two software packages for comprehensive hydrologic modeling of watersheds. These models should model water quantity and quality. Specify your search engine and key words. Do not include models discussed in the text. *Hint:* For key words, try *surface runoff, water quality models,* and variations of these.

24. A stream is 3 m deep with a velocity of 0.5 m/s and a flow of 6 m^3/s. The allowable BOD of a 0.5 m^3/s waste stream is to be constrained so the oxygen concentration in the channel will not fall below 5.5 g/m^3 (i.e., the maximum allowed deficit is 2.7 g/m^3). The oxygen uptake rate of a sample of waste to be discharged into the river is 0.15 day^{-1}. We may neglect the algal uptake rates. The benthic uptake rate is − 0.9 g/m^2-day. The initial oxygen concentration is 7.5 g/m^3, and the allowable deficit is 2.7 g/m^3. The BOD just above the point discharge is 10 g/m^3. The standard well-aerated oxygen concentration value is 8.2 g/m^3. The temperature is 30°C. Compute the maximum BOD of the waste before mixing and the distance downstream reached when the oxygen deficit is greatest. *Note:* Solution requires the use of the Streeter-Phelps model shown in Table 15.2. One may refer to texts such as Tchobanoglous and Schroeder (1985) for additional details.

Notes

1. Following Simons and Senturk (1991), these units are in kilograms-force, leading to the unusual unit of Newton-mass for density.
2. Personal communication from Dr. Judy Meyers, Institute of Ecology, University of Georgia, Athens.

References

Ambrose, R. B., Wool, T. A., and Martin, J. L. 1993. *The water quality analysis simulation program, WASP5. Part A, Model documentation. EPA/600/3-87/039.* Environmental Research Laboratory, EPA, Athens, GA.

Anon. 1999. *National Geographic* 195(2): insert map.

Bardach, J. 1962. *Downstream: A natural history of the river.* Harper & Row, New York, NY.

Bookchin, D., and D. Farnsworth. 1996. Environmental laws and regulations. In *Environmental engineers handbook,* 2d ed., ed. D. H. F. Liu, B. G. Liptak, and P. A. Bouis. Lewis Publishers, Boca Raton, FL.

Chow, V. T. 1959. *Open channel hydraulics.* McGraw-Hill, New York, NY.

Clesceri, L. S., A. E. Greenberg, and A. D. Eaton. 1998. *Standard methods for the examination of water and wastewater.* 20th ed. American Public Health Association, Washington, DC.

Cole, G. A. 1994. *Textbook of limnology.* 4th ed. Waveland Press, Prospect Heights, IL.

Correll, D. 1998. The role of phosphorus in the eutrophication of receiving waters: A review. *Journal of Environmental Quality* 27:261–66.

Crites, R., and G. Tchobanoglous. 1998. *Small and decentralized wastewater management systems.* WCB, McGraw-Hill, New York, NY.

diCenzo, P. D., and S. H. Luk. 1997. Gully erosion and sediment transport in a small subtropical catchment, South China. *Catena* 29(2):161–76.

FAO. 1993. *Guidelines for land-use planning.* UN Food and Agriculture Organization, Rome, Italy.

French, R. H. 1985. *Open channel hydraulics.* McGraw-Hill, New York, NY.

Graf, W. H. 1971. *Hydraulics of sediment transport.* McGraw-Hill, New York, NY.

Interagency Committee on Water Resources. 1963. *A study of methods used in measurement and analyses of sediment loads in streams.* Report no. 14. Minneapolis, MN.

Kundell, J. E., and T. C. Rasmussen. 1995. *Erosion and sedimentation: Scientific and regulatory issues.* Report by Georgia Board of Regent's Scientific Panel. Vinson Institute of Government, University of Georgia, Athens, GA.

Lahlou, M., L. Shoemaker, S. Choudhury, A. Hu, H. Manguerra, and A. Parker. 2000. *BASINS version 2 users manual.* USEPA, Washington, DC.

Maitland, P. S. 1990. *Biology of fresh waters.* 2d ed. Chapman & Hall, New York, NY.

McCutcheon, S. C., J. L. Martin, and T. O. Barnwell, Jr. 1993. Water quality. In *Handbook of hydrology,* ed. D. R. Maidment. McGraw-Hill, New York, NY.

Miller, B. I., L. C. Brown, E. W. Tollner, J. C. Hayes, T. G. Franti, and U. S. Tim. 1997. *Critical review of data from research on vegetative control systems.* ASAE Technical Paper no. 972058. American Society of Agricultural Engineers, St. Joseph, MI.

Mitsch, W. J., and J. G. Gosselink. 1993. *Wetlands.* 2d ed. Van Nostrand Reinhold, New York, NY.

Newbold, J. D. 1996. Cycles and spirals of nutrients. In *Rivers' flows and channel forms: Selected abstracts from the rivers handbook,* ed. G. Petts and P. Calow. Blackwell Science, Oxford.

NRCS. 1998. Stream corridor restoration: Principles, processes, and practices. USDA Natural Resources Conservation Service, http://www.usda.nrcs.gov, Washington, DC.

Perry, J., and E. Vanderklein. 1996. *Water quality: Management of a natural resource.* Blackwell Science, Cambridge, MA.

Schiechtl, H. 1980. *Bioengineering for land reclamation and conservation.* University of Alberta Press, Calgary.

Simons, D. B., and F. Senturk. 1992. *Sediment transport technology: Water and sediment dynamics.* Water Resources Publications, Littleton, CO.

Streeter, V. L. 1971. *Fluid mechanics.* 5th ed. McGraw-Hill, New York, NY.

Stumm, W., and J. J. Morgan. 1981. *Aquatic chemistry: An introduction emphasizing chemical equilibria in natural waters.* 2d ed. John Wiley & Sons, New York, NY.

Tchobanoglous, G., and E. D. Schroeder. 1985. *Water quality: Characteristics, modeling, and modification.* Addison-Wesley, Reading, MA.

Thomas, A. W., and R. Welch. 1988. Measurement of ephemeral gully erosion. *Transactions of the ASAE* 31(6):1723–28.

USEPA. 1994. *SWMM: Storm water management model.* EPA 823-C-94-001. Washington, DC.

van der Leeden, F., F. L. Troise, and D. K. Todd. 1990. *The water encyclopedia.* Lewis Publishers, Chelsea, MI.

Viessman, W., Jr., and G. L. Lewis. 1996. *Introduction to hydrology.* 4th ed. HarperCollins College Publishers, New York, NY.

Whitlow, J. R., and K. J. Gregory. 1989. Changes in urban stream channels in Zimbabwe. *Regulated Rivers* 4:27–42.

Yang, C. T. 1995. *Sediment transport: Theory and practice.* McGraw-Hill, New York, NY.

Yoo, K. H., and C. E. Boyd. 1994. *Hydrology and water supply for pond aquaculture.* Chapman & Hall, New York, NY

Appendix A

Ethics, Biotic Relationships, and Resource Economics

Most people believe that it is necessary to conserve natural resources but also that exploitation of resources for short-term economic gain may also be necessary. The conflict between conservation and exploitation creates ethical dilemmas leading to an examination of individual and societal values.

This appendix surveys selected topics that are not commonly in engineering curricula. Ethics, ecological design, resource economics issues, and criteria for a quality project are presented to stimulate understanding of and sympathy for diverse views.

The goal is not to define a "one size fits all" ethical position. Each person needs to be able to raise a conscientious defense for positions taken when ethical questions are involved. One hears constant reference to values and ethics in literature on total quality management (TQM). In this literature, which the author embraces, distinctions are made between values and ethics, as Table A.1 shows.

Morals are what others observe; values, like motives, are invisible until mapped into tangible forms by one's observed morals. Ethics provide written or spoken standards.

Ethics is the discipline concerned with the process for deciding what is good behavior and what is bad behavior in particular situations. In other words, what is one's moral obligation in particular instances? How do our values map into behavior in specific situations? Here the concern is with conservation values and ethics versus resource development values and ethics. Engineers have an interest in and can offer solutions from both points of view.

Review of Values and Ethics

Philosophers have proposed many theories concerning what a morally good action means.[1] Geisler (1989) summarized historical approaches to this question, shown in Table A.2. Approach 1 says that one who controls the most land, labor, capital, and/or technology is in the driver's seat. Technology is a recent addition to the land, labor, and capital list.

Approach 2 has been the subject of a vigorous debate in the civil rights movement and the modern political correctness movement. The tendency of police to charge alleged offenders from poverty-stricken neighborhoods but not to file charges for similar offenses against those from affluent neighborhoods illustrates possible racial tensions arising from number 2. Some have attributed racial motives to the tendency to site a potentially polluting animal production operation in a poor county rather than an affluent county. Responding to these situations, some have coined the term *environmental racism*. Ethics are subjective according to the approach 2 ethical position. Ethical subjectivism enables an open mind but fails to provide an adequate objective standard of behavior.

Table A.1. Correspondences between ethics, values, and morals.

Ethics	Values: ethical positions claimed by individuals or organizations	Morals: the "fruits" of ethics and values
Articulated statements, spoken or written, defining the standards for right and wrong behavior; the study of right and wrong.	Define the individual	Translate values into actions
	Are constant	Are changing
	Are internally derived	Are situationally determined
	Are concerned with virtue	Are concerned with justice
	Are general	Are highly specific
	Are stated morally	Are stated behaviorally
	Are judged as good or bad	Are judged as present or absent
	Set priorities	Set limits for appropriate behavior

Source: Adapted from Stamatis 1994.

Approach 3 in modern discussion represents the tension between the rugged individualist versus the collectivist. As society continues to integrate, individuals now call upon the courts to define constitutional rights in the face of pressure to act collectively. A modern outgrowth of approach 3 is the belief of the positivist school that the act of asserting makes the assertion true. The statement "If it is in print it is true; if it is in print on glossy paper it is definitely true" is a belief held by many that may relate to the approach 3 ethical position. Ethical subjectivism stems from position approach 3. In simple terms, ethical subjectivism states that man is infallible. According to ethical subjectivism, anyone may say anything about any subject if one sees a possibility of influencing another by doing so. This idea is deficient because it fails to consider reason.

Approaches 4a and 4b come to us in modern form via the animal rights movement. Society has largely discarded approach 4a. Approach 4b is a basic value of many in the environmental movement. Leopold (1966) wrote, "A thing is right when it preserves the integrity, stability and beauty of the biotic community." It is wrong otherwise. This position is the ethical basis for many environmental writers. During the past 50 years, it has been a major challenge to philosophers, sociologists, theologians, ethicists, ecologists, and many others to define and justify the place of a human being within the natural environment (Nash 1989; List 1993). The results are confusing and often contradictory, as one would expect from the variety of ethical positions available. Views range from anti-industrial ideas such as advocated by *Deep Ecology* and *radical environmentalism* (e.g., Devell and Sessions 1987) to the pro-industrial philosophy of *objectivism* (Rand 1971). The anti-industrial approach takes the view that the human being is an insignificant member of the biotic community. It grants him no right to interfere with the natural ecological processes for his benefit. Pro-industrial approaches consider the human being as the master of nature. They allow the right to explore, exploit, and alter the nature for the betterment of humanity. In the middle of this wide spectrum, many other bio- or anthropocentric philosophies exist. They narrow the human role to varying degrees of natural stewardship (Orr 1994; Leopold 1966). Ecology can become the "tail that wags the dog" when taken to extremes. On the other hand, a tragedy occurs when a species goes extinct that could contain the unique chemistry for saving a human's life. Responding to the issue of preserving biodiversity and conserving natural resources, ethicists are debating an ethical standing for plants and/or animals. Historical ethical theories have not addressed the standing of plants and animals compared with people. The usual approach for structuring the debate is to consider past issues and how philosophers handled them. There is much similarity in positions 3, 4a, and 4b in that they are inherently subjective.

Moderation, expressed in approach 5 of Table A.2, is usually appropriate in resource development questions. The "end justifies the means" (approach 8), basing questions solely on human pleasure (approach 6), and basing the meaning of *good* only on pleasure or beauty (approach 9) can lead to

Table A.2. Summary of historical ethical positions or worldviews.

Approach	Ethical approach	Leading proponent if known	Issues or difficulties
1	Might is right, where might is land, labor, capital, technology.	Thrasymachus (character in Plato's *Republic*)	It is possible to be good without power and powerful without being good; "possession is 9/10ths of the law" Subjective and/or relativistic ethics
2	Morals are mores.		Ethics are ethnic (e.g., racial), subjective, and/or relativistic ethics.
3	Man is the measure.	Protagoras (480?–411? BC) Hare (1952)	Each person's own will is the standard for defining right and wrong. The act of assertion makes it right (Positivist). Subjective.
4a	The Human Race is the basis of right.	See List 1993; Nash 1989; Leopold 1966	Does not allow for dissent; has in the past allowed child sacrifice, slavery. Who speaks for the human race? Who speaks for the biosphere?
4b	The biosphere is the basis of right		
5	Right is moderation.	Aristotle (384–322 BC)	Some virtues are inappropriately expressed in moderate amounts
6	Right is what brings pleasure (hedonism).	Epicurus (341–270 BC)	What kind of pleasure should be used for the test?
7	Right is the greatest good for the greatest number (utilitarianism).	J. S. Mill (1806–73)	Does not confront the issue of what is "good." Can be a modern version of hedonism.
8	Right is what is desirable for its own sake; rule based ethics (Kantianism; deontology).	Immanuel Kant (1724–1804)	Can lead to "goodness" being defined not by the means but by the ends; easy to confuse what is being desired with what ought to be desired. People regarded as an end, not a means to an end.
9	Right is indefinable "spirit" or consciousness-based ethics (contractualism).	Rousseau (1712–1768) G. E. Moore (1873–1958)	Attempts to define "good" fall prey to the "naturalistic fallacy" of assuming that since one can attribute pleasure to "good," they are identical.
10	Ethics is personalistic; virtue-based theory.	Martin Buber (1878–1965) Karl Jaspers (1883–1969)	Each person defines his or her own ethics; can lead to anarchy. Can allow psycho egoism or ethical egoism to unconsciously control. Can be subjective.
11	Good is what God wills (some of items 7–9); divine command; natural order.	Many religions (St. Augustine, Thomas Aquinas, and others)	Can lead to authoritarianism when people place themselves between people and God. Views of the nature of God vary widely even within the major religions. Can become rule based. Provides compelling ideas.

Source: Adapted from Geisler 1989.

Note: The shaded viewpoints are the modern predominant ethical positions.

significant difficulties. One person's pleasure is another person's pain. The fundamental canons of the National Society of Professional Engineers (NSPE) are most recognizable as virtue ethics. They are as follows:

- Hold paramount the safety, health, and welfare of the public.
- Perform services only in areas of competence.
- Issue public statements only in an objective and truthful manner.
- Act for each employer or client as a faithful agent or trustee.
- Avoid deceptive acts.
- Conduct oneself honorably, responsibly, ethically, and lawfully to enhance the honor, reputation, and usefulness of the profession.

The fundamental canons are followed by rules of practice and stated professional obligations. That these categories have evolved speaks to the difficulties with virtue ethics.

Approach 6 is the classical pleasure-based ethical system. "If it feels good, do it" typifies this approach. The time horizon is immediate instead of long term. The correct definition of *pleasure* is the difficult issue for this approach. Pleasure-based ethical systems sometimes form the basis for compromise, rightly or wrongly.

In a democratic society, utilitarianism (approach 7) seems to take precedence among items 1–9, perhaps because it defers the fundamental issue of what is good. Utilitarian approaches form the basis of modern public administration. The following may represent approach 7 in modern vernacular: *(1) the consensus within a diverse public is constructed around the best possible scientific understanding of the issue; and (2) the solution must be economically viable in the short and long term.* Approaches 1–4 (especially approach 4b) become foci for vigorous debate while trying to fulfill the utilitarian objective. Ethicists have criticized utilitarianism, claiming it is indifferent to individuals as it distributes economic satisfactions (e.g., pleasures and happiness) in society. What is the basis for pleasure? Is that all there is? Ethicists have also criticized it for not embodying an adequate description of human economic behavior. Utilitarianism has been called modern hedonism. In spite of these criticisms, society has accepted modernized utilitarianism as the primary society-based ethic in public administration (Perman et al. 1996).

The legal profession applies rule-based ethics (approach 8). Immanuel Kant articulated the rule of law for the sake of law. This idea lies at the heart of the concept of accountable standards and at the heart of modern bureaucratic organizations and government. The scientific management approaches of Frederick Taylor and management thinkers had a high rule orientation.

Spirit- or consciousness-based ethics (approach 9) is applicable when seeking to understand and act on intents arising from the general will of society. The reappearance of democracy in the eighteenth century was stimulated by Rousseau's contributions to the formation of the idea of a social contract between a government and the governed. The social contract of Rousseau was predicated on the governed waiving certain rights to take advantage of governmental protections. On the organizational level, humans agree to certain policies of organizations in order to receive certain benefits. The social contract approach evolved to counter natural behavior associated with egoism. Legislative sessions and team meetings are in a sense a continuing negotiation of general or specific social contracts. Radical libertarianism, or the total disdain of all government, is the opposite of spirit-based ethics (i.e., flashback to approach 3).

Personalistic ethics (approach 10) is, in a sense, not concerned so much with what is legal or what accords with community standards but with what is right. Virtues are the basis of this ethic. It borrows from Aristotle and other teachers. Noteworthy reformers (e.g., Gandhi, Martin Luther, M. L. King) appealed to this ethic. Current approaches to transforming leadership, as opposed to transactional leadership, appeal to personalistic leadership. In applying personalistic ethics, one must ensure that base motives do not enter unawares into the discussion. Rachels (1999) describes two terms relating to self-interest: psycho egoism and ethical egoism. Psycho egoism says that each individual knows what is best for himself. Humans are ultimately selfish (Rand 1971). This idea ultimately suggests the infallibility of selfish humanity. Motives for many behaviors can easily be reinterpreted to fit this ethic. It can also become irrefutable in that it blocks all signals indicating potential errors in paradigms. For example, once committed to a mental institution, proving the absence of mental disease is not possible under psycho egoism. Ethical egoism implies that altruism is an invasion of privacy, which creates unnatural dependencies among people, unless a real difference may be identified that justifies the special treatment. Not surprisingly, there is endless debate over the realness of this or that difference. Ethical egoism is also based on self-interest. Forms of ethical subjectivism may also enter into debates under the cloak of doing what is right. Vindication of personalistic ethics is usually experienced long after the fact. Approach 10 arises from the important role personal conviction plays in having and applying integrity to situations. Values, coupled with virtues, the author asserts, are essential. Challenges arise when those who claim spokesmanship for God legislate behaviors (ethics) often associated with particular values.

Many add approach 11 to the top (or bottom) of the hierarchy. The divine command and natural law ethics were articulated in the Middle Ages. Both build on the premise that to have ethical laws, there must be an ethical lawgiver. The idea of *divine command* implies there is a God and God's will is best for us. The idea of natural law implies that one can discern God and God's will from nature. Overreliance on natural law can result in the is-ought fallacy, a confusion of what ought to be and what is. Although *ethics* associated with approach 11 can be diverse, commonly held *values* are essential for a group to work together successfully.

Items 7, 8, 9, and 10 are normally applied in a hierarchical way (Hitt 1990, 1998). Aspects of approaches 1–6 figure into the 7–11 ethical hierarchy. Each individual orders the ethical hierarchies uniquely. It is interesting to study various cases to decide which values the various proponents held. Underwood (1998) summarizes many ecologically related controversies in this way.

Ethical Concepts and the Environment

Despite one's philosophical worldview, most people accept that human activities have interrupted or disturbed many ecological processes throughout history. Humans have altered many aspects of nature. Some believe, or even insist, that nature will restore itself if we stop intervening with its processes (List 1993). Without commonly held philosophy and belief systems, society must use various governmental problem-solving mechanisms to develop consensus. Values are not necessarily right or wrong. Achieving a consensus requires identifying and balancing values of the group.

The growing environmental and societal concern about engineered products creates a compelling need to consider these factors in the *design phase*. Direct and indirect impacts of modern technology on the environment, on the quality of natural resources, and on human society likewise create a compelling need to evaluate design methods. Today, customers' concerns go beyond mere technical specifications and economic affordability. Concerns include short- and long-term impacts of manufacturing, use/operation, and retirement of industrial products. University and government researchers are proposing design methodologies that attempt to capture key social elements (e.g., Esteghlalian et al. 1999; Bouchard 1999). These approaches build upon quality function deployment (QFD) ideas. These ideas will become more important as one advances into management.

The appearance of ISO 14000, an international standard for strategic environmental management, is perhaps the most comprehensive societal response to environmental stewardship needs (Tibor and Feldman 1997). ISO 14000 lays out an environmental management protocol, including comprehensive approaches for documenting initial review, processes, employee training, supplier management, record keeping, and audits. The growing realization that good environmental stewardship enhances profits over time furthers the adoption of ISO 14000. The act of forming ISO 14000 standards could be construed as beginning personalistically, progressing to debates about including this concept in the social contract, and finally leading to codification, all in the name of creating status for utilitarian distribution. Accomplishment of these goals in a pluralistic society is a continuing challenge. Nevertheless, similar standards are beginning to be incorporated in ecological design methodologies. Design methodologies considering fundamental environmental qualities and desirable societal attributes respond to such needs and concerns (Wann 1996; Orr 1994).

Principles of Ecological Engineering or Design

Ecology is the discipline describing plant-animal-human ecosystems and their interrelationships. The overarching goal of the ecologist is to model mass and energy flows within and among these large-scale systems. They apply models to learn how the system self-selects plant and animal species and populations. We are learning much about the ecology of plant-animal systems.

From the mathematical analogy mentioned in Chapter 1, the forcing function is essentially zero. Manipulations around the boundary are minimally invasive. Odum (1971) is regarded as the father of ecological modeling. Ideas of ecologists are invaluable for solving natural resource problems. Their knowledge of plant and animal biology and their interrelationships transcend the economically important plants and animals.

The following terms appear in the ecological literature:

Ecosystem: a soil-water-air-biotic system that exchanges water, soil, air, and biotic masses and energy with its environment.

Ecotone: a transition zone between two different ecological systems.

Time scale: time duration required for cyclical patterns to repeat; the time constant of a first-order response to an impulse or step function.

Space scale: distance bounding the effective range of key animal species within an ecosystem; distance required for boundary effects on fluxes not to be significant.

In its most natural state, the world ecosystem, often called *biota, biotic community, natural environment,* or simply *nature,* comprises animals, plants, bacteria, and their surroundings—namely, air, water, and soil—forming a *complex interdependent* system. Any change inflicted on any entity within this system will affect the whole community either directly or indirectly. In subsets of this large community, countless numbers of plants and animals, each with their own unique functions, growth patterns, and nutritional and environmental (both chemical and physical) needs, are in constant interaction with each other and with their surroundings, therefore creating a *diverse* and *dynamic* system. Each member plays many roles and connects many events and entities. Thus, each member is essential for maintaining the *integrity, stability,* and *functionality* of the system. The prey-predator food chains and symbiotic and synergistic living systems are examples of such interactions among living entities. The natural phenomena of nature support each other and reinforce certain behaviors. Although each of these events may be a linear chain of cause and effect, many of them have common elements, leading to large and interconnected networks. For instance, herbaceous plants extract minerals from soil. They serve as a food source to herbivores, which in turn feed small carnivores. Herbaceous plants provide a surface cover for the soil, therefore reducing the soil erosion. They also foster soil porosity and water infiltration because of their compact root systems.

A careful examination of different parts and various entities of nature and ecological processes reveals that each member, despite its instrumental value to humans, and apart from humans' reactions and judgments, is valuable and essential for system survival. Therefore, we must preserve the fundamental qualities of nature such as diversity, complexity, and nonlinearity. This is true not only from the human life viewpoint but for the sake of the whole system. The system adapts, feeds, and inspires the human race along with many other entities.

The multitude of animal and plant species, the various shapes and sizes of plants (bushes, shrubs, trees) and animals (birds, insects, mammals), and the variety of land use and landforms (forest, moorland, farms) create the diversity essential for the stability of the universal ecosystem and serve as a source of human inspiration, recreation, and aesthetics. We feel, sense, and appreciate the latter qualities. We categorize entities and settings with such attributes as beautiful, splendid, serene, exciting, or peaceful. However, the significance or the right-to-live of the smallest, or even the largest, ecosystem member does not stem from instrumental values. It stems from the role the member plays within the biotic community. Here we must bring to bear approaches 8 and 9 of Table A.2.

Although ecosystems are often described using the chain or building block metaphors, such systems are in reality *complex* and *nonlinear*. Therefore, the absence of a single member may affect the whole system through many *pathways* and *loops*. In any *nonlinear* system (either ecological or human behavior systems), even a minor change in one part may precipitate a series of unpredictable changes and adaptations (Bella 1997; Kosko 1993). The consequences may include formation of new loops (e.g., food chains) or development of new patterns or behaviors (e.g., growth rate) to maintain *balance* and *stability*. In other words, when a disturbance is not too "violent," a system can restore itself through multiple adjustments and emerge into a *new order*. Emergence of new orders is slow. Continuing to disturb the system during the renewal period could delay the attainment of the *new balance,* hence leaving the system at an unstable and unbalanced state for a longer period. In addition, if a system is violently disturbed, its restoration will require more than just a simple adjustment. Such alterations often result in a *systemic imbalance*. Imbalanced systems manifest strong tendencies to accommodate certain components and reject others. Systems continuously disturbed may not reach the disastrous level of *chaos* or *collapse*. However, they suffer from *lack of diversity* and, therefore, *low stability* (Odum 1971).

A fundamental tenet of ecology is self-organization (Odum 1988). The ecologist strives to determine a sustainable pattern with nature. The self-selected plant-animal-human system is the one that will be environmentally and economically productive. This requires that humankind accept the resulting water-soil-air-biotic states and fluxes. To say the least, this is problematic.

Odum (1989) advocates the discipline of ecological engineering. The ecological engineering paradigm is "you give us the problem and we will manage for a high environmental and economic performance." He compares this with the conventional engineering approach, reported as "you decide what you want and we will design and build it to specification." The Odum (1989) idea of the classical engineering approach is an accurate reflection of typical forcing functions imposed on a problem space by many nineteenth- and twentieth-century engineers. The functioning ecological engineer will begin to predominate in the twenty-first century. The lack of knowledge of the large, diverse biotic system hinders efforts to design our surroundings. The ecological engineer may relate to other engineers as do landscape and building architects. Architects and landscapers design based on aesthetic principles. Ecological designers design based on biological and ecological principles. In the future, architects, landscape designers, and ecological designers along with natural resources and other engineers will jointly plan projects.

Jorgensen and Mitsch (1989) summarized a group of principles dealing with problems of biological selection:

1. The forcing functions of the system determine ecosystem structure and function. Alterations of the forcing functions cause the most drastic changes in ecosystems. Forcing functions as used by Jorgensen and Mitsch (1989) are more like the set of relationships describing the problem space. The term in this text refers to the activities imposed by humankind or through natural catastrophe on the problem space.
2. Ecosystems are self-designing systems. The more one works with the self-designing ability of nature, the lower the costs of energy to maintain the system. For example, cattails grow in swampy areas and not other places because they are much more adapted to swampy areas than to uplands.
3. Ecosystems recycle elements. Models for calculating mass balances are most helpful in studying these issues. If one is operating a wetland, for example, one will need periodically to harvest biomass to keep nutrients from leaving.
4. Homeostasis, or constancy of long-term behavior, of ecosystems requires accord between biological function and chemical and physical composition. One may beneficially aid biological composition by maintaining chemical states and physical conditions known to foster diverse biological activity.

5. Processes in ecosystems have characteristic time scales varying over several orders of magnitude. One must adapt manipulation of ecosystems to the ecosystem dynamics. In certain regions, the deer population is now large due to the cumulative impact of wildlife management programs that responded to diminished populations 25 years ago. Natural resources engineers must learn to factor in long temporal scales into management issues. In the engineered wetland, time scale dynamics will no doubt be at work in unanticipated ways; intensive studies are required. Ecosystem components have characteristic space scales. Manipulation of ecosystems should take into account the appropriate size necessary to achieve the desired result. As humans, plants, and animals become ever more crowded, overlap of space scales will undoubtedly occur. In specific situations, the biologist can advise regarding space requirements for natural areas. They can tell us what may happen over time if one imposes specified biological treatments.
6. Chemical and biological diversity contribute to the buffering capacities of ecosystems. When designing ecosystems, one should introduce many parts from which the ecosystem's self-designing ability can choose. Biological diversity depends on chemical diversity. A wetland with high diversity will respond more readily to a one-time or a step increase of waste loading.
7. Ecosystems are most vulnerable at their geographic edges. Ecological management should take advantage of ecosystems and their biota in their optimal geographic range. One should select species with range in mind.
8. Ecotones, or transition zones, form at the transitions between ecosystems. The interfaces between humankind's activities and nature should be gradual, not abrupt.
9. Ecosystems are coupled with other biotic systems. One should maintain this coupling wherever possible. Ecosystems should not be isolated from their surroundings.
10. Ecosystems with pulsing or cyclical/repeating input patterns are often highly productive. One should recognize the importance of pulsing patterns. This principle underlies the need to specify rest periods for infiltration basins and engineered wetlands.
11. Everything is networked to everything else in the ecosystem. Managing one component of an ecosystem without affecting other parts is impossible. The development of modeling techniques for ecological systems is a thriving research area.
12. The feedback mechanisms, resilience, and buffer capacities of ecosystems are products of the ecosystem's preceding evolution. Existing ecosystems do not match well with manufactured synthetic chemicals. New emerging ecosystems can sometimes deal with them. This fact explains, for example, why many manufactured fibers may not easily decompose in a bioconversion system.

These principles appear to embody a moderate view of approach 4b in Table A.2. The natural resources engineer exercises creativity and judgment in the forcing function selection and boundary manipulations for problem spaces. As our knowledge of the biotic component develops and human acceptance of various fluxes and levels occurs, society will perhaps continue integrating ecological principles into the development of ever more precise forcing functions and boundary manipulations.

Resource Economics: Social Traps

Ecosystems provide scarce resources to humankind. Society imposes resource scarcity on the individual via income and wealth limits. Technological change eventually responds to market prices that in turn signal resource scarcity. These ideas are fundamental to traditional economics. They have explained how and why engineers work on particular problems. Valuation of the commodities and services of ecosystems occurs through pricing structures. One assumes perfect markets.

Many buyers and sellers, ease of marketplace entry and exit by buyers and sellers, an inability of individual buyers and sellers to set the price, and a product identical with respect to buyers and sellers define the perfect market (Seitz et al. 1994). Grain and livestock producers and buyers in the past were one of the closest approximations to the perfect market.

Resource management and environmental management have been historically treated as separate issues, but population pressures and scarcities are forcing increasing overlap of these issues. The uncertainty of where environmental management ends and resource management begins is a "fuzzy overlap" zone. Fuzzy overlap makes it difficult to draw crisp boundaries around the problem space.

For many ecosystem goods and services of human interest, price formation is limited due to missing markets, called "externalities." Subsidies and taxes distort prices (Maxwell and Costanza 1989). Problems are often larger than political boundaries. Markets may fail to encourage ecologically sound management when long time scales are involved. One cannot solely determine the ultimate value of a sustainable ecosystem from the economic value.

One may develop ecological designs in the presence of social traps. Social traps are situations where short-run, local reinforcements guiding individual behavior are inconsistent with the long-run global best interest of the individual and society. Traps can arise out of simple ignorance of the relevant reinforcements. They may arise when change of reinforcements occurs with time. Such short-term reinforcements may include monetary incentives, social acceptance, and physical pleasure and pain. Traps can arise due to externality conditions. For instance, an upstream landowner can lose soil from his cropland. This landowner never knows the cost of removing this sediment from the water downstream when others treat it for drinking purposes. Traps can arise in collective situations. Consider a hunting club composed of individuals who each decide to exceed the limit by one animal. For an individual, the cost seems small compared to the benefits of another animal. However, the cumulative cost to the hunt club is far more than any individual realizes.

Natural resources economists are researching the impact of social traps caused by erosion. Researchers estimated the onsite cost of wind erosion in New Mexico at $10 million annually. The offsite cost was an estimated $400 million annually. Documentation of the huge, off-site cost, or exposure of the social trap, now provides justification for increased research. This could lead to imposed controls, generating tension with regard to approach 3 in Table A.2. New dimensions to the meaning of "individual responsibility" continue to appear.

For turning some environmental traps into tradeoffs, some have proposed tax structures for altering behavior. The difficulties inherent in valuing an ecosystem suggest one will not soon realize the vision of the ultimate ecological designer. Sustained education about long-term impacts and superordinate authority such as legal systems and religion must occur. Converting of traps to tradeoffs by correcting the system to give accurate short-term reinforcement is helpful. One goal of ISO 14000 is to set up structures leading to improved identification of social traps through full cost accounting.

Economists have proposed two approaches for placing valuation on environmental problems. One is the decline in property values following a major environmental problem. For example, Palmquist et al. (1997) determined that proximity to a large hog operation caused a $5000+ decline in house values within 0.3 km (0.5 miles) of the operation in North Carolina. The other approach for placing valuation on environmental problems is the cost borne by nearby residents to remediate the effects of environmental problems. For example, in a situation with polluted water, the cost borne by nearby residents individually or collectively to remove the pollution is an environmental cost. One must factor in ethical considerations to evaluate these tradeoffs.

What Defines a Well-Designed Project?

The current emphases on ISO standards have clarified approaches to the question of what is a good or quality effort. One may adapt these criteria, given by Stamatis (1994), to natural resources engineering. A project is considered a quality effort if the following dimensions are present:

1. Function: Does the project provide the primary required function?
2. Features: Are the expected bells and whistles present?
3. Conformance: Are government agency and other client requirements satisfied?
4. Reliability: Will the project function for the design life with specified maintenance?

5. Serviceability: Can key components of the project be accessed if something fails prematurely?
6. Aesthetics: Are aesthetics considered and appropriate for the location and client?
7. Perception: Does the project enhance the reputation of those involved by being cost effective, employing leading or appropriate technology, and harmonizing with the above aspects of quality?

These dimensions of quality represent a good consensus among most corporate groups and government leaders. Corporate interests form the largest client base for natural resources engineers, second only to government.

All the ethical positions are usually evident in project debates. Addressing project function and other dimensions in an appropriate (to a democratic society) utilitarian manner requires the best in public administration. One must meet legal requirements and social acceptability needs. Economic realities limit excesses in each of the quality dimensions and may limit the dimensions themselves.

Case Study 1: Pitcher Plant Community Viability

The pitcher plant is an exotic endangered species found in the Georgia and North Carolina mountains. The Nature Conservancy is evaluating purchase of selected sites for preservation purposes.

Question: Has the hydrology of lands surrounding a pitcher plant (*Sarracenia oreophila*) community been altered to the point of endangering the continued viability of the pitcher plant community?

Background: The site is on the shore of Lake Chatuge, as shown in Figure A.1. The pitcher plant thrives in a nitrogen-poor soil and needs open canopy affording some exposure to direct sunlight. The pitcher plant thrives in areas with much sunlight and humidity coupled with protection from high winds. The soil appeared to have a high clay fraction. The plant's competitive advantage stems from the ability to capture and use N from entrapped insects. The plant community is within a zone owned by two property owners (A and C). Property A is now up for sale. The property C owner has agreed to limit development and will not apply nutrients. The canopy is open. A rock outcrop may indicate the presence of an aquitard or aquiclude. The aquiclude limits the deep-rooted species and provides steady water via seepage from nearby hills and mountains (not shown on the sketch). A flowing spring at the drain tile outlet suggested an underlying clay aquitard. The clay tile drain, 4 ft deep, was located near the aquitard depth. The rock feature did not seem to extend over to the drainage ditch, based on quick observation.

Many questions remain about how the plant community arrived at its present state. Ecologists have advanced three theories about how the pitcher plant survived in the site in years past: (1) burning and cultivation by the native Indian population, (2) native beaver populations, and (3) burning by lightning. These theories were tentative at best. The property owner burned the site annually in the recent past. The site was apparently not plowed or otherwise disturbed. This fortuitous burning worked to the competitive advantage of the pitcher plant community. Sorting out the spatial-temporal scale effects from hydrologic disturbance effects is hard.

There is another pitcher plant community nearby. This community was in a boggy area also. This region was either plowed in the past (by mule) or was situated near land plowed, based on recollection by local people.

Past Hydrological Alterations

The hydrological alterations began in 1941–42 with TVA construction of Lake Chatuge. The plant community lies within the TVA 500 yr easement. The lake to date has never reached the level of the pitcher plants.

The second major hydrological alteration was the construction of the tile drainage line (depth estimated at 4 ft) and a drainage ditch in 1965 on property A. This ditch was recently cleaned. Trees, 30–40 ft high, grew along the ditch. It is about 150–200 ft from the plant community. Assuming the clayey soils in the region have a low K_{sat} value, the influence of the drain is probably not too great. Capillary fringe would probably add 1 ft or so to the piezometric level. There was a low flow in the

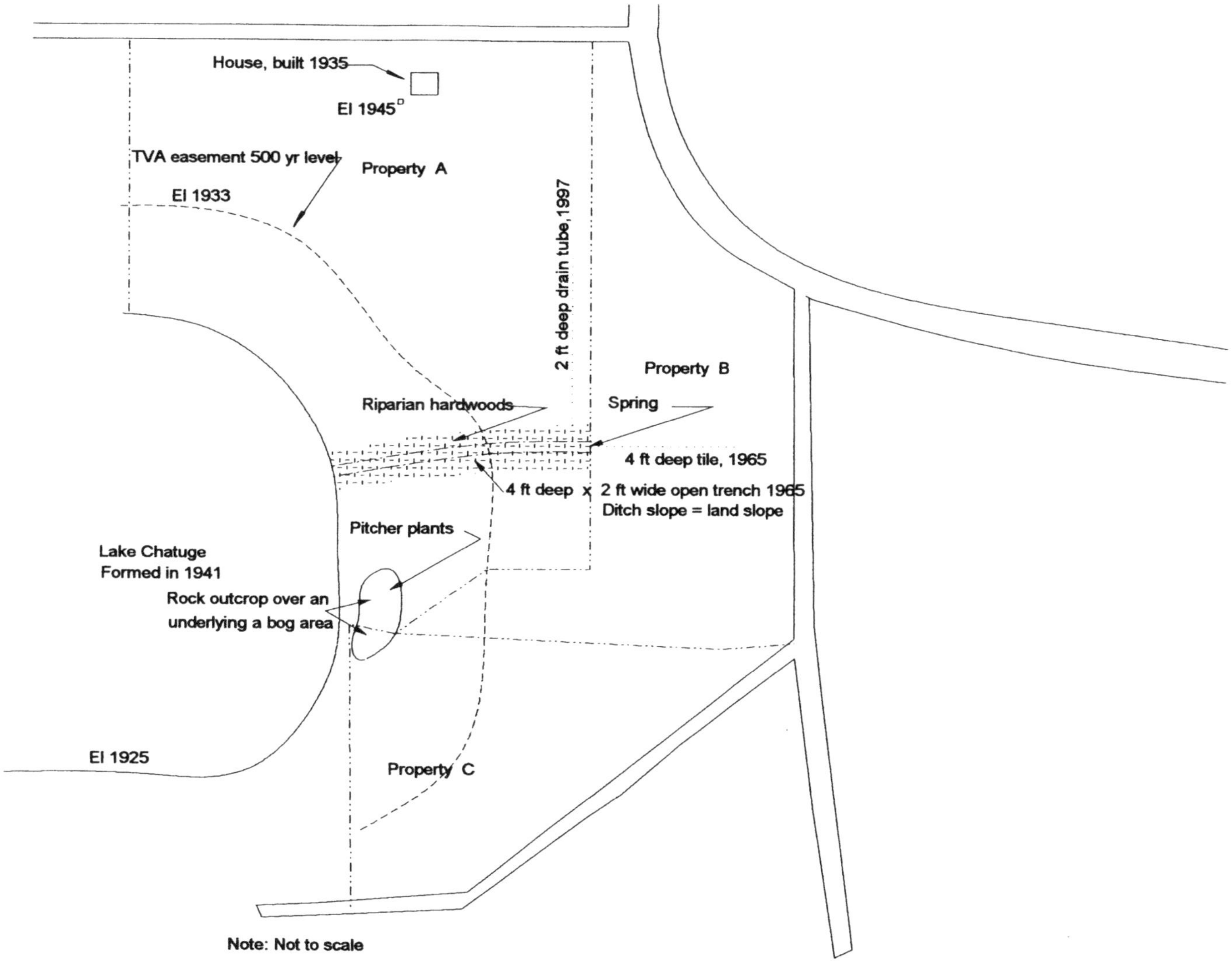

Figure A.1. Pitcher plant situation.

ditch welling up from a spring near the tile drain outlet. This spring may indicate a break in the aquitard on which the clay tile was installed. The 1965 tile drain was not discharging. If the spring is a recent development, the effect on the surface moisture regime is unknown. The spring could reduce upward water movement through the aquitard by functioning as a pressure relief. A 30–40 ft high growth of mixed hardwoods exists along the ditch, suggesting a constant presence of water in the ditch. Since the drain installation, the owners have managed the adjacent field as an improved pasture, implying occasional N applications.

The third major hydrological alteration was the construction of the 1997 perforated tube drain (estimated depth of 2 ft), coupled with cleanout of the 1965 drain channel. The owner evidently had not deepened the channel during cleanout. The 1997 tile was not discharging.

Potential Hydrological Alterations

An infrequent rising of Lake Chatuge is possible. Climatologists estimated that a 1994 rainfall event was a 500–1000 yr rainfall event. The impact of short-term (e.g., 1 month or less) flooding on the pitcher plant community is unknown.

Land values in the area indicate the potential of development all around the site. With development will come the possibility of aquifer drawdown due to well installation. More significantly, N levels in the soil may increase due to increased numbers of septic systems. Surface runoff may also increase due to reduced infiltration due to development.

The management of another adjacent property (property B) is a major unknown. This property seems to exert the dominant up-elevation influence on the pitcher plant community. If the area persists in pasture, surface runoff probably would not substantially increase. An intensification of cattle production could increase surface runoff and N concentrations to the disadvantage of the pitcher plant. It could also increase the groundwater N concentrations.

Analysis

The pitcher plant seems to be resilient to disturbances in nearby land cultivation, pasturage, and groundwater fluctuations. There is much uncertainty about exactly how the pitcher plant has survived in the natural state. This uncertainty clouds our ability to extrapolate future survival likelihood in areas undergoing nearby development.

The major surface hydrological alteration was the construction of Lake Chatuge in 1941–42, followed by the drainage ditch in 1965. Runoff from the improved pasture may also have occurred past and present. The lake, if anything, has elevated the water table somewhat, favoring the pitcher plant. Raising the level to the estimated 500 yr level would submerge the area. The 1965 tile line may be draining the top of the aquitard/aquiclude if such exists. The spring at the tile outlet may be relieving pressure, reducing possible upward movement of water through the clay aquitard if such exists. The 1997 tile line is across the channel from the site. It is 2 ft above from the channel bottom. This should not affect the site. Most of the modifications affecting the site happened 30 yr previously. The pitcher plant community may have reached into the pasture before drainage. The population should have equilibrated to the pasture, assuming pasture management has been almost constant over the years.

Development of nearby property, assuming the pitcher plant community would not be disturbed, would probably have minimal impact owing to the existing drainage ditch. The ditch would intercept surface and above-aquitard groundwater originating from development around the existing property residence.

Future development of the surrounding property could be disadvantageous to the viability of the pitcher plant community. Increased surface and subsurface N levels may jeopardize the current competitive advantage now enjoyed.

Conclusions

The pitcher plant may be quite resilient, considering past events. The hydrologic modification to date should not in itself jeopardize the pitcher plant community, based on its viability since the 30 yr drainage. The caveat here is the uncertainty about the spring's age. The spring's impact on the moisture regime in the pitcher plant community is uncertain. Lake Chatuge could inundate the area with large regional precipitation events. One cannot rule out inundations, though infrequent. The most significant threat to the pitcher plant community may be the impact of future development. This would increase potential for higher N levels.

In summary, there is a good likelihood that past owners have not altered the surface and subsurface hydrology sufficient to cause certain death for the pitcher plant community. Our inability to comprehend spatial and temporal scale effects at work here complicates matters. The greater risk lies in likely future development on surrounding property.

The pitcher plant case study was an exercise in reverse ecological engineering. Most ethical positions in Table A.2 play into the problem. The analysis was an attempt to understand how the community arrived at its present state using the guiding principles of ecological engineering. The task force contained members from the Nature Conservancy and technical experts from several science and engineering disciplines. The potential buyer of the property considered scientific soundness and economic affordability, with the additional value that one should save the plants at all cost if possible. Thus, approaches 4b and 7 of Table A.2 seemed to permeate most of this debate. Other debates would no doubt bring other items to the forefront.

Case Study 2: Clean Water Act Compliance, a Hypothetical Case

The state agency for Protection, Licensing, Investigation, and Compliance—Environment (P'LICE) has determined that a certain stretch of a river in the state exceeds water quality standards for dissolved oxygen (DO).[2] P'LICE is one of those state agencies to which EPA has delegated the authority to administer the Clean Water Act program within its state borders. As part of that delegation, P'LICE submits lists of so-called impaired waters not meeting water quality standards to the EPA every two years.

The City of Blissful operates a wastewater treatment plant that discharges into the stretch of river P'LICE has listed as impaired for DO. This discharge is pursuant to an NPDES permit issued by P'LICE. One pollutant regulated under the City's permit is biological oxygen demand (BOD). The permit is coming up for renewal and P'LICE is threatening to make the BOD limits in Blissful's permit much more stringent to improve the DO content of the listed stretch of the river. Blissful's consultant estimates that meeting the new DO limit proposed by P'LICE will require construction of a $10 million addition to the wastewater treatment plant.

Agriculture Is Us, Inc., operates an animal feeding operation (AFO) that discharges nutrients into the affected stretch of river. In addition, there are homes on both sides of the listed stretch of river having old septic tanks known to be leaking. P'LICE currently regulates neither of these activities.

The mayor of Blissful held a town meeting to discuss the anticipated tightening of permit limits by P'LICE and the proposed $10 million expenditure, which would be funded by increased taxes on users of the wastewater treatment system and the issuance of bonds. The mayor expressed regret for imposing these additional user fees. He said he had no choice because Blissful needed the permit to continue to operate the plant, and failure to comply with new permit limits could result in severe fines. During the meeting, some of those present suggested that Agriculture Is Us, Inc., might be responsible for part of the problem. Representatives of Agriculture Is Us, Inc., responded that they were utilizing BMPs to control any pollution from their operations and that, in any event, more likely sources of the problem were Blissful's treatment plant and leakage from old septic tanks utilized by homeowners. The meeting ended in a shouting match among the various factions present.

How can one resolve this issue?

1. Collect data to determine the relative contributions of Blissful, Agriculture Is Us, and other sources such as the leaking septic tanks.
2. Determine if control of sources other than the Blissful treatment plant would result in sufficient improvement of the DO content in the river to allow removal from the impaired waters list.

3. Determine if one could achieve significant reductions in discharges of oxygen-demanding substances from sources other than the Blissful treatment plant at costs significantly less than the $10 million proposed for the treatment plant addition.
4. If the answer to item 3 is yes, work out an agreement under which Blissful would fund part or all of the cost of controlling these other sources.
5. Present this agreement to P'LICE with the request that P'LICE issue an administrative order making the terms of this agreement enforceable.

Anon. 1999 presents additional case studies.

Ethics, Ecological Knowledge, and Natural Resources Problem Solution

For a long time, solar energy was the only source of energy available to humans. Their tools were primitive and simple, and they had no effective means of transportation. As a result, humans were nothing but simple members of their natural environment, with little control or dominion. The results of their interactions with nature (e.g., hunting, fishing, food collecting) were insignificant and the ecological processes of nature rapidly balanced the effect (Manner and Mikesell 1974). Moreover, humans could never take more than what they returned to the system, because they had no concentrated form of energy to power their tools.

Discovery, extraction, and use of fossil fuels provided humans with an abundant source of energy and enabled them to explore aggressively and exploit nature. Intensive agriculture, production and use of synthetic materials, development of urban areas, and emergence of new transportation means were among the results of this discovery. Consequently, humans created their own subsystem within the larger natural system and then began to expand rapidly. This subsystem is energy intensive and produces large quantities of waste. Moreover, the human-made subsystem is resource dependent. The human system consumes large quantities of natural raw materials but does not return to nature as much as it takes from it. Practices such as intensive pest control, intensive fishing and hunting by modern aids, deforestation, dam construction, intensive agriculture, and extraction of minerals have reduced nature's biodiversity, interrupted many ecological interactions (loops), and blocked or altered many natural channels of energy.

This section raises questions on ethics and points out the vagueness in our knowledge of ecological principles and spatial-temporal scales. How does one balance the time value of money and the apparent vagueness in realizing various goals associated with environmental preservation? The section suggests that the market system is inadequate for defining the total scope of economic impact. In a pluralistic society with diverse belief systems and imperfect markets, debate of issues structured around scientific foundations is the best approach for consensus building. The fuzzy boundary between resource considerations and environmental management further complicates resolution of conflicts and suggests the likelihood of social traps. The pitcher plant situation discussed exemplifies the difficulty of perceiving the fuzzy boundary between resource management and environmental management. Does this plant have "standing" compared to animals and humankind? Some think it is deserving of "standing," and they allocate resources based on a perceived "standing." Years ago one would have perceived the boundary between saving the pitcher plant and economic activity as clear and distinct. Now, increasing numbers of people are perceiving value to the community because of its rarity. This value is hard to assign a monetary value, although for the pitcher plant the per hectare development value approaches one-quarter million dollars. Yet, because we do not clearly understand how the community succeeded in surviving to this point, we are uncertain about how development will affect the community. Thus, the fuzzy overlap appears. The fuzzy overlap condition can breed many ethical questions and social traps.

The dilemma faced by the hypothetical city of Blissful in case study 2 is identical to that anticipated by many cities on streams with mandatory TMDL requirements. The creative solution proposed may or may not work in the real environment. One must identify the fuzzy boundaries involved, the nature of expense (e.g., capital vs. variable costs), and the personalities of all the potential disputant individuals and groups. One must identify the boundary fluxes and sources/sinks of materials and information flow. Solutions may then emerge.

The emerging science of ecological design helps us appreciate the key role of time and space scales. The advantage of pulsing comes from ecological design. Chemical and biotic diversity are important for productive ecosystems. Special care to protect an ecosystem at its boundaries, or ecotones, is important. Resource economics will play an increasingly important role in allocating materials and in selecting appropriate forcing functions. Social traps will continue to blind us as to the nature of many problems and the benefits possible. Economists will no doubt continue to expose social traps (e.g., the wind erosion example discussed above) in the future.

The diverse spectrum of individual and collective values and of the resulting ethical positions explains why conflicts arise. Add to this the inherent vagueness of ecological principles and imperfections in the marketplace. Assuming society maintains the view that science-based solutions are best, opportunities for the natural resources engineer and other environmentalists will grow. Thucydides (c. 455–400 B.C.), commenting on the Athenian mood during the Peloponnesian Wars, said: "The ability to understand a question from all sides meant one was totally unfit for action. Fanatical enthusiasm was the mark of the real man." Francis Bacon (1561–1626) stated: "There is no great concurrence between learning and wisdom." Although a detailed study of ethical issues can be paralyzing if taken to extreme, wisdom demands we go forward, appreciating the importance of ethical considerations. Opportunities inspire hope!

Notes

1. This section was developed with the aid of Dr. Alireza Esteghlalian, postdoctoral associate, Department of Biological and Agricultural Engineering, Driftmier Engineering Center, University of Georgia, Athens.
2. Presented courtesy of Harvey A. Rosenzweig, TroutmanSanders LLP, Atlanta, GA.

References

Anon. 1999. National Agricultural Biotechnology Council BioEthics Institute Handbook. Distributed at the North Carolina State University BioEthics Institute Workshop, 22–27 May 1999, Raleigh, NC.

Bella, D. 1997. Class notes: Contemporary technology. Department of Civil Engineering, Oregon State University, Corvallis.

Bouchard, C. E. 1999. A model for successful community resource planning. ASAE Technical Paper no. 992260. American Society of Agricultural Engineers, St. Joseph, MI.

Devell, B., and G. Sessions. 1987. Deep ecology: Living as if nature mattered. Gibbs Smith, Layton, UT.

Esteghlalian, A., B. P. Verma, T. Foutz, and S. Thompson. 1999. A step-wise procedure for incorporating environmental and societal consideration into engineering design. ASAE Technical Paper no. 992257. American Society of Agricultural Engineers, St. Joseph, MI.

Geisler, N. L. 1989. *Christian ethics: Options and issues*. Baker Books, Grand Rapids, MI.

Hare, R. M. 1952. *The language of morals*. Oxford University Press, London. (A 1991 edition is available also.)

Hitt, W. D. 1990. *Ethics and leadership: Putting theory into practice*. Battelle Press, Columbus, OH.

———. 1998. *A global ethic: The leadership challenge*. Battelle Press, Columbus, OH.

Jorgensen, S. E., and W. J. Mitsch. 1989. Ecological engineering principles. In *Ecological engineering: An introduction to ecotechnology*, ed. W. J. Mitsch and S. E. Jorgensen. John Wiley & Sons, New York, NY.

Kosko, B. 1993. *Fuzzy thinking*. Hyperion, New York, NY.

Leopold, A. 1966. *A Sand County almanac*. Ballantine, New York, NY.

List, P. 1993. *Radical environmentalism: Philosophy and tactics*. Wadsworth Publishing Co., Belmont, CA.

Manner, I., and M. Mikesell, eds. 1974. *Perspectives on environment.* Association of American Geographers, Washington, DC.

Maxwell, J., and R. Costanza. 1989. An ecological economics for ecological engineering. In *Ecological engineering: An introduction to ecotechnology*, ed. W. J. Mitsch and S. E. Jorgensen. John Wiley & Sons, New York, NY.

Nash, R. F. 1989. *The rights of nature: A history of environmental ethics.* University of Wisconsin Press, Madison.

Odum, H. T. 1971. *Environment, power, and society.* John Wiley & Sons, New York, NY.

———. 1988. Self-organization, transformity, and information. *Science* 242:1132–39.

———. 1989. Ecological engineering and self-organization. In *Ecological engineering: An introduction to ecotechnology,* ed. W. J. Mitsch and S. E. Jorgensen. John Wiley & Sons, New York, NY.

Orr, D. 1994. *Earth in mind: On education, environment, and the human prospect.* Island Press, Washington, DC.

Palmquist, R. B., F. M. Roka, and T. Vukina. 1997. Hog operations, environmental effects, and residential property values. *Land Economics* 73(1):1114–24.

Perman, R., Y. Ma, and J. McGilvray. 1996. *Natural resource and environmental economics.* Longman Press, New York, NY.

Rachels, J. 1999. *Elements of moral philosophy.* 3d ed. McGraw-Hill, New York, NY.

Rand, A. 1971. *The new left: The anti-industrial revolution.* Penguin Press, New York, NY.

Seitz, W. D., G. C. Nelson, and H. G. Halcrow. 1994. *Economics of resources, agriculture, and food.* McGraw-Hill, New York, NY.

Stamatis, D. H. 1994. *Total quality service: Principles, practices, and implementation.* St. Lucie Press, Delray Beach, FL.

Tibor, T., and I. Feldman. 1997. *Implementing ISO 14000.* McGraw-Hill, New York, NY.

Underwood, L. 1998. *Case studies in environmental sciences.* Saunders College Publishing (Harcourt Brace), New York, NY.

Wann, D. 1996. *Deep design: Pathways to a livable future.* Island Press, Washington, DC.

Appendix B

Solutions for Trapezoidal Channels with Constrained Velocity

One may design channels with known flow rates to not exceed set velocities by constraining the hydraulic radius or slope. Hydraulic radius is the most often used approach. Knowing the velocity and flow rate, one can solve for area from the continuity equation. One can solve for the required hydraulic radius from Manning's equation knowing the slope.

Substituting values for area (Avl) and hydraulic radius (Rvl) into the hydraulic elements, one can solve for depth d and bottom width b. Side slope z is assumed known. One may use the quadratic formula to solve for depth:

$$d = \frac{-Avl\,(-z - 2\sqrt{1 + z^2}) + \sqrt{4\,Avl\,Rvl^2\,(4 + 3z^2)\,(-z - 2\sqrt{1 + z^2}) + Avl^2\,(-z - 2\sqrt{1 + z^2})^2}}{2\,Rvl\,(4 + 3z^2)},$$

$$d = \frac{-Avl\,(-z - 2\sqrt{1 + z^2}) - \sqrt{4\,Avl\,Rvl^2\,(4 + 3z^2)\,(-z - 2\sqrt{1 + z^2}) + Avl^2\,(-z - 2\sqrt{1 + z^2})^2}}{2\,Rvl\,(4 + 3z^2)} \tag{B.1}$$

The solution to choose is usually the smallest real value for depth d. One may substitute this into the hydraulic elements relationship for bottom width b:

$$b = -\frac{-Avl + d^2 z}{d} \tag{B.2}$$

where Avl and d are as previously defined.

Sometimes slope is used to control the velocity when the bottom width is determined (e.g., by available construction equipment). One may use the continuity equation to compute area as before. One may then use the definition of hydraulic radius and area to write a quadratic equation with the solution shown as

$$d = \frac{-b\sqrt{Vpset} + \sqrt{b^2 Vpset + 4QZ}}{2\sqrt{Vpset}\,Z} \tag{B.3a}$$

$$d = \frac{-b\sqrt{Vpset} - \sqrt{b^2 Vpset + 4QZ}}{2\sqrt{Vpset}\,Z} \tag{B.3b}$$

where

b is the preset bottom width;
Q is the flow;
Vpset is the permissible velocity;
z is the side slope;
d is the depth.

One normally chooses the shallowest depth. With a selected depth, one may compute the hydraulic elements, substitute them into the Manning equation, and solve for slope.

Appendix C

Structure Installation and Soil-Imposed Forces

This appendix discusses several topics related to soils that in themselves are not natural resources engineering topics per se. Many natural resources engineering problems require soils engineering. This appendix addresses cut/fill computations, techniques for structure/channel installation, and loads imposed on culverts and tiles by soils.

Impoundment and Embankment Volume Calculations

Computations for pond cuts and fills take maximum advantage of geometric regularity. One may visualize a pond dam as a series of trapezoids as shown in Figure C.1. Tables C.1 and C.2 organize the data needed for pond and dam volume calculations, respectively. The contour can also be a basis for geometric regularity, given the area enclosed by each contour along with the distance (interval) between the contours. This is shown for a water body in Figure C.2 and for a hilltop in Figure C.3.

The average end area formula is useful for calculating water and/or earth volumes. The average end area formula is

$$Vol = \frac{d_{inte}}{2}(A_1 + A_2) \qquad \textbf{(C.1a)}$$

where

Vol is volume of cut (m^3);
A_i (i = 1 or 2) is the end area (L^2);
d_{inte} is the distance between the end areas (L).

This formula is useful for computing water and soil volumes. With water, A_i represents the area enclosed by adjacent contours and d_{inte} is the distance interval between the adjacent contours. With embankment dam volumes, A_i represents the cross-sectional area of the dam at adjacent points separated by d_{inte}. Table C.1 summarizes the calculations needed for water volume calculations. One could also apply the approach in Table C.1 to soils when contour areas were available.

To calculate the soil volume from a borrow area, one may use identical approaches used in calculating water volumes. The required information would be the area surrounded by contours describing the proposed excavated area.

The cross-sectional area of the dam cross section is well approximated by a trapezoidal shape. The volume of a trapezoidal dam is calculated by applying the average end area formula to a

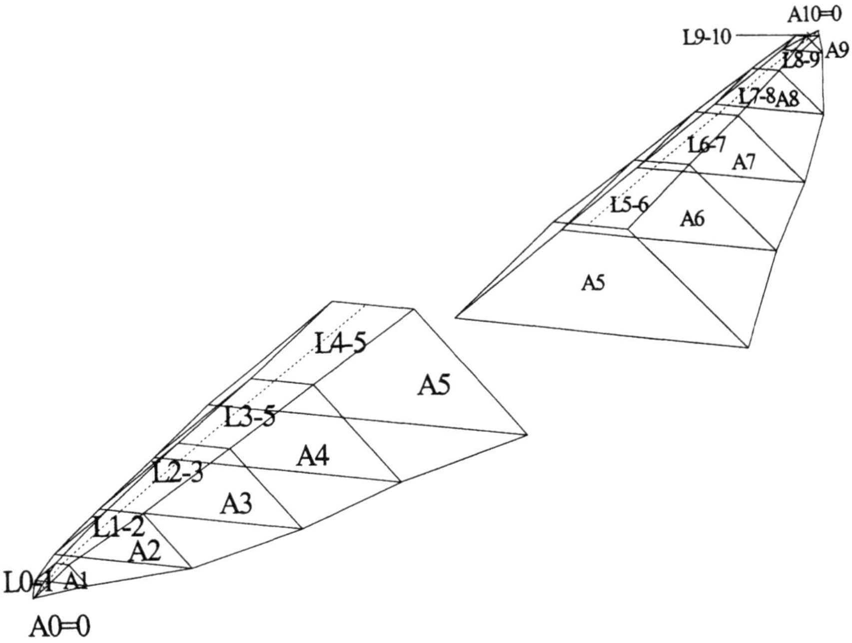

Figure C.1. Perspective sketch of hypothetical pond dam shown in two halves. Note the trapezoidal cross sections. Volume can be found knowing the area of each trapezoid and the length along the dam between the trapezoids and applying the average end area formula, as indicated for the dam in the figure. The number of sections depends on the particular situation.

Table C.1. Calculation template for computing water volumes.

Stage Inc = d (Col A) (m)	Interval (Col B)	Water surface planimeter reading (units) (Col C)	Area (Col D) (m^2)	Volume in slab "I" (Col E) (m^3)	Cum. volume (Col F) (m^3)
Contour elev 1		0	0	0	0
Contour elev 2	Col A(i)– Col A($i-2$)	. . .	. . .	Equation C.1a where d_{inte} = Col B and areas A1 and A2 given in D($i-1$) and D(i)	Col F($i-1$) + Col E(i)
Contour elev 3		. . .	. . .	. . .	
. . .		. . .	. . .	. . .	
Max. contour elev.		. . .	. . .	. . .	

Col A: Predetermined depths at which volume increments will be computed. Available contours are used.
Col B: Planimeter reading or computed using a computer graphics package.
Col C: Convert Col B to area units if necessary.
Col D: Apply average end area formula to previous area and current area to get the volume increment.
Col E: Sum the volume increments to arrive at the total volume.

Note: For a levee or excavated pond, one may use equation C.1d or C.1e instead of this tabular computation.

Table C.2. Calculation template for earth embankment volume.

Station along center line of dam[a] (Col A)	Height of dam at this station[b] (m) (Col B)	Cross-sectional area[c] (m^2) (Col C)	Avg cross-sectional area (m^2) (Col D)	Length of section (m) (Col E)	Volume of section (m^3) (Col F)
0 (measured) Station $i = 0$	0 (measured)	0 use equation C. 4b or C. 4c	—	—	—
. . . Station i	. . .	. . .	[Col C($i-1$) + Col C(i)]/2	(Col A(i)– Col A($i-1$))	(D(i) × E(i))
. . . Station $i = i_{max}$	. . .	. . .	. . .	. . .	. . .
					$\sum$ = Sum Col F (1)... F(i_{max}) + volume in core

Note: Soil volume in core trench = average depth × average width × average length.
[a]Measured based on scaled topographic map.
[b]Determine the elevation of the top of the dam from the contour map. Then determine the elevation of the bottom of the dam at the station. The dam height is the difference.
[c]Assumption: $Z = 3$ inside the dam and $Z = 2$ on the back of the dam if equation C.4c is used.

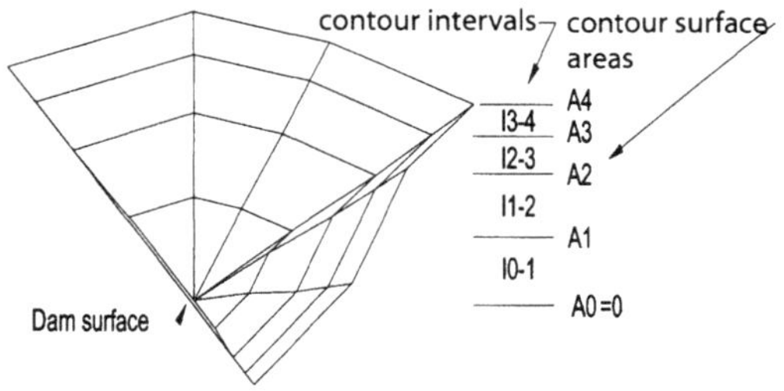

Note: Measure area enclosed by each contour as viewed from top

Figure C.2. Perspective view of water body with contours denoting water level as a function of elevation. The water volume can be computed with the average area formula knowing contour intervals and surface area associated with each contour.

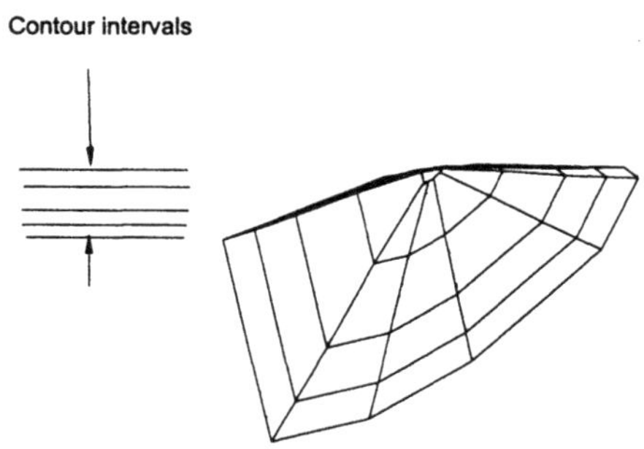

Note: Measure area enclosed by each contour as viewed from top.

Figure C.3. Perspective view of small hill showing contour lines for soil volume calculation with the average end area formula.

series of trapezoidal cross sections of the dam as shown in Figure C.1. The area of a trapezoid (A_i) is given by

$$A_i = H_{emb} \frac{W_{botw} + W_{topw}}{2} \quad \textbf{(C.1b)}$$

where

W_{botw} is the bottom length of the trapezoid as read from the topographic map;
W_{topw} is the top width of the trapezoid as determined from equation 12.8;
H_{emb} is the height of the embankment as read from the topographic map.

Equation C.1c provides an alternative approach for computing the area of the trapezoidal cross section:

$$A_i = \left(\frac{Z_{front} + Z_{back}}{2} \right) H_{emb}^2 + W_{topw} H_{emb} \quad \textbf{(C.1c)}$$

where

Z_{front} is the side slope ratio for the submerged dam slope (typically $Z_{front} = 3$);
Z_{back} is the side slope ratio for the back of the dam (typically $Z_{back} = 2$).

Table C.2 shows an embankment volume calculation.

An alternative volume formula, the Culp (1948) cone frustum approximation (as modified for general units by Schwab et al. 1993) for watershed-type ponds, is

$$Vol = A_s d + \frac{177 d^2 A_s^{1/2}}{S} \quad \textbf{(C.1d)}$$

where

A_s is the area (L^2) of the pond surface at a lower reference contour;
d is the depth of water above the reference contour;
S is the average slope of the reservoir sides through the range of d in percent;
Vol is the volume (L^3).

The cone frustum approximation was originally developed for quantifying flood storage, and the reference contour was originally the area of water surface with the stage at the principal spillway crest.

The volume of levees and similar structures that are used for storage, lagoons, and aquaculture and that have regular side slope values may be described using an implementation of the prismoidal formula (Crites and Tchobanoglous 1998):

$$V = [LW + (L - 2Zd)(W - 2Zd) + 4(L - Zd)(W - Zd)]d/6 \quad \textbf{(C.1e)}$$

where

V = cell volume (L^3);
L = levee length (L);
W = levee width (L);
Z = slope factor ($Z:1$ is horizontal unit distance to 1 vertical unit);
d = depth of levee (L).

When moving soil from one state to another, the volume of solids is a conserved quantity. The soil volume can change due to varying compaction between the borrow site and the fill location. Example 12.5 gives an example of dam and water volume calculations. One can visualize the relevant contour areas and intervals on a contour map. Similarly, one can visualize the cross-sectional trapezoids including respective bottom widths, top widths, and embankment heights for each trapezoid from topographic maps. Visualization comes with practice.

Example C.1: Suppose an earthen dam with a volume of 1000 m^3 and void ratio of 0.5 was under construction with soil from a borrow area with a bulk density of 1.3. What borrow volume is required?

The void ratio of the borrow area soil is computed by first getting the porosity, then computing the void ratio:

$$n_p = 1 - BD/PD = 1 - (1.3 \text{ Mg/m}^3)/2.5 \text{ Mg/m}^3 = 0.48$$

$$e = n/(1 - n) = 0.48/(1 - 0.48) = 0.92$$

Get the solids volume:

$$Vt = Vs + eVs$$

$$1000 \text{ m}^3 = Vs + 0.5Vs = 1.5Vs$$

$$Vs = 1000 \text{ m}^3/1.5 = 666.7 \text{ m}^3$$

We take the same solids from the borrow area. Therefore,

$$Vb = eVs + Vs$$

$$Vb = (0.92 + 1)(666.7 \text{ m}^3) = 1280 \text{ m}^3$$

Surveying skills are essential to doing quality land-forming calculations and installations.

Example C.2: Suppose a soil volume of 144.45 m^3 existed in a borrow area with a void ratio of 1.1. Further, the volume fills a depression to be compacted to a void ratio of 0.6. What depression volume would this soil occupy?

The volume of soil solids in the borrow pit must equal the soil solids volume in the compacted structure. First, compute volume of solids:

$$V = 144.45 \text{ m}^3 = Vs + Ve = Vs + 1.1Vs = 2.1Vs$$

This yields $Vs = 68.79$ m^3. The final volume is given by

$$V = 68.79 \text{ m}^3 + 0.6(68.79 \text{ m}^3) = 110.06 \text{ m}^3$$

Setting Slope Stakes for Cuts and Fills

Slope stakes are helpful for guiding construction equipment for ponds and channels. One may estimate slope widths using the slope stakes equation:

$$Dis = \Delta El\, Z + \frac{b}{2} \qquad \textbf{(C.2)}$$

where

Dis is the distance from centerline (CL) to edge of fill or cut (L);
ΔE is the difference in elevation from the stake to the bottom of ditch or top of fill (*L*);
Z is the side slope ratio;
b is the bottom width of the channel (in the case of a dam, replace *b* with W_{TW}, the top width of the dam).

Figure C.4 shows the slope stakes equation nomenclature. First, stake the fill centerline (CL). Slope stakes are then set for a cut by determining distance from the surface to cut bottom at the centerline, tentatively placing a stake, then determining the elevation of the position above or below the stake. Combine the distance to bottom and tentative slope stake elevation to get elevation from cut bottom to slope stake. Measure horizontal distance to the stake from the centerline. Compute the distance with the slope stakes equation. If the computed distance exceeds the measured distance,

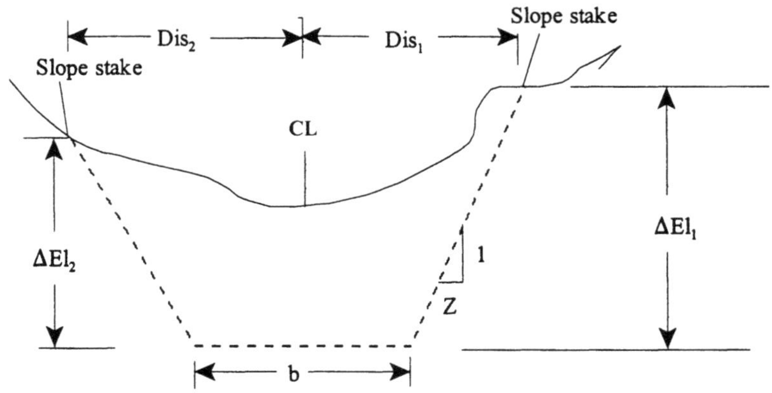

Figure C.4. Slope stake definition sketch for a cut. For a fill such as a dam, the sign reverses on ΔEl_1 and ΔEl_2.

move the slope stake farther out from the centerline. If the computed distance is not enough, move the slope stake inward. Redo the elevation measurements and calculation until they agree within 5% or so. For a fill such as a dam, Figure C.5 illustrates the required measurements.

TECHNIQUES FOR INSTALLATION OF CHANNELS AND OTHER STRUCTURES

Stake the centerline before construction. Tentative slope stakes are set and elevations established from the design fill height to the tentative slope stakes. Measure and compare horizontal distance from the centerline to the tentative stakes with the calculated value. If calculated value exceeds measured value, move the stake outward. If the calculated value is less than the measured value, move the stake inward. After each move, redo the elevations and recalculate. Stop when measured and calculated values are within 5%.

Plots of the original valley floor along the proposed fill centerline at stations of 15–30 m are helpful in determining first estimates of slope stake positions. In addition, one can make volume calculations from these plots.

Layout of Circular Curves

With channels designed with a constant slope, changes in direction will be necessary, requiring gradual curves. One specifies the curve sharpness with a degree of curvature. Curvature may be as high

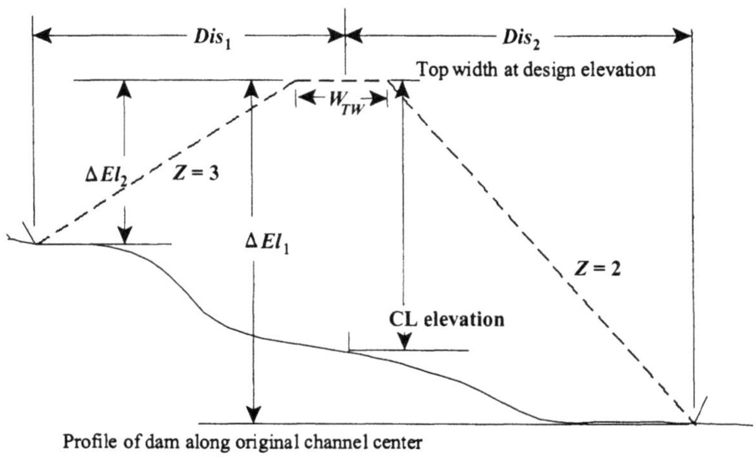

Figure C.5. Nomenclature for setting slope stakes for pond dam.

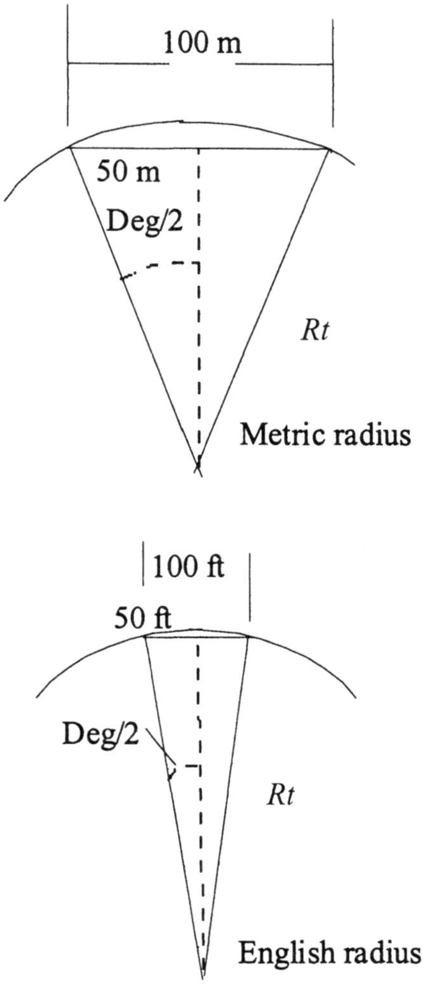

Figure C.6. Schematic for defining radius of curvature for English and metric units.

as 60 degrees for small channels and lower for larger channels. See Chapter 8 for additional discussion of curvature. Curvature is defined for English and metric units as shown in Figure C.6. One may compute the radius of curvature to degree of curvature using

$$Rt = 50/\sin(Deg/2) \tag{C.3}$$

where

Rt is the radius of curvature in m or ft;
Deg is the degree of curvature in English or metric units.

In laying out a curve of known degree and computed radius of curvature, set up the transit or total station at the point of intersection as shown in Figure C.7.

Calculate the tangent distance using

$$Tt = Rt \tan\frac{I_i}{2} \tag{C.4}$$

where I_i is the angle of intersection.

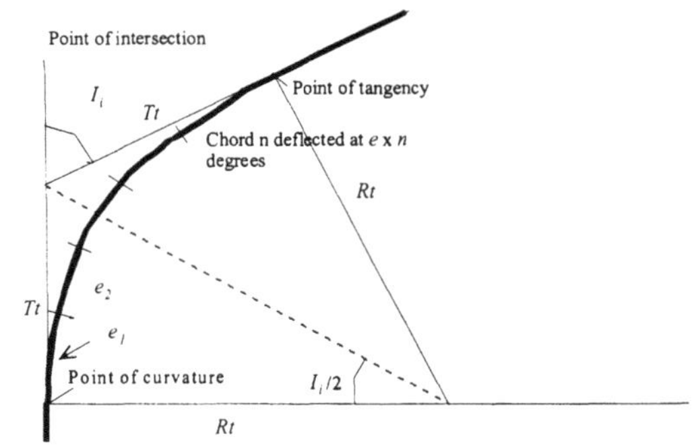

Figure C.7. Definition sketch for laying out a circular curve.

Move the instrument to the point of curvature (or point of tangency), which is T away from the point of intersection. Next, establish chord distances to points along the curve. Establish a chord length of 50 or 100 units. Calculate the deflection angle (e_1) to position the other end or the chord using

$$e_1 = c_1 Deg/200 \quad \textbf{(C.5)}$$

where

Deg is the degree of curvature as defined before;
c_1 is the chord length;
e_1 is the deflection angle to point marking the end of the first chord.

For the second chord deflection, double the angle and chain from point e_1. For successive deflections, add e_1 to the previous angle and chain from the previous point. There will be 2 deflections if the chord length is 100 units, and 4 deflections if the chord length is 50 units. The total length of the curve is given by

$$L = 100I/Deg \quad \textbf{(C.6)}$$

Nearly flat topography is assumed in equation C.6.

Example C.3: Design a 30 (SI) degree curve if the angle of intersection is 50 degrees. Applying equations C.3, C.4, and C.6 gives $R = 50/\sin(30/2) = 193.18$ m; distance $T = 190.18 \tan(30/2) = 90.08$ m; curve length $= 100(50/30) = 166.67$. Select a chord length of 50 units. We show the chord length and deflections in Table C.3.

Table C.3. Chord length calculations.

Deflection	Degrees (SI) (from eq. C.5)	Unit chord length (m)	Cumulative chord length from point of curvature or prior point (from eq. C.6) (m)
e_1	7.5	50	41.67
e_2	15	50	83.33
e_3	22.5	50	125
e_4	30	50	166.67

Small Channels or Minor Topography Modification

Grade establishment in the field is important for channel installation and for profile modification. Figures C.8 and C.9 show methods for grade establishment. Cut depths are shown on stakes. Placing lime in an auger hole along with the hub stake is a way to convey to equipment operators when the cutting blade has approached the channel bottom. Laser grading techniques also help equipment operators grade accurately. Fill stakes are left as long as possible, piling soil in the vicinity. For channels or dams, the slope stakes are left undisturbed through construction.

Land cut-and-fill design operations should be conducted with frequent interaction between the engineer and the contractor. Laser grade control is now the state of the art for jobs other than small projects. Subsurface drainage installation depends heavily on laser grade control.

Lay out the channel using the following steps:

1. Get an accurate topographic map and soils characterization of the field site or, at a minimum, a profile survey of the channel route.
2. Locate the channel on the map and determine soils characterization.
3. Plot channel bottom with respect to distance along channel and compute the desired slope.
4. Plot surface elevation with respect to distance along channel.
5. Plot the channel cross section 15–30 m apart; get distance from centerline to slope edges.
6. Plot slope stakes on the map (plan view).
7. Use average end area or prismoidal formula for computing volume of cuts.
8. If one anticipates using part of the spoil for filling a depression, one must know the void ratio of the existing material and the void ratio of the material after placement in the depression. Void ratios of existing material in agricultural soils range from 0.9 to 1.2. Void ratios of compacted material may be lower than 0.4.

Be careful to control inflow to a newly constructed vegetated channel. The channel is most vulnerable at construction time. Before vegetation emerges, one should recommend some sort of mulch and fiber such as jute or coconut mesh for stabilization. The greatest risk of failure is during initial cover establishment. Seeding should occur at optimum time for selected vegetation. Cost depends largely on earthmoving costs. Maintenance costs are commonly 5–10% of construction cost per year. One must take precautions to avoid failure after establishment due to livestock-induced breaching; farm implement-induced breaching; rodent-induced failure by leaking; breaks in sod cover; and sediment accumulation.

Terrace Construction and Layout

A tripod level and application of surveying techniques are required to lay out a terrace system. Stakes are normally set every 15 m along the proposed channel (Schwab et al. 1993). Appropriate staking provides grade in the channel. One calculates the difference in vertical interval along each proposed terrace needed to achieve the design terrace channel slope. A 15 m interval is suggested. In the field, one sets the channel stakes every 15 m along points corresponding to the calculated vertical interval measurements. Terrace channel bottoms should intersect the outlet waterway bottom. One must design the outlet after layout of the terrace system.

The terrace depth of cut for a uniform-grade terrace is given by

$$\text{Depth of cut} = 0.5(h + SW/100) \tag{C.7}$$

where

h is the height of the terrace;
S is the land slope (%);
W is the specified width of the frontslope, often a specified machine width.

Terraces may be made parallel by adjusting the location of the stakes set for the uniform grade. This may require deeper cuts and more fills. Beasley et al. (1984) suggest plotting each channel

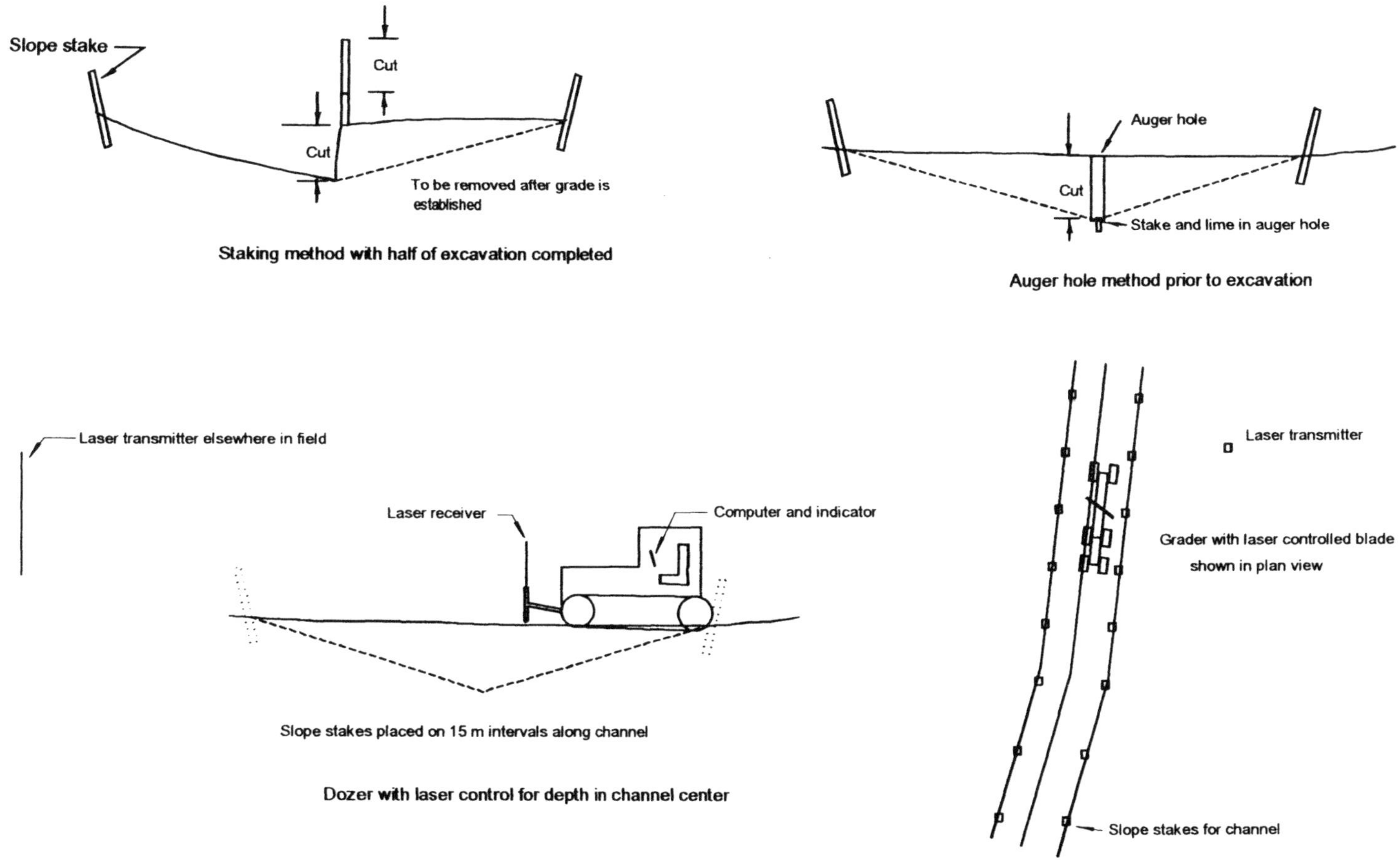

Figure C.8. Traditional and modern methods of establishing grade in the field.

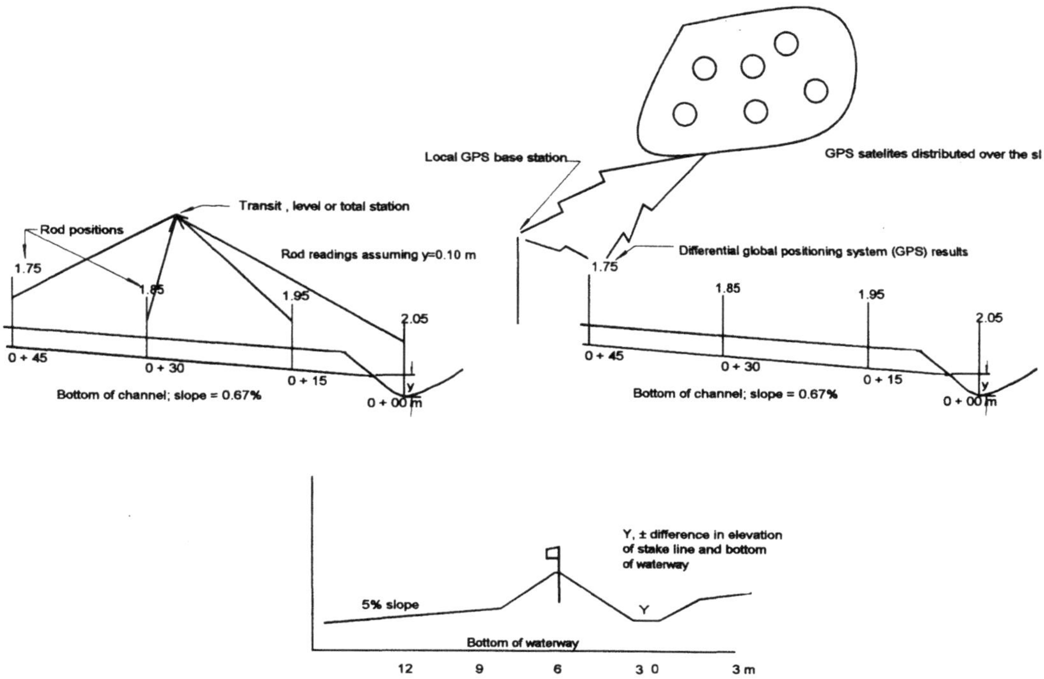

Figure C.9. Layout procedure for graded terraces with a grassed waterway outlet using conventional surveying equipment or global positioning system (GPS) equipment. (Modified from Schwab et al. 1993.)

bottom elevation, ground surface elevation, and the profile elevation after construction. Alterations in the profile determine the depth of cuts and fills for the design. One should try to balance the cuts and fills for each terrace. Erosion and equipment width specifications govern spacing of parallel terraces, as discussed by Beasley et al. (1984).

Projects greater than the field scale (i.e., farm scale or larger) require detailed consideration of construction techniques for economic and engineering reasons. Channel design must factor in available construction resources and how they mesh with site topography. Available construction resources often determine how excavation will occur and how spoil is deposited. Engineers on site during construction routinely check soils, grades, and compaction.

Begin the design by breaking the problem down into an excavation analysis and then transport/spreading analyses. Equipment listed in Table 8.7 is useful for ponds in addition to channels. Figures C.10–C.13 show typical pond and channel construction equipment. Works such as Stephens 1991 and Harris 1994 present extensive discussion on ground engineering and construction techniques. Local offices of the NRCS may also provide information about the availability of various types of construction equipment in a region. Associations such as the Land Improvement Contractors Association (LICA, http://ourworld.compuserve.com/homepages/delmarva/staff.htm.), International Commission on Irrigation and Drainage (ICID, http://cirrus.hrwallingford.co.uk/g7icid.htm), ASAE (http://www.asae.org), and ASCE (http://www.asce.org) also link to helpful information. The Internet site http://www.drainpro.com/resource.htm provides excellent links to earthmoving equipment dealers and selected contractors in regions of the United States.

Materials Specifications for Channels and Structures

Hydraulic structures may be fabricated using polyethylenes (PE), polyvinylchlorides (PVC), various epoxies and other plastics, metal, or concrete. Most of the PVC and related plastic materials are prefabricated and available off the shelf. Manufacturers can provide much information on

Figure C.10. Crawler shovel with pivoting scoop for finish work.

corrosion resistance and field durability. They should provide test results (American Society of Testing and Materials, http://www.astm.org) on request. One should write ASTM or similar test results and expectations into contracts related to structure installation. Corrosion, durability in the presence of sunlight, temperature (especially with PVC and PE materials), pressure requirements, brittleness, expansion-contraction (with temperature), and crushing resistance are important. One must specify minimum allowable hydraulic conductivity for pond liners (e.g., clay blankets, HDPE, other geosynthetic materials). Relevant ASTM material specifications should be included. Metals used in structures range from various steels to galvanized metal products. One can purchase prefabricated weirs and flumes made of metal off the shelf. Vendors and manufacturers can provide ASTM test results. One should clearly write expectations of these results into project contracts. The American Steel

Figure C.11a. Installation of a geosynthetic liner in a structure. Surface is prepared by compacting with a smooth-wheeled roller.

Figure C.11b. Placement of soil cover over the geosynthetic liner. Note the laser guide pole on the dozer blade for grade control purposes.

and Construction Institute (ASCI, http://www.aisc.org) provides excellent publications relating to reinforcing steel design and materials specification.

Concrete specification for structures built on site should specify air content and consistency, mix design, inspection and testing of concrete, specification for mixing, conveying, placing, consolidating and curing, preparation and removal of forms and fabrication and placing of reinforcement. The American Concrete Institute (http://www.aci-int.org) and Portland Cement Institute (http://www.portcement.org) provide excellent publications and guidance for specifying concrete.

Standards organization such as ASAE (http://www.asae.org) and ASCE (http://www. asce.org) provide excellent resources on hydraulic structure installation and maintenance. Local offices of the NRCS also provide excellent assistance. Works such as Harris 1994 and Stephens 1991 contain

Figure C.12. Construction of a drainage ditch demonstrating the types of obstacles commonly found in urban areas. Backhoes and shovels are frequently used, although a small crane-dragline is in use here.

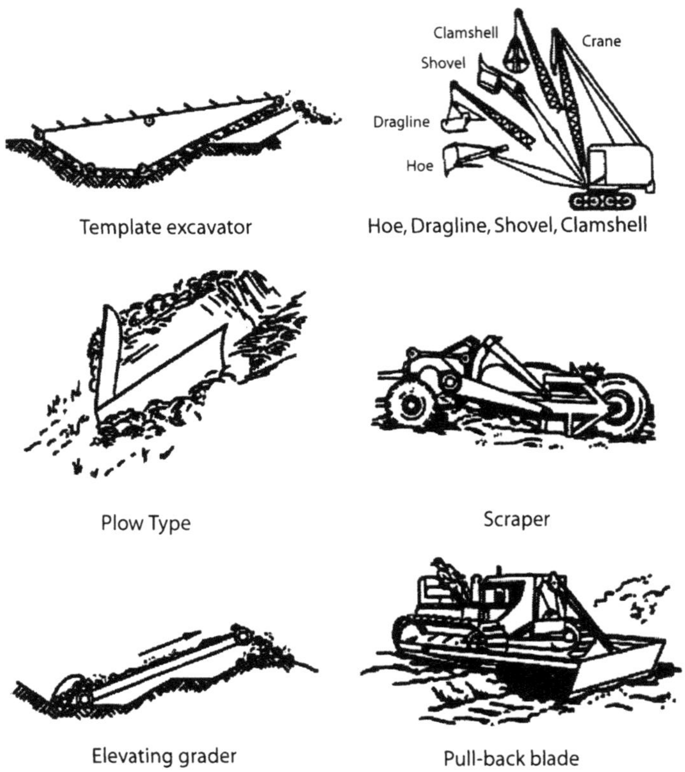

Figure C.13. Miscellaneous construction equipment for open channels. (From Schwab et al. 1993; used with permission of John Wiley & Sons.)

useful information on construction and siting. One may also contact local NRCS offices for additional construction guidelines.

Rural Road Construction

One should construct access roads wide enough to accommodate all equipment and personnel transport needs. One should minimize switchbacks. If the road is narrow, there should be turnouts designed as needed. Speeds for rural roads are often limited to 20–25 km/hr (30–40 mph). Lane width may vary from 3 to 4 m depending on traffic and right-of-way constraints. Shoulders should be 2 m or more. Slope along the line of travel should not be greater than 20% if possible. Speeds should not exceed 15 km/hr on steep slopes and otherwise should not exceed 60 km/hr. Further details on road construction may be found in Garber and Hoel 1999. Figure C.14 shows a typical road profile.

The key to minimum maintenance is to provide adequate drainage. Drainage ditches should be at least 2 ft below the road edge. This ensures the road base will not be in a sustained wet, plastic condition

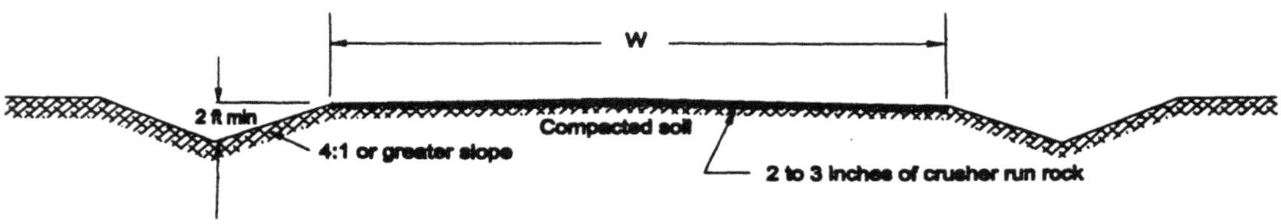

Figure C.14. Typical rural road profile with camber for two-sided drainage. Surfaces may be sloped to drain in one direction also.

having low bearing strength. Slightly crown the surface to enable surface drainage. Crushed rock provides additional surface durability for heavily traveled roads. The width W is determined from equipment width and passage requirements. The 4 : 1 slope requirement may be relaxed in cases with low traffic.

In addition to drainage, the possibility of frost, substratum composition, and traffic loading patterns also influence road construction specifications. Van Lier et al. (1999) give a nomograph relating axle weight and California bearing ratio test results to a strength index. The strength index correlates to various pavement compositions. Traffic, precipitation, gradient, and sun exposure also correlate to the need for pavement versus unpaved surface.

Promising chemical amendments (e.g., various polyacrylamids) are under investigation for use on forest logging roads. Amendments reduce susceptibility to erosion and increase surface strength until chemical breakdown occurs.

Buried Pipe Loading

Most flow and drainage structures are on or near the soil surface. Most structures will not have surface-imposed loads due to vehicular traffic. The small pipe culvert is an exception. Drainage and subirrigation pipe (Chapter 14) is placed at least 0.6 m below the surface to avoid crushing by vehicular traffic. In instances of large culverts or underground containments, one should consult a soil mechanics text for additional techniques for loading evaluation.

The purpose of presenting the pipe loading discussion is to show the importance of using acceptable quality materials and to stress using proper installation techniques. Improper materials and poor installation practice can result in physical failure. Physical failure in turn leads to hydraulic failure. This failure progression holds for all natural resources engineering structures.

The load on a buried pipe has two components: static loading due to the soil plus the static-dynamic loading possible by surface forces. One may adjust the surface force to account for dynamic effects (e.g., moving machinery). The soil static loading is computed using equations C.8, C.9, or C.10 (Schwab et al. 1993):

$$W_c = C_d \gamma_s B_d^2 \quad \textbf{(C.8)}$$

$$W_c = C_c \gamma_s B_c^2 \quad \textbf{(C.9)}$$

$$W_c = C_d \gamma_s B_d B_c \quad \textbf{(C.10)}$$

where

B_d is the width of the trench (m);
B_c is the outside diameter of a projecting pipe;
γ_s is the unit weight of the soil (N/m^3);
C_d is the load coefficient in trench conditions;
C_c is the load coefficient in projecting conditions.

Equation C.8 is applicable when the nonflexible pipe is placed in a trench less than 3 times the pipe diameter. Projection loading is applicable when trench side walls do not effectively support the weight of the backfill. Projecting conditions (equation C.9) are assumed when the trench for a pipe is greater than 2 or 3 pipe diameters. Projecting conditions apply when the pipe is placed on the surface (i.e., infinitely wide trench), then covered with a soil mass, with or without a projecting end. Equation C.10 is used for flexible tubing. Obtain the load coefficients for equations C.8–C.10 from Figures C.15 and C.16.

Surface static/dynamic loads are determined using

$$W_t = \frac{1}{L} I_c C_t T \quad \textbf{(C.11)}$$

where nomenclature is as defined on Figures C.15 and C.16. The total load on a pipe is the sum of the relevant soil static loading plus the surface static-dynamic component.

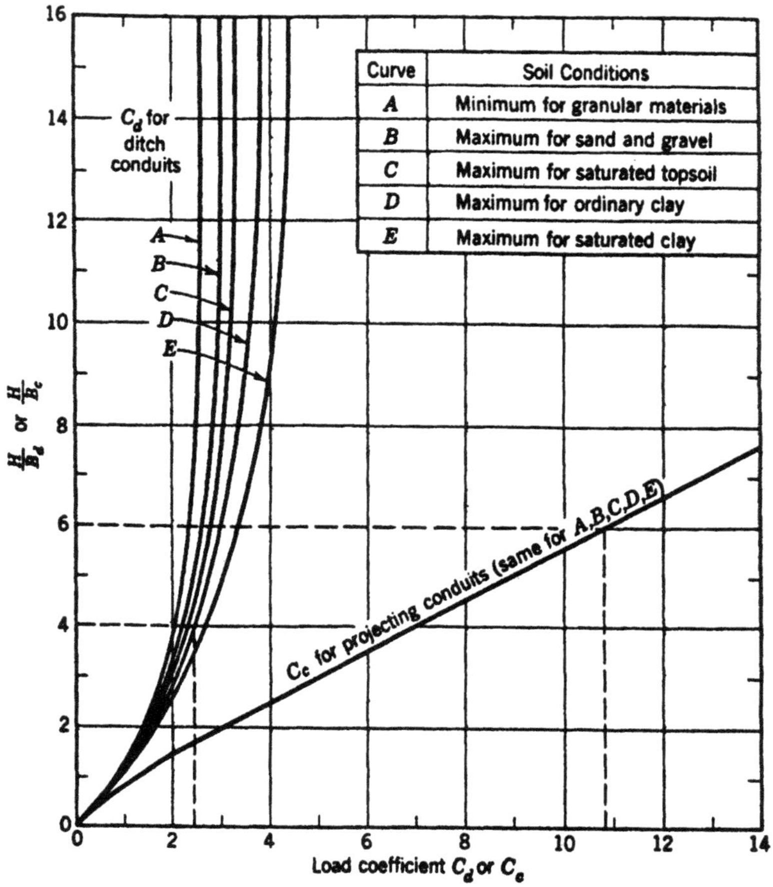

Figure C.15. Load coefficients for underground tubing due to overburden weight for indicated soil conditions. (From Schwab et al. 1993; used with permission of John Wiley & Sons.)

Pipe strength of flexible pipes is evaluated using a parallel plate deflection test. Deflection is the failure criterion. A section of pipe is placed between two parallel plates and loaded. One measures deflections as a function of load until pipe collapse or deformation exceeds a preset value. Design safety factors (load factors, L.C.) come into play when installing the pipe in a trench conforming to the tile shape. Figure C.17 shows accepted load factors resulting from indicated installation approaches.

An alternative approach for estimating the contributions of surface loads is the Boussinesq equation:

$$p = \frac{3d^3P}{2\pi\sqrt{x^2 + y^2 + d^2}} \qquad \textbf{(C.12)}$$

where

d is distance below surface (L);
x, y are offsets from directly over the point (L) (see Figure C.18);
P is the surface load (F);
p is the unit pressure due to P at point d (F/L^2).

The Boussinesq equation is useful when the load P is not directly over point d. One should use safety factors of 2 or 3 with the Boussinesq equation. Surface and soil loadings calculated from equation C.12, divided by the load factors in Figure C.17 should not exceed bearing capacity values in Table C.4 for the particular pipe. This condition is easily satisfied in most culvert and drainage situations.

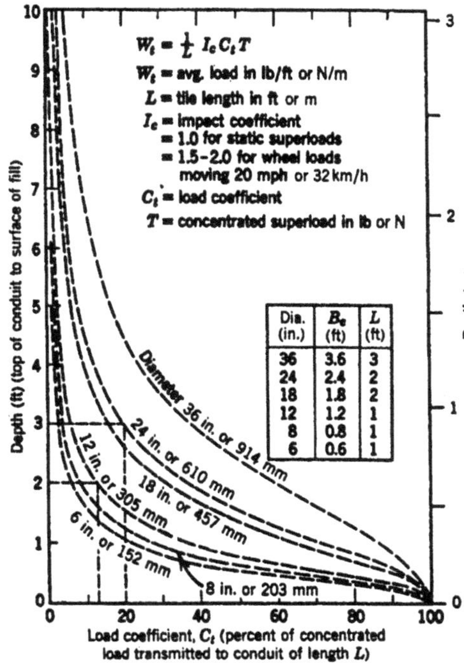

Figure C.16. Load coefficients for underground tubing due to concentrated surface loads. (From Schwab et al. 1993; used with permission of John Wiley & Sons.)

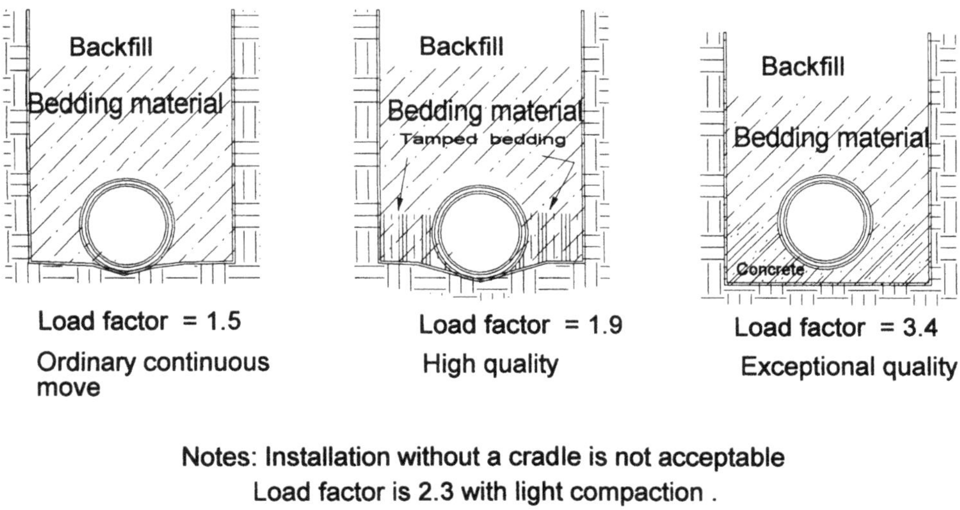

Figure C.17. Schematic showing installation load factors resulting from various acceptable pipe installation approaches. The design bearing strength must be greater than or equal to the computed design load after dividing the design strength by the load factor. Most tile drainage installations have a load factor of 1.5. Properly installed culvert installations have load factors of 1.9 or higher. (Redrawn from Spangler 1947.)

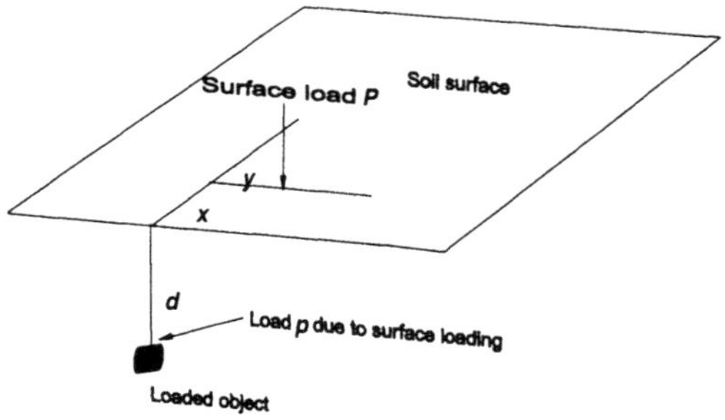

Figure C.18. Definition sketch showing nomenclature for Boussinesq surface load equation, useful for loads offset x and y from the point directly over the loaded object.

Table C.4. Representative loads or strength moduli causing failure for indicated pipe or material types. The design loads divided by the installation load factor must not exceeed the design bearing load tabulated below.

Pipe type	Bearing load, rigid pipes (N/m)[a]
Standard clay drain tile	11,680 for dia up to 305 mm, 12,700 dia = 381 mm; not available any larger
Extra-quality clay drain tile	16,060 for dia up to 305mm; increases to 23,360 (dia. = 610 mm)
Heavy-duty clay drain tile	20,440 (dia 152mm or less) to 43,800 for dia = 610 mm
Standard concrete drain tile	11,680 for dia up to 305 mm; not available for larger diameters.
Extra-quality concrete drain tile	16,060 (dia up to 381 mm) to 23,360 (dia = 610 mm)
Reinforced concrete culvert pipe	140,000 to 166,000
16 Ga corrugated steel pipe	244,550 (8 mm rivits single) to 419,020 (9.5 mm rivits double)
12 Ga corrugated steel pipe	341,640 (9.5 mm rivits single) to 774,000 (8 mm rivits single)
8 Ga corrugated steel pipe	374,000 (9.5 mm rivits single) to 1,032,000 (9.5 mm rivits single)
Geocomposite materials	24,000 to 290,000 per meter of width
Aluminum pipe 101 mm diameter, 1.58 mm thick	32,000
	Stiffness modulus $E_{s10\%}$ (N/m^2)[b]
Standard-duty corrugated tubing	130,000
Heavy-duty corrugated tubing	175,000
Plastic pipe[c]	ASTM D3033 275,600 to 351,000 ASTM D3034 454,740 to 565,000 ABS Composite D2680 1,364,000 to 1,736,000 Smooth PE 110,000

Sources: Schwab et al. 1993 (corrugated tubing, clay and concrete values); AISI 1967 (steel pipe values); Koerner 1990 (geocomposites); Spangler and Handy 1989 (reinforced concrete); Chambers et al. 1980.

[a]Use loading formulas only for preliminary failure evaluation. The Figure C.17 load factor divided by the total of surface and soil loading should not exceed the bearing loads tabulated below.

[b]Use relevant loading formulas and the Iowa formula with soil modulus and pipe modulus to check if deflection exceeds 10%.

[c]Installation technique is critically important here. Pipe supported by small aggregates will fail.

Notes: Failure mechanism is joint failure for steel pipes and brittle shattering for clay and concrete pipe. The values for corrugated steel pipe hold for spotwelded and helical lock seam fabrication also. See AISI 1967 for the effect of deflection on failure of steel pipes.

References

AISI. 1967. *Handbook of steel drainage and highway construction products*. American Iron and Steel Institute. New York, NY.

Beasley, R. P., J. M. Gregory, and T. R. McCarty. 1984. *Erosion and sediment pollution control*. Iowa State University Press, Ames, IA.

Chambers, R. O. E., E. G. McGrath, and F. I. J. I. Hager. 1980. *Plastic pipe for subsurface drainage of transportation facilities*. National Cooperative Highway Research Program Report 225. Transportation Research Board, National Research Council, Washington, DC.

Crites, R., and G. Tchobanoglous. 1998. *Small and decentralized wastewater management systems*. McGraw-Hill, New York, NY.

Culp, M. R. 1948. The effect of spillway storage on the design of upstream reservoirs. *Agricultural Engineering* 29 : 344–46.

Garber, N. J., and L. A. Hoel. 1999. *Traffic and highway engineering*. 2d ed. PAWS Publishing, New York, NY.

Harris, F. 1994. *Modern construction and ground engineering equipment and methods*. 2d ed. Longman Publishing Group, White Plains, NY.

Koerner, R. M. 1990. *Designing with geosynthetics*. Prentice Hall, Englewood Cliffs, NJ.

Schwab, G. O., D. D. Fangmeier, W.J. Elliot, and R. K. Frevert. 1993. *Soil and water conservation engineering*. 4th ed. John Wiley & Sons, New York, NY.

Spangler, M. G., and R. L. Handy. 1989. *Soil engineering*. 4th ed. Addison-Wesley, New York, NY.

Stephens, T. 1991. *Handbook on small earth dams and weirs: A guide for siting, design, and construction*. Paul & Company Publishers Consortium, Concord, NH.

van Lier, H. N., L. S. Pereira, and F. R. Steiner. 1999. *CIGR handbook of agricultural engineering*. Vol. 1, *Land and water engineering*. American Society of Agricultural Engineers, St. Joseph, MI.

Appendix D

Land-Forming Computations for Surface Drainage

Many farm and nonfarm land uses require or benefit from land forming. Slope modification is the primary objective. Land forming is important for some practices such as surface and level basin irrigation, drainage, and farming specialized crops such as rice. Preparation of a field for level basin irrigation is shown in Figure D.1.

In some applications, land forming is done to create a uniform slope while reducing cut/fill transport. In other applications land forming is done to create a predetermined slope value at a given location. Cut-and-fill balancing via a survey usually precedes the actual land-leveling operation. Cut/fill calculations with the goal of creating a uniform slope proceed as shown in Example D.1.

Figure D.1. Land leveling with laser-controlled scraper for precise slope modification. Laser control may be interfaced with conventional graders, bulldozers, and drain tube laying equipment also.

Example D.1: Using cut/fill computation, estimate the distribution of cuts and fills for the scenario shown in Figure D.2.

1. Get an accurate topographic map of the field being modified. The Global Positioning System (GPS) is a popular source for obtaining elevation information. The USGS 7.5 minute quadrangles (with a 20 ft contour interval) are usually not sufficient. GPS provides more accurate and more accessible data.

2. Set up a grid over the field (100 ft or 30 m maximum size) as shown in Figure D.2. Fit a plane through the points using regression analyses, defining the plane as

$$Elevation = A + S_x X + S_y Y \qquad \textbf{(D.1)}$$

One may compute S_x and S_y from equations D.2a–b, which arise from linear regression theory:

$$(\Sigma X^2 - nX_c)S_x + [\Sigma(XY) - nX_cY_c]S_y = \Sigma(XE) - nX_cE_c \qquad \textbf{(D.2a)}$$

$$(\Sigma Y^2 - nY_c)S_y + [\Sigma(XY) - nX_cY_c]S_x = \Sigma(YE) - nY_cE_c \qquad \textbf{(D.2b)}$$

where

n is the total number of grid intersections;
X_c is the X distance to centroid of the field;
Y_c is the Y distance to the centroid of the field;
E_c is the elevation of the centroid.

The elevation centroid is the simple average of all elevations. The X_c can be found by following these steps:

A. Number the lines running in the Y direction intersecting X, counting the stakes on each line and forming a cumulative sum of the (X distance for each line along $Y \times$ number of stakes in the respective line) products.

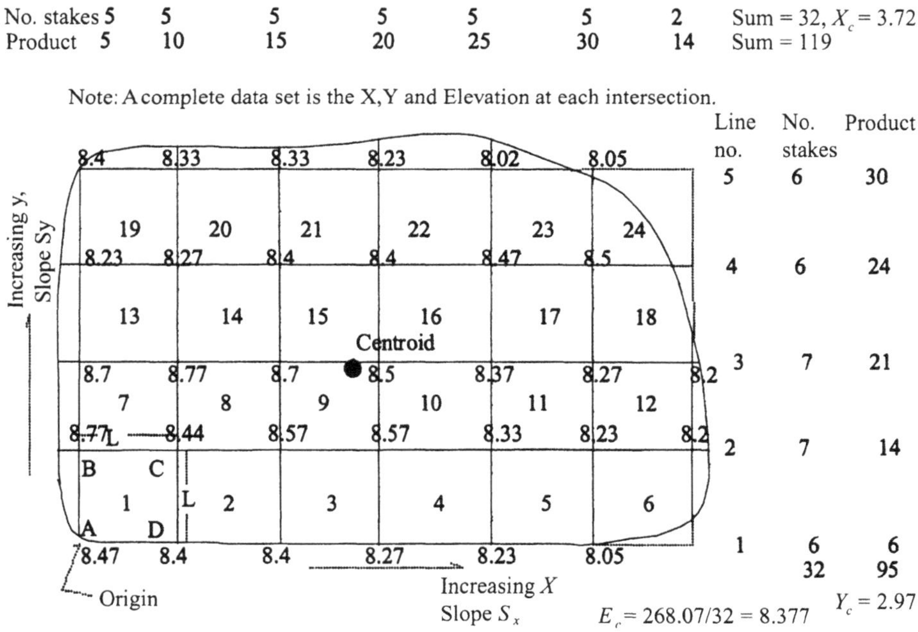

Figure D.2. Example land-forming situation showing measured elevations in the grid and centroid calculations.

B. Divide the total by the total intersections in the field.

C. Repeat for the Y centroid.

One can then solve the above equations for S_x and S_y, the slopes along X and Y respectively. For the above example, $S_x = -0.0579$ and $S_y = -0.02579$. Substituting the centroid coordinates and the centroid elevation into the equation and solving will give the intercept. In this example, the intercept is 8.669 m. Solve the relationship for the predicted elevation at each X-Y intersection. Elevations below the actual elevation require cuts and elevations above the actual elevation require fills.

3. Compute the volume of cuts and fills using the 4-point formula in equations D.3a–b:

$$V_c = \sum_{i=1}^{N=no.\ of\ squares} \frac{L^2(\Sigma C)^2}{4(\Sigma C + \Sigma F)} \quad \textbf{(D.3a)}$$

$$V_f = \sum_{i=1}^{N=no.\ of\ squares} \frac{L^2(\Sigma F)^2}{4(\Sigma C + \Sigma F)} \quad \textbf{(D.3b)}$$

where

V_c is the volume of cut (L^3) for the field under grid;
V_f is the volume of fill (L^3) for the field under grid;
L is the grid spacing (L);
C is the cut on the grid corners (L);
F is the fill on the grid corners (L).

The summations within each square refer to points on the corner of each square. For boundary points with incomplete squares, estimate missing points based on nearby points. Adjust the area (L^2) downward to reflect actual area within the given partial grid.

Adjust the cut/fill ratio by varying the intercept and recomputing predicted elevations until the cut/fill ratio is satisfactory. Set a stake at each intersection denoting cut or fill, noting the respective amount. Experience has shown cut/fill ratios should range from 1.1 to 2.0, with 1.3–1.5 being most common. One must cut more soil areas than are filled to compensate for compaction by the heavy machinery. Thus, one should lower the design elevation by 0.003–0.05 m as a first guess.

Make the cut/fill calculations using the above equation for volume of cuts, then fills. For example, identify the 4-point grids in the lower left of the field. In doing cuts calculations, move around the grid and sum the cuts at each grid. Do the same for fills in a separate calculation. Move to the next grid (e.g., to the right), and for these 4 points do the same thing. Continue until all grids are processed. For partial squares such as those at field boundaries, reduce the area (L^2) by an appropriate amount and assume boundary points to be similar to the nearest measured point.

A simple test of cuts and fills is to do a cumulative sum of cuts and one for fills. As expected, cuts equal fills. To get an estimated cut/fill ratio as desired Schwab et al. (1993) suggest a deviation of -0.05 m. For the example, this resulted in an estimated cut/fill ratio of 2.25, which is high. A deviation of -0.03 m would be more appropriate. Application of the 4-point method would give a more accurate cut/fill ratio.

Calculate the cut/fill ratio. If the ratio is too low, lower the design plane and redo. If the ratio is too high, lower the plane a smaller amount. Schwab et al. (1993) give further examples of land-leveling calculations. Laser-controlled grading equipment (Dedrick 1979; Schwab 1988; Spectra-Physics 1981) aids accurate land-forming work. Software packages are available from the NRCS for surface drainage computations.

References

Dedrick, A. I. R. 1979. *Land leveling precision attained with laser controlled drag scrapers.* ASAE Technical Paper no. 79-2565. American Society of Agricultural Engineers, St. Joseph, MI.

Schwab, G. O. 1988. Land leveling. In *Handbook of engineering for agriculture,* vol. 2, ed. R. H. Brown. CRC Press, Boca Raton, FL.

Schwab, G. O., D. D. Fangmeier, W. J. Elliot, and R. K. Frevert. 1993. *Soil and water conservation engineering.* 4th ed. John Wiley & Sons, New York, NY.

Spectra-Physics, Inc. 1981. *Land leveling design system software for use with Apple II computers.* Construction and Agricultural Division, Dayton, OH.

Appendix E

Centrifugal Pumps and Pumping System Design

Centrifugal pumps are widely used devices for pumping water and many other fluids up grades. Pumps add energy and raise pressure. Most water distribution strategies for irrigation and bioremediation require pumps. This appendix focuses on centrifugal pumps; however, with the appropriate relationship between pump flow rate and pressure head, one can generalize most of the ideas to other pump types. Irrigation and drainage texts more thoroughly cover pump selection and pumping systems (e.g., Cuenca 1989; NRCS 1968). Pump impeller design is beyond the scope of this text. Refer to works such as Wilcox 1997 for additional pump design discussion. Figures E.1 and E.2 show two examples of centrifugal pumps.

Pump Fundamentals

The centrifugal pump operates by centrifugal force. One may describe the flow through the pump, referring to Figure E.3 (Henderson and Perry 1976), by

$$q = 2\pi r_1 w_1 v_{norm1} = 2\pi r_2 w_2 v_{norm2} \tag{E.1}$$

Figure E.1. Pump power units and centrifugal multistage deep well submersible pump.

Figure E.2. Electric-powered centrifugal pump.

where

q is the flow rate through the impeller;
r_x is the radius at radius 1 and radius 2;
w_x is the impeller width at radius 1 and radius 2;
v_{norm1}, v_{norm2} are the velocities normal to the impeller at r_1 and at r_2.

The torque (Tq) exerted on a fluid element is

$$Tq = \frac{dq}{d\theta}\frac{\gamma}{g}[r_2 v_2 \cos(a_2) - r_1 v_1 \cos(a_1)] \quad \textbf{(E.2)}$$

Integrating around the entire wheel and multiplying both sides by the angular velocity ω and sub-

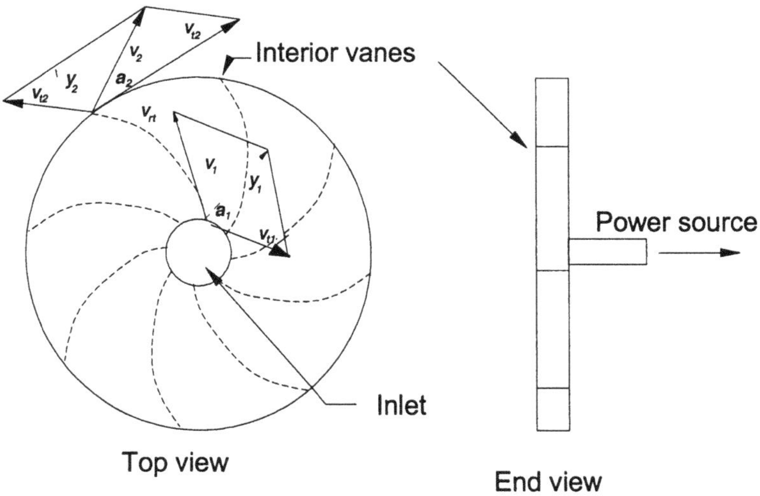

Figure E.3. Schematic of centrifugal pump impeller showing velocity vectors.

stituting v_x for ωr_x allows us to write

$$Pwr = T\omega = \frac{q}{g}\gamma[v_{t2}v_2\cos(a_2) - v_{t1}v_1\cos(a_1)] = q\gamma H \tag{E.3}$$

where

Pwr is power;
H is static head;
v_x is the velocity at point x;
γ is the unit weight of water.

Solving equation E.3 for head H, one can then write

$$H = \frac{v_{t2}v_2\cos(a_2) - v_{t1}v_1\cos(a_1)}{g} \tag{E.4}$$

Equation E.4, which assumes no hydraulic losses, relates the impeller configuration to the pump theoretical total head. Hydraulic engineers also know equation E.4 as the Euler equation (Henderson and Perry 1976). This is the theoretical head possible from the impeller. Velocity relationships may be written from the impeller geometry (applying the law of cosines) as shown:

$$v_{r2}^2 = v_2^2 + v_{t2}^2 - 2v_{t2}v_2\cos(a_2) \tag{E.5}$$

and

$$v_{r1}^2 = y_1^2 + v_{t1}^2 - 2v_{t1}v_1\cos(a_1) \tag{E.6}$$

where v_{r1} and v_{r2} represent the tangential velocity to the impeller vanes at the entrance (1) and the exit (2) locations. Substituting these into the Euler equation (equation E.4) results in

$$H = \frac{v_2^2 - v_1^2}{2g} + \frac{v_{t2}^2 - v_{t1}^2}{2g} + \frac{v_{r1}^2 - v_{r2}^2}{2g} \tag{E.7}$$

Rearranging the second and third terms to categorize them leads to (Henderson and Perry 1976)

$$H = \frac{v_2^2 - v_1^2}{2g} + \frac{v_{t2}^2 - v_{r2}^2}{2g} - \frac{v_{t1}^2 - v_{r1}^2}{2g} \tag{E.8}$$

The first term of equation E.8 represents kinetic energy gain as the flow goes through the pump. The second and third terms jointly represent static head increases from the inlet to the outlet.

Centrifugal Pump Type Selection

One can categorize pumps based on specific speed, calculated as follows:

$$n_{SpecS} = \frac{C_{ucon}N_{rpm}q^{1/2}}{H^{3/4}} \tag{E.9}$$

where

n_{SpecS} is specific speed in rpm;
N_{rpm} is the impeller speed in rpm;
q is the discharge (L^3/T);
H is total head in meters (L);
C_{ucon} is a units constant equaling 51.65 (q in m^3/s; H in m), 1.63 (q in l/s; H in m), or 1.0 (q in gal/min; H in ft).

The radial-flow centrifugal pump is usually used in irrigation and in other applications with high head requirements. Axial- and mixed-flow pumps are used in high-flow/low-head applications such as drainage pumping.

Specific speeds of radial-flow centrifugal pumps for sprinkler irrigation range from below 500 to 1500. Mixed-flow and propeller axial-flow pumps are commonly used in drainage (high volume/low head). They have specific speeds ranging from 1500 up to 15,000. Radial-flow pumping efficiencies range from 60 to 75%. Efficiencies for mixed-flow and axial-flow pumps can be as high as 90%.

Centrifugal Pump Performance Curves

The candidate pump for a given application will have available performance curves supplied by the manufacturer. They are similar to the curves shown in Figure E.4. One may plot discharge versus total pumping head (elaborated upon below). Pumping efficiency, defined as the ratio of water power to mechanical power, relates to head. Pumping efficiencies should range from 60 to 75% over the flow range with well-designed pumps. The NPSH (net positive suction head, head required to move water into the eye of the impeller) relates to discharge. NPSH is a pressure head condition on the suction side. One must satisfy the NPSH to avoid cavitation. The suction line design discussion gives further details. A cavitation condition results in excess noise, impeller wear, and premature failure.

Friction Losses in Pipes

In practice, the vast majority of water flows in pipes have a Reynolds number greater than 2000, implying turbulent flows. Friction losses in turbulent flows can be expressed using the Moody diagram and the Darcy-Weisbach equation (Streeter 1971 or other fluids text). We discuss the Moody diagram below. For water at near room temperatures flowing through smooth pipes more than 25 mm in diameter, the Scobey equation (Scobey 1930), shown in equation E.10, provides a simpler approach to computing friction losses:

$$H_f = \frac{K_{sc}\, L\, q^{m_{sc}}}{D^{m_{sc}+3}}\, 4.1 \times 10^6 \tag{E.10}$$

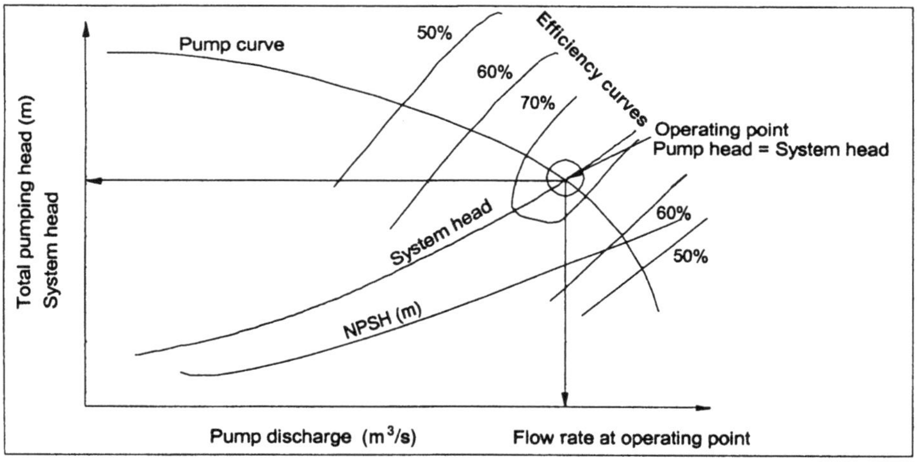

Figure E.4. Performance curve for a typical centrifugal pump showing a hypothetical system head curve intersecting the pump curve at an operating point of optimal efficiency, the ideal condition.

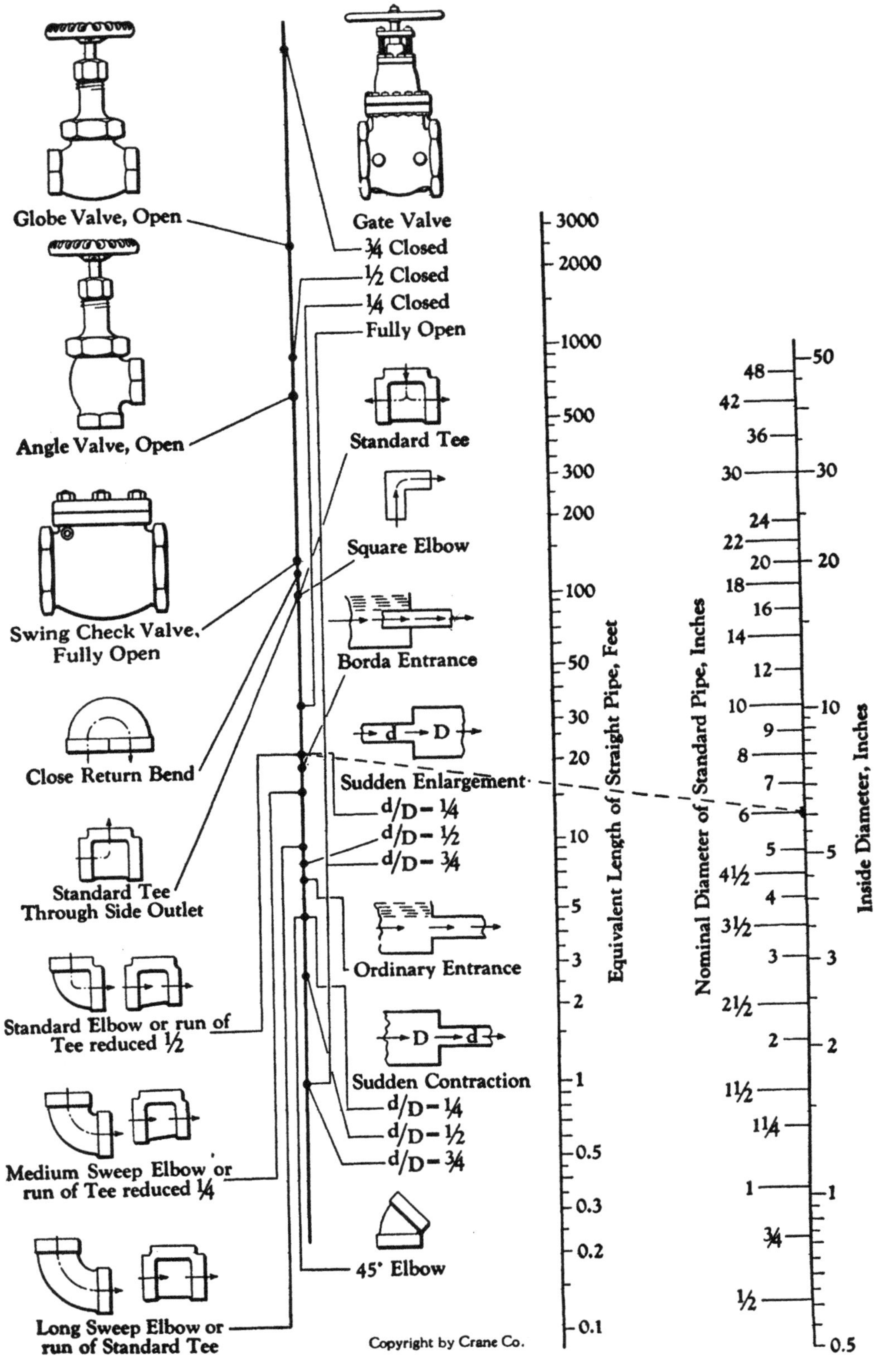

Figure E.5. Equivalent lengths for various minor losses in typically rough pipe transporting water. The swing flap valve may be considered similar to a check valve for many applications. An intake screen may be considered similar to a one-quarter-closed gate valve. Conversions: 25.4 mm per in; 3.28 ft per m. (From Chicago Pump Co., FMC Corporation, Chicago, IL.)

where

H_f is the total friction loss in line;
K_{sc} is the Scobey coefficient (0.42 for portable galvanized steel pipe with couplers; 0.4 for steel pipe; 0.4 for portable aluminum pipe with couplers; 0.32 for PVC pipe with couplers);
q is the discharge in l/s;
D is the inside pipe diameter in mm;
L is the actual plus equivalent length of minor losses (m);
m_{sc} is 1.9 for most irrigation pipe; m_{sc} is 1.75 for PVC and other smooth pipe.

One can estimate the minor losses from Figure E.5. *Be mindful of units for the Scobey equation!*

The Scobey equation presumes water at near room temperature. Fluids behaving as Newtonian but with different viscosity values should use the Moody diagram and the Darcy-Weisbach equation for head loss calculations. The Darcy-Weisbach equation is further discussed below. Figure E.5 is useful for nonviscous Newtonian fluids in rough pipe. One may also apply Figure E.5 to PVC and aluminum pipe with reasonable accuracy.

Example E.1: What are the line losses with a 10 l/s flow through a 101 mm, 50 m long, portable aluminum pipe having two 90 degree elbows?

The equivalent length of the 90 degree elbows is 2(3 m) = 6 m from Figure E.5 for a 56 m friction loss length. The line loss from equation E.10 is then

$$4.1 \times 10^6\ (0.4)(56\ \text{m})[(10\ \text{l/s})^{1.9}]/[(101\ \text{mm})^{4.9}] = 1.0697\ \text{m}$$

Minor losses can add much equivalent length to a pipe. For example, consider the effect of adding a globe valve to this system!

Suction Line Design Considerations

Design of the suction line is important with pumps operating above the water source. One determines maximum practical suction lift when the net positive suction head is within allowable levels. The NPSH value from the pump performance chart is the minimum allowable value of NPSH required to avoid cavitation.[1] In practice, $NPSH_{Perf.\ Curve}$ is the velocity head plus the losses through the pump proper. $NPSH_{actual}$ is the value of static head available to lift the pumped fluid. The actual value must exceed the baseline minimum or chart NPSH. This condition can be expressed as

$$NPSH_{table,\ chart} => NPSH_{calculated} = BP - VP_w - H_{1s} - M_{1s} - S_1 - D_D - F_S \qquad \textbf{(E.11)}$$

where

$NPSH_{table,chart}$ is the value, not to be exceeded, read from the manufacturer-supplied table or performance chart for the pump (m);
$NPSH_{calculated}$ is the computed value above (m);
BP is the barometric pressure (m);
VP_w is the vapor pressure of water at the given temperature (m);[2]
H_{1s} is the head loss in suction pipe;
M_{ls} is the minor loss on the suction side (m) (the actual and equivalent lengths are added; use the sum of the lengths in the Scobey equation to compute the total head loss);
S_l is the suction lift (m);
D_D is the drawdown (m) with a well;
F_s is a factor of safety, usually 0.6 m (Schwab et al. [1993] suggest 0.6 m).

Any lengths unit is appropriate for NPSH calculation. Failure to satisfy the NPSH requirement will result in cavitation, leading to impeller failure, poor performance, and noisy operation. Pumps are usually located to minimize suction line length. Suction lines are frequently 1 or 2 sizes larger than discharge due to cavitation prevention concerns.

One may compute maximum suction lift using the above equation and the chart value of NPSH. Substitute the chart value of NSPH for the computed value and solve for S_l. Suction line design is not required with submerged pumps.

Example E.2: Determine the maximum practical suction lift if NPSH versus discharge is as shown in Table E.1.

Table E.1. Example relationship between NPSH and discharge for Example E.2.

Discharge (m^3/s)	NPSH (m) not to be exceeded
0.01	1
0.02	2.5
0.03	3.5
0.04	5.5
0.05	8

The discharge is 0.038 m^3/s, temperature is 20°C, total suction line friction losses are 1.58 m, and the pump sets at 300 m altitude. Rearrange the suction lift equation to solve for suction lift.

Interpolate to find NPSH = 4.82 m. Standard atmospheric pressure may be approximated using $Patm = 101.325$ kPa − 300 m × 0.01055 = 98.16 kPa, which converts to 10.0 m. From a steam table or by using equation 4.21, one may estimate a vapor pressure at 20°C to be 2.35 kPa or 0.24 m water. Note that 0.1023 m H_2O = 1 kPa. Then, from equation E.11 one can solve S_l, if $D_D = 0$ m, $F_s = 0.6$ m.

$$S_L = 10 \text{ m} - 1.58 \text{ m} - 0.24 \text{ m} - 4.82 \text{ m} - 0.6 \text{ m} = 2.26 \text{ m}$$

Suction line design and NPSH considerations are not relevant with submerged pumps and jet pumps.

Total Dynamic Head and Power

Compute the total dynamic head (H_s) using

$$H_s = S_L + D_L + D_D + H_{1s} + H_{1d} + M_{1s} + M_{1d} + H_o + VH \qquad \textbf{(E.12)}$$

where

H_s is total system head;

S_L is suction side lift;

D_L is discharge side lift;

$D_D(t)$ is the time- and flow-dependent drawdown; one may use a known or estimated maximum value;

H_{ls}, M_{ls} are the pipe friction losses and minor losses on the suction side; often computed together by adding actual pipe length and minor loss equivalent length and then inserting the total into the Scobey equation;

H_{ld}, M_{ld} are the pipe friction losses and minor losses on the discharge side;
H_o is the required pressure at the end of the discharge line;
VH is the velocity head.

All heads are expressed in units of length. One can show the total system head as in Figure E.6. Compute the power requirement using

$$Pwr = \frac{9.8qH_s}{E_p E_d E_m} \tag{E.13}$$

where

Pwr is input power to the power plant in kW;
q is flow in m^3/s (if the discharge q is given in l/s; recall that 1 l/s = 0.001 m^3/s; assume the density is 1000 kg/m^3 here);
H_s is total dynamic head in m;
E_p is the pump efficiency (see pump performance curve);
E_d is the efficiency of the drive;
E_m is the overall energy efficiency of the power plant (use E_m when computing fuel or electrical usage; consider E_m = 100% when estimating the pump load to be matched with the rated load of the power plant).

Power plant selection depends on the pattern of use and on local economies. Amount of power required, initial cost, availability and cost of fuel or electricity (e.g., 3 phase if power is greater than 4 kW or so), annual use, and duration and pumping frequency are some factors to consider. One prefers electric power unless other factors preclude it. Diesel engines are justified if annual use exceeds 800–1000 hr; otherwise, using gasoline may be practical. Propane and natural gas are also options. Efficiencies of diesel engines are about 80%; water-cooled gasoline, 70%; and air-cooled gas engines, 60%. Efficiency of the electric motor is 80–90%. The electric motor is also the easiest power source to control with automation.

Drive efficiencies range from near 100% in the case of direct drive, 95% for gear drives, 90–95% for V belts, to 85–90% for flat belts.

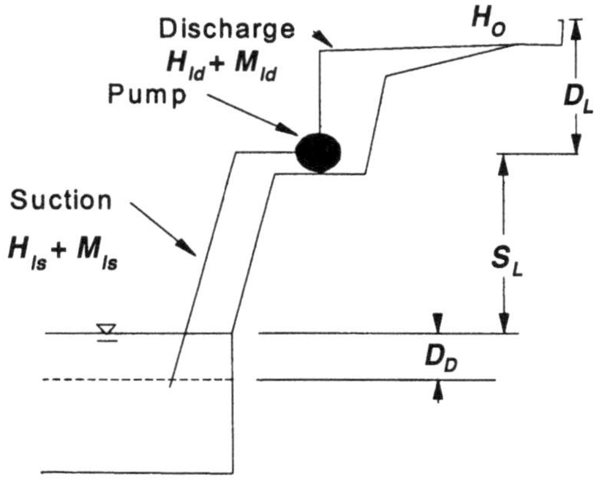

Figure E.6. Schematic of pressure heads used in pump design problems. Discharge and suction friction and minor losses may be related to length, diameter, and flow via the Scobey equation. Drawdown estimation depends on the nature of the source.

Determination of the System Head-Flow Curve

The system head curve reflects all static and dynamic head components required to raise water from one level/pressure to another level/pressure. The static suction lift is a measure of the initial elevation difference between the pump centerline and the water surface. The static discharge lift is the elevation difference between the point of use and the pump centerline. The friction loss is the sum of pipe loss and minor loss in the suction and discharge lines.

Operating head is the system head as a function of flow rate. For example, if a sprinkler nozzle were the end use, the orifice equation would describe the pressure-flow relation. If the end use were a sprinkler main, analyzing the sprinkler system pressure and losses at several representative flow rates would give the flow-pressure relation. If the system were a drainage pump, the operating pressure would be low for all flows assuming free discharge.

The well drawdown attempts to account for the lowering of a well level as a function of water removal. For well drawdown estimation, Longenbaugh (1980) suggests the following series relationship:

$$Drawd = \frac{q}{4\pi T_r}\left[-0.5772 - \ln\left(\frac{r^2S}{4T_rt}\right) + \frac{r^2S}{4T_rt} - \frac{r^2S}{4T_rt(2)(2!)} + \frac{r^2S}{4T_rt(3)(3!)} \cdots\right] \quad \textbf{(E.14)}$$

where

q is the discharge in m^3/day;
r is the well radius in m;
S is the dimensionless storage coefficient (volume of water yielded per unit, per unit drop in head, per unit of surface area);
T_r is the well transmissivity (product of aquifer height and conductivity) in m^2/day;
t is the time since pumping started (day);
Drawd is the drawdown (m).

The time t is usually no longer than 1 or 2 days. When pumping from a replenished pond, the drawdown term is zero or could be a linear function indicating the lowest level of the pond. For a

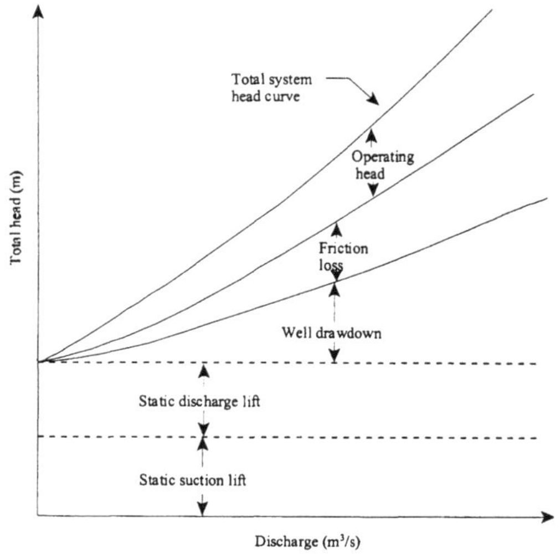

Figure E.7. Components of the total system head curve.

given nozzle or orifice size, one can estimate pressure required to produce various flow rates with the orifice equation. Suction and discharge friction losses (including minor losses) may be estimated for a range of flow rates using the Scobey equation. Summing these components and adding discharge and suction lifts results in the total system head curve as shown in Figure E.7.

Selecting and Matching Pumps to a System

Matching a pump to a system is required for efficient operation. The analysis requires knowledge of the flow versus total head for the system. Figure E.4 shows a system curve. The system operation point is at the intersection of the well capacity and pump curves.

The system curve shown in Figure E.4 intersects the pump curve in the zone of optimal efficiency. What if this point does not occur at an optimal efficiency for an available pump? One may select another pump from a manufacturers catalog. If the match is close, the existing pump impeller may be trimmed (up to 10% of diameter). Shaft speed may be varied within 20%. The pump affinity laws, useful for exploring these questions, are shown in equations E.15–E.17 (Henderson and Perry 1976; Longenbaugh 1980):

$$\frac{q_1}{q_2} = \frac{D_1}{D_2}\frac{Nrpm_1}{Nrpm_2} \tag{E.15}$$

$$\frac{H_1}{H_2} = \left(\frac{D_1}{D_2}\right)^2\left(\frac{Nrpm_1}{Nrpm_2}\right)^2 \tag{E.16}$$

$$\frac{Pwr_1}{Pwr_2} = \left(\frac{D_1}{D_2}\right)^5\left(\frac{Nrpm_1}{Nrpm_2}\right)^3 \tag{E.17}$$

where

Pwr is power;
D is impeller diameter;
Nrpm is rpm.

One can arrive at the equations for adjusting NSPH, speed, and impeller diameter (Henderson and Perry 1976):

$$\frac{(NPSH_r)_1}{(NPSH_r)_2} = \left(\frac{D_1}{D_2}\right)^2\left(\frac{Nrpm_1}{Nrpm_2}\right)^2 \tag{E.18}$$

$$N_2 = N_1\left(\frac{q_1^{0.5}}{H_1^{0.75}}\right)\left(\frac{H_2^{0.75}}{q_2^{0.5}}\right) \tag{E.19}$$

$$D_2 = D_1\frac{H_1^{0.25}}{q_1^{0.5}}\frac{q_2^{0.5}}{H_2^{0.25}} \tag{E.20}$$

When changes are small, assume the efficiency of the pump is constant.

One can use the above relations to develop pump curves at different impeller diameters and/or operating speeds. Use the following procedure:

- Plot the system head-capacity curve.
- If a well is the source, get the drawdown at each flow rate assuming steady-state conditions. Add the drawdown amount to the initial lift to get final lift. For estimating well drawdown, use the steady-state relationships given above.
- Determine the point on the system curve resulting in optimal operation.
- Select a pump delivering required flow at the desired head, preferably at or near the point of optimal pump efficiency.

If a pump or drive modification is the goal:

- Select a new impeller diameter or speed (smaller unless pump is to change).

- Holding speed (or diameter in the case of varying speed) constant, get new flow ratio, head ratio, and power ratio.
- Compute the new flow, head, and power for several points and plot the new curves.
- Vary diameter and/or speed until deriving an acceptable operating point.
- These new points bear the same relation to the efficiency curve as did the old points when D or *Nrpm* varies by less than 10%. Greater changes have impacts on efficiency and on specific speed (e.g., optimal impeller shape).
- Develop the NPSH curve for the alternate pump using the affinity relations. Check to see that calculated actual NPSH for the alternate speed or diameter exceeds the system NPSH.

Example E.3: One is given the pump performance data in the first two columns of Table E.2. Suppose the speed was decreased by 20% from 1740 rpm. The new speed (rpm_2) is 1392 rpm. The impeller diameter is unchanged. Find the modified performance curve. The solution is shown in Figure E.8 and Table E.2.

Table E.2. Relationships between original and modified heads and discharges for Example E.3.

Head H_1 (m)	Discharge q^1 (m^3/s)	Modified head H^2 (m) where the ratio of head is proportional to the ratio of rpm^2	Modified discharge q_2 (m^3/s) where the ratio of flow is proportional to the ratio of rpm
88	0.01	56.3	0.008
87	0.02	55.68	0.016
80	0.03	51.2	0.024
60	0.04	38.4	0.032
30	0.05	19.2	0.04

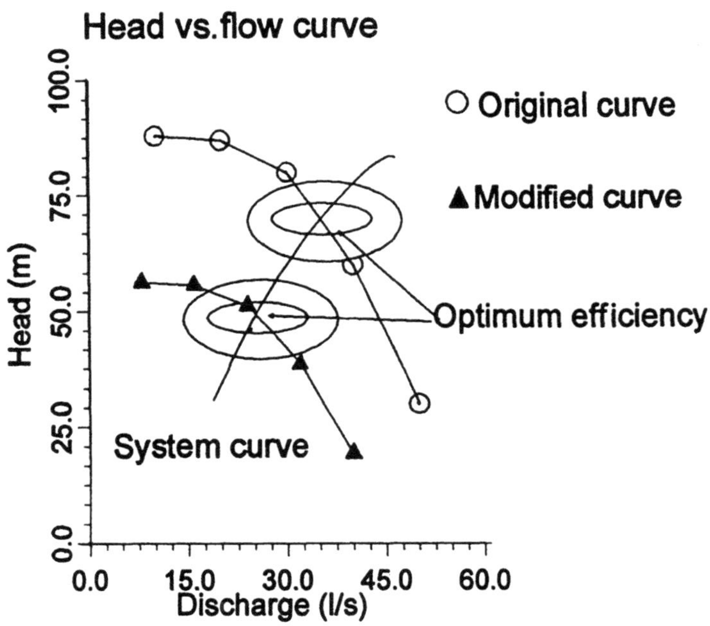

Figure E.8. Original and modified head vs. flow curves.

Similar calculations can be done for power and NPSH. The point of optimum efficiency shifts down and to the left along with the modified curve. The goal is to cause the system curve and the pump curve to coincide at the optimum efficiency point.

Further complicating the pump-system matching issue is the need to match the torque-speed curve of the power plant to the pump. Motors and engine speeds vary in response to load, based on a shaft torque-speed relationship. Most motors have a flat torque-speed curve over most of the operating region (i.e., reasonably constant torque over a range of shaft speed). Four-cycle engines also have a reasonably flat torque-speed performance curve. Governors further control engine speeds. Two-cycle engines are not popular for pumping power plants. Electric motors and diesel engines with governors are most often used for pumping power plants.

Example E.4: Suppose we desire to deliver 0.005 m^3/s against a 19.8 m head. A curve is available for a 178 mm impeller running at 760 rpm pumping 0.01 m^3/s against 11 m at 1760 rpm, at 2.17 kW. What combination of diameter and speed will meet the flow requirement?

Using equations E.19 and E.20,

$$D_2 = 178\text{ mm }[(11\text{ m})^{0.25}/(0.01\text{ m}^3/\text{s})^{0.5}][(0.005\text{ m}^3/\text{s})^{0.5}/(19.8\text{ m})^{0.25}] = 109\text{ mm}$$

$$N_2 = 1760\text{ rpm }[(0.01\text{ m}^3/\text{s})^{0.5}/(11\text{ m})^{0.75}][(19.8\text{ m})^{0.75}/(0.005\text{ m}^3/\text{s})^{0.5}] = 3868\text{ rpm}$$

From equation E.17, the power requirement is

$$Pwr = 2.17\text{ kw }[(3870\text{ rpm}^3)/(1760\text{ rpm})^3][(109\text{ mm})^5/(178\text{ mm})^5] = 1.98\text{ kW}$$

Pumps in Series and Parallel

Pumps can be used in series (e.g., a booster pump) or in parallel. For two pumps operating in series, the pump curves can be added vertically to get a combined pump curve if the two pumps are approximately the same size. Pumps in parallel can be added horizontally if the pumps are near the same size. One should not operate dissimilar-sized pumps in parallel due to the likelihood of flow looping through the pumps. This looping arises because the pumps have different pressure-flow relationships. Longenbaugh (1980) discusses these issues in detail and shows how to estimate the efficiency of series and parallel pump stations.

Storage Requirements at Pumping Plants

From an economic perspective, using a small, properly matched pump for extended periods is better than operating a large pump intermittently. However, in practice this is not always possible. Variation of flow rates or difficulty in finding a suitable low-flow pump may necessitate intermittent operation.

Experience has shown (ASAE 1997) that one should try to avoid starting an engine more than twice per day. Electrical pumps may be operated with pressure switches and float valves. Experience has shown that electrical motors should not cycle on and off more than 10–15 times per hour (Larson and Manbeck 1961). The relationship describing storage, inflow, and outflow relationships is

$$\frac{3600}{Ncyc} = \frac{Sv}{q - q_i} + \frac{Sv}{q_i} \tag{E.21}$$

where

$Ncyc$ is the on-off cycles per hour;

Sv is the storage volume in m^3;
q is the pumping rate in m^3/s;
q_i is the inflow rate (or, in the case of a pressurized reservoir, the water use rate) in m^3/s.

One can show the storage to be optimum when the inflow rate is twice the pumping rate, as shown in equation E.22:

$$Sv = \frac{900q}{Ncyc} \quad \textbf{(E.22)}$$

This discussion did not consider utility rate structures, which becomes a significant issue for large (e.g., drainage) pumping instalations.

Example E.5: A drainage pump has a design capacity of 0.04 m^3/s and is installed in a 2.5 m diameter sump with a maximum cycling frequency of 10 cycles/hr. What is the minimum difference in elevation between the start and stop levels of the sump?

From equation E.22,

$$Sv = 900(0.04\ m^3/s)/10\ cycles/hr = 3.6\ m^3$$

Given a sump diameter of 2.5 m,

$$3.6\ m^3 = H[3.14(2.5)^2/4]$$

or

$$H = 0.734\ m$$

Example E.6: Suppose a trickle system is applying water at a rate of 0.004 m^3/s. What is the optimum pumping rate and storage for intermittent operation, 10 cycles/hr cycle frequency?

The pumping rate is twice the inflow (or outflow) rate. Thus,

$$q = 2(0.004) = 0.008\ m^3/s = 8\ l/s$$

From equation E.22,

$$Sv = 900(0.008\ m^3/s)/(10\ cy/hr) = 0.72\ m^3$$

Pump and Power Plant Fittings and Protection

For maximum system life, pump installations must be fitted with standard safety features and properly operated. Centrifugal pump installations should include a check valve to eliminate backspin at shutoff. Manual or electric priming pumps should be included. Low-pressure positive prime valves as an aid for maintaining prime are sometimes used. Bourdon pressure gauges are routinely used for discharge pressure measurement. A pressure relief valve is placed to protect discharge piping from excessive pressure. Filters passing only what subsequent system components will pass should be included. Flow meters are sometimes required. Globe and/or gate valves are located to control the discharge. Globe valves work best for systems that are turned on and off often. If nutrient injection is involved, nutrients should be injected beyond other discharge fittings. All valves and fittings necessary for proper winterization should be included.

Precautions should be taken when charging the high-pressure side of the system to prevent water hammer from damaging components on the discharge side. After turning the pump on, one should "crack" the valve, maintaining very low pressure until the line is filled. Once the line is filled, the discharge side may be slowly advanced to full open.

Loss-of-discharge pressure cutoffs and loss-of-suction cutoffs are important safety precautions. Other protections for the internal combustion power plant include cutoffs for loss of oil pressure, excessive engine speed, and excessive engine temperature. Protections for electric motors include pump temperature and low-voltage cutoffs. Because the liquid is an important lubricant for the pump, it should not be operated dry. Nutrient injection pumps should be automatically cut off when the main pump is turned off. NEMA III all-weather electrical components should be used for all controls. Regulations often specify safety precautions for specific operations.

Safety features for noncentrifugal pump installations are similar to those for centrifugal pumps. Self-priming pumps are more forgiving from the point of view of maintaining prime. Noncentrifugal pumps require more maintenance. Otherwise, power plant protections and operation are similar.

Problems and Questions

1. You are given the following situation:

Water temperature = 15°C
Elevation of pump above sea level = 350 m
Elevation of discharge end above pump = 30 m
Elevation of pump above water surface = 2.5 m
Drawdown = 0 m
Fittings on suction side are projecting entrance, 90 degree bend, and swing flap valve; fittings on the discharge side include two 45 degree bends
Required flow = 0.05 m^3/s
Required pressure at end of discharge = 20 m water
Discharge line diameter = 100 mm
Suction line length = 25 m
Suction line diameter = 150 mm
Discharge line length = 50 m
All pipes are galvanized schedule 40
Pump NPSH = 5 m

Will this pump work as configured? Why?

Change the suction line diameter to 250 mm, shorten suction line length to 10 m, and lengthen discharge line to 65 m and resolve.

What would you recommend regarding cavitation?

What is the total dynamic head required with the reconfigured suction line diameter, suction, and discharge lengths?

What are the power requirements with a combined water, drive, and power source efficiency of 0.50?

2. Suppose an irrigator wanted to pump 250 gal/min through a traveling gun sprayer. The pump was situated 2 m above the water surface with the suction and discharge fittings of Problem 1, except the following: flow is 250 gal/min; the discharge pipe above was hooked to 1000 ft hose having friction losses of 4.5 ft head per 100 ft hose. A nozzle at the hose end requires 50 psi. Sketch the situation. Size the pump. Will cavitation occur with the 10 m long, 250 mm diameter suction line?

3. A sprinkler main with a maximum length of 400 ft is to be operated. The main lies upslope at 0.5%.

Required flow = 250 gpm = 15.77 l/s
Required pressure at lateral connection = 60 psi

Table E.3. Pump characteristics for Problem E.5.

Head (ft.)	gpm	Eff. (%)
220 ft	100	40
200	130	50
190	160	60
180	180	65
170	200	70
160	230	75
140	250	70
120	230	65
110	200	60

What pipe diameter for the main is needed to keep the loss within accepted loss tolerance? What pressure must be delivered to the main?

4. A lateral is attached to the end of the main in Problem 3. The lateral has 16 sprinklers, each applying 16 gpm. The lateral is 800 ft long. The lateral lies up a 1% slope and operates sprinklers at 50 psi. Riser height is 1 m. Size the lateral.

5. Develop a system curve covering flow rates ranging from plus and minus 50% of the flow in Problem 3. Assume the operating head is operating an orifice delivering 250 gpm at 60 psi.

A pump has the characteristics shown in Table E.3. Determine the operating point of the system. Find an alternative pump curve to optimize efficiency.

6. Develop prompt sheets describing customer requirements and technical specifications for pump selection and design.

7. A swine waste slurry of 25% solids is pumped 300 m. Estimate the diameter of pipe needed to keep the pumping power to about 200 kW. Two standard elbows are in the system.

Notes

1. Cavitation is what happens when pumping situations reduce the actual pressure at the pump inlet below atmospheric pressure. Pumps experiencing cavitation vibrate excessively and perform poorly.
2. Some useful conversions are standard atmospheric pressure = 101.3 kPa = 33.93 ft water = 10.36 m water = 14.7 psi; 0.1023 m water = 1 kPa. Vapor pressure as a function of temperature may be found using equation 6.21. Convert the vapor pressure to units of head.

References

ASAE. 1997. *Design of agricultural drainage pumping plants*. EP369.1 DEC94. American Society of Agricultural Engineers, St. Joseph, MI.

Cuenca, R. H. 1989. *Irrigation system design: An engineering approach*. Prentice-Hall, Englewood Cliffs, NJ.

Henderson, S. M., and R. L. Perry. 1976. *Agricultural process engineering*. 3d ed. AVI Publishing Co., Westport, CT.

Larson, C. L., and D. M. Manbeck. 1961. Factors in drainage pumping efficiency. *Agricultural Engineering* 42:296–97, 305.

Longenbaugh, R. A. 1980. Farm pumps. In *Design and operation of farm irrigation systems*. ASAE Monograph no. 3. American Society of Agricultural Engineers, St. Joseph, MI.

NRCS. 1968. *Drainage of agricultural land*. (Reprint of NEH-16, Drainage.) Water Information Center, Syosset, NY.

Scobey, F. C. 1930. *The flow of water in riveted steel and analogous pipes*. USDA Technical Bulletin 150, Washington, DC.

Schwab, G. O., D. D. Fangmeier, W. J. Elliot, and R. K. Frevert. 1993. *Introduction to soil and water conservation*. 4th ed. John Wiley & Sons, New York, NY.

Streeter, V. L. 1971. *Fluid mechanics*. McGraw-Hill, New York, NY.

Wilcox, D. C. 1997. *Basic fluid mechanics*. DCW Industries, La Canada, CA.

Appendix F

Plant Water Potential

How does a plant or cell operate against high negative soil matric potentials? Alternatively, how does a microbial cell react to various solutes in its environment? Physiologists originally held that plant tissue acted like a wick. For these capillary forces to be sufficient to transport water in a large tree, the capillary tubes would have to be about 150 nanometers (nm) in diameter (Canny 1998). Most xylem tubes are larger than 150 nm. The typical resistance generated by moving observed daily transpiration rates through tubes this small is unrealistic. Thus, other mechanisms are involved.

The most commonly held explanation for plant water movement within tissue is the cohesion theory. Cohesion theory ignores matric potential. Physiologists partition the total water potential in a cell into two dominant components: osmotic and turgor. The various potentials are related by

$$\text{Diffusion pressure deficit (total potential)} = \text{osmotic potential} - \text{turgor potential} \qquad \textbf{(F.1)}$$

where the diffusion pressure deficit (DPD) is the algebraic sum of the osmotic and turgor potentials. Turgor potential is measured using the pressure bomb apparatus. In making the DPD measurement, a freshly cut plant stem is placed in a chamber, which is sealed around the stem. One end of the stem is in the pressurized chamber and the other end is exposed to atmospheric pressure. Pressure in the chamber is slowly raised until free liquid is visible on the surface of the stem. The pressure at incipient liquid appearance is recorded as the turgor potential.

Osmotic potentials stem from semipermeable membranes. The plant acts as a semipermeable membrane. It can therefore create concentration gradients, accounting for high (negative) osmotic pressure levels. With semipermeable membranes, pressure relates to molar anion and cation concentration by

$$\Psi_{os} = -cR_{gas}Te_{abs} \qquad \textbf{(F.2)}$$

where

Ψ_{os} is osmotic potential (atmospheres);
c is the concentration of anions and cations (mole/liter);
R_{gas} is the gas constant (0.08205 liter-atm/K gram mole);
Te_{abs} is absolute temperature (K).

Turgor potential is the potential that overcomes wilting. The turgor potential, measured using the pressure bomb, and the osmotic potential, calculated using equation F.2, lead to the diffusion pressure deficit determination, or total potential, via equation F.1. These potentials relate as shown in Figure F.1.

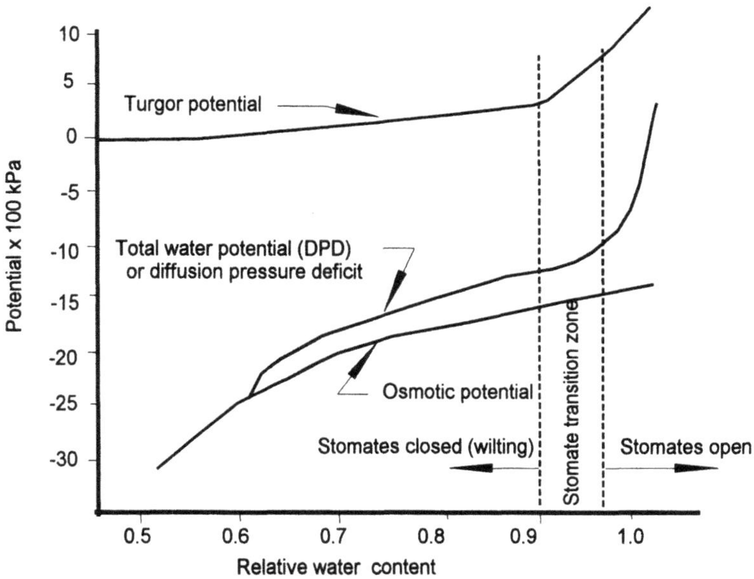

Figure F.1. Plant cell water relations for a bell pepper. The pepper plant wilts at −1.2 MPa. The typical wilting point is about −1.5 MPa. Xerophilous (e.g., cactus) plants wilt at −3 MPa. (Redrawn from Baver et al. 1972.)

Knowing the DPD, one may estimate concentrations of solutes and relative water content, the osmotic potential, and turgor potential for a cell. Figure F.1 shows relations between total, osmotic, and turgor potential for bell peppers. Total potential and osmotic potential are negative over the range of practical relative water content values. Turgor potential is between −0.5 and −1 MPa for functioning cells. A −1 MPa pressure corresponds to approximately 9,800 cm water suction. As relative water content decreases, turgor potential reduces toward zero. This manifests as plant wilt or, with microbial activity, cessation. The DPD enables water to move against soil matric potential forces and root resistance.

The plant responds to more negative soil pressures by losing some turgor pressure and relative cell water content. This leads to osmotic pressures that are more negative. The osmotic potential becomes lower than the surrounding soil matric potential. Thus, water flows into the root. This causes the plant to gain water and thus results in increased osmotic potential, moderating the water flow.

The degree of plant stomatal opening correlates to the relative water content. For most plants, significant closure of stomata does not occur until more than 60% of the available water is gone (Larcher 1980; Pearcy et al. 1989). This corresponds to soil potential levels of −0.05 to −0.05 MPa. Plant cell walls lose rigidity at the wilting point. At potential levels approaching wilting, stomata are completely closed. For irrigation design purposes, one strives to keep enough moisture in the soil to prevent stomatal closure. The management-allowed deficit (MAD) approach is based on using no more than 50–65% of the available water, thus ensuring water stress does not close stomata.

Available water from the water release curve is on a per unit volume or unit depth basis. One must know the effective depth of the root zone. A rule of thumb is to set the bottom of the soil water management zone at one-half the maximum rooting depth or to the depth of a compacted layer. Irrigation guides provide these values for important crops. For bioremediation applications, much work remains to be done to develop comparable tabular information. In summary, osmotic forces account for large potentials.

References

Baver, L. D., W. H. Gardner, and W. R. Gardner. 1972. *Soil physics*. 4th ed. John Wiley & Sons, New York, NY.

Canny, M. J. 1998. Transporting water in plants. *American Scientist* 86:152–59.

Larcher, W. 1980. *Physiological plant ecology*. 2d ed. Springer-Verlag, Berlin.

Pearcy, R. W., J. Ehleringer, H. A. Mooney, and P. W. Rundel. 1989. *Plant physiological ecology: Field methods and instrumentation*. Chapman & Hall, New York, NY.

Appendix G

Unit Conversions

Table G.1 summarizes unit conversions used in this text.

Table G.1. Selected unit conversion factors.

From	To	Multiply By	From	To	Multiply By
acre	ft^2	43560	cal/g	MJ/kg	0.004184
acre	m^2	4049	csm/in	M^3/ha mm s	4.3055E-6
acre	ha	0.4049	dyne/cm	mN/m	1
ac-in/hr	ft^3/s	0.9992	ft	cm	30.48
acre ft	ha-m	0.1233	ft	in	12
acre in	ha-m	0.010275	ft	m	0.304878
atmosphere	ft head	34	ft lb	n m	1.389
atmosphere	m head	10.3	ft head	psi	0.43
atmosphere	psi	14.696	ft/s	m/s	0.304878
atmosphere	kg/m^2	10352	ft^2	m^2	0.0929
atmosphere	mm Hg	760	ft^3	m^3	0.0283
atmosphere	Torr	760	ft^3	l	28.32
atmosphere	kPa	101.3	ft^3/s	gal/min	448
atmosphere	mbar	1013	ft^3/s	m^3/s	0.0283
bar	atmosphere	1.0128	ft^3/s	l/min	1699
bar	kPa	100	ft^3/s	l/s	28.32
bar	cm head	10128	ft^3/s ft	m^3/s m	0.093
bar	psi	14.5	ft lb	J	1.3556
bar	mm Hg	760	ft^3/mi^2 in s	m^3/ha mm s	4.3055E-6
Btu	J	1055.056	100 ft tn in/ac yr	MJ mm/ha yr	17.02
Btu	cal	251.99	gal	l	3.785
Btu/hr	W	0.293	gal	m^3	0-0037886
Btu/ft^2 min	W/m^2	189.276	gal/min	l/min	3.785
Btu/ft^2 hr	W/m^2	3.1546	gal/min	l/s	0.06309
Btu/lb	kJ/kg	2.326	ha	m^2	10000
Btu/ft^2	MJ/m^2	0.011357	ha-m	m^3	10000
cal/cm^2 day	W/m^2	0.4845	ha-m	ft^3	353198
cal/cm^2 day	MJ/m^2 day	0.04181	hp	kW	0.746
cal/g	J/g, kJ/kg	4.1868	in	cm	2.54

Table G.1. (*Continued*).

From	To	Multiply By	From	To	Multiply By
in	mm	25.4	mi/hr	m/s	0.447
in Hg (15C)	kPa	3.377	mm Hg	kPa	0.133
in water (15C)	kPa	0.2488	n/m^3	lb/ft^3	0.006
in^3	l	0.016387	Ounce (fluid)	m^3	2.9574E-5
J/s	W	1	Ounce (fluid)	l	0.029573
kPa	mm Hg	7.501	Pa	n/m^2	1
kg/cm^2	kPa	98.07	psi	kPa	6.89
kg/cm^2	cm Head (4C)	1000	psi	mm Hg	51.73
kg water (4C)	l water (4C)	1	psi	ft head	2.31
km/day	m/s	0.0116	psi	m- head	0.7042
l	m^3	0.001	rod	m	5.029
l	cm^3	1000	ton (English short,		
l	ft^3	0.0352	2000 lb)	t, MG	0.9072
l/s	l/hr	3600	ton ac hr/	t ha hr/	
Langley	cal/cm^2	1	100 ac ft tonf in	ha MJ mm	7.59
Langley	J/m^2	41868	W	cal/hr	860
Langley/min	W/m^2	697.8	W/m^2	MJ/m^2 day	0.08648
Langley/min	MJ/m^2 day	60.29	yd	m	0.907
lbf	kgf	0.4536	yd^3	m^3	0.746
lb	n	4.384	yd^3	l	746
lb/ft	dyne/cm	175126.8	yd^3	ft^3	27
lb/ft^2	kPa	0.04788	Temperature conversions		
lb/ft^3	Mg/m^3	0.016	$F = 1.8C + 32$		
m^3	l	1000	$F = R - 460$		
m head	kPa	9.81	$C = K + 273$		
mbar	kPa	0.1	$C = (F - 32)*0.56$		
mho	S	1	Viscosity conversions		
mmho/cm	dS/m	1	lb s/ft^2	centipoise	47880.3
mi	ft	5280	lb s/in^2	centipoise	6894760
mi	km	1.60976	centipoise	Pa-s	0.001
mi^2	m^2	2591315	centistokes	mm^2/s	1
mi^2	ha	259.13	ft^2/s	centistokes	92903
mi^2	acre	640	m^2/s	mm^2/s	1000000
mi/hr	km/hr	1.609			

Note: Head here implies water at 20°C unless otherwise indicated. Hg is referenced at 0°C unless otherwise indicated.

Index

Page numbers given in *italics* refer to tables.

Abrasion resistance, 441
Acid mine drainage, 482
Acid rain, 197
Adsorption
  reversible, 483
  surface, 192
AGNPS, 98, 168, 487, 488
Agriculture
  erosion and, 139
  extensive, 2
  intensive, 2, 199
  water consumption from, 3
Air
  displacement of, 22
  heated, 22
  masses, 22, 23
  moisture level of, 80
  quality, 12
Alfalfa, 81
Algae, 475
  in irrigation water, 415
  photosynthetic, 477
Algal blooms, 454
American Concrete Institute, 390, 523
American Institute of Steel Contractors, 389
American Society of Civil Engineers (ASCE), website, 170, 253, 398
American Society of Testing and Materials (ASTM), 388, 389, 522
  soil classification, 334
American Steel and Construction Institute (ASCI), materials specifications, 522-523
Anaerobic conditions, streams, 477
Animal production, 185
  CNMPs for, 181
  pollution and, 177, 185
  water contamination and, 8
Antecedent moisture conditions, *1-3*
Anti-industrial approach, to ethics, 494
Antiseep collars, 344
Application rate,
  center pivot irrigation system, 438
Appropriation, 423-426
Aquaculture, 291, 322
Aquatic species, endangered, 475
Aquicludes, 391, 396
Aquifers, 390, 391
ARCVIEW, 486
Area, channels, 206
Artesian well, 391
ASAE
  drainage ditch design flow rate, 116-117, 133
  1983, 66
  1985, 87
  20-40 rule, 374-375
  website, 170, 398
ASCE, website, 170, 253, 398
Ash, 182
Asphalt, 345
  liners, 227
  runoff, 483-484
Atmosphere, carrying capacity, 70
Atmospheric pressure
  evaporation and, 71, 72
  precipitation and, 29
  standard, 74
Atterburg limits, 330, 332, 334, 345
Available water, 409
Average, weighted, isoheytal plot, 45
Average end area formula, 511
Axial-flow centrifugal pump, 537

Backflow, prevention, 445
Backslope, 265

Bacteria
  in soil, 440
  in streams, 475
Bacteriophages, 183
Baffling, 358
BASINS, 200, 486, 488
Basins
  sedimentation, 356-362, 384
  settling, 357
  stilling, 294, 298
Bearing load, pipe, *528*
Bedding channels, 372
Bed load, 464, 466
Bennett, H.H., 137
Benthic habitat, 474-476, 481, 482
Bentonite clay, 332
Berms, 168
  height of, tile outlet terrace, 276-281
Best Management Practices (BMPs), 6, 179
  crop and livestock production, *194*, *195*
  ecology and, 195-197
  fundamentals of, 189
  nomenclature, 190-191
  pollution control
    nonstructural, *193*
    structural, *192*
  stream maintenance, 473
  structural vs. nonstructural approaches, 191-194
Biochemical cycling, 198
Biochemical oxygen demand (BOD), 183, 414
  streams, 482
Bioconversion, 2
  odor and, 12
Biogeochemical cycles, 179
Biological selection, principles of, 499-500
Biomass yield, 69
Bioremediation, 2, 11-12
  irrigation and, 402
  and microirrigation, 442
  and sprinkler systems, 439
  terracing and, 258, 282
  wells for, 394-399
Biota, 498
Biotic community, 498
Blackflies, 475
Blaney-Criddle method,
    evaporation prediction, 70, 83-84
Blaney-Criddle term, 85
Blasius equation, 443
BMPs. *See* Best Management Practices (BMPs)
BOD. *See* Biochemical oxygen demand
Bogs, 330
Boring, 396
Boron, 415
Bottom width, 254
  channels, 253
Boundaries, 460
Boundary limits, equations for, 459-461
Bourden pressure gauges, 546
Boussinesq equation, 526
Bowen ratio method, 70, 73-74, 86-87
Box-inlet drop spillway, 297
Breathers, 384, 386
Brush layers, 471
Buffers
  definition of, 190
  natural, 473
Buffer strips, 166
Buried pipe loading, 525-528

Cable tool, 396
Caddis flies, 183, 474, 475
Calcium, 414, 415
Canal, irrigation, 227
Canopy cover, 151
Capacity designs, vegetated waterways, 247-249
Capillary fringe, 412
Capillary potential, 53, 409.
    See also Matric potential
Capillary rise, 54-55
Capillary tubes, 550
Carbon, 179, 485
  in streams, 477
Carbon dioxide, dissolved, 482
Catchment, definition of, 96-98
Catchment area
  diversions, 270
  diversion terrace, 270
Cations, soluble, 482
Cavitation, 540
Census of Agriculture, 402
Center pivot irrigation systems, 405, 436-439
Certified Conservation Compliance Plan, 185
Channel design, 215-225, *284-285*
  construction, 227-231
  customer requirements, *230*
  engineering factors, *230*
  flow transition analyses, 225-226
  hydrograph routing, 226-227
  seepage losses, 227
Channels
  circular, 219
  continuous flow, 205
  cross-sections, 207, *208*, 217-225
  curvature of, 228, *229*, 516-518
  depth of, 461

designed, 205
deterioration of, 231
diversion, cross-section, 270-274
earthen, 249
economical, 223-225, 228
elements of, 205-207
ephemeral flow, 205
erodible, 219-222
erosion control terraces, 264-269
flow path, 107-108
formation of, 140-142
gradient, 206, 519
installation techniques, 516-524
intermittent flow, 205, 238, 240. *See also* Waterways, vegetated
linings for, 223, 227
maintenance/restoration of, 471-473
materials specifications, 521-524
natural, 205
occasional flow, 223
open, velocity profiles, 458-461
parabolic, 219, 241-242, 247, 248, 252
perennial flow, 205
rectangular, 212-213
side slope, 244-245
slope of, 227
small, 519
spoil, 228
stability of
permissible velocity method, 220-222
tractive force method, 220
trapezoidal, 217-219, 241-242, 246, 249-254
with constrained velocity, 509-510
side slope values, *224*
triangular, 219, 241-242, 247, 248
velocity values, 244, *247*
Chemical oxygen demand (COD), 182
Chemigation, 403
Chezy coefficient, 215
Chezy uniform flow equation, 214-215
Chord deflection, 518
Christiansen's factors, *431*, 444
Chutes, 292
Circular curves, layout of, 516-518
Clarity, of water, 453
Clay
activity, 332
bentonite, 332
blanket, hydraulic conductivity of, 56-57
flocculated, 345
kaolinite, 332, 334
montmorillonite, 332, 334
soils, 52, 94, 410
types of, 332
Clean Water Act, 6, 181, 183, 185, 186, 452
hypothetical case, 505-506
section 319, 186
Climate, and water erosion, 138
Cloudiness, percent, 90
CNMPs. *See* Comprehensive nutrient management plans
Coalescence, of water droplets, 21
Coastal Zone Management Act (1990), 185
Codes, storm sewer design, 321
Coefficient of variation, probability distributions, 28
Cohesion theory, 550
Coliforms, 183
Colloids, in soil, 52
Color, of water, 182
Composite conductivity, 55-56
Compost berms, 168
Composting, 190
odor and, 12
Comprehensive Environmental Response, Compensation, and Liability Act (CERCLA), 186
Comprehensive nutrient management plans (CNMPs), 177, 181, 186
Concentrated Animal Feeding Operations, 185
Concrete, 521
drainage structures, 390
specifications for, 523
Condensation, 3, 21-22
Condensation nuclei, 21
Conductivity, composite, 55-56
Conduit hydraulics, 292
Conduit loss factor, 308
Conduits, 292
closed, 305
Cone frustum approximation, 514
Cone of depression, 391, 394
Cone of injection, 391, 394
Confidence intervals
precipitation data, 30
and record length, 33
Confined aquifer equation, 393
Conservation, of wetlands, 371, 398
Conservation practice factor, 151, 155
Conservation tillage. *See* Tillage, conservation
Construction
erosion control and, 7, 157-158, 258, 274-276
sediment control in, 168
site planning, 162
vegetated waterways, 253

Contact cooling, 22
Continuity equation, 209
  economical channel, 223-225
  terrace channels, 267
  with velocity constraints, 219-222
  without constraints, 217-219
Contour, 190
Contour farming, 5
Contouring, 165-166
Contouring-practice factor, 151, *154*
Convection coefficient, 75
Conveyance, 215
Conveyance efficiency, 416-417
Conveyance factor, 308, 309
Cooling
  contact, 22
  dynamic, 21
Corrosion resistance, 441
Coshocton wheel, 148
Cover
  crops, 166, 190
  hydrograph shape and, 111, 114
  management factor, 151, *152*, *153*, 154-158
  vegetal, retardance classes, 240, *246*
CREAMS, 168, 169, 170-171
Crest boards, 386
Cribbing, 472
Critical depth, hydraulic structures, 294, 296
Critical flow, 210
Critical flow model, coefficients for, *304*
Crop coefficients, 70, *79*, 81
Crop growth index, 58
Cropland
  irrigating, 9
  and soil erosion, 7, 137
Cropping, 192
Crop production, and pollution, 177, 185-186
Crop residues, infiltration and, 51
Crops
  biomass yield, 69
  cover, 166, 190
  rotation of, 151
  salt tolerance, *416*
Crop water use coefficient, 421
Cross-sections
  channels, 207, *208*, 217-225
  diversion channels, 270-274
  terrace channels, 259, 265
  vegetated waterways, 241-242
  weirs, 298
Crustaceans, bottom-dwelling, 183, 474, 475
Culverts
  flow hydraulics, 305-316
  pipe flow equation, 308, 463
  small pipe, 525
  stage-discharge curves, 316-321
Curb gutters, 321
Curb inlets, 321
Curvature
  channel, 228, *229*, 516-518
  radius of, 517
Curve number equation, 98-104, 133
Curve numbers
  agricultural, *99*
  antecedent condition conversion factors, *103*
  nonagricultural, *100*
Cut-and-fill, 519-521, 530-532
Cutoffs
  loss-of-discharge pressure, 547
  loss-of-suction, 547
Cut off walls, 293
Cutslope, 265, 269
Cyanobacteria, 454
Cycle time, lateral, 427
Cycling, 191

Dalton coefficient, 71-72
Dalton's law, 71-72
Dams, straw bale, 362
Daphnia, 183, 474
Darcian flux, 52, 375
Darcy's law, 52-57, 59, 70
  and subsurface drainage, 375
Darcy-Weisbach equation, 537
Darcy-Weisbach friction factor diagram, 307
Decomposition, 46
Deep Ecology, 494
Deflection test, 526
Deforestation, 7, 15, 199
Delivery ratio, 164-165, 361
Denitrification, 440
Dense nonaqueous phase liquids, 439
Dependence, of hydrologic data, 28
Depth
  channel, 461
  critical, 294, 296
Design, and ethics, 497
Design storm
  peak flow, 122
  runoff estimates, 108
Dew, formation of, 20, 21, 22
Dewatering, sedimentation basins, 360
Dew point temperature, 79-80
Diesel engines, 541
Diffusion pressure deficit, 550-551
Dike, 190
Discharge
  permits, 185
  storm sewers, 321

Discharge coefficient, 298
Dispersants, 345
Dissolved oxygen, 182
Distribution, hydrologic data, 27-33
Distribution boxes, flow, 322
Distribution uniformity, irrigation, 418
Ditches
  cross-slope, 374
  diversion, 168
  drainage, 227, 371-375
    trapezoidal, 217
  excavated, 372
  flow travel time, 107-108
Diversion
  definition of, 257
  design summary, 286-287
Diversion channels, 260
  design of, 271-274
Diversion ditches, 168
Diversions, 238, 270-274
  catchment area, 270
  shape of, 258
Diversion terraces, 283
  agricultural or urban applications, 270
  for construction site erosion control, 274-276
  design summary, 286
  uses, 258
Diversity, species, 481
Divine command, 497
Domain of Hinlet, 320
Doppler techniques
  sediment sampling, 468
  stream survey, 463
Downstream integration, 211
Drag coefficient, 214, 219, 459
Drainage
  materials specifications, 388-390, 398
  non-steady-state, 378-379
  pipe, 376-390, 525.
    See also Drainpipe; Pipe
  rates, 380-381
  and salinity control, 380-381
  steady-state, 376-378
  streams, 396-397, 454-456, 473
  subsurface, 10-11, 365-376, 371
    function of, 376
  surface, 10-11, 371-375
    land-forming computations, 530-532
Drainage coefficients
  arid regions, 381
  humid areas, *380*
Drainage systems
  accessories, 386
  design and installation, 383-386
Drain line, installation of, 384-385
DRAINMOD, 385, 398
Drainpipe, flow velocities, 279, *280*, 281.
  See also Pipe
Drains, size of, 381-383
Draw, natural, 276
Drilling, 396
Drinking water, standards, 183, *184*
Drip irrigation systems, 403, 442
Drive efficiencies, 541
Drizzle, formation of, 20, 21
Drop inlets, 292
Drop spillway, 297, 298
Dry bulk density, 331, 409
Dupuit-Forcheimer assumptions, 371, 376, 392-398, 398
Dynamic cooling, 21
Dynamic heating, 21

Earth Info- National Climatic Data Center (NCDC) Summary of the Day, 33
E.coli, 183
Ecological design, 507
Ecological engineering, 12-15
  paradigm, 499
  principles of, 498-500
  reverse, 505
Ecologist, stream, 474
Ecology, 498
  best management practices and, 195-197
  lentic, 475, 479, 481
  lotic, 475, 479, 481
Economics, resource, 507
Ecosystems
  biological principles of, 499-500
  diversity of, 498-499
  restoration of, 471
  world, 498
Ecotones, 473, 498, 500
Eddy correlation, 87
Effective available water, 433
Egoism, ethical, 497
Electrical conductivity, 182
Elevation
  centroid, 531-532
  energy, 210
  information, 531
  invert, 306
  measuring, 81
  potential, 53
Ellipse equation, 376-378
Embankment construction
  soil classification, 332-335
  soil properties, 330-332
  soil survey, 330

Embankments
design summary, 365
filters for, 347-348
function of, 329
height/slope of, 343-344
profile of, 348-349
safety issues, 354-355, 363
seepage control, 345-348, 350
slope of, 343-344
soil limitation ratings, *344*
volume of, 351, 511-515
Emitter, pressure-compensated, 442-445
Energy dissipaters, 213
Energy dissipation, in outlets, 323
Energy equation, 210-211
Energy loss rate, 211
English units, 158-159
Enrichment ratio, 198
Environment, 1, 2
Environmental assessment, 452-454
Environmental management, versus resource management, 502-505, 506
Environmental racism, 493
ENWATBAL, 87, 91
EPA. *See* U.S. Environmental Protection Agency
Ephemeral gully erosion model (EGEM), 471
Epilimnion, 479
Epoxies, 521
Equipment, and channel design/construction, 228, *229*, 231, 265, 268-270
Erosion
average annual, 151
catastrophic, 7
cost of, 501
gullies, 468-471
modeling software for, 114
pollution and, 168
soil. *See* Soil erosion
streambanks, 468-471
structure design and, 293
types of, 137
upland, 148-153, 453
water. *See* Water erosion
Erosion control, 7-8, 137
in agriculture, 165-168
during construction, 7, 157-158, 258, 274-276
mats, 471-472
modeling, 162-165
principles of, 200
terracing for, 257-258, 261, 283, 286
Erosivity, 219
Error analysis, runoff assessment, 132
Estuarine wetlands, 477
Ethical egoism, 497
Ethical subjectivism, 493-494, 497
Ethics
definition of, 493
and the environment, 497-498
historical summary of, *495*
personalistic, 496
pleasure-based, 496
rule-based, 496
spirit-based, 496
values and morals, *494*
Euler equation, 536
Eutrophication, 454
lakes, 480, *481*
Evaporation, 3
definition of, 69, 90
factors affecting, 70
from free water surfaces, 71-72
measuring, 4, 89-90
and pollutants, 90
and runoff volume, 94
Evaporation coefficients, 70
Evaporation pan, 4, 70
Evaporation prediction
approaches to, 70
Blaney-Criddle method, 83-84
Bowen ratio ET method, 86-87
Bowen ratio method, 73-74
Dalton's law, 71-72
energy balance based, 73-74
FAO modified Blaney-Criddle method, 85-86
Jensen-Haise method, 70, 82-83
mass-transfer based, 71-72
Penman-Monteith method, 74-82, 90
Priestly-Taylor method, 86
soil-canopy simulation, 87
standard evaporation pan, 87-89
Thornthwaite method, 87
Evapotranspiration, 69, 70
computing requirements, 421-423
daily, 419, 420
measuring, 4, 89-90
peak, 420
and water stress, 91
Exit velocity, 225
Exit water, 305, 323
pipe flow and, 315-316
Explosives, 231
Externalities, 501

Factory scale, 2
Falling raindrop energy, 138
FAO modified Blaney-Criddle method, evaporation prediction, 85-86

Farm Bill (1996), 185
Farming, contour, 5, 190, 261
Farmland, prime, 5
Farm scale, 2
Federal Farm Security Act (1985), 8
Federal Insecticide, Fungicide, and
  Rodenticide Act, 185-186
Feedlots, 181
Fencing, for embankments, 363
Fertigation, 403
Fertilizers, excessive application of, 194
Fick's law, 53, 64, 70
Field capacity, 409
Field scale, 2
Filters
  drain line, 384-385
  gravel, 347
  ponds and embankments, 347-348
  pump, 546
  rock toe, 347
Filter sock, 385
Filter strips, 166, 190
Filtration, 191
Fines, 332
Flap valves, 384
Fleas, water, 183
Flies, 183, 474, 475
Float, 461, 463
Flooding, 6-7
Flood routing, 335
  NRCS TR55 method, 337-338
  storage routing, 338-342
Flow
  clarity of, 453
  depth, channels, 206
  gravity well, 391-392
  measurement devices, 322
  meter, 63
  modeling, 226-227
  overland, 107-111, 198
  peak, 96
  pollution transport and, 198
  potential, 53
  sediment-laden, 301, 304
  sheet, 143
  specific energy of, 294, 296
  subcritical, 210
  subsurface, 375-376
  supercritical, 210
  time, horizontal, 358
  transitions, 211, 296
  turbulent, 68
  velocity, channel, 220
Flow-controlling mechanisms, 291
Flow hydraulics, culverts, 305-316
Flow rate
  center pivot irrigation system, 438
  channel, 207
  and drain size, 381-383
Flow transitions, 211
  subcritical-supercritical, 296
Fluid
  boundary layers, 459-461
  energy, 210
  pressure, 208-209
  surface tension of, 54
Fluids, multiphase, 439
Fluid statics, 53
Fluid statistics, 208-215
Flumes, 132, 292, 322, 522
  formless, 297
  H, 301-304
  Parshall, 301
  stage-discharge relationships, 297-304
Flush valves, 445
Flux, 59
  Darcian, 72, 375
  Green-Ampt, 59, 60
  pollutant transport modeling, 64-65
Fog, 20, 21, 22
Food chains, 498
Food Security Act (1985), 11
Forage, infiltration and, 52
Force, hydrostatic, 208
Forestry, and water quality, 177
Forests, in erosion control, 166
Fossil fuels, 506
Fourier's law of heat conduction, 53, 70
Freeboard, 206-207
  channel, 227
  terrace channels, 268-269
  vegetated waterways, 249-253
Free outfall, 293
Frequency analysis, hydrologic, 27-33, 47
Frequency factors,
  lognormal probability distribution, *29*
Freshwater, evaporation from, 72
Friction
  skin, 459
  velocity, 459-460, 465-466, 487
Friction losses, 542
  irrigation lines, 430, 432-433
  in pipe, 537-539
Frontal systems, 22
Frontslope, 265
Froude number, 209-210
  weir formula, 297-298, 303
Fungi
  in soil, 440
  in streams, 475

Gabions, 190
Ganges River, 7
Gas injection, 395-396
Gauges
  precipitation, 23-24, 43
  pressure, 445
  rain, 3-4
  snow, 3-4
  stream, 3
Gauging station, 96, 97
Geologic zones, 412
Georgia, floods, 7
Geotextiles, 472
Gills, pollutant diffusion across, 481
Global Positioning System, 531
Gradation, assessing, 333
Grading, channel, 206, 519
Grasses
  infiltration and, 52
  retardance classes, 240, *246*
Grates, 386
Gravel filter, 347
Gravimetric analysis, sediment concentration, 468
Gravimetric water content, 331
Gravitational constant, 298
Gravity equation, 392
Gravity well, 391-392
Green-Ampt
  equation, 59-62, 65-66
  infiltration parameters, *57*
  runoff calculation, 125, *126*
Groundwater, 4
  Dupuit-Forcheimer assumptions, 371, 376, 398
  pollution of, 11
  regulation of, 423-426
  seeps, 468
  table, 390, 391
Gullies
  control structures, 190
  erosion of, 137, 141-142, 143, 468-471
  headwalls, 471
Gumbel equation, 27
Gutters, 321
Gypsum, 345

Hail, 20, 21
Head, 210
  cutting, 141
  equation, 309
  loss, 444
  submerged inlets, 305-312
  submerged outlets, 312-313
  wall, 293
Heating, dynamic, 21
Heat islands, 44
Heat of vaporization, latent, 80
Heat transfer, 22, 75-76
  evaporation and, 73-74
Heavy metals, in streams, 482, 483
Hedge layers, 471
Herschfield atlas, 35-36, 37
High water marks, 461
Holtan equation, 58
Homeostasis, 499
Horton equation, 57-58
Hose-tow irrigation system, 403
Humidity, relative, 70, 80
Hurricanes, 23
Hydraulic conductivity
  constant, 52
  definition of, 53
  measuring, 53-57
  saturated, 59, 60
Hydraulic depth, 206
Hydraulic elements, vegetated waterways, 245, *248*
Hydraulic jumps, 210, 211, 294
  rectangular channel, 212
Hydraulic radius, 206
  constrained, 509-510
  triangular channels, 267
Hydraulics
  conduit, 292
  fundamentals of, 205-*208*
  inlet, 292
  outlet, 292
  structures, 294-296
Hydrocarbons
  bioremediation of, 12
  polycyclic, 484
Hydrographs
  basic runoff, 124
  computer modeling, 169
  definition of, 96
  generalized, 125
  incoming, 337
  NRCS triangular, 111-116
  NRCS unit, 124-125
  rational, 196
  runoff, 123-132
  synthetic, 113
  unit, 96, 123-125
Hydrologic cycle, 1, 3-4
Hydrology, 2, 20, 46
Hydroseeders, 168
Hyetographs
  precipitation, 23, 26
  rainfall, 123

and storm type, 38
Hypolimnion, 479

Ice crystallization, 21
Impeller, 536, 540, 543
Impoundments
cost of, 329
function of, 329
storage in, 334
volume calculations, 511-515
Independence, of hydrologic data, 28
Indicator organisms, 474
Indicator species, water pollution, 183
Infiltration
basin, 190
cumulative, 61
curve, 61
Darcy's Law, 52-57
definition of, 51, 65
factors affecting, 51-52
measurement of, 62-63
models, 57-62, 65-66
parameters, Green-Ampt, *57*
and pollutant transport, 64-65
uniform, 396-397
Infiltration rate
Green-Ampt, 60
Holtan, 58
Horton, 58
soil, 94
steady-state, 58
Infiltrometers, 62-63
Inflow rate, pumps, 546
Infrastructure, creation/expansion of, 5
Initial abstraction, 99, 100
Injection wells, 394-395
Inlet head, weir equation, 301-303
Inlet hydraulics, 292, 323
Inlets, 292
blind, 386
drop, 292
flow control, 305
submerged, 312-313
free discharge, 305-312
surface, 386
Intake rate, soil, 413-414
Integration, upstream, 211
Intensity-Duration-Frequency (IDF) Maps, 37
Interagency Committee on Water Resources, 467
Interflow, 3
Intermediate water management zone, 412
International Erosion Control Association (IECA), website, 170
International Soil Classification System, 334
International Standards Organization (ISO), 16, 196, 497, 501
Interrill erosion, 137, 140
Intrinsic permeability, 53
Invert elevation, 306
Ion exchange processes, in streams, 483
Irrigation, 8-10
agricultural applications, 402
bioremediation projects, 402
canal, 227
closed surface, 403
drip, 8-9
efficiency of, 418-419
flow control, 322
intake rate, 413-414
interval, 427
level basin, 530
managed, 8, 190
nonagricultural applications, 402
open surface, 403
Parshall flume in, 301
pumping rate, 423
requirements
predicting, 74-82
seasonal, 421-423
scheduling, 91, 419-420
soil intake rate, 421
sprinkler, 9
urban, 439
and wastewater disposal, 402
water measurement structures, 322
and water quality, 414-416
water table, 406
Irrigation Association, 445, 446
Irrigation systems
application rate, 447
center pivot, 436-439
closed conduit, 405
continuous, 405
intermittent, 405
maintenance of, 445-447
materials specifications, 445, 447
nonagricultural applications, 439-442
overview of, 403-406
safety of, 445-447
soil intake rate and, 447
sprinkler lateral design, 426-436
summary of, *408*
ISO 14000, 16, 196, 497, 501
Isoheytal method, precipitation analysis, 45

Jensen-Haise method, evaporation prediction, 70, 82-83
Jetting, 396

Jump zone, 294
Junction boxes, 386

Kaolinite clay, 332, 334
Kentucky mining reclamation approach, 162-165
Kinematic viscosity, 460
Kirpich formula, 106-107, 110
KYSPILL, 65

Lacustrine wetlands, 477
Lagoons, objectives for, 365
Lake Chatuge, 502-505
Lakes
  biological processes, 479-481
  dystrophic, 480
  equatorial, 480
  eutrophic, 480, *481*
  oligotrophic, 480, *481*
  oxbow, 476
  temperate, 480
  tropical, 480
Land
  ownership, 4-5
  slope, terracing and, 261
  transition, impact on pollution, 187
  use
    changes in, 4-6, 15
    regulation of, 6
Land Improvement Contractors Association (LICA), website, 170, 230, 253, 398
Lanes' law, 456-457
Laser grading techniques, 519
Laterals. *See* Sprinkler laterals
Leaching fraction, 376, 416
Leaching requirement, 415
Ledge rock, 344, 345
Length factor, USLE, 263-264
Lentic ecology, 475, 479, 481
Less dense nonaqueous phase liquids, 439
Levees, soil limitation ratings, *344*
Light penetration, measuring, 474
Linear move irrigation systems, 405
Liners
  asphalt, 227
  clay, 345, 347
  plastic, 227, 345
  pond, 522
Linings
  channel, 227
  diversion terraces, 270
  erodible, 217
  minimizing costs, 223
  nonerodible, 217
  temporary, 253
  waterway, maintaining, 253
Liquid limits, 330, 331, 332
Littoral habitat, 474, 479, 481
Log cribbing, 472
Lognormal distribution, hydrologic data, 27-33
Loss coefficient, 211
Loss-of-discharge pressure cutoffs, 547
Loss-of-suction cutoffs, 547
Lotic ecology, 475, 479, 481
Lysimeters, 4, 88, 89-90

Machine width, and channel design, 265, 268-270
Magnesium, in irrigation water, 414, 415
Main lines, sprinkler lateral systems, 426
Maintenance, vegetated waterways, 253
Managed soil water zone, 412
Management-allowed deficit (MAD), 406, 551
Manning equation
  economical channel, 223-225
  roughness and particle size, 460
  roughness, 215, *216*
  streams, 460-461
  for tile line sizing, 381
  uniform flow, 215
  vegetated waterways, 238, 239-240
  with velocity constraints, 219-222
  without constraints, 217-219
Manufacturing, environmentally conscious, 5
Mapping units, 330
Marine wetlands, 477
Mass transfer, 76-77
  evaporation prediction, 70
Matric potential, 52, 53, 54-55, 59
  computing, 409-410
  values of, *57*
Matrix porosity, 64
Mats, erosion control, 471-472
Mavis plots, orifice flow, 310
Maximum-minimum allowable slope, terrace channels, 265
Mayflies, 183, 474
Meadows, to reduce soil loss, 155
Mean, probability distributions, 28
Meandering, of streams, 456, 458, 473
Measurements
  hydrologic, 3-4
  units of, 158-159
Mechanical impedance, plant root growth, 412
Media, porous, 191, 192
Membranes, semipermeable, 550
Metals, 521

in drainage structures, 389
removal of, 194
Meteorology, 20
Meyer equations, 71, 72
Meyer-Peter-Mueller relationship, 466
Microirrigation systems, 403, 439
design of, 442-445
Microorganisms
in bioremediation, 11-12
indicator, 474
as pollutants, 181
Microsprinklers, 403
Minerals, in soil, 330
Mining, site planning, 162
Minnows, 183
MIT, soil classification system, 334
Mixed-flow centrifugal pump, 537
Mockus equation, 29-33
Modeling, infiltration, 57-62, 65-66
Modeling software
constituent transport, 65
drainage, 385, 486
erosion, 114
flood routing, 342
ponds, 366
runoff, 133, 168
sedimentation basins, 360
streamflow, 486
stream water quality, 479
Modified universal soil loss equation (MUSLE), 144-145
Moisture conservation, and terracing, 261
Moisture flux, computer simulation of, 87
Mollusks, 475
Momentum equation, 211-213
Monitoring wells, 4
Monsoon, 21
Moody diagram, 537
Morals, 493
ethics and values, *494*
Motors
electric, 541, 547
speed of, 545
Mountainside cultivation, and flooding, 7
Mulches
evaporation and, 91
infiltration and, 51, 52
for sediment control, 168
Multiple outlet factors, *431*
Muskingum-Cunge storage approach, 227
MUSLE, 144-145

National Ground Water Association, website, 398
National Pollutant Discharge Elimination System, 181, 186
National Society of Professional Engineers (NSPE), canons of, 496
National Storm Water Management Program, 191
Natural draw, 276
Natural environment, 498
Natural forces, 2
Natural gas, 541
Natural law ethics, 497
Natural Resource Conservation Service (NRCS), 4. *See also* NRCS
Natural resources engineer, function of, 16
Natural resources engineering
definition of, xv, 2
purpose of, xv
scope of, 6-17
Nature, 498
NCSS, isoheytal analysis software, 45
Nematodes, in streams, 475
Net positive suction head, 537, 539-540
Nettings, 253
Nitrate control, 398
Nitrogen
cycle, 192, 485
loss, due to drainage, 386
removal, 198-199
samples, 484-485
in soil, 179
in water, 454
Nonpoint source pollution. *See* Pollution, nonpoint source
Normal distribution, hydrologic data, 27
Nozzles
spacing of, 428
V-jet, 163
NPDES permits, 452
NRCS curve numbers, for runoff computation, *127*
NRCS effective runoff calculation, 127-131, *132*
NRCS 1983, 438
channel restoration, 471
embankment safety, 355
NRCS runoff curves, 118-121
NRCS runoff volume method, 98-104
NRCS storm coordinates, *41-42*
NRCS storm magnitude computation, 37-43
NRCS TR55 method
peak runoff rate estimates, 117-123
reservoir routing, 337-338
NRCS triangular hydrograph
version 1, 111-112
version 2, 112-116
NRCS waste management handbook, 200

NRCS website, 133, 200
Nutrients
  cycling, 485-487
  in ground water, 8
  irrigation and, 403
  management, 12, 191
  removal of, 192
  retention mechanisms, 46
  sampling, 484-485
  transport mechanisms, 46

Oasis effect, 72
Odor
  animal-feeding facilities, 12
  composting, 12
  of water, 182
Ohm's law of electrical current, 53, 70
Oils, biodegradable, 168
100 yr storm, determining, 28-33
On-off cycle, pumps, 545-546
Open channel flow, 314-315, 319
Open channels, 217-225
Operating head, 542
Organic compounds
  in groundwater, 8
  in irrigation water, 414
  in soil, 179
  solubility of, 439
  in water, 197
Organisms, indicator, 474
Orifice
  coefficients, 304, 305
  discharge relations, 304-315
  equation, 304, 430
  flow, 305, 315
    Mavis plots for, 310
    versus pipe flow, 309
  hydraulic relationships, 298
  plate, 278-279
Osmotic potential, 550
Outlet hydraulics, 292
Outlets, 292
  design of, 293
  energy dissipation in, 323
  pipe, 276
  submerged, 312-313
  terrace, 276
  underground, 277-278
Overgrazing, 6, 15
Overland flow, time of concentration, 107-111
Oxbow lake, 476
Oxidation-reduction potentials, streams, 482
Oxygen, in streams, 477-479, 482

Palustrine wetlands, 477
Pan coefficient, 88
Pan evaporation, 70
Parallel field drains, 371, 372, *374*
Parallel plate deflection test, 526
Parshall flume, 301
Particle density, 331
Particle size
  distribution, 163-164
  and Manning roughness, 460
  parameter, 148
Partition coefficients, 483
Pastures
  in erosion control, 166
  management of, 190
Pavement, porous, 190
Peak flow, 96
  frequency analysis of, 122-123
  NRCS TR55, 118
Peak runoff rate, estimating
  NRCS TR55 graphical method, 117-123
  rational method, 105-117, 133
Peak stage, record of, 123
Pelagic habitat, 474-475, 479, 481
Penman-Monteith method,
    evaporation prediction, 70, 74-82, 90
Penman transformation, 77-78
Percent passing curve, 332-333
Percolation, 51
  deep, 381, 422
  definition of, 65
Percolation prediction models, 57-62
Permissible velocity method,
    channel stability, 220-222
Personalistic ethics, 496
Pesticides
  adsorption of, 483
  excessive application of, 194
  in groundwater, 8
  irrigation and, 403
  in water, 183
pH, 182
  streams, 482
Phillip equation, 58-59
Phosphorus
  control, 196
  cycling, 485
  removal, 198-199
  in soil, 179
  in water, 454
Photometers, 474
Piezometric surface, 391
Pipe
  circular, 308

depth of, 379
diameter, 207, 279-281, 315
drainage, 376-390, 525
drainage rates, 380-381
flexible, 526
friction losses in, 537-539
horizontal, 319
loading, 525-528
materials specifications, 525-528
perforated, 279, 376
rodent exclusion, 386
sealed, 279
seepage collars, 294
size, storm sewers, 321
slope drain (chute), 190
slope of, 379
spacing of, 376-379
vertical, 316-319
Pipe flow, 304-316
versus orifice flow, 309
Piping, 350
Pitcher plant, community viability study, 502-505
Pitot tubes, 463
Pivot length, 437-438
Plant growth, in streams, 454
Plant water potentials, 550-551
Plasticity index, 330, 332
Plastic limits, 330, 332
Plastics, 521
as drainage materials, 388-389
Pleasure-based ethics, 496
Point source pollution.
See Pollution, point source
Pollutants
evaporation and, 90
freshwater, *46*
industrial, 8
sediment as, 137
in streams, 482
transport by subsurface flow, 64-65
wetlands removal of, 477
Pollution, 15-16
biological, 181
chemical, 181
definition of, 181
erosion and, 168
groundwater, 11
industrial, 186
measures of, 182-183
nonpoint source, 6, 46
abatement, 189-197
characteristics of, 181-182
generators of, 177
reducing, 200
removal/renovation mechanisms, *197*
scale effects, 186-189
sources of, 185-186
transport of, 198
physical, 181
point source, 11, 181, 185
regulation of, 6
regulation of, 185-186
Polyethylene, 388, 521
Polypropylene, 472
Polyvinylchloride, 388, 521
Pond design, 342-351
summary of, 364-365
Ponds, 190
agricultural, 355-356
biological processes in, 479-481
embankment height, 343
excavated, 354
filters for, 347-348
fish, 291
flow control, 322
function of, 329
hillside, 354
infiltration of, 56, 57
liners for, 522
objectives for, 365
safety issues, 362
sedimentation, 276, 362
seepage control, 345-348
side slope, 348
sites for, 342-344
sizing, 337
soil limitation ratings, *343*
temperate, 480
types of, 342
water volume computations, 511-512
Population, increases in, 4, 199
Porosity, soil, 331
Portland Cement Institute, 390
Postharvest handling facilities, 5
Potential, 53
Poultry operations, odor and, 12
Power plant, selection of, 541
Power requirement, computing, 541
Precipitation
characteristics of, 21, 27
convective, 22
cyclonic, 22-23
data analyses, 25-26
definition of, 20-21
duration of, 27
effective, 420-421, 422

Precipitation (*continued*)
frequency of, 27
gauge, 23-24, 43
intensity of, 25-26, 27
local-scale analyses, 43-46
measurement of, 3, 23-25
mechanisms of, 21-22
orographic, 22
pollution transport by, 46
return period analysis, 27-33, 47
seasonal effects, 23
storm magnitude calculations, 33-43
in streams, 482
U.S. average, 3
and water erosion, 138
Precipitation equation
duration factors, *34*
return period factors, *34*
Precipitation estimates, Weiss method, 34-37
Precipitation hyetograph, 23, 26
Precipitation maps, Herschfield, 35-36, 37
Pressure
differentials, 22
fluid, 208-209
gauges, 546
regulators, 445
sprinkler nozzles, 429
Pressure-compensated emitter, 442-445
Pressure depth energies, 210
Pressure head, 53
components, 432
Pressure/vacuum relief valve, 441
Prey-predator food chains, 498
Price current meter, 463
Priestly-Taylor evaporation prediction method, 70, 86
Primary water management zone, 412
Principal spillways, design of, 352
Prior use, 151
Prismoidal formula, 514
Probability distributions, hydrologic data, 27-33
Probability plots, precipitation data, 30-33
Probable maximum precipitation (PMP), 43
Problem space, 2
Proctor density, 331, 345
Production
environmentally conscious, 5
sustainable, 15
Profundal habitat, 474, 479, 481
Pro-industrial approach, to ethics, 494
Projection loading, 525
Projects, well-designed, 501-502
Propane, 541
Protozoans, in streams, 475
Psychrometric constant, 80, 81
Pump affinity laws, 543
Pump curves, 543-544
Pumping, bioremediation projects, 395-396
Pumping plants, storage requirements, 545-546
Pumping rate, 546
irrigation, 423
Pumps
capacity of, 433
centrifugal
flow through, 534-536
performance curves, 537
type selection, 536-537
efficiency of, 537, 543
impeller, 536
matching to a system, 543-545
in parallel, 545-546
safety features, 546
in series, 545-546
Pump stations, 375

Qtot, 320
QUAL2E, 479, 486, 487, 488
Quality function deployment, *230*, 231

Radial-flow centrifugal pump, 537
Radiation
heat transfer, 22, 73-74
net, 80, 81
Radionuclides, 181
Radius of influence, 393
Rain, 20, 21. *See also* Precipitation
acid, 197
gauges, 3-4
Raindrop dislodgement, 142-143
Raindrop erosion, 140, 142-144
Raindrops
size of, 21, 142
terminal velocity of, 21, 142-143
Rainfall, 20. *See also* Precipitation
effective, 125, *126*, 420-421, 422
excess, 95. *See also* Runoff
increments, 125
mean annual, 138
and runoff erosivity index, 145-146, *147*
simulating, 63, 148
Rainfall erosivity cover factor, weighted, 154-158
Rational method, estimating peak runoff rate, 105-117, 133
Rational runoff coefficient, 105
field terraces, 266
selection of, 108, *109*, *110*
Readily available water, 410-411

Recession limb, 96
Recession limb to rising limb ratio, 112-116
Record length, 28-29
  and confidence intervals, 33
  Mockus equation for, 29-33
Recurrence interval, 27
  grassed waterways, 239
Redox potentials, streams, 482
Reflection coefficients, 74
Reflectivity, 81
Reflectivity coefficients, *75*
Regime theory, 457
Relief pipes, 384, 386
Relief valves, 441-442
Remaining water, 410
Reoxygenation, streams, 477-479
Reservoir routing, NRCS TR55 method, 337-338
Reservoirs
  and downstream sediment load, 457
  evaporation from, 71-72
Resource economics, 507
Resource management, versus environmental management, 502-505, 506
Retardance classes, grasses, 240, *246*
Retardation factors, *65*
Retention, maximum, 99
Return period
  analysis, 27-33, 47
  calculating, 27-30
  definition of, 27
Revised universal soil loss equation (RUSLE), 144-154, 169-170
Reynolds number, 307
Richards equation, 58
Rill erosion, 137, 140-141, 143, 169
Riparian buffers, 190, 473
Riparian water rights, 423
Riparian zones, 8, 200
Riprap, 190
Rising limb, 96
Rising stage sampler, 467
Riverine wetlands, 477
Rivers, water quality requirements, 453
Roads, 524-525
Rock, decorative, 472
Rock gabions, 472
Rock toe filters, 347
Rodent exclusion, from pipes, 386
Rohwer equation, 71, 72
Rollers
  sheepsfoot, 347
  smooth, 347
Rooting depth, 410, *411*
Roots, plant, 376, 412
Root zone thickness, 58
Rotation interval, center pivot irrigation system, 438
Rotifers, in streams, 475
Roughness coefficient, 215
  vegetation and, 238
Roughness elements, 459
Roughness factor, Manning, 215, *216*
Routing, hydrologic, 338
Routing coefficient, 163, 164
Routing term, 338, *339*
Rule-based ethics, 496
Runoff
  asphalt, 483-484
  assessment, 132
  definition of, 94, 95
  direct, 51
  effective, 125, 127-131, *132*
  evaporation and, 94
  frequency analysis of, 122-123
  hydrograph, 123-132
  impoundment, 98
  measuring, 3
  NRCS curves, 118-121
  peak, 94
  rate, 132
  vegetated waterways, 241
  velocity control, 168
Runoff coefficient, *108*
  ASAE drainage, *116*
  excess rain, 112
Runoff curve numbers
  agricultural, *99*
  nonagricultural, *100*
Runoff volume, 96, 104
  estimates, compact watershed, 98-104
  and storm duration, 122
  tile outlet terrace, 276-279
Rural areas, flood damage, 6-7
Rural-urban interface, 16
RUSLE, 144-154, 169-170
  terrace spacing method, 262-263, 269

Safe Drinking Water Act, 6
Safety precautions
  motors, 547
  pumps, 547
Salinity
  control, drainage and, 380-381
  excess, 371
  of irrigation water, 414-416
  remediation, 11
  vapor pressure and, 72
  of water, 70, 181, 183
Salmon streams, 471

Saltation, 464
Salt tolerance, crops, *416*
Samples, water, 183
Sand, 332, 347
Santa Ana winds, 21
Scale, 2
Scobey equation, 430-431, 443, 537, 539
Scour, structure design and, 293
Sealants, 344
  artificial, 342
  natural, 342
  soil, 345-346
Sealing potential, soils, 102, 345-346
Season, precipitation and, 23
Secchi disk, 474
Sedigraphs, 165, 361
  computer modeling, 169
Sediment
  barriers, 362
  bouncing, 464
  concentration,
    gravimetric analysis for, 468
  control
    in construction, 168
    principles of, 200
  management, 168
  mass, 198
  mass flow rate of, 165
  as pollution, 181
  ponds, 362
  retention basin, 190
  routing, modeling, 198-199
  sampling, streams, 467-468
  size distribution of, 164
  streams, 464-467, 482
  traps, 362
  wash load, 476
  in water, 197
Sedimentation, and spillway design, 354
Sedimentation basins, 356-362
  dewatering, 360
  in drainage systems, 384
  surface area requirements, *359*
Sedimentation pond, 276
Sediment delivery ratio, 159-162
  and pollution transport, 198
Sediment prediction potential, 454
Sediment yield, 137
  prediction, 159-165
  versus soil erosion, 140
Seepage collars, 294, 350
  embankments, 350
Seepage control, ponds/embankments, 345-348
Seepage losses, channels, 227
Seepage rates, 345
  lined pond, 56
Selenium, in irrigation water, 415
Self-design, system, 14, 15
Self-organization, in ecology, 499
Self-renovation, of water bodies, 185
Sensible heat, 75
Serial correlation, hydrologic data, 28
Set time, lateral, 428
Settling, 191
Settling basins, 357
Sewer networks, design of, 321
Shaft speed, 543
Shear stress, 220
Sheet flow, 143
Shields curve, 465-467
Shrimp, brine, 183
Shrinkage limit, 332
Shrink-swell potential, 330
Side roll irrigation system, 403
Side slope
  channels, 244-245
  trapezoidal channels, *224*
Side walls, 293
Signage, 363
Silt basins, 356
Silt fences, 168, 362
Silviculture, and nonpoint source pollution, 199
Simple average method,
    precipitation analysis, 44
Siphon, 316
SI units, 158-159
Skin friction, 459
Sleet, 20, 21
Slope
  channel, 206, 227
  concave, 139-140, 150
  constrained, 509-510
  convex, 139-140, 150
  diversion channel, 270, 272
  diversion terraces, 270
  embankment, 343-344
  modification, 530
  of pipe, 379
  shapes of, 139-140, 150
  side, 206
  standard, 150
  of streams, 457-458
  surface, 211
  terrace channels, 267-270
  tile drains, *380*
  uniform, 139-140, 150
  vegetated waterways, 221
Slope length factor, 148-150

Slope stakes
equation, 348, 515-516
tentative, 516
Sluice gates
analysis, 304
hydraulic relationships of, 298
Snow
formation of, 20, 21
measuring, 3-4, 24-25
surveys, 3
Social traps, 501
Societal forces, 2
and changes in land use, 4-6, 15
Soda ash, 345
Sodium, in irrigation water, 414, 415
Sodium tetra phosphate (STP), 345
Software, modeling.
See Modeling software
Soil
aeration, 376
analysis, for seepage control, 345
artificial, 148
bioengineering, 471-472
buffer capacity of, 197
cements, 345
classification of, 332-335
clay in, 94
compaction, 166
composition of, 346-347
conductivity, 81
conservation practices, 5
consolidation, 345
detachability, 139
formation, 140
fractions, mobility of, 199
groups, hydrologic, *102*-104
heating/cooling, 81, 90
infiltration rate, 94
intake rate, 413-414, 421
irrigation and, 437, 447
layered, hydraulic conductivity of, 55
loss tolerance, 140
minerals in, 330
moisture, 151, 420
montmorillonite, 332, 334
organic, 330
percent passing curve, 332-335
permeability code, 148
and pollution adsorption, 483
pore size, 51, 52, 412
properties, 330-332
summary of, *411*
water well design problems, *393*
residues, 179
salinization of, 11, 53
sandy, water retention of, 410
sealing potential of, 345-346
solids in, 330
stability analysis, 348
stabilizers, 7
static loading, 525
structure code, 148
surface sealing, 51-52, 102, 166
surveys, 330
transportability, 139
vadose zone, 412, 439
volume, 511-515
water content of, 52
water erosion and, 138
water management zone, 410, 412
water potential, 420
water retention, 410
water storage reservoir, 406, 409, 410
Soil-canopy simulation, evaporation prediction, 87
Soil detachment
predicting, 144-158
reducing, 165-168
Soil erodibility factor, 146-148, 155
Soil erosion
in streams, 467
versus sediment yield, 140
Soil limitation ratings
levees and embankments, *344*
ponds, *343*
Soil loss
construction site, 157-158
equation, 145
rates, global, 8
ratio, *152*
reducing, 155
Soil suction head, 59
Solar energy, 506
and evapotranspiration, 70
Solar radiation, mean, *76*
Solids
filterable, 182
fixed, 182
in soil, 330
suspended, 191
volatile, 182
Sorption, in streams, 482-483
Space scale, 498
Spacing tolerance ratio, 263
Spatial scales, nonpoint source pollution, 187
Specialty-crop production, 9
Specific energy, 210-211
structure flow, 294, 296

Spillways
  box-inlet drop, 297
  configuration of, 338
  design of, 352-354
  drop, 297, 298
  emergency, 321, 338, 348-349, 352-353
  principal, 352
  sizing, 337
  stage-discharge curves, 316-321
Spirit-based ethics, 496
Spoil bank, 228
Sprinkler irrigation equipment, comparison of, *428*
Sprinkler laterals, 403, 405
  agricultural applications, 426-436
  design summary, 433-436
  inlet pressure, 432
  number of sprinklers per, 430
  size of, 430-431
  spacing of, 428
Sprinklers
  capacity of, 429
  drip, 403
  nozzles, 428-430
Sprinkler systems, 9, 406
Sprinkler wetting radius, 438
Stability
  channel design, 217-225
  vegetated waterways, 244, 245-247
Standard deviation, probability distributions, 28
Static discharge, 542
Static loading, 525
Static suction lift, 542
Stefan-Boltzmann coefficient, 81
STELLA, 342
Stiff diagram, 159-162
Stilling basins, 294, 298
Stone flies, 183, 474
Storage routing, 338-342
  curve calculations, *339*
Storage volume, spillways, 352, 353
Storm duration, 122, 125
Storm magnitude, calculating, 33-43
Storm prediction, return period analysis, 27-33, 47
Storm type, and hyetographs, 38
Storm water collection, closed conduit systems, 321
Storm water discharge
  management of, 12
  regulation of, 186, 191
Storm Water Management Model (SWMM), 170, 200, 325, 486, 488
Streamflow
  effect of slope and velocity on, 458-461
  measuring, 461-464
Stream gauges, 3
Stream routing, 338
Streams
  bank erosion, 142, 469-471
  bed degradation, 457
  biological processes in, 474-476
  braided, 456, 476
  channel erosion, 137, 142
  channel maintenance/restoration, 471-473
  chemical processes in, 482-484
  deltaic, 456
  dendritic, 455, 476
  depositing (aggrading), 464
  ephemeral, 456, 476
  eutrophic, 454
  flow measurements, 461-464
  impaired, 454
  intermittent, 456, 476
  large, 463
  meandering, 456, 458, 473
  morphology of, 454-458, 473, 481, 487
  multibasin patterns, 455
  nutrient cycling, 485-487
  nutrient sampling, 484-485
  order of, 455
  oxygen levels, 482
  oxygen transport, 477-479
  perennial, 456
  pH of, 482
  radial, 455, 476
  sediment chemistry, 482
  sediment sampling, 467-468
  sediment transport, 464-467
  soil erosion in, 467
  total maximum daily load (TMDL), 453, 487
  trellis, 455, 476
  velocity of, 459-461, 473
  water quality requirements, 453
Streeter-Phelps equations, 478-479
Streptococcus, fecal, 183
Strip-cropping, 5, 166, 190
Strip-crop practice factor, 151, *154*
Structures
  conservation, 291, 324
  customer requirements, *324*
  design and selection summary, 323
  discharge relations, 304-315
  foundations for, 294
  function of, 291
  hydraulic concepts, 294-296
  materials specifications, 521-524
  resources on, 523

stage-discharge relationships, 297-304
technical requirements, *324*
types of, 292-294
Subcritical-supercritical flow transitions, 296
Subirrigation, 386-387, 406
Subirrigation pipe, loading, 525
Subsurface drainage, 365-376
Subsurface flow, 51
Suction lift, 540
Suction line, design considerations, 539-540
Sulfide deposits, 482
Sulfur cycle, 179, 192, 485
Sumps, 386
Sunshine-temperature factor, 85
Superfund, 186
Surface cover, 151
Surface drainage, 371-375
land-forming computations, 530-532
Surface roughness, 151, 159
Surface runoff factor, 159-160
Surface sealing, 166
Surface water
regulation of, 423-426
standards, 183
Surfer, 45
Swales, 190, 362
Swamp factor, *114*
Swamps, 330
Swine operations, odor and, 12
SWMM, 170, 200, 325, 486, 488
Synthetic storm approach,
design storm prediction, 27-33, 47
System head curve, 542-543

Tackifiers, asphalt-based, 168
Temperature
dew point, 79-80
precipitation and, 29
and water erosion, 138
Temporal scales, nonpoint source pollution, 189
Tennessee-Cumberland River, 474
Tensiometer, 420
Terrace channels, 264-265
capacity of, 267-269, 270
cross-section, 265
depth of, 267-269, 270
grades, 265-267, *268*
length of, 265
Terrace outlets, 261
selection of, 276
Terraces, 7, 167, 190, 257
bench, 259, 261
broadbase, 259, 261
construction and layout, 519-521
definition of, 257, 259
design of, 261-270
erosion control, 155, 283, 286
function of, 261
grassed backslope, 259
key, 282
nonparallel, 258, 259
parallel, 258, 259, 281-283, 521
shape of, 258
spacing, 261-264, 269, 521
temporary, 274
tile outlet, 276-281, 287
types of, 259-261
unbalanced, 282-283
and vegetated waterways, 249
water conservation, 261, 282, 283
Terracing factor, 151
Tetra sodium pyrophosphate (TSPP), 345
Thermal stratification, 480
Thermocline, 479
Thiessen polygon method, precipitation analysis, 44-45
Thornthwaite method, evaporation prediction, 87
Tile drains, minimum slopes for, *380*
Tile line, sizing, 381
Tile outlet terraces, 276-281
design summary, 287
Tillage
conservation, 7, 155, 167-168, 190, 257, 258
conventional, 167-168
and rill erosion, 140-141
for weed control, 194
Time, cover management factor and, 151
Time of concentration
compact watersheds, 106-107
NRCS, 118
overland flow, 107-111
Time scale, 498
TMDL. *See* Total maximum daily load
Toe walls, 293
Topography
hydrograph shape and, 111, 114
and water erosion, 139-140
Top width, 206
Torque, 535
Torque-speed curve, 545
Total dissolved solids, 182
Total dynamic head, 540-541
Total hydraulic potential, 53
Total Kjedahl nitrogen, 183
Total load, 466
Total maximum daily load (TMDL)
definition of, 182
streams, 453, 487

Total solids, 182
TR20, 325
Tractive force method, channel stability, 220
Transition flows, analyzing, 211
Transition loss coefficients, *225*
Transpiration, 3, 69
Transport, convective, 65
Transport parameters, 78
Trapezoid, area of, 514
Trapping efficiency, 361
Trapping efficiency coefficient, 358
Traps
  sediment, 362
  social, 501
Traveling guns, 405
Tropical forests, loss of, 6
Trout streams, 471
Tube radius, 384
Tubing, corrugated, 381
Turbidity
  streams/lakes, 480
  of water, 182
Turbulence
  in settling basins, 358-359
  in streams, 477
Turbulent transfer coefficients, 74
Turgor potential, 550
Turnover time, nutrients, 486

Ultraviolet (UV) light, 484
Unified Soil Classification (USC) System, 332, 334
Uniform flow equations
  Chezy, 214-215
  Manning, 215
Uniformity coefficient, 417-418
Unit conversions, *553-554*
United Nations Food and Agriculture Organization (FAO), 69
  environmental assessment, 453
U.S. Bureau of Reclamation, 227
U.S. Department of Agriculture (USDA)
  National Resources Inventory website, 5
  NRCS soil survey, 330
  Soil Conservation Service, 4
U.S. Environmental Protection Agency (EPA)
  drinking water standards, 183, *184*
  environmental assessment, 452
  streamflow modeling, 486
  TMDL, 453, 487
  website, 66, 133
U.S. Forest Service, sediment delivery ratio, 159-162
U.S. Geological Survey
  groundwater aquifers, 390
  streamflow measurement stations, 463
  website, 4, 66
Unit hydrograph
  definition of, 123-124
  NRCS, 124-125
  NRCS effective runoff, 127-131, *132*
Unit plot, 147
Units
  confusing, 81
  English, 158-159
  SI, 158-159
Universal soil loss equation (USLE), 144-158, 169-171, 263-264
Uplands
  erosion of, 148-153, 453
  nonpoint pollution, 177-200
Upstream integration, 211
Urban areas, erosion from, 137
Urbanization, 4, 5, 6
  definition of, 3
  erosion and, 139
  and nonpoint source pollution, 199
  temperature and, 43-44
  unplanned, 15
  water quality and, 177
USLE, 144-158, 169-171
  terrace spacing methods, 263-264
Utilitarianism, 496

Vacuum breakers, 445
Vadose zone, 412, 439
Values, 493
  ethics and morals, *494*
Valves
  backflow prevention, 445
  flush, 445
  relief, 441-442, 546
Van Schilfgaarde equation, 378-379
Vapor convection coefficient, 77
Vapor flux rate, 76
Vapor movement, 75
Vapor pressure, 70
  of air, 79
  evaporation and, 71, 72, 79
Vegetation
  erosion reduction by, 139
  selection of, 242-244
Vegetation strips, 8, 473
Velocity
  channel, 207
  exit, 225
  point, 463

streams, 461-464
terrace channels, 268
vegetated channels, 244, *247*
Velocity control
hydraulic radius restrictions for, 220-222
slope restrictions for, 220
Vertical interval method, terrace spacing, 262, 264, 269
Vertical riser, spillways, 316, *317-319*
Virtue ethics, 496
V-jet nozzle, 162
Void, ratio, 331
Volatile solids, 182
Volatilization, 46, 182, 192, 198
Volumetric water content, 409
von Karman coefficient, 459

Walls, structure, 293
Wash load, 464
Waste
biological, 179
industrial, 481
in water, 197-198
Waste application systems
and pipe drainage, 376
sprinklers, 439
Waste lagoons, 8, 363
Wastewater treatment
BOD and, 414
irrigation and, 402
and spillway design, 354
Water, 3
adjustment table, 440
application, 417, 429
balance, 419
conservation, 261, 282
drinking, 183, *184*
electrical conductivity of, 415
exit, 305, 315-316, 323
hammer, 433
harvesting, 282
hydrologic cycle, 3-4
intended use, 177
nutrient constituents, 197
pesticides limits, 183
plant available, 410-412
and pollutant transport, 64-65
regulation of, 423-426
requirements, agricultural ponds, *355*
rights, 423-426, 433
salinity of, 70, 181
specific gravity of, 409
supply, public, 445
surface, 183, 461
surface tension of, 54-55
unit weight of, 208
volume, 511-512
Water bodies
assimilative capacity of, 187
self-renovation of, 185
Water erosion
computer modeling, 170
factors affecting, 138-140
flow-induced, 143-144
mechanics of, 142-144
prediction, 169
rainfall erosivity cover factor, 154-158
types of, 140-142
Water Erosion Prediction Project (WEPP), 168, 169, 170-171, 471
Water management structures. *See* Structures
Water management zone, soil, 410, 412
Water potential, soil, 420
Water quality, 183-186
global, 177
Internet database browser, 200
irrigation water, 414-416
management, 8
and nonpoint pollution, 46
Water release curve, 406, 409
Watershed
assessment, 453
compact
definition of, 95
runoff rate, 105-123, 133
runoff volume, 98-104
time of concentration, 106-107
complex, 96
definition of, 96-98
divide, 97
gradient, 107, 108
inch, 337
precipitation distribution, 44-46
upstream, 6
Watershed ponds, 355-356
design of, 343-344
Watershed-stream assessment, 12
Water stress, 551
evapotranspiration and, 91
Water tables
confined, 391
lowering, 375, 378
Water use
efficiency, 69, 417. *See also* Transpiration ratio
seasonal, 421-423

Waterways, grassed, 5, 190. *See also* Waterways, vegetated
Waterways, vegetated, 238
  channel cross-sections, 241-242
  characteristics of, 239
  construction, 253
  design of, 245-255
  maintenance of, 253
  planning, 240-245
  vegetation selection, 242-244
Water well design problems, soil properties, *393*
Weed control, 194
Weibull plotting position equation, 27
Weir, 132, 210, 292, 322
  constant, 298-300
  cross-section of, 298
  equation, 297-303, *304*
  flow, 313-314, 315, 319, 321
  inlets, stage-discharge relationships, 297-304
  prefabricated, 522
Weiss method, precipitation estimates, 34-37
Wells
  Artesian, 391
  drawdown equation, 542-543
  drilling, 396
  gravity, 391-392
  hydraulics of, 390
  injection, 394-395
  shallow, 390
WEPP, 168, 169, 170-171, 471
Wetlands, 8, 190, 476
  biological processes, 476-477
  conservation, 371, 398
  pollution and, 197, 199
  types of, 477
Wetted perimeter, 206, 462
Wetting front, depth of, 59-60
Wilting, 409, 410, 550
Wind, 3
  direction, 4
  effect on sprinkler patterns, 429
  erosion control, 5, 7-8
  function, 78
  precipitation and, 20
  speed, 4, 71, 81
Wing walls, 293
Wyoming bentonite, 345

Xylem tubes, 550

Yield, 2, 410

Zhang model, conductivity, 63

www.ingramcontent.com/pod-product-compliance
Ingram Content Group UK Ltd.
Pitfield, Milton Keynes, MK11 3LW, UK
UKHW050949290926
15470UKWH00005B/1734

9 780813 818474